THE SALT MINE

Salt mining, Ethiopia

Bureau of Economic Geology Udden Book Series No. 5

AAPG Memoir 99

THE SALT MINE
A Digital Atlas of Salt Tectonics

Michael R. Hudec and Martin P. A. Jackson

Illustrated by
Nancy Cottington

Modeling by
Bruno C. Vendeville
Daniel D. Schultz-Ela
Tim P. Dooley

Published jointly by

Bureau of Economic Geology
Scott W. Tinker, Director

and

American Association of Petroleum Geologists

BUREAU OF ECONOMIC GEOLOGY

Jackson School of Geosciences, The University of Texas at Austin
Austin, Texas 78713-8924, U.S.A.

AMERICAN ASSOCIATION OF PETROLEUM GEOLOGISTS
Tulsa, Oklahoma 74101, U.S.A.

First published 2011
Printed in the United States of America

ISBN 978-0-615-51836-7

Illustrations:

Front cover: Salt glaciers spreading from the summit of Kuh-e-Namak, Bushehr Province, Zagros Ranges, Iran (Martin Jackson).

Frontispiece: Afar and Tigray salt miners digging salt slabs from the shore of Lake Assal, Danakil Depression, Afar, Ethiopia (Paul Nicholson).

Table of Contents facing: Adit into the historic Muntanya de Sal salt mine, which reached 1,300 m depth in the Cardona diapir, Ebro Basin, Spain (Martin Jackson).

Table of Contents: Laden with hand-carved slabs of salt, a camel caravan crosses the flooded shores of Lake Assal in the Danakil Depression at the start of a 3-day trek up the rift escarpment to markets in the Ethiopian highlands (Paul Nicholson).

Back cover: Seismic image courtesy of CGGVeritas.

This publication should be cited as:

Hudec, M. R., and M. P. A. Jackson, 2011, The salt mine: a digital atlas of salt tectonics: The University of Texas at Austin, Bureau of Economic Geology, Udden Book Series No. 5; AAPG Memoir 99, 305 p.

This book is available from:

U.T. Bureau of Economic Geology Bookstore
University Station, Box X
Austin, TX 78713-8924, U.S.A.
Phone: 512.471.7144
Toll-free phone: 1.888.839.4365 (USA)
Toll-free fax: 1.888.839.6277 (USA)
E-mail: pubsales@beg.utexas.edu
http://www.beg.utexas.edu/pubs/pubsales.php

The AAPG Bookstore
P.O. Box 979
Tulsa, OK, 74101-0979, U.S.A.
Tel.: +1.800.364.2274AAPG (USA, Canada)
or +1.918.584.2555
Fax: +1.800.898.2274 (USA)
or +1.918.560.2652
E-mail: bookstore@aapg.org
http://bookstore.aapg.org

Geological Society Publishing House
Unit 7, Brassmill Enterprise Centre
Brassmill Lane, Bath, BA1 3JN, U.K.
Tel.: +44.1225.445046
Fax: +44.1225.442836
www.geolsoc.org.uk
E-mail: dawn.angel@ geolsoc.org.uk
http://bookstore.geolsoc.org.uk

The American Association of Petroleum Geologists Books Refereeing Procedures

The Association makes every effort to ensure that the scientific and production quality of its books matches that of its journals. Since 1937, all book proposals have been refereed by specialist reviewers as well as by the Association's Publications Committee. If the referees identify weaknesses in the proposal, these must be addressed before the proposal is accepted.

Once the book is accepted, the Association Book Editors ensure that the volume editors follow strict guidelines on refereeing and quality control. We insist that individual papers can only be accepted after satisfactory review by two independent referees. The questions on the review forms are similar to those for the *AAPG Bulletin*. The referees' forms and comments must be available to the Association's Book Editors on request.

Although many of the books result from meetings, the editors are expected to commission papers that were not presented at the meeting to ensure that the book provides a balanced coverage of the subject. Being accepted for presentation at the meeting does not guarantee inclusion in the book. More information about submitting a proposal and producing a book for The American Association of Petroleum Geologists can be found on its web site: www.aapg.org.

Compilation and production of this atlas were funded
by the Applied Geodynamics Laboratory consortium
with supplementary funds from the
Jackson School of Geosciences
(The University of Texas at Austin).

Established in 1988, the Applied Geodynamics Laboratory is dedicated to producing innovative new concepts in salt tectonics. This research comprises a mix of physical and mathematical modeling and seismic-based mapping and structural-stratigraphic analysis of some of the world's most spectacular salt basins. The laboratory is widely considered the world's premier institute for research on salt tectonics. Concepts and terminology pioneered by the Applied Geodynamics Laboratory have had a profound influence and are widely disseminated throughout the oil industry.

The following companies supported construction of this atlas as members of the Applied Geodynamics Laboratory consortium.

Anadarko Petroleum Corporation
BHP Billiton Petroleum
BP
Burlington Resources
CGGVeritas
Chevron Corporation
Cobalt International Energy
ConocoPhillips
Devon Energy
Ecopetrol
Encana Natural Gas
Eni S.p.A
Enterprise Oil
ExxonMobil Corporation
Fugro
Global Geophysical Services
Hess Corporation
INEXS
ION Geophysical Corporation
Instituto Mexicano del Petróleo
Korea National Oil Corporation
Maersk Oil
Marathon Oil Corporation
Mariner Energy
McMoRan Exploration
Murphy Oil Corporation
Nexen
Noble Energy
Petrobras
Petróleos Mexicanos
Petroleum Geo-Services
Repsol YPF
Samson
Saudi Aramco
Schlumberger
Shell Oil Company
Statoil
TGS-NOPEC Geophysical Company
Total S.A.
Woodside Energy

Cardona salt mine, Spain

Contents

Chapter 1 Salt Pillows and Salt Anticlines........................ 1
Chapter 2 Salt Stocks and Salt Walls.............................31
Chapter 3 Triangular Diapirs.......................................85
Chapter 4 Indented Diapirs.. 109
Chapter 5 Extensional Salt Detachments.................... 121
Chapter 6 Contractional Salt Detachments................ 141
Chapter 7 Salt Sheets.. 161
Chapter 8 Minibasins.. 217
Chapter 9 Roho Systems.. 235
Chapter 10 Counterregional Systems........................... 245
Chapter 11 Location Maps.. 253
Glossary.. 269
References and Figure Credits................................ 285
Subject Index.. 295
Geographic Index.. 303

Salt caravan, Ethiopia

Preface

Salt tectonics is the process of deformation involving salt: how and why salt structures evolve. Because salt is weak, it lubricates and intensifies deformation in a sedimentary basin. The effects of salt tectonics are therefore of vital interest to exploration for oil, gas, and some minerals. Salt tectonics forms abundant hydrocarbon traps around salt bodies. Moreover, salt canopies form vast shrouds that affect geophysical imaging of deeper hydrocarbon traps. Salt tectonics has modified more than 130 basins around the world, including some rich in hydrocarbons such as the Gulf of Mexico, offshore West Africa, offshore Brazil, the North Sea, the Middle East, and the North Caspian area.

Successful hydrocarbon exploration has produced a surge of interest in salt tectonics in the last 20 years. This attention, combined with advances in seismic data acquisition and processing, has revolutionized our understanding of salt tectonics. However, the wealth of information is also a glut. More than 3,800 papers related to salt tectonics have been published since 1990. It is difficult for interpreters newly assigned to salt basins to learn the basics or for experienced salt interpreters to keep up with advances in the field.

Given this glut of information, what is the best way to gain and maintain a working knowledge of salt tectonics? Traditional methods of learning, such as reading papers and attending oral presentations, are time intensive. Furthermore, a long interval commonly separates the time when information is received and when it is applied. Finally, the information is typically buried in a mass of other data that are less relevant to the particular problem at hand. We need a system that quickly sorts through the entire spectrum of knowledge of salt tectonics, isolates relevant information, and provides pathways to more-detailed information.

The DVD in the back of this book illustrates such a system in the form of an interactive digital atlas. *The Salt Mine* DVD is an HTML-based atlas of salt structures and associated sediment geometries. The atlas contains annotated images of salt structures with detailed captions that discuss key principles. Animations illustrate many of the physical and conceptual models. All images are grouped into structural styles on the basis of a geometric classification rather than on their mode of origin, which is often contentious or ambiguous. Images include field exposures (outcrop views, geologic maps, aerial photographs, and satellite images), seismic sections, geologic cross sections, conceptual sketches, and animations. In addition, the atlas showcases hundreds of the best examples from the Applied Geodynamics Laboratory's collection of physical and numerical models.

Designing an atlas should take into account its users, the information they seek, and how long they are willing to spend searching. *The Salt Mine* is aimed at novices and experienced geoscientists in salt tectonics. We assumed that both types of users wish to learn more about salt tectonics, look for analogs or alternative interpretations, or deepen their understanding of a specific type of salt structure. Finally, we assumed that users are willing to spend, at most, 5 minutes looking for answers to their questions.

As evident from the preceding description, the core of *The Salt Mine* is the DVD in the back pocket, which contains all 1,349 images and more than 40 animations. Images in the DVD can be found using four different methods. The DVD is packaged with a printed book containing 400 of the most interesting and visually appealing of these images.

We are grateful to Bruno Vendeville, Dan Schultz-Ela, and Tim Dooley for providing the results of their largely unpublished modeling at the Applied Geodynamics Laboratory, which make this atlas especially distinctive. We also selected a few model images from other experimenters, who are acknowledged in captions. The following companies allowed us to reproduce their proprietary images: Anadarko, BHP Billiton, BP, CGGVeritas, Chevron, ConocoPhillips, ExxonMobil, Reading & Bates, Shell, Vanco, WesternGeco, and Total. The following scientists sent material for the atlas: Ian Davison, Chris Harrison, Charles Kluth, Lou Liro, Paul Nicholson, Mark Rowan, Delphine Rouby, and Chris Talbot. We are indebted to our countless colleagues in industry for influencing our ideas and for encouraging this atlas.

Production Team

This atlas would never have left the ground without
the efforts of a talented production team
at the Bureau of Economic Geology.

Drafting: Nancy Cottington, with contributions by Bill Bergquist

HTML Authoring: Sarah Clough, with contributions by Xuejiao Liu and Rong Li

Book Design and Layout: Jamie Coggin

Animations: Tim Dooley, Dan Schultz-Ela, Nancy Cottington, Xuejiao Liu, and Bruno Vendeville

Geographic Information Systems: Cari Breton, with contributions by John Andrews and Scott Rodgers

Multimedia Design and Supervision: Scott Rodgers

Editing: Lana Dieterich and Susie Doenges

Indexing: Michael Hudec, Martin Jackson, and Amanda Masterson

Copyright Release: Nancy Cottington, with contributions by Cathy Brown

Search-Engine Programming: Joseph Su

Book Production Manager: Cathy Brown

Administration: Jay Kipper

FEATURES OF THE BOOK

Chapter Introductions. We classified 10 types of salt structures, assigning each a chapter on the basis of geometry, as explained in the Introduction. Each chapter introduction summarizes the structural style and presents examples of that style in map and cross section.

Chapter Contents. Images in each chapter are grouped by location in alphabetical order and then by physical, mathematical, and conceptual model analogs. Each figure has a detailed caption that highlights features of interest related to salt tectonics.

Text includes a definition of the structural style, significance and occurrence in salt basins, and key aspects of structural evolution.

A

Chapter introduction

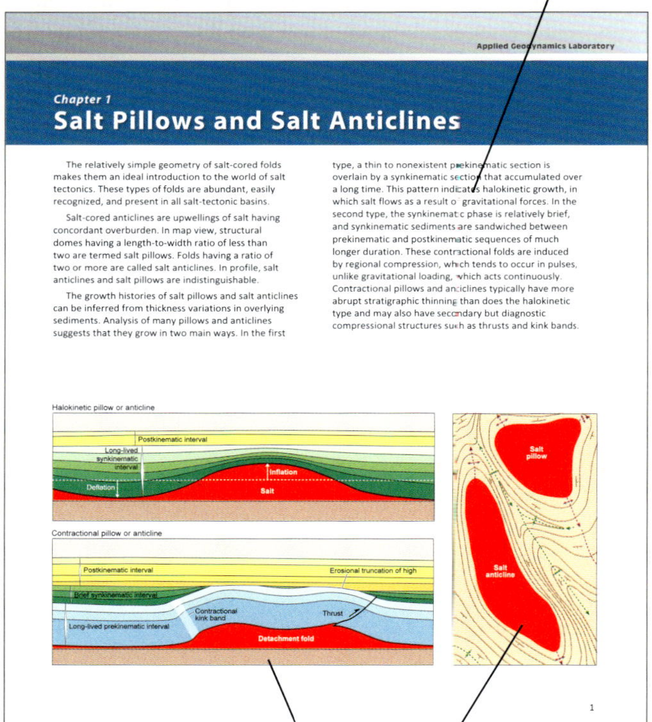
Applied Geodynamics Laboratory

Chapter 1

Salt Pillows and Salt Anticlines

The relatively simple geometry of salt-cored folds makes them an ideal introduction to the world of salt tectonics. These types of folds are abundant, easily recognized, and present in all salt-tectonic basins.

Salt-cored anticlines are upwellings of salt having concordant overburden. In map view, structural domes having a length-to-width ratio of less than two are termed salt pillows. Folds having a ratio of two or more are called salt anticlines. In profile, salt anticlines and salt pillows are indistinguishable.

The growth histories of salt pillows and salt anticlines can be inferred from thickness variations in overlying sediments. Analysis of many pillows and anticlines suggests that they grow in two main ways. In the first type, a thin to nonexistent prekinematic section is overlain by a synkinematic section that accumulated over a long time. This pattern indicates halokinetic growth, in which salt flows as a result of gravitational forces. In the second type, the synkinematic phase is relatively brief, and synkinematic sediments are sandwiched between prekinematic and postkinematic sequences of much longer duration. These contractional folds are induced by regional compression, which tends to occur in pulses, unlike gravitational loading, which acts continuously. Contractional pillows and anticlines typically have more abrupt stratigraphic thinning than does the halokinetic type and may also have secondary but diagnostic compressional structures such as thrusts and kink bands.

Map and cross-sectional examples of an ideal structure.

Title provides information on type of structure, location, or type of model.

B

Image page

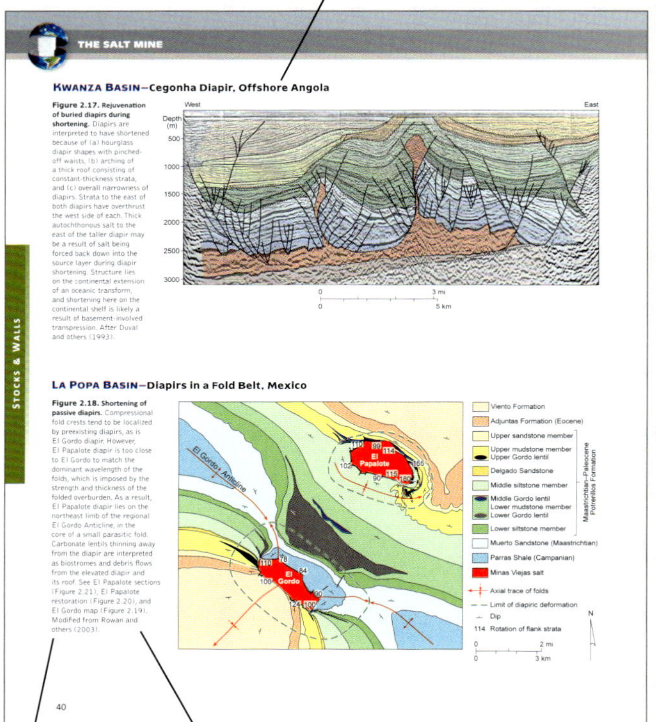
THE SALT MINE

KWANZA BASIN—Cegonha Diapir, Offshore Angola

Figure 2.17. Rejuvenation of buried diapirs during shortening. Diapirs are interpreted to have shortened because of (a) hourglass diapir shapes with pinched-off waists, (b) arching of a thick roof consisting of constant-thickness strata, and (c) overall narrowness of diapirs. Strata to the east of both diapirs have overthrust the west side of each. Thick autochthonous salt to the east of the taller diapir may be a result of salt being forced back down into the source layer during diapir shortening. Structure lies on the continental extension of an oceanic transform, and shortening here on the continental shelf is likely a result of basement-involved transpression. After Duval and others (1993).

LA POPA BASIN—Diapirs in a Fold Belt, Mexico

Figure 2.18. Shortening of passive diapirs. Compressional fold crests tend to be localized by preexisting diapirs, as is El Gordo diapir. However, El Papalote diapir is too close to El Gordo to match the dominant wavelength of the folds, which is imposed by the strength and thickness of the folded overburden. As a result, El Papalote diapir lies on the northeast limb of the regional El Gordo Anticline, in the core of a small parasitic fold. Carbonate lentils thinning away from the diapir are interpreted as biostromes and debris flows from the elevated diapir and its roof. See El Papalote sections (Figure 2.21), El Papalote restoration (Figure 2.20), and El Gordo map (Figure 2.19). Modified from Rowan and others (2003).

Caption ends with acknowledgments or credits for the image.

The caption discusses aspects of the geometry or structural evolution that are particularly well illustrated.

Map Chapter. The final chapter shows map locations of all images in the book for which the location is known and is not proprietary.

Glossary. At the back of the book a glossary contains 243 terms related to salt tectonics. We intensively revised the glossary of Jackson and Talbot (1991) to reflect current understanding of salt tectonics and expanded it to include new terms from the last 20 years.

C

Map page

Maps presented alphabetically by geographic name.

Regional map shows location of area.

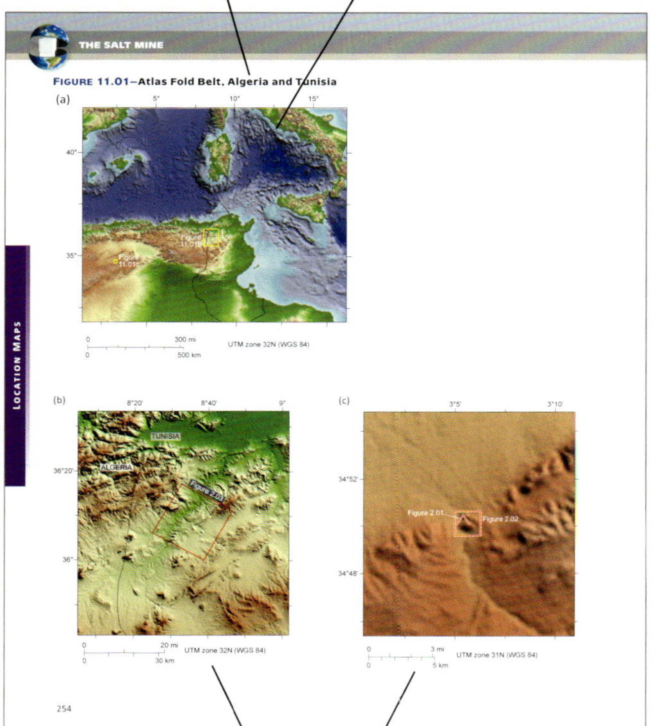

Detailed maps show locations of figures.

D

Glossary page

Italic type indicates cross reference to another glossary entry.

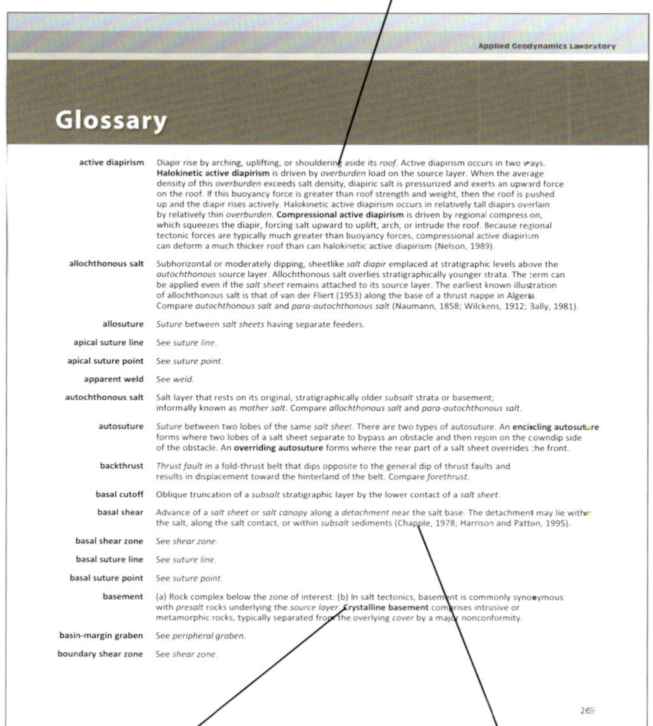

Applied Geodynamics Laboratory

Glossary

active diapirism Diapir rise by arching, uplifting, or shouldering aside its *roof*. Active diapirism occurs in two ways. **Halokinetic active diapirism** is driven by *overburden* load on the source layer. When the average density of this *overburden* exceeds salt density, diapiric salt is pressurized and exerts an upward force on the roof. If this buoyancy force is greater than roof strength and weight, then the roof is pushed up and the diapir rises actively. Halokinetic active diapirism occurs in relatively tall diapirs overlain by relatively thin *overburden*. **Compressional active diapirism** is driven by regional compression, which squeezes the diapir, forcing salt upward to uplift, arch, or intrude the roof. Because regional tectonic forces are typically much greater than buoyancy forces, compressional active diapirism can deform a much thicker roof than can halokinetic active diapirism (Nelson, 1989).

allochthonous salt Subhorizontal or moderately dipping, sheetlike *salt diapir* emplaced at stratigraphic levels above the *autochthonous* source layer. Allochthonous salt overlies stratigraphically younger strata. The term can be applied even if the *salt sheet* remains attached to its source layer. The earliest known illustration of allochthonous salt is that of van der Fliert (1953) along the base of a thrust nappe in Algeria. Compare *autochthonous salt* and *para-autochthonous salt* (Naumann, 1858; Wilckens, 1912; Bally, 1981).

allosuture *Suture* between *salt sheets* having separate feeders.

apical suture line See *suture line*.

apical suture point See *suture point*.

apparent weld See *weld*.

autochthonous salt Salt layer that rests on its original, stratigraphically older *subsalt* strata or basement; informally known as *mother salt*. Compare *allochthonous salt* and *para-autochthonous salt*.

autosuture *Suture* between two lobes of the same *salt sheet*. There are two types of autosuture. An **encircling autosuture** forms where two lobes of a salt sheet separate to bypass an obstacle and then rejoin on the downdip side of the obstacle. An **overriding autosuture** forms where the rear part of a salt sheet overrides the front.

backthrust *Thrust fault* in a fold-thrust belt that dips opposite to the general dip of thrust faults and results in displacement toward the hinterland of the belt. Compare *forethrust*.

basal cutoff Oblique truncation of a *subsalt* stratigraphic layer by the lower contact of a *salt sheet*.

basal shear Advance of a *salt sheet* or *salt canopy* along a *detachment* near the salt base. The detachment may lie within the salt, along the salt contact, or within *subsalt* sediments (Chapple, 1978; Harrison and Patton, 1995).

basal shear zone See *shear zone*.

basal suture line See *suture line*.

basal suture point See *suture point*.

basement (a) Rock complex below the zone of interest. (b) In salt tectonics, basement is commonly synonymous with *presalt* rocks underlying the *source layer*. **Crystalline basement** comprises intrusive or metamorphic rocks, typically separated from the overlying *cover* by a major nonconformity.

basin-margin graben See *peripheral graben*.

boundary shear zone See *shear zone*.

Boldface type indicates primary reference for a term.

Lead author indicates earliest known reference.

Features of the DVD

Main Menu. The digital version of *The Salt Mine* in the back pocket is small but mighty. It contains 3 times as many images as the book and more than 50 animations. These images and animations can be accessed through the main menu page, which provides links to three types of search and three forms of help.

Keyword Search. *The Salt Mine*'s search engine allows examples to be extracted from the database and assembled in user-defined groups by using keywords for geographic location, salt geometry, basin geometry, tectonic setting, geomorphic province, sediment geometry, fault style, or image type.

A

Salt Mine main menu

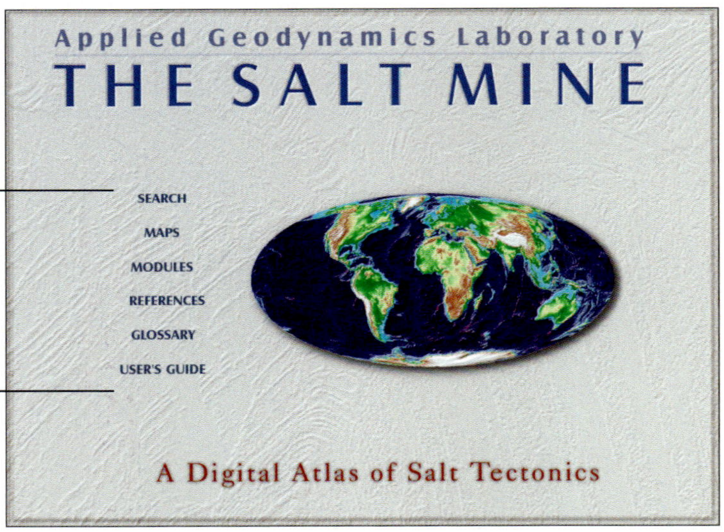

- Search by keyword
- Search by geographic location
- Search by structural style
- List of references cited
- Salt-tectonics glossary
- Directions for using *The Salt Mine*

B

Keyword search

Boolean searches for *and*, *or*, and *and not* are permitted.

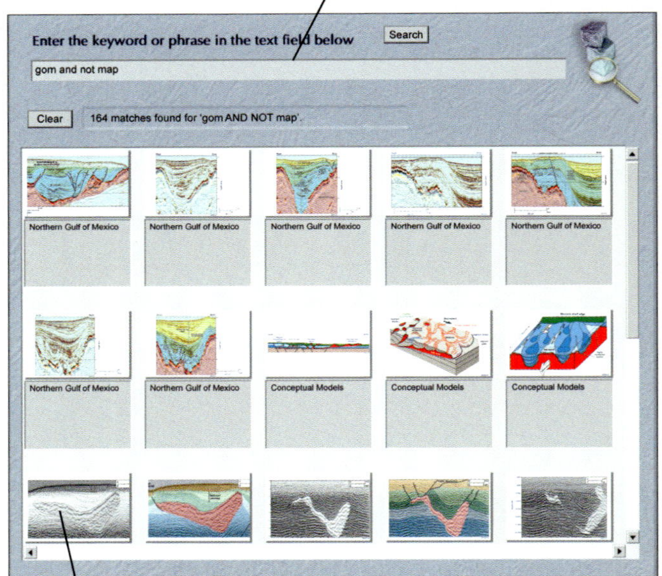

Each search returns a scrollable page of thumbnails. Clicking on a thumbnail links to the full-sized image and caption.

Map Search. Every image for which we have releasable geographic coordinates has been entered into a GIS database. Locations of these sections and maps have been plotted on basin maps and are linked to the full-sized images.

Module Search. Images in *The Salt Mine* are grouped by structural style (modules), equivalent to chapters in the book. Users already familiar with *The Salt Mine* can jump straight to the appropriate part of the atlas using the module search function.

C

Geographic search

Regional map shows location of index map.

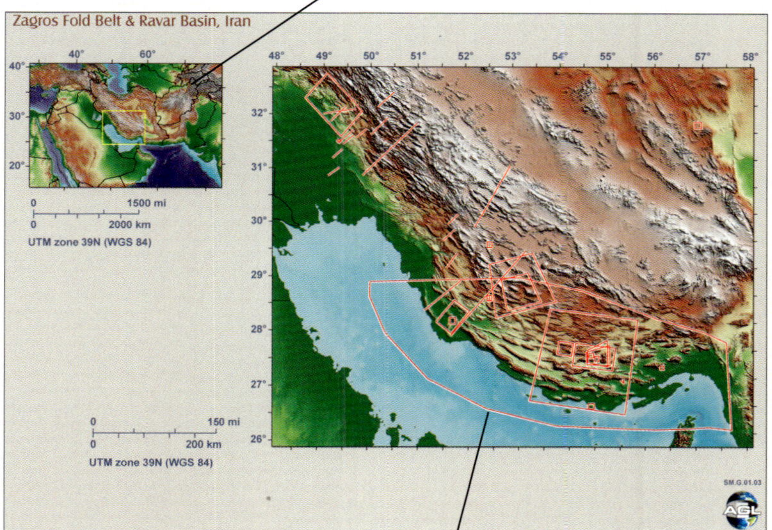

On the index map a section line or map outline may be selected. These link to a full-sized image.

D

Module search

Each module is illustrated by a simplified section.

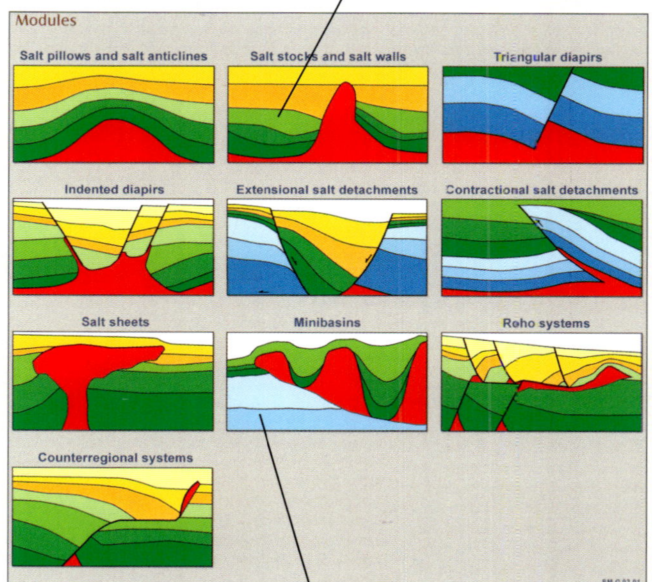

Clicking on a module links to the module page.

FEATURES OF THE DVD

Module Page. The module page is the gateway to all images illustrating a particular structural style. Users can link to pages of thumbnails showing module contents sorted by geographic area or model type.

User's Guide. *The Salt Mine* User's Guide is a graphical tour to navigate the DVD. Instructions are simple and brief, with minimal text.

E

Module page

Summary of structural style.

Module contents organized by geographic location.

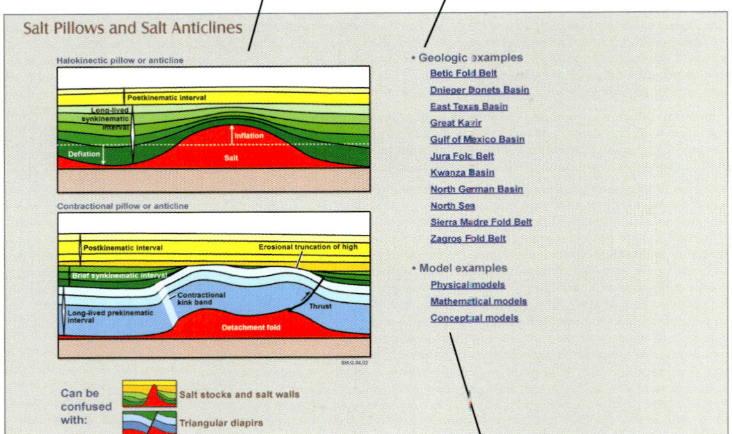

Links to other structural styles similar enough to cause confusion.

Module contents organized by model type.

F

User's Guide main menu

Clicking on a page type links to an enlarged version that explains how each link on the page works.

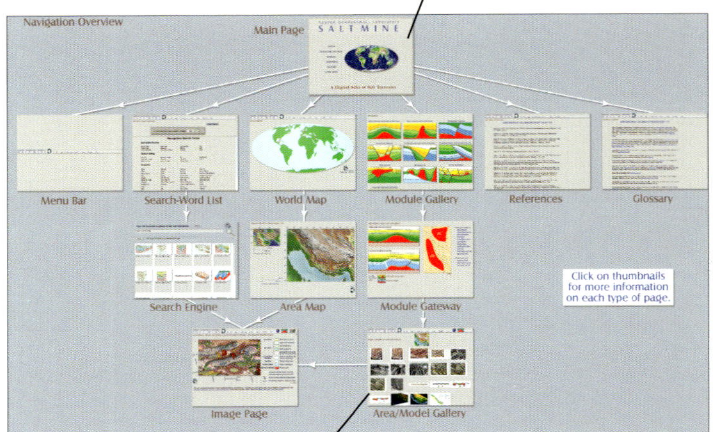

Main menu shows links between all types of pages in the DVD.

INTRODUCTION

Three large challenges face any compilation like *The Salt Mine*: selection, classification, and annotation.

In selection, the problem is how to decide which images to include and exclude. Any such project inescapably reflects the experience and background of the compilers. For example, our consortium has in the last decade conducted major research projects in the Gulf of Mexico and offshore West Africa, and the image selection reflects this focus. We have tried to widen the scope of *The Salt Mine* and include images drawn from published literature in other basins. However, scientists with more expertise and familiarity with the literature in these basins could no doubt improve our choices. Furthermore, we compiled chapters one at a time and did not revisit them after completion, apart from adding material from our own research group. Some chapters are therefore more up-to-date than others. For example, recent books by Mohriak and others (2008) and Brun and Fort (2008) were published after the relevant chapters in *The Salt Mine* were finished, so they are not included in our atlas. *The Salt Mine* is thus a personal survey of salt tectonics and would have a different mix had it been compiled by other authors.

In classification, the challenge is how to organize the selected pieces into manageable groups. As shown in the glossary, salt-structure nomenclature is a tangle of genetic and nongenetic terms, proposed by dozens of workers over more than 100 years. Because we are unaware of any published classifications covering all types of salt structures, we had to experiment for this atlas. Our goal was to try to keep classification entirely geometric so that the novice would easily classify a form without needing to know how it evolved kinematically or why it formed mechanically.

We were only partly successful in achieving this goal, while facing three problems inherent in geometric classification of images in *The Salt Mine*. First, individual images commonly feature more than one type of structure. In some images, we ignored a dominant structure and focused on a rare one. Second, individual structures commonly evolve through different stages and structural styles. For example, a diapir can be initiated as a triangular diapir, then pierce to the surface to form a salt wall, then extrude to form a salt sheet, which can then be reactivated to form a network of subsiding minibasins and rising salt walls. How should such a structure be classified? Third, many structures are hybrids because either successive growth stages are superposed or two processes act together. For example, a roho system can evolve into a counterregional system or a passive diapir can continue to grow passively after regional extension begins. Because of these three problems, we ended up classifying individual images by using a mixture of geometric and genetic criteria. This classification is naturally inconsistent and idiosyncratic, but we found it to be the most practical.

Finally, after selecting and classifying the images, we needed to annotate images and write captions about the structures. A decision here was what to focus on in each figure. Rather than attempting a complete analysis of every structure, we wrote about what we thought was most interesting, unique, or particularly well illustrated in each figure. Some captions focused merely on geometries; others include advanced insights on kinematics and dynamics. Because *The Salt Mine* is an atlas and not a textbook, its information is not presented in a logical sequence. However, the range of topics covered in figure captions is extremely broad, and many of the topics that would be covered in a salt-tectonics textbook are presented here. We hope that novices and experts alike will find *The Salt Mine* a useful source of analogs that stimulate new insights into the geometry, origin, and evolution of salt structures.

Chapter 1

Salt Pillows and Salt Anticlines

The relatively simple geometry of salt-cored folds makes them an ideal introduction to the world of salt tectonics. These types of folds are abundant, easily recognized, and present in all salt-tectonic basins.

Salt-cored anticlines are upwellings of salt having concordant overburden. In map view, structural domes having a length-to-width ratio of less than two are termed salt pillows. Folds having a ratio of two or more are called salt anticlines. In profile, salt anticlines and salt pillows are indistinguishable.

The growth histories of salt pillows and salt anticlines can be inferred from thickness variations in overlying sediments. Analysis of many pillows and anticlines suggests that they grow in two main ways. In the first type, a thin to nonexistent prekinematic section is overlain by a synkinematic section that accumulated over a long time. This pattern indicates halokinetic growth, in which salt flows as a result of gravitational forces. In the second type, the synkinematic phase is relatively brief, and synkinematic sediments are sandwiched between prekinematic and postkinematic sequences of much longer duration. These contractional folds are induced by regional compression, which tends to occur in pulses, unlike gravitational loading, which acts continuously. Contractional pillows and anticlines typically have more abrupt stratigraphic thinning than does the halokinetic type and may also have secondary but diagnostic compressional structures such as thrusts and kink bands.

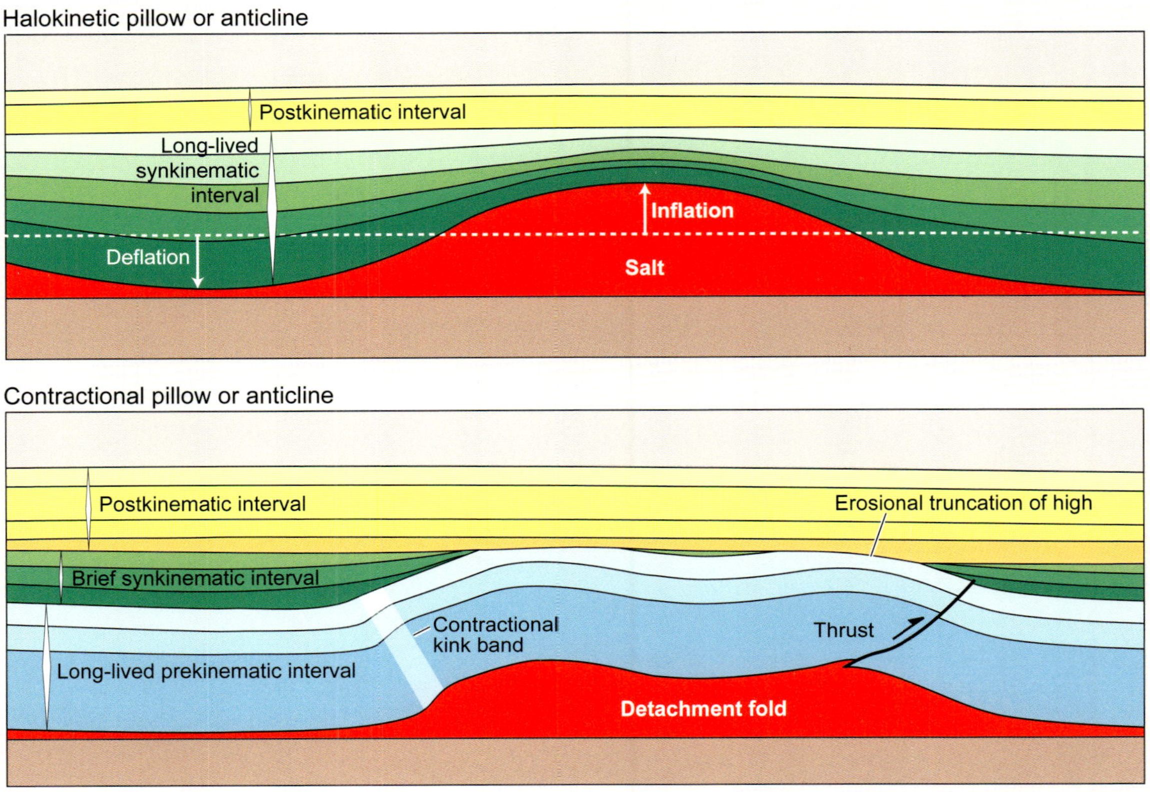

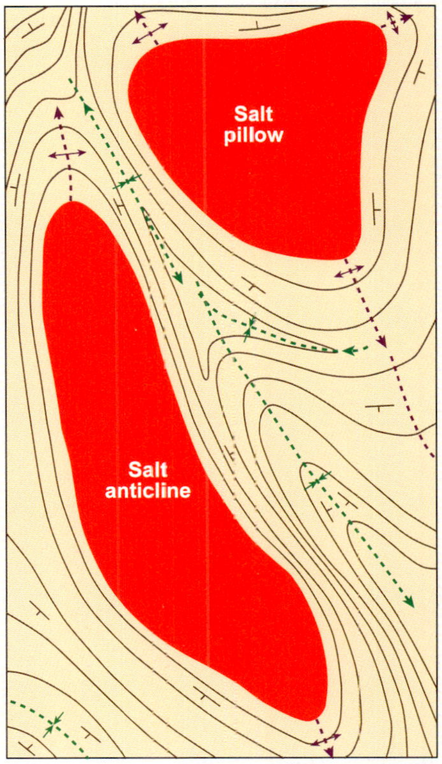

DNIEPER-DONETS BASIN—Compressional Salt Anticline, Ukraine

Figure 1.01. Growth of salt anticline during shortening. This salt anticline initiated during Early Permian shortening, and amplified during a second phase of shortening in the Late Cretaceous. Evidence of a compressional origin of both deformational episodes includes (a) crestal uplift and erosion, (b) a thick prekinematic roof, and (c) brief deformation events. The small normal fault over the crest of the structure is interpreted to result from crestal bending. Modified from Stovba and Stephenson (2003).

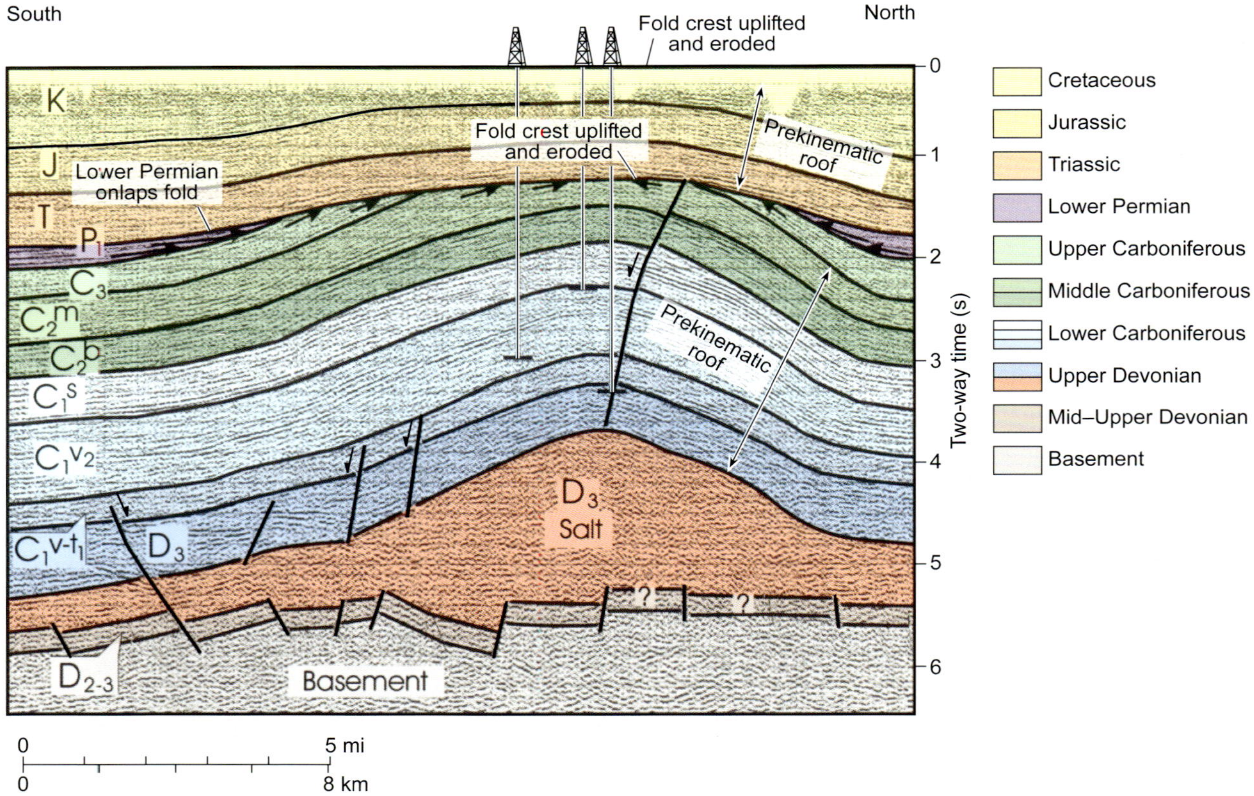

DNIEPER-DONETS BASIN—Breached Salt Anticline, Pripyat Trough, Ukraine

Figure 1.02. Growth of salt anticline during possible shortening. Early thinning onto the crest of the structure suggests that the diapir originated as a low-relief salt anticline shortly after Upper Devonian salt deposition. This anticline was then buried beneath a thick constant-thickness package (interkinematic section) before being reactivated and breaching its roof later in the Devonian. Signs that this reactivation was compressional are (a) the uplifted and eroded crest, (b) a thick interkinematic roof, and (c) a short-lived uplift. However, any shortening in the Late Devonian is counter to the consensus that the Late Devonian was a time of extension, volcanism, and basin formation. This anticline suggests a previously unrecognized tectonic event in the Pripyat Trough. Alternatively the bedding geometries are inaccurate. A second episode of shortening in the Permian further amplified the structure and caused renewed crestal erosion. Modified from Stovba and Stephenson (2003).

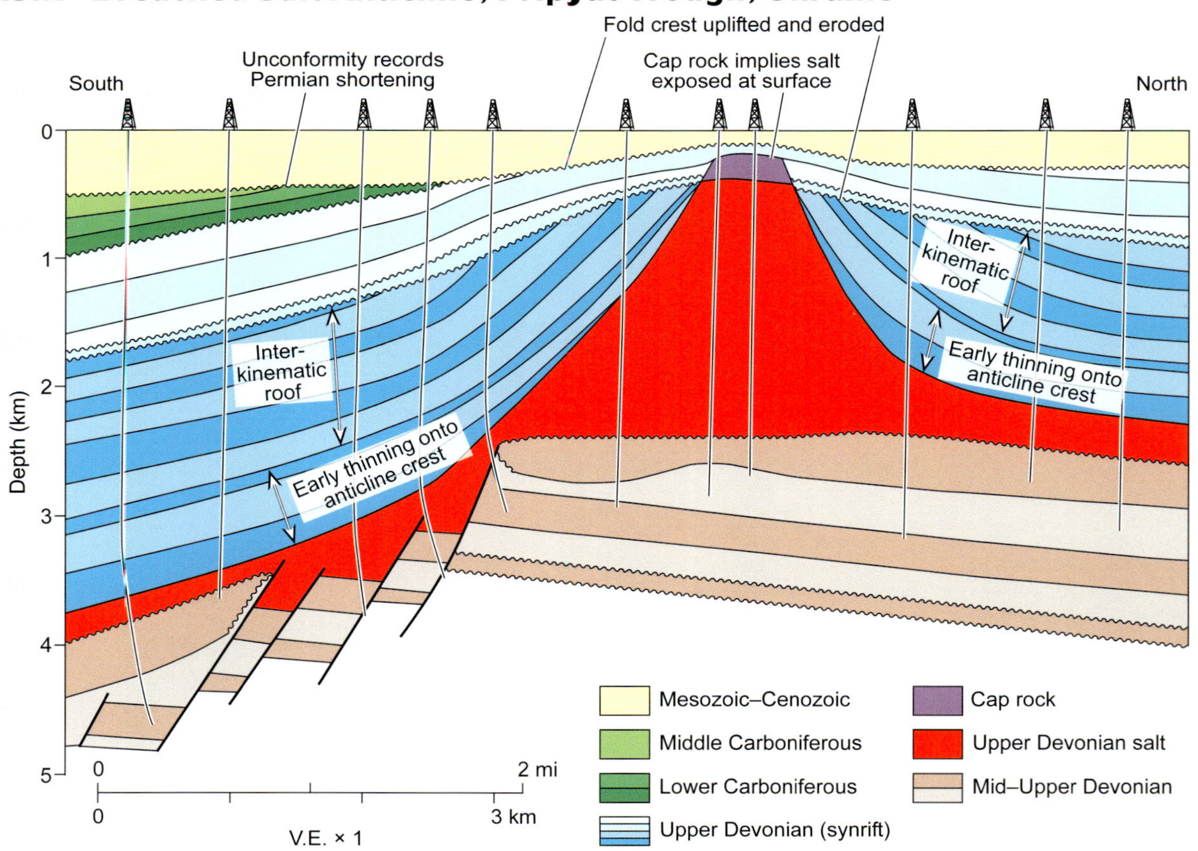

GREAT KAVIR—Domes and Basins, Iran

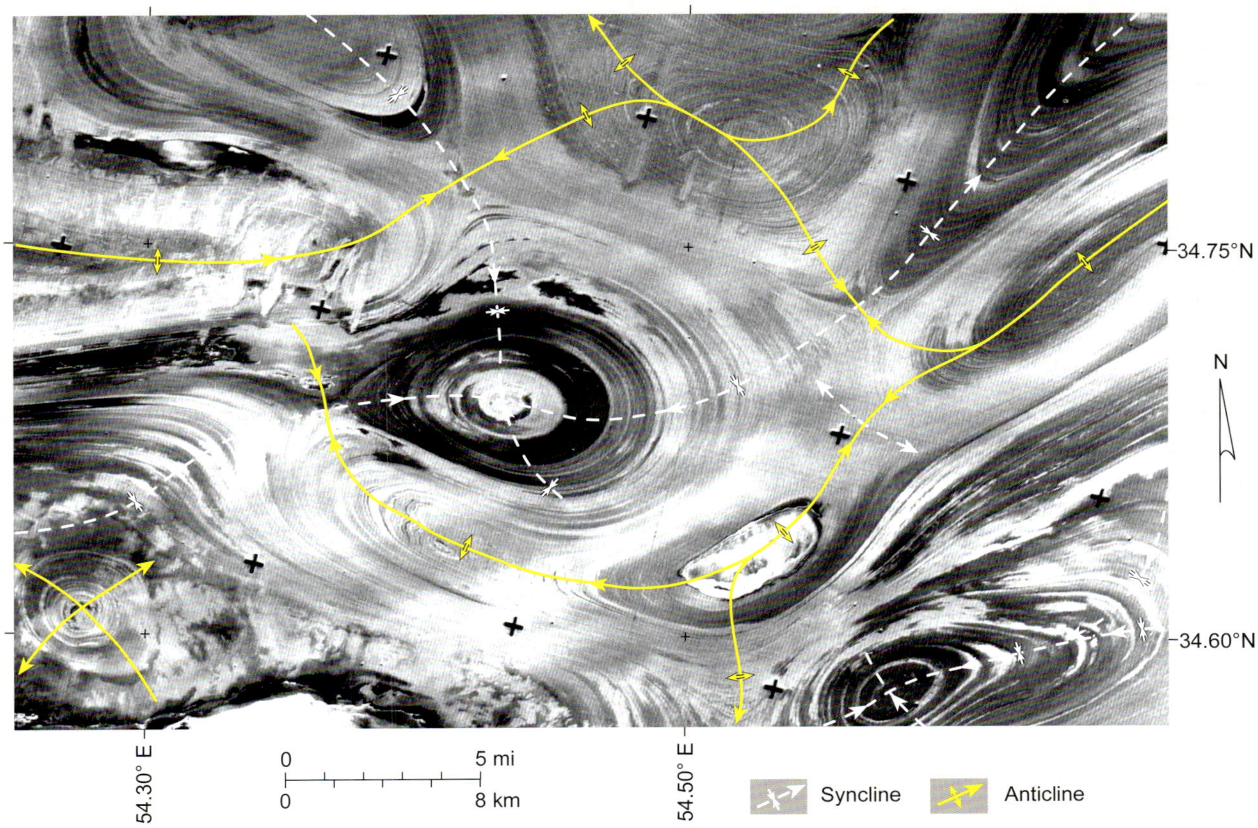

Figure 1.03. Fold interference patterns above salt. The interference structures in the Miocene Upper Red Formation probably detach concordantly on Miocene evaporites. Four domes are linked in an anticlinal ring around a central basin, a pattern atypical of dome-and-basin interference. The white ellipse in the south-center is the erosionally breached core of one of the structural domes. X-shaped saddles are exposed at the quadruple junction of two domes and two basins. Jackson and others (1990) interpreted this pattern as interference between NE-trending Alborz compressional folds and an earlier phase of halokinetic anticlines. Return Beam Vidicon image 83117706160XB supplied by NOAA. From Jackson and others (1990).

GREAT KAVIR—Domes and Basins Adjacent to a Transpressional Fault, Iran

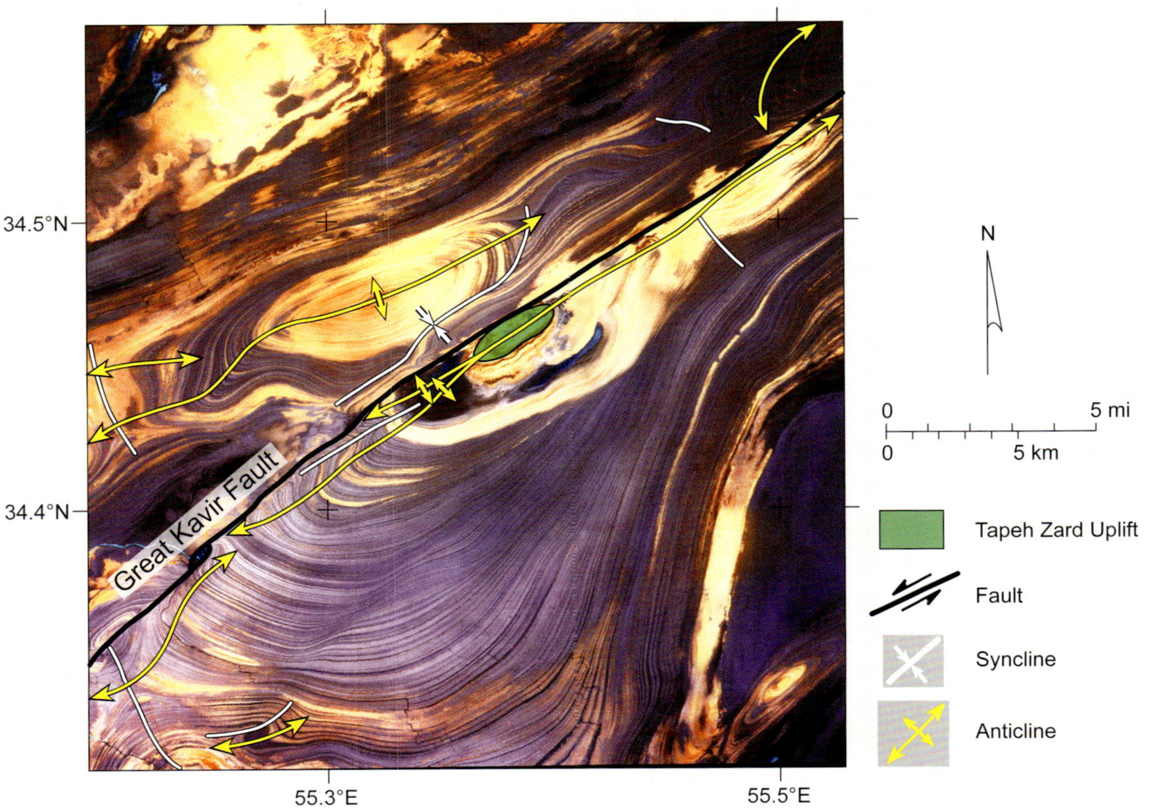

Figure 1.04. Transpressional salt anticlines. The 700-km-long transpressional Great Kavir Fault slices through folded Upper Red Miocene strata thought to detach on thin Miocene evaporites. The fault scarp to the southwest is 10 m high and upthrown to the north. Tapeh Zard [Yellow Hill] is cored by schist, representing the only pre-Cretaceous basement exposed in the central Kavir. The slight obliquity between the fault and salt-cored folds on its northwest side suggests left-lateral shear on the fault. Interpretation based on Jackson and others (1990). Landsat image courtesy of ConocoPhillips Remote Sensing Laboratory.

GREAT KAVIR—West-Central Part of West Dulasian Anticline, Iran

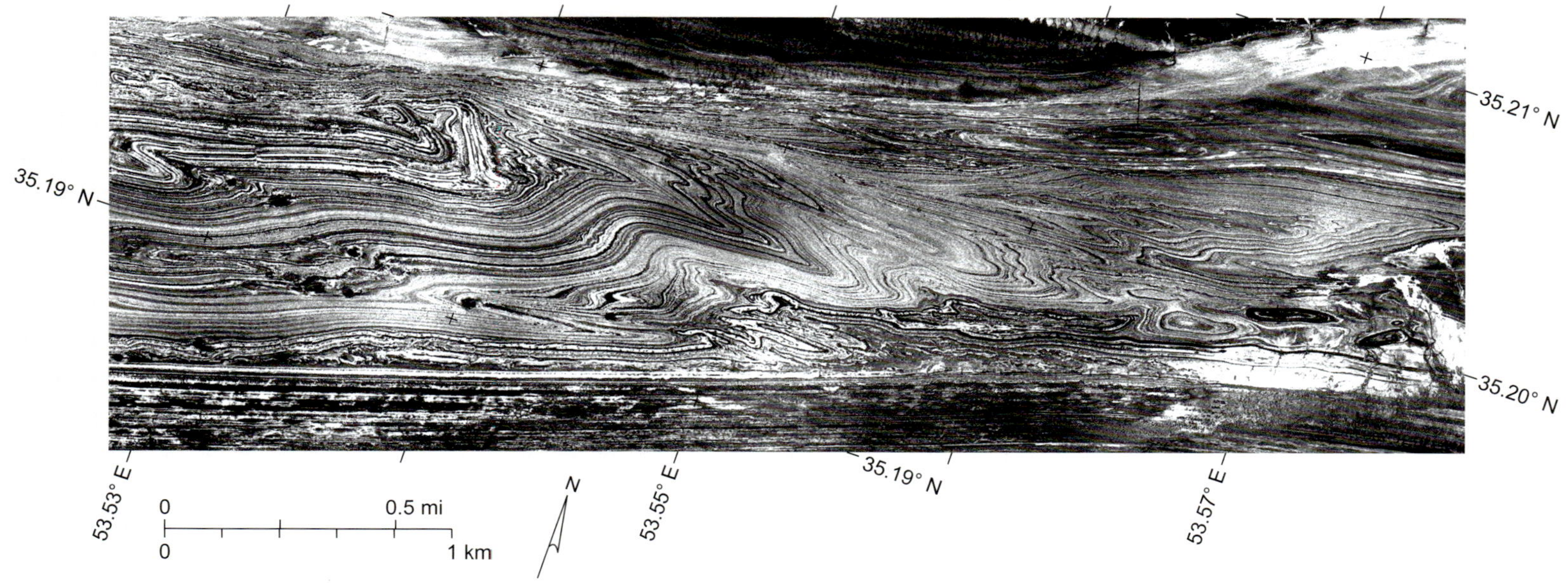

Figure 1.05. Internal folding in a compressional salt anticline. Interbedded halite, gypsum, and mudstone give Miocene evaporites a strong banding, which defines the internal structure of exposed evaporites in the core of the West Dulasian anticline. Similar-style flow folds include elliptical closed patterns in the southeast, which are probably sheath folds produced by extreme shear within the fold core. Farther west, the southern contact is concordant to the overburden but discordant to isoclinal folds within the anticline. The northern contact of the anticline core is obscured by modern playas. See structures farther east in Figure 1.06. From Jackson and others (1990).

GREAT KAVIR—East End of West Dulasian Anticline, Iran

Figure 1.06. Internal folding in a compressional salt anticline. Interbedded halite, gypsum, and mudstone give Miocene evaporites a strong banding, which defines the internal structure of exposed evaporites in the core of the West Dulasian anticline. Rounded folds are visible in oblique profile at the blunt east end of the evaporitic fold core. These folds die out upward and outward toward the contact shear zone of the fold core. Faults offset the Upper Red overburden in the southern fold limb. Black patches in the east are cloud shadows. Diagonal white lines are negative scratches. See structures farther west in Figure 1.05. Photograph by Worldwide Aerial Surveys, Inc.

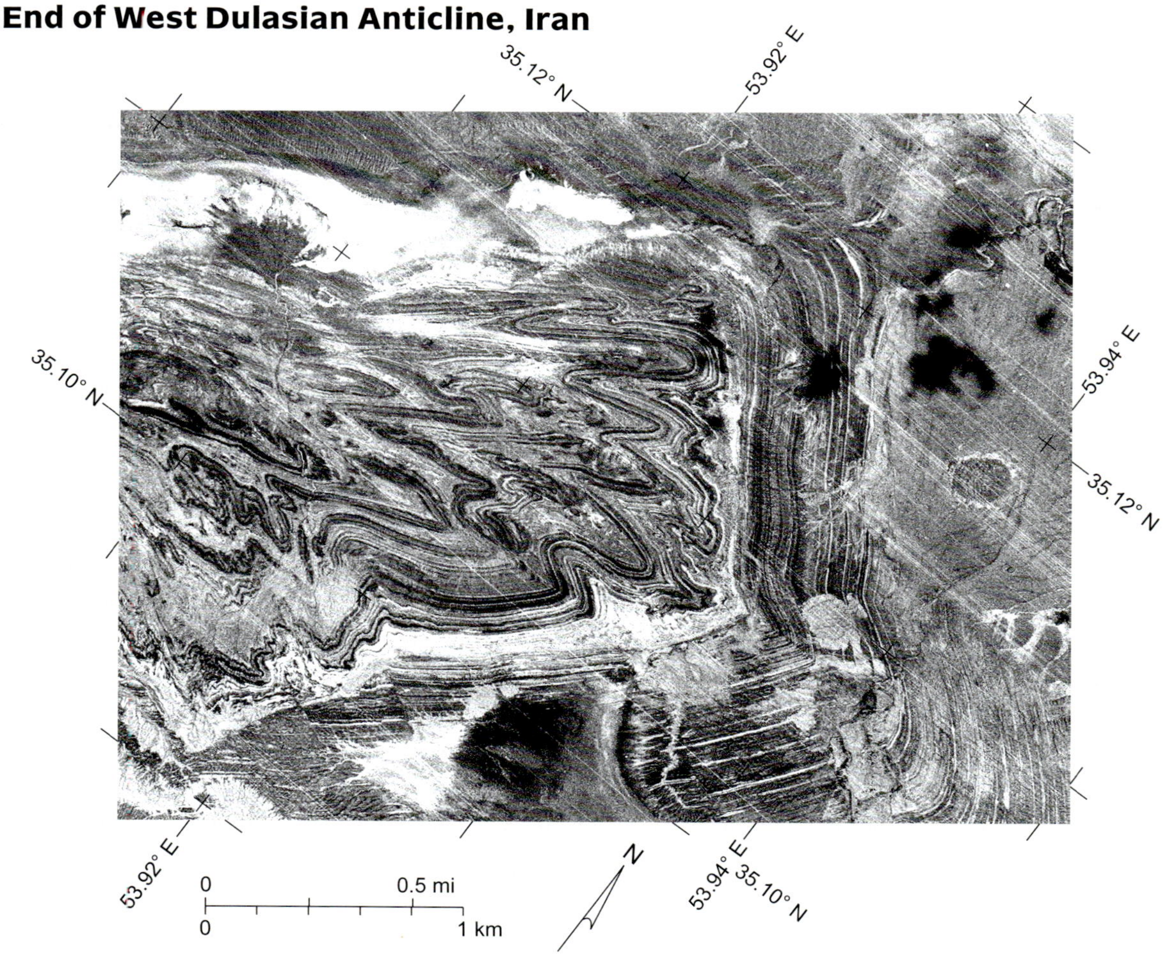

GULF OF MEXICO BASIN—Atwater Fold Belt, Offshore USA

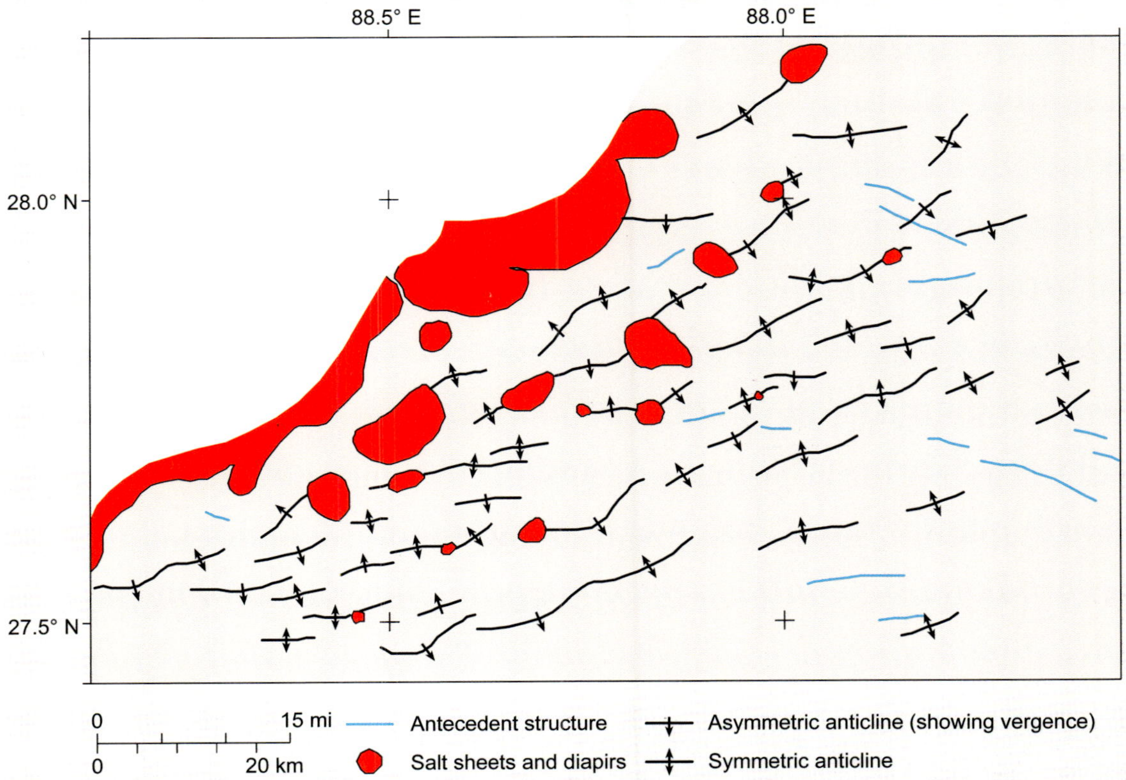

Figure 1.07. Growth of compressional toe structures. The Miocene-Pliocene Atwater fold belt formed at the downdip end of the northern Gulf of Mexico salt basin, near the seaward termination of salt. The fold belt projects under the complex of salt sheets at its northwest edge. Diapirs tend to occur along anticlinal axial traces, commonly near fold terminations or at changes in fold trend. The diapirs predated shortening and are interpreted to have nucleated many of the folds. As is typical over salt detachments, fold vergence is both landward and seaward. From Rowan and others (2000).

GULF OF MEXICO BASIN—Mad Dog & K2 Salt Anticlines, Atwater Fold Belt, Offshore USA

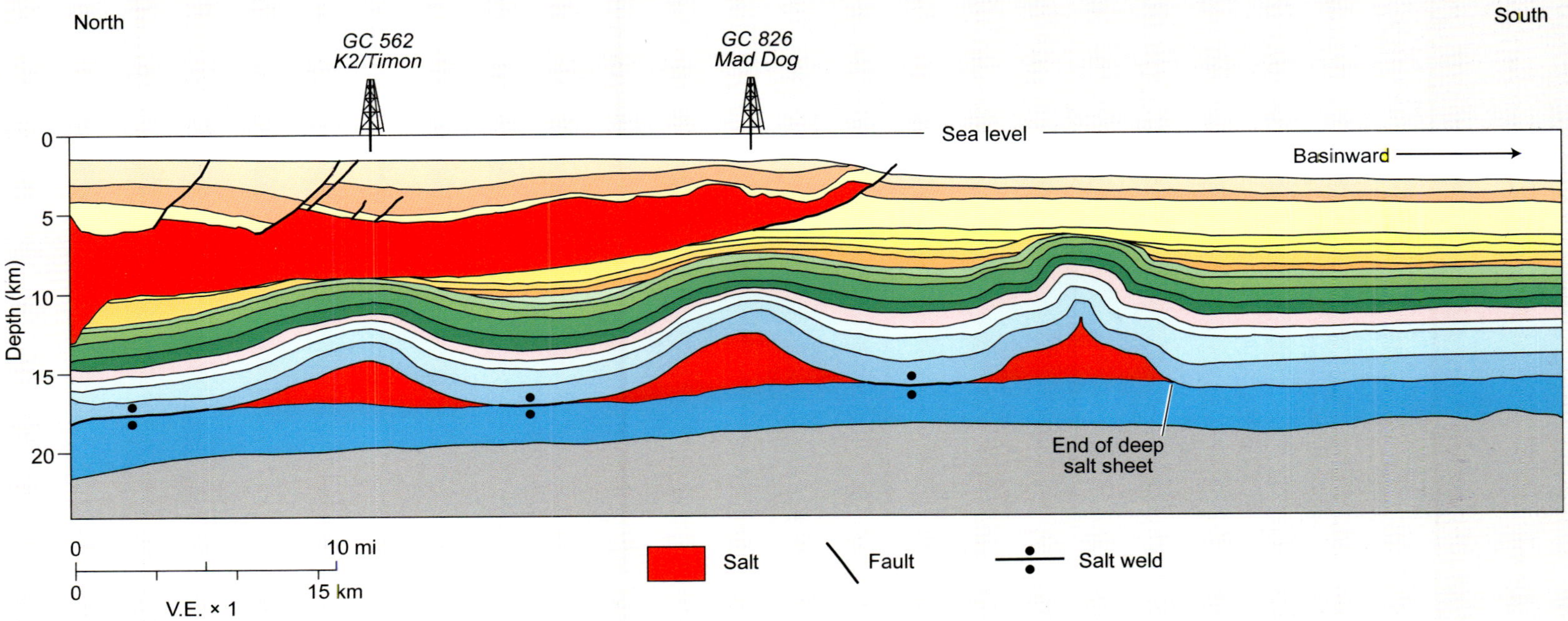

Figure 1.08. Compressional growth of salt anticlines. The Atwater Fold Belt is interpreted to have formed above a deep Mesozoic allochthonous salt sheet at the seaward end of the Louann salt basin (Peel, 2001). The frontal (southern) anticline overlies the seaward pinch-out of the deep salt sheet. This frontal fold continued shortening after the others had stopped, producing a box fold at the crest of an otherwise smoothly curved anticline. The fold belt is overridden by a shallower allochthonous salt sheet that ends in a roof-edge thrust. Modified from Rowan and others (2004).

GULF OF MEXICO BASIN—Compressional Salt Anticlines in Perdido Fold Belt, Offshore USA

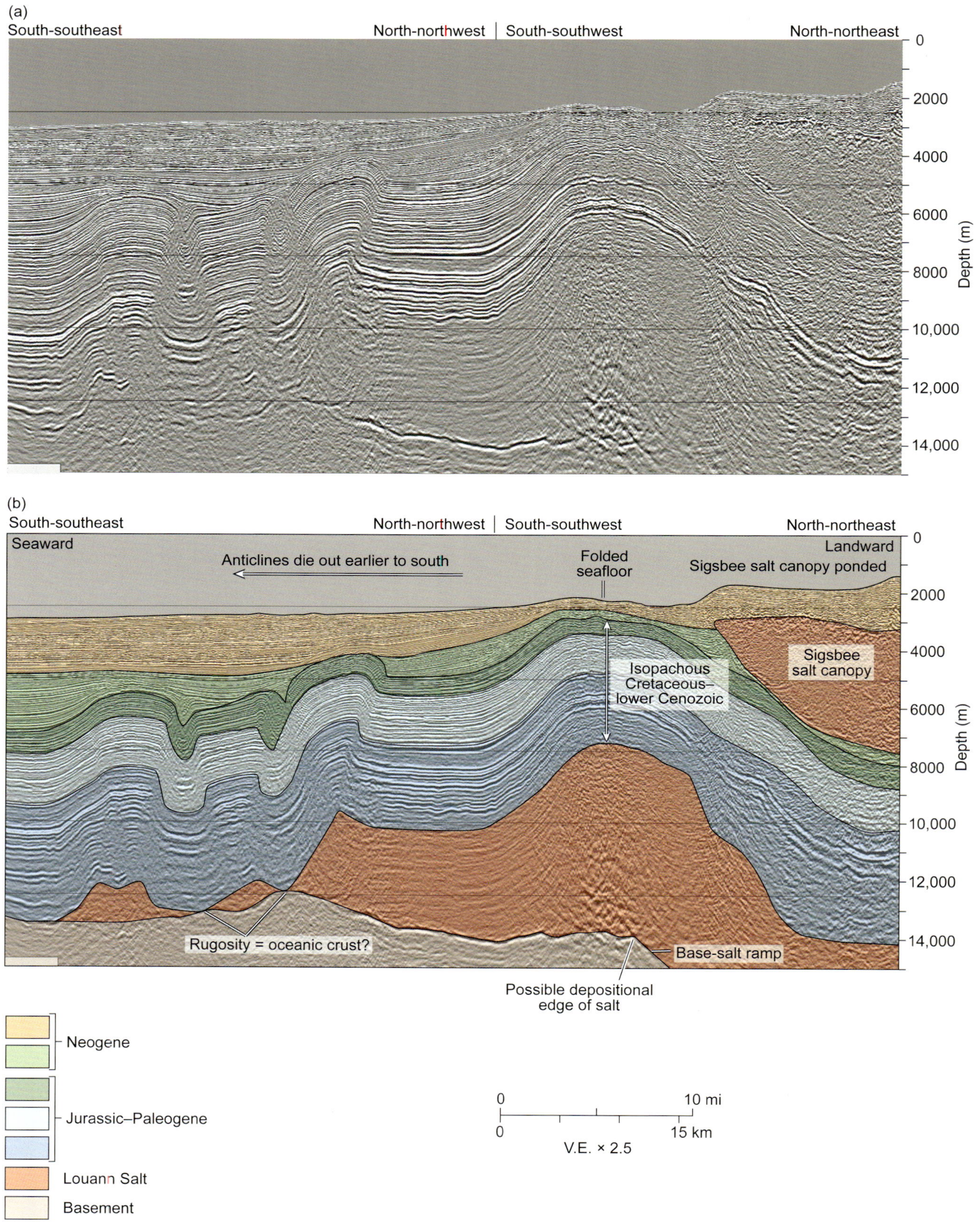

Figure 1.09. Compressional growth of salt anticline. The isopachous Cretaceous–lower Cenozoic section in the Perdido fold belt indicates that shortening did not begin until the Oligocene. Folds near the seaward end of the belt are lower amplitude and are cored by less salt, and they died out earlier than landward folds to the north. Folded seafloor indicates that shortening continues near the north end of the fold belt. The Sigsbee salt canopy was emplaced during growth of this northern fold and appears to have ponded against the back limb. A major base-salt ramp beneath the landward end of the fold belt may mark the original seaward pinch-out of the salt basin. If so, the salt seaward of the ramp was emplaced as an early allochthonous sheet. The base of salt beneath the Perdido fold belt is rugose, consistent with interpretation as oceanic or volcanic crust. Seismic courtesy of CGGVeritas.

GULF OF MEXICO BASIN—Halokinetic Salt Anticline, St. Malo Discovery, Offshore USA

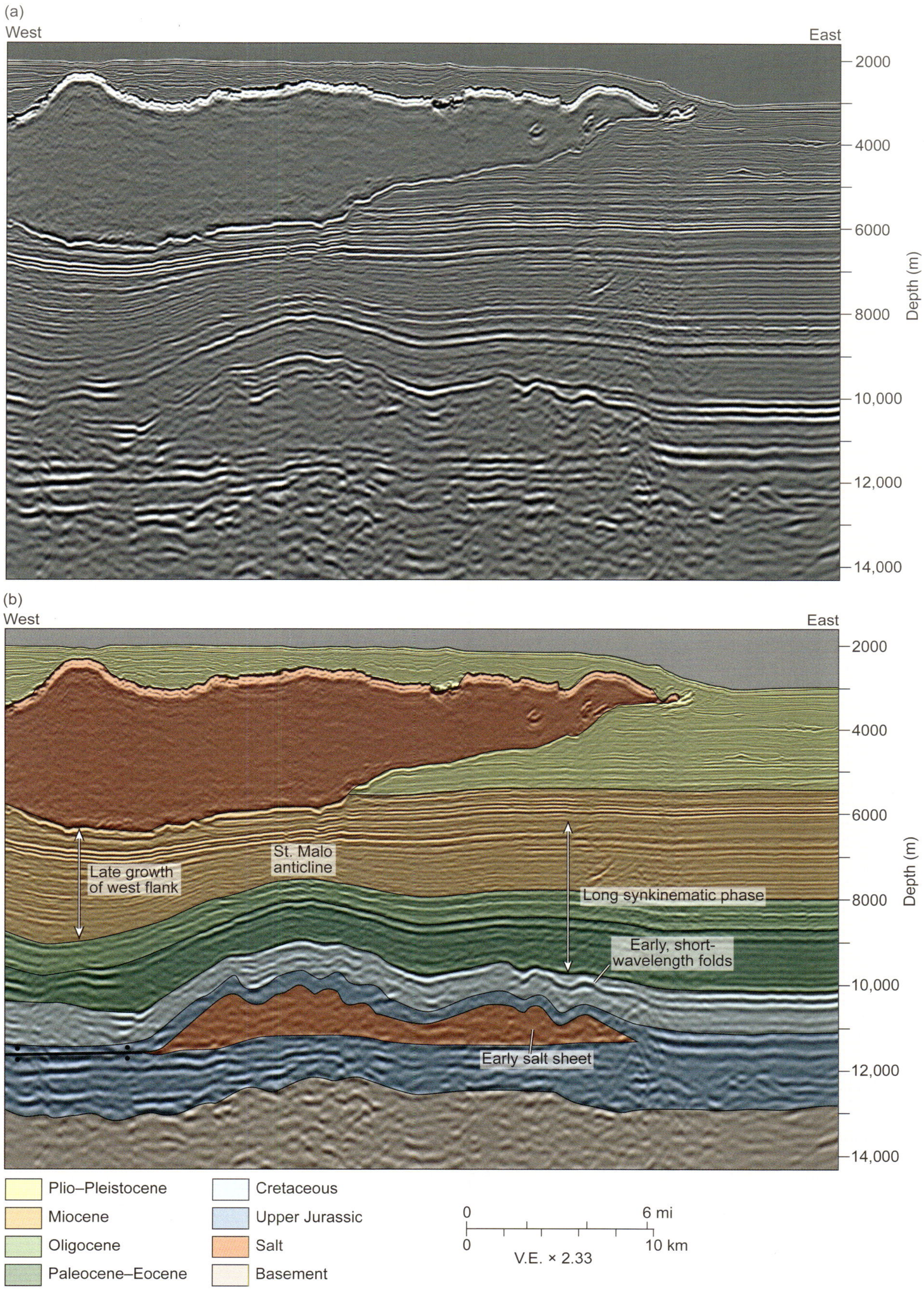

Figure 1.10 Halokinetic growth of salt anticline. The St. Malo salt anticline overlies an allochthonous salt sheet that extruded from the seaward end of the Louann salt basin. This sheet originally formed the compressional toe of the Gulf of Mexico gravity-spreading system. Shortening deformed the sheet roof into folds having a short wavelength because of the thin Mesozoic roof. Shortening stopped near the end of the Cretaceous, when the structure began long, slow growth. This later folding had a much longer wavelength than the Mesozoic folds because the roof had thickened in the Paleogene. Such a long synkinematic phase (Paleocene–Miocene) is more characteristic of halokinetic deformation than shortening, so the second phase of anticline growth is interpreted as halokinetic. Consistent with this interpretation is the fact that limb rotation was diachronous. The west flank subsided after deformation on the east flank had mostly ceased. Seismic courtesy of CGGVeritas.

GULF OF MEXICO BASIN—Compressional Salt Anticline on Allochthonous Sheet, Kansas Discovery, Offshore USA

Figure 1.11. Compressional growth of salt anticline. After seafloor spreading began in the Late Jurassic, tilting of the basin caused salt to flow southeast, extruding a thin salt sheet over the abyssal plain. After an early phase of shortening that formed small thrust faults above this sheet, the Kansas anticline remained inactive during the Paleogene, leading to deposition of a thick isopachous section over the future fold. This isopachous section was folded in the late Miocene, when renewed translation along the base-salt detachment produced the Kansas anticline at the toe of the system. Modified from Mount and others (2007).

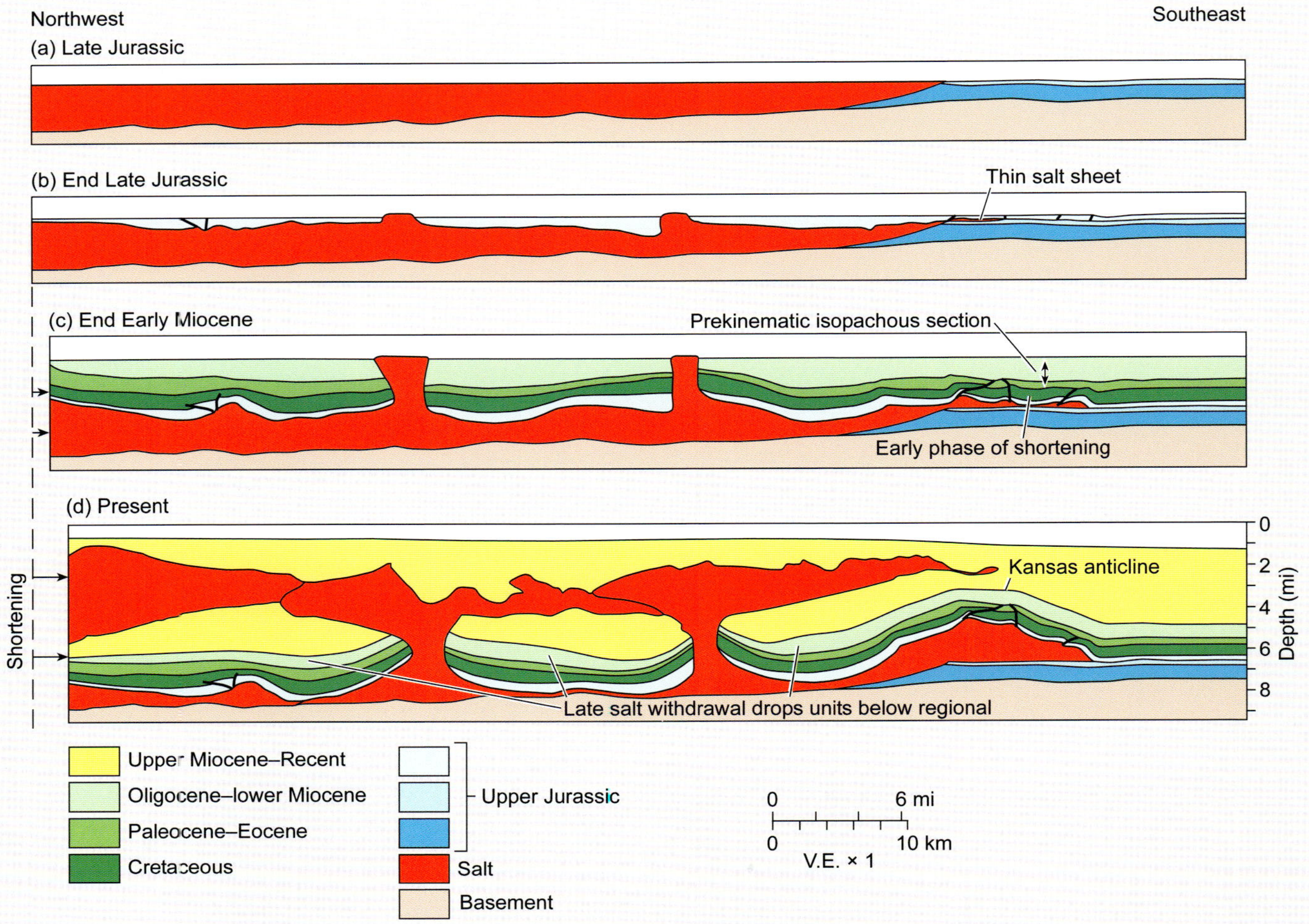

JURA FOLD BELT—Weissenstein Anticline, Switzerland

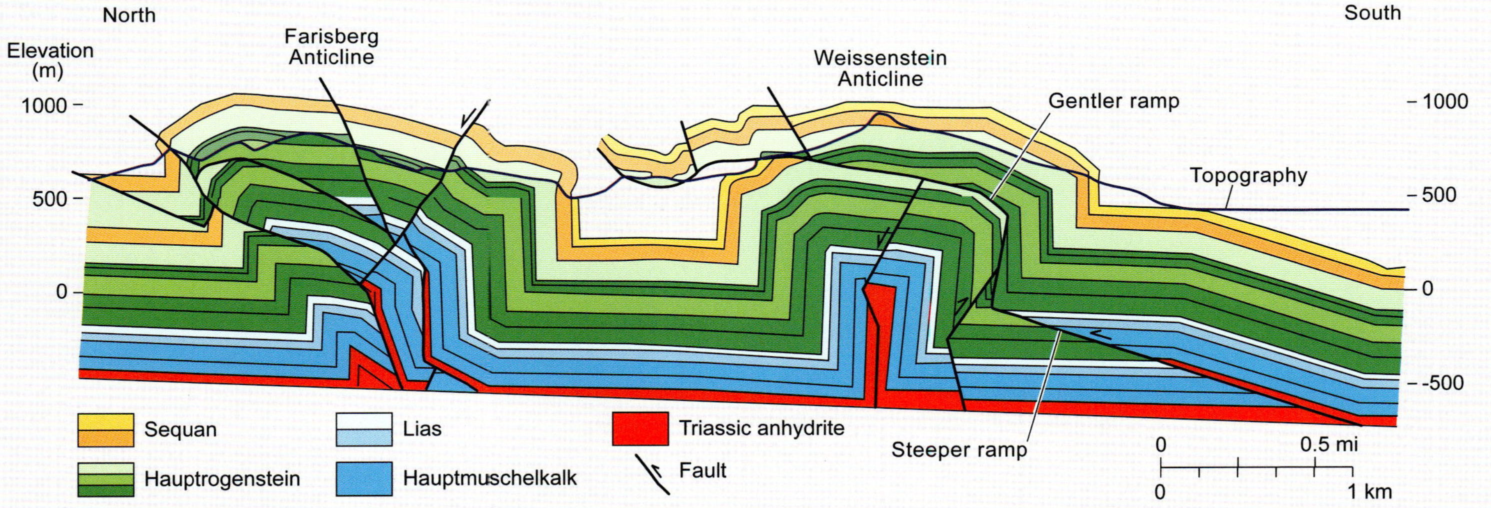

Figure 1.12. Shortening above thin salt. The classic box-fold profile of the Weissenstein anticline is offset by a large folded thrust, which ramps upward from the detachment of thin anhydrite to the Lias level then flattens to a much gentler ramp upward to the Sequan level. The same folded thrust is shown in a nearby cross section in Figure 1.13. This section is based on the box-fold concept of Meier (1977). From Bitterli (1990).

JURA FOLD BELT—Weissenstein Anticline (Dunnern Gorge), Switzerland

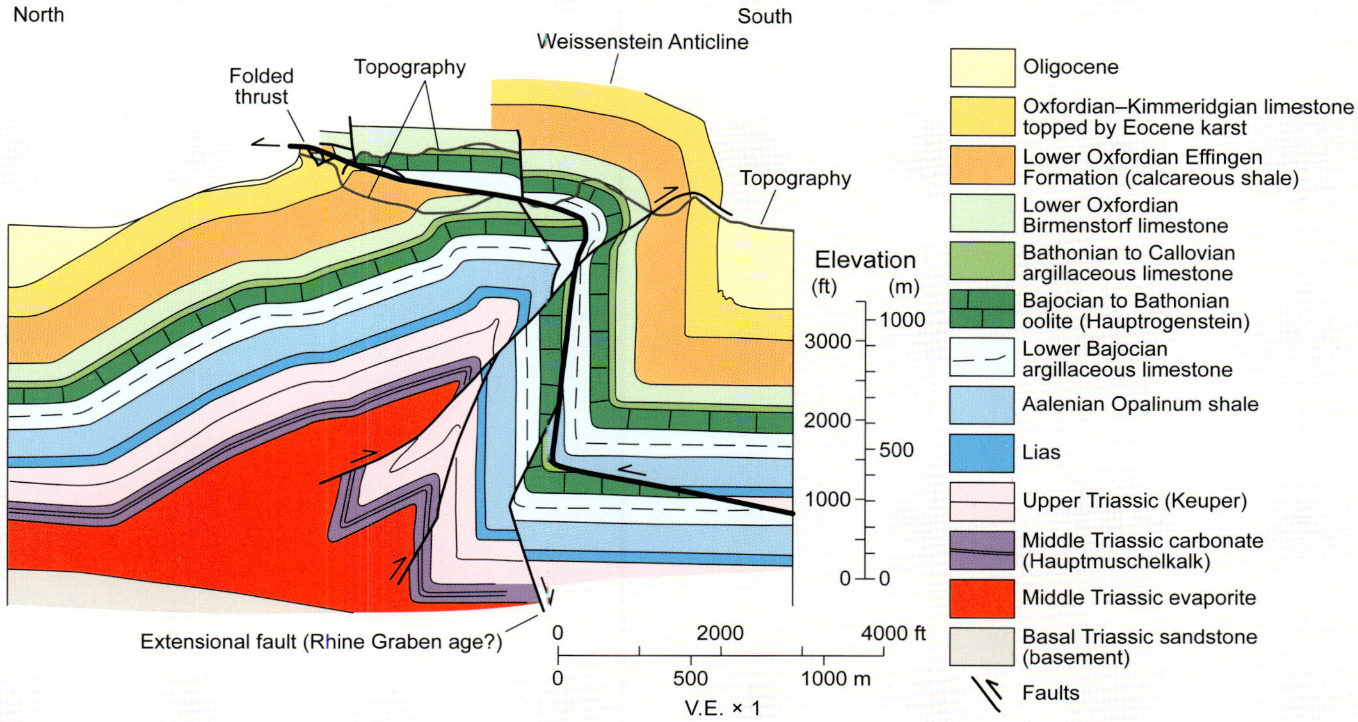

Figure 1.13. Shortening above thin salt. This southward-verging anticline is cut by (1) folded thrusts created during Jura shortening and (2) younger extensional faults partly reactivated by sinistral strike slip. The main folded thrust is the same as that in the nearby cross section in Figure 1.12. From Meier (1977) and Laubscher and Bernoulli (1980).

KWANZA BASIN—Polyphase Salt Anticlines, Offshore Angola

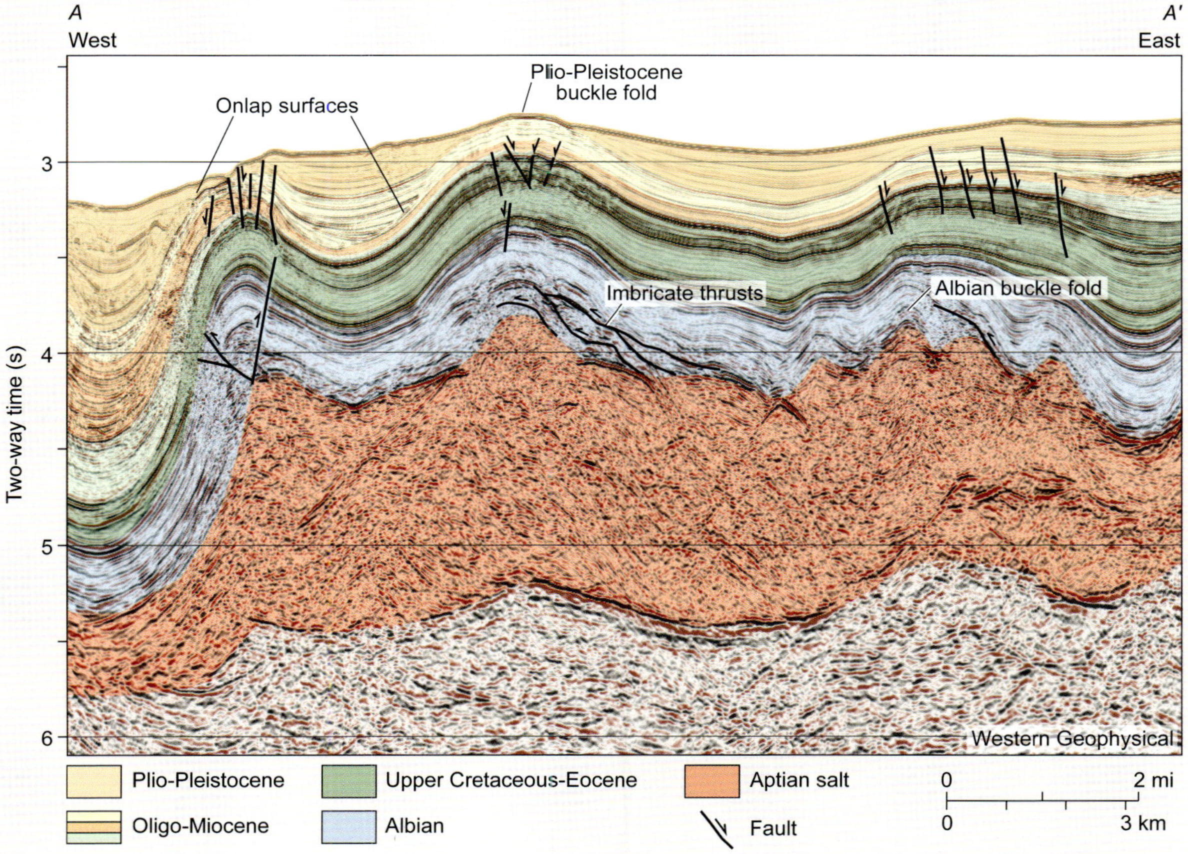

Figure 1.14. Shortening above thick salt. Albian contraction is recorded by small buckle folds and imbricate thrust faults. After subsequent tectonic quiescence, Cenozoic growth folds detached on Aptian salt inflated by internal flow. The Cenozoic section is condensed above this salt plateau. Neogene timing of buckling is recorded by onlap and thinning against the anticlines. Seismic image courtesy of WesternGeco. Interpretation by Martin Jackson and Mike Hudec.

KWANZA BASIN—Reactivated Salt Anticlines, Offshore Angola

Figure 1.15. Shortening above thick salt. Albian contraction is recorded by small buckle folds. Cenozoic growth folds detached on Aptian salt inflated by internal flow. The Cenozoic section is condensed above this salt plateau. At right, a reactive diapir grew throughout the Late Cretaceous before being inverted by compression in the Plio-Pleistocene. At left is a long-lived minibasin. Seismic image courtesy of WesternGeco. Interpretation by Martin Jackson and Mike Hudec.

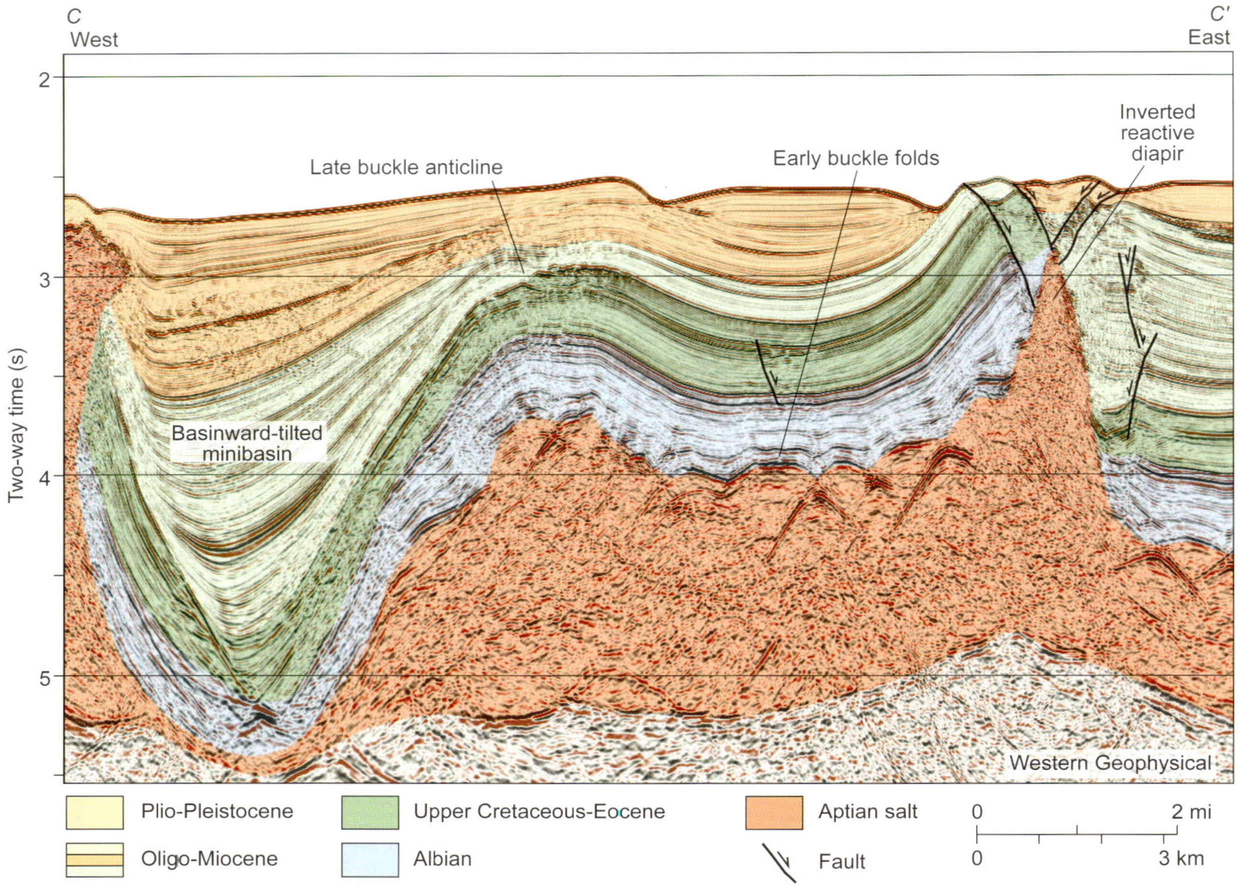

KWANZA BASIN—Salt Anticlines, Angola

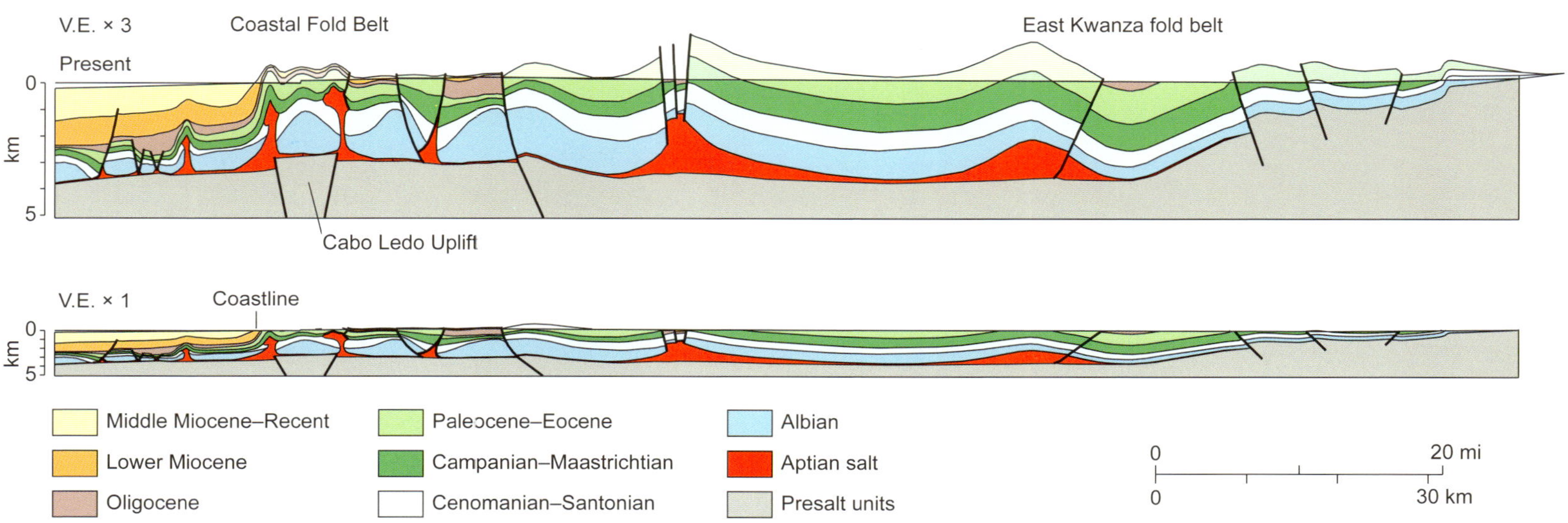

Figure 1.16. Shortening related to basement inversion. Tertiary basement inversion is recorded by basement-involved thrusting in the East Kwanza fold belt and by Tertiary rise of the Cabo Ledo Uplift. Above the salt, the same shortening episode produced gentle contractional folds in the center of the section and squeezed diapirs in the Coastal fold belt. After Hudec and Jackson (2002).

NORTH GERMAN BASIN—Growth of a Turtle Structure next to a Salt Anticline, Düste Oil Field, Germany

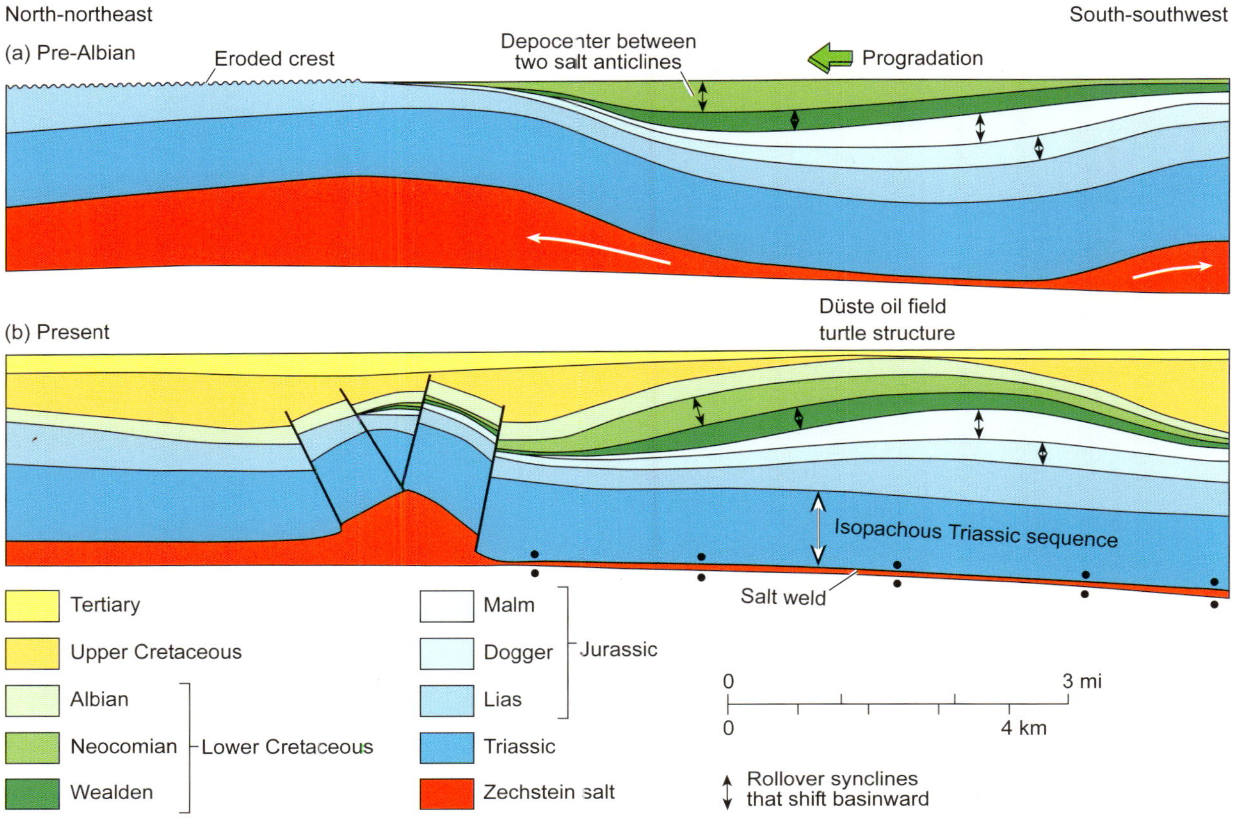

Figure 1.17. Subsidence of salt anticlines. The pre-Albian restoration shows an Upper Jurassic–Lower Cretaceous depocenter between two salt anticlines. Welding of salt below this depocenter caused subsequent salt evacuation to shift to the two flanking anticlines, which subsided in the Late Cretaceous. This subsidence left the older depocenter as a structural high, forming an anticlinal turtle structure. The turtle structure rests on a thick, nearly isopachous Triassic sequence. The delayed initiation of the turtle probably reflects Jurassic progradation of sediments, which loaded the underlying salt. This differential load caused the uneven evacuation that eventually formed the turtle structure. Modified from Trusheim (1960).

NORTH GERMAN BASIN—Salt Anticline Formed during Shortening, Barrien Salt Anticline, Germany

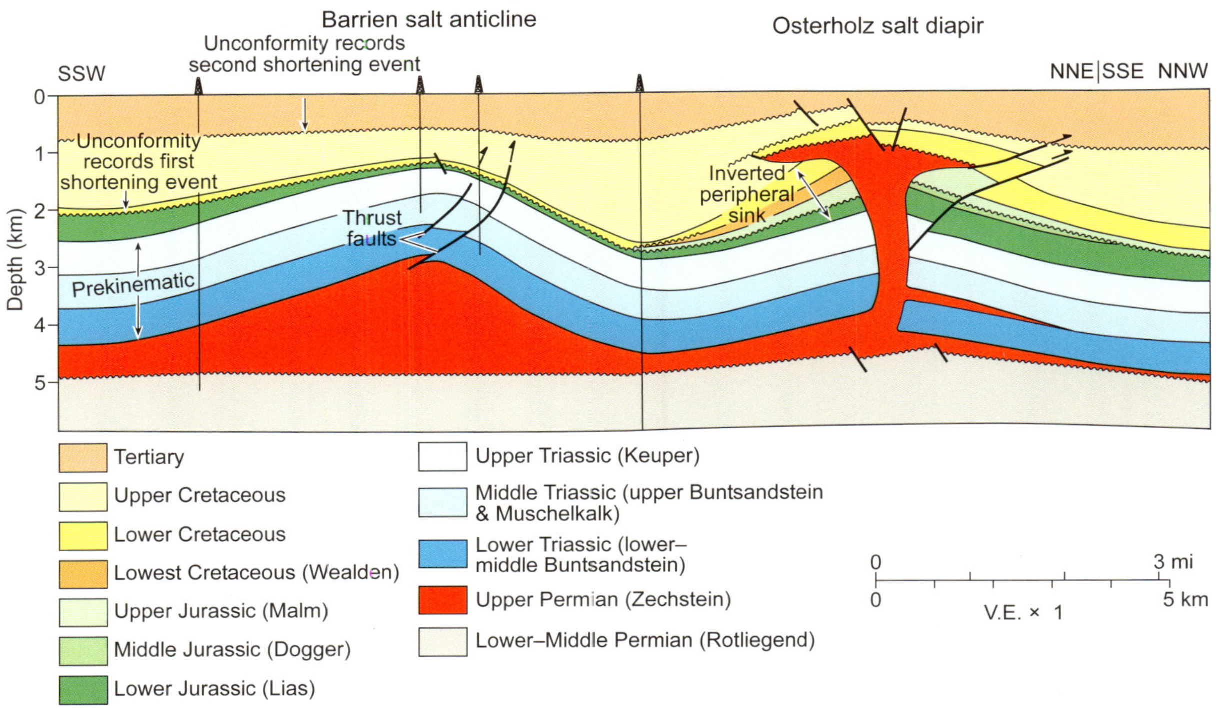

Figure 1.18. Compressional growth of salt anticline. Strata are isopachous near the Barrien and Osterholz salt structures throughout the Triassic and probably into the Middle Jurassic (the age of a major unconformity). These formerly flat lying beds indicate that the salt anticline probably did not begin growing until the Middle Jurassic. Folding of a thick, strong, prekinematic section at Barrien suggests that the Middle Jurassic initiation of the anticline was compressional (as in the Jerstedt salt anticline in Figure 1.19). A compressional origin is consistent with the small thrust faults in the crest of the anticline. Further shortening during the Late Cretaceous–early Cenozoic amplified both the anticline and diapir and formed thrusts on the flank of the Osterholz diapir. Modified from Baldschuhn and others (2001).

NORTH GERMAN BASIN—Salt Anticline Formed during Shortening, Jerstedt Salt Anticline, Germany

Figure 1.19. Compressional growth of salt anticline. Structures adjacent to the Harz basement uplift formed during two phases of shortening during the Middle Jurassic and Late Cretaceous–early Cenozoic. The isopachous Triassic–Middle Jurassic units indicate that neither the Jerstedt salt anticline nor the Salzgitter diapir existed prior to the Middle Jurassic. Middle Jurassic shortening created the Jerstedt anticline, as recorded by erosion of Triassic and Lower Jurassic units over its crest (compare with the Barrien salt anticline in Figure 1.18). Continuity of Lower Jurassic units across the crest of the Salzgitter diapir indicates that this structure was only slightly shortened then. By contrast, Late Cretaceous shortening preferentially arched the roof of the Salzgitter diapir, with only slight reactivation of the Jerstedt anticline. Modified from Baldschuhn and others (2001).

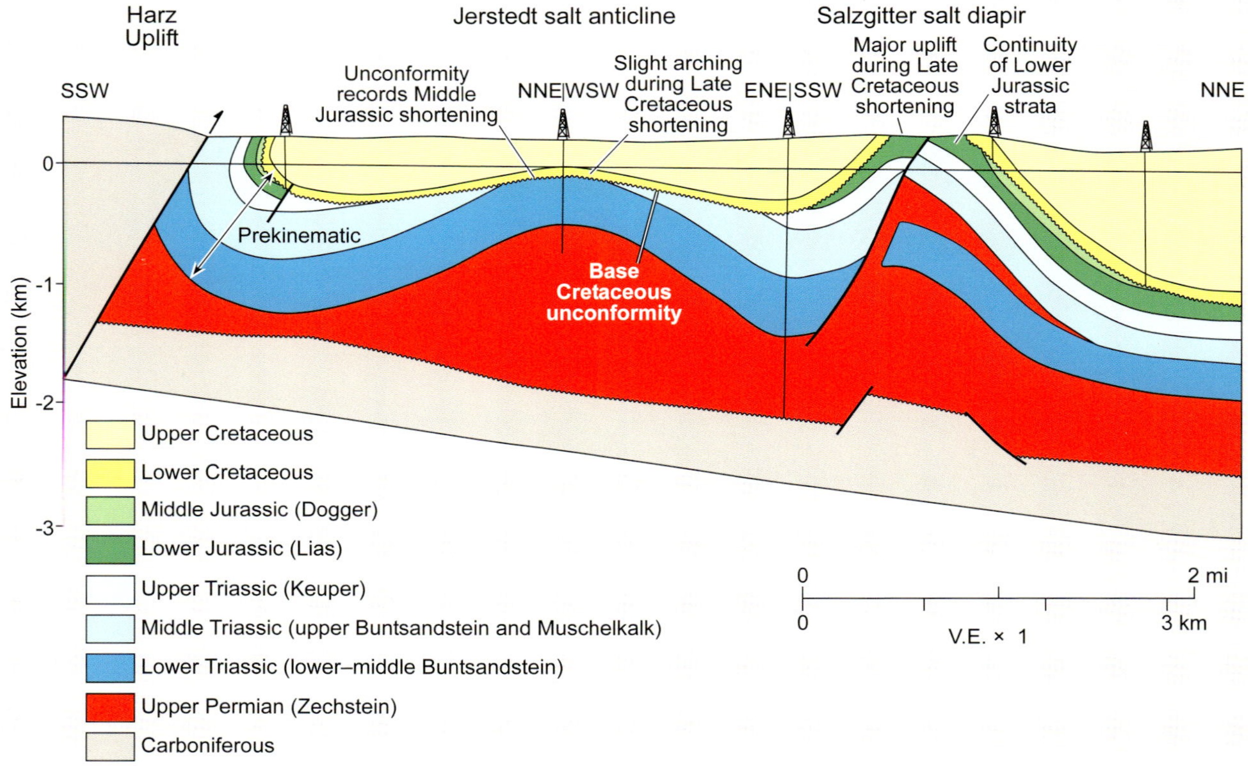

NORTH SEA—Salt Anticline, Sole Pit

Figure 1.20. Repeated reactivation of a salt anticline. Where preserved, Triassic–Jurassic units are isopachous across the area, indicating deposition during structural quiescence. The base Cretaceous unconformity constrains the end of the main phase of shortening and uplift. Gravity-driven extension off the northeast flank of the anticline detached on a thin Triassic salt layer. The anticline overlies an interpreted basement horst, a relationship common in the North Sea. A second phase of shortening occurred in the late Tertiary, as indicated by another unconformity centered on the same salt anticline. From Owen and Taylor (1983).

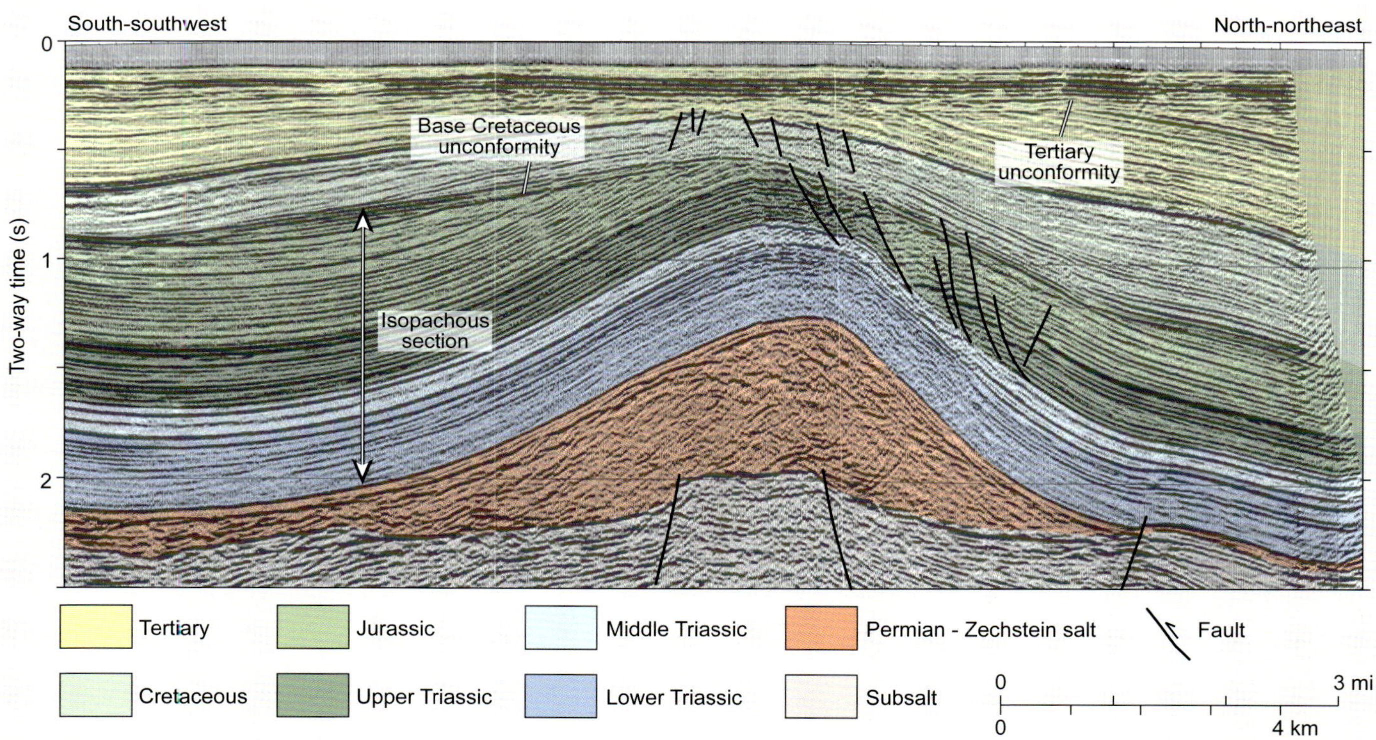

NORTH SEA—Salt Anticline above Basement Reverse Fault, Offshore Norway

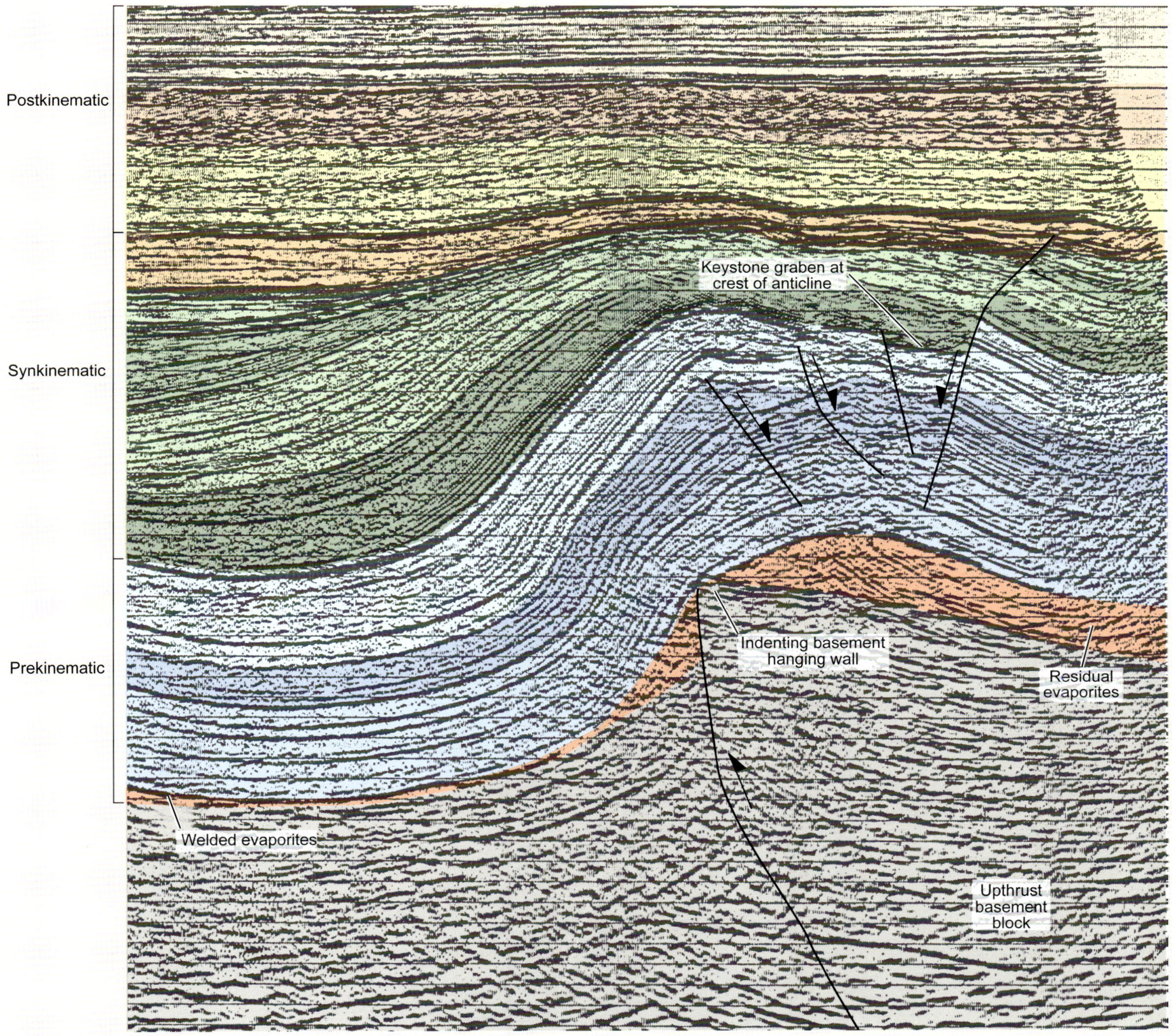

Figure 1.21. Folding above thin salt during basement shortening. Reverse slip on an ill-defined fault zone in the basement formed an asymmetric, salt-cored drape fold in the overlying cover. The cover detaches on evaporite, probably coeval with the Zechstein (Upper Permian) salt. The time of shortening is recorded by the synkinematic interval. The synkinematic depocenters migrated progressively farther (to the left) from the basement fault. As the cover thickened, its flexural strength increased, so the fold flexed on a longer wavelength. The crestal graben in the prekinematic interval formed by local stretching in the outer arc of the draped cover. The apex of the basement upthrown block impinged against the base of the prekinematic overburden and slightly kinked it. The base of the postkinematic interval records the end of major contraction. Exact location and scales are unknown. Interpretation by Martin Jackson. Seismic data courtesy of Total.

SIERRA MADRE FOLD BELT—Salt Anticlines, Monterrey Salient of Sierra Madre Oriental Fold Belt, Mexico

Figure 1.22. **Shortening above thin salt.** Curved forested ranges in the Monterrey salient define compressional anticlines. The east and west edges of the mountain belt coincide with the depositional edges of the anhydrite-gypsum layer on which the folds are detached. Landsat TM, band combination R(7), G(4), B(1), which approximates natural colors. Image courtesy of BHP Billiton.

ZAGROS FOLD BELT—Regional Geology, Iran

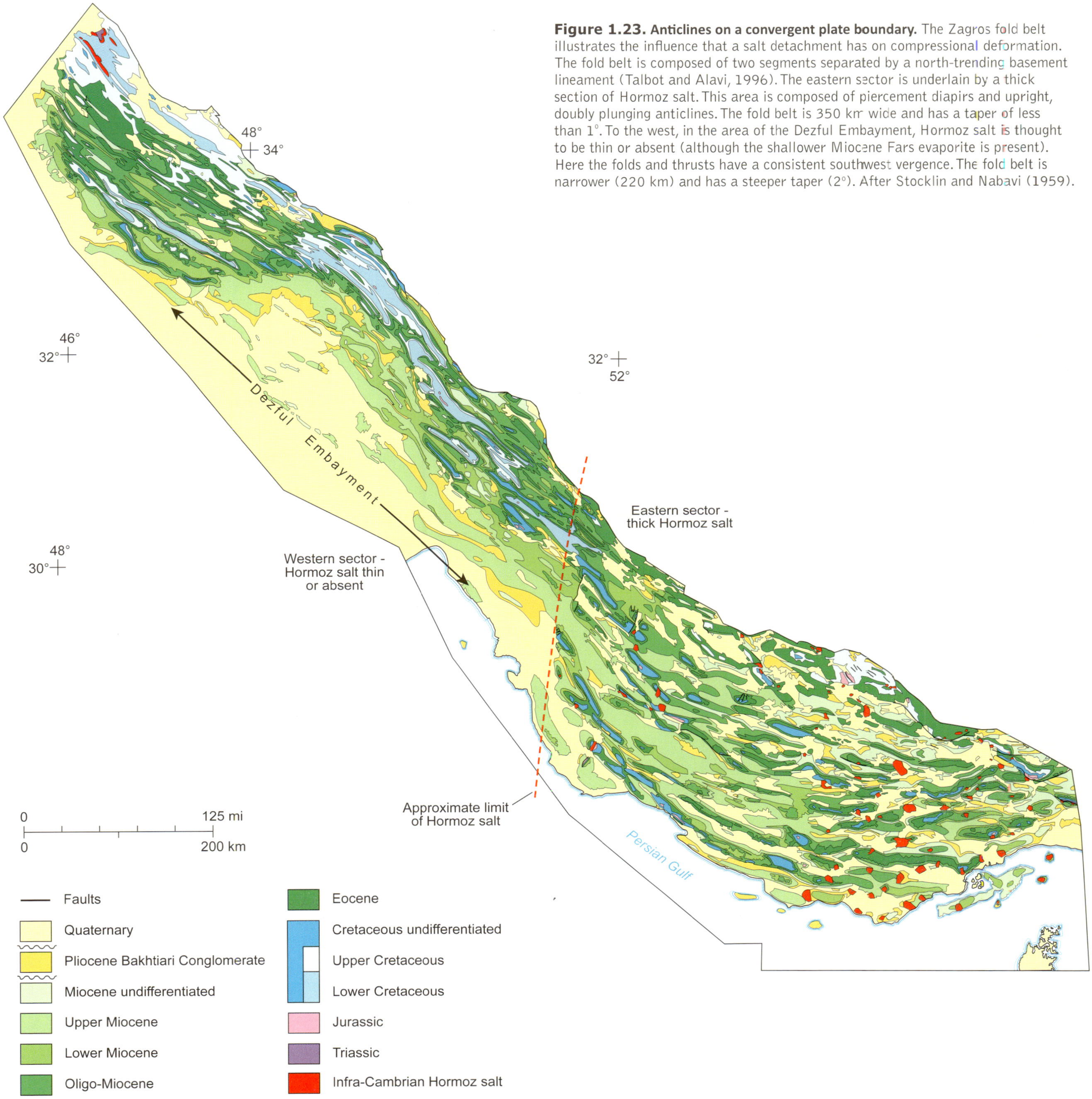

Figure 1.23. Anticlines on a convergent plate boundary. The Zagros fold belt illustrates the influence that a salt detachment has on compressional deformation. The fold belt is composed of two segments separated by a north-trending basement lineament (Talbot and Alavi, 1996). The eastern sector is underlain by a thick section of Hormoz salt. This area is composed of piercement diapirs and upright, doubly plunging anticlines. The fold belt is 350 km wide and has a taper of less than 1°. To the west, in the area of the Dezful Embayment, Hormoz salt is thought to be thin or absent (although the shallower Miocene Fars evaporite is present). Here the folds and thrusts have a consistent southwest vergence. The fold belt is narrower (220 km) and has a steeper taper (2°). After Stocklin and Nabavi (1959).

Zagros Fold Belt—Satellite Image of Folds and Diapirs, Iran

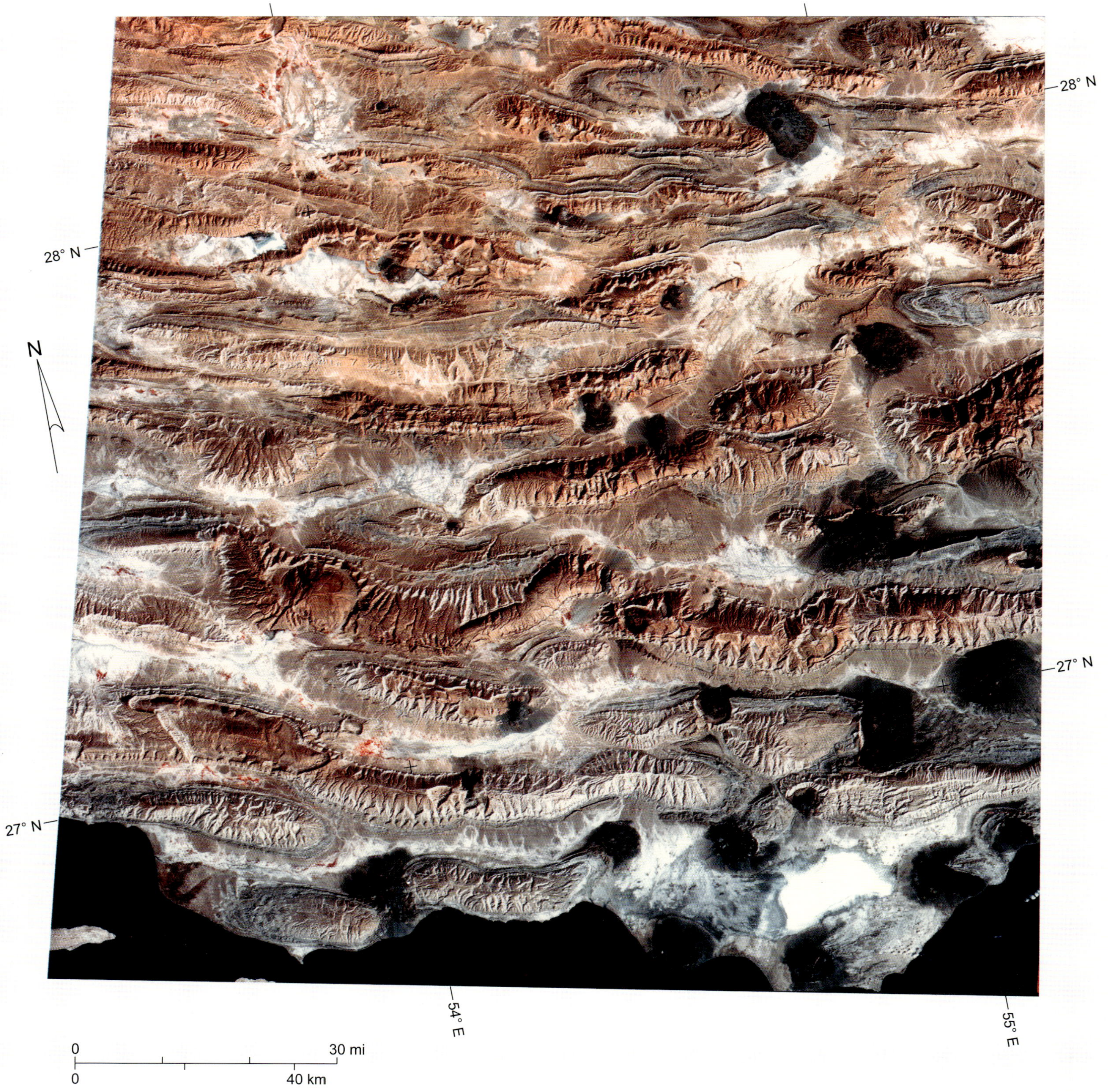

Figure 1.24. Shortening of salt diapirs. The Zagros fold belt formed as the Arabian and Eurasian plates collided. Anticlines are broad and widely spaced owing to the thick overburden being shortened. Because precollisional diapirs (black in the image due to insoluble residue on top of the salt) localized shortening, they are located along the crests or limbs of the anticlines. See interpreted image (Figure 1.25) and detailed area (Figure 1.26). Image courtesy of ExxonMobil Upstream Research Company.

ZAGROS FOLD BELT—Interpreted Satellite Image of Folds and Diapirs, Iran

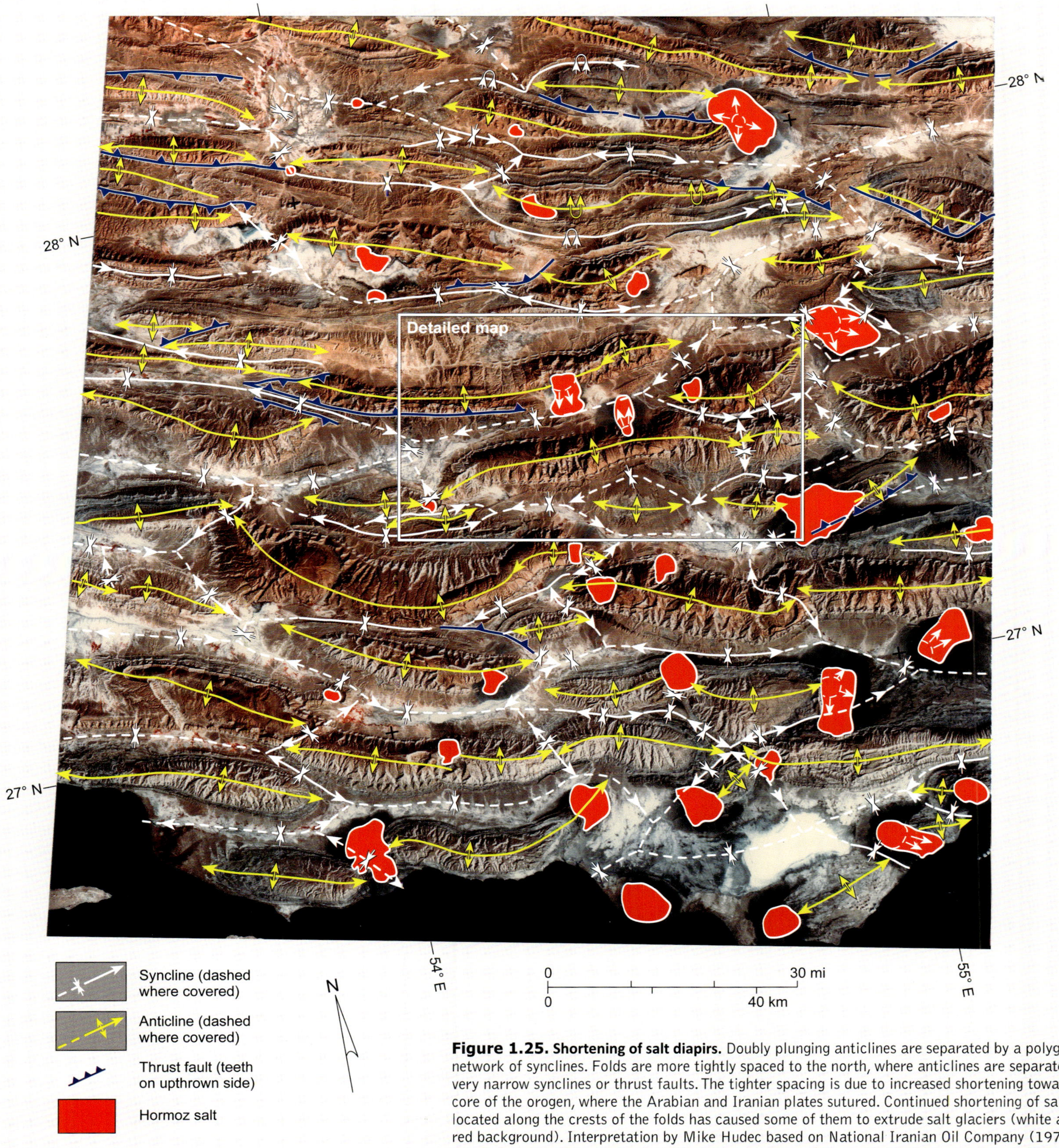

Figure 1.25. Shortening of salt diapirs. Doubly plunging anticlines are separated by a polygonal network of synclines. Folds are more tightly spaced to the north, where anticlines are separated by very narrow synclines or thrust faults. The tighter spacing is due to increased shortening toward the core of the orogen, where the Arabian and Iranian plates sutured. Continued shortening of salt domes located along the crests of the folds has caused some of them to extrude salt glaciers (white arrows on red background). Interpretation by Mike Hudec based on National Iranian Oil Company (1975b). See detailed map in Figure 1.26. Satellite image courtesy of ExxonMobil Upstream Research Company.

ZAGROS FOLD BELT—Detail of Interpreted Satellite Image of Folds and Diapirs, Iran

Figure 1.26. **Shortening of salt diapirs.** Hills are composed primarily of folded Tertiary limestones and dolomites. Shortening was localized by preexisting salt diapirs. Continued shortening of these diapirs caused many of them to extrude salt glaciers (see satellite image draped on topography in Figure 1.27). Shortening is dated as Neogene by the synkinematic Agha Jari Formation and the Bakhtiari conglomerate. Interpretation by Mike Hudec based on National Iranian Oil Company (1975b). Image courtesy of ExxonMobil Upstream Research Company.

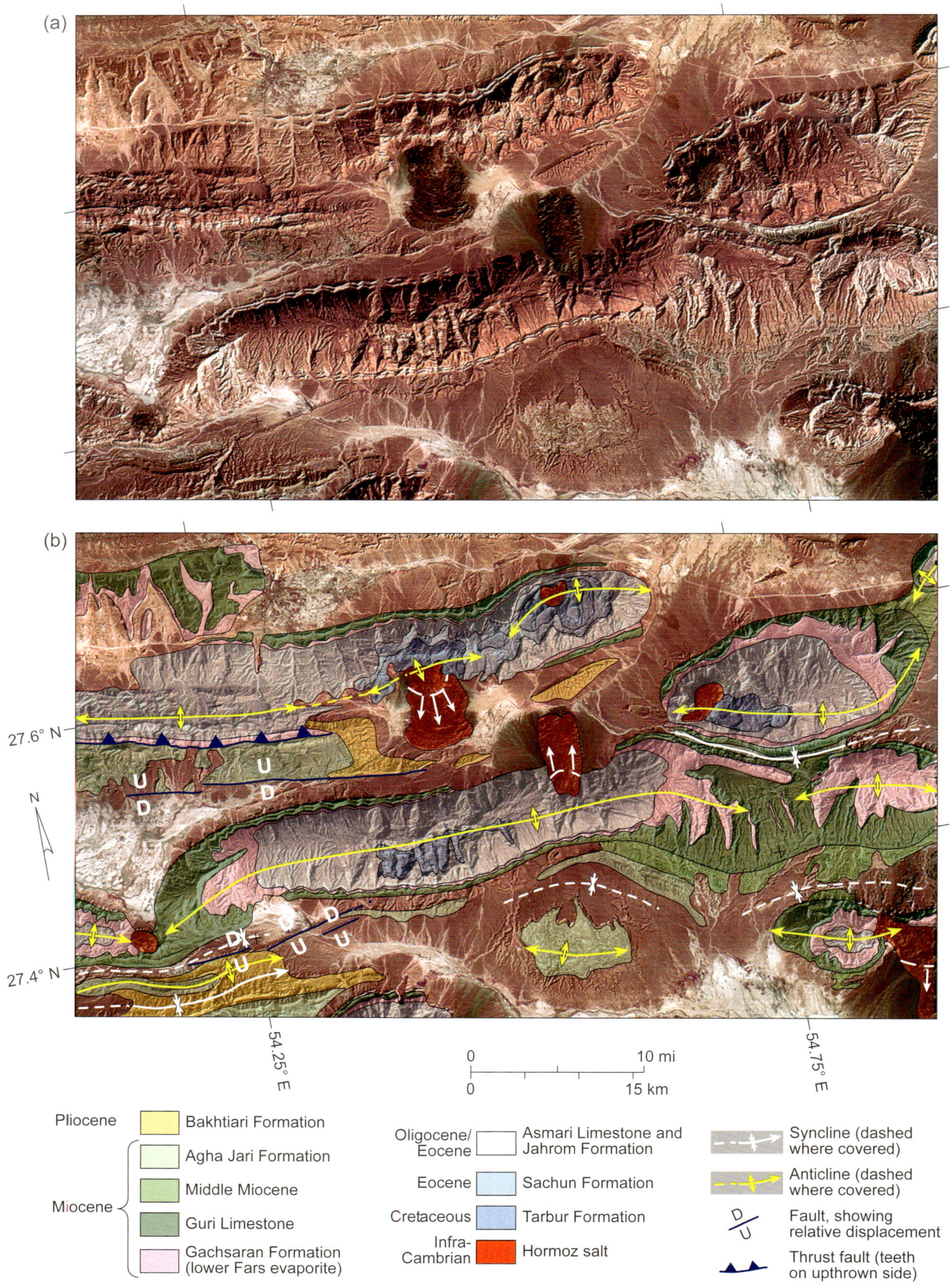

ZAGROS FOLD BELT—Salt Diapirs in the Cores of Compressional Folds, Iran

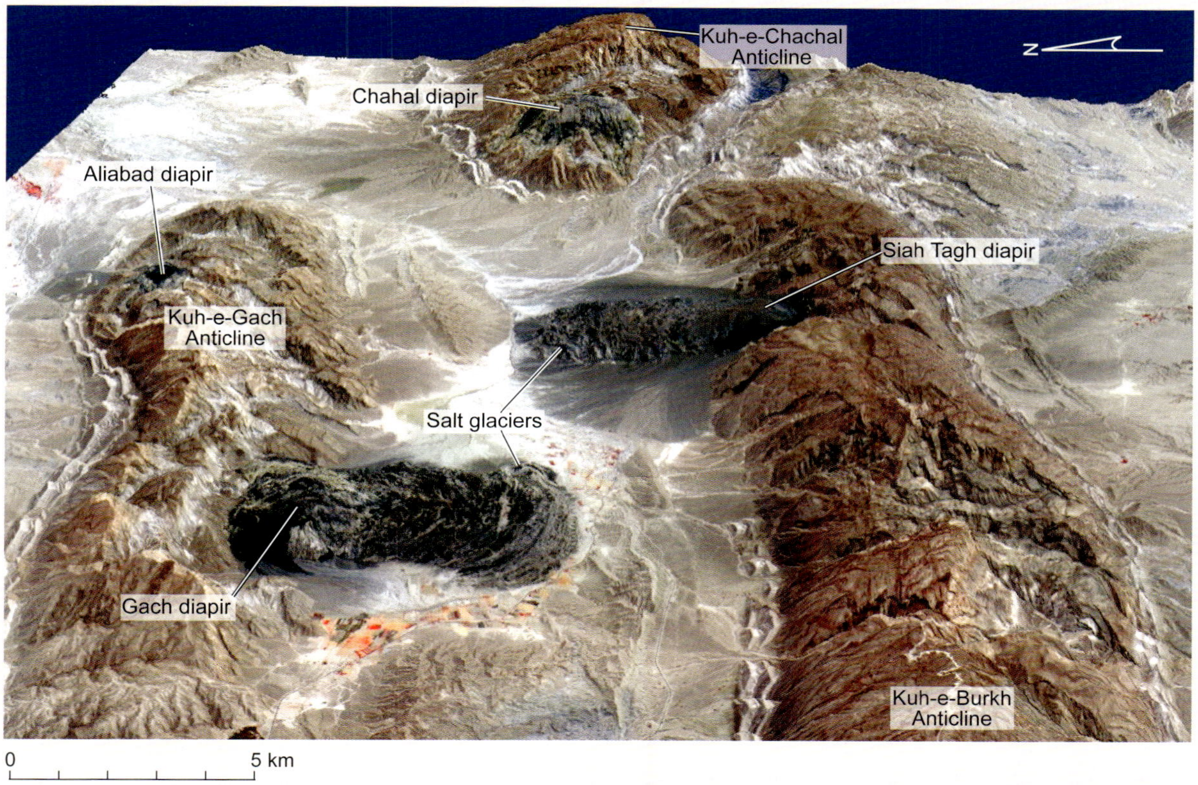

Figure 1.27. Shortening of salt diapirs. Salt diapirs are weaker than their indurated wall rocks, so they shorten preferentially in compression. Preshortening diapirs nucleate compressional folds, which grow to connect the diapirs, as here with the Aliabad and Gach diapirs. Shortening of diapirs in the cores of compressional folds may expel salt upward to extrude as salt glaciers. Salt appears black because of an insoluble mantle of fragments of limestone originally interbedded with the Hormoz evaporites. Spacing between Kuh-e-Burkh and Kuh-e-Gach Anticlines is about 15 km. See interpreted satellite image of the same area in Figure 1.26. This oblique view with 2× vertical exaggeration was created by draping a band 3-2-1 (RGB) image over an ASTER-derived digital elevation model acquired on August 10, 2001. Image from NASA/GSFC/METI/ERSDAC/JAROS and U.S./Japan ASTER Science Team.

ZAGROS FOLD BELT—Shuttle Photograph of Salt Anticlines, Iran

Figure 1.28. Shortening of thick overburden. Anticlines are broad and widely spaced owing to the thick overburden being shortened. Anticlines form doubly plunging whaleback structures, separated by polygonal patterns of synclines. Photograph provided by NASA, image number STS61A-42-36.

ZAGROS FOLD BELT—Salt Diapir along the Crest of a Compressional Anticline, Kuh-e-Namak (Bushehr), Iran

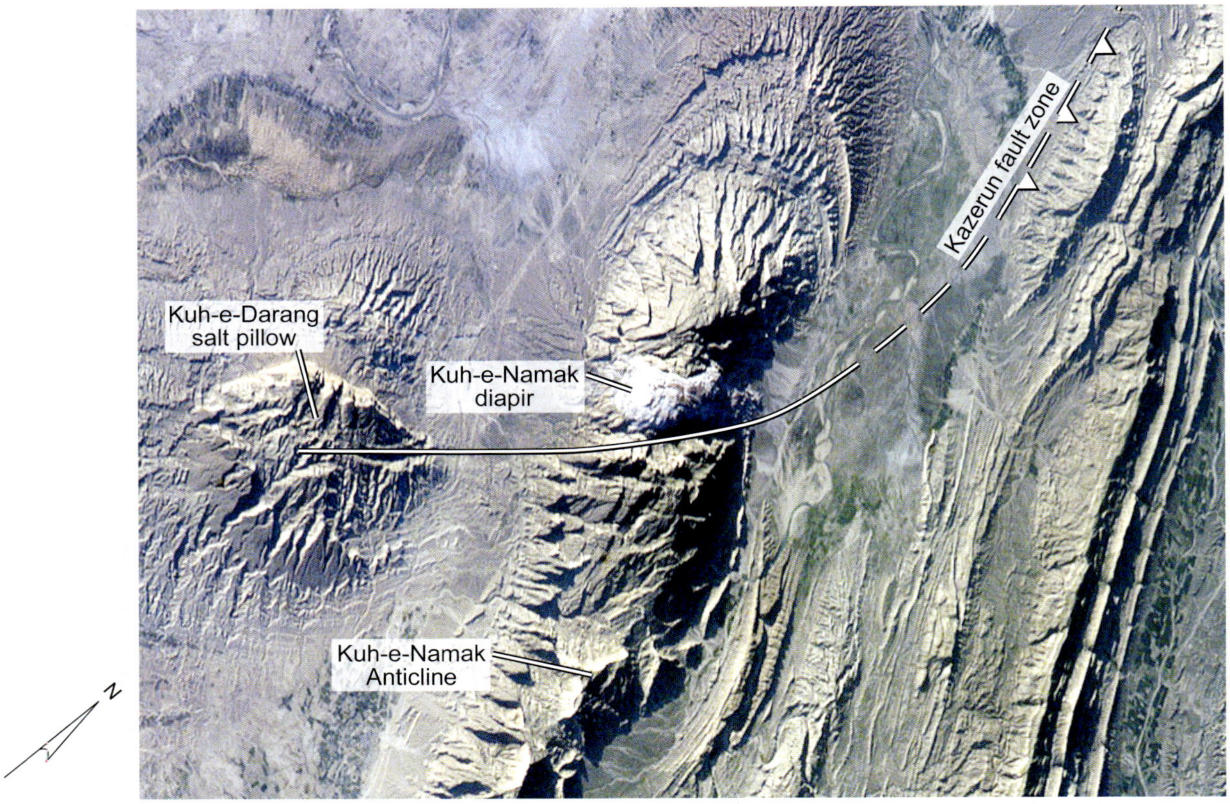

Figure 1.29. Shortening of salt diapirs. The Kuh-e-Namak (Bushehr) diapir lies at the intersection of the Kuh-e-Namak Anticline with the Kazerun fault zone. The diapir has been squeezed by shortening associated both with folding and with a restraining bend in the fault zone. A dynamic bulge of salt at the top of the diapir has produced relief on the central dome that is 400 m above the eroded crest of the anticline and 1400 m above the surrounding plain. Several salt glaciers extruded down the flanks of the anticline. To the southeast of the Kuh-e-Namak diapir, also along the Kazerun fault zone, lies the Kuh-e-Darang salt pillow. The Kuh-e-Darang pillow does not lie along a major compressional fold, perhaps because its intact roof was stronger than the salt-filled Kuh-e-Namak piercement to the northeast. Shuttle image courtesy of NASA.

ZAGROS FOLD BELT—Salt Diapir near a Compressional Anticline, Do-Au Diapir, Iran

Figure 1.30. Shortening of salt diapirs. Outcrops in the foreground are part of the Do-Au diapir, exposed along the crest of a compressional anticline. White rocks are mostly halite, covered by a mottled yellow and red gypsum residual soil. Black rocks to the left are limestones that were originally interbedded within the Hormoz evaporite sequence and that were brought up to the surface as part of the diapir. From the middle distance to the skyline is the south limb of the Kuh-e-Baviyun Anticline, which extends along strike for a remarkable 120 km. Flatirons at the near edge of the anticline are composed of the Miocene Guri Limestone. Topographic relief on the Anticline is roughly 1500 m. View to the northwest. Photograph by Martin Jackson.

ZAGROS FOLD BELT—Halft Kel Anticline, Iran

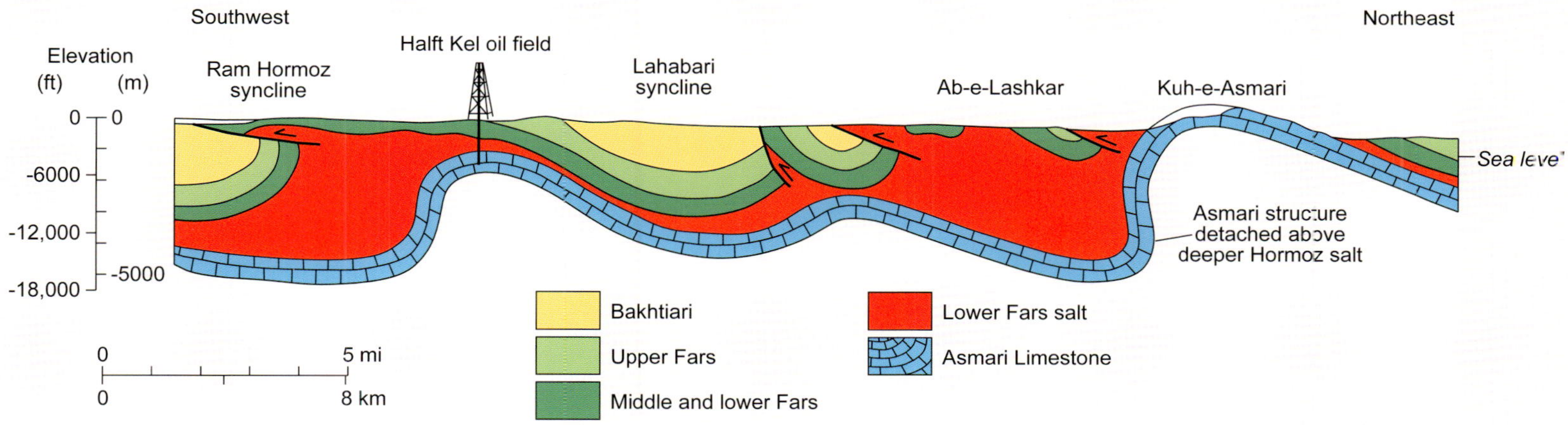

Figure 1.31. Shortening above two salt layers. Here both the Infra-Cambrian Hormoz salt and the Miocene Gachsaran (lower Fars) salt act as detachments. The productive Asmari limestone is asymmetrically folded above the deep Hormoz detachment (below section). Folds detached above the shallower Gachsaran salt are generally discordant to Asmari folds. From O'Brien (1957).

PHYSICAL MODEL—Salt Pillow Formed during Sediment Progradation

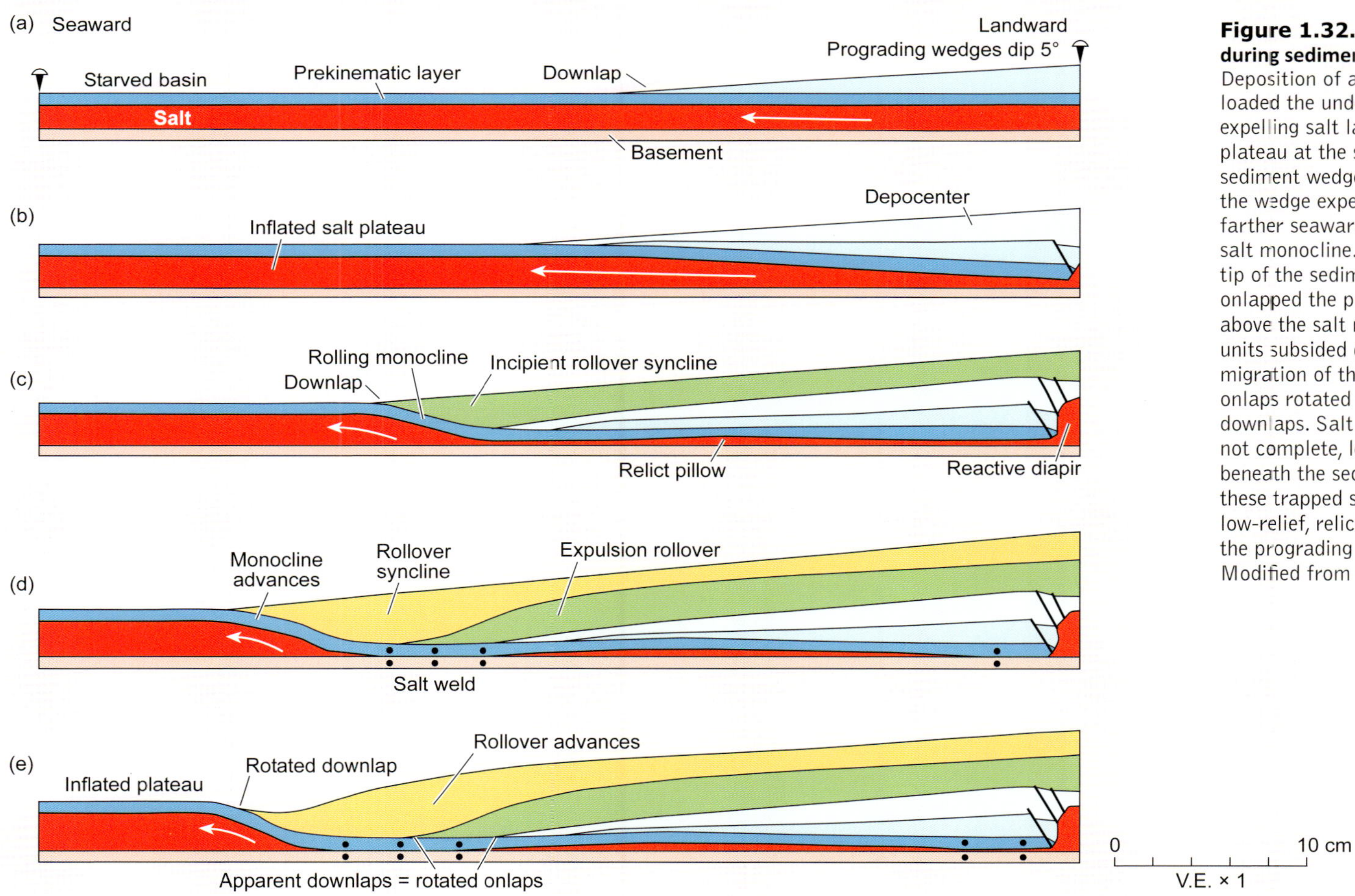

Figure 1.32. Differential loading during sediment progradation. Deposition of a sediment wedge loaded the underlying salt layer, expelling salt laterally into an inflated plateau at the seaward end of the sediment wedge. Progradation of the wedge expelled salt progressively farther seaward, forming a rolling salt monocline. Strata at the seaward tip of the sediment wedge initially onlapped the prekinematic layer above the salt monocline. As these units subsided during seaward migration of the monocline, these onlaps rotated to form apparent downlaps. Salt evacuation was not complete, leaving salt trapped beneath the sediment wedge. One of these trapped salt bodies formed a low-relief, relict salt pillow beneath the prograding sediment wedge. Modified from Ge and others (1997).

PHYSICAL MODEL—Shortening by Flexural-Slip Folding

Figure 1.33. Shortening of multilayered stratigraphy. Multilayers were produced by building of a microlaminate of alternating silicone putty (pale) and modeling clay (dark-red or black). This layering allowed bedding-parallel shear (flexural slip) during shortening. Fold profiles change upward from narrow, sinusoidal anticlines and broad synclines to broad, flat-topped box anticlines and tight, narrow synclines. Salt (pink) was simulated by silicone putty. Model by Hemin Koyi.

PHYSICAL MODEL—Cylindrical Folding

Figure 1.34. Shortening of multilayered stratigraphy. Rounded anticlines are highly cylindrical because of laterally uniform layers and uniform regional shortening. Grid lines on the upper surface were prekinematic, before 44% shortening (stretch = 0.56). Basal salt detachment layer (pale pink) was simulated by silicone putty. Prekinematic overburden was simulated by a microlaminate of alternating weak silicone putty (pale) and strong modeling clay (dark). From Johns and Mosher (1996).

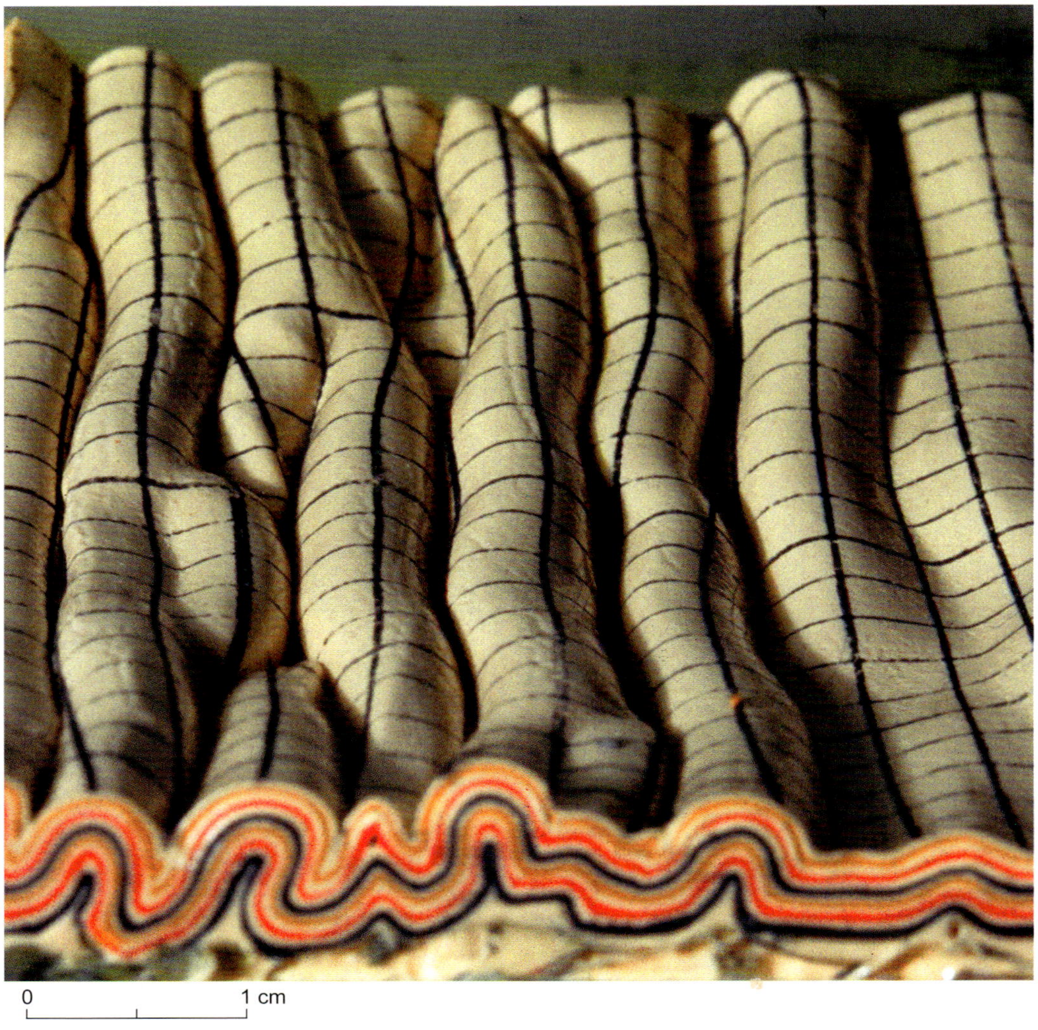

PHYSICAL MODEL—Salt Anticline Formed above Basement Normal Faults

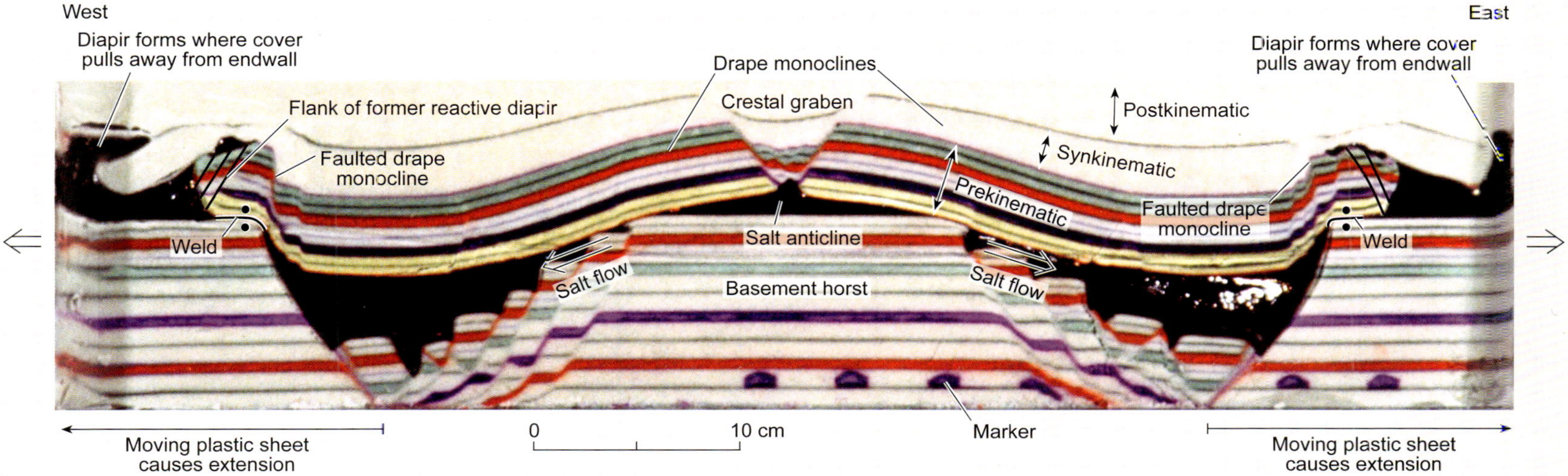

Figure 1.35. Basement extension below salt. Basement, overburden, and salt all extended by the same amount, but they deformed very differently. Salt flowed to partly fill the basement grabens. The overburden draped over the basement blocks, forming an anticline over the crest of the basement horst. Draping was especially pronounced after the salt layer thinned enough to weld the overburden against the footwall corners (see restoration in Figure 1.36). Crestal grabens formed at the crests of the drape monoclines and were later filled by reactive diapiric walls. Overburden extension was mostly at the ends of the model, where the overburden pulled away from the endwalls. Salt rose as reactive diapirs at the endwalls and eventually became emergent. Model by Bruno Vendeville.

PHYSICAL MODEL—Restoration of Salt Anticline Formed above Basement Normal Faults

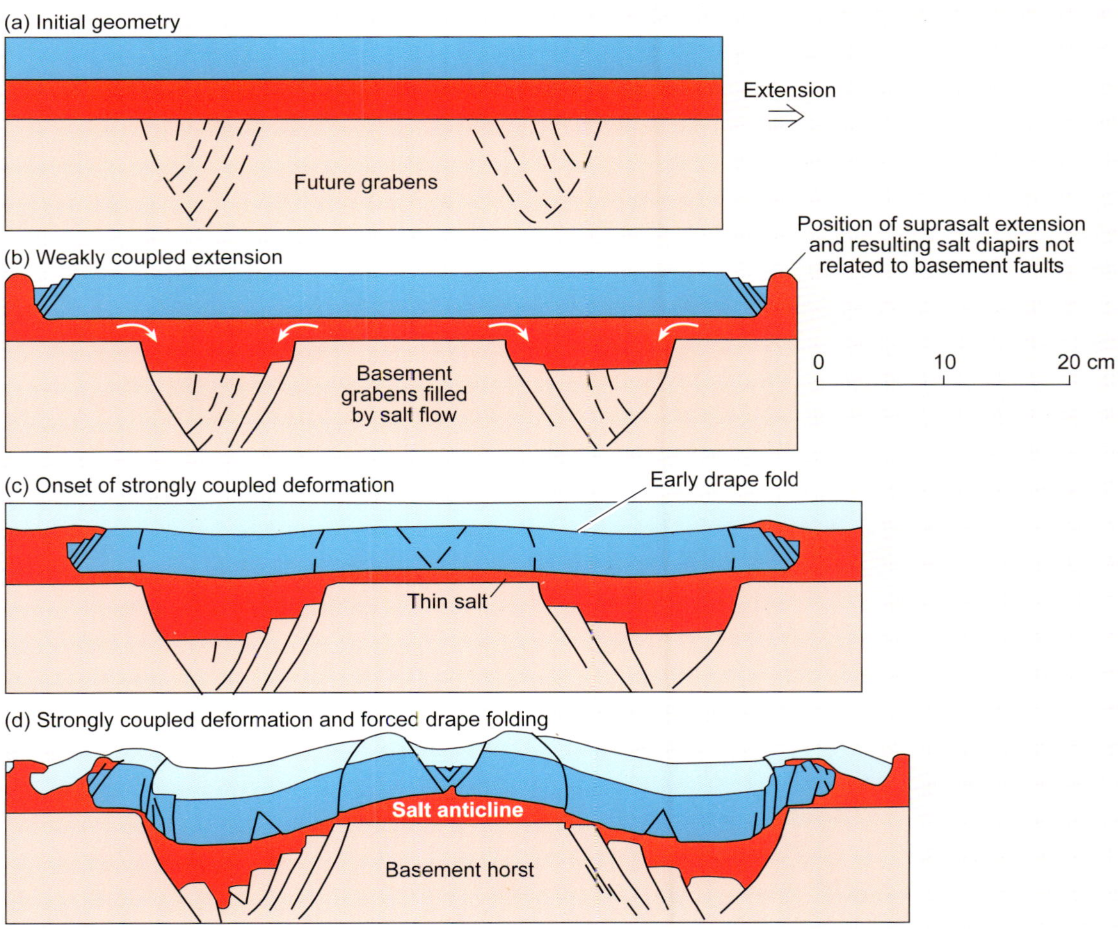

Figure 1.36. Basement extension below salt. While the salt layer was thick, suprasalt extension was only weakly coupled to subsalt extension. Most subsidence into basement grabens was filled by salt flow, and the location of suprasalt extension was unrelated to basement faults. As the experiment progressed, however, the salt over basement high blocks thinned, increasing the degree of coupling. Later, the overburden draped and stretched over the edges of basement blocks. This formed a salt anticline above the basement horst (see section from experiment in Figure 1.35). Modified from Vendeville and others (1995).

PHYSICAL MODEL—Shortening above Two Detachments during Slow Shortening

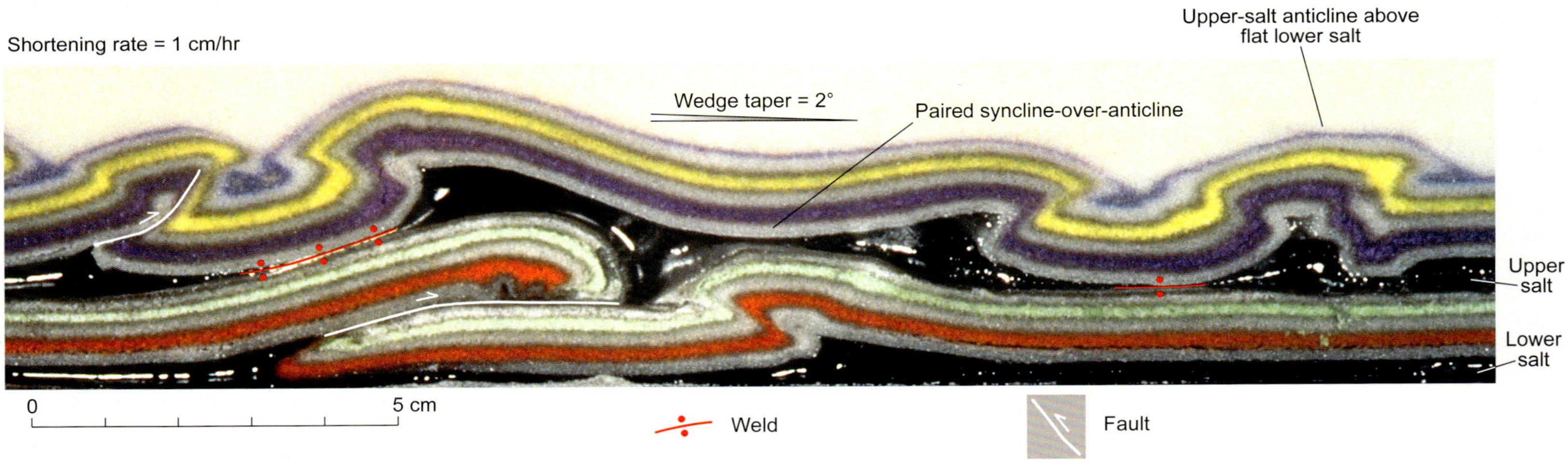

Figure 1.37. Shortening above two detachments. During slow shortening there was little coupling across the upper salt detachment because the upper salt had time to adjust by flow to the evolving structural relief above the lower folds. The upper surface of the model had a gentle overall taper of 2°. Compare with the equivalent rapidly shortened model in Figure 1.38. Left-hand endwall moved slowly to the right. Total shortening was 18% (stretch = 0.82). Model by Brent Couzens-Schultz and Bruno Vendeville.

PHYSICAL MODEL—Shortening above Two Detachments during Rapid Shortening

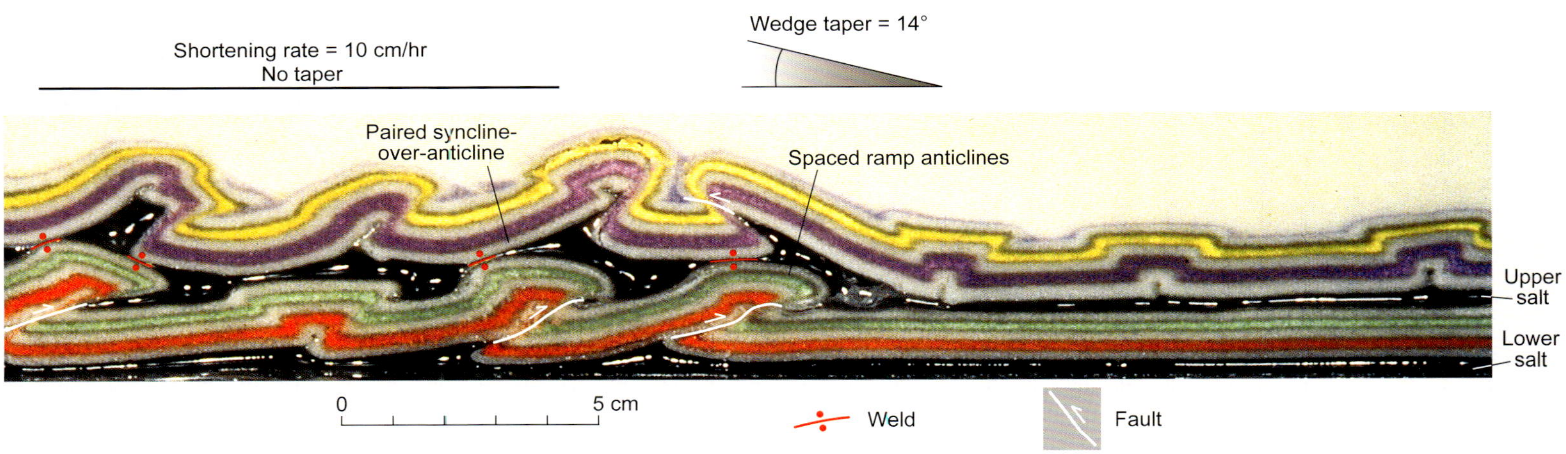

Figure 1.38. Shortening above two detachments. During rapid shortening there was a high degree of coupling across the upper salt detachment (compare with slow shortening model in Figure 1.37). The mountain front of the upper fold belt had a steep surface taper of 14°, up to a comparatively flat plateau in the left half of the model. Left-hand endwall moved rapidly to the right. From Couzens-Schultz and others (2003).

PHYSICAL MODEL—Box Folds above a Salt Detachment

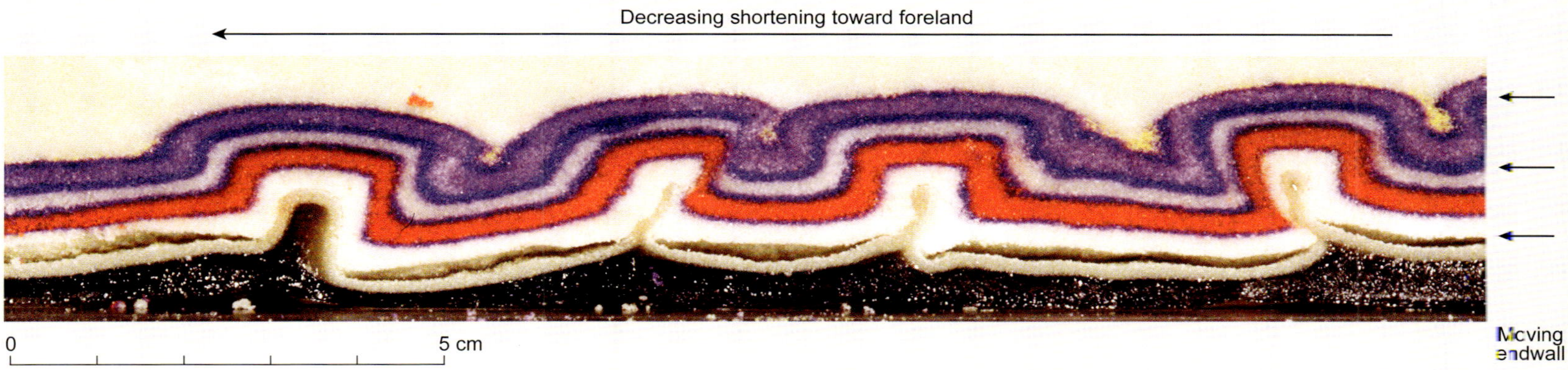

Figure 1.39. Folding above thick salt. A moving endwall at the right edge of the model caused deformation to decrease into the foreland. Undeformed layers are at the extreme left. Shortening from the right produced symmetric box folds and extremely low wedge taper, both typical of very weak detachments like salt. The youngest (leftmost) anticline is cored by the silicone detachment (black). All other anticlines expelled their salt cores and welded isoclinally shut. The rightmost fold has a small thrust weld in its core. Lowermost skin of white sand delaminated and sagged during model sectioning, creating shadowed cavities. Model by Bruno Vendeville.

NUMERICAL MODEL—En Echelon Salt Anticlines with Mechanically Layered Stratigraphy

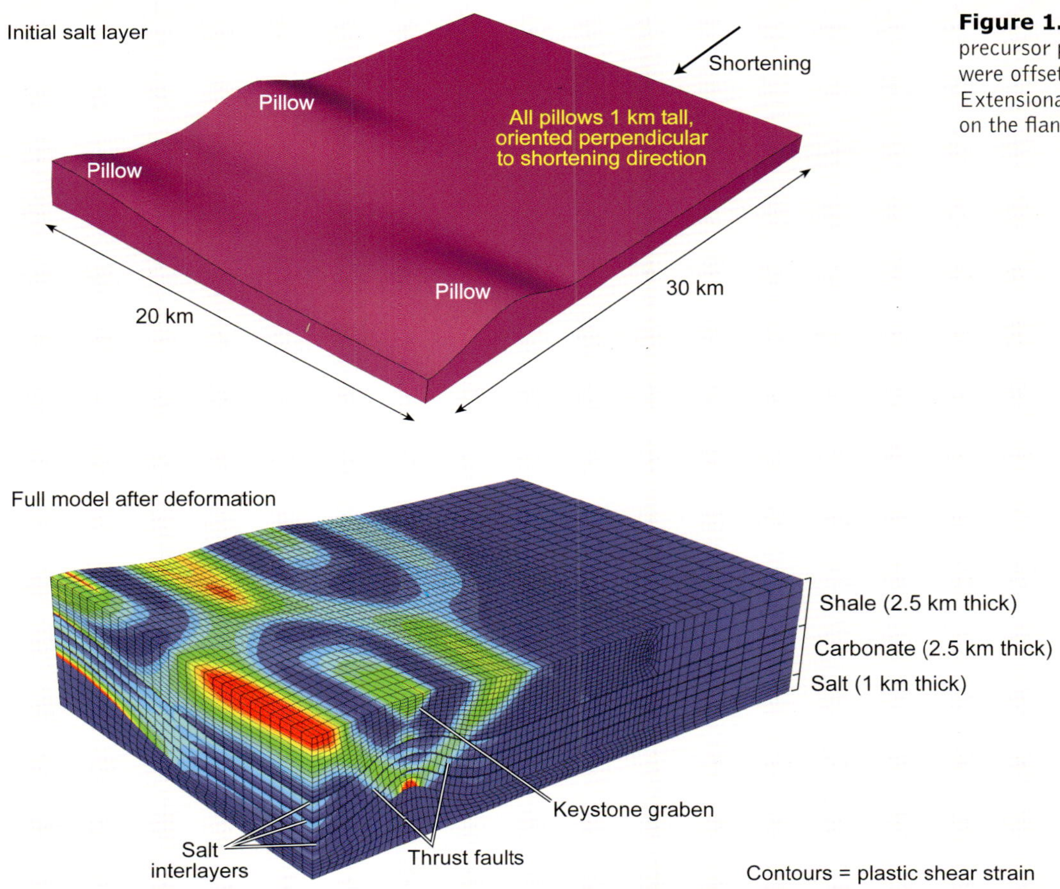

Figure 1.40. Shortening of preexisting salt pillows. All three precursor pillows localized folds during shortening. Because the pillows were offset from one another, the resulting folds also stepped laterally. Extensional grabens formed over fold crests while thrust faults formed on the flanks. Abaqus finite-element model by Dan Schultz-Ela.

NUMERICAL MODEL—Growth of Salt Anticlines

Figure 1.41. Shortening of multilayered overburden. Folds formed adjacent to a fixed endwall at the downdip end of a tilted salt layer. Salt initially flowed into the cores of anticlines but was later expelled as the fold tightened. The anticline originally grew by limb rotation, and later by limb lengthening. Steep limbs underwent layer-parallel extension. See a plot of pressures in Figure 1.42. GeoSim finite-element model by Dan Schultz-Ela.

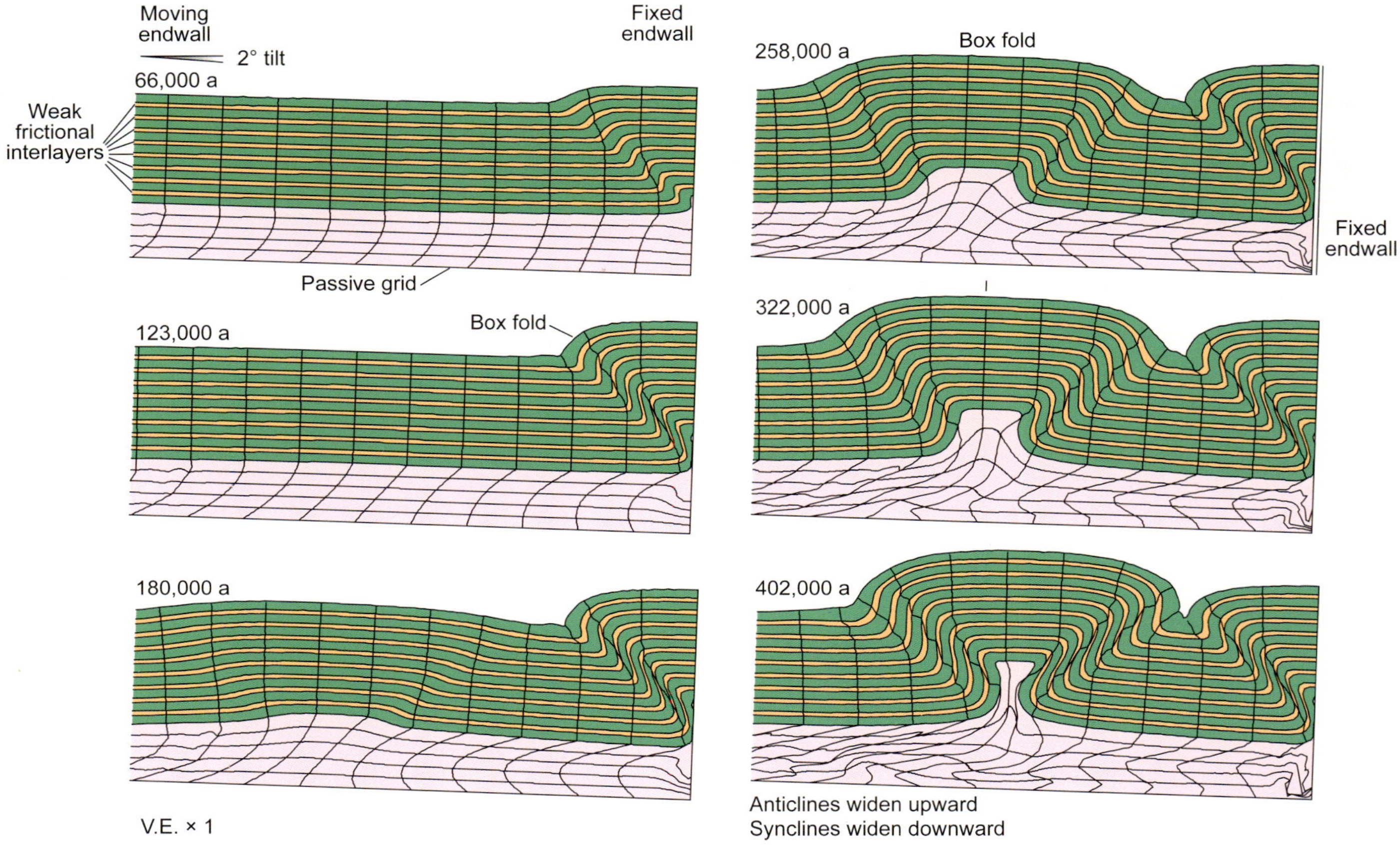

NUMERICAL MODEL—Pressure Distribution around Growing Salt Anticlines

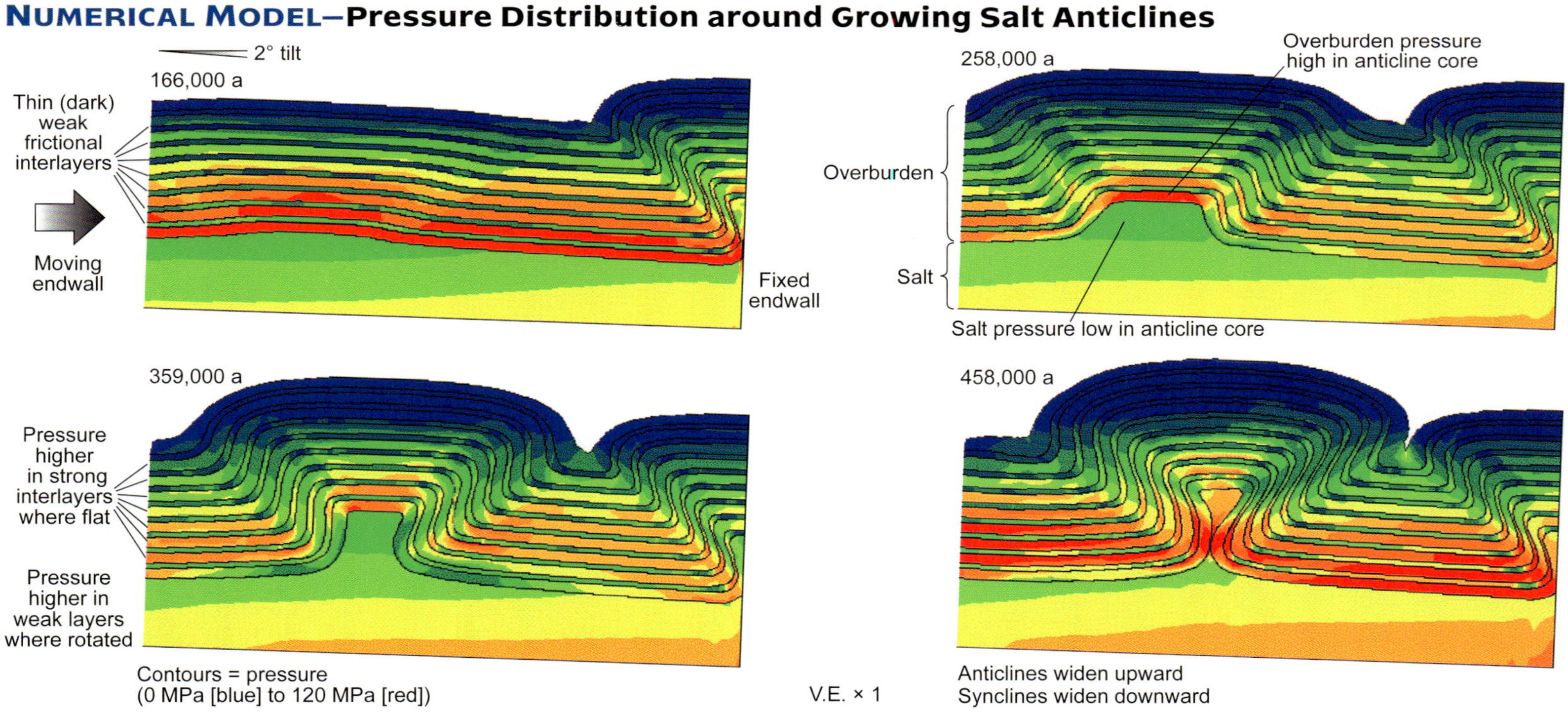

Figure 1.42. Shortening of multilayered overburden. Folds formed adjacent to a fixed endwall at the downdip end of a tilted salt layer. Low salt pressure in the immature anticline core shows that salt flowed passively into a space created by buckling of the overburden. After the fold core pinched shut (at 458,000 a) salt in the core was pressurized, which could hydraulically fracture its overburden. See the evolution of structures in the model in Figure 1.41. GeoSim finite-element model by Dan Schultz-Ela.

NUMERICAL MODEL—Salt Anticline with Mechanically Layered Stratigraphy, No Strain Weakening

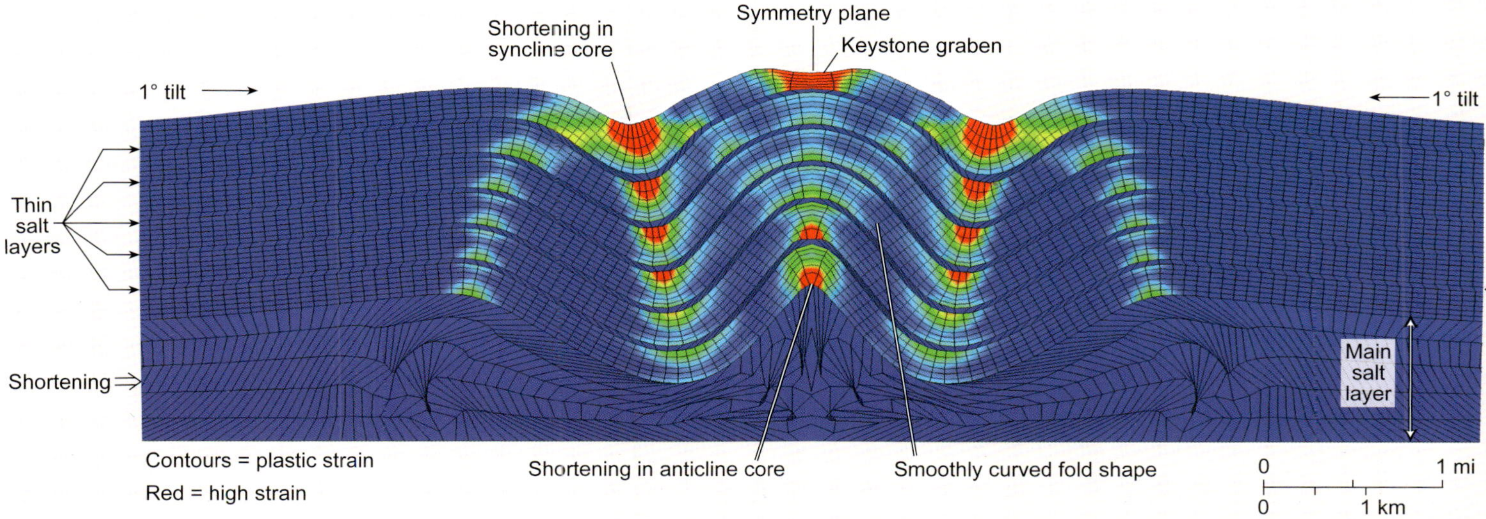

Figure 1.43. Folding of mechanically layered stratigraphy. Deformation was initiated in the model by tipping the left and right halves of the model inward by 1°. The main anticline formed at the intersection of these two dip panels. The absence of strain weakening inhibited the formation of shear zones, resulting in a very smooth fold form. Salt initially flowed into the core of the anticline. As shortening continued, the flanks of the anticline closed in and salt movement reversed to flow back out of the core. Compare with a similar model with strain weakening in Figure 1.44. Abaqus finite-element model by Dan Schultz-Ela.

NUMERICAL MODEL—Salt Anticline with Mechanically Layered Stratigraphy, Strain Weakening

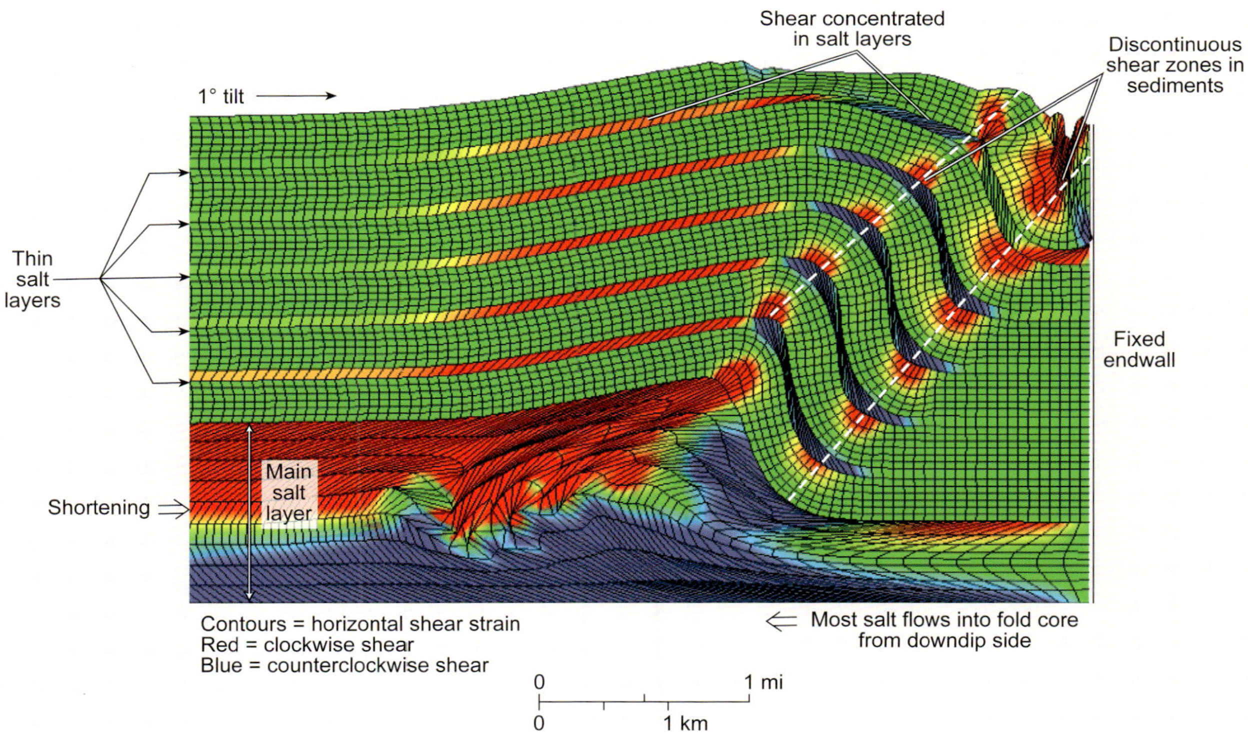

Figure 1.44. Shortening above thick salt. Shortening occurred at the downdip end of a model that was tipped 1° to the right. The presence of salt interlayers allowed the beds to slip past one another in flexural slip. The presence of strain weakening in the model allowed two discontinuous shear zones (red) to form in the sediments over the crest of the fold (compare with a similar experiment without strain weakening in Figure 1.43). The contours show that most of the shear was concentrated strongly in the salt layers. Abaqus finite-element model by Dan Schultz-Ela.

NUMERICAL MODEL—Contractional Drape above Normal Fault Inverted at 50,000 a, Isotropic Roof

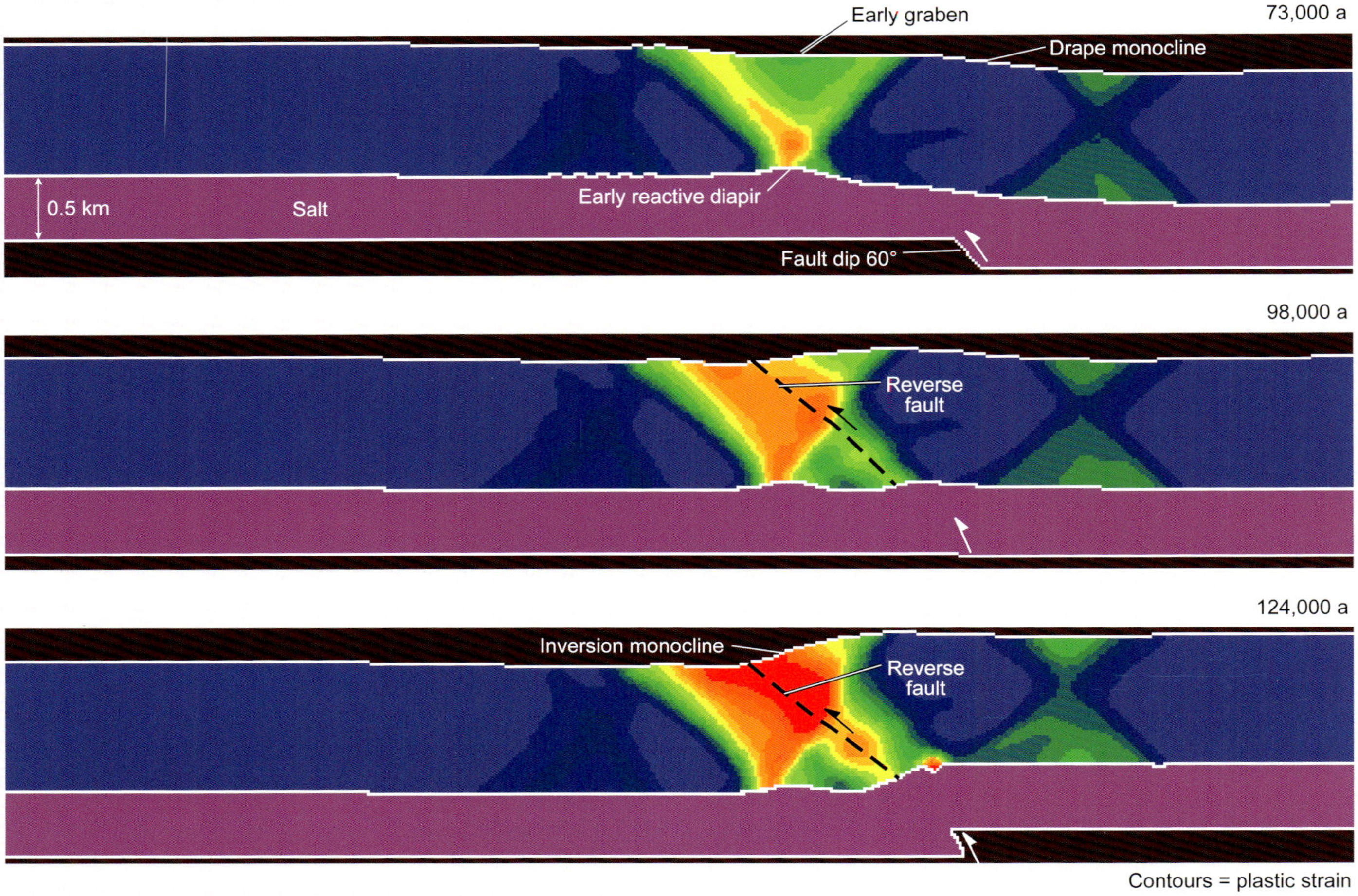

Figure 1.45. Subsalt inversion. Structures formed during the early extensional phase were little modified for most of the inversion; it is difficult to see the effects of inversion even after slip was almost completely reversed (98,000 a). In the absence of synkinematic sediments, which would show that the former basin was uplifted, it might be difficult to recognize shortening until the inversion was quite advanced (124,000 a). GeoSim finite-element model by Dan Schultz-Ela.

NUMERICAL MODEL—Contractional Drape above Reverse Fault, Free-Slip Roof Layers

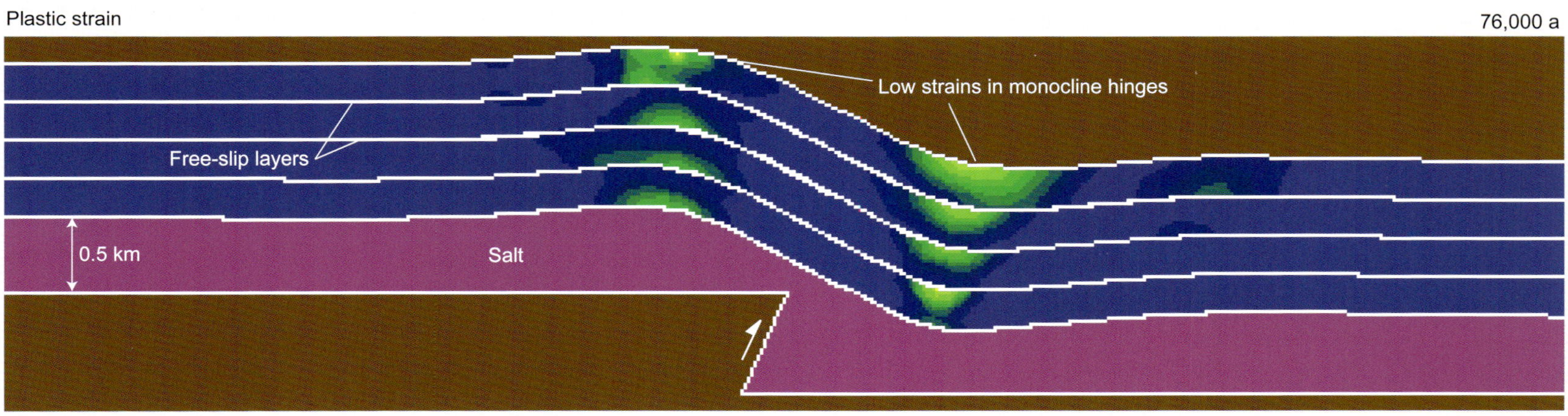

Figure 1.46. Subsalt reverse faulting. The overburden was forced to shorten by the same amount as that of the basement. Free-slip layers enhanced the ability of rocks to fold, so basement offset was accommodated in the overburden by a drape fold without significant faulting (compare with the same model using frictional layering in the roof in Figure 1.47). If deformation continued, however, the basement high block would eventually impinge on the bottom of the overburden layer, and deformation would become much more strongly coupled. GeoSim finite-element model by Dan Schultz-Ela.

Numerical Model—Contractional Drape above Reverse Fault, Frictional Roof Layers

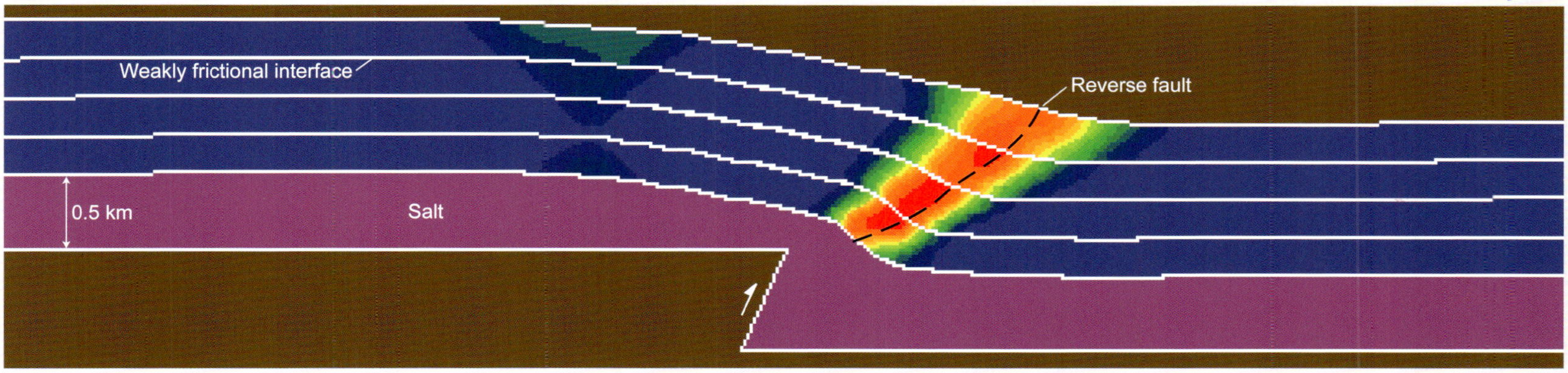

Figure 1.47. Subsalt reverse faulting. The overburden was forced to shorten by the same amount as that of the basement. Weak friction between layers was enough to resist flexural slip, so folding was inhibited relative to the same model using free-slip layers (Figure 1.46). Most of the overburden shortening was accommodated in a thrust fault that formed at the lower bend of the drape monocline. Friction angle between layers = 10°. GeoSim finite-element model by Dan Schultz-Ela.

Numerical Model—Contractional Drape above Multiple Reverse Faults, Isotropic Roof

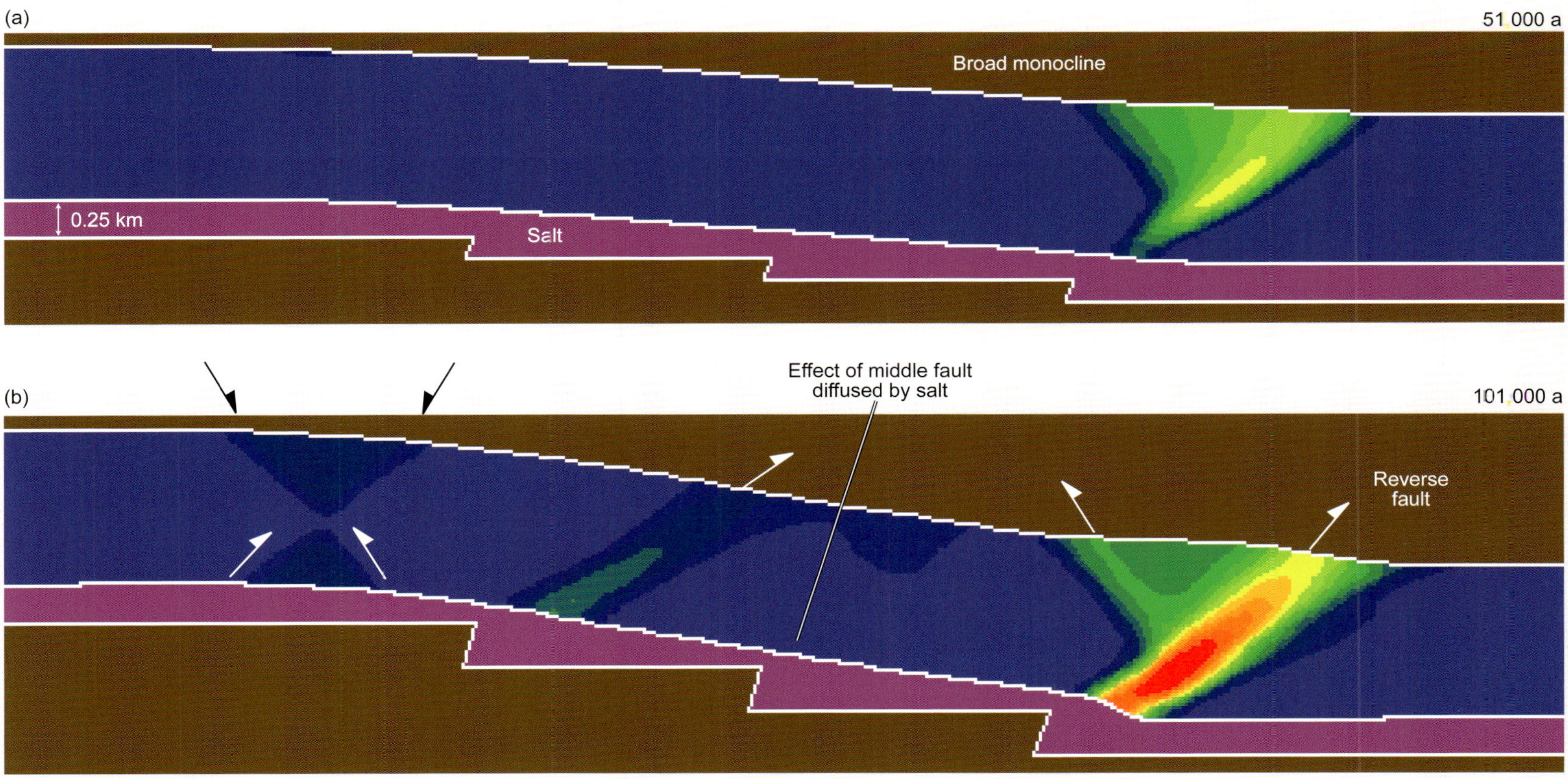

Figure 1.48. Subsalt reverse faulting. Salt was thin enough that deformation was partly coupled above the upper and lower thrust faults. The thrust at the bottom of the monocline had more slip because it also accommodated gravity sliding down the flank of the monocline. At the top of the ramp, the overburden acted as a beam, with inner-arc shortening and outer-arc extension separated by a neutral surface. GeoSim finite-element model by Dan Schultz-Ela.

Chapter 2

Salt Stocks and Salt Walls

Salt stocks and salt walls are the most classic of salt structures, rising as mountainous pillars from the basin floor. They are remarkable for their size (some stocks are more than 8 km tall, and some walls are more than 100 km long) and for their abundance (the Precaspian basin alone contains about 1800 diapirs). Despite their prominence in salt-tectonic literature, many aspects of their growth have not been understood until recently.

Salt stocks and salt walls are plugs or ridges of salt having discordant contacts with postsalt strata. Salt walls having triangular profiles are covered in Chapter 3 because of their unique evolution; other types are presented here. In map view, diapirs having a length-to-width ratio of less than two are *salt stocks* (sometimes called *salt domes*). Diapirs having a ratio of two or more are *salt walls*. In profile, salt stocks and salt walls are indistinguishable.

Nontriangular salt stocks and salt walls grow in two main ways: passively and actively. In passive diapirism (also known as *downbuilding*) the diapir crest remains at or near the sediment surface as surrounding sediments aggrade and the basin floor sinks. Any roof that accumulates is thin and ephemeral, being eroded or shouldered aside by the rising diapir. In active diapirism the rising diapir arches a thicker, more competent roof. The force required to arch the roof may be from buoyancy, differential compaction, or regional shortening.

Passive diapir

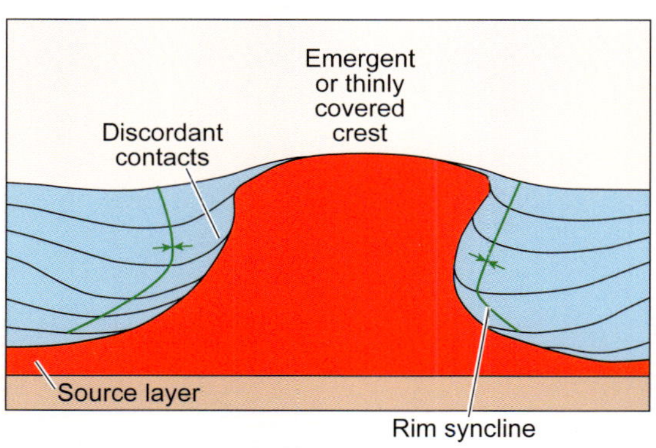

Active diapir

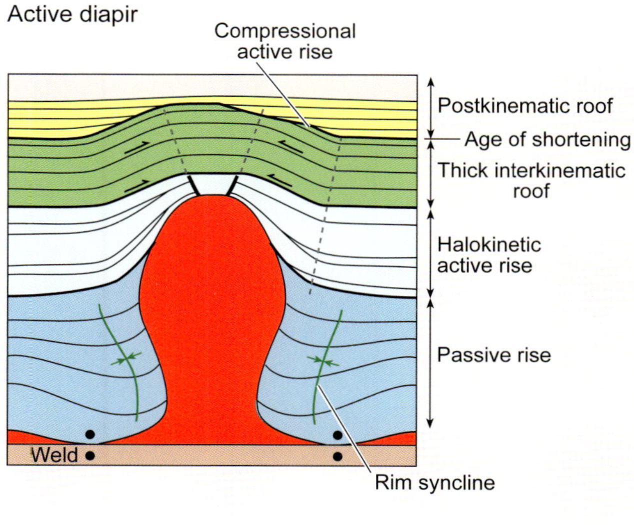

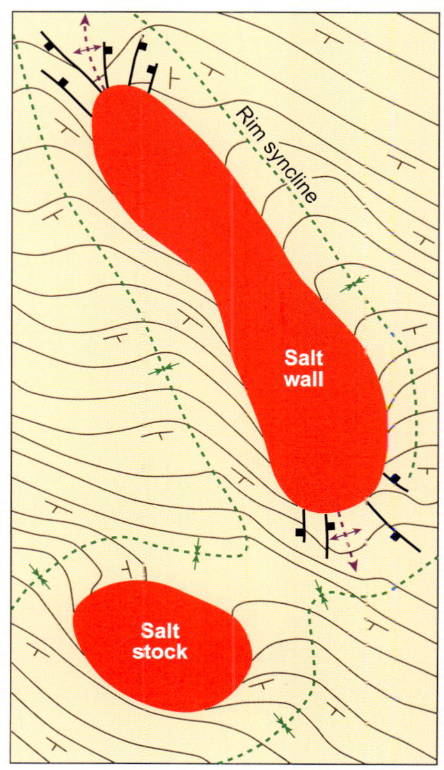

ATLAS FOLD BELT—Field Photograph of Ran el Melah Diapir, Algeria

Figure 2.01. Passive diapirism. The karstic surface of the diapir results from rainfall-induced dissolution to form sinkholes. Despite this weathering, the diapir forms a hill, suggesting that diapiric salt is still rising. Halite dissolved from the diapir is redeposited in an ephemeral stream in the foreground, fed by salt springs within the diapir. The inward-dipping contact on the east side of the diapir indicates that the diapir is flaring outward. Ran el Melah (also known as Rocher de Sel de Djelfa) is historically significant as the first salt diapir described in the geologic literature (Ville, 1856). See map in Figure 2.02. Photograph by Martin Jackson.

ATLAS FOLD BELT—Map of Ran el Melah Diapir, Algeria

Figure 2.02. Passive diapirism. In the Holocene the actively rising Ran el Melah diapir extruded a small salt glacier to the north. The salt in this glacier has now dissolved, leaving a remnant carpet of brecciated Jurassic limestone, which is exotic to the country rocks and must have been shed from the diapir. This limestone is younger than the Triassic salt but was probably a foundered roof carried within the rising salt, together with lesser amounts of Cretaceous sandstone. Blocks of these units remain inside the main body of the diapir. Contorted layering within the diapir reflects complex curtain folding as a result of lateral flow within the source layer, followed by vertical flow up the diapir. Ran el Melah (also known as Rocher de Sel de Djelfa) is historically noteworthy as the first salt diapir described in the geologic literature (Ville, 1856). See photograph in Figure 2.01. Modified from Kulke (1991).

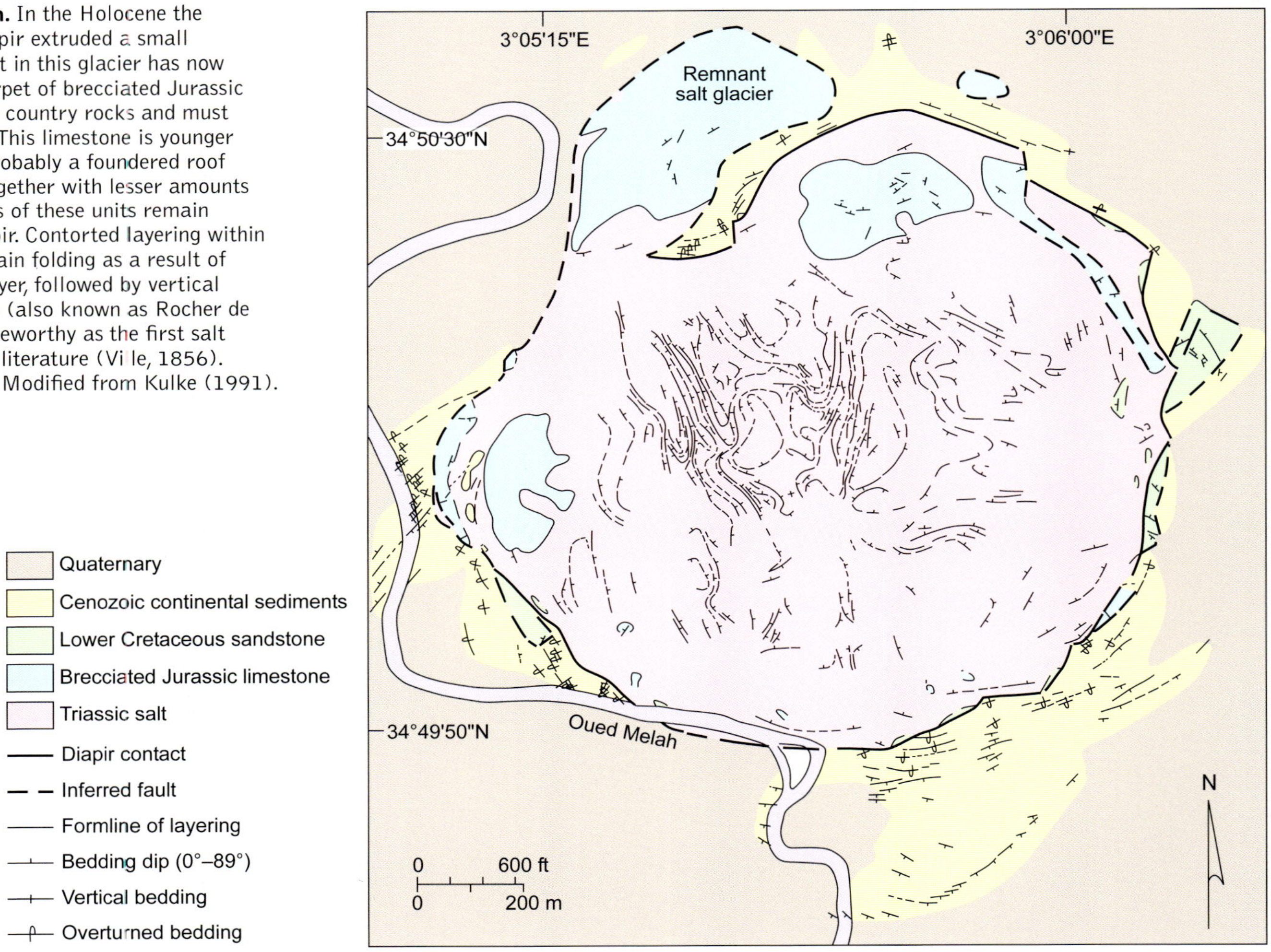

ATLAS FOLD BELT—Djebel Debadib Diapir, Tunisia

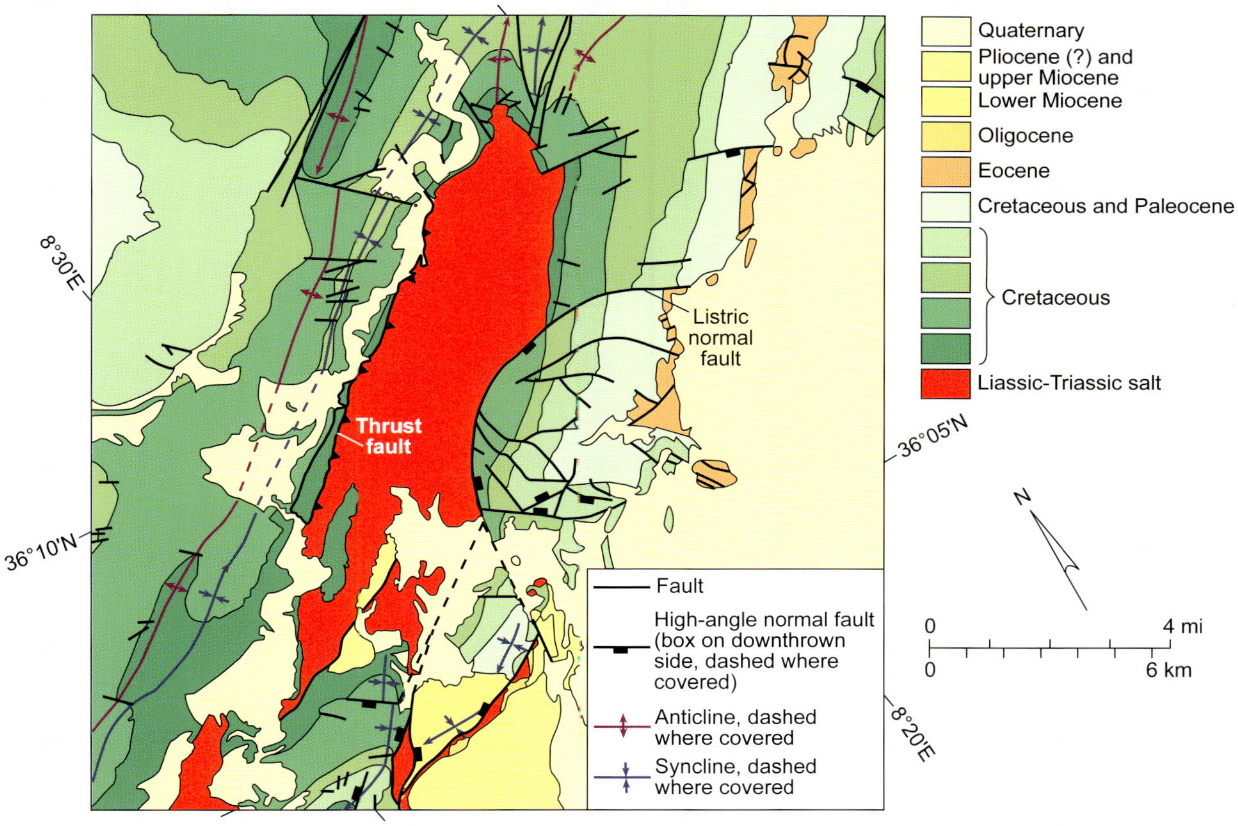

Figure 2.03. Rejuvenation of buried diapir during shortening. The diapir roof was arched during NW-SE-directed regional shortening. Three types of faults formed during arching. First, a thrust fault offset the NW flank of the diapir. Second, radial faults accommodated circumferential stretching during arching. These have a slightly preferred orientation, striking NE-SW, perpendicular to the shortening direction; some offset the diapir flank. Finally, one large listric normal fault appears to accommodate slumping off the uplifted SE side of the diapir. After Snoke and others (1988).

EAST TEXAS BASIN—Salt Structures, Texas, USA

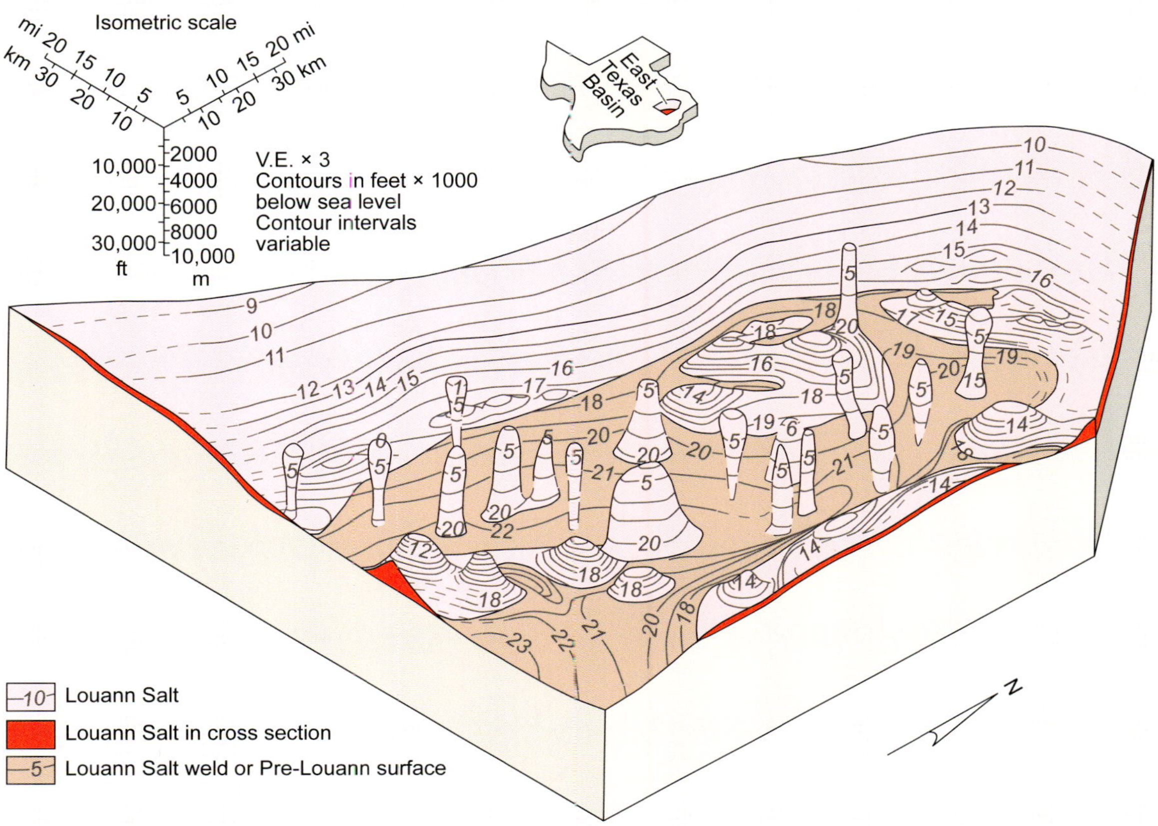

Figure 2.04. Diapir growth. Salt diapirs form a north-trending belt through the center of the East Texas Basin, with flanking belts of salt pillows (see diapir map in Figure 2.05). This pattern probably reflects the original salt isopach, which was thickest in the middle of the basin and thinned toward its margins. Diapirs stopped growing when salt supply from the source layer was exhausted, as indicated by a widespread weld on the basin floor. Modified from Jackson and Seni (1983).

EAST TEXAS BASIN—Salt Structures and Turtle Structures, Texas, USA

Figure 2.05. Diapir growth. Salt diapirs form a north-trending belt through the center of the East Texas Basin, with flanking belts of salt pillows (see block diagram in Figure 2.04). This pattern probably reflects the original salt isopach, which was thickest in the middle of the basin and thinned toward its margins. The Mexia-Talco and Elkhart Graben–Mt. Enterprise Fault Zones are extensional breakaways for minor slip into the basin off the structural highs around the basin. Because there is limited slip into the basin center, the pillows probably formed by halokinesis rather than shortening. A halokinetic origin is consistent with the near-circular planform of many of the pillows. Many of the turtle structures between salt domes are composite structures, formed by expulsion of salt into as many as eight salt structures. Modified from Jackson and Seni (1984).

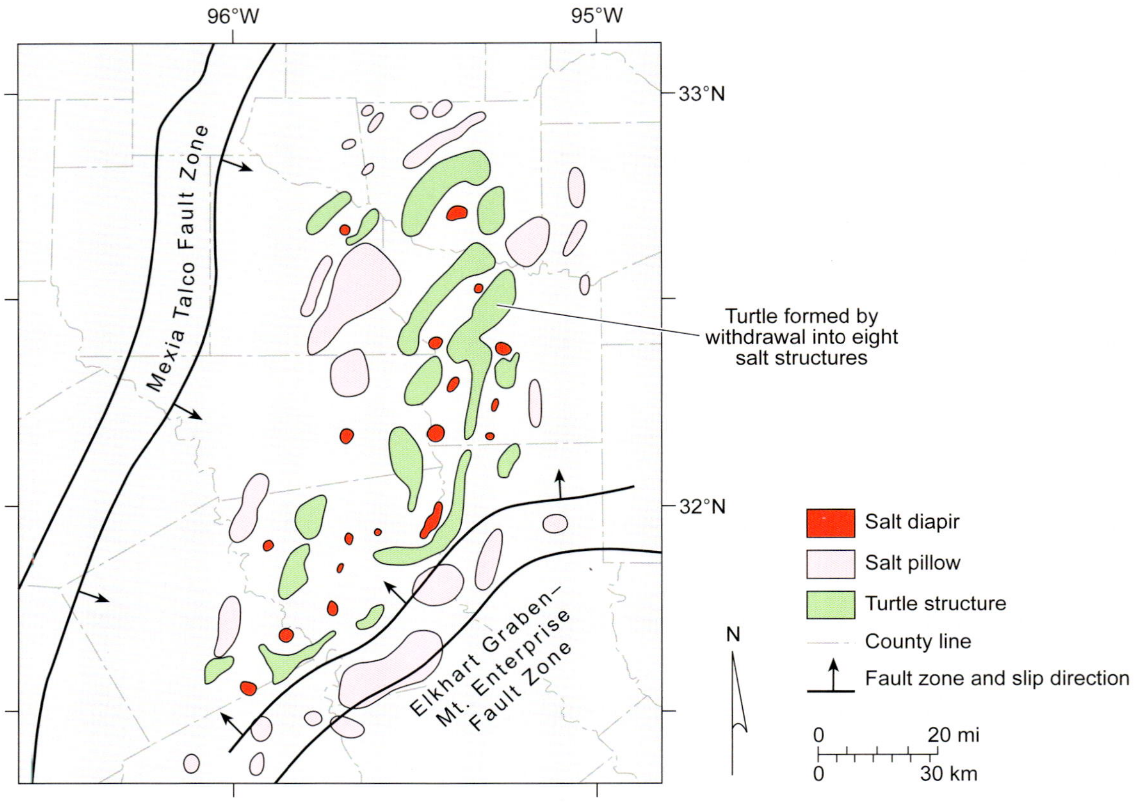

EAST TEXAS BASIN—Reef Buildup between Withdrawal Basins, Fairway Field, Texas, USA

Figure 2.06. Salt withdrawal. The James Limestone reservoir at Fairway field is composed of reef and associated facies, deposited on a saddle formed as a remnant high between two salt-withdrawal basins. Localization of the reef on the saddle indicates that salt withdrawal was affecting bathymetry because sedimentation was slower than subsidence. Modified from Seni and Jackson (1983).

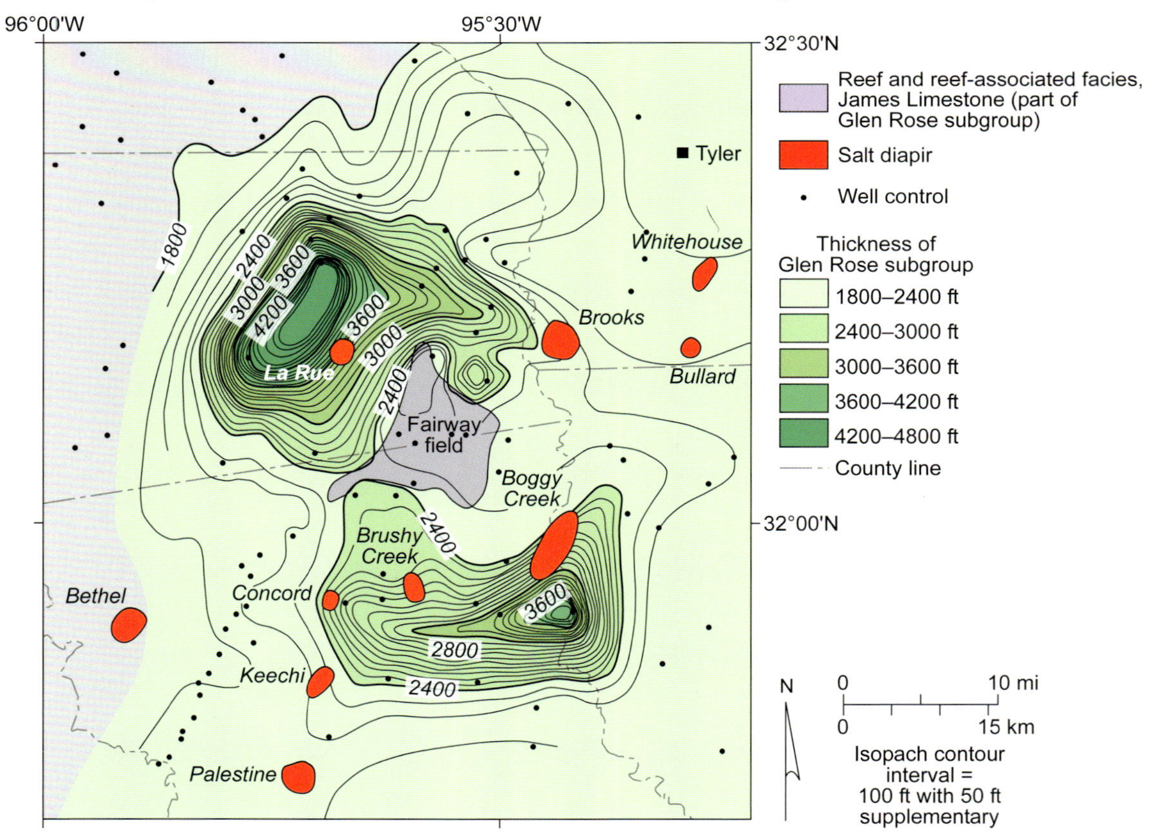

EAST TEXAS BASIN—Hourglass-Shaped Salt Stock, Hainesville Dome, Texas, USA

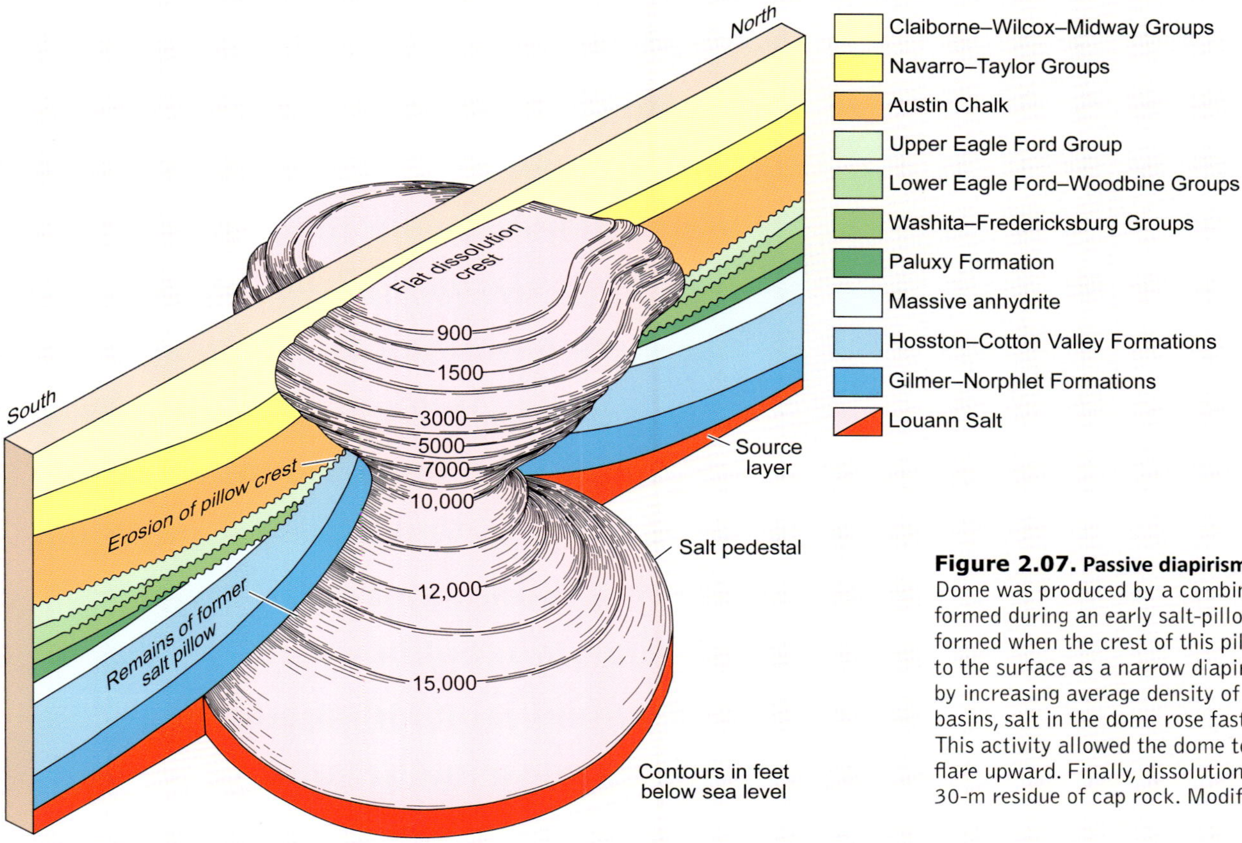

Figure 2.07. Passive diapirism. The hourglass shape of Hainesville Dome was produced by a combination of factors. The pedestal at the base formed during an early salt-pillow stage. The narrow waist in the middle formed when the crest of this pillow was eroded, allowing salt to pierce to the surface as a narrow diapir. The bulb at the top formed when, driven by increasing average density of sediments in the adjacent withdrawal basins, salt in the dome rose faster than the local sediments aggraded. This activity allowed the dome to extrude outward, causing the diapir to flare upward. Finally, dissolution planed the diapir crest flat, leaving a 30-m residue of cap rock. Modified from Jackson and Seni (1984).

EAST TEXAS BASIN—Overhanging Salt Stock, Oakwood Dome, Texas, USA

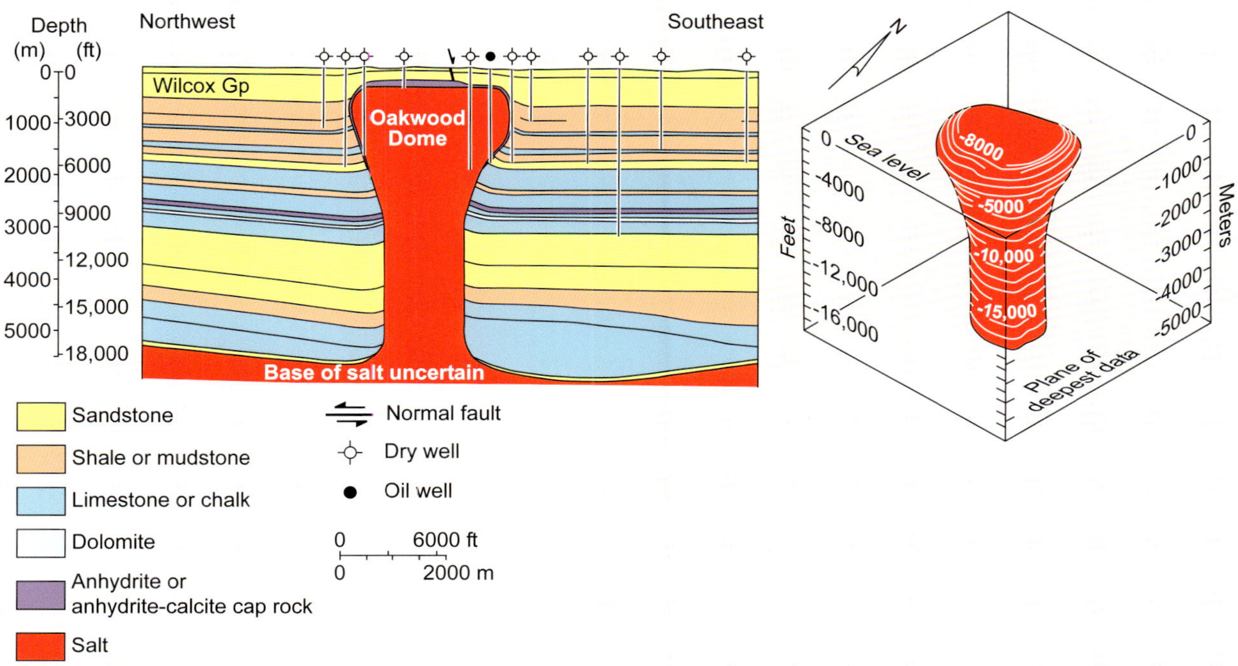

Figure 2.08. Passive diapirism. The overhang near the top of Oakwood Dome formed when, driven by increasing average density of sediments in the adjacent withdrawal basins, salt in the dome rose faster than the local sediments aggraded. This activity allowed the dome to extrude outward, causing the diapir to flare upward. The dome continued to rise slightly after burial in Wilcox time, arching the roof and diverting Wilcox sands around the dome. Despite the slight late arching, the dome retains the relatively flat top imposed by dissolution and erosion. Seni and Jackson (1984) calculated that 70% of the salt in the basin was dissolved at diapir crests. The cap rock at Oakwood Dome is 30 m thick, which is too thin to account for the missing salt, so the authors concluded that the missing salt and its less soluble minerals were dispersed by extrusion and erosion. Modified from Jackson and Seni (1984).

EAST TEXAS BASIN—Crestal Anticline above Butler Dome, Texas, USA

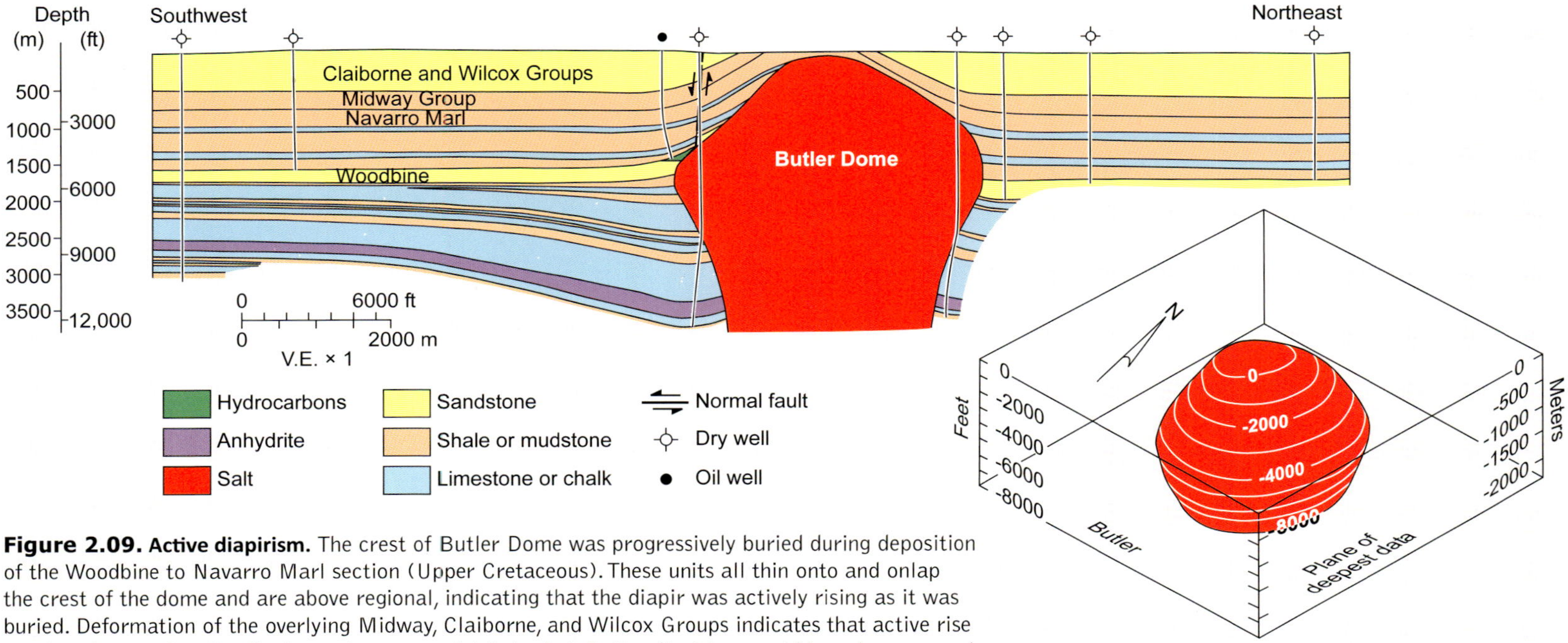

Figure 2.09. Active diapirism. The crest of Butler Dome was progressively buried during deposition of the Woodbine to Navarro Marl section (Upper Cretaceous). These units all thin onto and onlap the crest of the dome and are above regional, indicating that the diapir was actively rising as it was buried. Deformation of the overlying Midway, Claiborne, and Wilcox Groups indicates that active rise continued even after the diapir crest was completely buried. Active diapirism could have been caused by either (a) diapir inflation by pressurized salt from the Louann source layer or (b) diapir shortening. Either way, roof uplift was accompanied by normal faulting. One of these normal faults trapped a small pool of oil on the southwest side of the dome. Modified from Jackson and Seni (1984).

EBRO BASIN—Detail of Surface Structure, Cardona, Muntanya de Sal, Spain

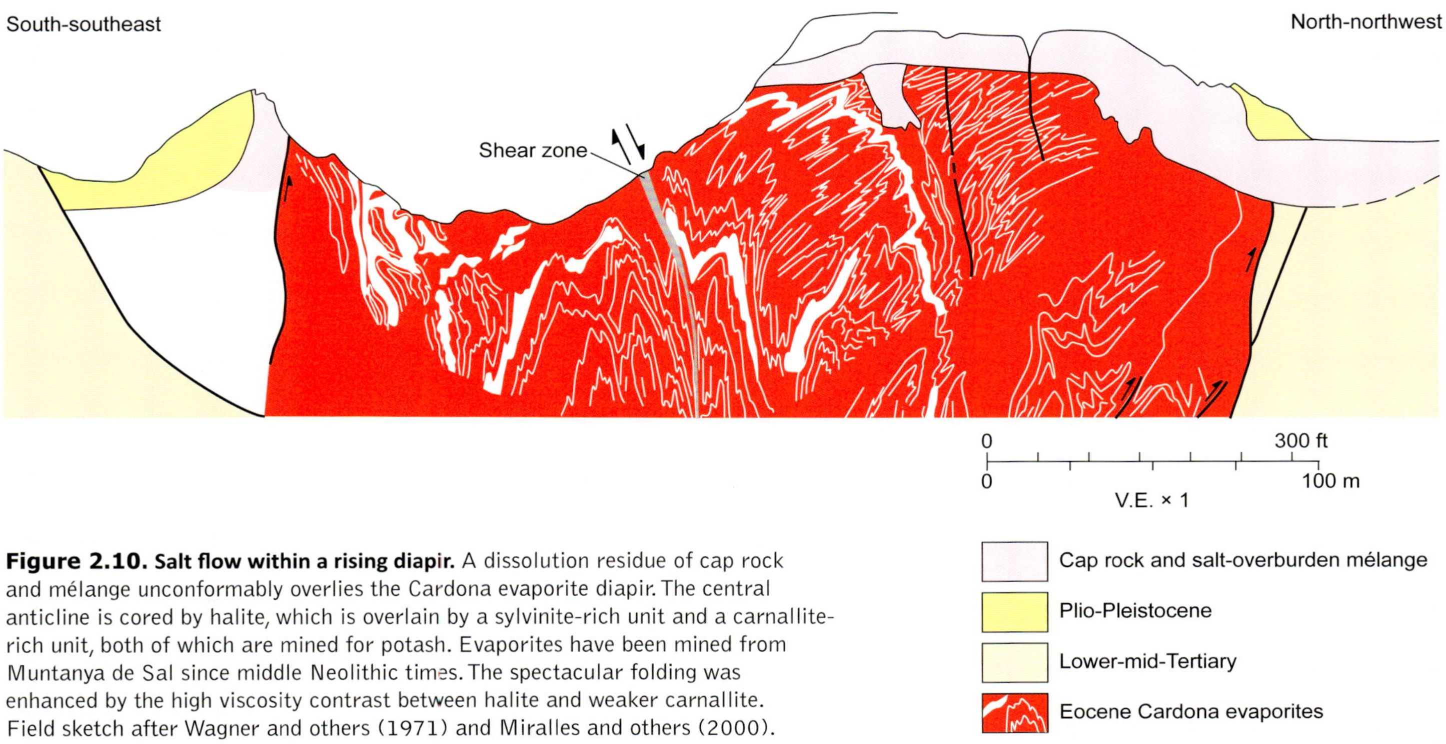

Figure 2.10. Salt flow within a rising diapir. A dissolution residue of cap rock and mélange unconformably overlies the Cardona evaporite diapir. The central anticline is cored by halite, which is overlain by a sylvinite-rich unit and a carnallite-rich unit, both of which are mined for potash. Evaporites have been mined from Muntanya de Sal since middle Neolithic times. The spectacular folding was enhanced by the high viscosity contrast between halite and weaker carnallite. Field sketch after Wagner and others (1971) and Miralles and others (2000).

GREAT KAVIR—Satellite Image of Diapirs 10 and 5, Central Iran

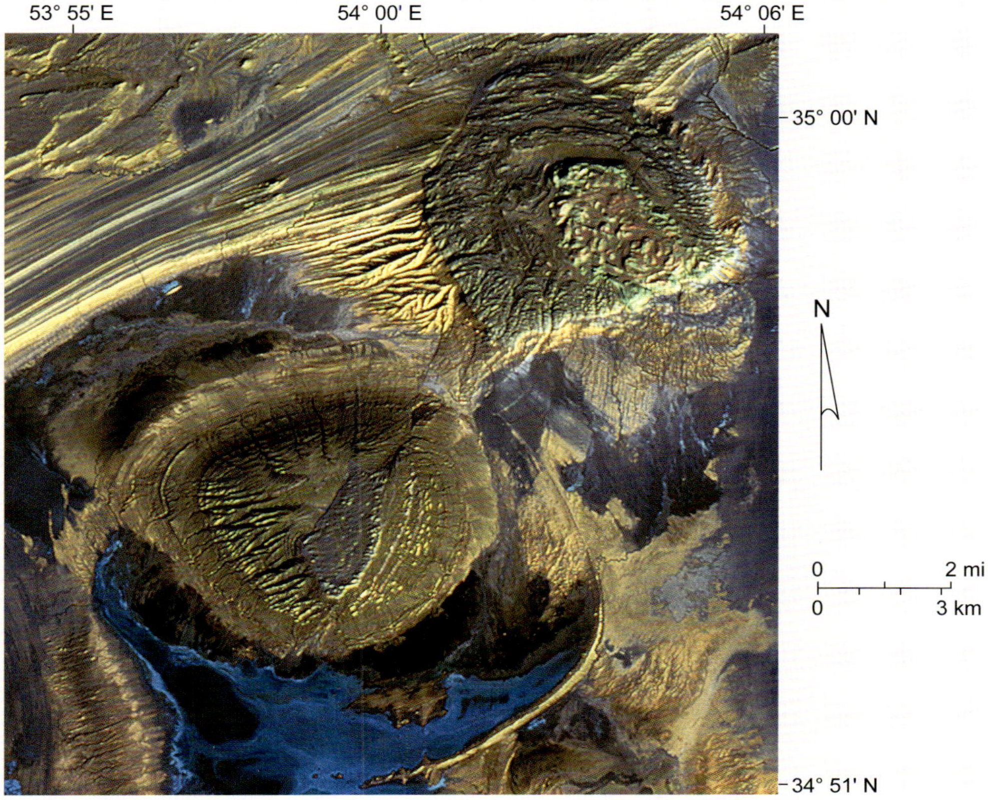

Figure 2.11. Active diapirism. Dome 10 (southwest) lies largely in the subsurface. The topography intersects its largely concordant roof (see aerial photograph in Figure 2.12). Dome 5 has pierced to a higher level, exposing strongly discordant sediment truncations against the sides of the diapir. The peripheral sink to the south of Dome 5 contains an ephemeral playa lake (blue). Image courtesy of ConocoPhillips.

GREAT KAVIR—Aerial Photograph of Diapir 10, Central Iran

Figure 2.12. Active diapirism. Photograph shows an erosional section across the uplifted roof of an arched diapir. The exposed evaporites (white residual gypsum soil overlying Eocene salt) in the center are the tip of a broader bulb in the subsurface. Roof sediments are broadly concordant to the evaporite core but locally discordant where faults intersect the salt. Black and white surrounding plain is a Holocene salt desert (kavir) in different stages of desiccation. See satellite image in Figure 2.11. After Jackson and others (1990); photograph by Worldwide Aerial Surveys.

GULF OF MEXICO BASIN—White Castle Dome, Louisiana, USA

Figure 2.13. Active diapirism. All units in the roof are stratigraphically thin, suggesting a long period of continuous diapir rise. Continuous rise is consistent with active diapirism. In this instance, all structures in the roof were extensional. After Smith and Reeve (1970) and Schultz-Ela and others (1994).

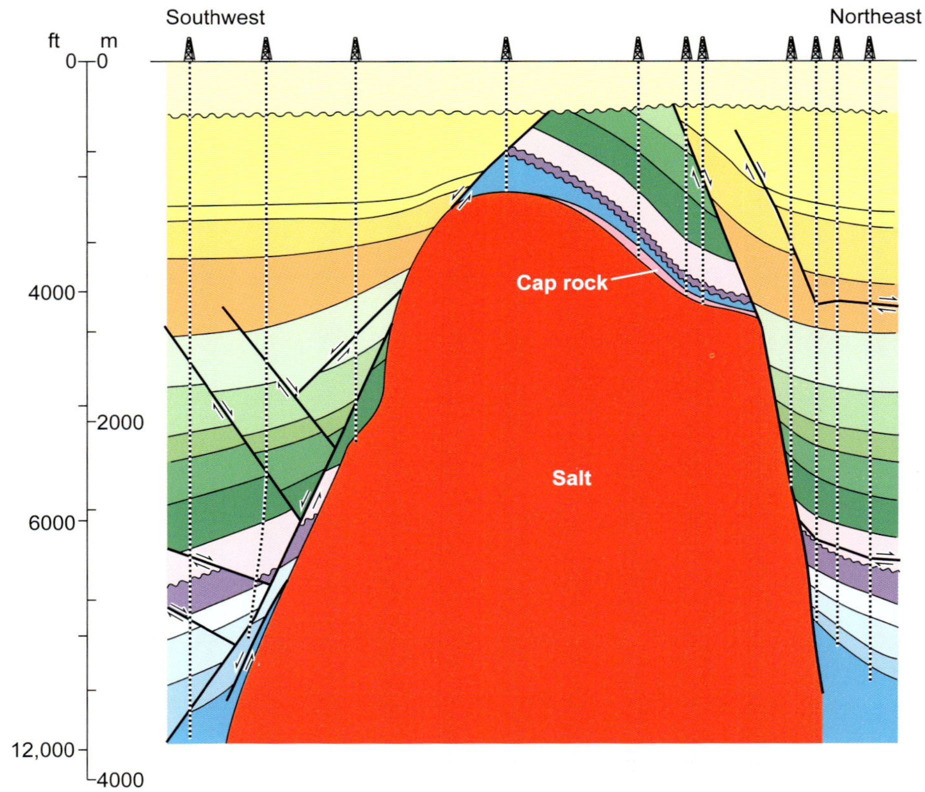

GULF OF MEXICO BASIN—Diapir in Green Canyon Area, Offshore USA

Figure 2.14. Active diapirism. All units in the roof thin over the crest, which is typical of halokinetic active diapirism. However, thinning is most extreme in the shallowest units, which may indicate recent acceleration in roof uplift. This acceleration could reflect a phase of diapir shortening or a sedimentary hiatus. After Seni (1994).

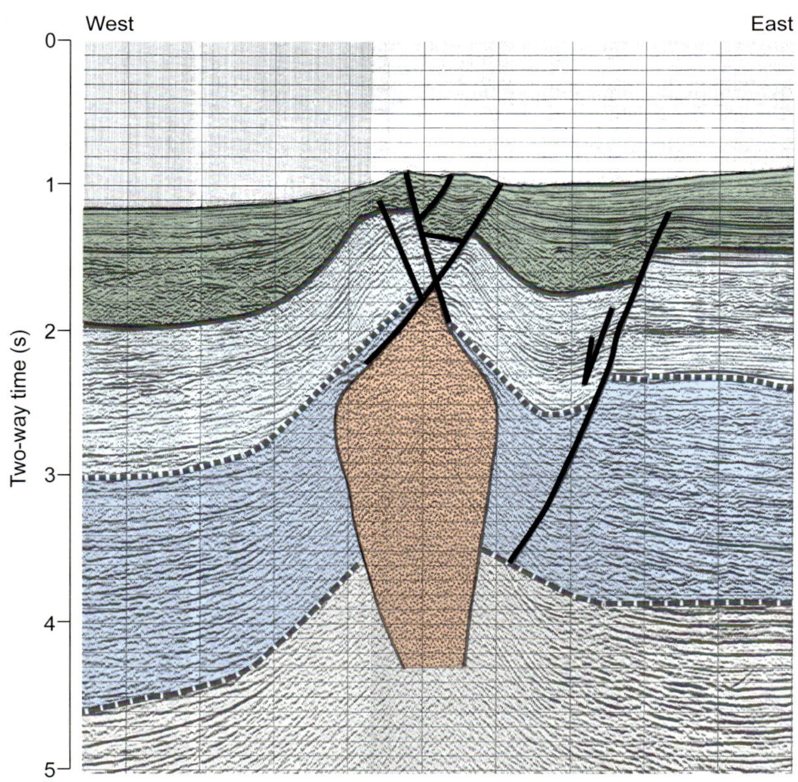

GULF OF MEXICO BASIN—Upturned Strata on Diapir Flank, Offshore USA

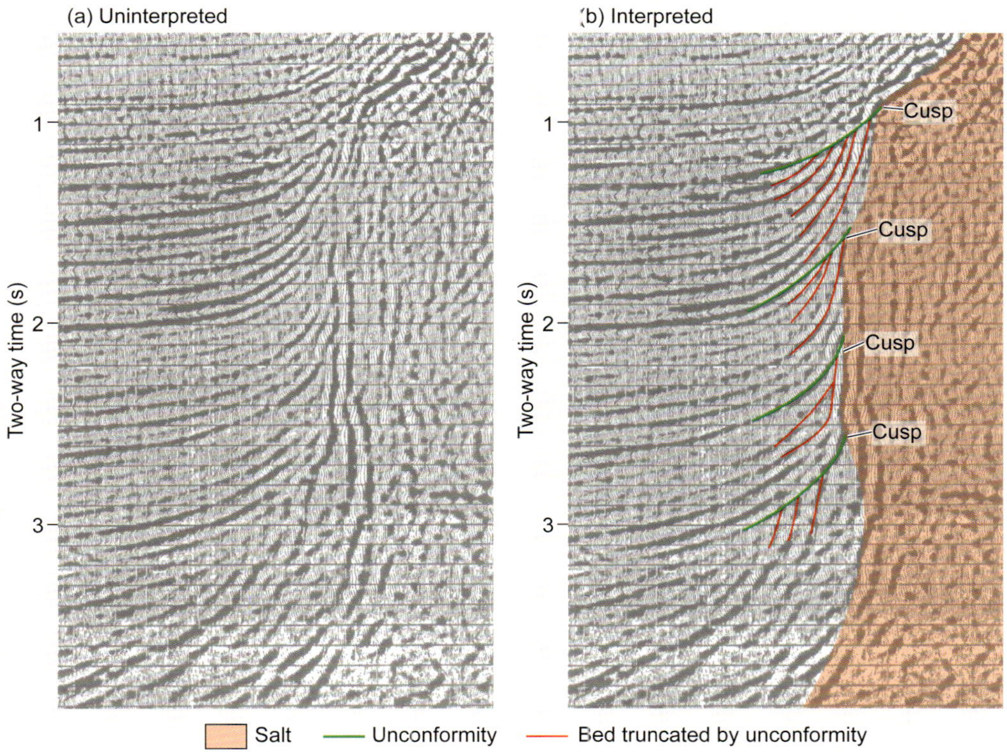

Figure 2.15. Upturn of beds against a diapir flank. Several unconformity-capped stratal packages are upturned against the flank of the dome. These halokinetic sequences form when a diapir uplifts and rotates its roof. Any roof strata not eroded off remain as upturned flaps against the diapir flank. The intersection of a halokinetic-sequence boundary with the diapir typically produces a cusp in the salt face. Modified from Ratcliff and others (1992).

KWANZA BASIN—Inverted Diapirs, Offshore Angola

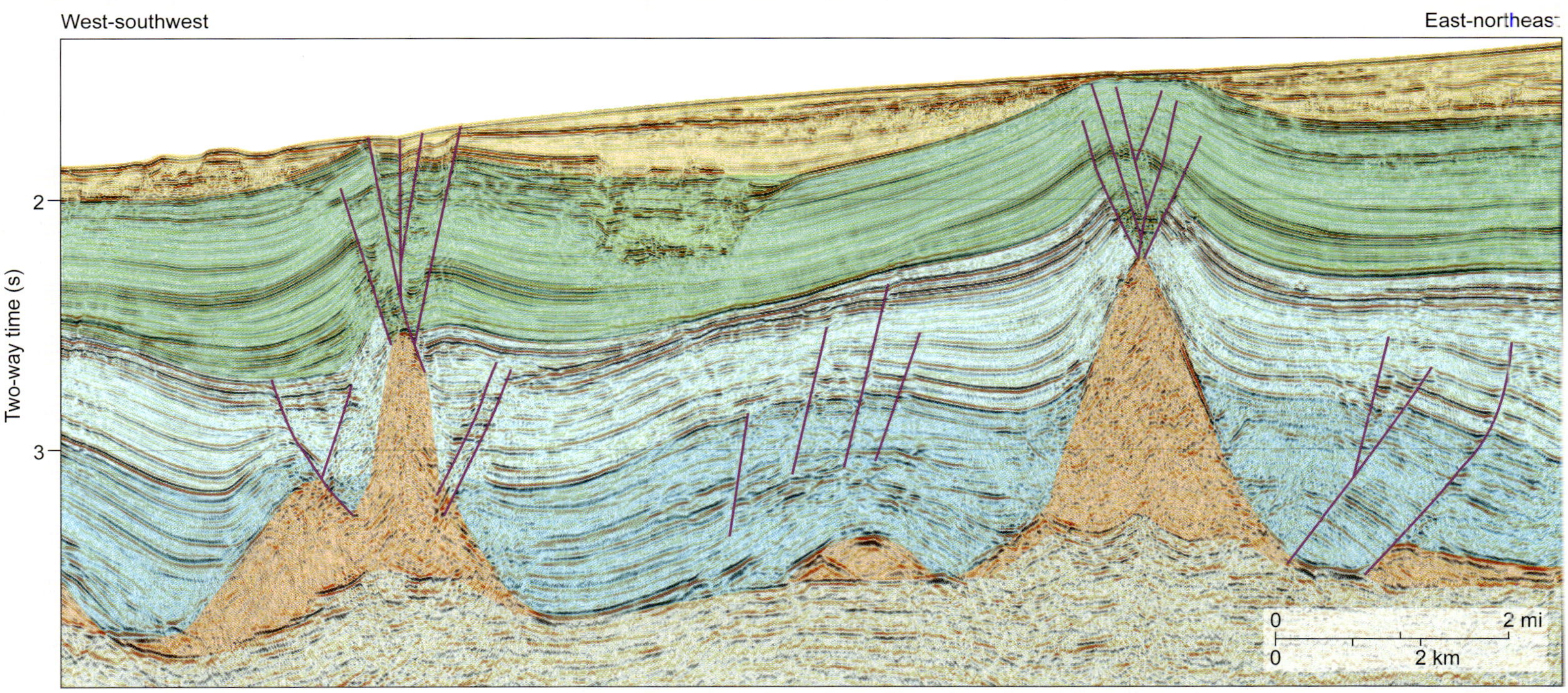

Figure 2.16. Rejuvenation of buried diapirs during shortening. Diapirs are interpreted to have shortened because of thick arched roofs consisting of constant-thickness or thickened strata. Timing of shortening is indicated by onlap of shallow orange unit. Seismic image courtesy of WesternGeco. Interpretation by Mike Hudec.

KWANZA BASIN—Cegonha Diapir, Offshore Angola

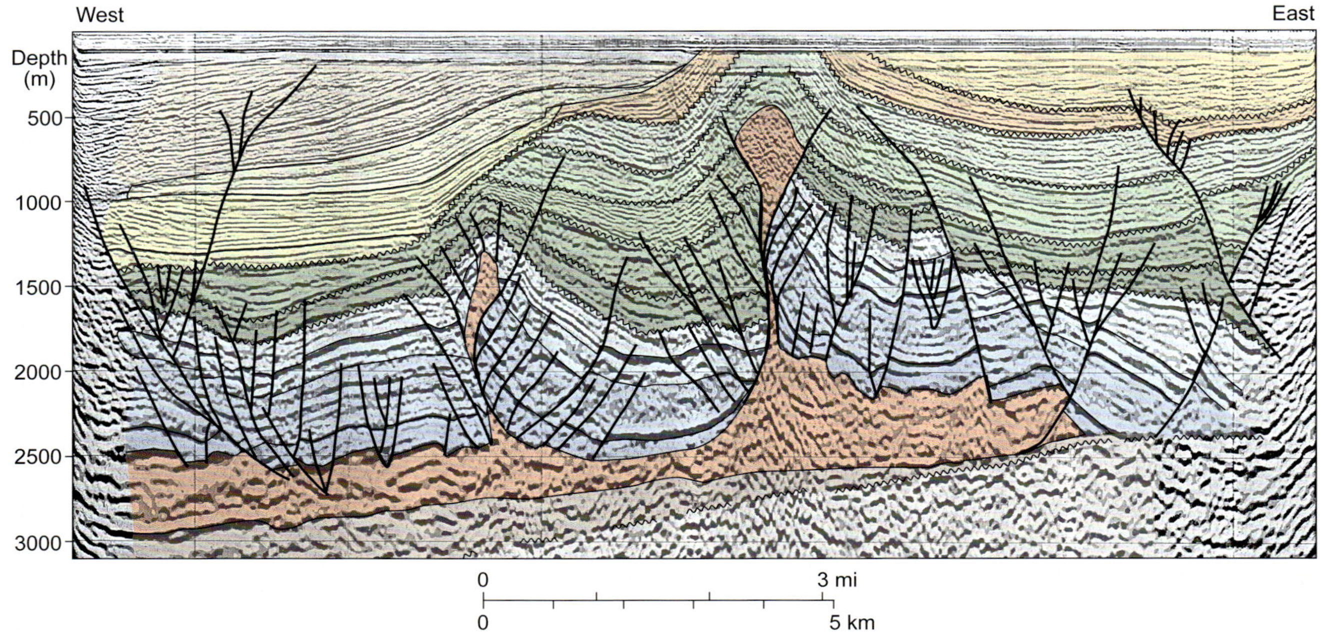

Figure 2.17. Rejuvenation of buried diapirs during shortening. Diapirs are interpreted to have shortened because of (a) hourglass diapir shapes with pinched-off waists, (b) arching of a thick roof consisting of constant-thickness strata, and (c) overall narrowness of diapirs. Strata to the east of both diapirs have overthrust the west side of each. Thick autochthonous salt to the east of the taller diapir may be a result of salt being forced back down into the source layer during diapir shortening. Structure lies on the continental extension of an oceanic transform, and shortening here on the continental shelf is likely a result of basement-involved transpression. After Duval and others (1993).

LA POPA BASIN—Diapirs in a Fold Belt, Mexico

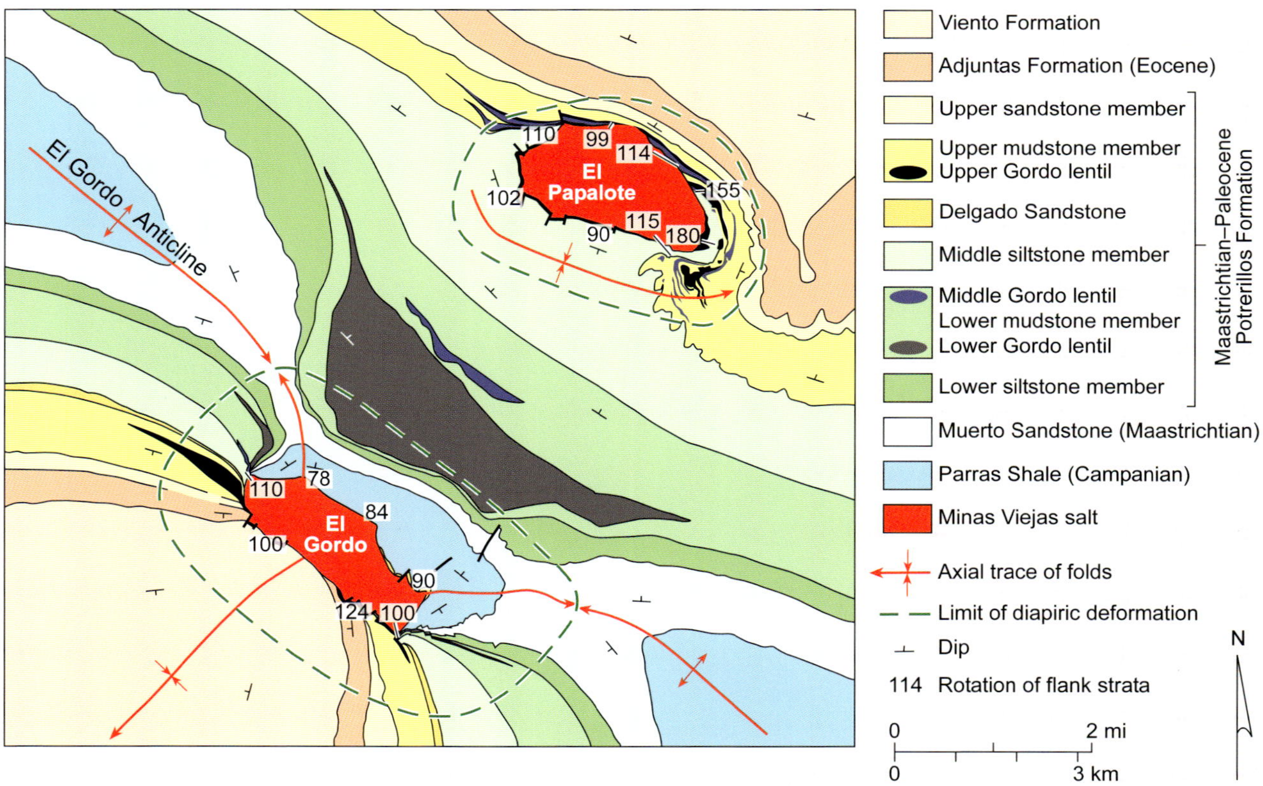

Figure 2.18. Shortening of passive diapirs. Compressional fold crests tend to be localized by preexisting diapirs, as is El Gordo diapir. However, El Papalote diapir is too close to El Gordo to match the dominant wavelength of the folds, which is imposed by the strength and thickness of the folded overburden. As a result, El Papalote diapir lies on the northeast limb of the regional El Gordo Anticline, in the core of a small parasitic fold. Carbonate lentils thinning away from the diapir are interpreted as biostromes and debris flows from the elevated diapir and its roof. See El Papalote sections (Figure 2.21), El Papalote restoration (Figure 2.20), and El Gordo map (Figure 2.19). Modified from Rowan and others (2003).

La Popa Basin—Halokinetic Sequences adjacent to El Gordo Diapir, Mexico

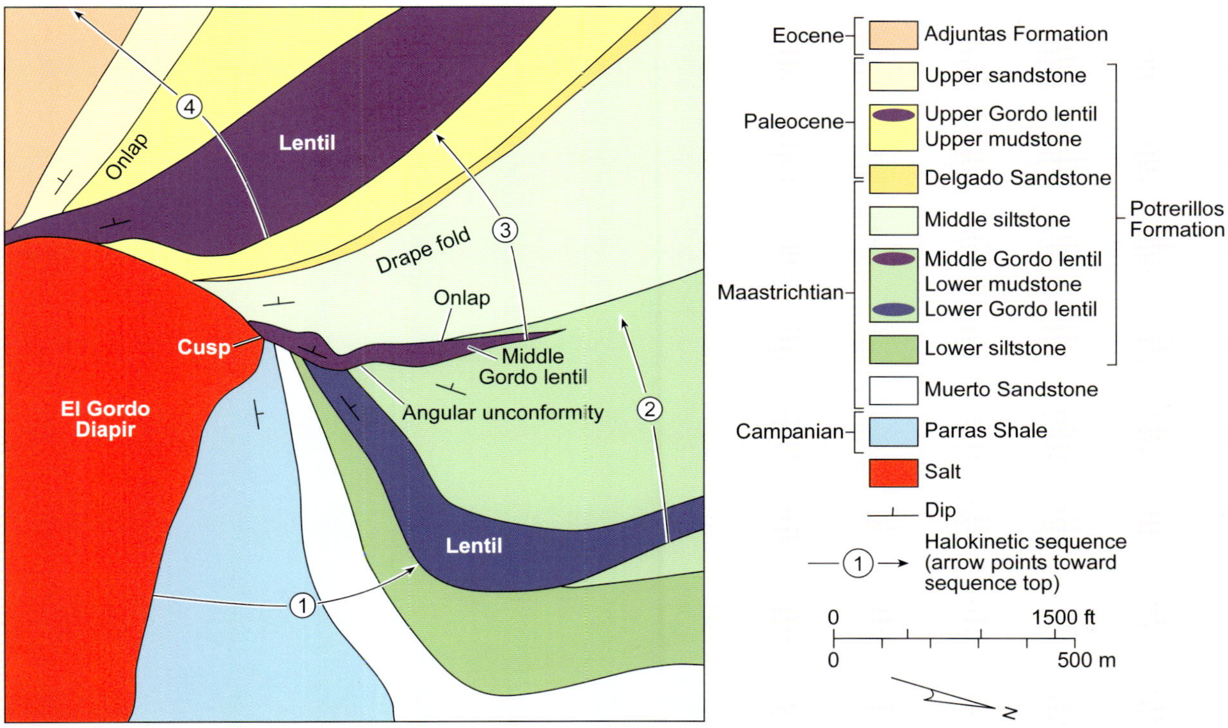

Figure 2.19. Formation of halokinetic sequences. The map is oriented to show a down-plunge view, or oblique cross section. Four stacked halokinetic sequences crop out around the northwest corner of El Gordo diapir. The upper three sequences have carbonate lentils resting on the basal unconformity. These lentils thin away from the diapir and pinch out (see El Papalote sections in Figure 2.21). The unconformity beneath the Middle Gordo lentil is strongly angular, truncating upturned beds from the underlying two sequences. This geometry (a "hook halokinetic sequence") forms when the diapir roof is lifted and shouldered aside by the rising diapir. The upturned roof erodes, forming the base of the next halokinetic sequence. Continued erosion of the diapir crest leads to deposition of the lentil on top of the unconformity. The cusp in the salt face at the intersection of the diapir and the unconformity is formed by shear along the unconformity during rotation. Modified from Rowan and others (2003).

La Popa Basin—Hook Halokinetic Sequences adjacent to El Papalote Diapir, Mexico

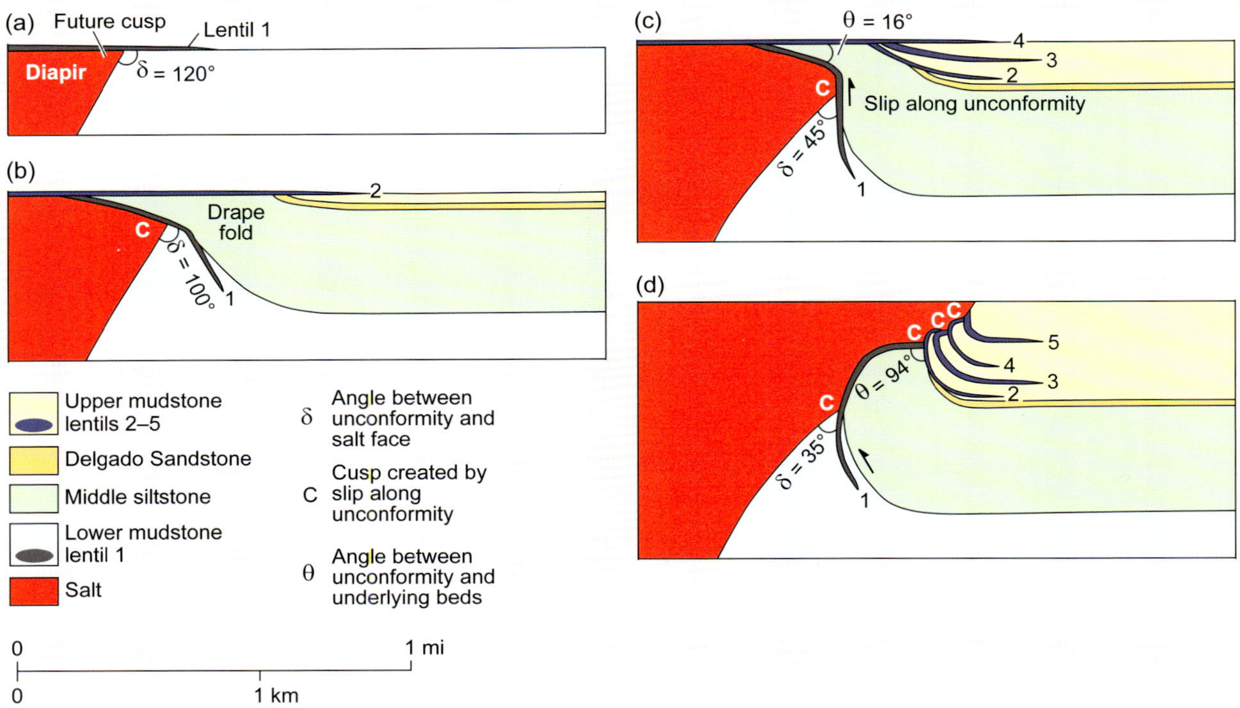

Figure 2.20. Formation of halokinetic sequences. Restoration of a section across the southeast flank of El Papalote diapir (see map in Figure 2.18 and sections in Figure 2.21) shows how halokinetic sequences form and stack. Each halokinetic sequence is marked at the base by a carbonate lentil composed of biostromes and debris flows from the diapir and its roof. Each steeply dipping sequence is formed when the rising diapir lifts, rotates, and shoulders aside its thin roof during times of reduced sedimentation, forming a "J-hook unconformity." As beds fold, they slip along unconformities, which changes angles between bedding and the salt face, and between unconformities and underlying beds. Cusps in the salt face are formed at intersections between the diapir flanks and roof. Modified from Rowan and others (2003).

LA POPA BASIN—Hook Halokinetic Sequences adjacent to El Papalote Diapir, Mexico

Figure 2.21. Formation of halokinetic sequences. Beds are steeply upturned to overturned against the southeast flank of El Papalote diapir. The upturned strata are grouped into unconformity-bounded packages called "halokinetic sequences." Each halokinetic sequence is marked at the base by a carbonate lentil, composed of biostromes and debris flows from the diapir and its roof, including exotic meta-igneous inclusions formerly entrained in salt (see map in Figure 2.18). Each steeply dipping sequence forms when the rising diapir lifts, rotates, and shoulders aside its thin roof when sedimentation slows (see restoration in Figure 2.20). Stacking of six halokinetic sequences produces diagnostic "J-hook unconformities." Modified from Giles and Lawton (2002) and Rowan and others (2003).

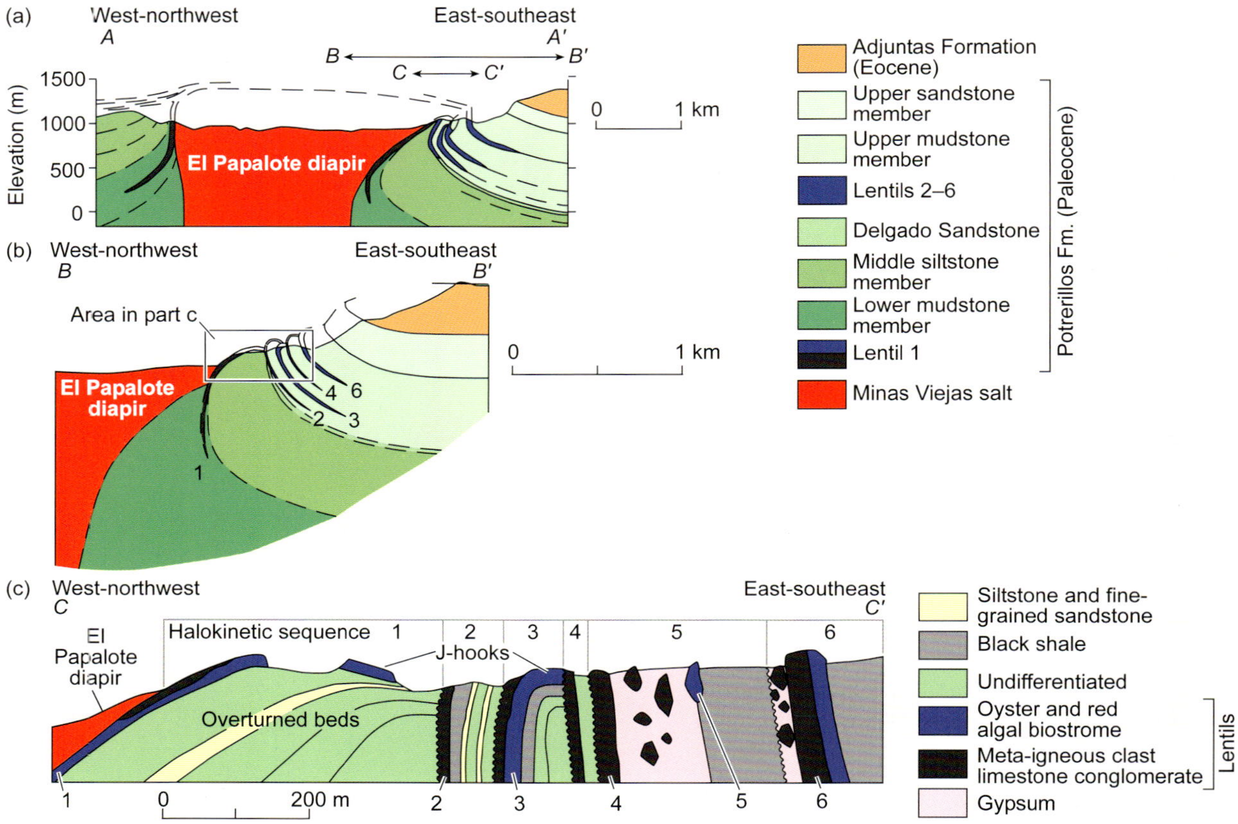

LOWER CONGO BASIN—Pinched-Off Diapir, Offshore Gabon

Figure 2.22. Rejuvenation of buried diapir during shortening. The pinched-off diapir stem and northwest-directed thrusting of the hanging wall on the steep weld suggest large shortening of the diapir. The roof is much less squeezed than are the deeper parts, suggesting that most of the shortening took place before the roof was deposited. The roof was arched in two stages. Early arching is shown by thinning of pale-green and lower orange units onto the diapir crest, possibly during waning phases of the shortening. The later stage is younger than any of the units shown on the section (all of which are arched) and may represent later compression or differential compaction. Interpretation by Martin Jackson, Mike Hudec, and Dave Jennette. Seismic data courtesy of Total.

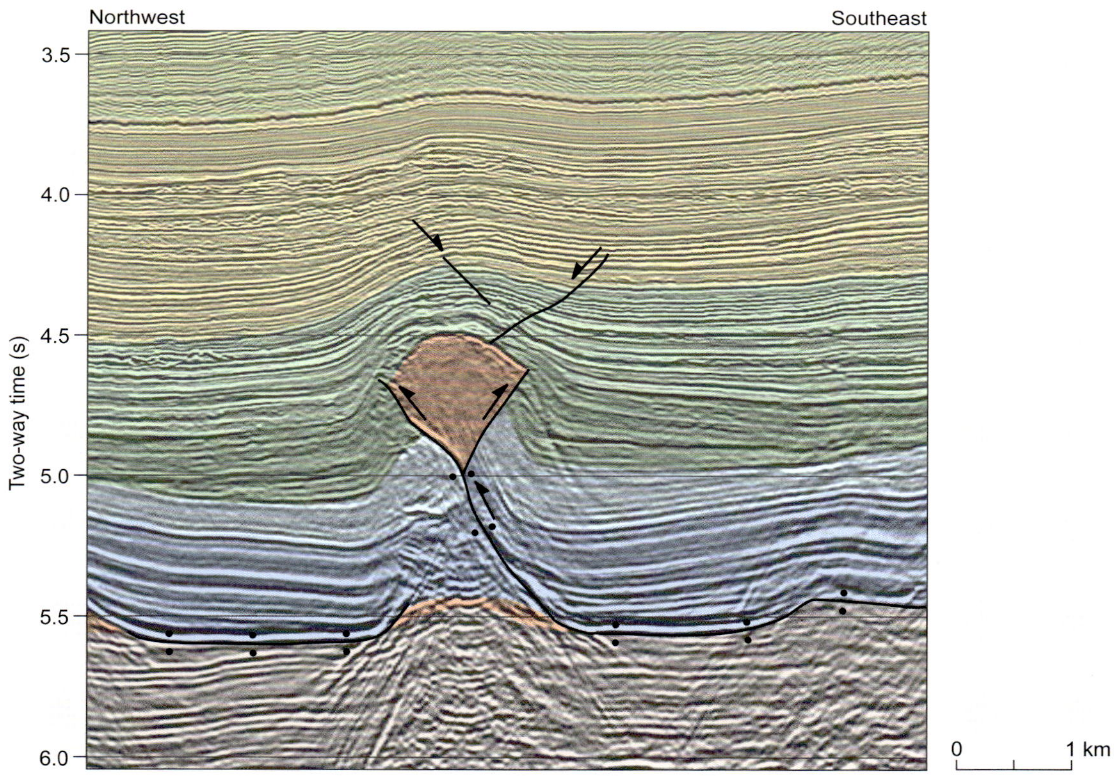

STOCKS & WALLS

MARITIME ALPS—Overthrust Diapir, Puget-Théniers, France

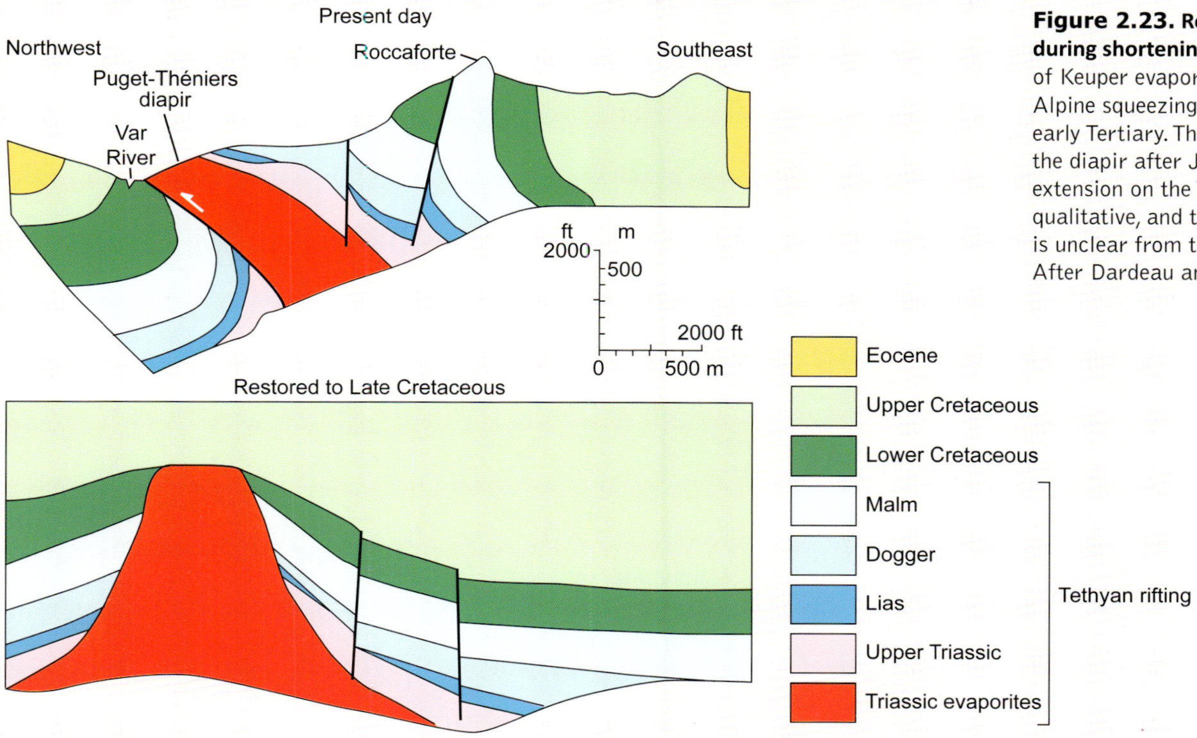

Figure 2.23. Rejuvenation of a buried diapir during shortening. The already-formed diapir of Keuper evaporites localized overthrusting during Alpine squeezing in the Late Cretaceous and early Tertiary. The restoration below shows the diapir after Jurassic and Early Cretaceous extension on the Tethyan margin. The restoration is qualitative, and the original geometry of the diapir is unclear from the outcrop information alone. After Dardeau and de Graciansky (1990).

MARITIME ALPS—Overthrust Diapir, La Cime, France

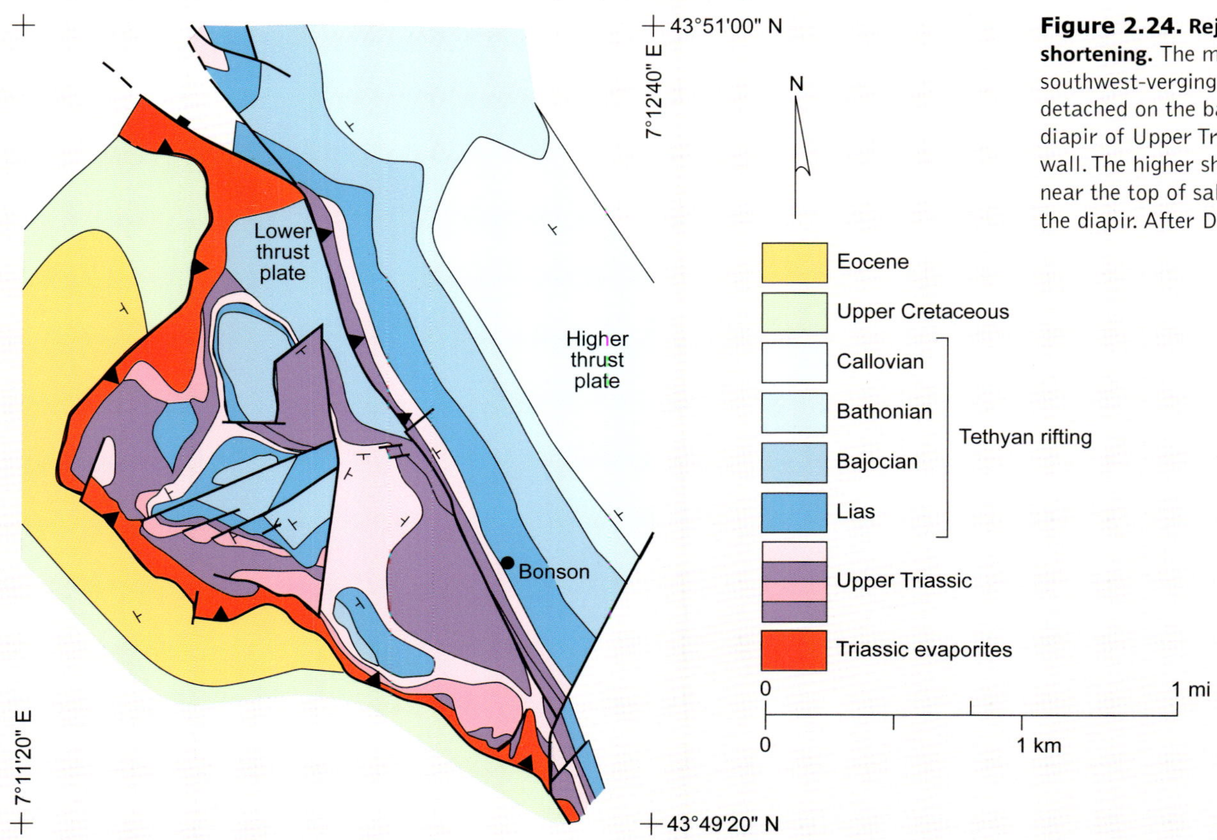

Figure 2.24. Rejuvenation of a buried diapir during shortening. The map shows two major thrust sheets in a southwest-verging thrust complex. The lower thrust sheet detached on the base of salt and carried the ribbonlike diapir of Upper Triassic Keuper evaporites in its hanging wall. The higher sheet detached in Upper Triassic sediments near the top of salt and overrode the northeast flank of the diapir. After Dardeau and de Graciansky (1990).

MARITIMES BASIN—Diapir Carapace, Broad Cove, Cape Breton Island, Nova Scotia, Canada

Figure 2.25. Shortening in a diapir roof. The inward-younging limbs of E1 limestone define a syncline in the crumpled roof strata. Overlying white gyprock was disharmonically squeezed into an antiform. Width of syncline at sea level is about 20 m. Location is 1 km SW of Broad Cove. Photograph by Martin Jackson. Interpretation after Brown (1998).

NORDKAPP BASIN—Squeezed Diapir, Barents Sea, Offshore Norway

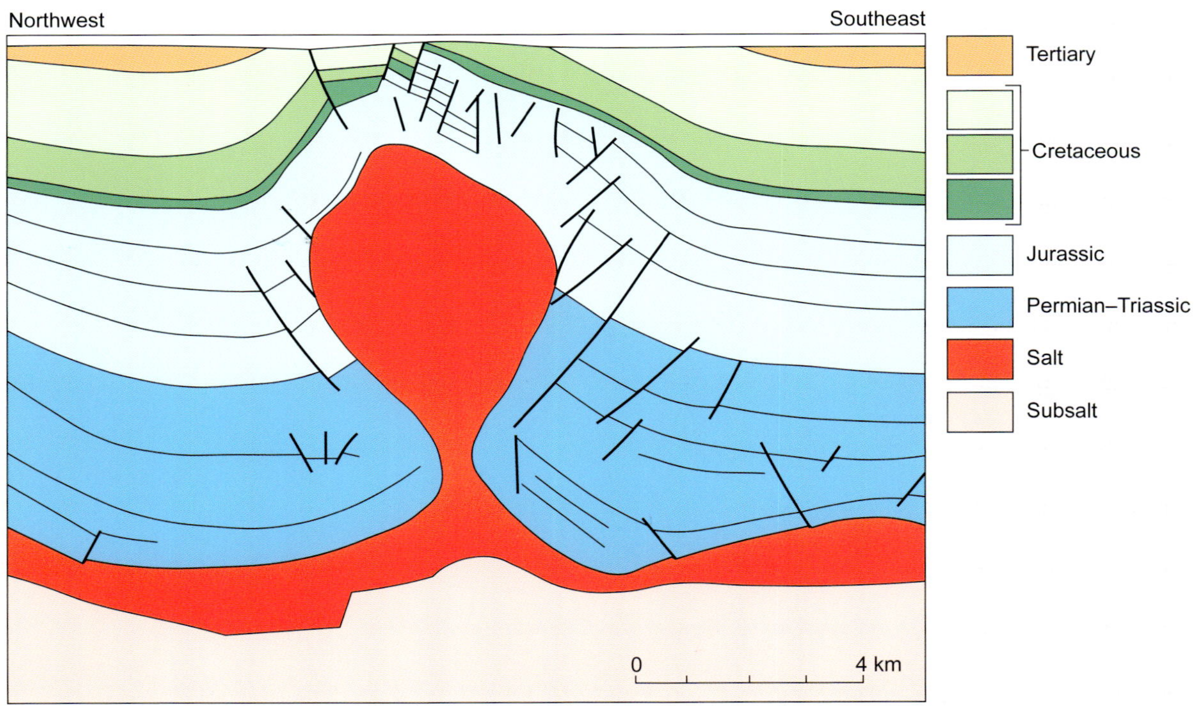

Figure 2.26. Rejuvenation of buried diapir during shortening. Two lines of evidence point to Tertiary shortening: (a) hourglass diapir shape with the neck nearly pinched off and (b) uplift and arching of a thick roof consisting of constant-thickness strata. Crestal normal faults accommodate stretching in the outer arc of the buckle fold. After Nilsen and others (1995).

NORTH GERMAN BASIN—Zwischenahn Diapir, Germany

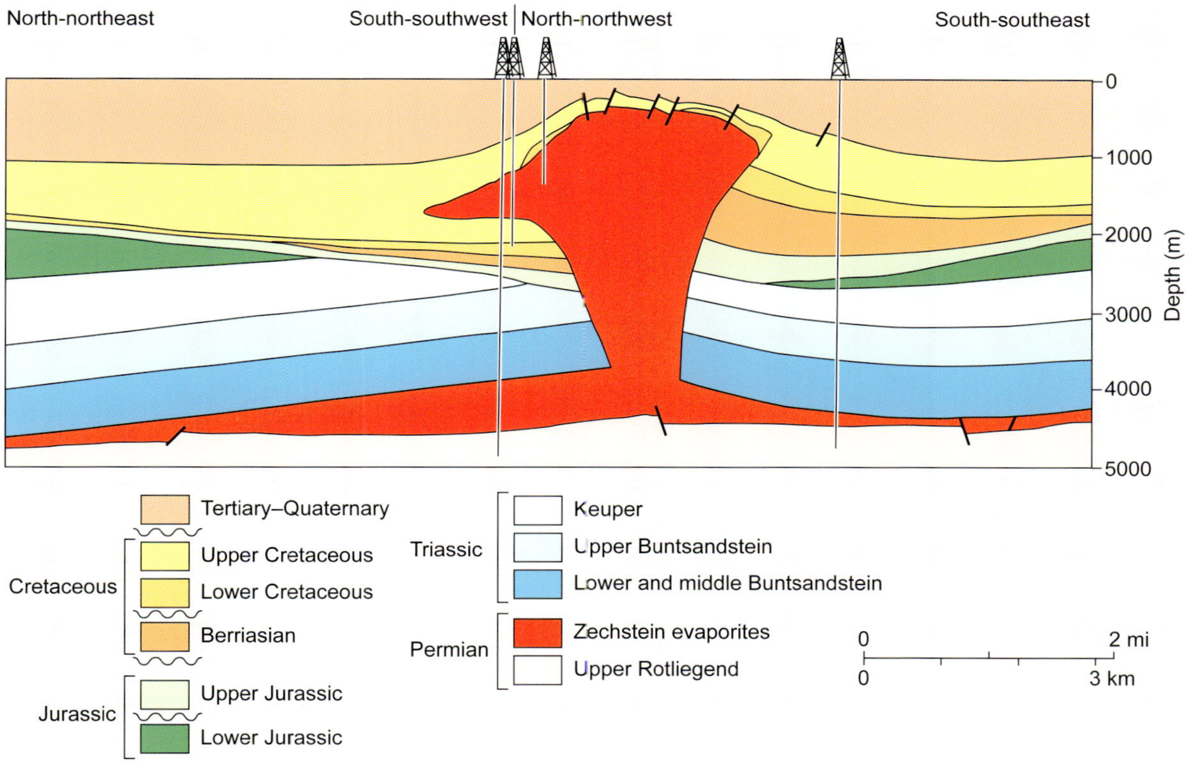

Figure 2.27. **Active diapirism.** The isopachous, deep stratigraphy shows that no salt structure existed prior to the Late Jurassic, when regional shortening created a salt-cored anticline. The unconformity at the top of the Lower Jurassic records decapitation of this anticline, which thinned the overburden enough to permit salt to pierce and start growing as a passive diapir. A second phase of shortening in the Late Cretaceous squeezed the diapir, forcing salt to extrude into the salt sheet preserved on the northeast side of the dome. Arching of the crest could be related to another phase of shortening or active diapirism during burial. After Kockel (1998) and Baldschuhn and others (1996, 2001).

NORTH GERMAN BASIN—Breached Salt Anticline, Husum Diapir, Germany

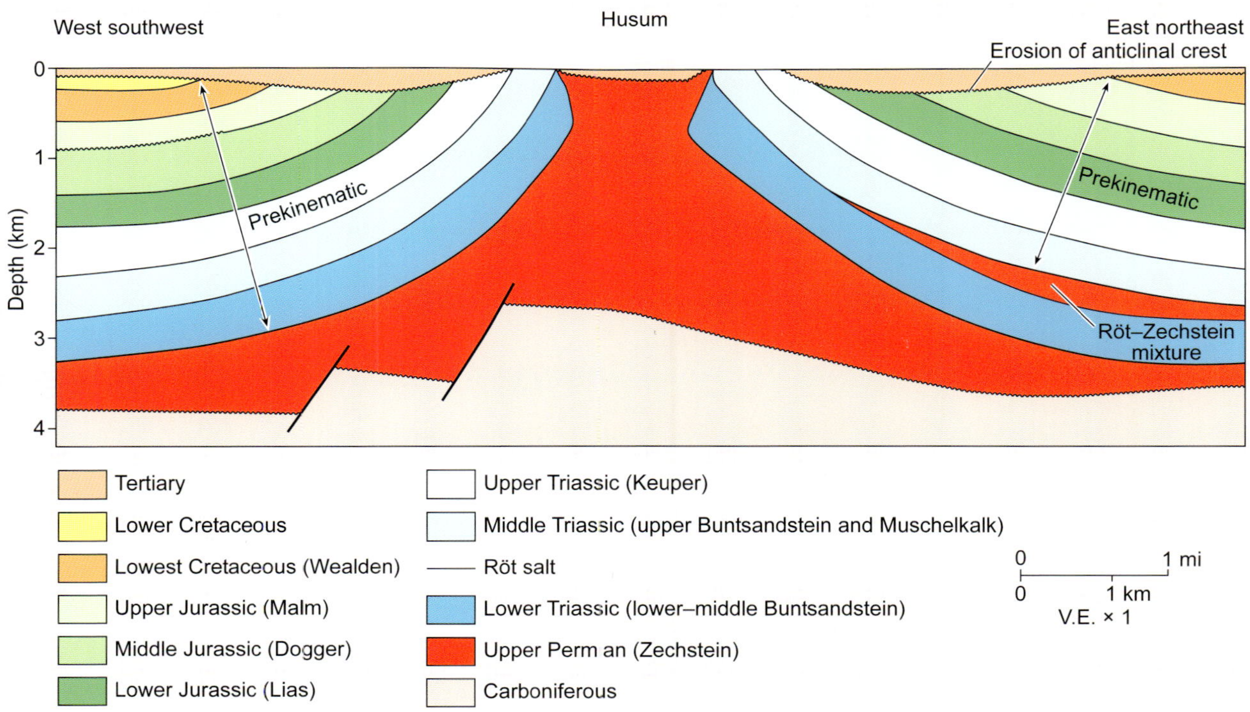

Figure 2.28. **Erosional breaching of a salt anticline.** The uniform thickness of Triassic–Lower Cretaceous units indicates that the Husum diapir did not exist prior to Late Cretaceous–Tertiary inversion of the Northwest German Basin. Shortening formed a large anticline, whose crest was eroded away. Eventually erosion was deep enough for pressurized salt to actively break through its thinned and weakened roof, forming the Husum diapir. The Röt-Zechstein salt mixture in the northeast is presumably a salt wing from a nearby diapir. After Baldschuhn and others (2001).

NORTH GERMAN BASIN—Inverted Peripheral Sinks adjacent to the Gifhorn Diapir, Germany

Figure 2.29. Transpressional basement uplift beneath a salt diapir. Primary to secondary peripheral sinks record shifting patterns of salt expulsion during early growth of the Gifhorn diapir. The diapir was buried in the Early Cretaceous then rejuvenated during Late Cretaceous–Tertiary inversion of the Northwest German Basin. The diapir overlies the Allertal Lineament, which was transpressionally reactivated during inversion. Basement uplift inverted secondary sinks on both sides of the diapir. This inversion would be difficult to explain without basement uplift because there is no thick salt beneath the sinks. Nor is it plausible that there was late salt withdrawal from beneath the primary sinks because the cessation of growth indicates that their salt was exhausted during the Jurassic. The south side of the lineament rose more than the north, as indicated by the near-complete erosion of the Lower Cretaceous to the south. After Baldschuhn and others (2001).

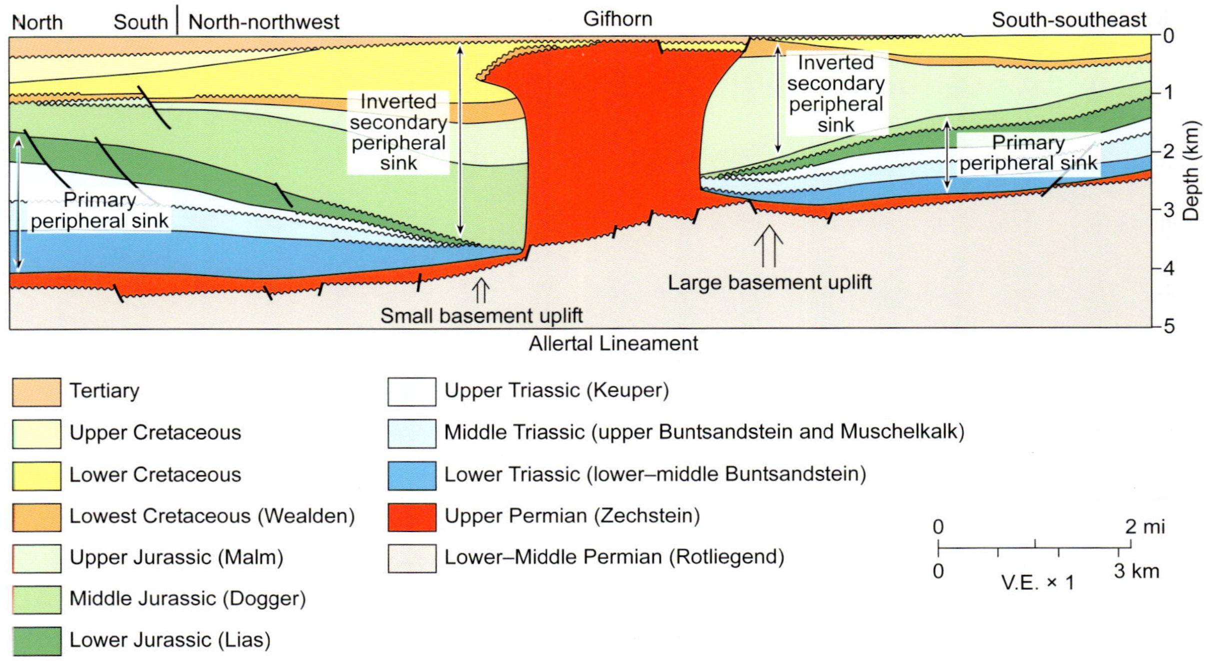

NORTH GERMAN BASIN—Diapir at Toe of Expulsion Rollover, Gorleben Diapir, Germany

Figure 2.30. Diapirism due to sediment progradation. Sediment progradation during the Triassic and Jurassic expelled salt to the northwest. A ramp in the base of salt restricted its flow and deflected salt flow upward, forming a salt anticline at the top of the ramp. Active rise of this anticline uplifted overlying strata, leading to erosion and then salt piercement. After surfacing, salt continued to rise diapirically as the Gorleben salt wall. Continued expulsion of salt from beneath the southeast side of the diapir formed a topographic low in the Cretaceous into which a salt glacier flowed. After Ge and others (1997).

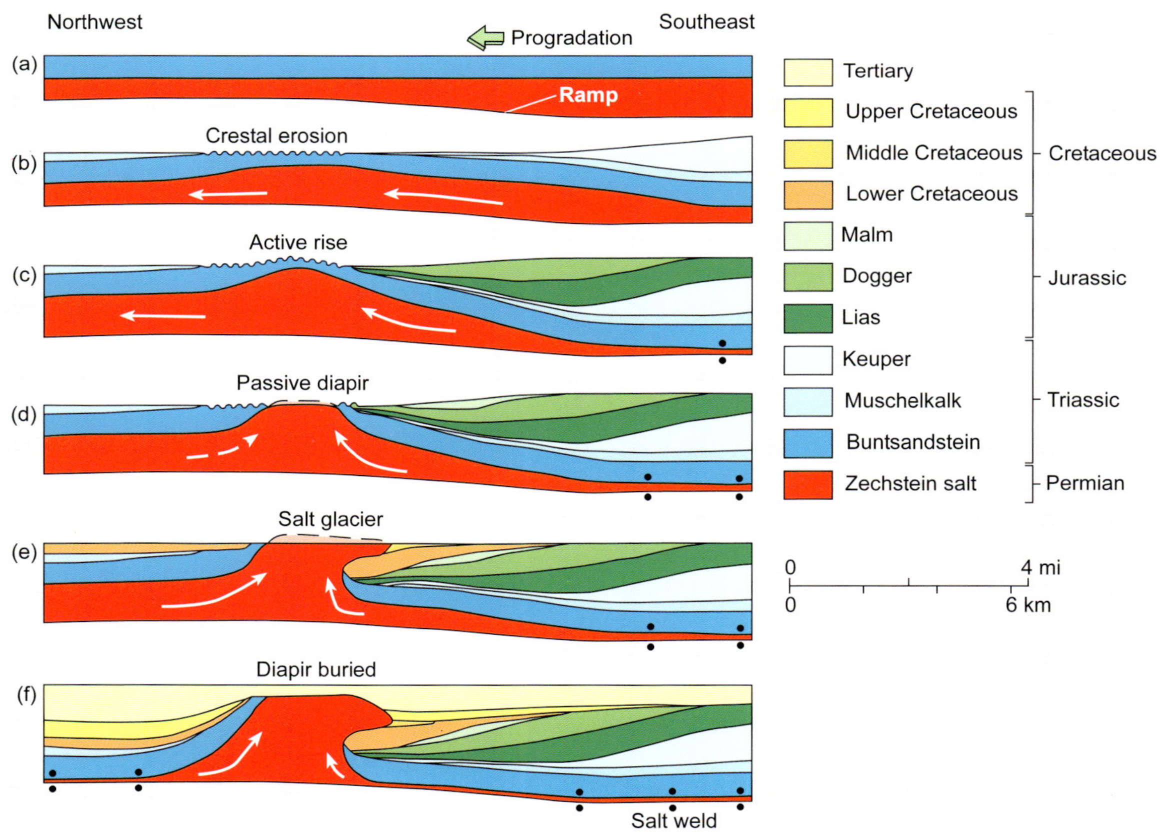

NORTH GERMAN BASIN—Hohne Turtle Structure, Germany

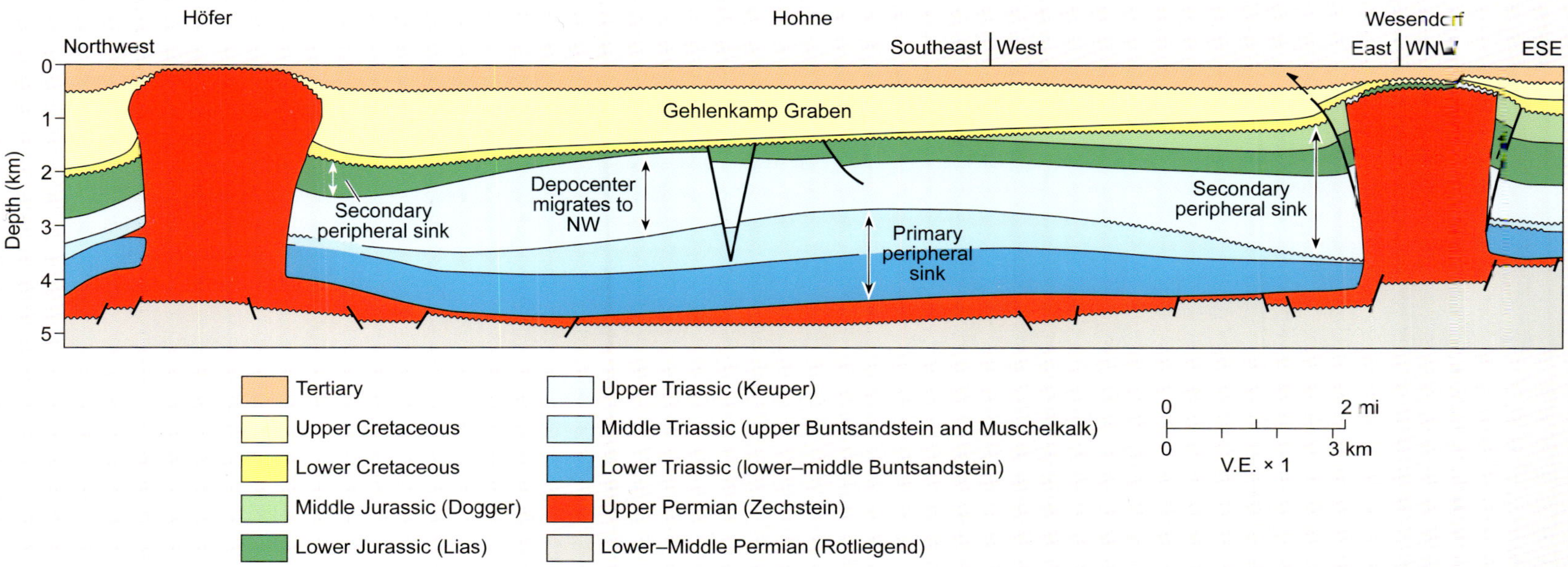

Figure 2.31. Turtle formation. Turtle structures form between two salt structures when salt expulsion shifts from a primary sink midway between the structures to secondary sinks located adjacent to the diapirs. Subsidence of the secondary sinks lets down the flanks of the primary depocenter, leaving the depocenter as an inverted structural high (turtle structure). Arching of the primary sink during this inversion can produce extensional faults over the crest of the turtle, as in the Gehlenkamp Graben. Other diapirs must be located out of the plane of section for a four-way turtle closure to form. After Baldschuhn and others (2001).

NORTH GERMAN BASIN—Heide Diapir Formed from Multiple Salt Layers, Germany

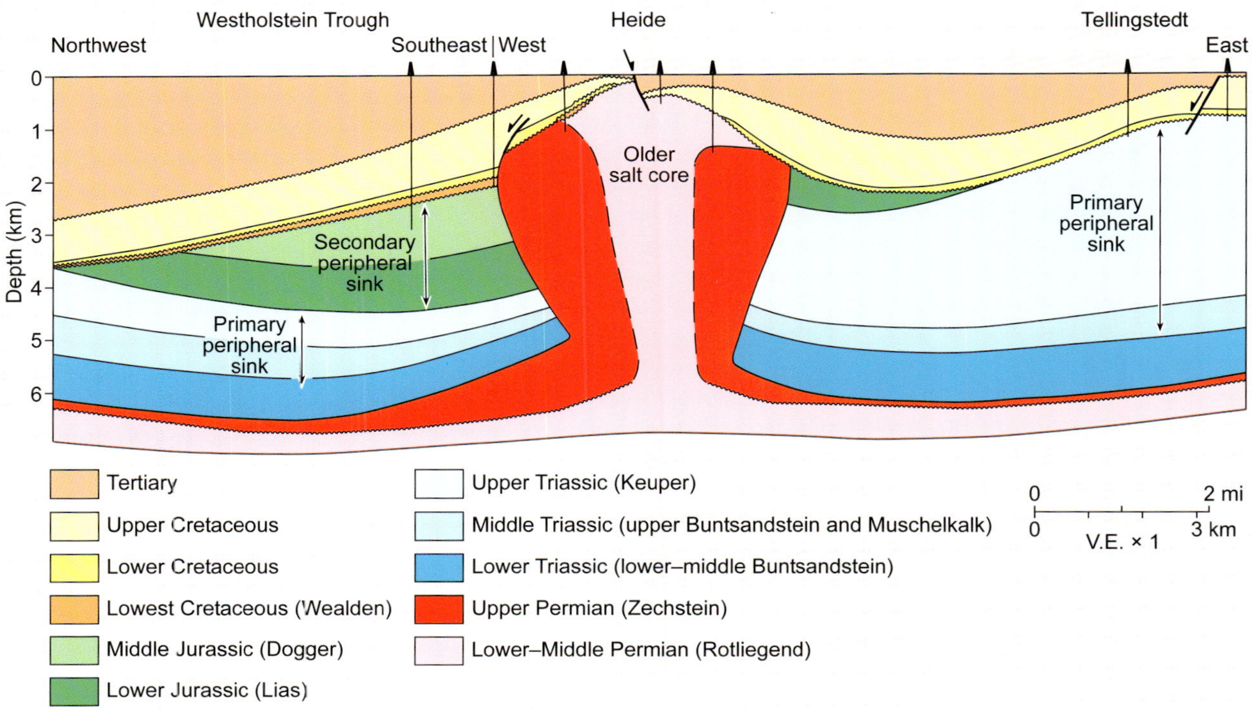

Figure 2.32. Diapiric mixing of salt layers. In parts of the Northwest German Basin, both Rötliegend and Zechstein Groups contain mobile evaporites. The deeper Rotliegend salt intrudes Zechstein salt, forming diapirs with Rotliegend cores and Zechstein rims. Normally mixing of the two salt layers would leave at least a thin Zechstein salt layer at a diapir's crest. The absence of such a layer (confirmed by well data) argues that upper parts of the diapir have been lost to dissolution. After Baldschuhn and others (2001).

NORTH GERMAN BASIN—Salt Wings adjacent to the Hope, Thören, and Meissendorf Diapirs, Germany

Figure 2.33. Intrusion of salt wings. Late Cretaceous–Tertiary basement inversion shortened all three diapirs. Some salt displaced by convergence of the squeezed diapir walls escaped as Late Cretaceous salt sheets (Hope and Meissendorf diapirs). Some salt arched diapir roofs (Thören and Meissendorf diapirs), and some rafted the displaced roof of the diapir upward (Meissendorf diapir). Yet other displaced salt intruded laterally into the Triassic Röt salt near the base of the diapir. This intrusion was usually allowed by compressional arching of units above the Röt salt as detachment folds. However, on the northeast side of the Meissendorf diapir the units above the Röt salt remained flat while the underlying Lower Triassic section sank into Zechstein salt. After Baldschuhn and others (2001).

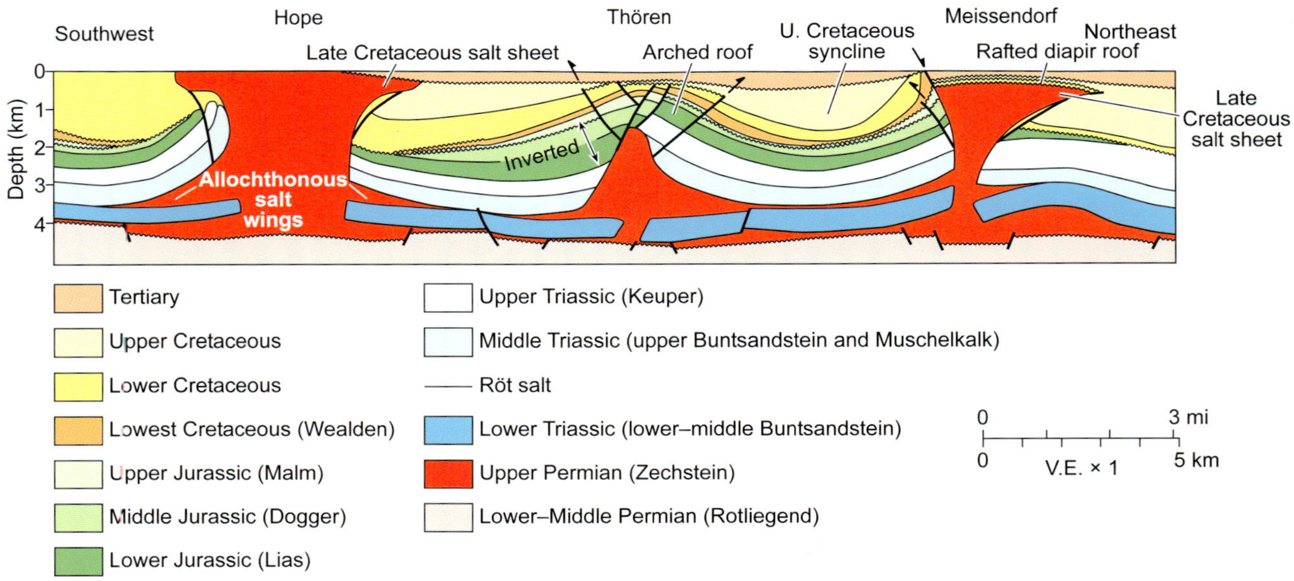

NORTH GERMAN BASIN—Restoration of Niendorf, Roshe-Thondorf, and Gülden-Brandel Diapirs, Germany

Figure 2.34. Inversion of salt-withdrawal basins. Upper Cretaceous–Tertiary shortening of the Northwest German Basin caused salt-related structures to invert in a variety of ways. Three are illustrated here. (1) Intrusion of a salt wing into the Röt salt on the west flank of the Niendorf diapir accompanied its asymmetric uplift of the roof. (2) The roofs of the Rosche-Thondorf and Gülden-Braudel diapirs arched as salt was displaced upward during diapir shortening. (3) The Upper Triassic (Keuper) withdrawal basins flanking the Gülden-Braudel diapir were inverted, apparently by uplift of the underlying basement. Basement uplift is indicated because (a) the uplifted basins are not underlain by thickened salt and (b) there is no withdrawal basin of post-Keuper age at Drawehn that could help explain the structure as a half-turtle. After Baldschuhn and others (2001).

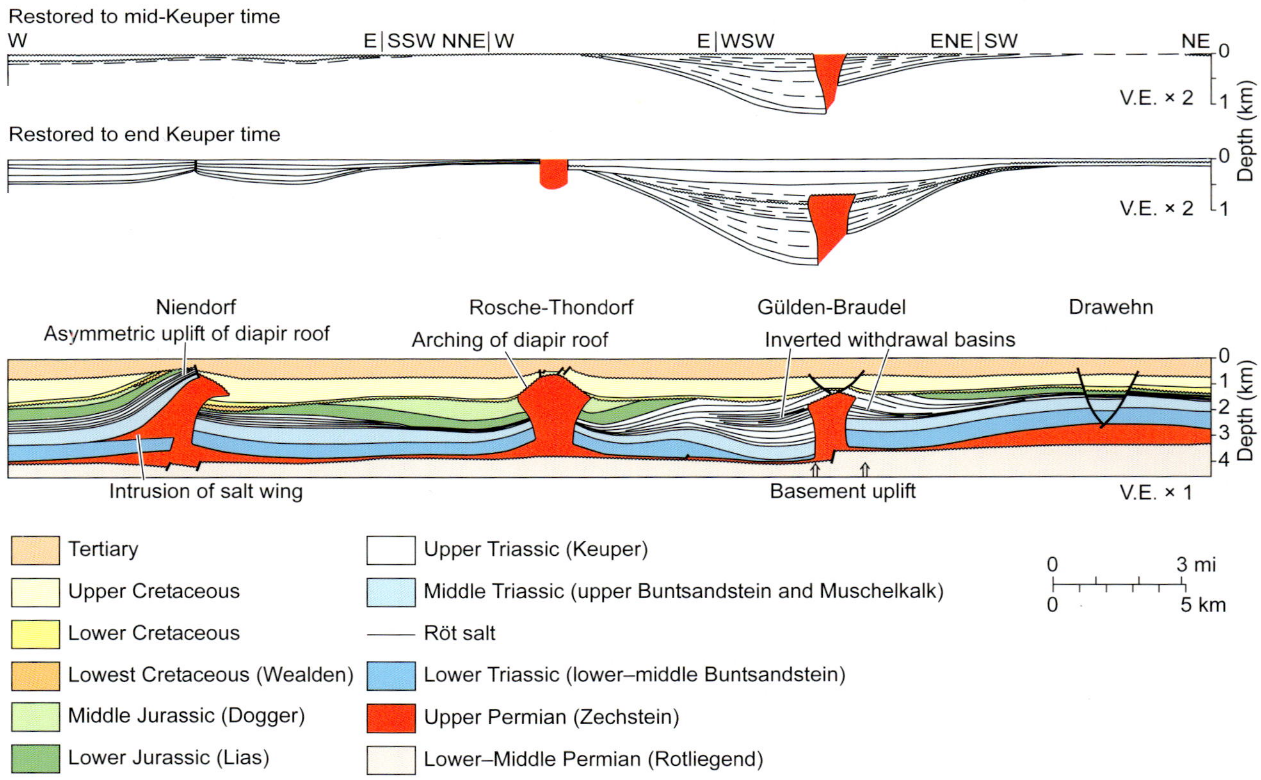

NORTH SEA—Passive Diapirs on West Central Shelf, Offshore UK

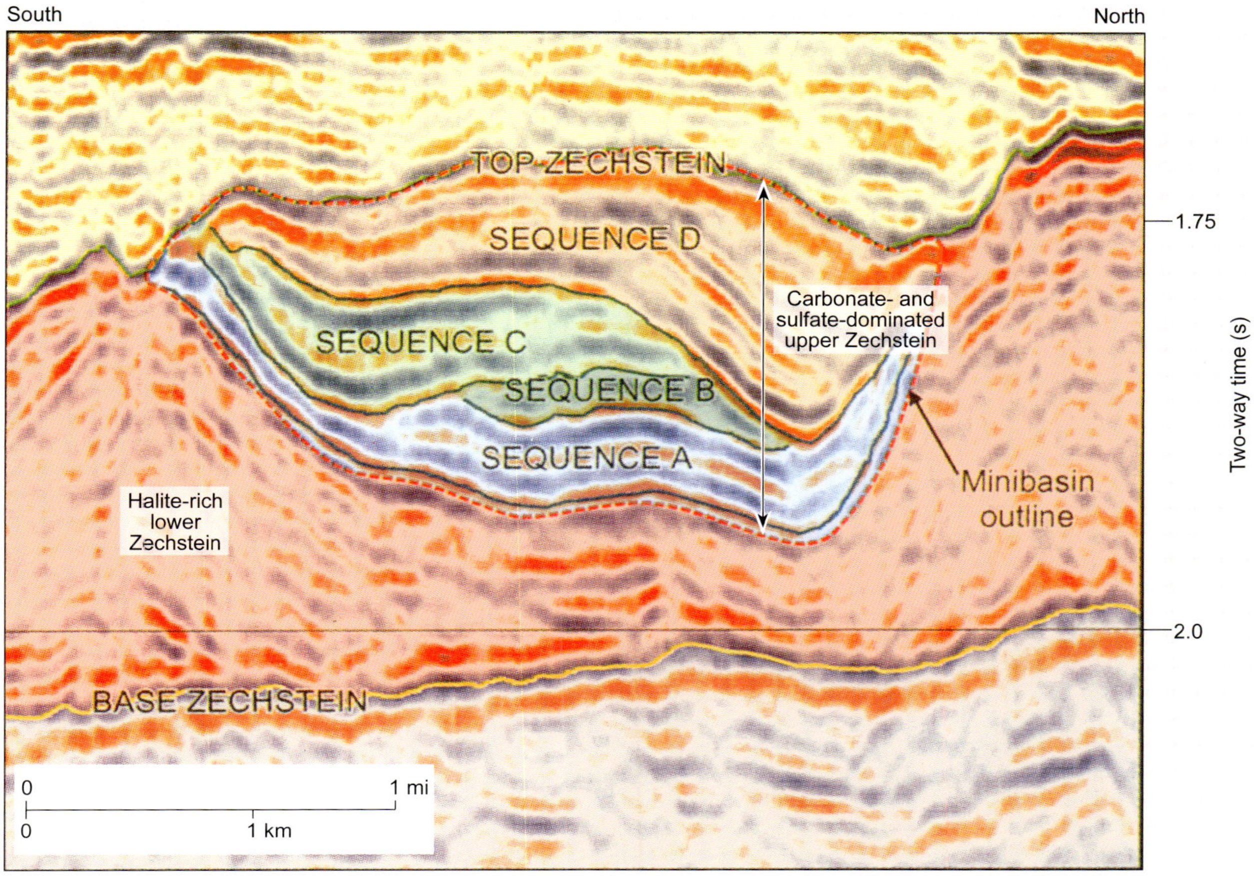

Figure 2.35. Subsidence of a dense minibasin. The upper Zechstein evaporites (sequences A–D) consist largely of carbonates and sulfates so are denser than the halite-dominated lower Zechstein. The minibasin may have subsided purely because of this density inversion, with possible assistance from regional extension and reactive rise of the triangular halite diapirs on either side. Modified from Stewart and Clark (1999).

NORTH SEA—Arched Diapirs, Central Graben, Offshore UK

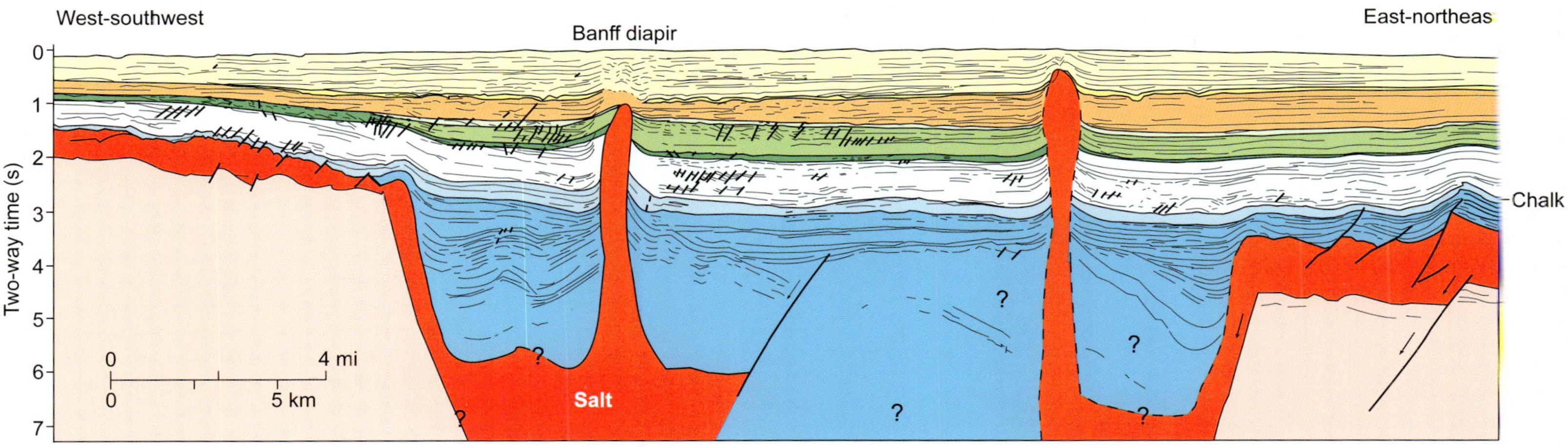

Figure 2.36. Active diapirism. Where arched roof thickness is more than 20% of diapir height, then the arching most likely requires shortening. Here the roofs are thin compared with diapir height, so the arching could be due to either shortening or active diapirism. Figures 2.37– 2.39 show other images of the Banff diapir. After Davison and others (2000a).

NORTH SEA—Seismic Section of Banff Diapir, Central Graben, Offshore UK

Figure 2.37. Active diapirism. The main criterion for interpretation as a shortened diapir is hourglass shape with stem nearly pinched off. Thrust faults adjacent to diapir shoulders suggest shortening, but they could also be rotated normal faults. Similarly, large-scale uplift of the roof (1.4 km at top Paleocene level) could be evidence of shortening. However, the roof was fairly thin during uplift, and thinning of beds onto the crest was continuous from Paleocene to early Pliocene. Both observations are consistent with halokinetic active diapirism. We interpret diapir burial beginning in the Paleocene, followed by halokinetic active diapirism until the Miocene. Pronounced onlap of Miocene strata suggests accelerated roof uplift, probably due to regional shortening. Other images of the Banff diapir include a regional section (Figure 2.36), perspective view (Figure 2.38), and structure map (Figure 2.39). After Evans and others (1999) and Davison and others (2000a).

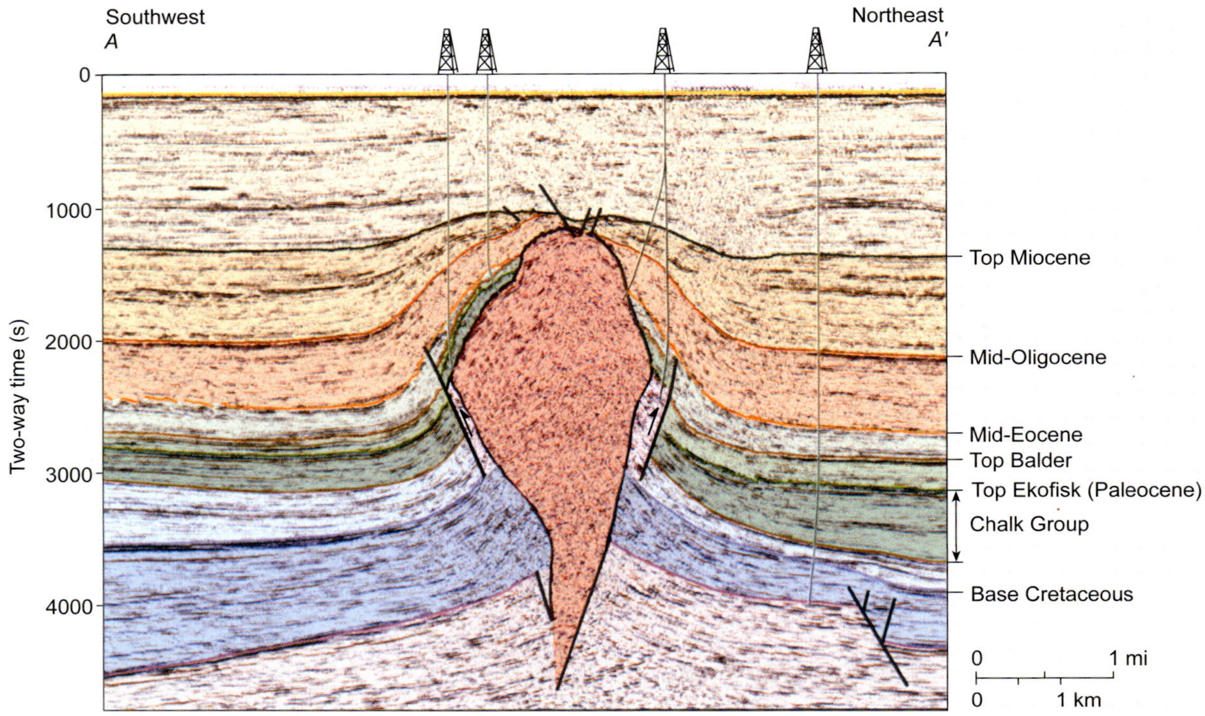

NORTH SEA—Structure Over Crest of Banff Diapir, Central Graben, Offshore UK

Figure 2.38. Radial faulting. Radial faults dominate as much as 2 to 3 km from the diapir. These faults were a result of circumferential stretching caused by arching of the elliptical diapir roof. The surrounding polygonal fault pattern is probably related to fluid expulsion. Inset map shows generalized structure map of this horizon. Faults dip consistently clockwise around the diapir, except at the northwest end. Perspective view is of Oligocene–Miocene horizon. See also regional section (Figure 2.36), seismic section (Figure 2.37), and structure map (Figure 2.39) of the Banff diapir. After Davison and others (2000a).

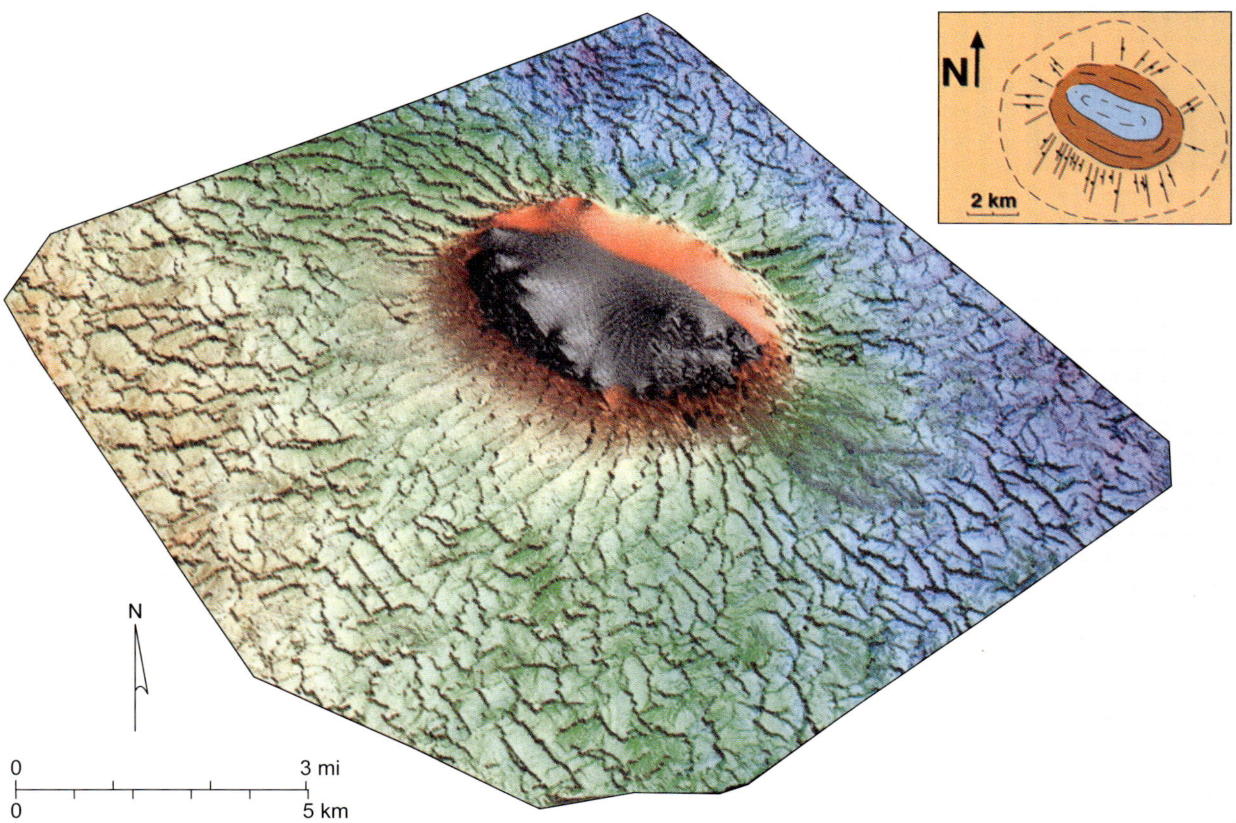

NORTH SEA—Map of Flanks of Banff Diapir, Central Graben, Offshore UK

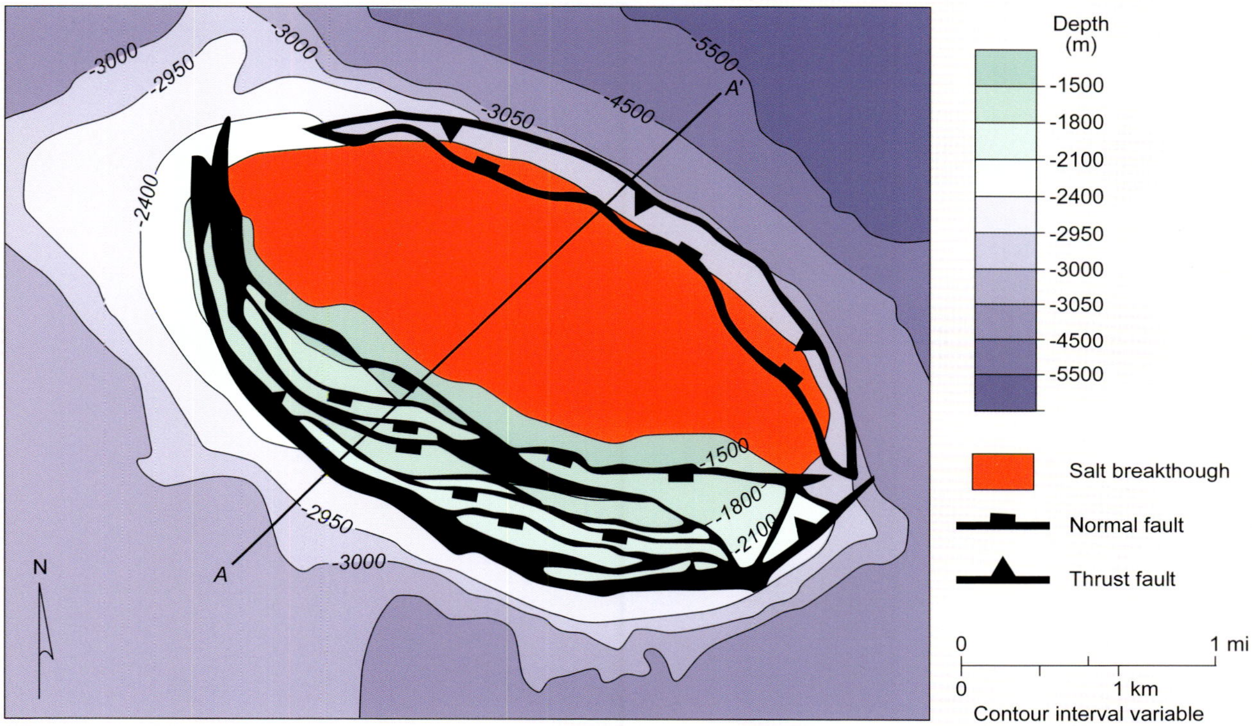

Figure 2.39. Concentric faulting. The concentric fault pattern at this structural level (top Paleocene sandstone) is very different from the radial pattern over the dome crest as shown in the perspective view (Figure 2.38). The radial pattern over the crest reflects stretching produced by arching of the diapir roof. Arching was negligible this far down the flank of the dome. Instead, the faulting accommodates the differential rise of the strata adjoining the dome. A pair of inward-dipping reverse faults bound the fault zones, although these may have initiated as normal faults prior to rotation of bedding. Line A-A′ shows location of seismic section (Figure 2.37). See also regional section of the Banff diapir (Figure 2.36). After Davison and others (2000a).

NORTH SEA—Crestal Structure of Machar Diapir, Central Graben, Offshore UK

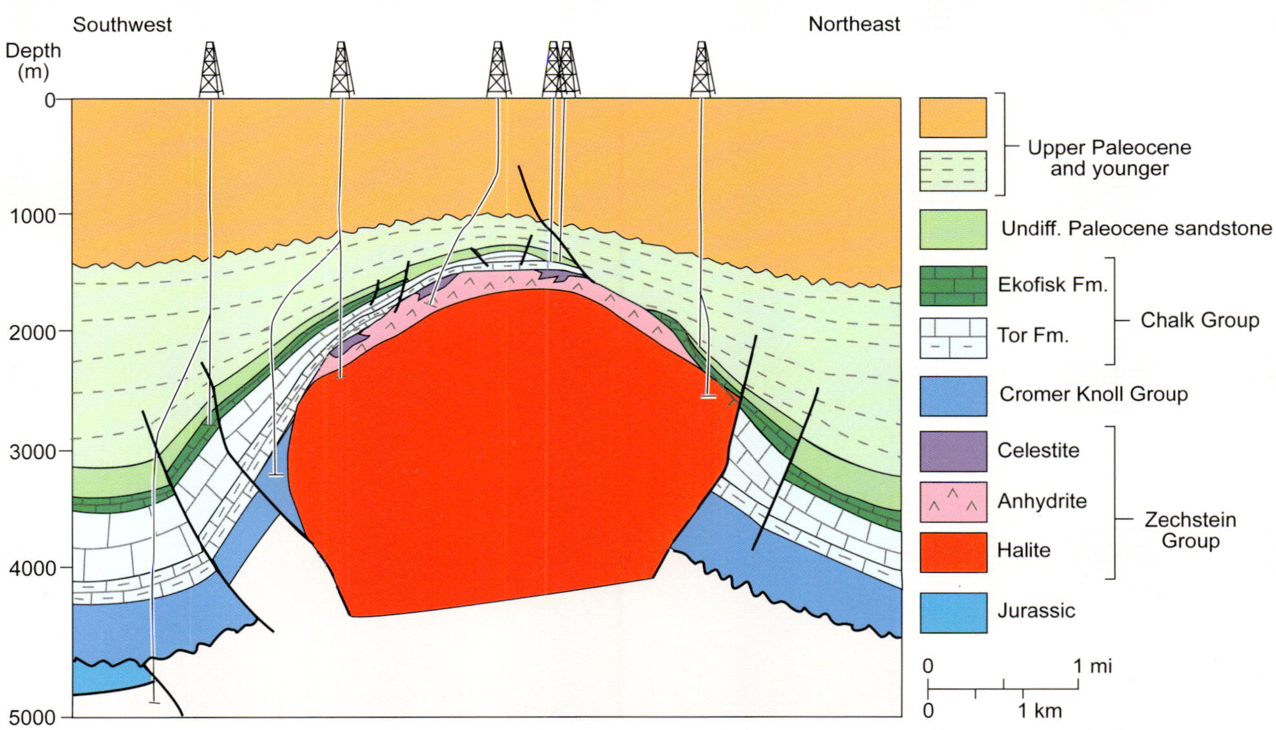

Figure 2.40. Active diapirism. Drill-hole cores of roof strata show extreme tectonic thinning in the chalk, recorded by normal faults, tensile fractures, and pressure solution seams. The overlying Paleocene sandstones are also extensively fractured, indicating that arching of the diapir roof was accompanied by much tectonic thinning below seismic resolution. Had the diapir roof been thinner or the sedimentation slower, piercement of the roof and reexposure of the diapir on the seafloor could have resulted. See also time sections of the Machar diapir (Figure 2.41). After Davison and others (2000b).

NORTH SEA–Stratigraphic Patterns above Machar Diapir, Central Graben, Offshore UK

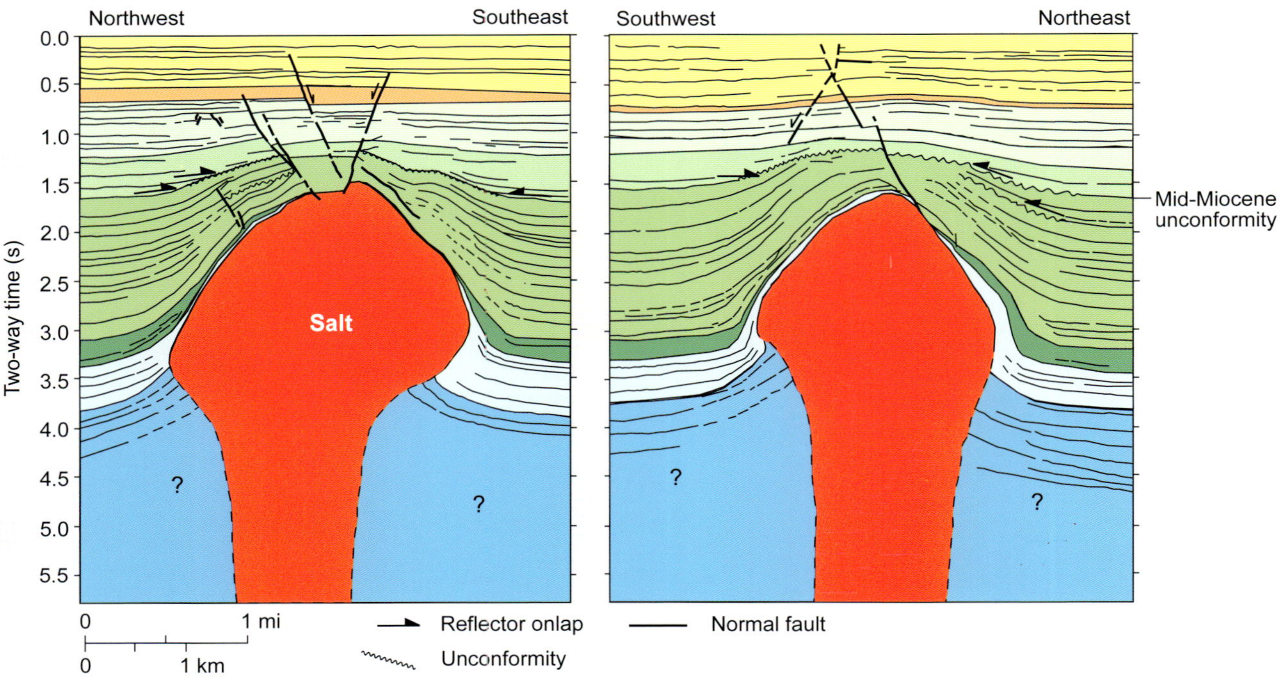

Figure 2.41. Active diapirism. No unambiguous criteria on these sections indicate origin of the crestal arch. All green units thin progressively onto the diapir crest, so uplift appears to have been continuous over a long period of time. This continuous uplift is consistent with either halokinetic active diapirism or a prolonged period of gradual shortening. See also depth section of the Machar diapir (Figure 2.40). After Davison and others (2000a).

NORTH SEA–Stratigraphic Patterns above Monan Diapir, Central Graben, Offshore UK

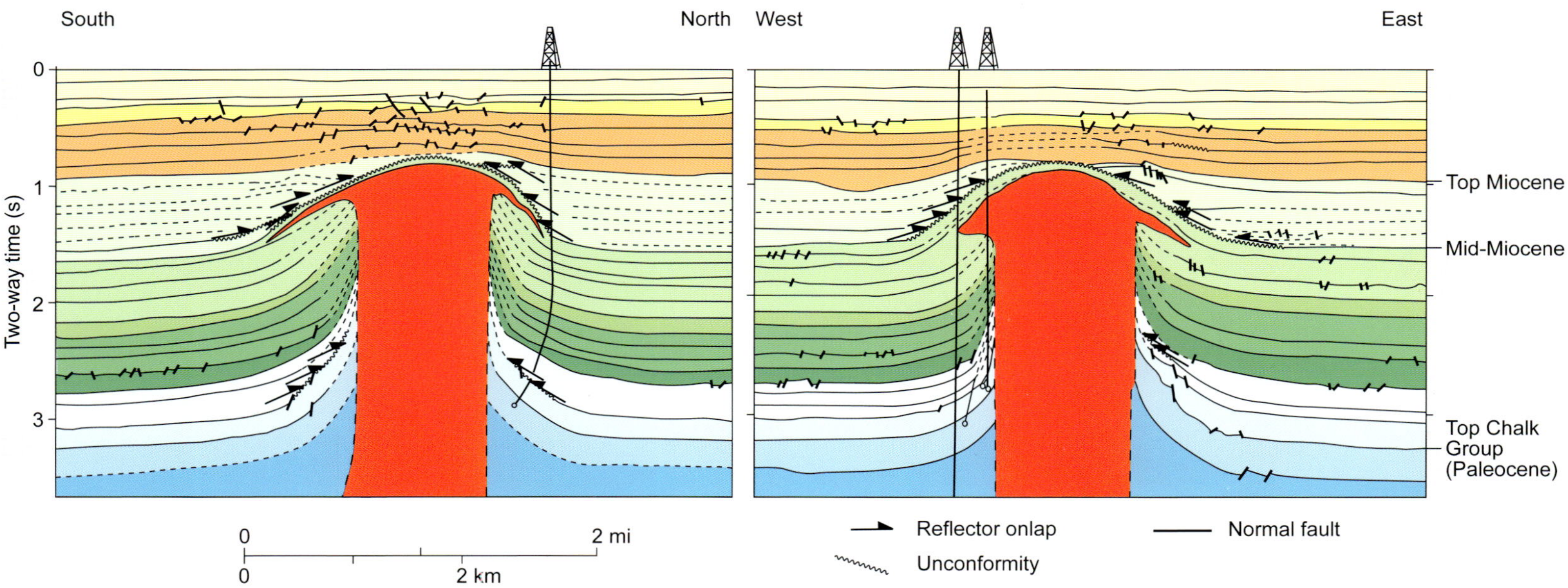

Figure 2.42. Diapir shortening. The mid-Miocene unconformity is onlapped by 0.7 s (~700 m) of strata on the diapir flanks. Comparison with modern diapirs (Davison and others, 1996) suggests that the maximum surface relief that can be generated by buoyancy forces alone is about 300 m, so the extreme relief shown by the mid-Miocene unconformity was probably a result of diapir shortening. Miocene shortening may also have caused rapid salt flow, leading to extrusion of thin salt sheets near the top of the diapir. After Davison and others (2000a).

NORTH SEA—Crestal Structures above Mungo Diapir, Central Graben, Offshore UK

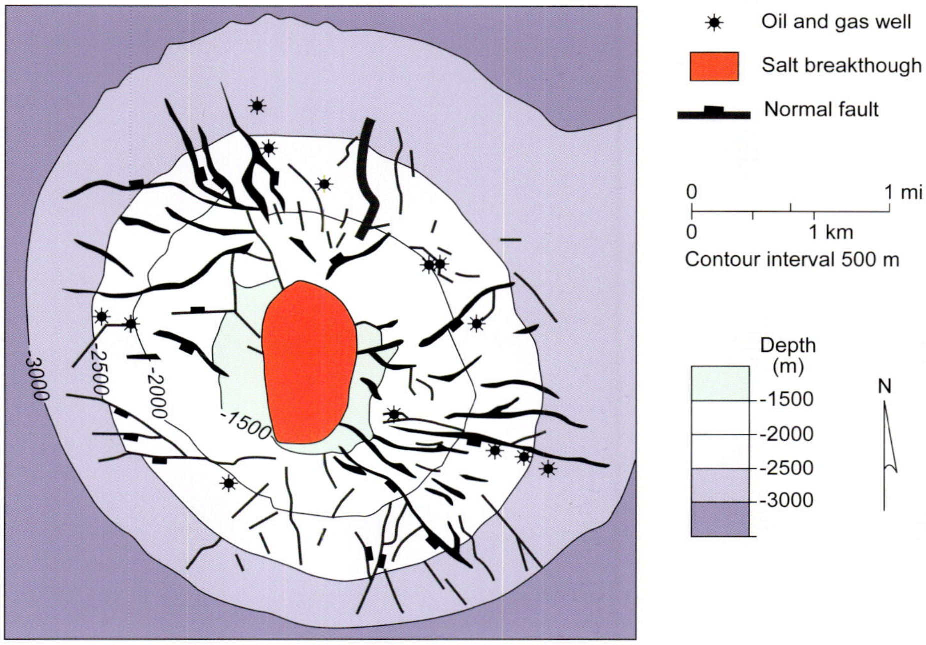

Figure 2.43. Radial faulting. Map is on the top of the Paleocene sandstone reservoir, which truncates underneath a nearly detached bulb of salt at the top of the diapir. Dominant fault pattern is radial, resulting from circumferential stretching as the diapir uplifted and pierced. Faults have slightly preferred orientations parallel to a regional NW-striking fault set. After Davison and others (2000a).

PARADOX BASIN—Uplifted Roof above Onion Creek Diapir, Utah, USA

Figure 2.44. Rotation of beds against a rising diapir. This westward view shows erosional remnants of the Plio-Pleistocene roof above the rising Onion Creek salt diapir. Rise of the west part of the diapir has been supplied from beneath a minibasin to the east. Beds in the minibasin thin onto the rising salt in a series of angular unconformities. See a close-up of the diapir flank in Figure 2.45. Photograph by Mike Hudec.

PARADOX BASIN—Upturn against the Flank of a Rising Dome, Onion Creek Diapir, Utah, USA

Figure 2.45. Upturn of beds against a diapir flank. Permian Cutler strata are upturned to near vertical against the north flank of the Onion Creek diapir, as marked by a thin, white limestone bed near the contact. Upturned beds are interpreted as former roof strata pushed up and away from the salt by Permian rise of the dome. The steep upturn is one of a series of tight folds exposed in Cutler rocks near the dome flanks. The sharpness and short wavelengths of these folds suggest that they are shallowly detached, perhaps on a salt wing just below the land surface. See also regional view of the dome in Figure 2.44. Photograph by Mike Hudec.

PARADOX BASIN—Erosion of a Pinched-Off Salt Dome, Upheaval Dome, Utah, USA

Figure 2.46. Salt-dome pinch-off. Upheaval Dome is a structural culmination whose crest has been eroded to form a crater, drained to the west by Upheaval Canyon (foreground). The preferential erosion is due to the uplift and extensive fracturing of the resistant sandstones that cap the mesa in the area. The uplifted and fractured sandstones were more easily eroded than their surrounding equivalents, allowing exposure and erosion of the less competent units below. The origin of the fracturing and uplift is hotly debated. Some claim that the structure is a deeply eroded meteorite impact structure, but Jackson and others (1998, 2001) argued that it represents the neck of a pinched-off salt dome. See also map (Figure 2.47), close-up photos (Figure 2.48), and restoration (Figure 2.49). Modified from Jackson and others (2001).

Paradox Basin—Constrictional Strain Associated with Diapir Pinch-Off, Upheaval Dome, Utah, USA

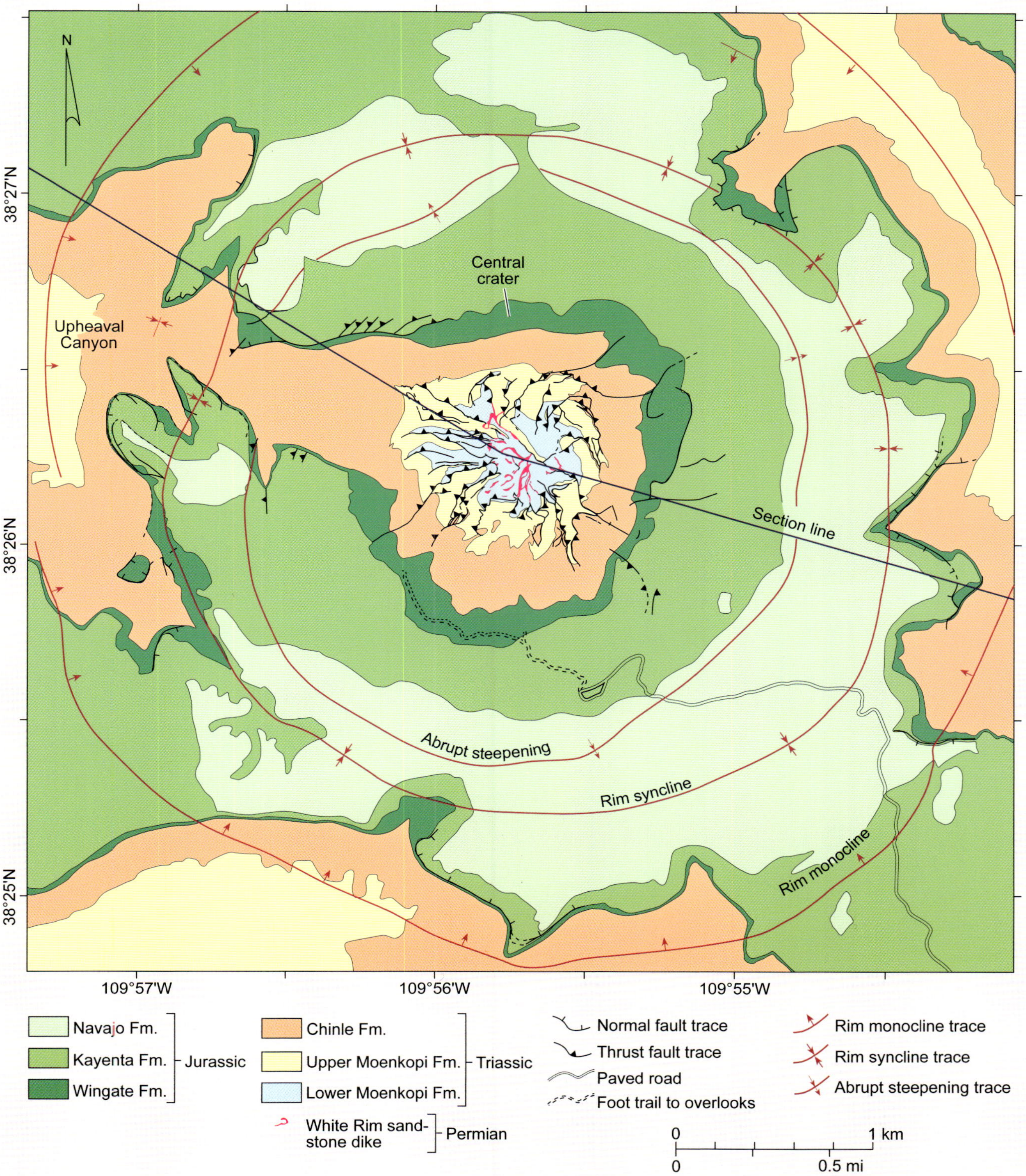

Figure 2.47. Salt-dome pinch-off. Upheaval Dome is a structural dome whose culmination has been eroded to form a crater, drained to the west by Upheaval Canyon. Thrust faults radiate out from the center of the structure, formed when rocks moved radially inward toward the center of the dome. This constrictional strain is analogous to closing the leaves of a camera diaphragm, where circumferential shortening increases inward. Jackson and others (1998) argued, on the basis of growth stratigraphy in the canyons surrounding the dome, that constriction occurred during pinch-off of a salt dome that occupied the position of the central crater (see restoration, Figure 2.49). This structure illustrates the marked inward increase in deformation during pinch-off of a diapir that was originally circular in plan view. See also oblique aerial photo (Figure 2.46) and close-up photos (Figure 2.48). Modified from Jackson and others (1998).

PARADOX BASIN—Thrust Faults Formed during Salt-Dome Pinch-Off, Upheaval Dome, Utah, USA

Figure 2.48. Salt-dome pinch-off. Radial inward motion of wall rocks to pinch off the Upheaval Dome salt structure (see restoration, Figure 2.49) produced constrictional strain. A major manifestation of this strain is a pattern of radial thrust faults, which allowed rocks to shorten in a manner analogous to closing the leaves of a camera iris diaphragm (see map, Figure 2.47). Strain intensity increases markedly toward the center of the structure, as wall rocks crowded from all sides into a progressively smaller space. See also oblique aerial photo (Figure 2.46). Modified from Jackson and others (2001).

(a)

(b)

PARADOX BASIN—Salt Dome Pinch-Off, Upheaval Dome, Utah, USA

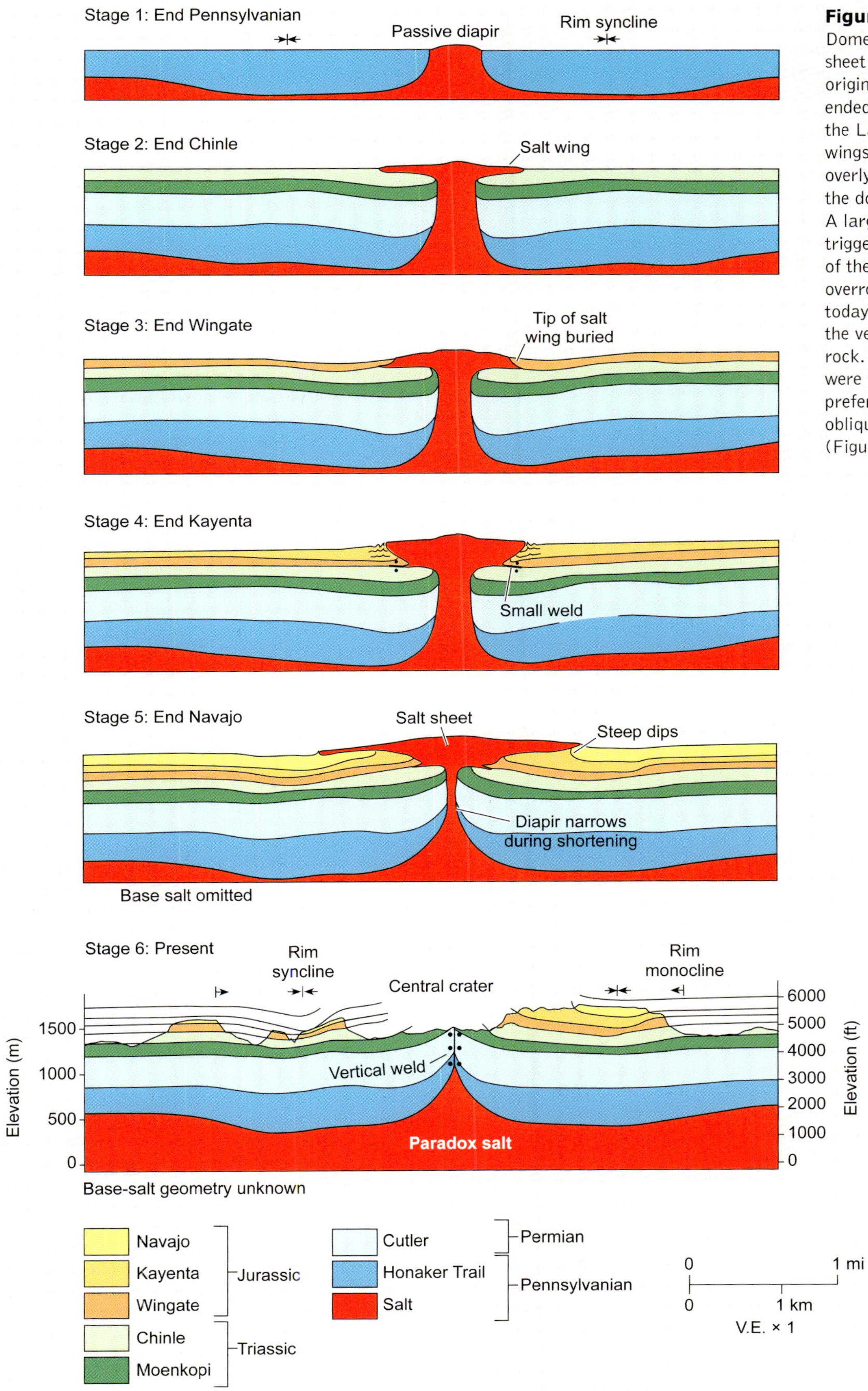

Figure 2.49. **Salt-dome pinch-off.** Structures at Upheaval Dome have been interpreted as the pinched-off root of a salt sheet that has now been eroded. In this hypothesis, the dome originated as a passive diapir shortly after salt deposition ended in the Pennsylvanian. Diapir rise continued through the Late Triassic, when small salt wings were extruded. These wings were buried during the Jurassic, and the weight of the overlying sediments expelled salt from the wings back into the dome, forming small salt welds on the flanks of the dome. A larger salt sheet spread during the Late Jurassic, perhaps triggered by expulsion of salt from shortening and pinch-off of the dome. Outward-spreading salt in the sheet rotated and overrode beds in its footwall, forming the steep dips preserved today. Diapir pinch-off formed a dense network of faults near the vertical weld (see map, Figure 2.47), which weakened the rock. Fractured rock and soluble evaporites in the dome center were more easily eroded than surrounding units, allowing preferential erosion to form the present crater. See also oblique aerial photo (Figure 2.46) and photos of thrust faults (Figure 2.48). Modified from Jackson and others (1998).

PRECASPIAN BASIN—Stacked Halokinetic Sequences along a Diapir Flank, Kazakhstan

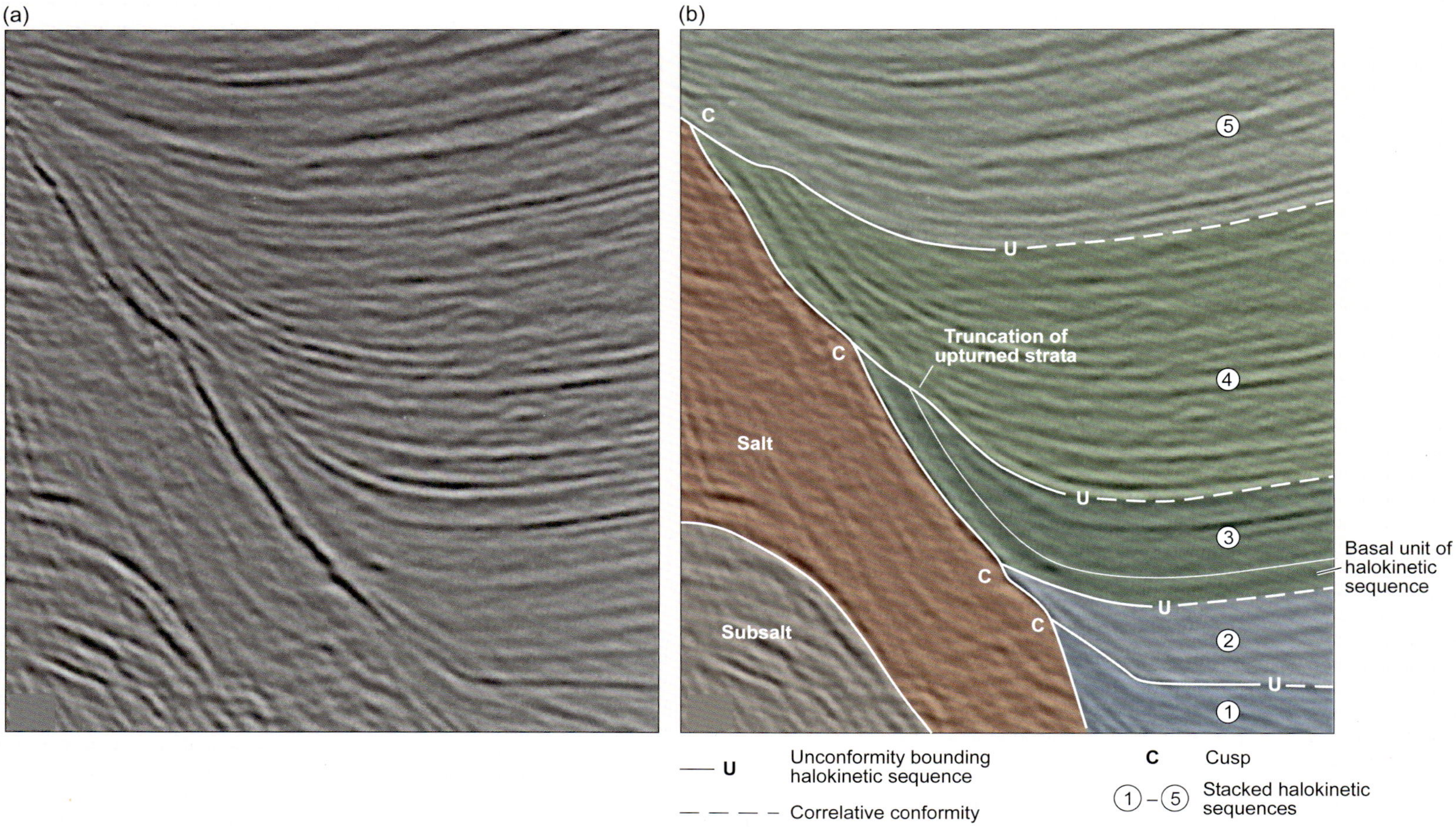

Figure 2.50. Formation of halokinetic sequences. Beds flanking a salt diapir are upturned when the rising diapir lifts and shoulders aside its roof. The steep roof may fail gravitationally, producing an angular unconformity near the diapir that truncates the upturned strata beneath (a "J-hook unconformity"). Shear along this unconformity may produce a cusp in the salt face where the unconformity intersects the salt flank. Modified from Rowan and others (2003).

RAVAR BASIN—Map of Diapir VII, Eastern Iran

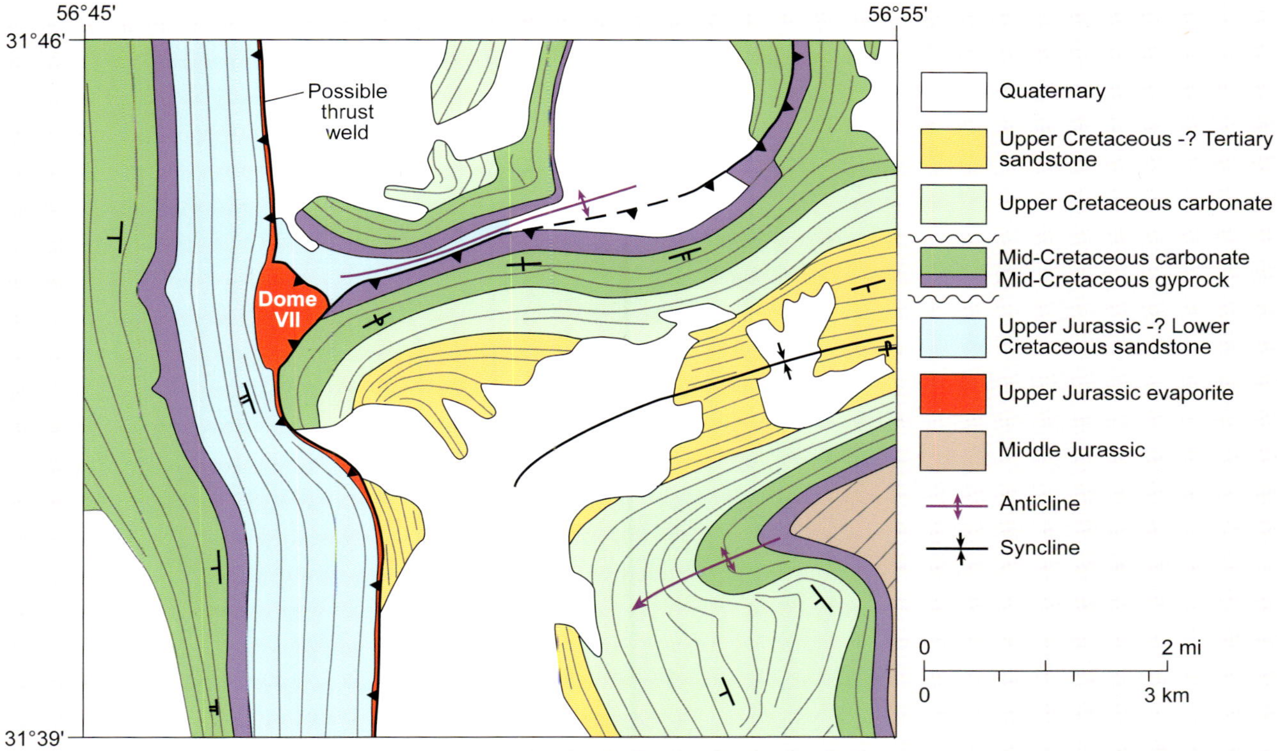

Figure 2.51. Diapir rise during thrusting. Dome VII is one of five major diapirs lying along a north-striking lineament more than 80 km long, which we interpret as a probable thrust weld. The thrust emplaced a concordant slab of Jurassic-Cretaceous sediments, which rest on the autochthonous salt, discordantly across older folds in its footwall. We infer that the diapirs formed early and controlled the location of the thrust ramp. The juxtaposition of the five diapirs along the east side (inferred footwall) of this lineament suggests that the thrust nappe smeared the underlying diapirs and expressed them ahead of it. If so, the diapirs are not vertically rooted but, rather, are allochthonous blobs that have been pushed eastward ahead of their smeared feeders. See oblique aerial photograph (Figure 2.52) and outcrop view (Figure 2.53). After Stöcklin (1961).

RAVAR BASIN—Oblique Aerial Photograph of Diapir VII, Eastern Iran

Figure 2.52. Diapir rise during thrusting. The prominent topographic bulge of soluble evaporites indicates continuing rise of the diapir. See geologic map (Figure 2.51) and outcrop view (Figure 2.53). Photograph by Jovan Stöcklin. From Jackson and others (1990).

RAVAR BASIN—Outcrop Photograph of Diapir VII, Eastern Iran

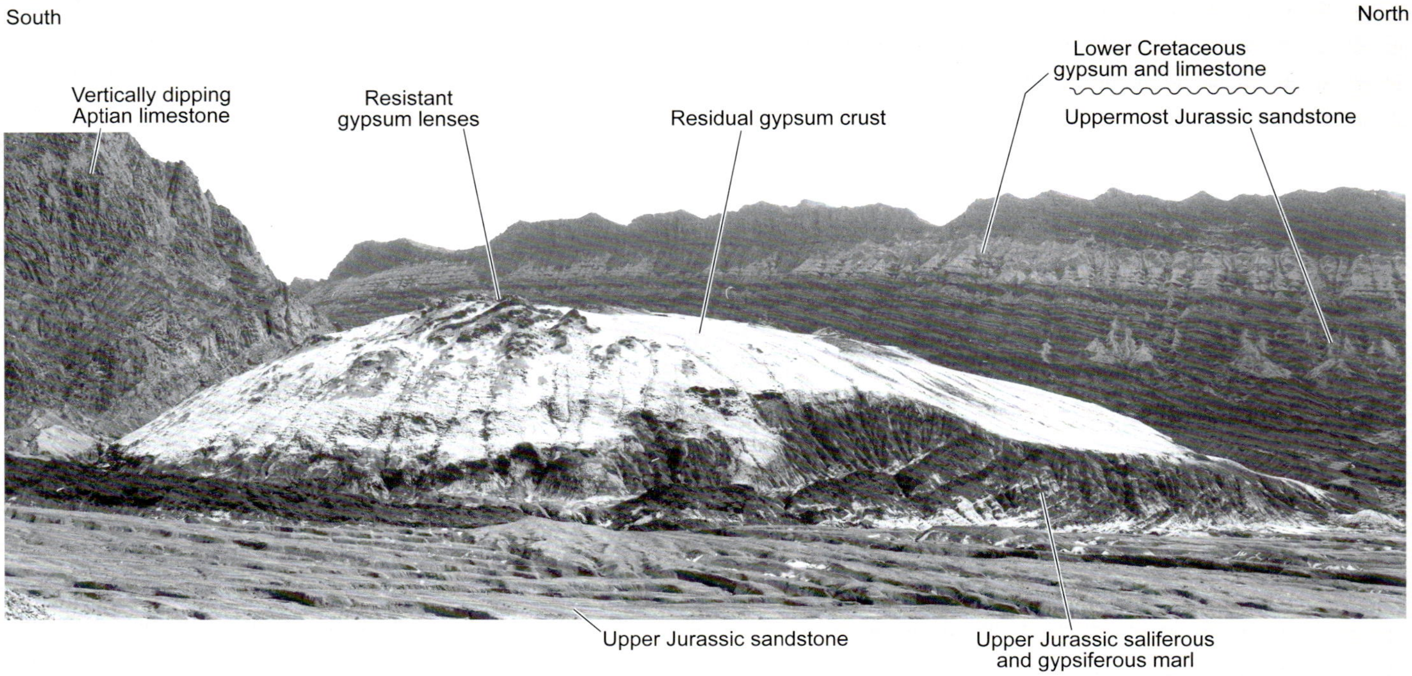

Figure 2.53. **Diapir rise during thrusting.** The prominent topographic bulge indicates continuing rise of the diapir. Flanking marls on the north arch upward toward the diapir until truncated by erosion. South of the diapir, the flanking strata are vertical to overturned. See oblique aerial view (Figure 2.52) and geologic map (Figure 2.51). Photograph by Jovan Stöcklin. After Jackson and others (1990).

PHYSICAL MODEL—Extensional Drape over Basement Grabens during Detached Translation

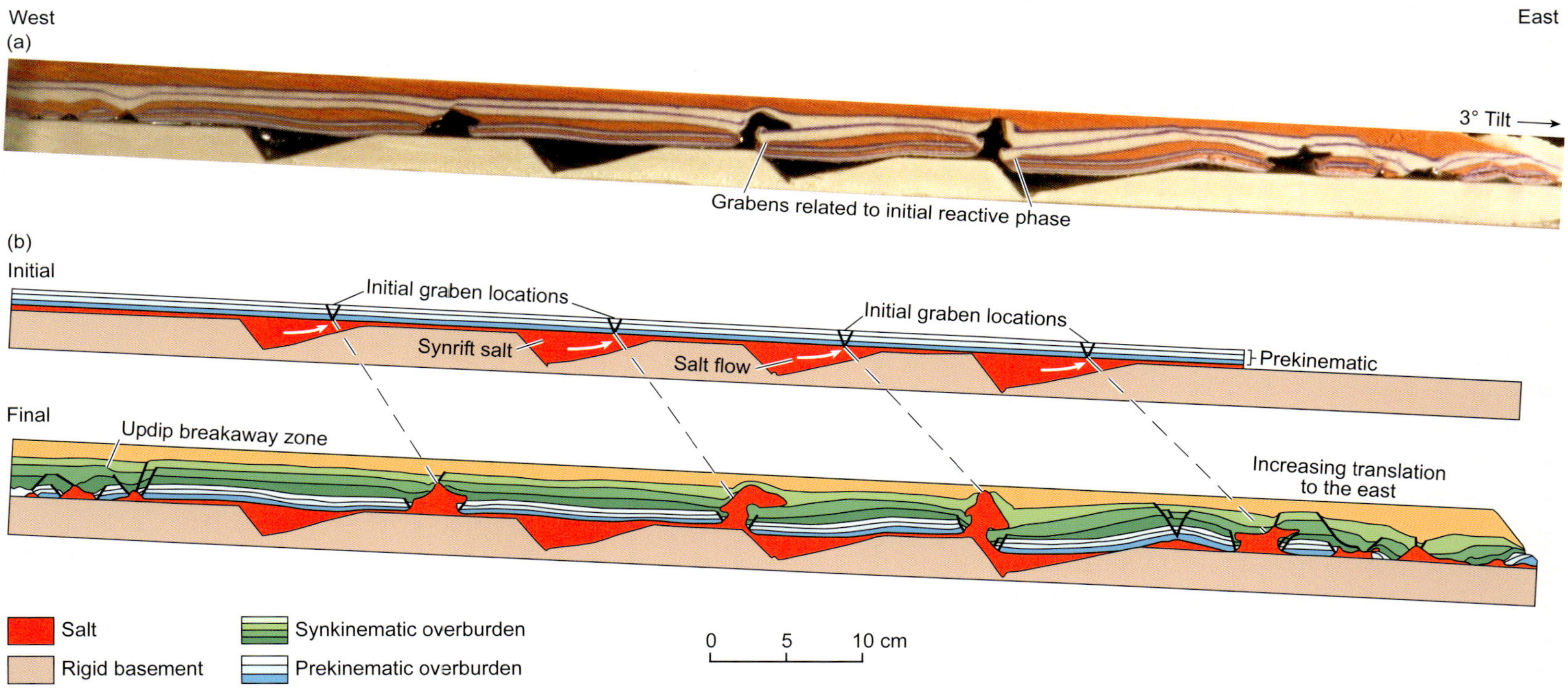

Figure 2.54. Translation over basement faults. Translation across a synrift salt layer produced four reactive diapirs that evolved into passive walls. Salt flowed downdip within each basement half-graben, forming a salt-cored bulge at its downdip edge. Stretching of the cover over the crest of each bulge nucleated the diapirs. Subsequent translation of these diapirs moved them from where they initially formed. Model by Bruno Vendeville.

PHYSICAL MODEL—Diapir Growth during Weakly Coupled Extension

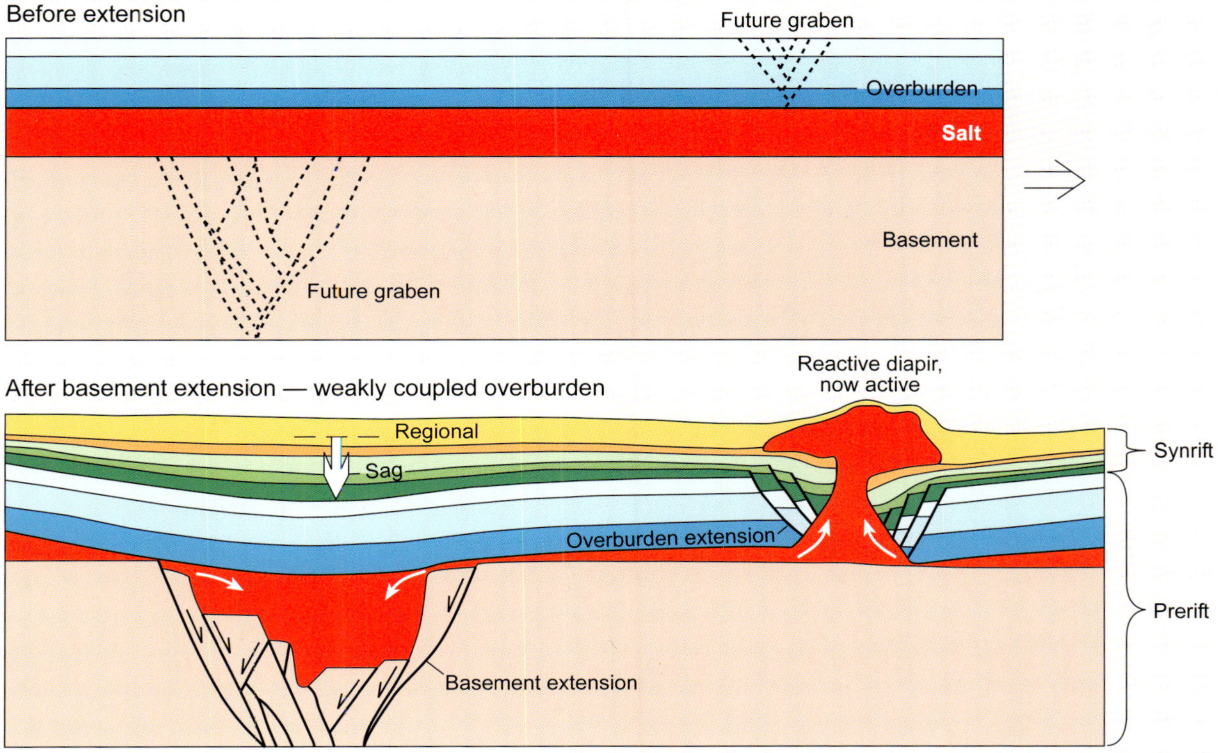

Figure 2.55. Basement extension below salt. Extension above and below salt was weakly coupled in this experiment. Most subsalt extension was accommodated by flow of salt into the basement graben. Equivalent extension in the overburden initiated a diapir in a position unrelated to the basement graben. Weak coupling is encouraged by thick salt, slow extension, and low-viscosity evaporites. Modified from Jackson and Vendeville (1994).

PHYSICAL MODEL—Extensional Drape over a Basement Graben, Northern Section

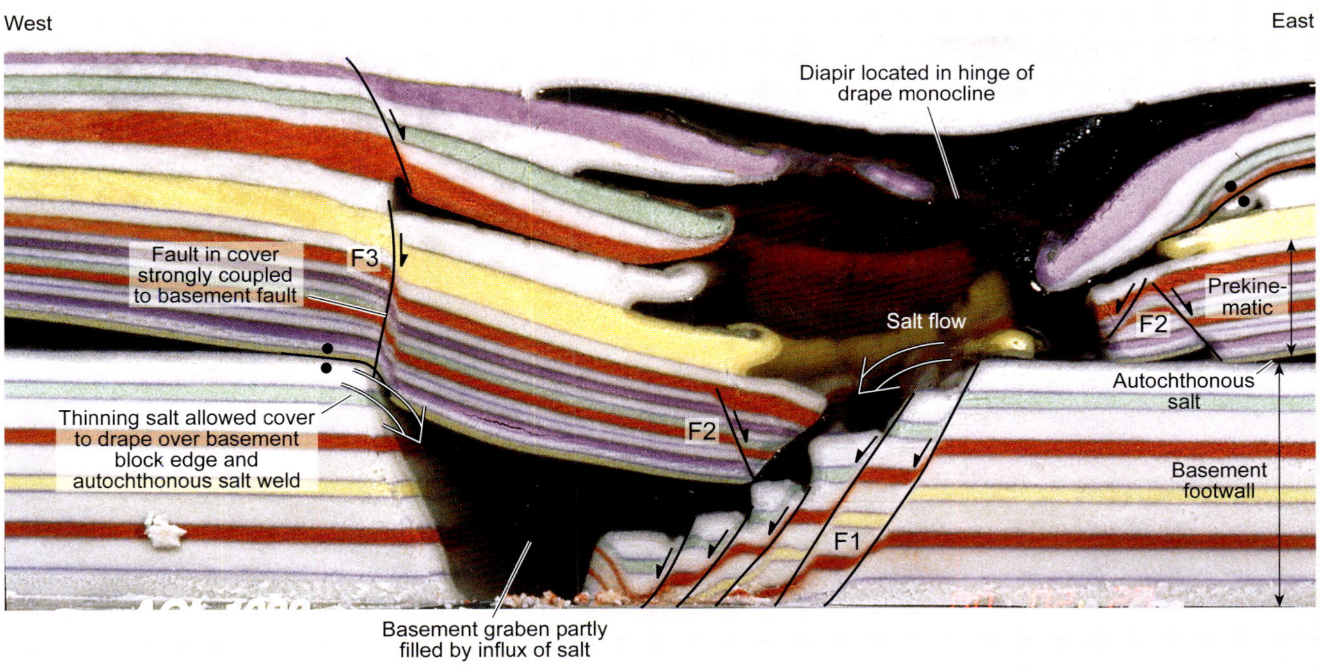

Figure 2.56. Basement extension below salt. Two types of coupling between basement faults and suprasalt deformation are illustrated. On the west side of the graben, autochthonous salt flowed down into the graben until the overburden impinged on the edge of the basement footwall. Fault F3 propagated up into the overburden from this corner, so that suprasalt deformation is here strongly coupled to basement faulting. Coupling was weaker in the east, where suprasalt extension localized in the crest of a drape monocline. A reactive diapiric wall rose along the axis of this extensional system, where it eventually pierced to the surface and extruded several generations of salt sheets. See also southern section of the same model (Figure 2.57). Model by Bruno Vendeville.

PHYSICAL MODEL—Extensional Drape over a Basement Graben, Southern Section

Figure 2.57. Basement extension below salt. The autochthonous salt layer was evacuated early in this part of the model because salt was flowing along strike to feed diapirs extruding over the sediment surface. The loss of autochthonous salt draped overburden over the edge of the western basement block much earlier than elsewhere in the model (compare with northern section, Figure 2.56). Rotation of this monocline is recorded by sediment thickening in the yellow through purple synkinematic layers in the rollover hanging wall of a large west-dipping normal fault. Coupling to basement structure strengthened near the end of the experiment, when fault F3 broke upward from the basement fault through the hinge of the monocline. Extensional separation of the prekinematic sequences allowed a diapir to rise above the basement graben. Model by Bruno Vendeville.

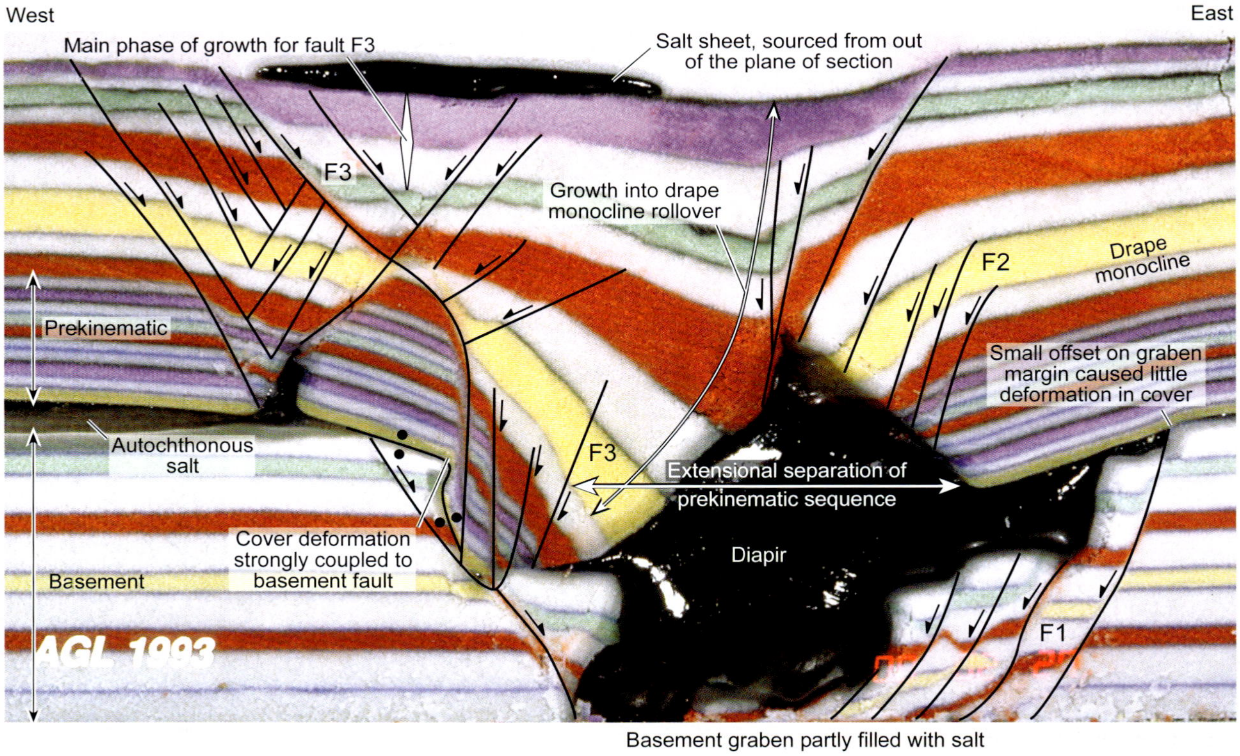

PHYSICAL MODEL—Centrifuge Model, Active Diapirs 2 and 3

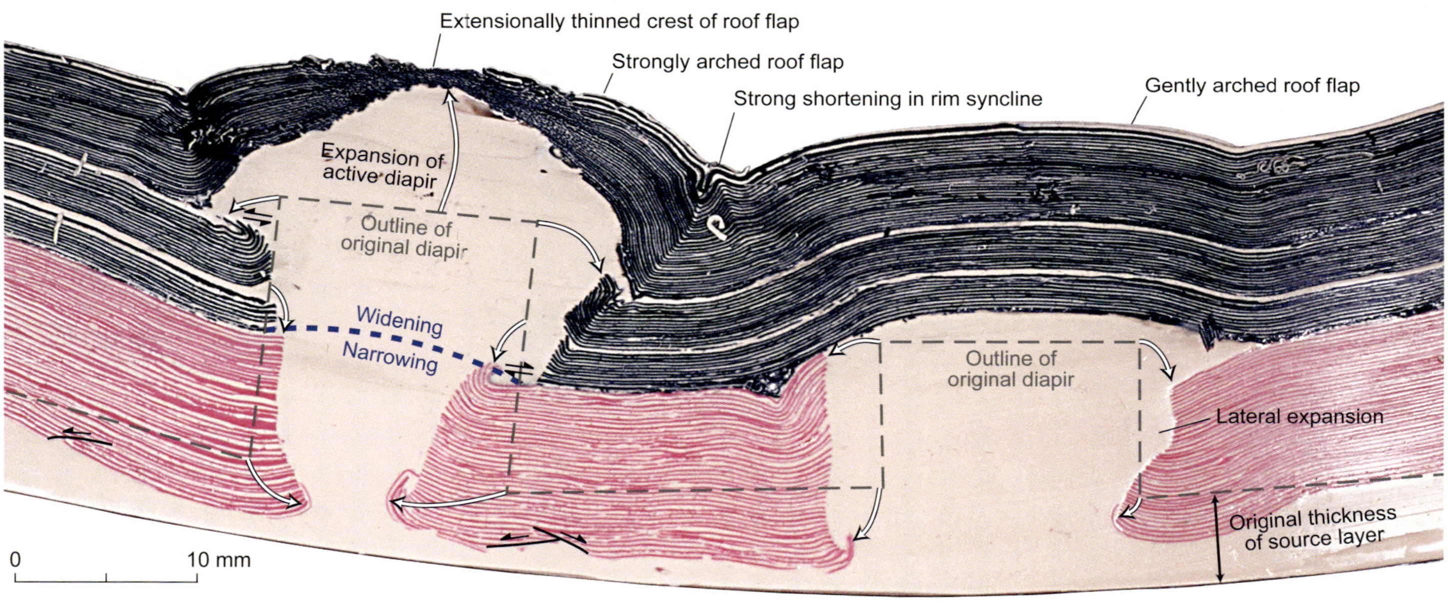

Figure 2.58. Stem pinch-off during halokinetic active diapirism. Diapirs can pinch off in the absence of regional shortening, simply by having the wall rocks spread laterally into the space occupied by the diapir. The low diapir on the right rose only slightly because its thick roof was too strong to undergo more than minor arching; the diapir expanded laterally at the expense of its neighbor on the left. Conversely, the tall diapir on the left expanded upward because its roof was thin and therefore weak enough to be rotated, extended, and thinned by the actively rising diapir. The upper part of the vigorously active diapir widened, whereas its lower part (below the dotted line) narrowed but did not pinch off entirely. This strain boundary coincided with the lower regional detachment (top of red unit, level with the crest of the lower diapir). The upper regional detachment was midway in the black unit, level with the crest of the taller diapir. No regional extension or shortening took place during diapirism. Moreover, widening of the lower diapir was balanced by narrowing of the lower part of the taller diapir. Dashed line shows the original geometry of the two diapirs and their source layer. Compare with diapir 1 (Figure 2.59), which pinched off entirely. Model by Martin Jackson.

PHYSICAL MODEL—Centrifuge Model, Active Diapir 1

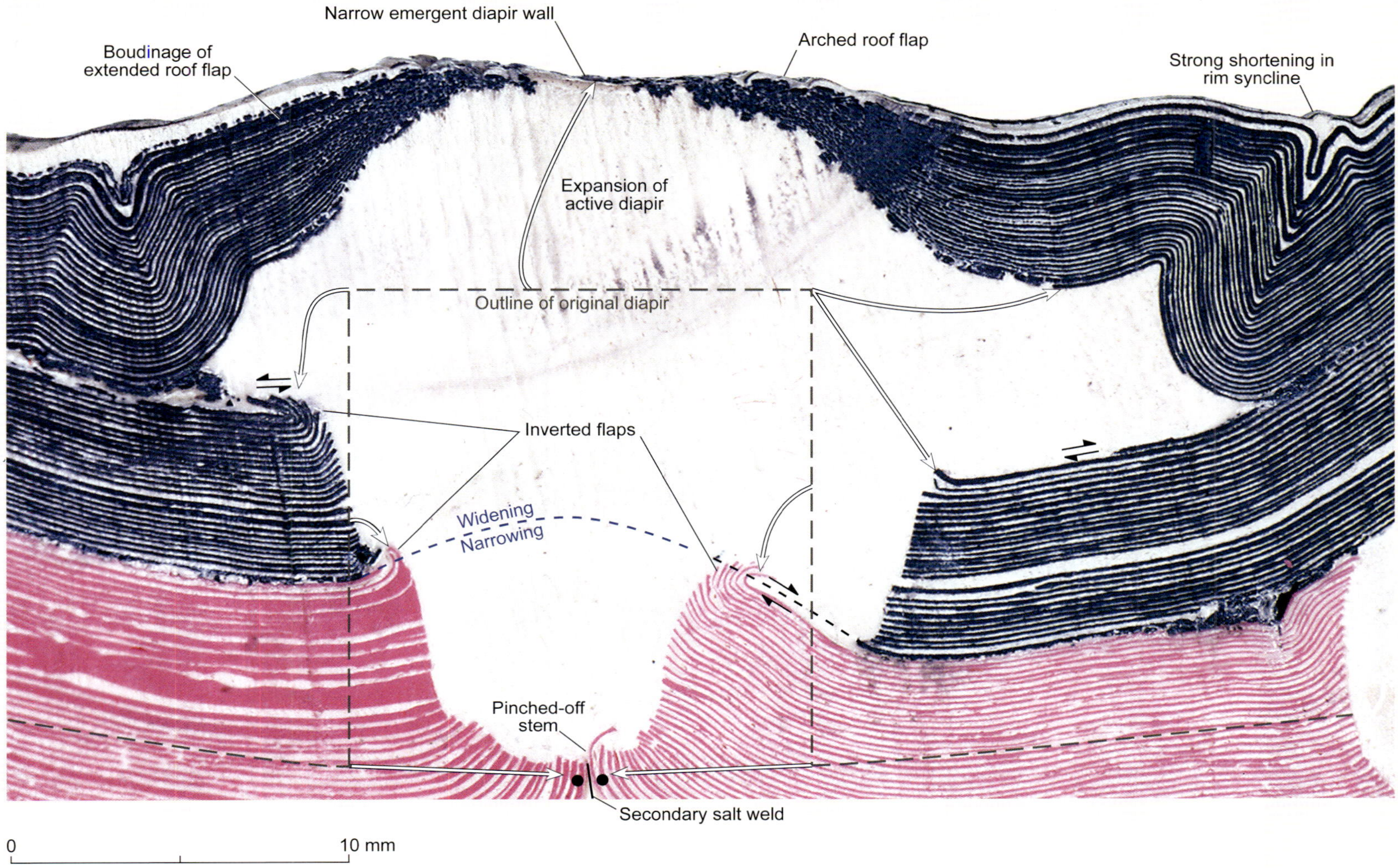

Figure 2.59. Stem pinch-off during halokinetic active diapirism. Diapirs can pinch off in the absence of regional shortening simply by having the wall rocks spread laterally into the diapir. Below the dotted line, the diapir narrowed until it pinched off entirely, forming a secondary salt weld. But the upper part of the diapir rose and widened upward because its roof was thin and weak enough to be rotated, extended, and thinned by the rising diapir. This strain boundary between narrowing and widening of the diapir coincided with the lower regional detachment (top of red unit). The upper regional detachment was midway in the black unit, level with the original crest of the diapir. Along this upper detachment, the diapir intruded, and the roof peeled back. No regional extension or shortening took place during diapirism. Thus, extension above the diapir was balanced by shortening in its rim syncline. Narrowing of the lower part of the diapir was balanced by extensional faulting in the red unit and by expansion of the lower diapir off section (Figure 2.58). Dashed line shows the original geometry of the diapir and its source layer. Model by Martin Jackson.

PHYSICAL MODEL—Squeezed Diapir

Figure 2.60. Rejuvenation of buried diapir during shortening. After exhausting the supply of salt from its source layer, the diapir stopped growing and was buried. Shortening caused the diapir to rise again. Criteria for diapir squeezing are (1) the thick, arched, isopachous roof; (2) the straight, shallow-dipping backthrust; and (3) diapir rise after the source layer was exhausted. The diapir stem pinched off in this plane of section, but it extends down to the source layer along strike. Model by Bruno Vendeville.

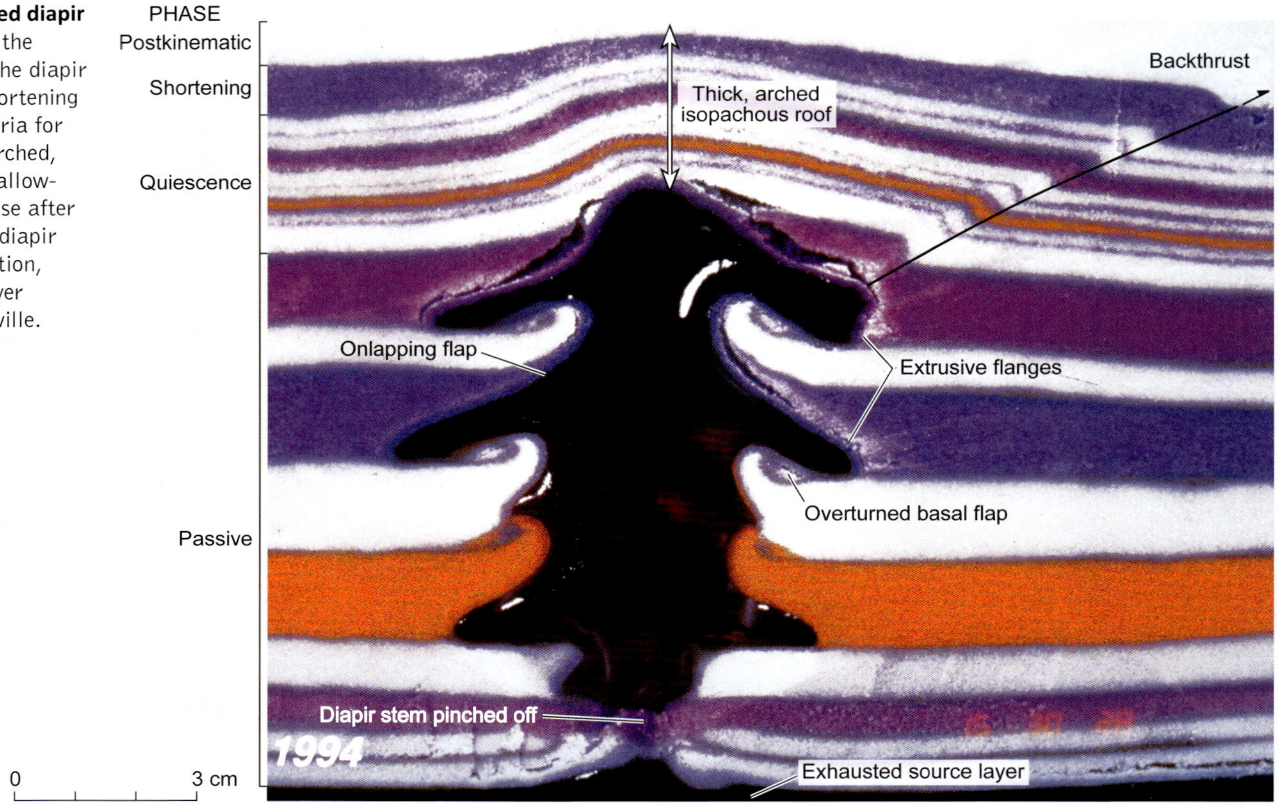

PHYSICAL MODEL—Inverted Graben above Diapir

Figure 2.61. Rejuvenation of buried diapir during mild regional shortening. The reactive diapir grew during regional extension but remained deeply buried. After extension and a brief tectonic quiescence, squeezing rejuvenated the diapir, even though it had virtually exhausted the supply of salt from its source layer. The signs of diapir squeezing were subtle: (1) the inverted depocenter, which accumulated after extension ended and which arched during shortening; (2) a single thrust fault; and (3) the welded tip of the diapir. Model by Bruno Vendeville.

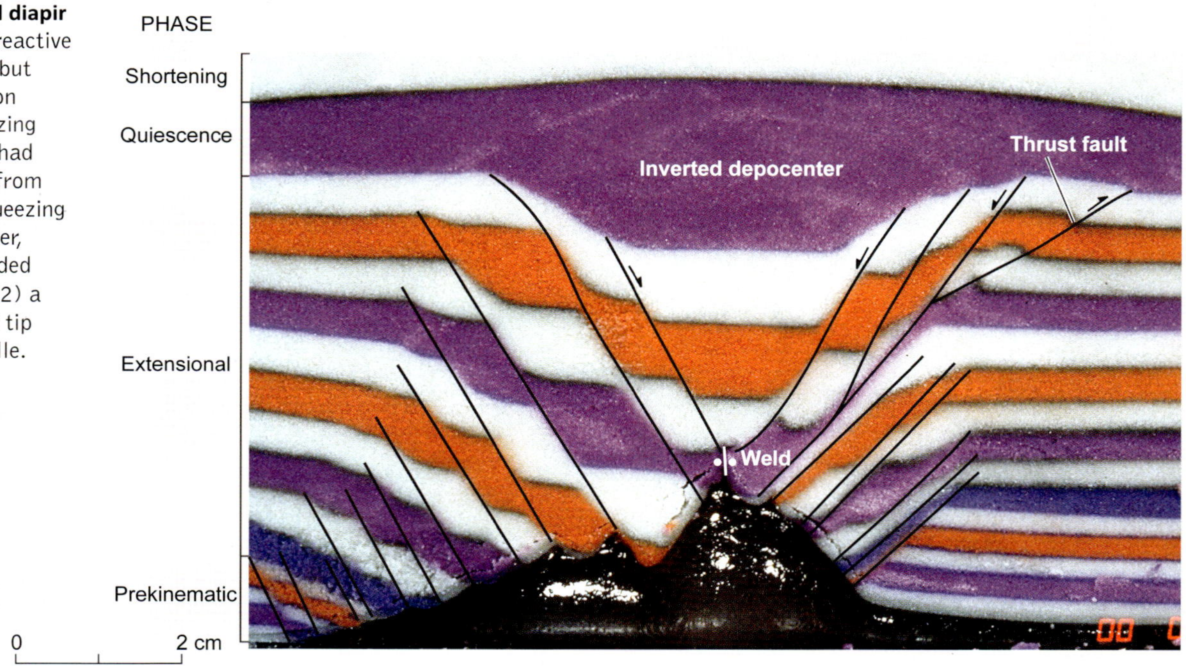

PHYSICAL MODEL–Salt Flow in Source Layer of Squeezed Salt Stock

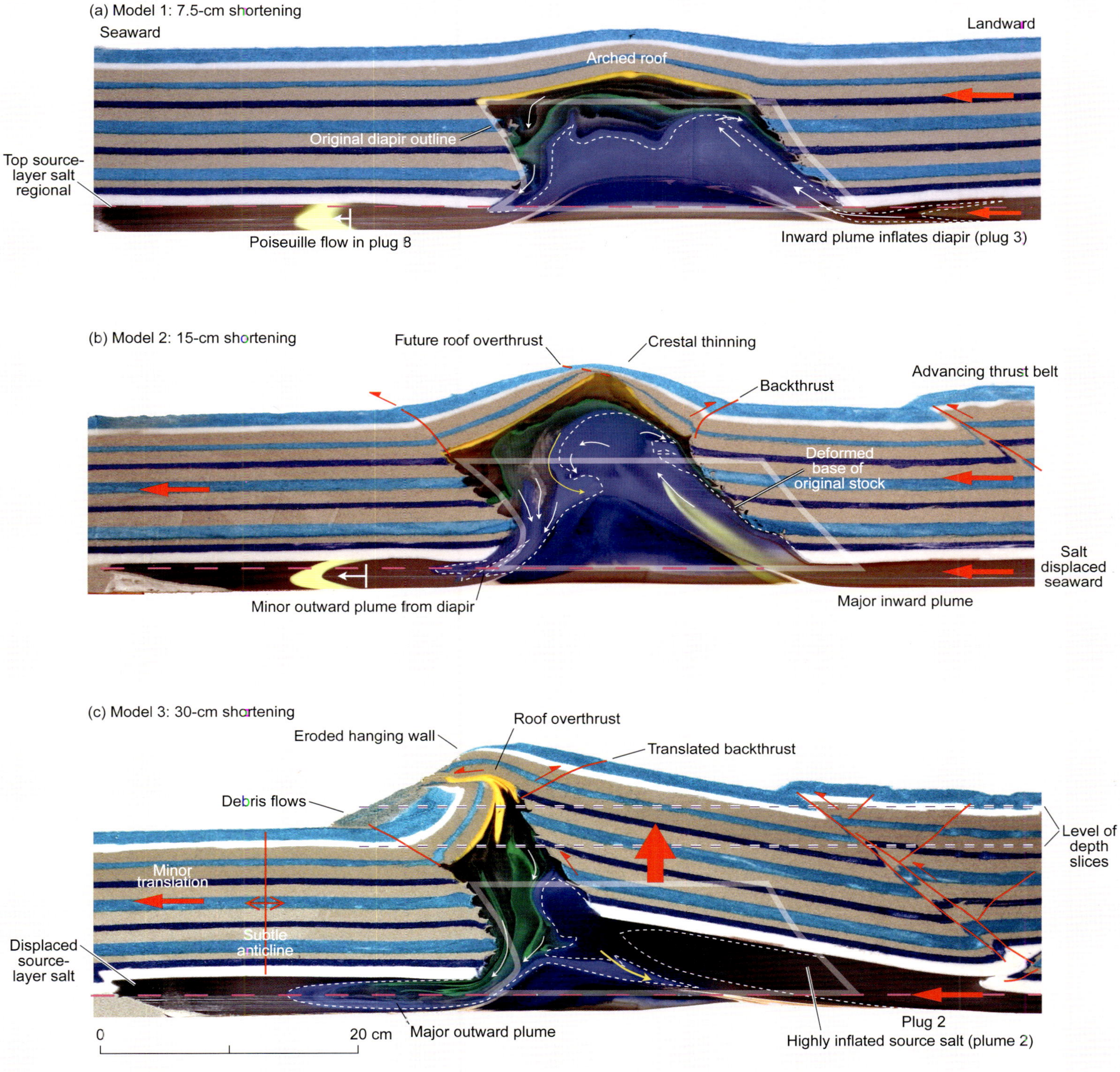

Figure 2.62. Shortening of preexisting diapirs. These sections are from three identical inclined salt stocks shortened by different amounts. Colored passive marker plugs and layers track the flow of salt into and out of the diapir (plug numbers refer to underside views in Figure 2.64). Until 15-cm shortening, the diapir inflated as inward plumes entered the base of the diapir. This inflation arched and structurally thinned the diapir roof. By 30-cm shortening, the landward side of the roof overthrust the seaward side, greatly decreasing the cross-sectional area of the diapir. Salt was expelled as a blue outward plume into the source layer on the seaward side of the diapir, inflating the salt into a subtle salt-cored anticline. See a so overhead views (Figure 2.63). Modified from Dooley and others (2009).

PHYSICAL MODEL—Evolution of a Squeezed Salt Stock

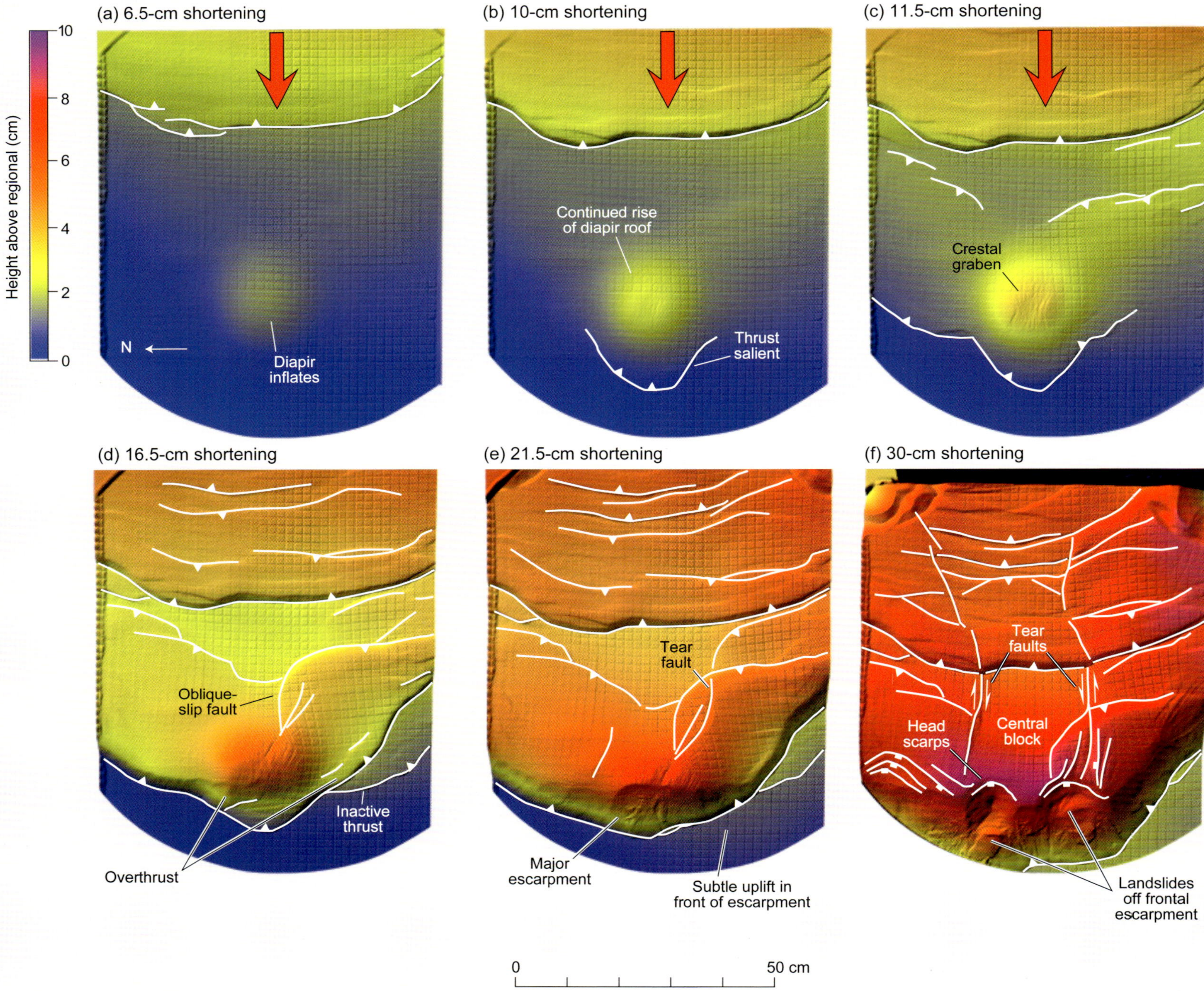

Figure 2.63. Shortening of preexisting diapirs. A stock in the foreland of an advancing thrust front originally inflated as salt from the compressed source layer flowed into the stock (see cross-section evolution in Figure 2.62). This inflation stretched the diapir roof, forming a crestal graben. Later, after the regional thrust front advanced to the diapir, tear faults separated a central block containing the diapir from adjoining country rock. Depth slices show that this central block moved south relative to the rocks on either side to indent the weak diapir. See also underside views (Figure 2.64). Modified from Dooley and others (2009).

PHYSICAL MODEL—Salt Flow in Source Layer of Squeezed Salt Stock

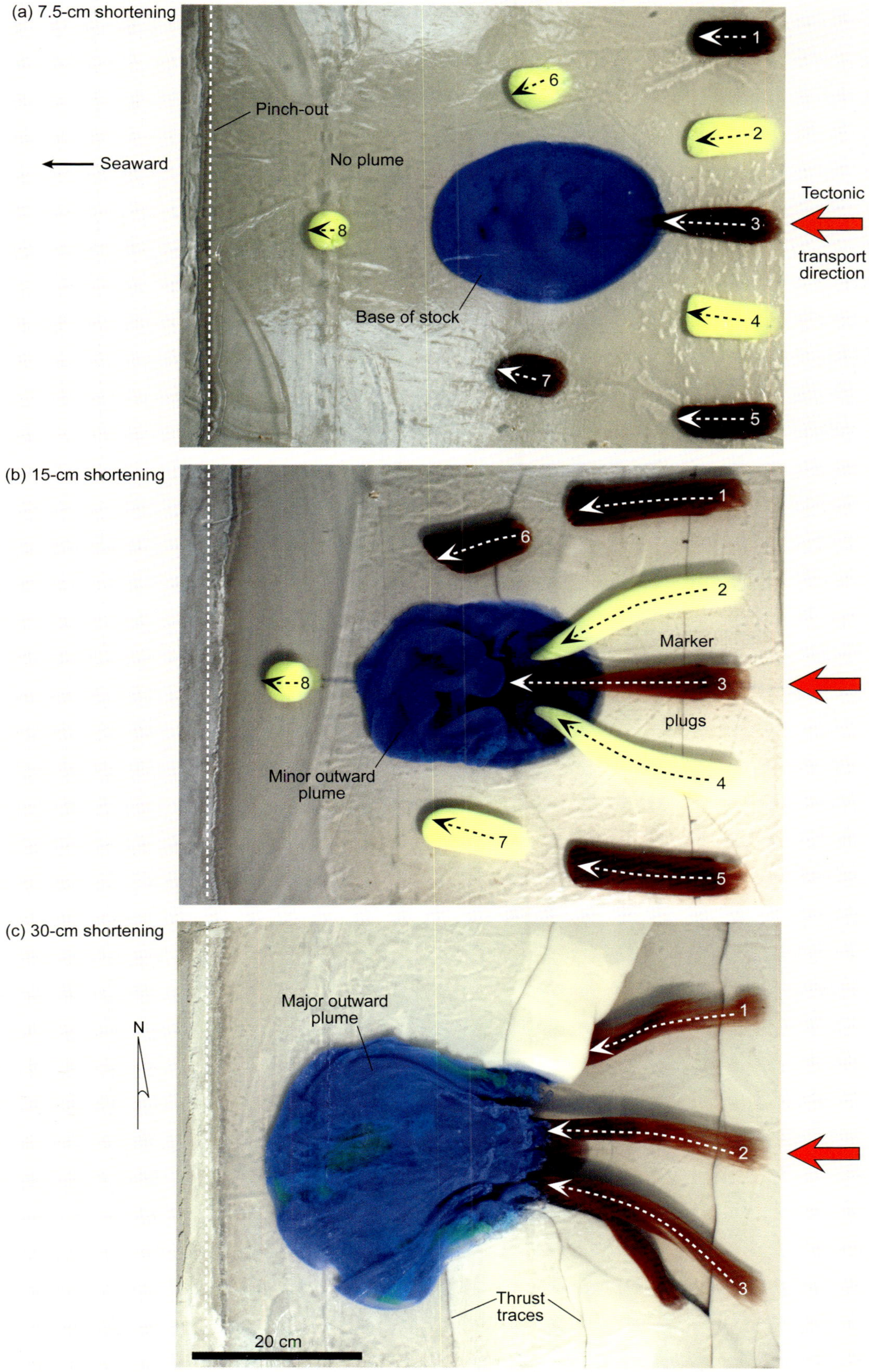

Figure 2.64. Shortening of preexisting diapirs. These photographs show the underside of three identical models of a shortened salt stock that were stopped at different amounts of shortening. Most of the underside of the model is clear silicone in the source layer, through which the base of the lowermost grayish sand bed is visible. Brown and yellow passive marker plugs (originally circular) show flow in the source layer. Until 15-cm shortening, salt flows into the diapir as the stock inflates (see cross-section evolution in Figure 2.62). After 15-cm shortening, an outward plume of blue silicone injected into the source layer from the seaward side of the stock. This outward plume of salt was expelled from the base of the diapir by inward movement of an indenter block (see overhead views in Figure 2.63). Modified from Dooley and others (2009).

PHYSICAL MODEL—Oblique Shortening of Salt Walls with Thin Source Layer

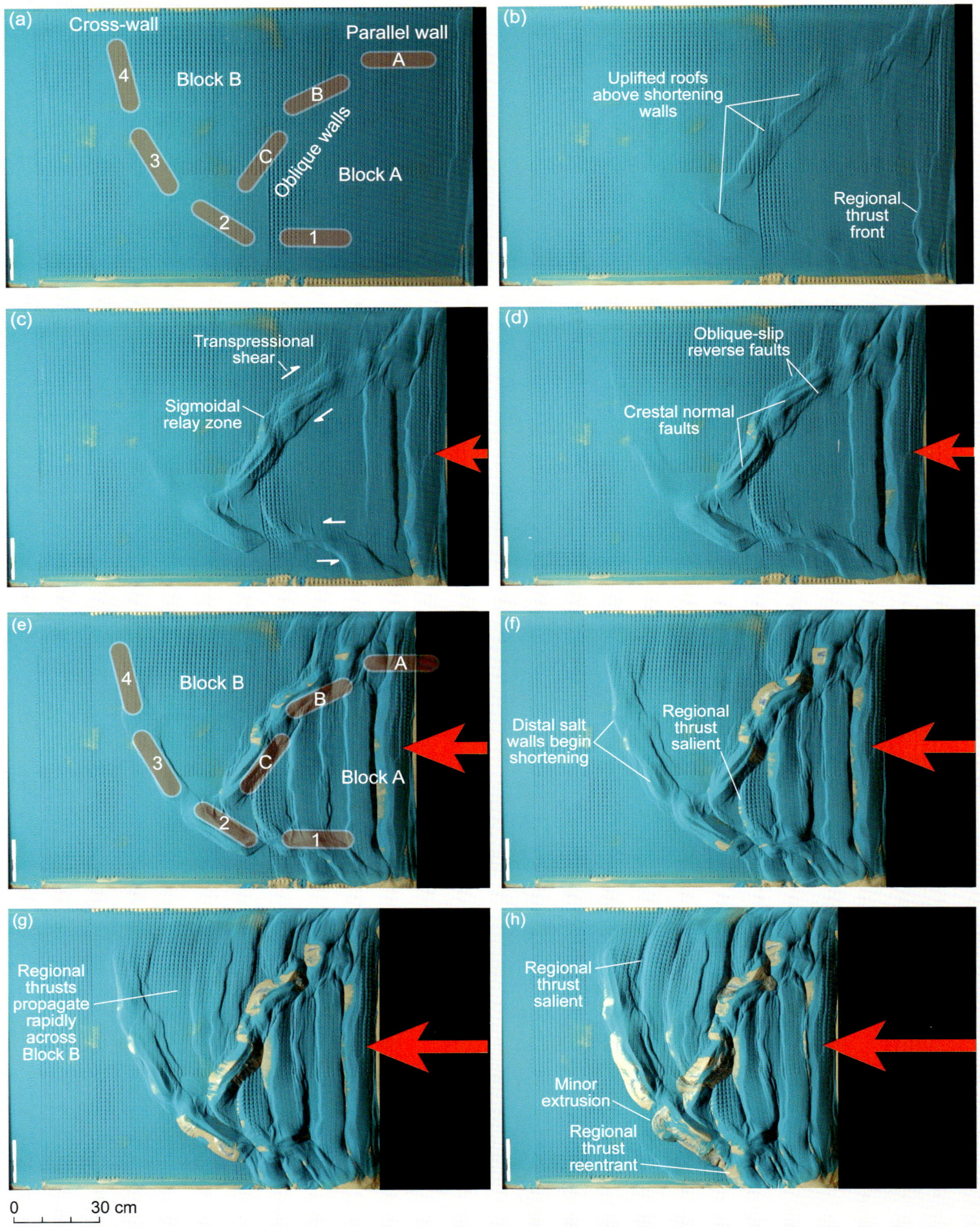

Figure 2.65. Shortening of preexisting diapirs. Preexisting salt walls were constructed at varying orientations to the regional shortening direction. Because the walls were weaker than surrounding overburden, they began shortening long before the regional thrust front propagated to them. Oblique walls shortened in transpression so that uplifts above diapirs were bounded by oblique-slip reverse faults (see section view in Figure 2.66). As diapir roofs arched by compression, extensional grabens formed at their crests. Sigmoidal relay zones connected the diapirs (see depth slice and laser scan in Figure 2.67). Regional thrusts propagated most rapidly in blocks bounded by shortening diapirs, forming first a salient in Block A (stage f) then in Block B (stage h). Little salt extruded despite extreme shortening (see section view [Figure 2.66] and depth slice [Figure 2.67]) because limited salt was available in the autochthonous source layer (compare with thick-source model [Figure 2.68]). Model by Tim Dooley.

PHYSICAL MODEL—Welding of Salt Wall with Thin Source Layer during Regional Shortening

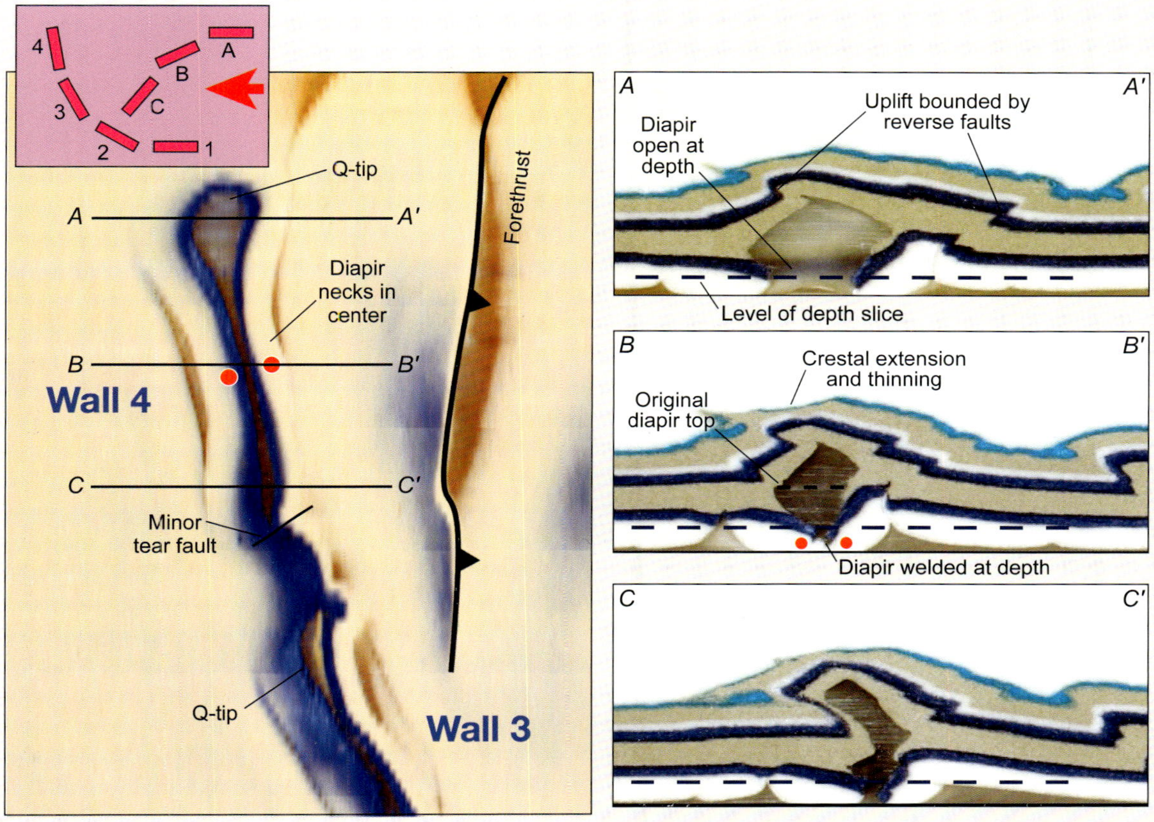

Figure 2.66. Shortening of preexisting diapirs. Shortening of a salt wall oriented at a high angle to the shortening direction (see overhead views of model in Figure 2.65) produced a deep weld that opened at its north end, forming a Q-tip® structure. This type of structure could form if the wall had been originally wider at its ends or if shortening had decreased toward the ends of the structure, as the tip of the wall was held open by the strength of the adjacent wall rocks. Salt within the wall was forced upward below the arched roof, allowing lower parts of the diapir to weld. Here there was too little salt to rupture the roof and extrude salt (compare with thick-source model [Figure 2.68]). See also depth slice of oblique walls in Figure 2.67. Model by Tim Dooley.

PHYSICAL MODEL—Welding of Oblique Salt Wall with Thin Source Layer during Regional Shortening

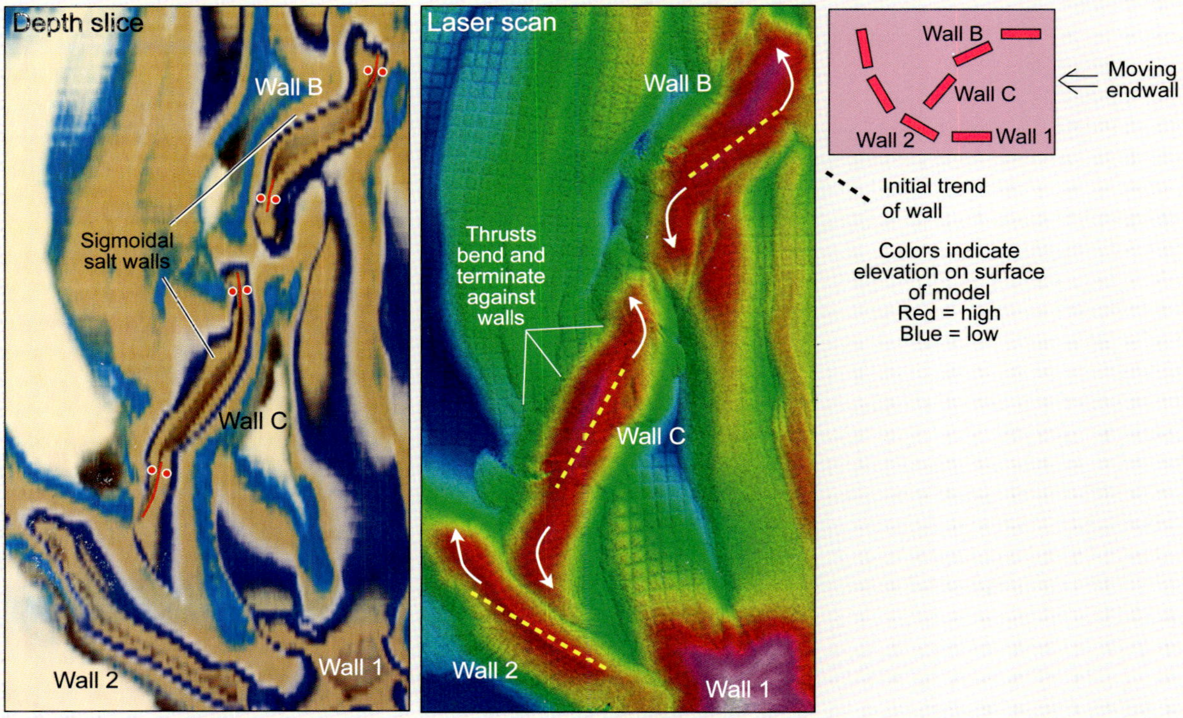

Figure 2.67. Shortening of preexisting diapirs. The depth slice shows that the shortened walls are sigmoidal, having welded tips rotated into a north-south orientation by interaction with regional thrusts in relays between salt walls. The laser scan of topography shows how thrusts bend and terminate into the walls. Overhead views of deformation (Figure 2.65) show that the thrusts each side of walls B and C are of different ages, so the walls acted as oblique boundaries to two independent thrust systems. See also sections (Figure 2.66) and model with thicker source layer (Figure 2.68). Model by Tim Dooley.

PHYSICAL MODEL—Oblique Shortening of Salt Walls with Thick Source Layer

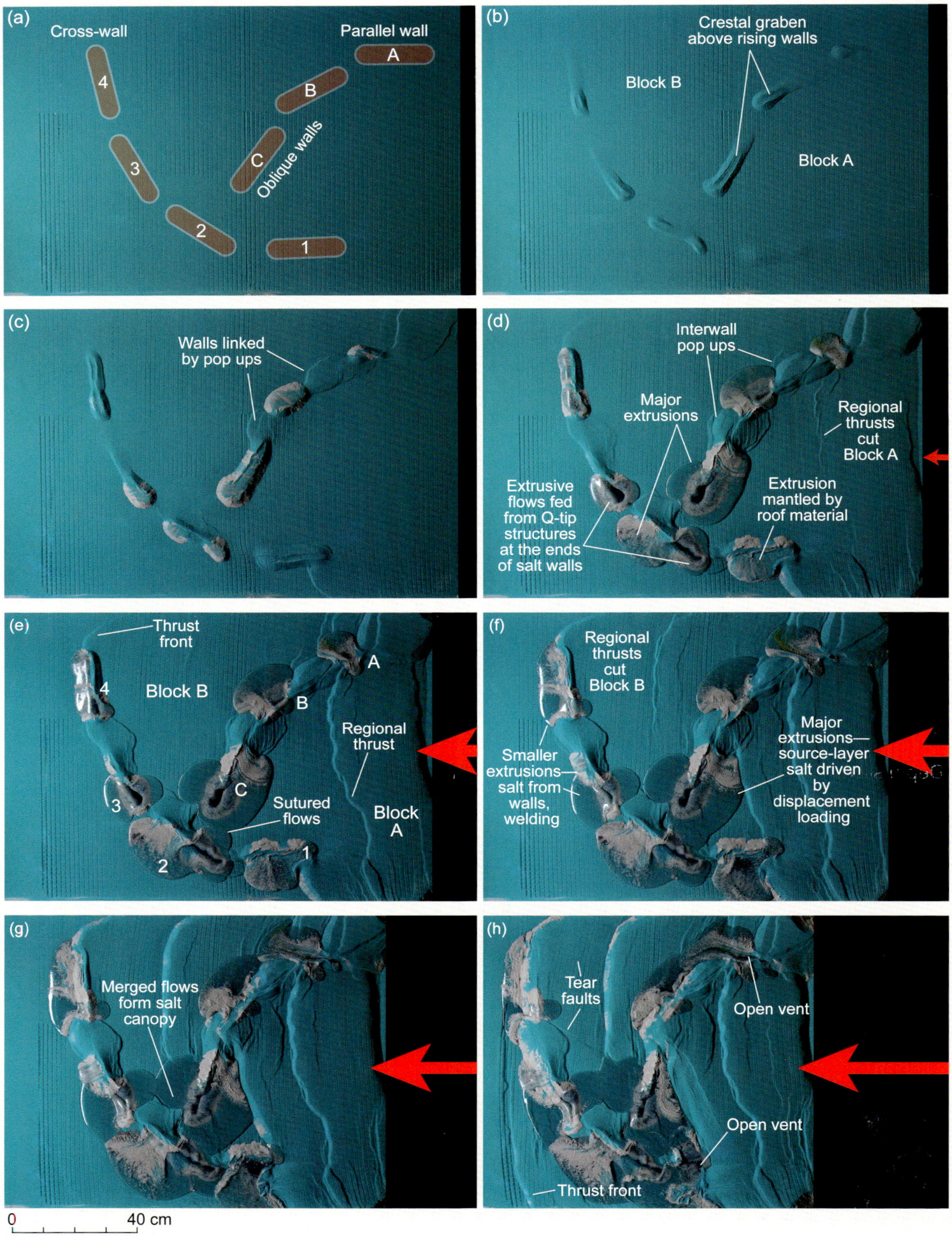

Figure 2.68. Shortening of preexisting diapirs. Salt walls were constructed at variable orientations to the regional shortening direction. Because the walls were weaker than surrounding overburden, they began shortening long before the regional thrust front propagated to them. Pop-up structures formed between shortening salt walls. Unlike the thin-source-layer models (Figure 2.65), these diapirs were supplied by thick salt, which fed major salt extrusions. These extrusions were sourced mostly from the bulbous ends of the walls, which formed Q-tip® structures (Figure 2.66). Extrusive flows merged to form salt canopies. The volume of extrusive salt was much greater than in the thin-source-layer model (Figure 2.65), suggesting that the salt canopies in this model were fed by salt displaced from the thick source layer by the advancing thrust front (see sections in Figure 2.69). Thrust-related topography strongly influenced salt-canopy shape. Model by Tim Dooley.

PHYSICAL MODEL—Welding of Salt Walls with Thick Source Layer during Regional Shortening

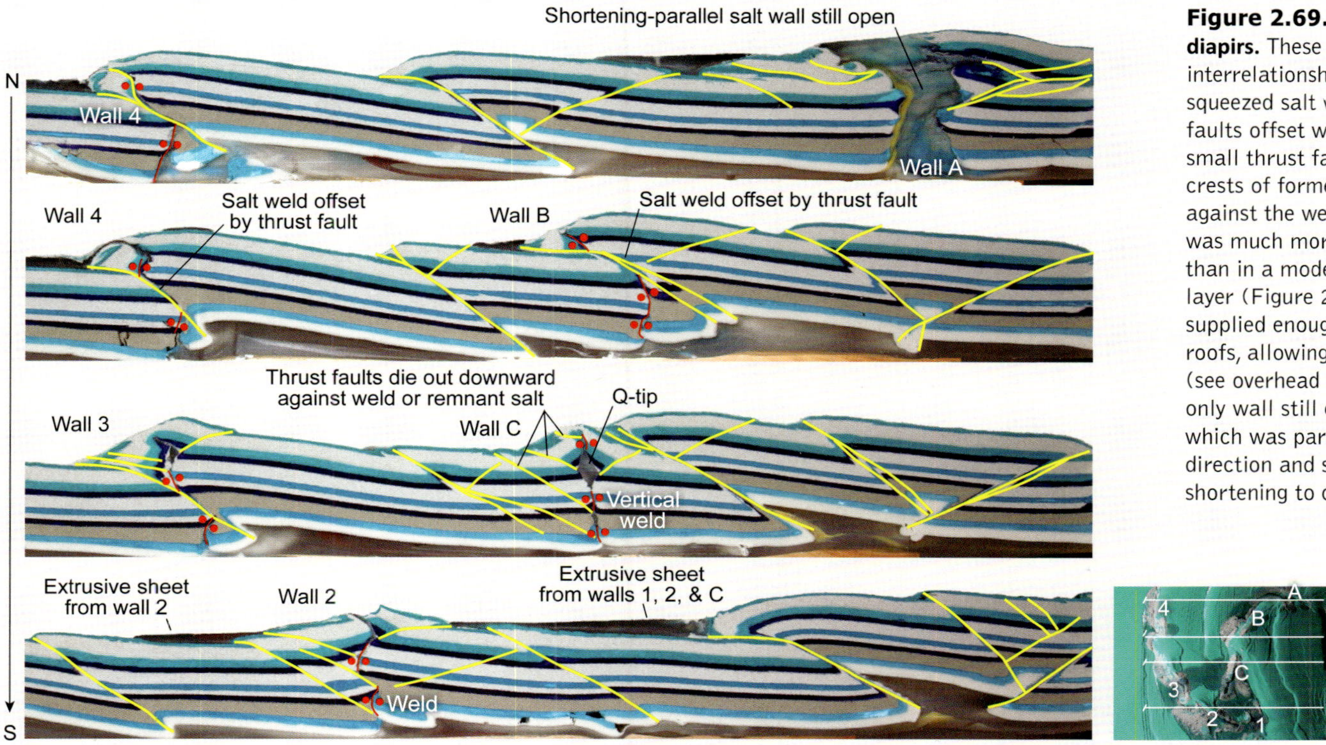

Figure 2.69. Shortening of preexisting diapirs. These sections show the complex interrelationships of thrust faults with squeezed salt walls. Large regional thrust faults offset welded salt walls. However, many small thrust faults, especially in the arched crests of former salt walls, die out downward against the weld or remnant salt. Welding was much more complete in these sections than in a model having a thinner source layer (Figure 2.66). The thicker source layer supplied enough salt to breach the diapiric roofs, allowing salt to escape as extrusions (see overhead views in Figure 2.68). The only wall still containing salt was wall A, which was parallel to the shortening direction and so required much more shortening to close. Model by Tim Dooley.

PHYSICAL MODEL—Squeezed Allochthonous Sheet

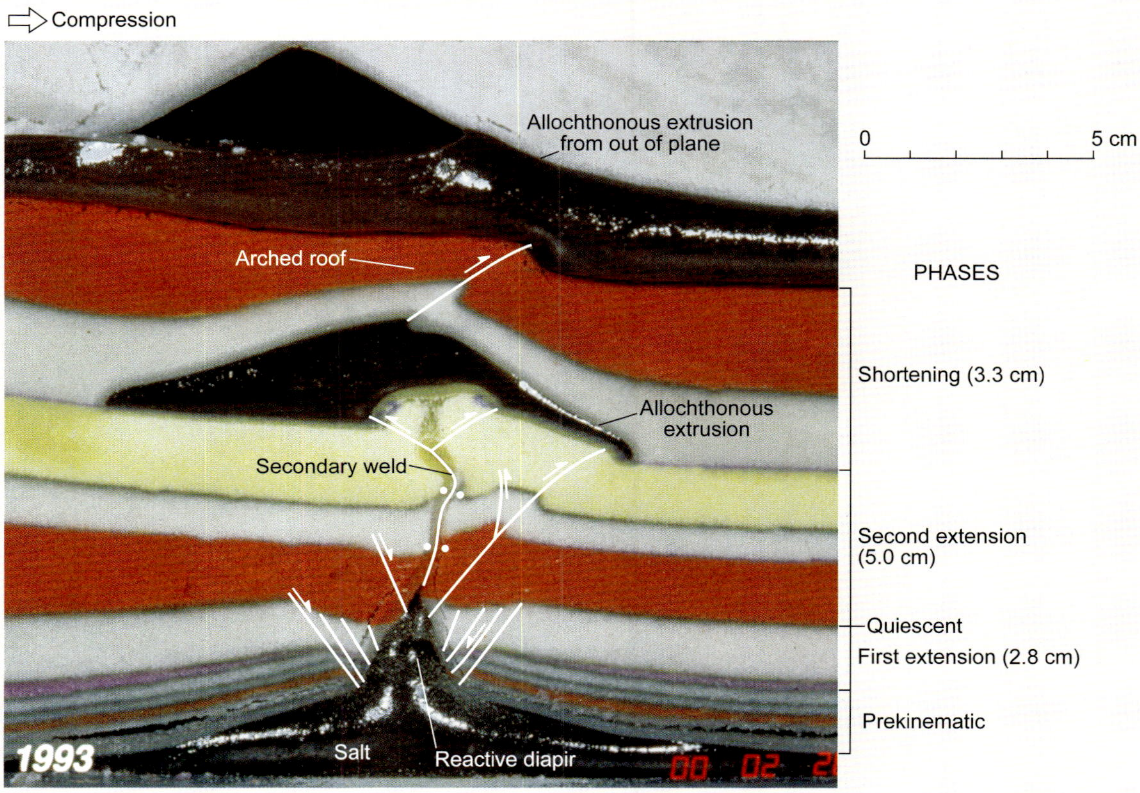

Figure 2.70. Rejuvenation of buried diapir during strong shortening. Because salt is much weaker than surrounding sediments, shortening tends to be absorbed by salt diapirs, whereas the wall rocks remain relatively unaffected. Prior to shortening, an originally reactive diapir extended almost to the paleosurface (halfway up this plane of section). Shortening raised the diapir's roof as a pop up and also laterally squeezed and evacuated the salt to form a secondary salt weld. Compression displaced the salt and forced it to extrude allochthonously from a nearby vent off the plane of section. Sediments deposited during continued shortening formed a roof to the diapir. The roof became arched and offset by a thrust. Displacement values shown are for the model's entire width of 66 cm. Model by Bruno Vendeville.

PHYSICAL MODEL—Deformation of Preexisting Diapirs during Strike Slip

Figure 2.71. Strike slip of preexisting diapirs. Thirteen salt stocks above a continuous salt source layer were located at variable distances from a dextral strike-slip fault in the basement. Laser scans of the topography show that the crests of all diapirs rose early, before faults cut the surface. Shapes of deformed stocks indicate that the roofs rose because the diapirs were shortened by pervasive compression. Uplift and shortening decreased away from the basement fault. A compressional stepover formed around the diapir above the basement fault, producing strong uplift of the diapir roof. Model by Tim Dooley.

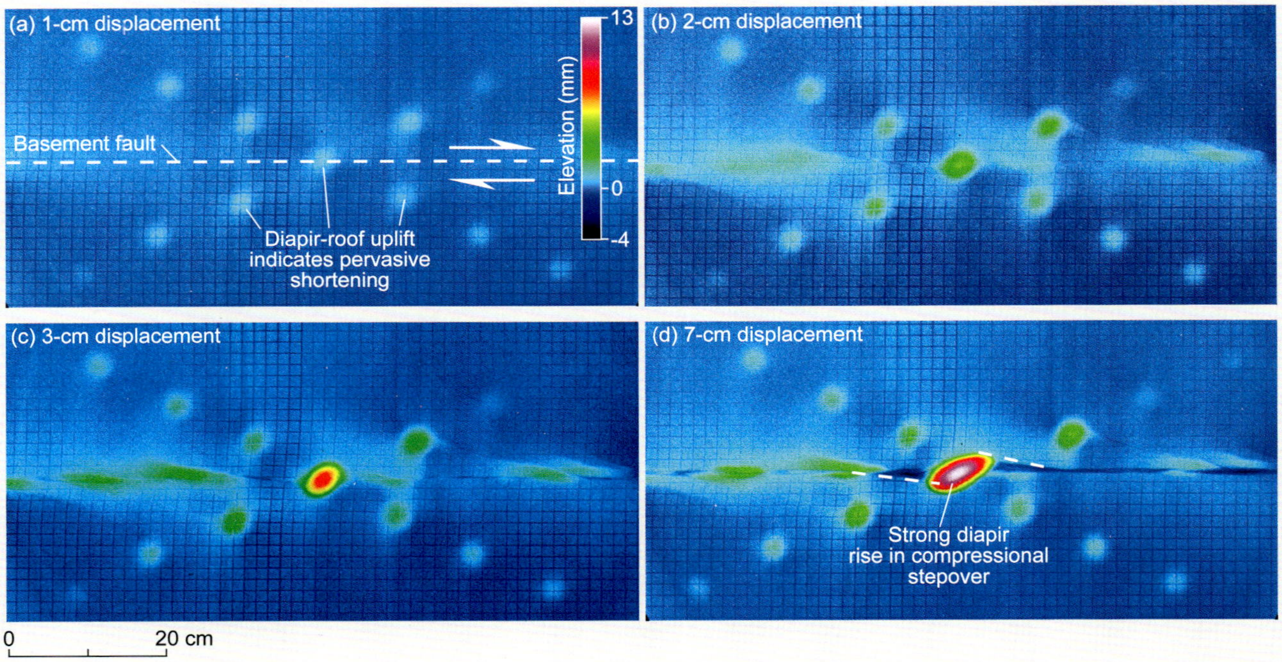

NUMERICAL MODEL—Growth of a Salt Pedestal at the Base of a Passive Diapir, Thick Layers

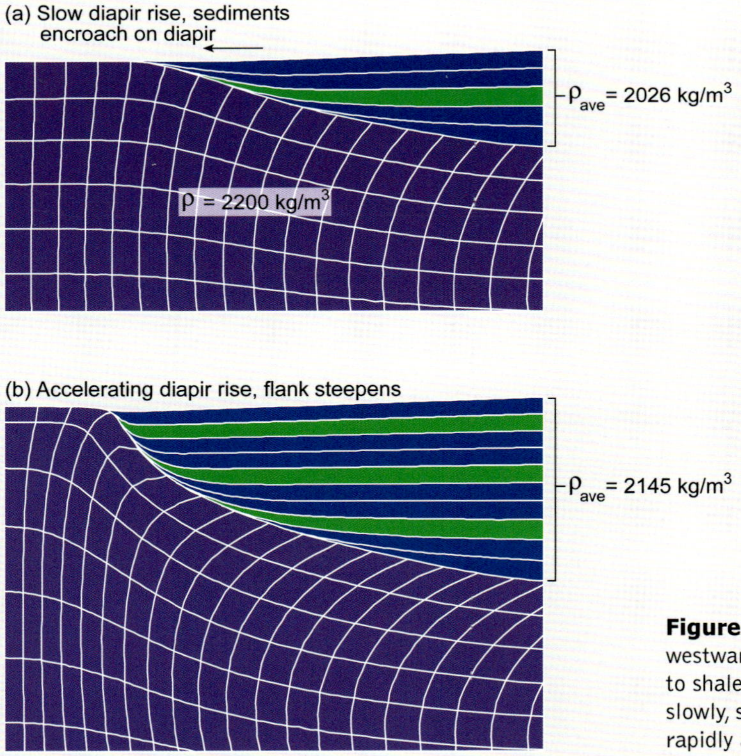

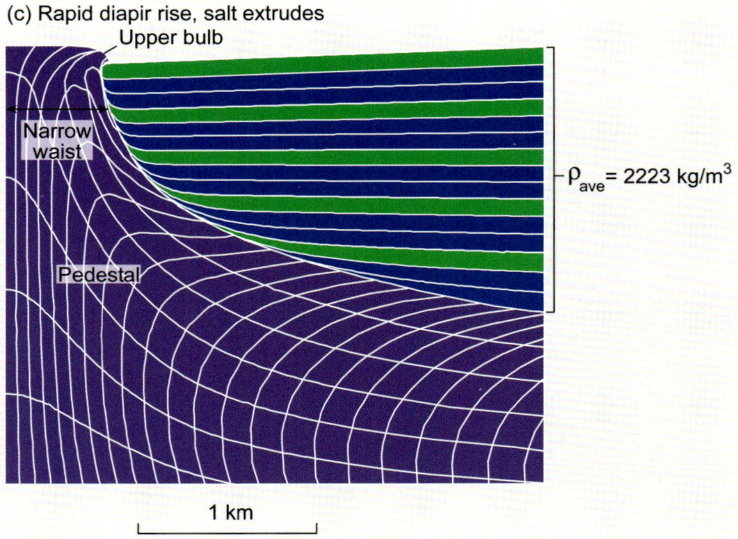

Figure 2.72. Passive diapirism. The diapir grew as a result of uneven loading produced by westward thinning of the overburden. Density of the overburden increased downward according to shale compaction curves so was less dense than salt for much of the time. The diapir first rose slowly, so overburden layers progressively onlapped and buried the diapir. Later the diapir rose more rapidly as the overburden thickened and compacted, steepening the diapir flank past vertical. The diapir eventually rose fast enough to extrude over the sediments. This history explains the hourglass cross-sectional shape of many passive diapirs. The wide base formed during slow diapir rise, the vertical waist formed when rates of diapir rise and sedimentation were nearly equal, and the wide bulb formed when the diapir began to spread by extrusion. Modified from Schultz-Ela (2003).

Numerical Model—Growth of Upright Diapir in Weak Sediments

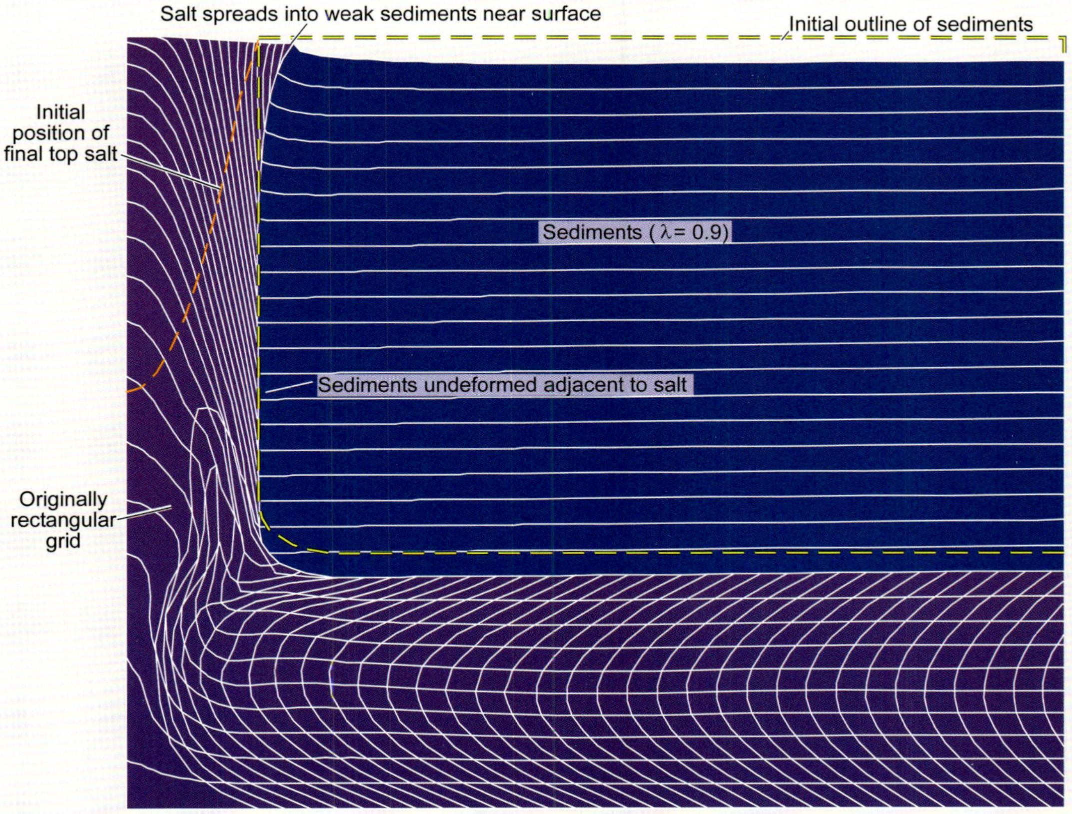

Figure 2.73. Upturn of beds against a diapir flank. Rising salt did not appreciably shear adjoining beds, even though these beds were very weak (equivalent to a pore-pressure coefficient of 0.9). This lack of shear reinforces the idea that beds adjacent to salt are only upturned if they are originally above the diapir. The shear zone accommodating relative rise of the salt is entirely within the diapir, leaving the sediment unaffected, as shown by deformed grid lines. Emergent salt was removed by simulated dissolution to keep it from spreading over the sediment surface. Modified from Schultz-Ela (2003).

Numerical Model—Halokinetic Sequences Produced by Rise of a Passive Diapir with Variable Sedimentation Rate

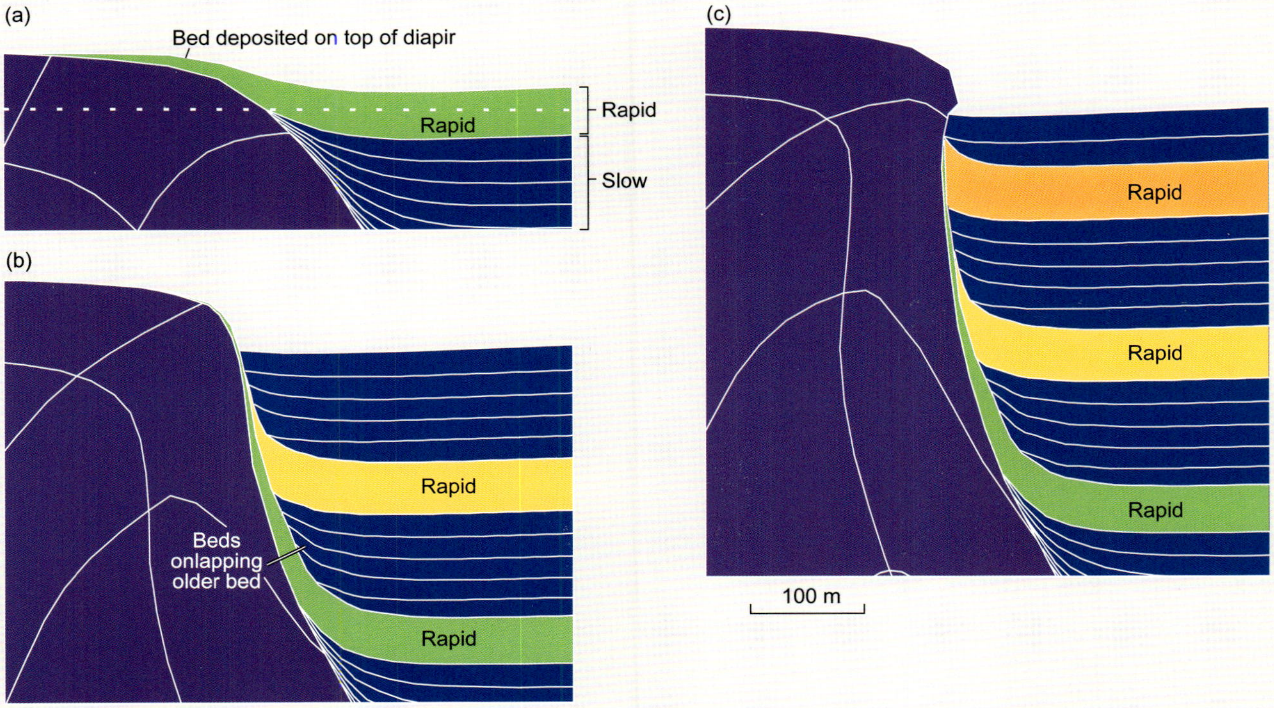

Figure 2.74. Upturn of beds against a diapir flank. The green layer was deposited over the top of the diapir during rapid sedimentation. Continued rise of the dome lifted and thinned this layer, which was rotated to near vertical as it became part of the diapir's flank. Beds deposited adjacent to salt during slower sedimentation onlapped the green layer, rather than the salt. Because these younger beds did not overlie the dome crest, they were less deformed by subsequent diapir rise. This process was less pronounced during later periods of rapid sedimentation (yellow and orange beds) because the diapir had more relief and so was not buried. Modified from Schultz-Ela (2003).

NUMERICAL MODEL—Arched Diapir, Isotropic Roof with Fixed Side Walls

Figure 2.75. Halokinetic active diapirism. Uplift of the diapir roof was driven by higher density sediments pushing down on the lower density salt source layer. Salt was forced out of the source layer and into the diapir with enough pressure to arch the roof. The roof deformed by outward rotation of two unstrained flaps. A graben formed where the two flaps join, and curved thrust faults propagated upward from the diapir shoulders. Sediment:salt density ratio = 1.09. Compare with active diapirism driven by shortening in Figure 2.76. GeoSim finite-element model by Dan Schultz-Ela.

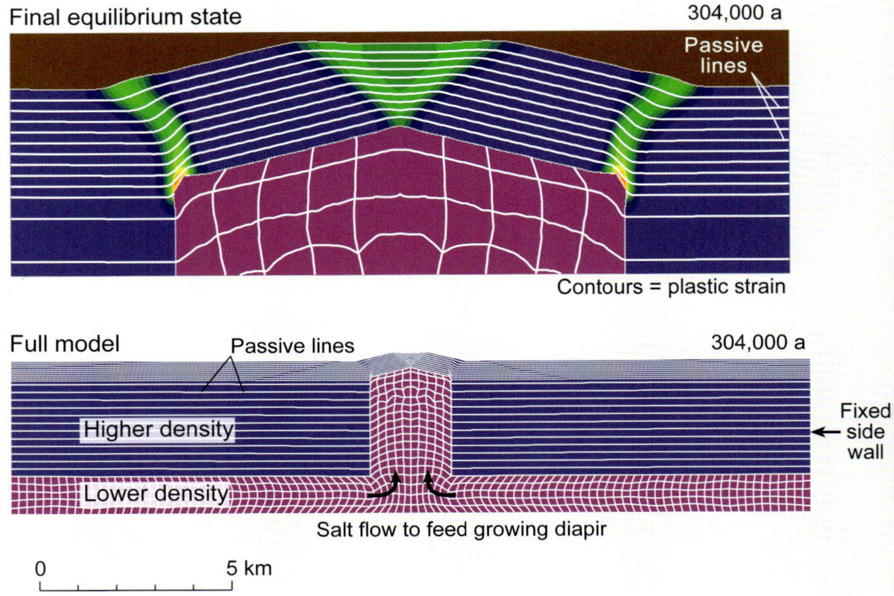

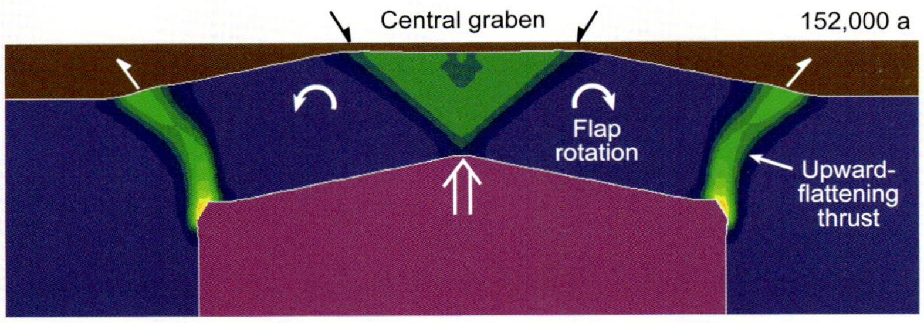

NUMERICAL MODEL—Arched Diapir, Isotropic Roof with Shortening

Figure 2.76. Active diapirism during shortening. Narrowing the diapir enhanced active rise. Some of the salt displaced by diapir shortening was forced back down into the source layer, inflating the salt and uplifting rocks adjacent to the diapir. The lower part of the roof shortened, forming a pop-down structure. Shortening rate = 1 mm/a. Sediment:salt density ratio = 1.09. Compare with active diapirism driven by density inversion in Figure 2.75. GeoSim finite-element model by Dan Schultz-Ela.

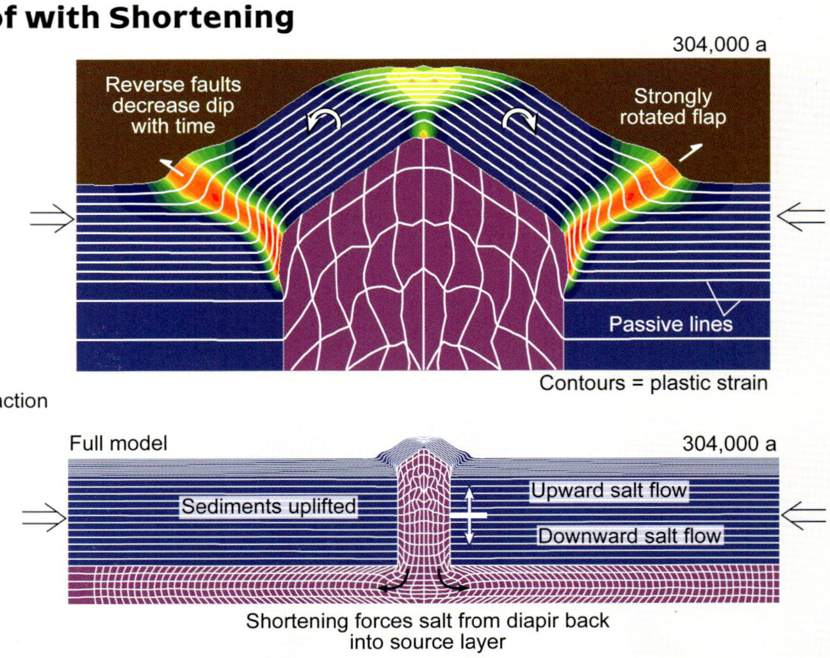

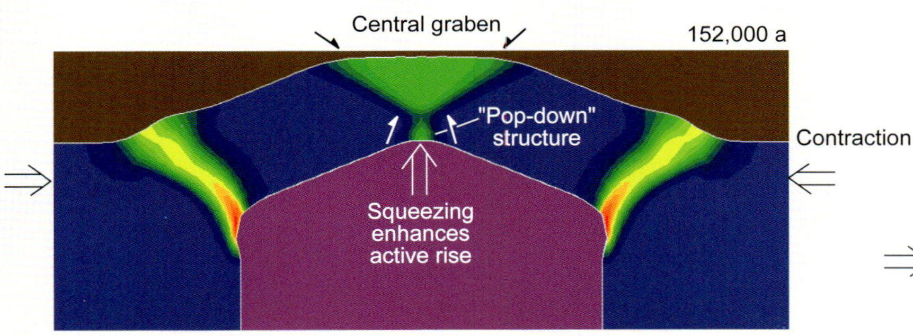

NUMERICAL MODEL—Arched Diapir with Steeply Sloped Top, Isotropic Roof with Fixed Side Walls

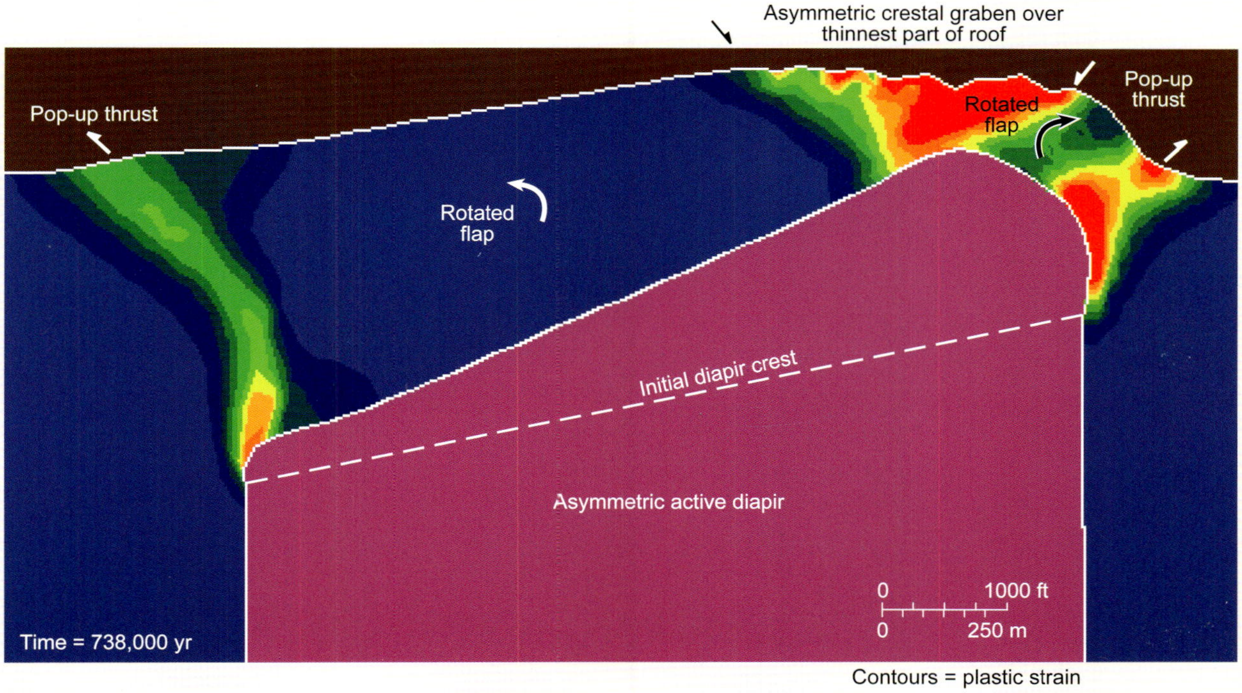

Figure 2.77. Halokinetic active diapirism. The diapir stopped growing at this stage, having reached equilibrium height for an active diapir with this geometry and sediment density. Further uplift would require roof erosion, increased load on the source layer adjacent to the diapir, or regional shortening. The crestal graben was strongly asymmetric, with a much steeper bounding fault adjoining the long flap. The graben formed near, but not at, the thinnest part of the roof. Sediment:salt density ratio = 1.09. GeoSim finite-element model by Dan Schultz-Ela.

CONCEPTUAL MODEL—Modes of Diapirism

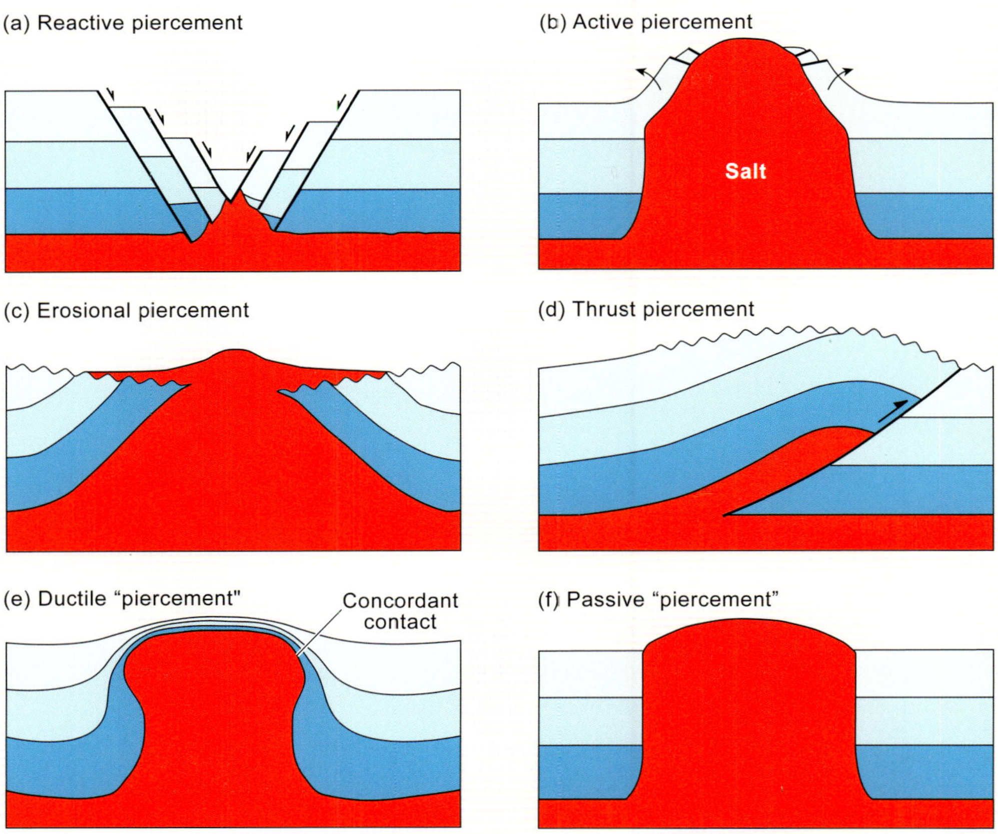

Figure 2.78. Rise of salt diapirs. In order for buried salt to be diapirically emplaced, the overburden previously occupying that space must be removed or displaced. This emplacement can occur in one of four ways for brittle overburdens. (a) The overburden may be extended, making room for a reactive diapir to rise between separated fault blocks. (b) Flaps of overburden may be lifted, rotated, and shouldered aside as the diapir forcibly breaks through them by active diapirism. (c) The roof may be removed by erosion. (d) Salt may be emplaced into its overburden in the hanging wall of a thrust fault. (e) Diapirs having fluid overburdens may rise by a fifth mechanism, ductile thinning of the diapir roof. These are not, strictly speaking, diapirs because they have concordant contacts with surrounding sediments. However, they are included here because they have risen with respect to flanking structures and because their shapes resemble those of true diapirs. (f) If a diapir completely pierces its overburden and is exposed at the sediment surface, passive diapirism or downbuilding becomes possible, and the diapir rises continually with respect to surrounding strata and remains exposed while sediments accumulate and sink around it. Modified from Hudec and Jackson (2007).

CONCEPTUAL MODEL–Three Modes of Diapirism

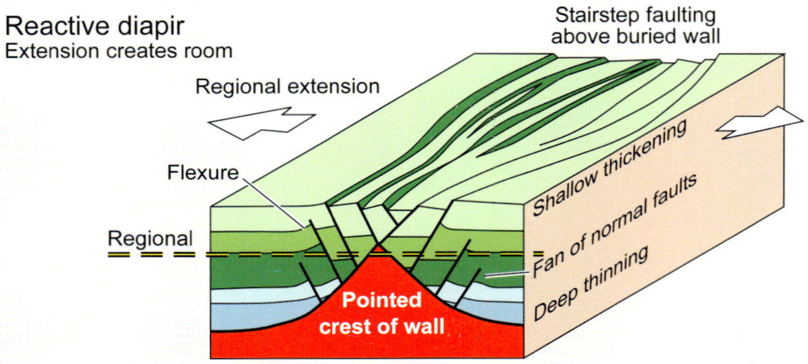

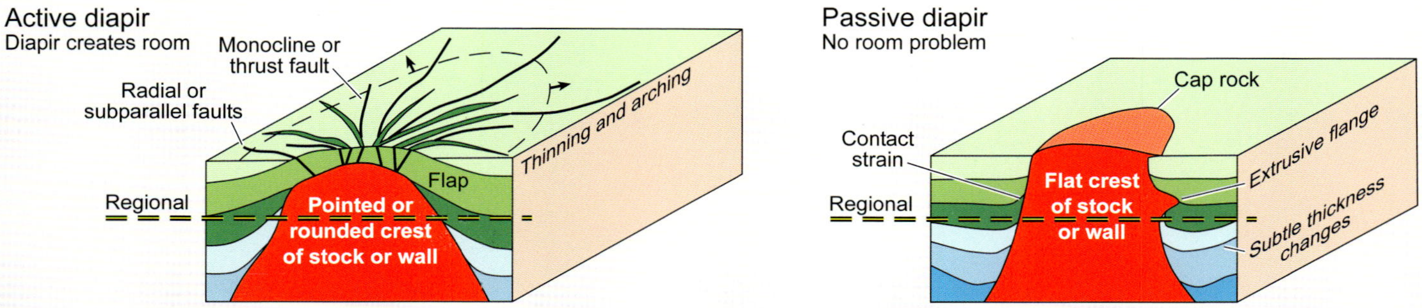

Figure 2.79. Growth of salt diapirs. For a salt diapir to pierce, there must be a solution to the room problem posed by overlying units: where is the overburden that was displaced by the diapir? The three primary modes of diapirism each solve this problem differently. In reactive diapirism, the overburden is rifted apart by extension. In active diapirism, overburden units are uplifted and pushed aside by the rising diapir. In passive diapirism, salt is already at the surface, so no overburden exists to be displaced. The diagrams show criteria to distinguish between the diapir modes. Modified from Jackson and others (1994).

CONCEPTUAL MODEL–Growth of an Hourglass-Shaped Passive Diapir

Figure 2.80. Passive diapirism. Hourglass-shaped diapirs have a wide bulb above a narrow waist above a broad pedestal. Most passive diapirs have an hourglass shape, which can be explained as follows. Initially, siliciclastic basins sinking into the salt layer are thin, weakly compacted, and less dense than the salt. These buoyant basins expel underlying salt slowly, producing only slow rise of nearby diapirs. Slowly rising diapirs tend to be buried, so the diapir narrows inward to build a pedestal. Continued aggradation thickens and densifies the withdrawal basins, accelerating flow of salt into the diapir. Faster diapir rise steepens the contact, and eventually the diapir widens outward, forming a wide crest. Halokinetic rise of the diapir ends when the source layer is exhausted, although an inactive diapir may be rejuvenated by later shortening (see Figure 2.81).

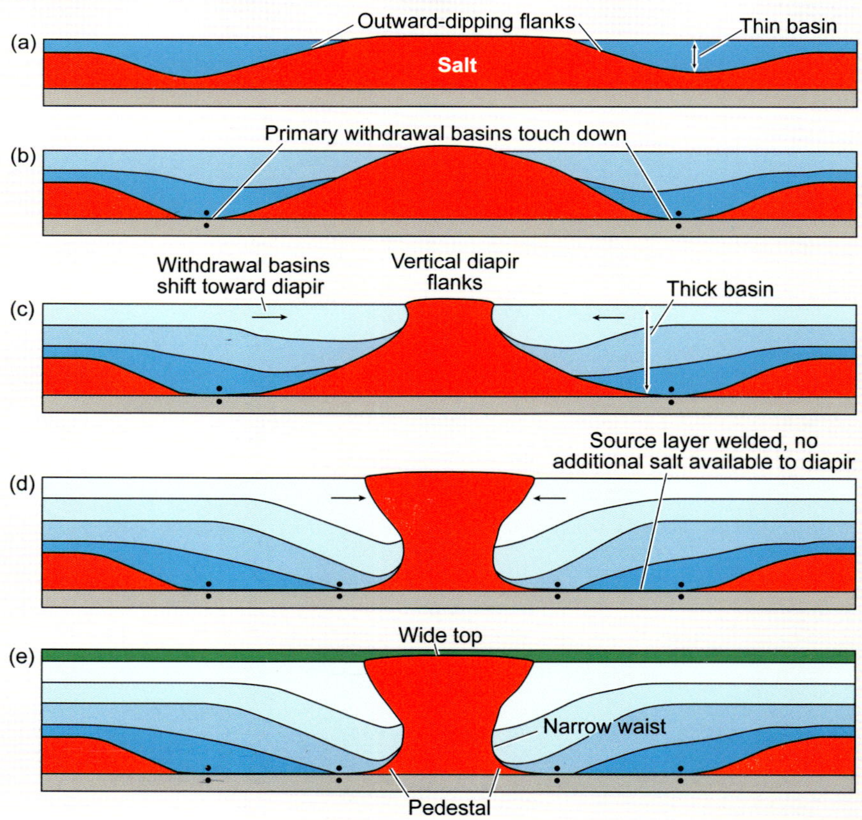

CONCEPTUAL MODEL—Hourglass-Shaped Diapir Rejuvenated by Shortening

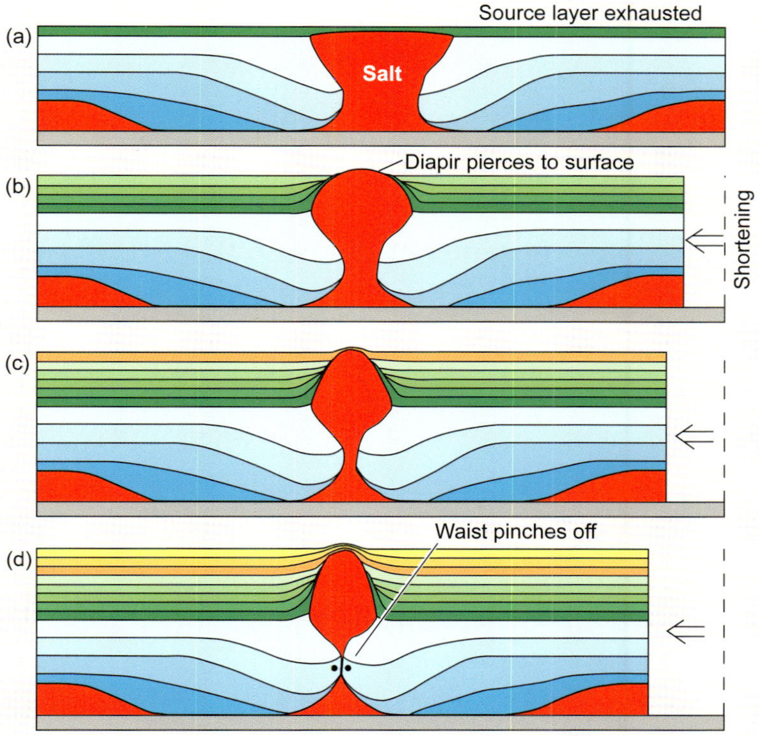

Figure 2.81. Shortening of a salt dome. Passive salt diapirs shaped like an hourglass are common (see hourglass evolution in Figure 2.80). Passive diapirs stop growing halokinetically after exhausting their source layers but may be rejuvenated by lateral shortening. Squeezing a diapir expels salt upward, which may allow it to pierce to the surface if the roof is thin enough. Squeezing also steepens the dip of flanking strata. The diapir continues to shorten until its waist pinches off, yielding a geometry resembling a balloon on a string. The wide top and base of this pinched-off diapir were inherited from the original hourglass shape.

CONCEPTUAL MODEL—Formation of Salt Diapirs by Sediment Progradation across Basement Steps

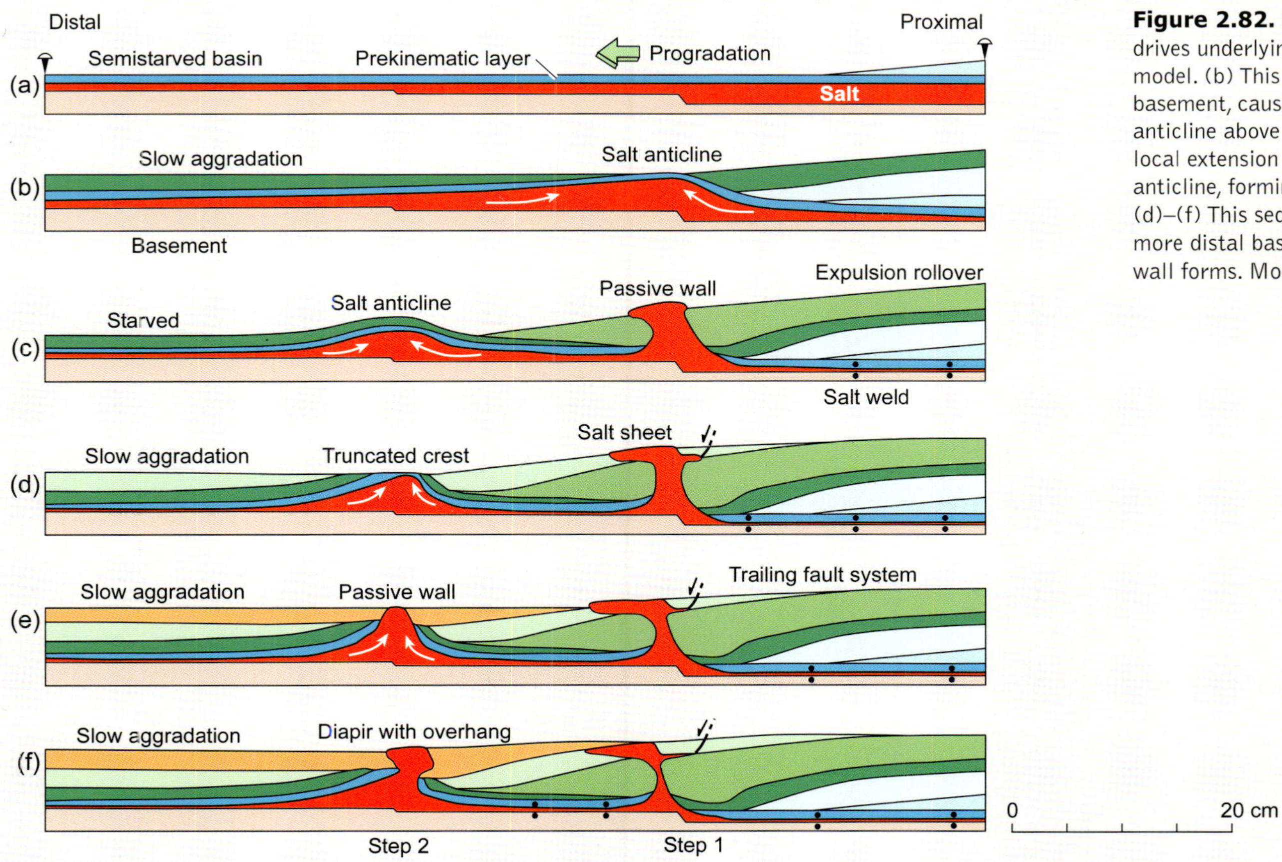

Figure 2.82. Passive diapirism. (a) Progradation drives underlying salt toward the distal end of the model. (b) This flow is resisted by a step in the basement, causing salt to back up and form a salt anticline above the basement step. (c) Erosion and local extension allow salt to pierce the crest of this anticline, forming a passive salt wall above the step. (d)–(f) This sequence of events repeats above a more distal basement step, where a younger passive wall forms. Modified from Ge and others (1997).

CONCEPTUAL MODEL—Growth of Hook Halokinetic Sequences during Overall Rapid Aggradation

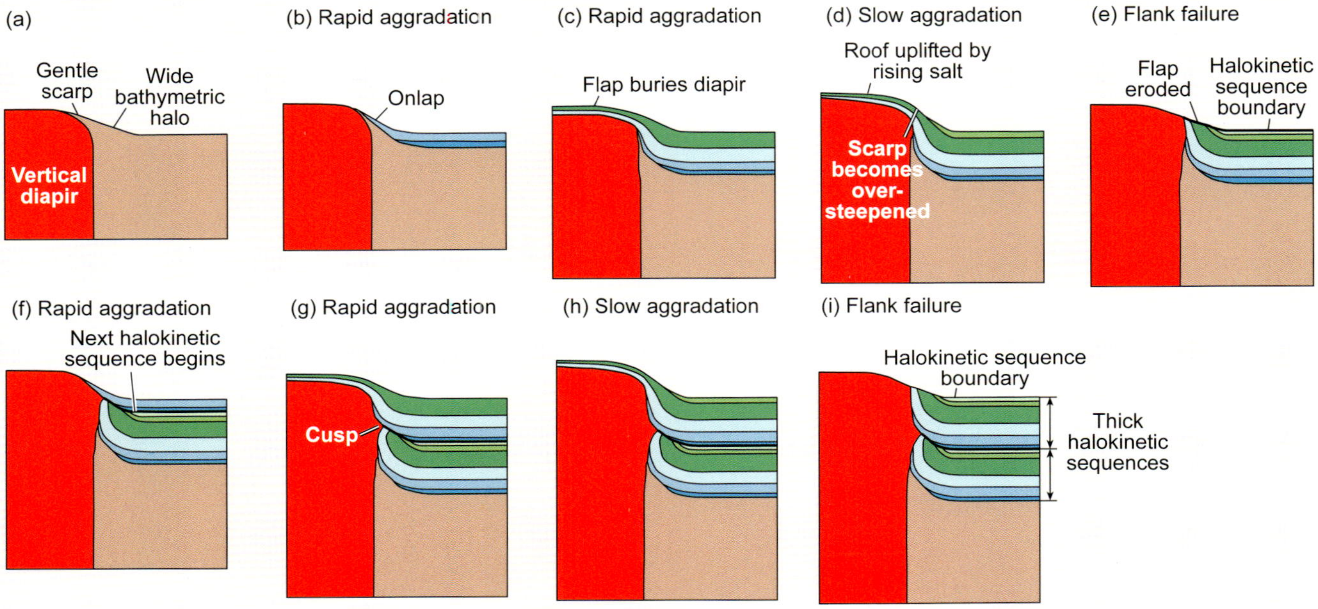

Figure 2.83. Formation of halokinetic sequences. Strata against a diapir flank are upturned and eroded when a wedge of sediment above the diapir is lifted and pushed aside by rising salt. The drape fold over the diapir flank periodically oversteepens during slow sedimentation, causing gravitational failure and erosional truncation of the upturned flap. These unconformity-bounded packages are termed halokinetic sequences. This process creates large, steeply upturned flaps from thick wedges deposited over the diapir crest and thinner, more gently upturned flaps from thin wedges deposited over the diapir crest (see slow aggradation model in Figure 2.84). Here, rapidly aggrading sediments bury the diapir crest, so flaps are thick and steeply upturned. Cusps of salt form where halokinetic sequence boundaries intersect the diapir flank. Modified from Rowan and others (2003).

CONCEPTUAL MODEL—Growth of Wedge Halokinetic Sequences during Overall Slow Aggradation

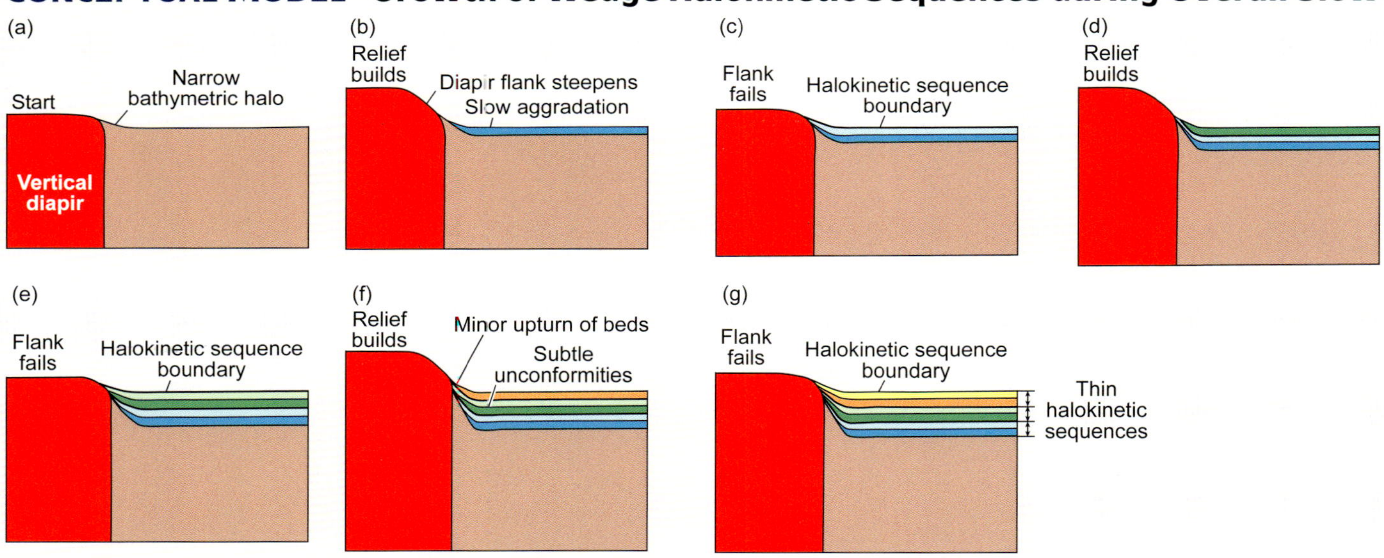

Figure 2.84. Formation of halokinetic sequences. Strata against a diapir flank are upturned and eroded when a wedge of sediment above the diapir is lifted and pushed aside by rising salt. The drape fold over the diapir flank periodically oversteepens during slow sedimentation, causing gravitational failure and erosional truncation of the upturned flap. These unconformity-bounded packages are termed halokinetic sequences. This process creates large, steeply upturned flaps from thick wedges deposited over the diapir crest (see fast aggradation model in Figure 2.83) and thinner, more gently upturned flaps from thin wedges deposited over the diapir crest. Here, slow aggradation of sediments is not able to bury the diapir crest, so flaps are thin and only moderately upturned. Modified from Rowan and others (2003).

Conceptual Model—Influence of Precursor Diapirs on Detached Extension

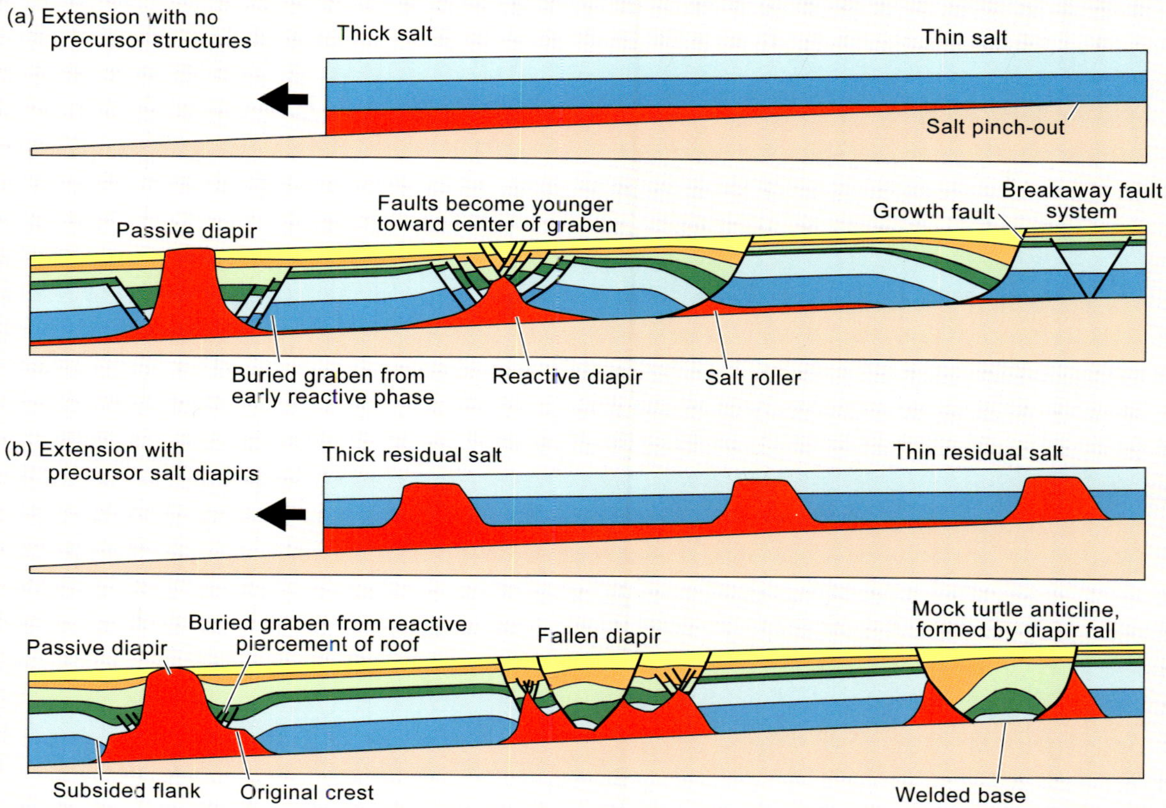

Figure 2.85. Extension of salt domes. (a) In the absence of precursor diapirs, the main control on extensional structural style is salt thickness. Thin evaporites localize the detachment but cannot form diapirs or large withdrawal basins. This structural style is dominated by listric growth faults and low-amplitude salt structures such as salt rollers. Above thicker salt, diapirs and adjacent withdrawal basins can grow larger. Some structures evolve through the reactive and active stages to become passive diapirs, which remain at the surface as long as there is salt to feed them. (b) Precursor diapirs widen during regional extension, so the most important factor controlling their evolution is the amount of salt available to maintain continued growth. If the source layer is exhausted, or if salt is imported from the surrounding source layer too slowly, diapirs will fall. If abundant salt is available, salt diapirs can continue to grow passively or reactively, although they may narrow upward because their older, deeper parts began widening earlier than their youthful diapir crests. Compare with influence of precursor diapirs on detached shortening in Figure 2.86. Modified from Hudec and Jackson (2007).

Conceptual Model—Influence of Precursor Diapirs on Detached Shortening

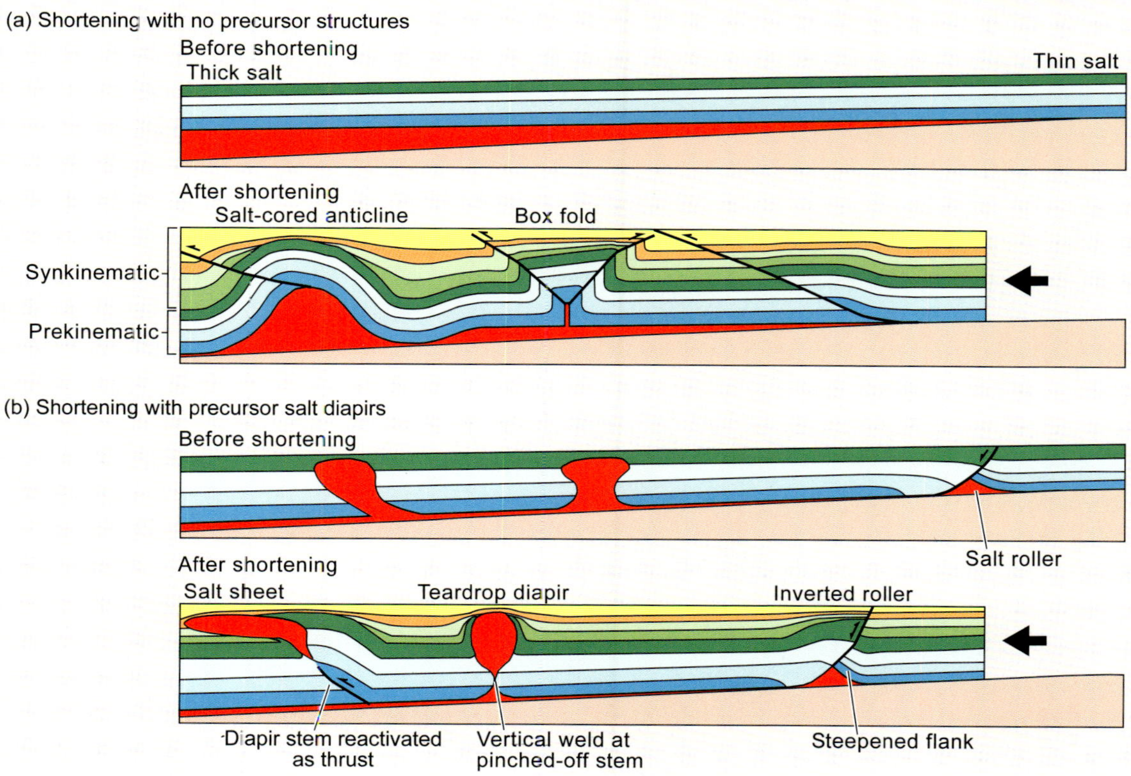

Figure 2.86. Shortening of salt domes. (a) Shortening thickens and therefore strengthens the overburden above salt, which inhibits new diapirs. In the absence of preexisting salt structures, salt functions mainly as a décollement. Above thin salt, structures are dominated by thrusts and narrow box-fold anticlines. More open, larger-amplitude detachment folds are possible where thicker salt can fill the cores of anticlines. (b) Preexisting diapirs are preferentially reactivated during shortening because they are mechanically weak. Salt bodies therefore shorten early and nucleate folds and thrust faults. Normal faults associated with early extensional diapirs can be inverted. Compare with influence of precursor diapirs on detached extension in Figure 2.85. Modified from Hudec and Jackson (2007).

STOCKS & WALLS

CONCEPTUAL MODEL—Rejuvenation of a Passive Diapir during Shortening

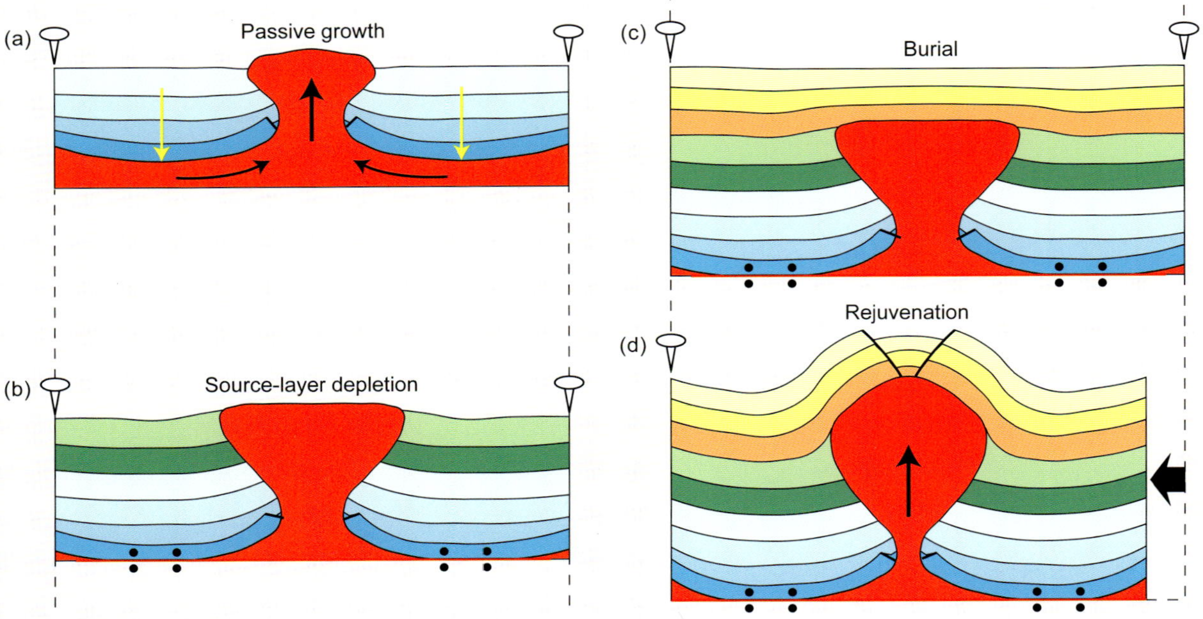

Figure 2.87. Rejuvenation of a buried diapir during shortening. As long as a diapir is being fed from its source layer, it has positive topographic relief and is difficult to bury. Once the source layer is depleted, the diapir becomes inactive and gets buried by layers that maintain constant thickness across the crest of the structure. If the diapir is later shortened, this constant-thickness roof buckles, and the diapir resumes its rise. After Vendeville and Nilsen (1995).

CONCEPTUAL MODEL—Characteristic Features of North Sea Diapirs

Figure 2.88. Active diapirism. Most North Sea domes evolved through several stages. The pillow stage is characterized by stratigraphic thinning onto the crest of the salt high. During the down-building stage, strata were truncated against salt at high angles. The rejuvenation stage started when the diapir first began to be buried. The layers deposited above the crest were continually being uplifted and stretched as deposition continued. The teardrop shape of the diapir suggests that shortening was involved in the rejuvenation process. After Davison and others (2000a).

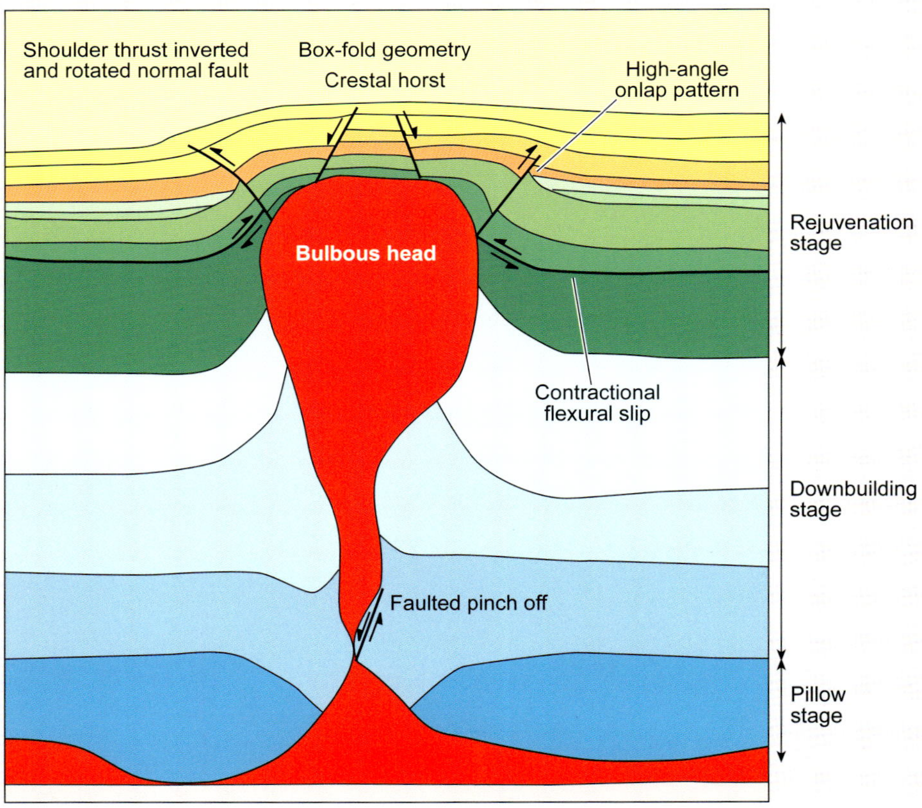

CONCEPTUAL MODEL—Influence of Precursor Diapirs on Strike-Slip Tectonics in a Releasing Stepover

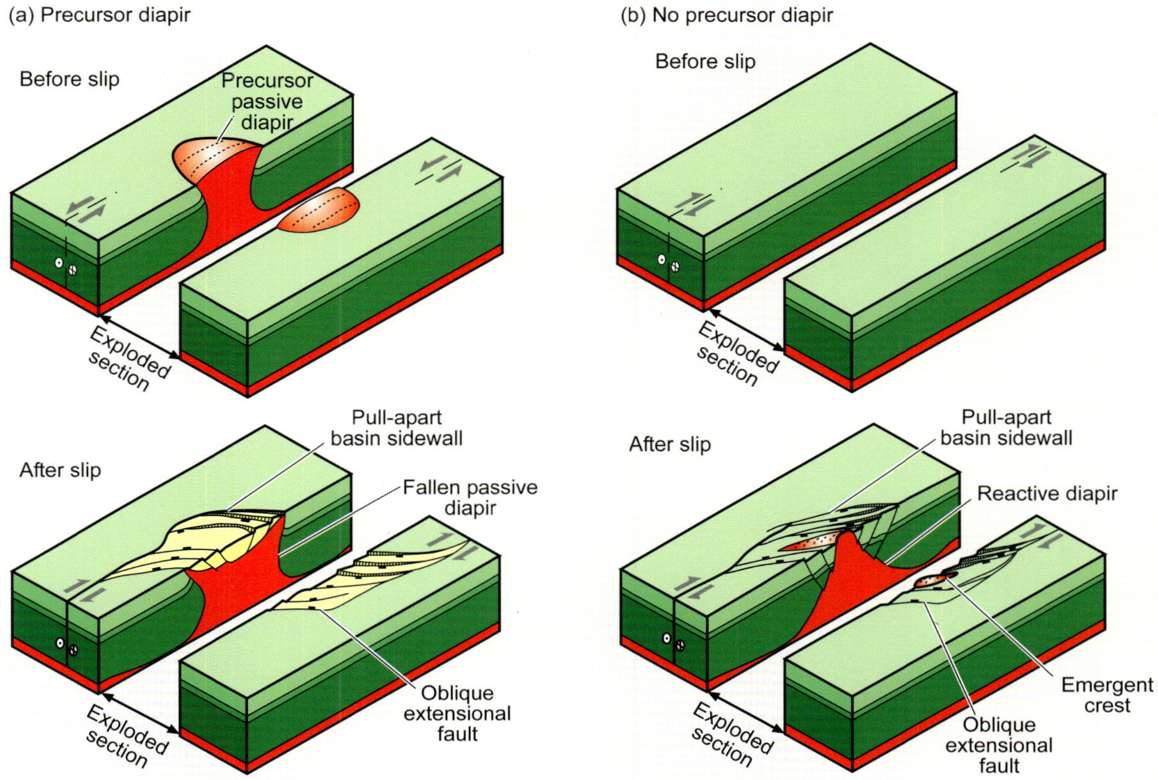

Figure 2.89. Strike-slip deformation of salt diapirs. Pure strike slip has little effect on a salt layer, but salt may flow if wrenching creates a component of tensional or compressional stress. Most salt-involved structures in strike-slip fault systems therefore occur at bends or stepovers. In these local strain zones, salt behaves much as in regional extensional and contractional provinces. Extensional stepovers or releasing bends form pull-apart basins. (a) Any preexisting diapir in one of these basins will widen and rise or fall depending on the rate of supply of new salt. (b) If no precursor structure exists, a reactive diapir may form in an extensional stepover. See influence of precursor diapirs in a restraining stepover in Figure 2.92. Modified from Hudec and Jackson (2007).

CONCEPTUAL MODEL—Influence of Precursor Diapirs on Strike-Slip Tectonics in a Restraining Stepover

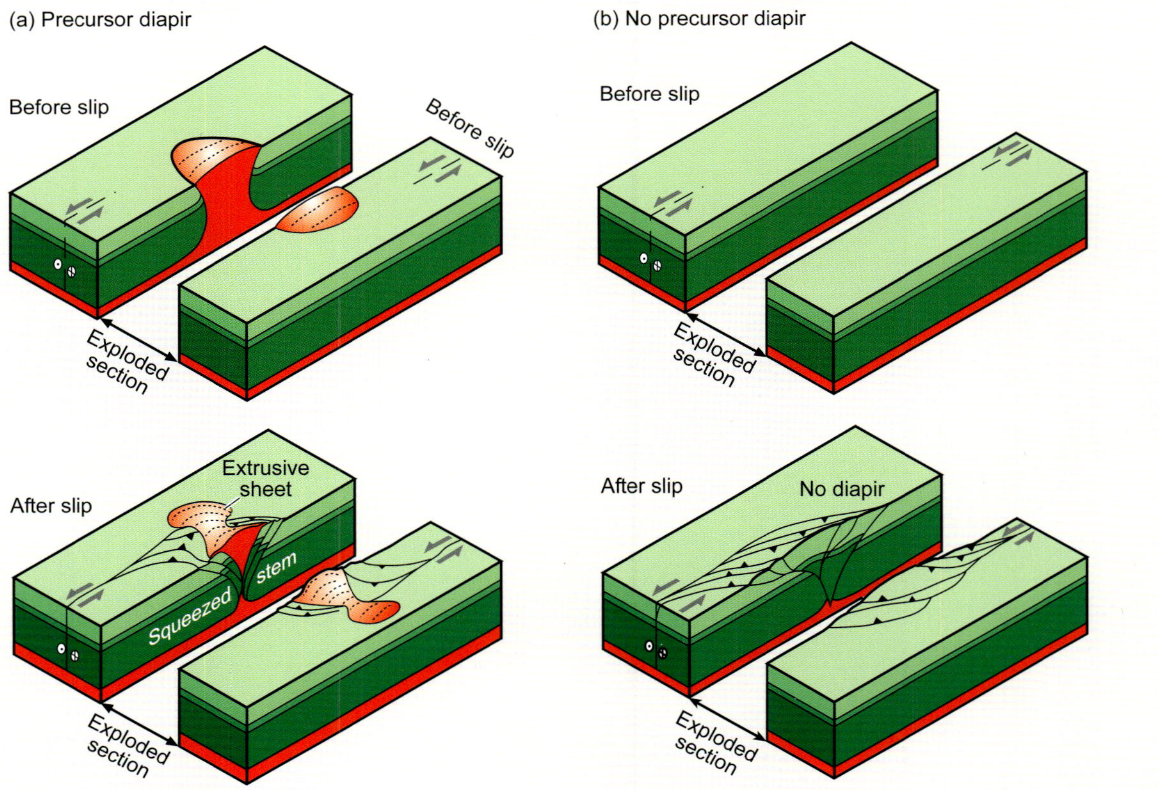

Figure 2.90. Strike-slip deformation of salt diapirs. Pure strike slip has little effect on a salt layer, but salt may rise if wrenching creates a component of tensional or compressional stress. Most salt-involved structures in strike-slip fault systems therefore occur at bends or stepovers. In these local strain zones, salt behaves much as in regional extensional and contractional provinces. Thrust systems form in restraining bends and stepovers, shortening any preexisting diapirs. (a) Where precursor diapirs exist, diapir stems are squeezed, and salt extrudes from the pinched-off stem to flow over the sediment surface. (b) In the absence of precursor structures, the salt acts as a décollement layer for thrusts. See influence of precursor diapirs in a releasing stepover in Figure 2.91. Modified from Hudec and Jackson (2007).

CONCEPTUAL MODEL—Internal Deformation of Salt in Source Layer and Diapir

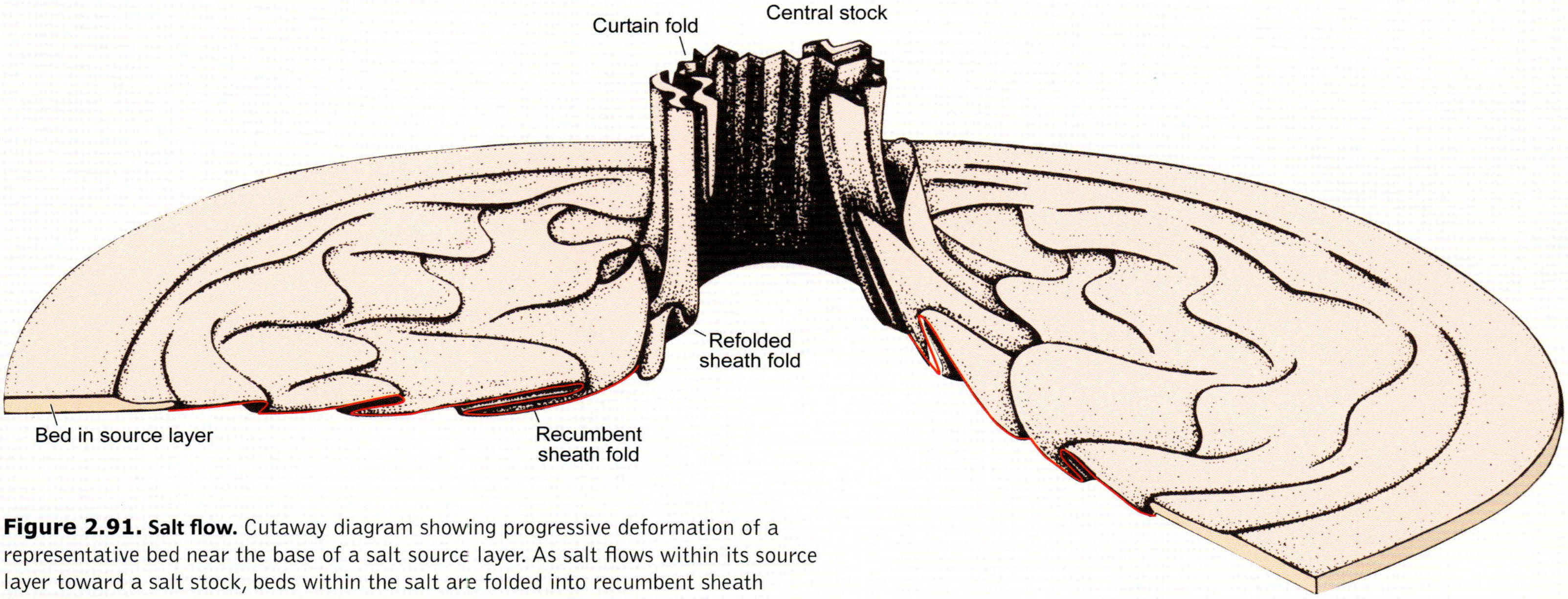

Figure 2.91. Salt flow. Cutaway diagram showing progressive deformation of a representative bed near the base of a salt source layer. As salt flows within its source layer toward a salt stock, beds within the salt are folded into recumbent sheath folds where stream lines cross bedding (typically because of irregularities on the base of salt). Radially inward constrictional flow into the stock produces vertically plunging curtain folds, which refold any sheath folds formed earlier in the source layer. Curtain folds have been likened to structures formed by pulling the center of a handkerchief through a small ring. Modified from Talbot and Jackson (1987).

CONCEPTUAL MODEL—Cross Sections of Mushroom Diapirs

Figure 2.92. Formation of mushroom diapirs by interference folding. A mushroom diapir has a crestal bulb fringed by one or more pendant peripheral lobes. If this lobe is entirely within the diapir, it forms an internal mushroom structure. More rarely the lobe encloses infolded country rock, as an external mushroom. Modeling suggests that mushrooms form during flow of two fluids having similar viscosities. Internal mushrooms therefore form in diapirs composed of a mixture of two evaporite layers, whereas external mushrooms can only form where the country rock acts as a fluid. Vortex structures (where the downward-pointing flange of the mushroom rolls back up) form by thermal convection within the diapir. See also block diagrams of mushroom diapirs in Figure 2.93. Modified from Jackson and Talbot (1989a).

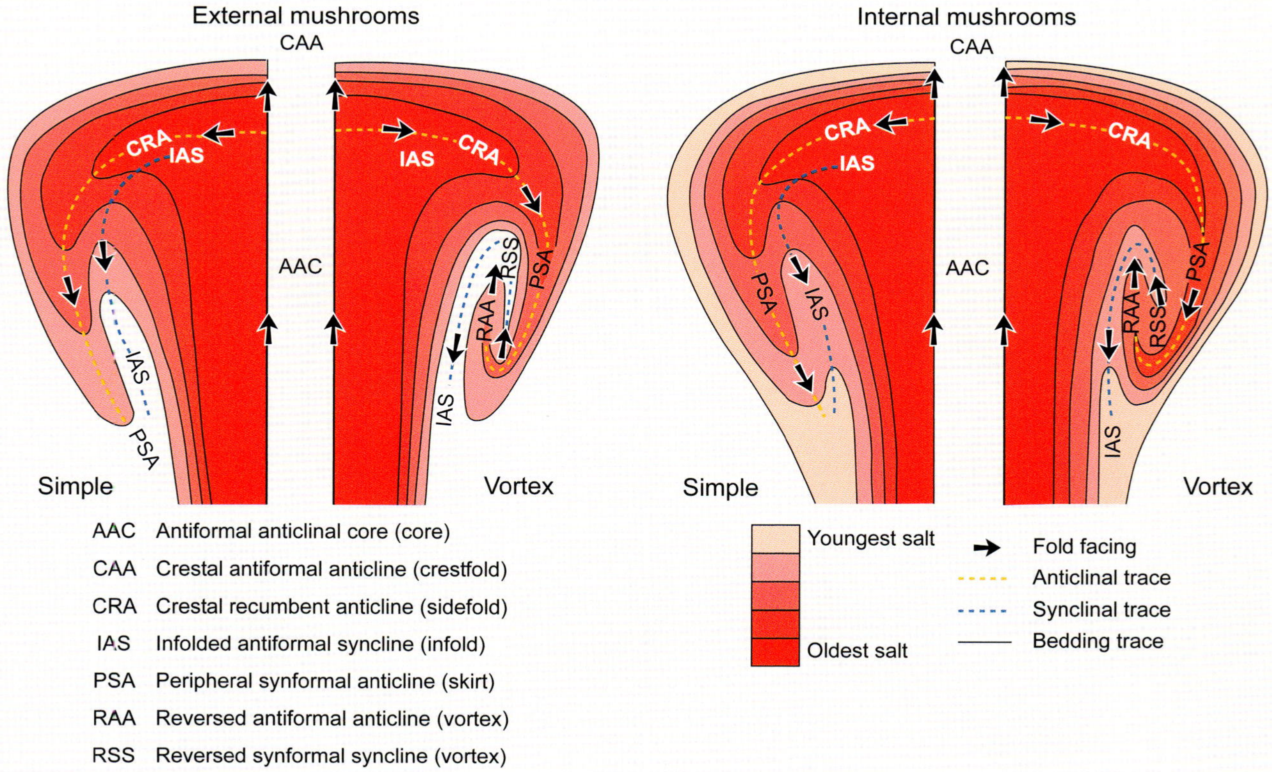

CONCEPTUAL MODEL–Block Diagrams of Mushroom Diapirs

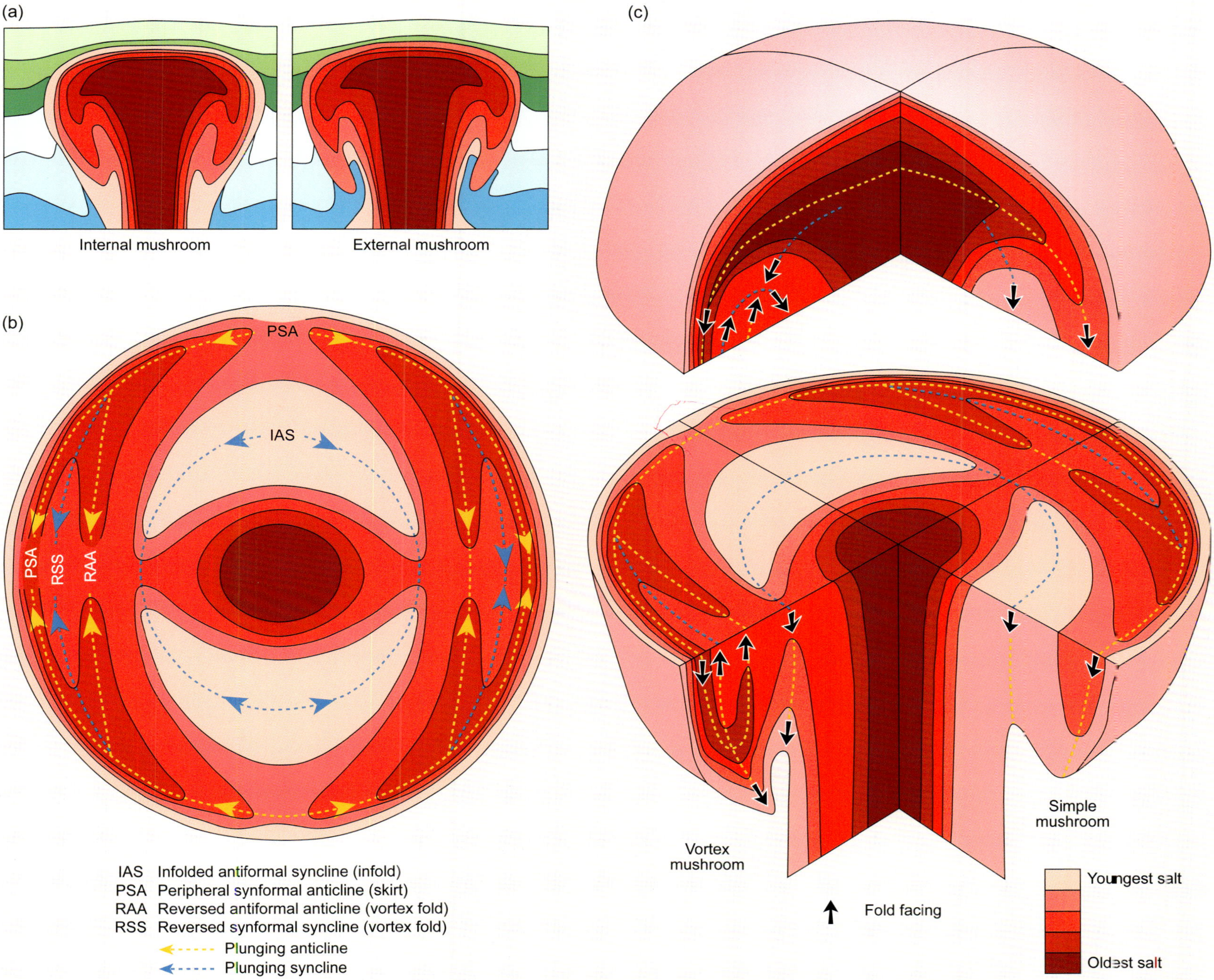

Figure 2.93. Formation of mushroom diapirs by interference folding. A mushroom diapir has a crestal bulb fringed by one or more pendant peripheral lobes, which are downward-facing folds. (a) This lobe is typically contained within the diapir (an internal mushroom) but more rarely encloses infolded country rock (an external mushroom, see cross sections of mushroom diapirs in Figure 2.92). (b) Horizontal section of the mushroom shows the axial traces of the internal crescentic folds. (c) Isometric cutaway diagram of the diapir shown in b, showing the difference between vortex and simple mushrooms. Vortex structures form by thermal convection. It is unlikely that a natural diapir would contain both vortex and simple mushrooms because it would imply that only half of the diapir was convecting or that convective overturn was more advanced on one side. Modified from Jackson and Talbot (1989a).

Chapter 3

Triangular Diapirs

In most salt basins, diapirism began during regional extension, often shortly after salt deposition ended. Extensional diapirs typically have triangular profiles, but this shape becomes rounded if the rising diapirs pierce to the sediment surface and begin to grow as passive diapirs.

Triangular diapirs are ridges of salt having triangular profiles and discordant contacts with their overburden. Triangular diapirs are one type of salt wall; nontriangular types that grow by different processes are presented in Chapter 2.

Most triangular diapirs grow during regional extension by reactive diapirism. In this process, the overburden sediments are thinned by normal faulting in response to regional extension. As the fault blocks are pulled apart, they make room for the diapiric salt to rise. Two types of triangular diapir are distinguished by geometries of the faults and overlying sediments. Symmetric reactive diapirs are overlain by a symmetric fan of inward-dipping normal faults that terminate against the salt face. Faults near the center of the fan are younger than those farther out. An asymmetric reactive diapir, by contrast, lies in the footwall of a single listric normal fault that flattens into the salt layer. The diapir has asymmetric flanks: one flank is discordant to the hanging wall of the fault, whereas the other salt flank is concordant to strata in the footwall.

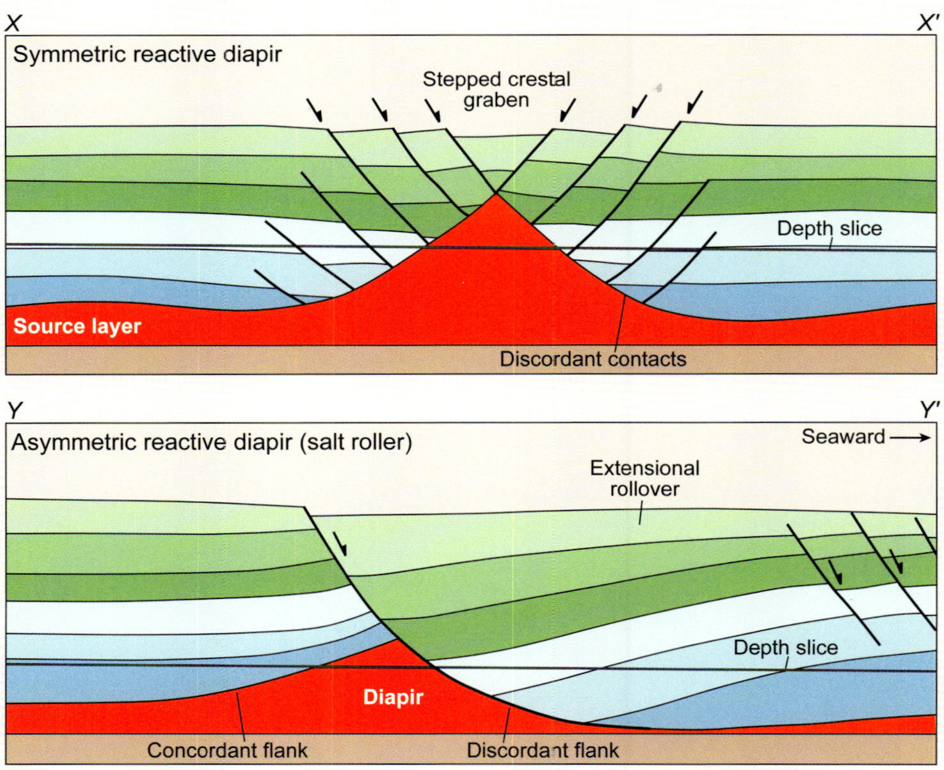

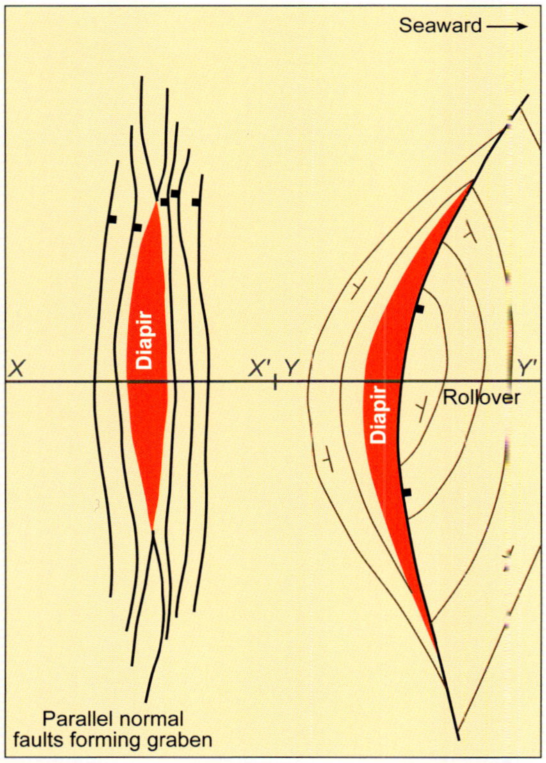

GULF OF MEXICO BASIN—Symmetrical Triangular Diapir above an Allochthonous Salt Sheet, Mississippi Canyon Area, Offshore USA

Figure 3.01. Reactive diapirism. The crestal faults above the diapir suggest only minor extension. However, the total offset of the deepest overburden reflectors indicates more than 5 km of extension, most of which took place by diapir widening. Offset on the outermost normal faults accompanied early reactive diapirism. These faults later became inactive as extension shifted inward. Salt is rising up the axis of the graben, driven by the difference in load between the full thickness of sediments adjacent to the diapir and the thinned roof. Modified from Rowan and others (1999).

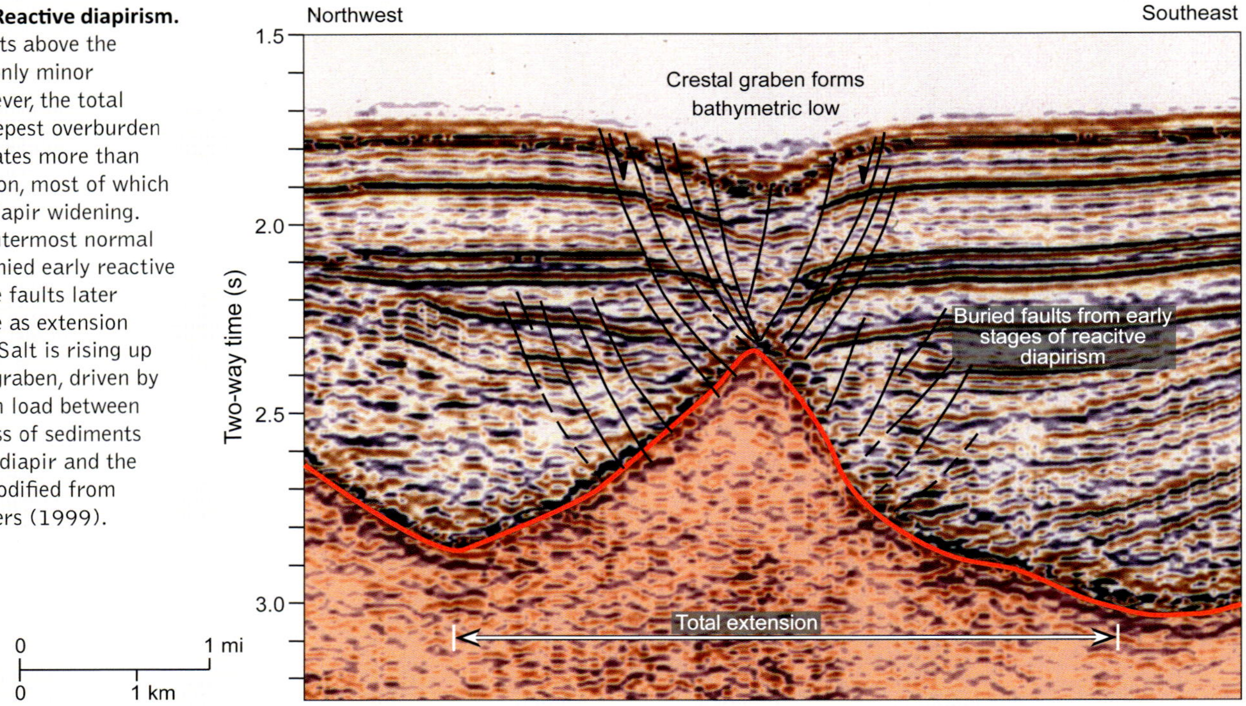

KWANZA BASIN—Raft Tectonics and Triangular Diapirs, Offshore Angola

Figure 3.02. Diapir fall. Two types of triangular diapirs occur in this restoration: reactive diapirs formed by diapir rise (end Albian stage) and relict structures resulting from later diapir fall (continuing to present day). The salt geometries are similar but can be distinguished by the age of nearby sediments. The oldest postsalt sediments are preserved on both sides of a reactive diapir. For a fallen diapir, however, the trough separating a relict triangle of salt from its twin lacks the oldest section at its base. The absent section represents the time when blocks on either side were separated by the parent diapir. Modified from Schultz-Ela (1992).

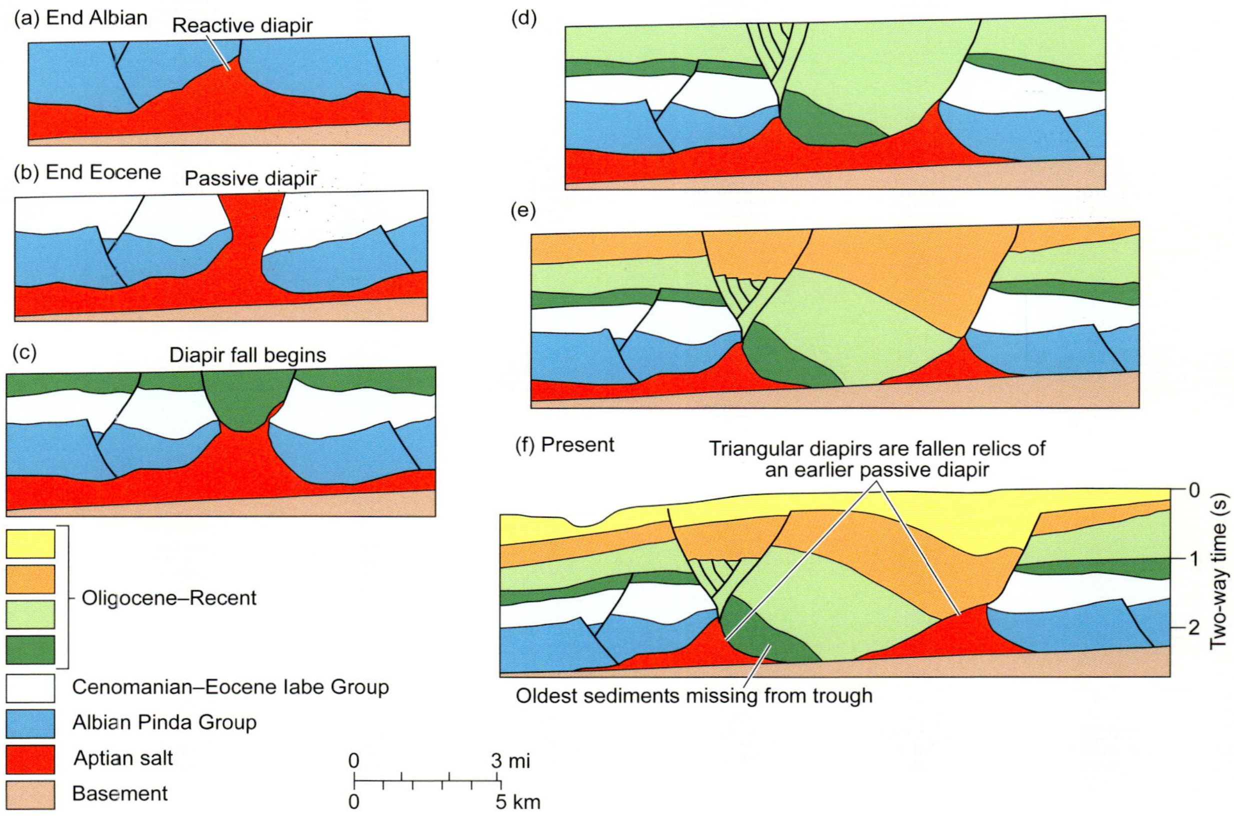

LOWER CONGO BASIN—Raft Tectonics, Offshore Angola

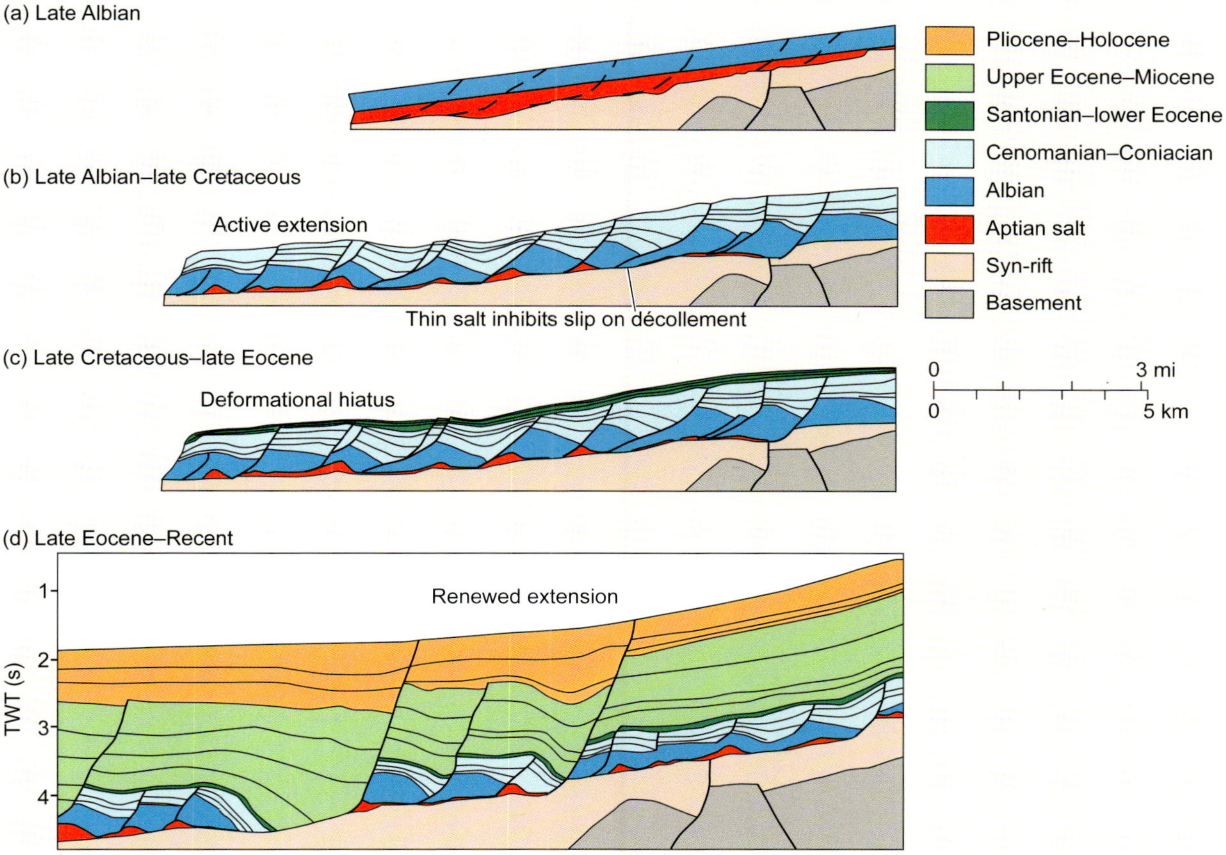

Figure 3.03. Raft tectonics. Extension had two main phases in the Lower Congo Basin. The Albian-Late Cretaceous phase was probably caused by gentle tilting of the margin during postrift thermal subsidence. It was probably terminated by increasing resistance to slip related to thinning of salt along the basal décollement. The second phase began in the late Oligocene, when rapid sedimentation caused progradation of the Congo delta here. This second phase comprises two subphases (28–21 Ma and 10–0 Ma). Reasons for the mid-Miocene (21–11 Ma) deformational lull are unclear. Modified from Valle and others (2001).

LOWER CONGO BASIN—Triangular Diapirs, Offshore Congo

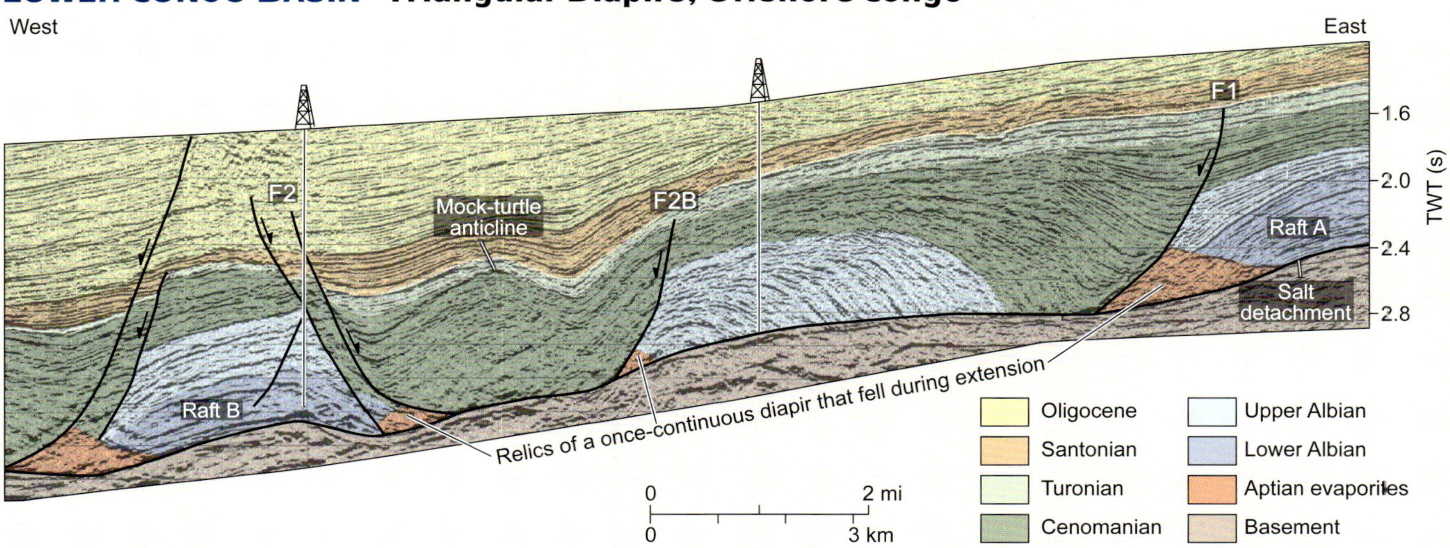

Figure 3.04. Raft tectonics. The two lower Albian rafts were once almost adjacent, separated only by a reactive diapiric ridge. This salt diapir then fell during continued extension. The scattered remains of this diapir are preserved in the footwalls of three normal faults. The lower Albian section is absent from the troughs between the dispersed diapir relics, indicating that extension transported Raft B more than 12 km downdip from its former neighbor, Raft A. Modified from Rouby and others (2002).

LOWER CONGO BASIN—Salt Rollers, Offshore Cabinda

Figure 3.05. Downdip translation above salt. Salt rollers are asymmetric reactive diapirs that form where salt-detached faults verge strongly (usually basinward). They tend to form in the updip, extensional parts of passive margins. In addition to the major roller-bounding faults, swarms of smaller faults accommodate stratal bending at the crests of rollover anticlines. See nearby section in Figure 3.06. Modified from Chimney and Kluth (2002).

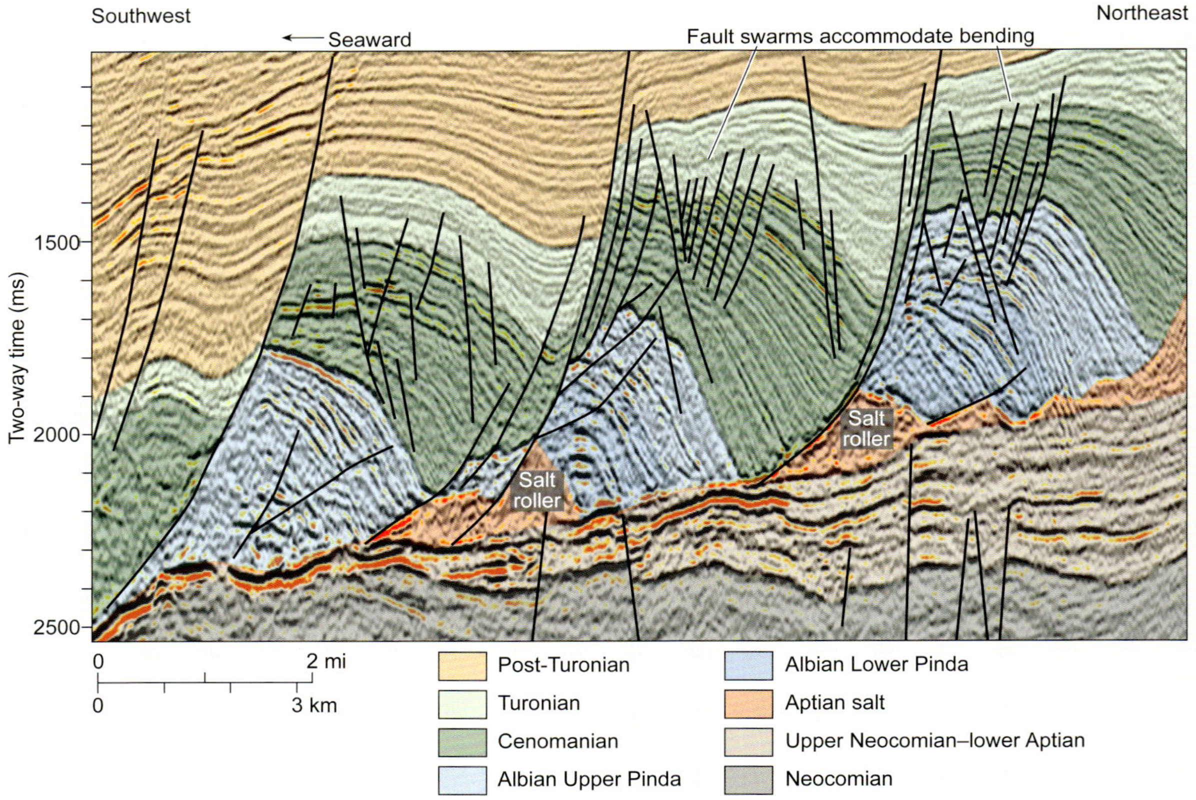

LOWER CONGO BASIN—Rotated Normal Fault above a Salt Roller, Offshore Cabinda

Figure 3.06. Downdip translation above salt. The low-angle fault in the center of the section is interpreted to have once been much steeper. The fault rotated to a lower dip, along with beds in the hanging wall of the next fault to the northeast. Once the fault dipped too gently for efficient slip, a new fault (Fault A) formed with steeper dip. See nearby section in Figure 3.05. Modified from Chimney and Kluth (2002).

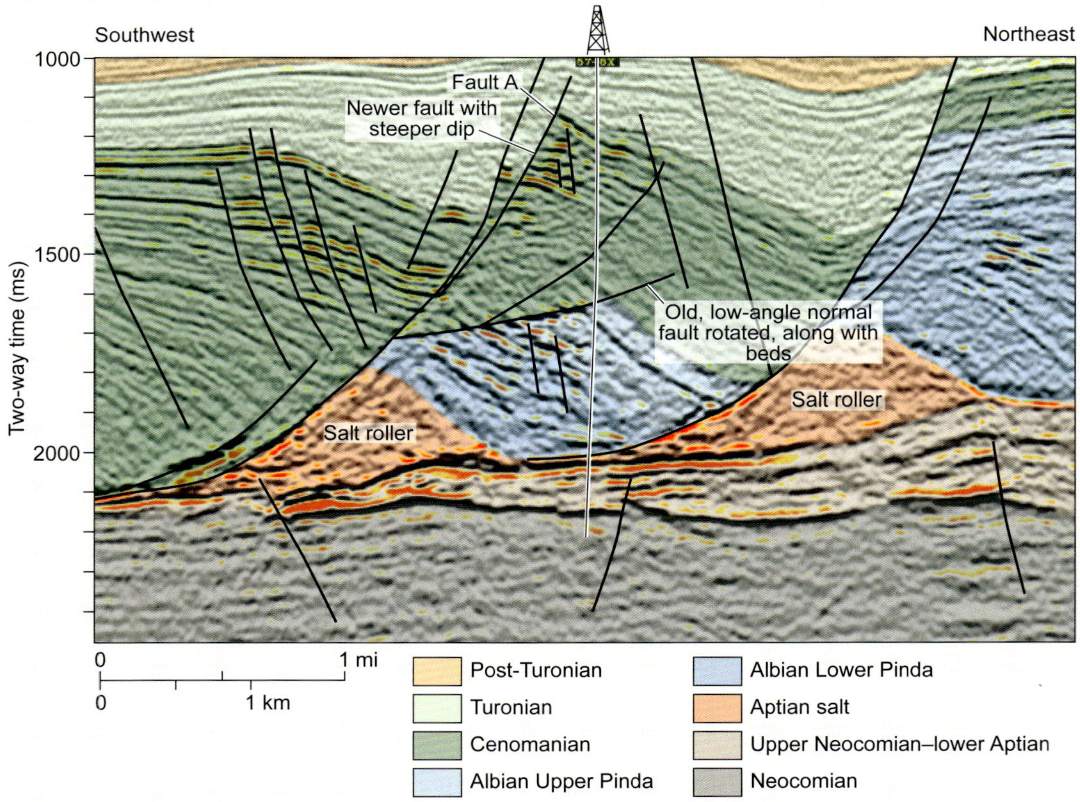

NORDKAPP BASIN—Extensional Drape above Basement Faults, Offshore Norway

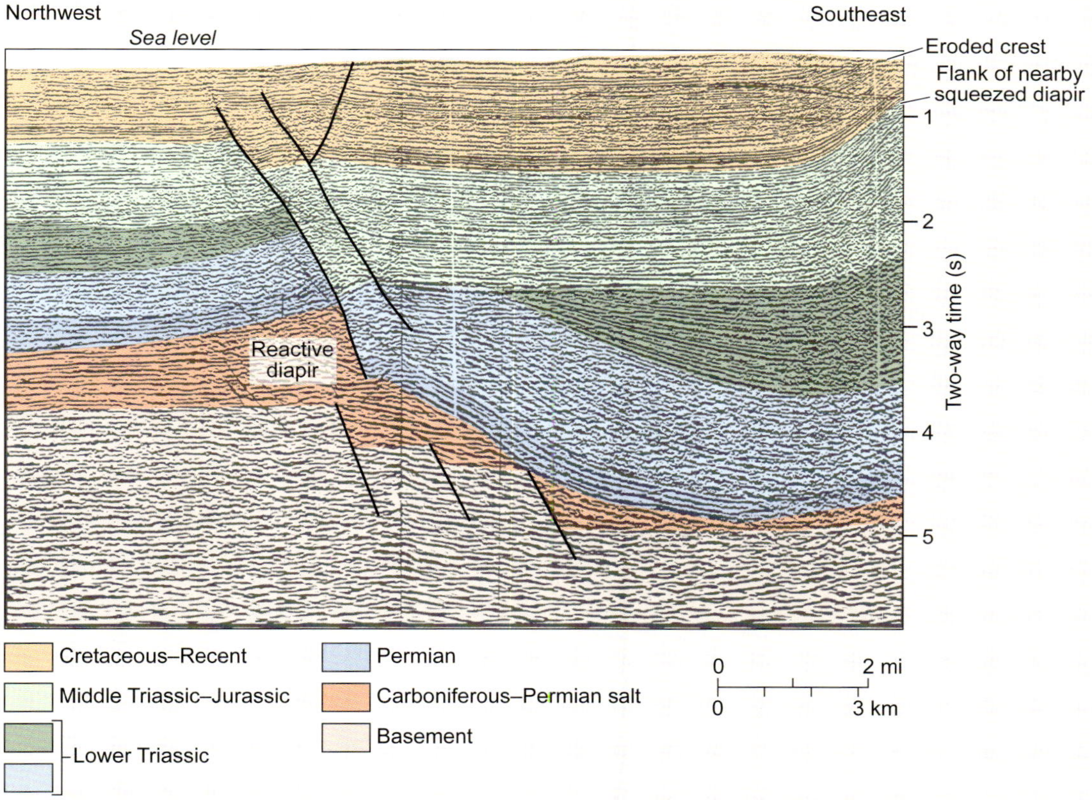

Figure 3.07. Basement extension below salt. A faulted monocline overlies basement-fault steps. All units after the Permian are thicker in the graben, suggesting that the drape is a product of long-term salt evacuation. This evacuation in turn suggests that the basement faults formed during salt deposition, so that the basement graben was filled with thick salt. Later evacuation of this salt produced the faulted drape fold (see nearby restoration, Figure 3.08). A later episode of postsalt extension may have led to extension in the overburden and growth of a reactive diapir above the basement faults. The diapir just off the southeast end of the section was shortened during the Cretaceous–Recent, uplifting and eroding its crest. Modified from Nilsen and others (1995).

NORDKAPP BASIN—Extensional Drape above Basin-Bounding Normal Faults, Offshore Norway

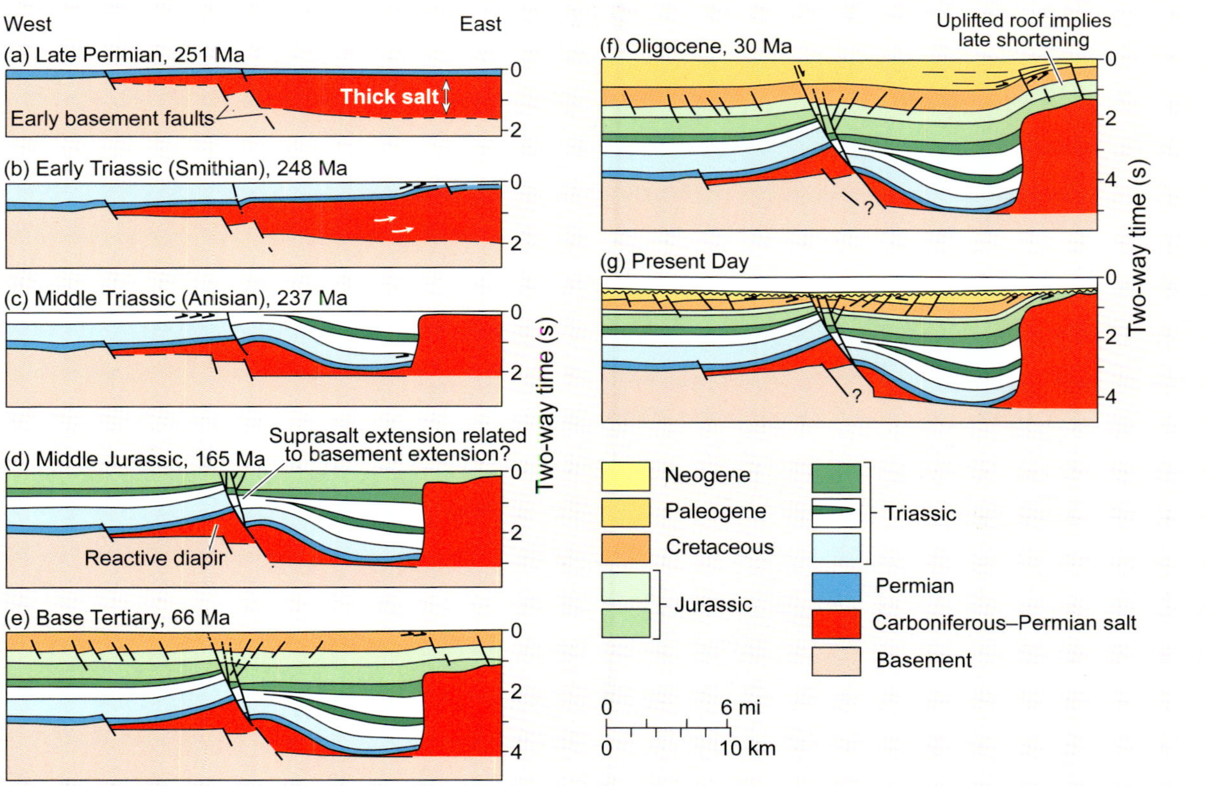

Figure 3.08. Basement extension below salt. This restoration suggests that basement faulting in the Nordkapp Basin ended by the close of Permian salt deposition, so that thick salt filled a basement graben. Younger structures then formed by salt flow in the absence of any regional lateral forces. Alternatively, some of the basement faulting occurred during the Triassic–Middle Jurassic. This interpretation would explain the Mesozoic extension forming a reactive diapir in the drape structure overlying the basement faults. Cenozoic uplift of the diapir's roof implies shortening after the extension on the adjacent triangular diapir. This shortening may have also been basement involved. See nearby seismic in Figure 3.07. Modified from Jensen and Sørensen (1992).

NORTH SEA—Salt Flow inside a Triangular Diapir, Anloo Diapir, Offshore Netherlands

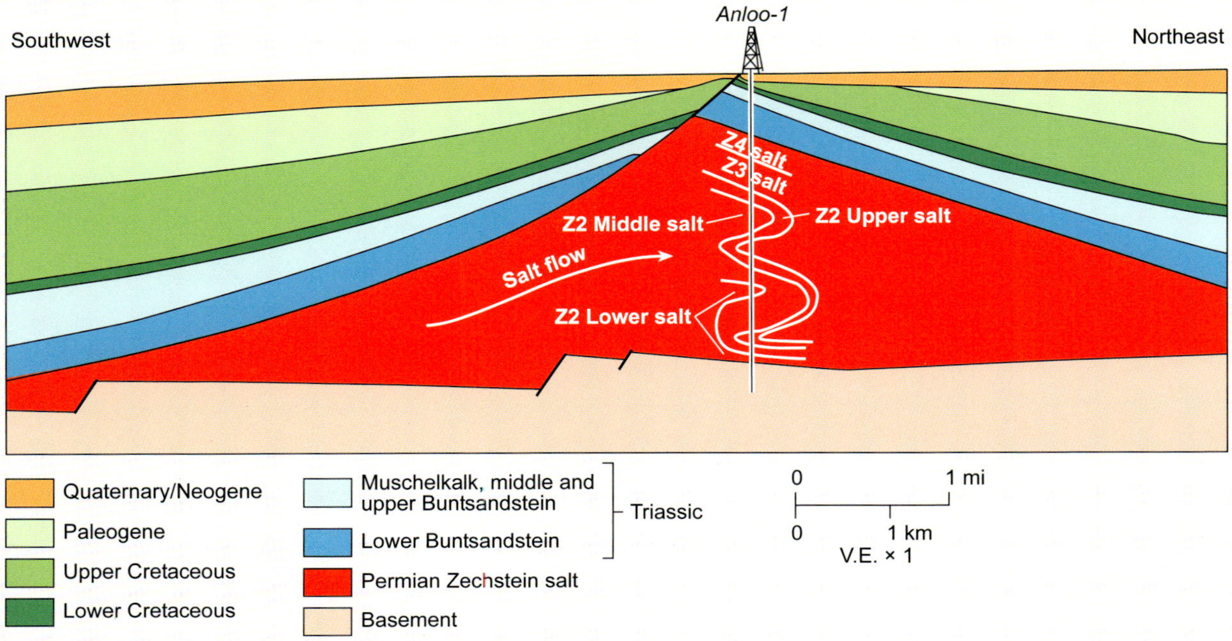

Figure 3.09. Flow of salt into triangular diapirs. The Zechstein salt contains five subunits, three of which are labeled. These are separated by thin marker beds of nonhalite material (white lines on section). The Anloo-1 well encountered repeats of these markers in the core of the Anloo diapir (some markers as many as five times). The stratigraphy of these flow folds suggests expulsion of salt mostly from the downthrown (southwest) flank of the diapir. Modified from Geluk (1998).

NORTH SEA—Triangular Diapir along the North Dogger Fault Zone, Offshore UK

Figure 3.10. Growth of a triangular diapir. The North Dogger Fault Zone lies along the north edge of the Southern North Sea salt basin. A faulted basement ramp allowed thicker, more mobile evaporites to accumulate southwest of the ramp. On the platform at the crest of the ramp, Zechstein units contained less halite, so were less mobile. The North Dogger Fault Zone formed at the break between mobile and immobile Zechstein facies, by accommodating southwestward flow of salt. A major unconformity at the base of the Cretaceous eroded most of the rollover into this fault, producing the present geometry. Modified from Allen and others (1994).

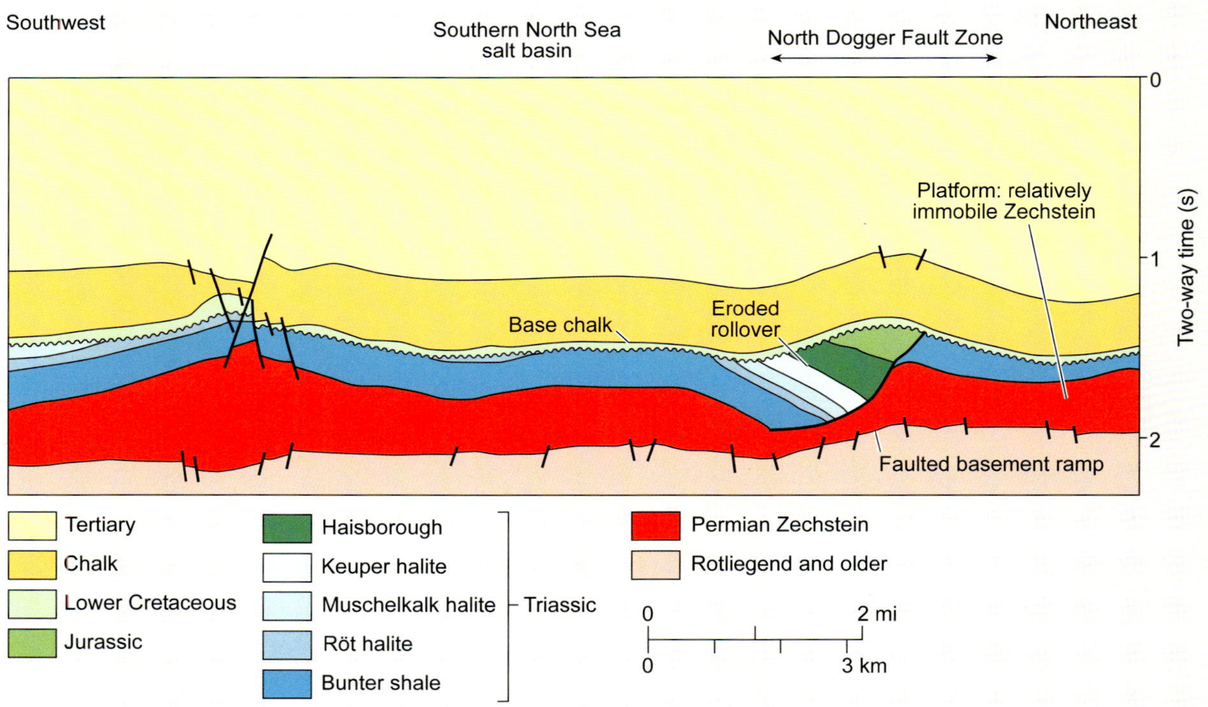

NORTH SEA—Growth of a Triangular Diapir, Offshore UK

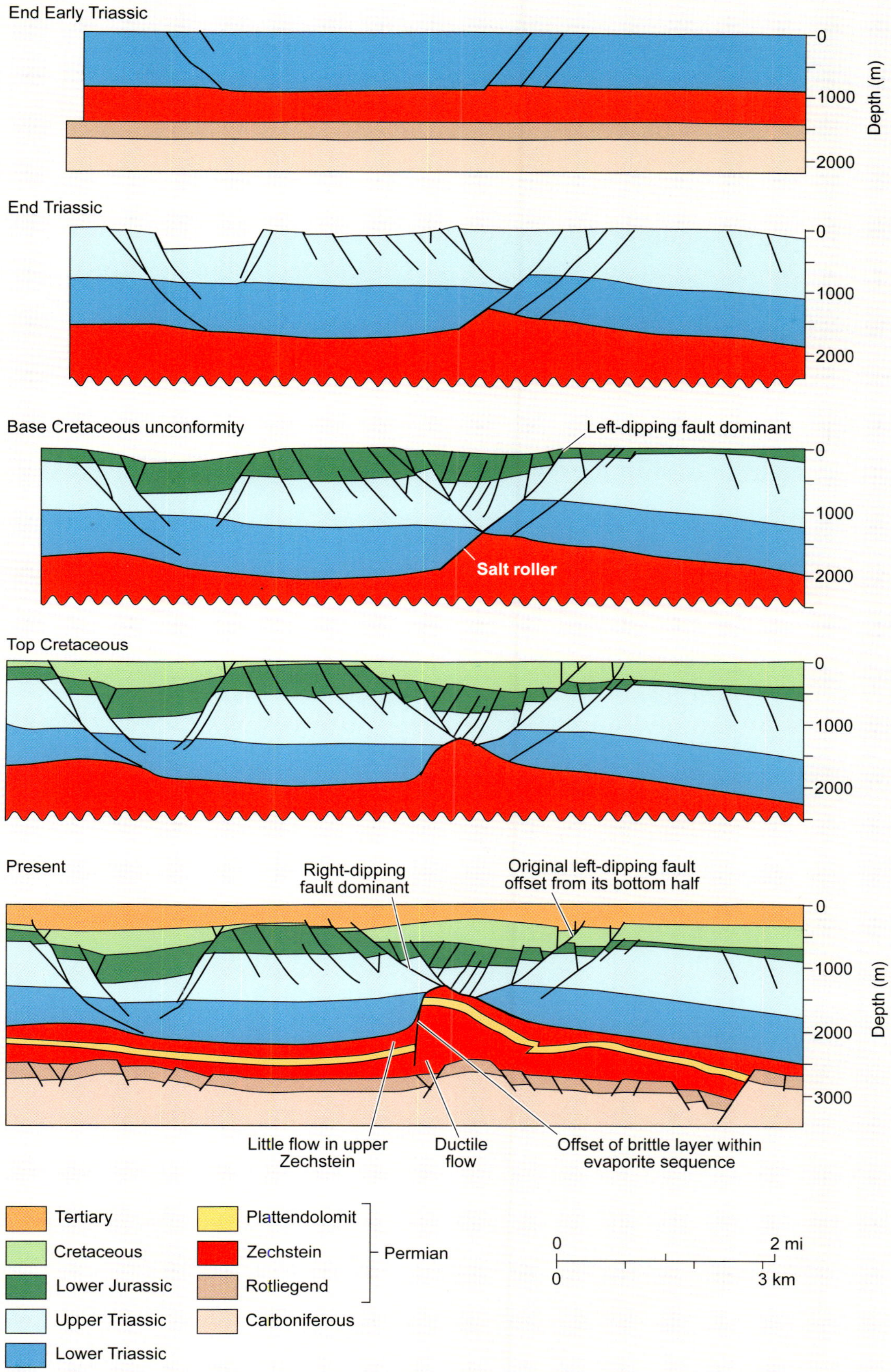

Figure 3.11. Growth of salt rollers. Polarity of faulting over the crest of a triangular diapir switched through time. The structure originally grew as a salt roller to the end of the Jurassic, mostly by slip on a left-dipping normal fault. During the Cretaceous, this left-dipping fault was offset by a major right-dipping fault system so that the final structure became more symmetric. Flow into the core of the salt structure mainly involved the lower part of the Zechstein evaporites. The upper Zechstein contains more nonhalite interbeds and was brittlely offset by the early left-dipping fault system. Modified from Hooper and More (1995).

NORTH SEA—Relationship between Salt Rollers and Fulmar Reservoir Deposition, North Sea, Offshore UK

Figure 3.12. Growth of salt rollers. Fulmar coastal facies sands were stacked in the hanging walls of large down-to-the-basin growth faults associated with salt rollers. Rising sea level during the Late Jurassic caused facies belts generally to retrograde, causing sand packages to step landward through time. The load of coastal sedimentation may also have driven faulting. Modified from Fraser and others (2003).

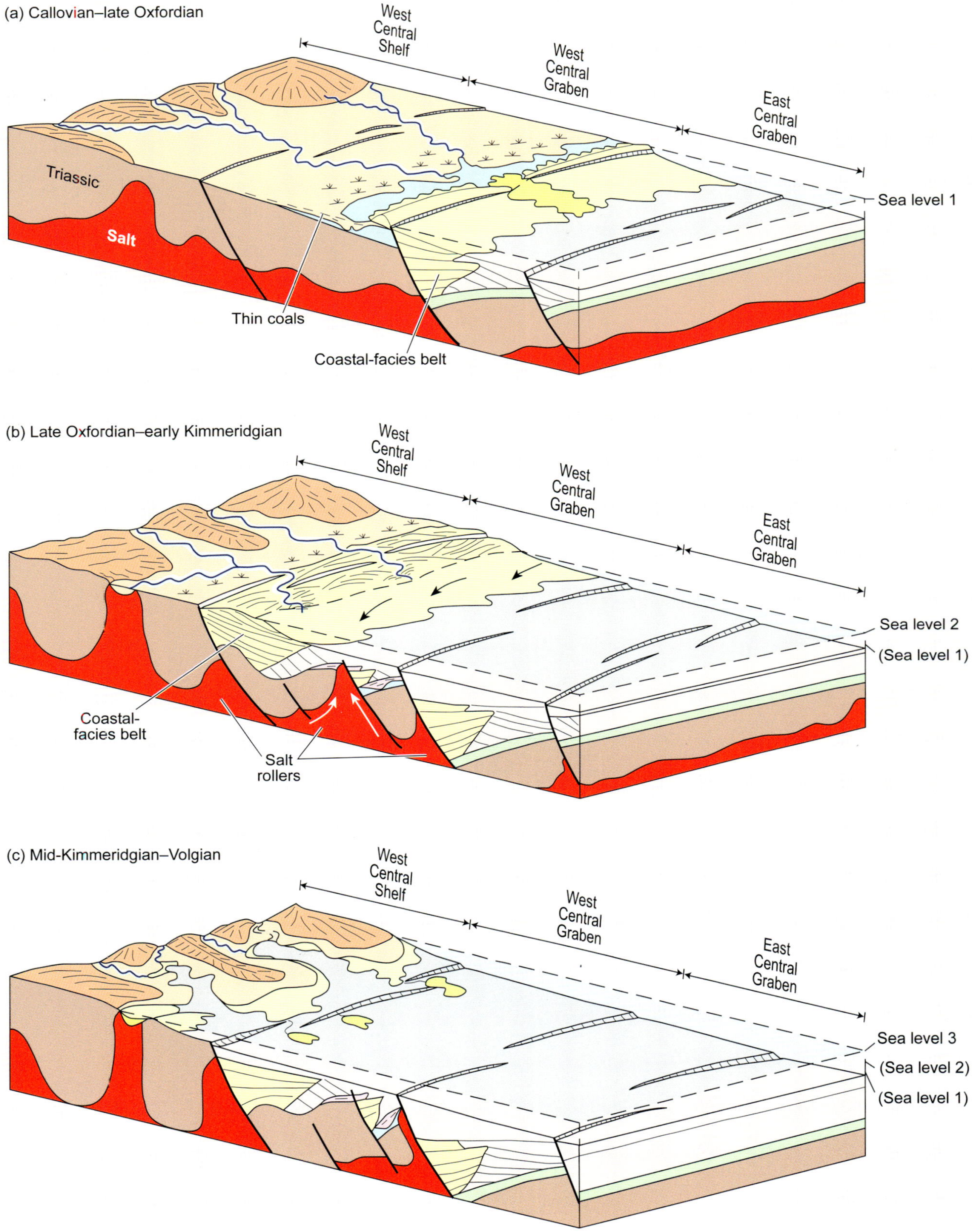

PARADOX BASIN—Anticlines Formed near a Salt Roller, Lisbon Valley Structure, Utah, USA

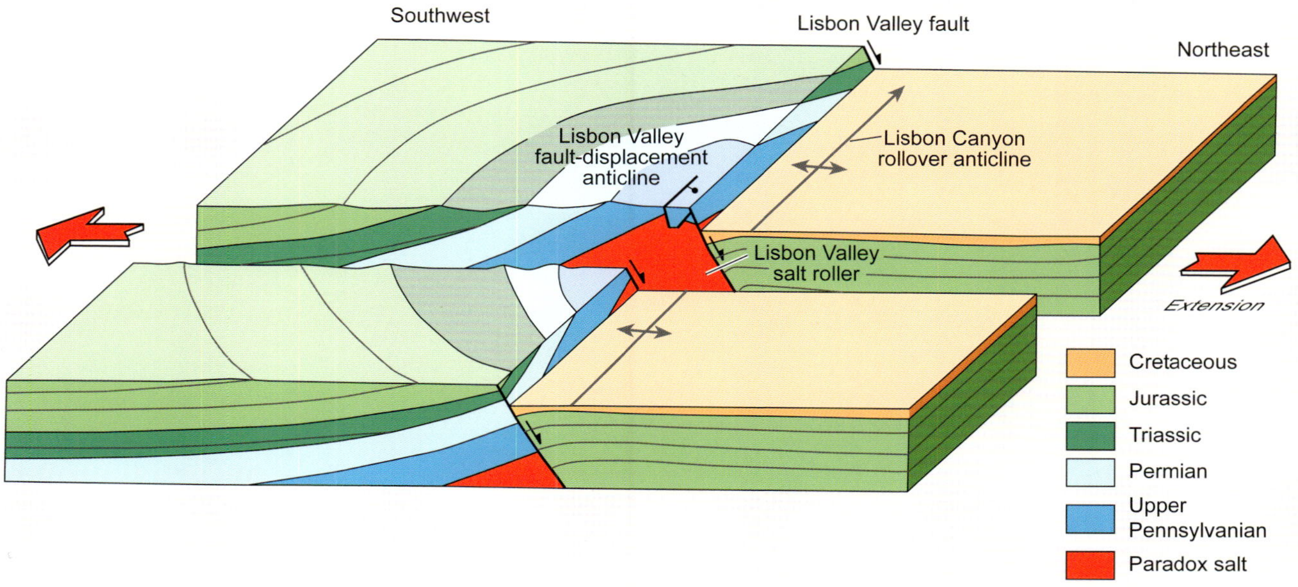

Figure 3.13. Uplift above a salt roller. Two types of anticlines formed near the Lisbon Valley salt roller. A domal fault-displacement anticline bulges the footwall of the Lisbon Valley fault. This type of fold is produced by variable footwall uplift during normal faulting and culminates near the maximum fault displacement. Here, the anticline has probably been enhanced by upwelling of buoyant salt in the underlying salt roller. A rollover anticline deforms the hanging wall of the Lisbon Valley fault. This type of fold is produced by slip down a listric fault surface. Modified from Ge and others (1995).

PHYSICAL MODEL—Growth of a Reactive Diapir

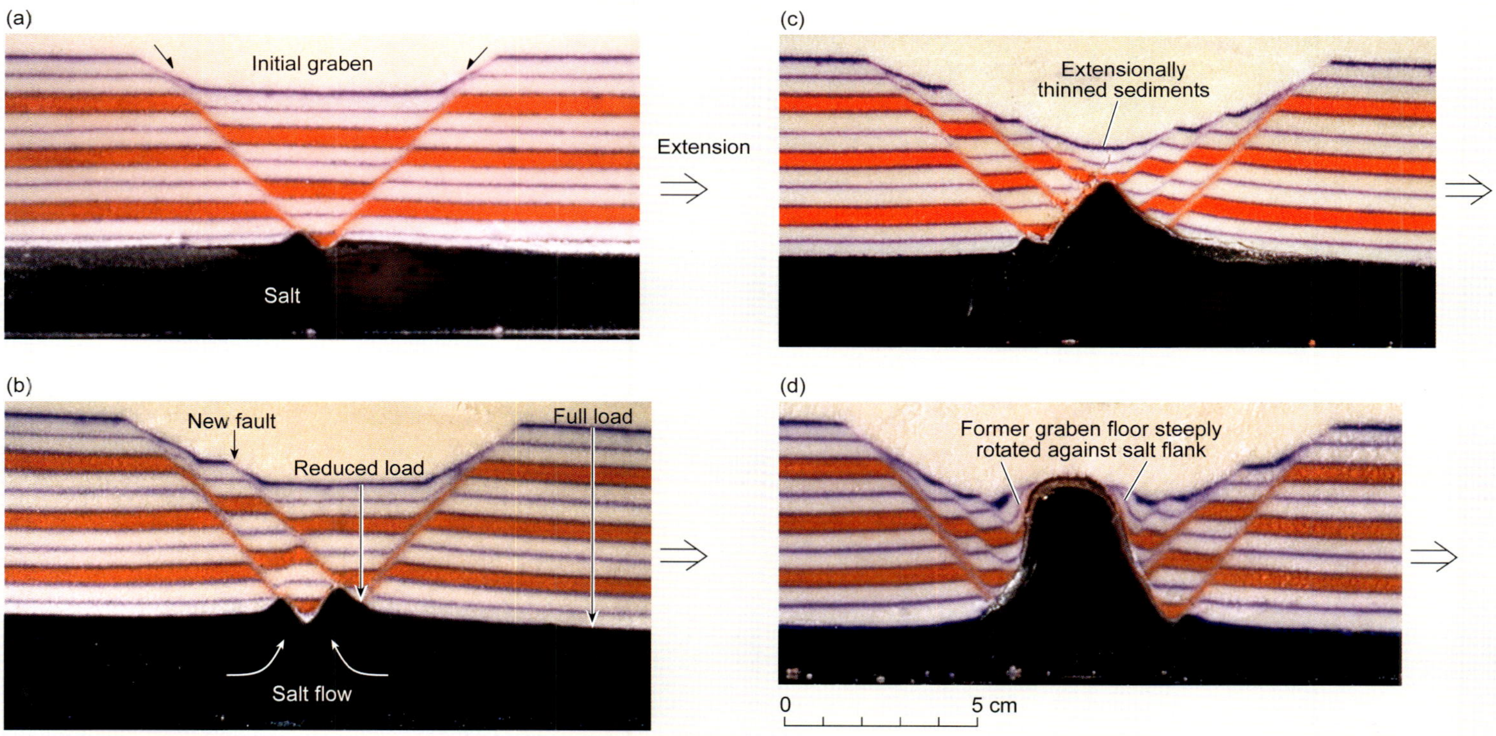

Figure 3.14. Reactive diapirism. Extension thins the sediments above salt. This thinning has two consequences, both of which favor diapir initiation. First, a load imbalance expels salt from beneath the unthinned graben flanks toward the extensionally thinned graben center. Second, faulting weakens the overburden, lowering its resistance to piercement. As a result, salt rises up the graben axis to form a reactive diapir. In this example, the reactive diapir eventually became active, uplifting and breaking through the graben floor to emerge at the surface. All units were prekinematic and originally tabular and horizontal. These images are from four identical experiments that were stopped and sectioned at different stages of evolution. Modified from Vendeville and Jackson (1992a).

PHYSICAL MODEL—Internal Strain in a Reactive Diapir

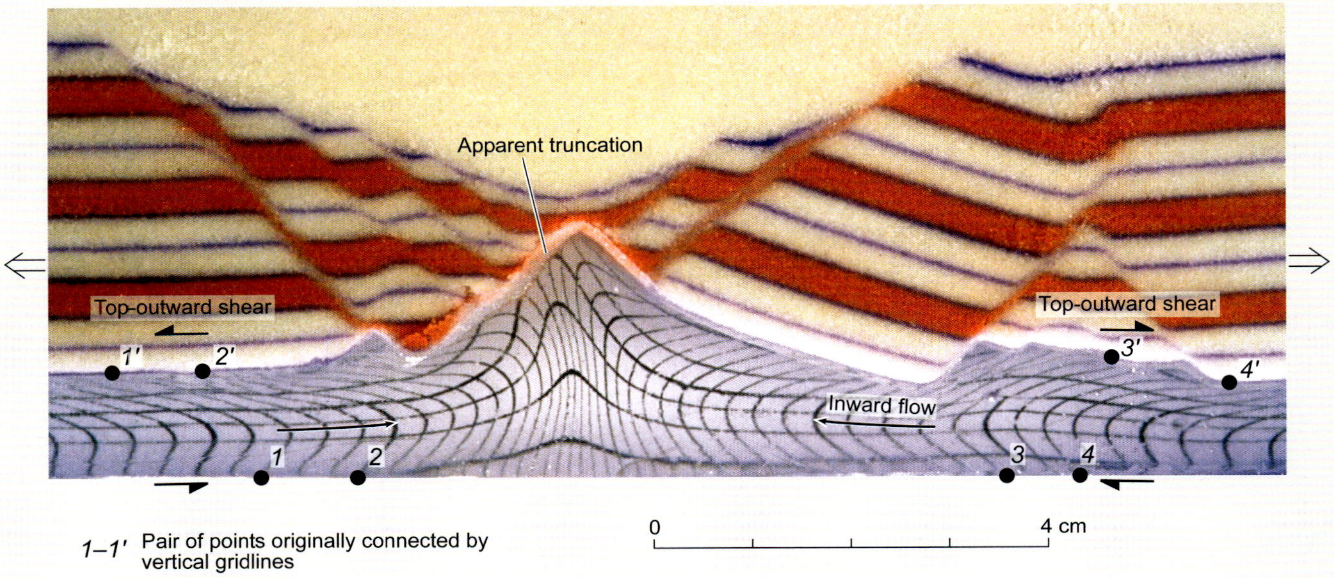

Figure 3.15. Reactive diapirism. A vertical, 4- × 4-mm rectangular grid imprinted within the salt illustrates the kinematics of salt flow during reactive diapirism. Both model endwalls moved outward during the experiment, causing the sedimentary cover to move away from the central graben. The sediment dragged salt with it as it moved outward, producing a top-outward shear in the salt on both sides of the diapir. The divergent flow at the top of salt was offset by a return inward flow near the center of the salt layer. This inward flow fed salt to the rising diapir. The grid was imprinted into the clear silicone some distance back from the section plane. The perspective of this offset and optical refraction produced apparent truncations of the grid against the top and base of salt. Model by Bruno Vendeville.

PHYSICAL MODEL—Variations in Reactive Diapirism along Strike

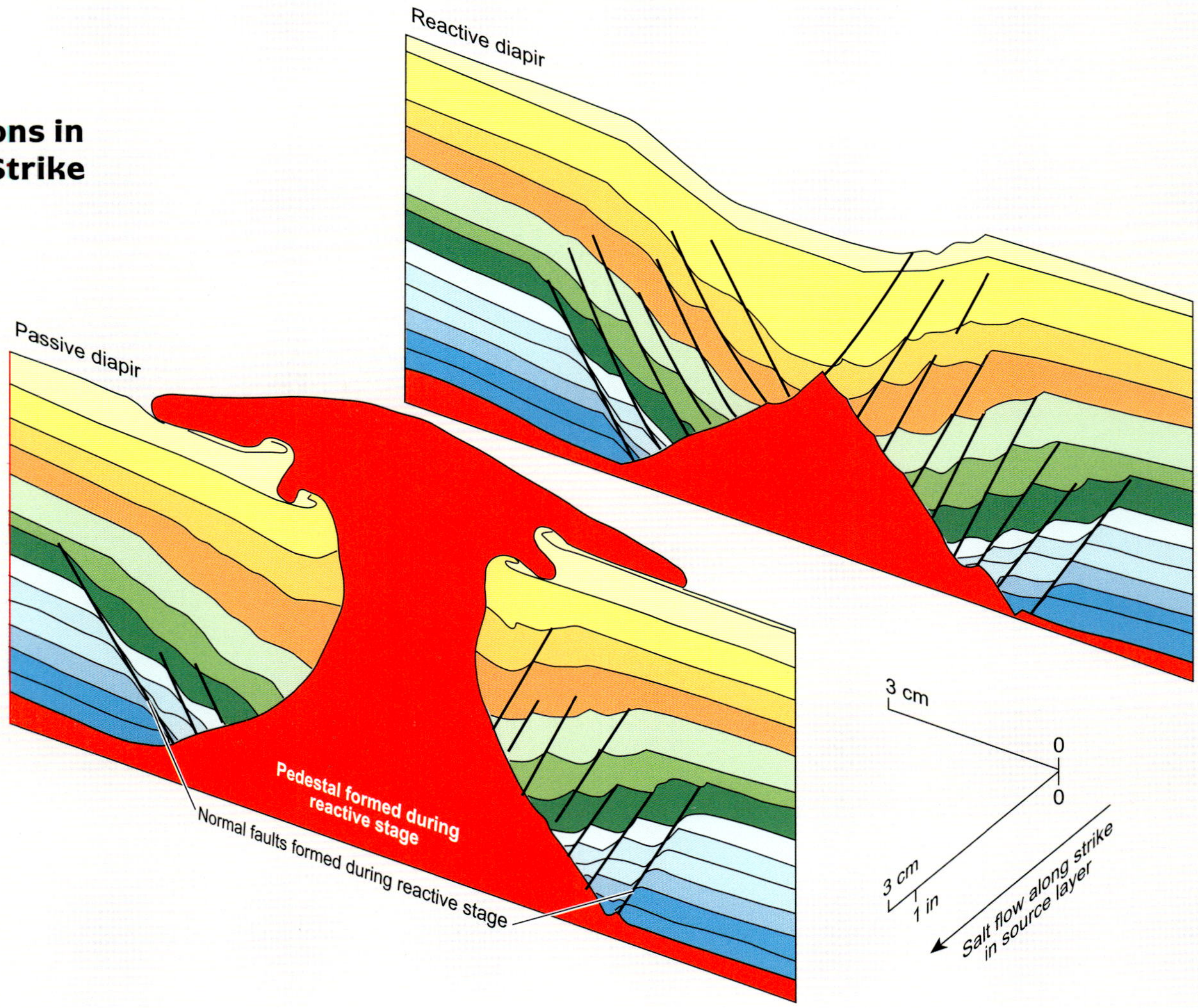

Figure 3.16. Along-strike variations in extensional diapirism. The front panel shows part of a diapiric wall that evolved from reactive through active and passive stages and finally extruded at the surface. The rear panel shows part of the diapiric ridge that remained in reactive stage because its source layer was tapped during the front diapir's vigorous growth (see section detail in Figure 3.17). Modified from Vendeville and Jackson (1992a).

PHYSICAL MODEL—Reactive Diapirism

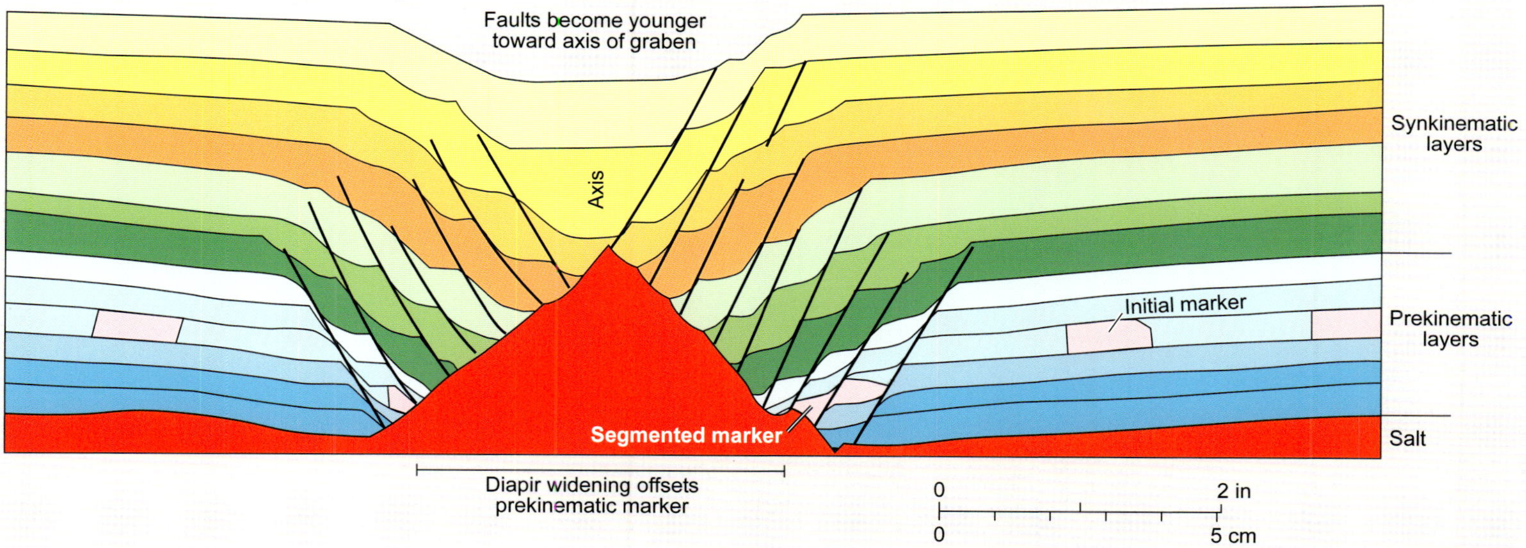

Figure 3.17. Widening a reactive diapir. The reactive diapir pierced its prekinematic cover but is still buried beneath synkinematic strata (see along-strike variations, Figure 3.16). Once a diapir has pierced a given layer, faults offsetting that layer become inactive because further extension is accommodated mostly by diapiric widening. Offset of two halves of a prekinematic marker block illustrates this widening. This process is one reason that faults became younger toward the axis of the graben; older, deeper faults farther from the graben axis became inactive once deeper beds had been pierced. Additionally, new faults tend to nucleate in the weaker, thinner overburden over the crest of the diapir. Modified from Vendeville and Jackson (1992c).

PHYSICAL MODEL—Reactive Diapirism during Synkinematic Sedimentation

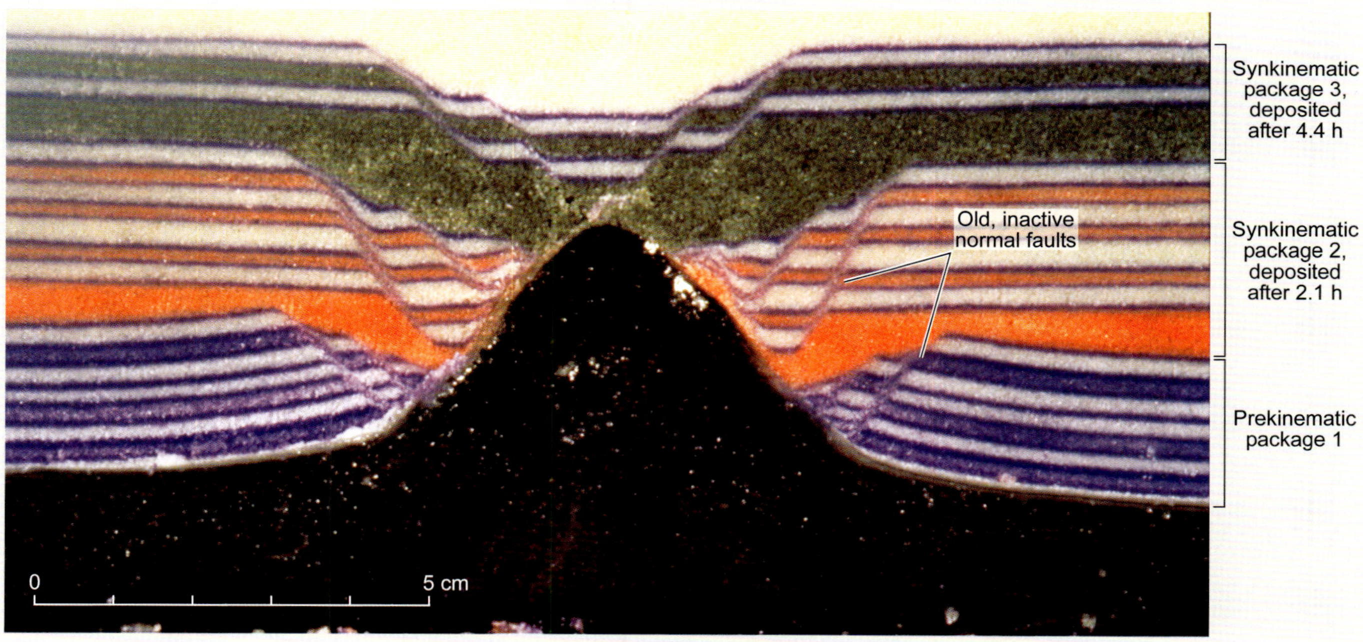

Figure 3.18. Reactive diapirism. The salt was initially buried beneath prekinematic package 1. Salt rising as a reactive diapir completely pierced this layer and overlying synkinematic package 2. However, the salt rise rate was lower than the aggradation rate, so salt never pierced to the surface (compare with reactive diapirism without synkinematic sediments in Figure 3.14). Piercement through each sediment package formed new normal faults. Older, inactive faults are preserved against the deeply buried flanks of the structure. Model by Bruno Vendeville.

PHYSICAL MODEL—Alternative Models for Evolution of Triangular Diapirs

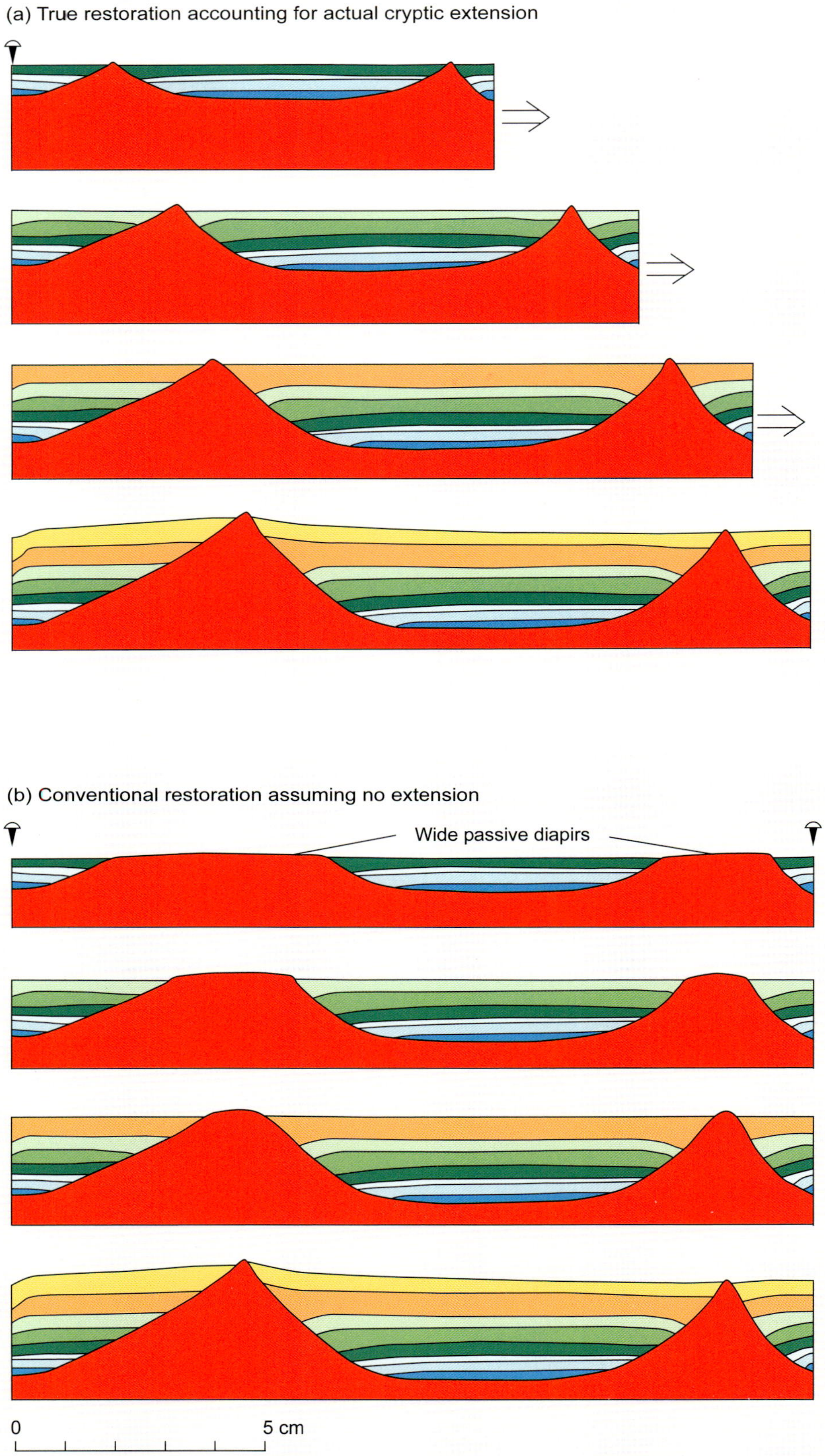

Figure 3.19. Extensional diapirism. Some triangular diapirs can be restored as either (a) extensional or (b) nonextensional structures. Nonextensional models restore to wide passive diapirs, whereas extensional models typically restore to passive diapirs that remained triangular throughout their growth. Both models are kinematically valid. In nature, an extensional model may be favored if the structure lies in an extensional province or if a normal fault extends beyond the edge of the salt structure in strike or dip view. Overhead views of this model during its growth indicate that the extensional interpretation is correct here. Modified from Vendeville and Jackson (1992a).

PHYSICAL MODEL—Evolution of Triangular-Diapir Relays

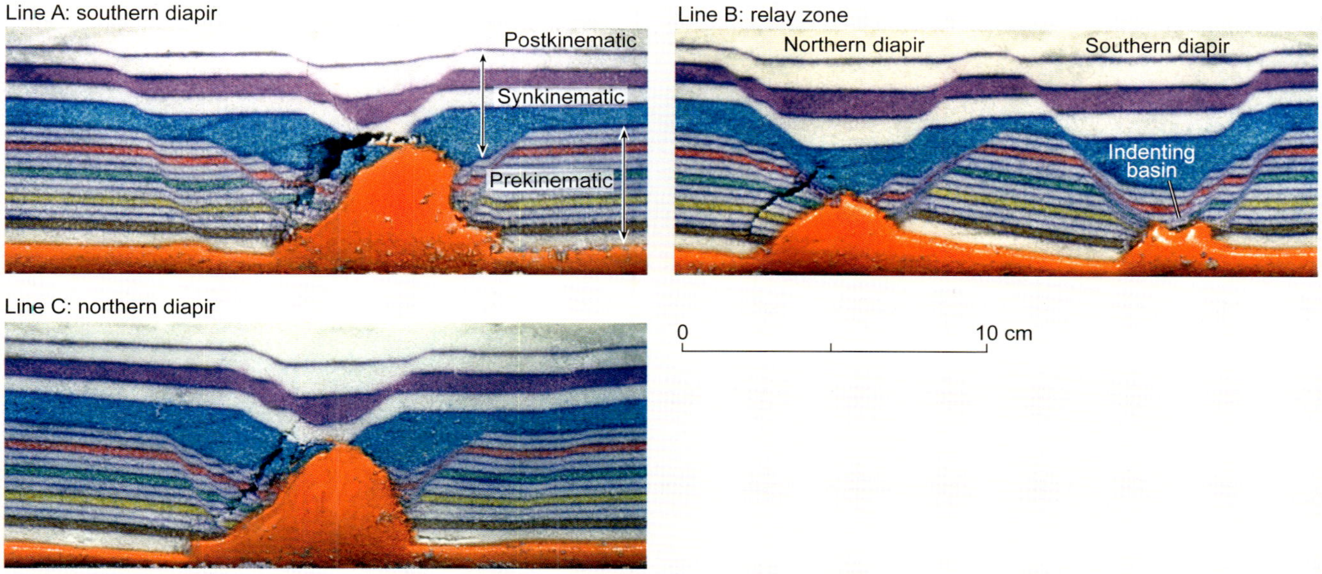

Figure 3.20. Extensional diapirism. Reactive diapirs formed above two preformed salt ridges to form a diapir relay (see overhead view in Figure 3.21). At the north and south ends of the model, abundant salt allowed the diapirs to keep rising. Within the relay zone, however, two diapirs shared the source-layer salt, which retarded their growth. A crestal graben indented the southern diapir in the relay zone, indicating diapir fall. The northern diapir may also have fallen, but its crestal graben was less diagnostic. Model by Bruno Vendeville.

PHYSICAL MODEL—Evolution of Triangular Diapir Relays before Synkinematic Sedimentation

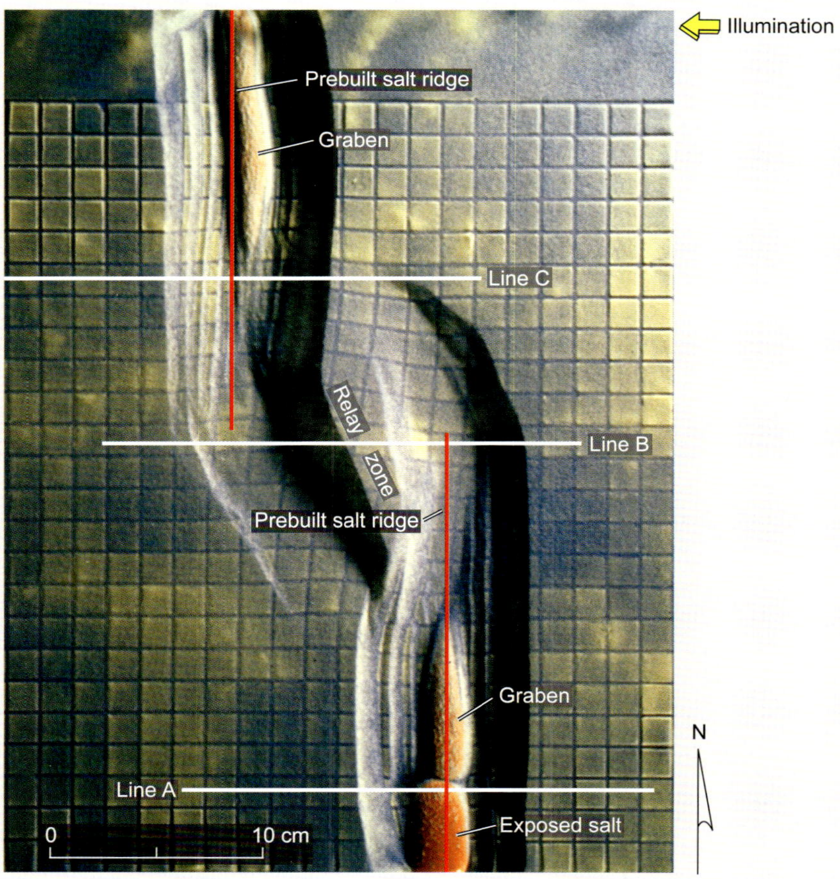

Figure 3.21. Extensional diapirism. Extension above two laterally offset salt ridges formed a relay system. Both ridges nucleated reactive diapirs, which locally pierced to the surface. Extensional faults and reactive diapirs extended beyond the ends of the prebuilt ridges and curved toward one another in the relay zone. The position and number of reactive diapirs varied with the location of cross sections (Figure 3.20). Model by Bruno Vendeville.

PHYSICAL MODEL—Overhead Views of Extension Caused by Tilting and Progradation across Salt

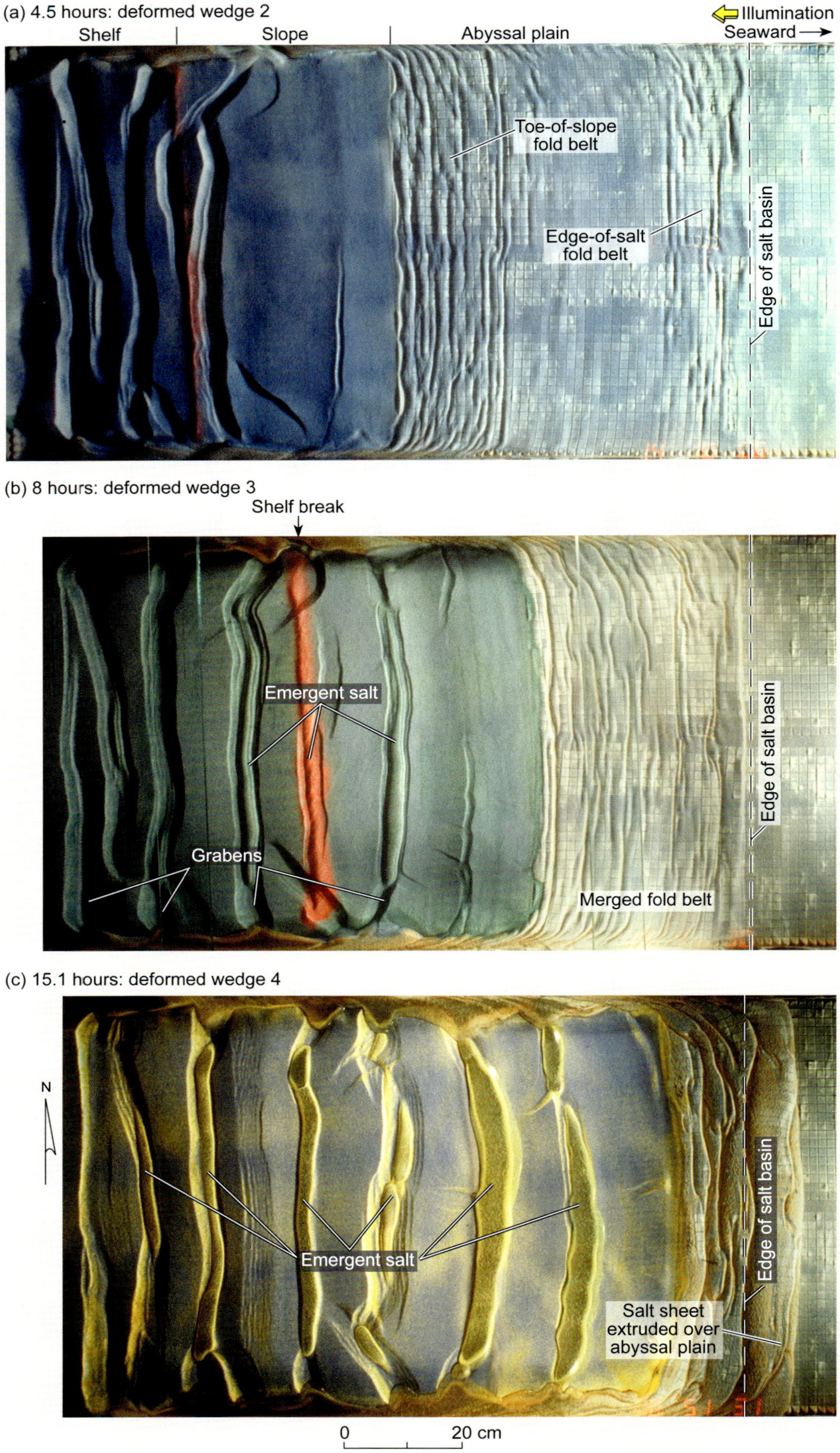

Figure 3.22. Extensional diapirism. A sedimentary wedge prograding across salt extended, forming grabens trending perpendicular to the slope direction. Reactive diapirs pierced these grabens, eventually reaching the surface (see restoration in Figure 3.23). Extension was concentrated on the shelf and upper slope, as in nature. (a) Fold belts initiated at the base of the slope and at the edge of the salt basin. (b) These later merged to form a single fold belt. During progradation the extensional province shifted seaward, until the entire salt basin was extending, accommodated by salt extrusion across the abyssal plain. Modified from Rowan and others (2004).

PHYSICAL MODEL—Restoration of Extension Caused by Tilting and Progradation across Salt

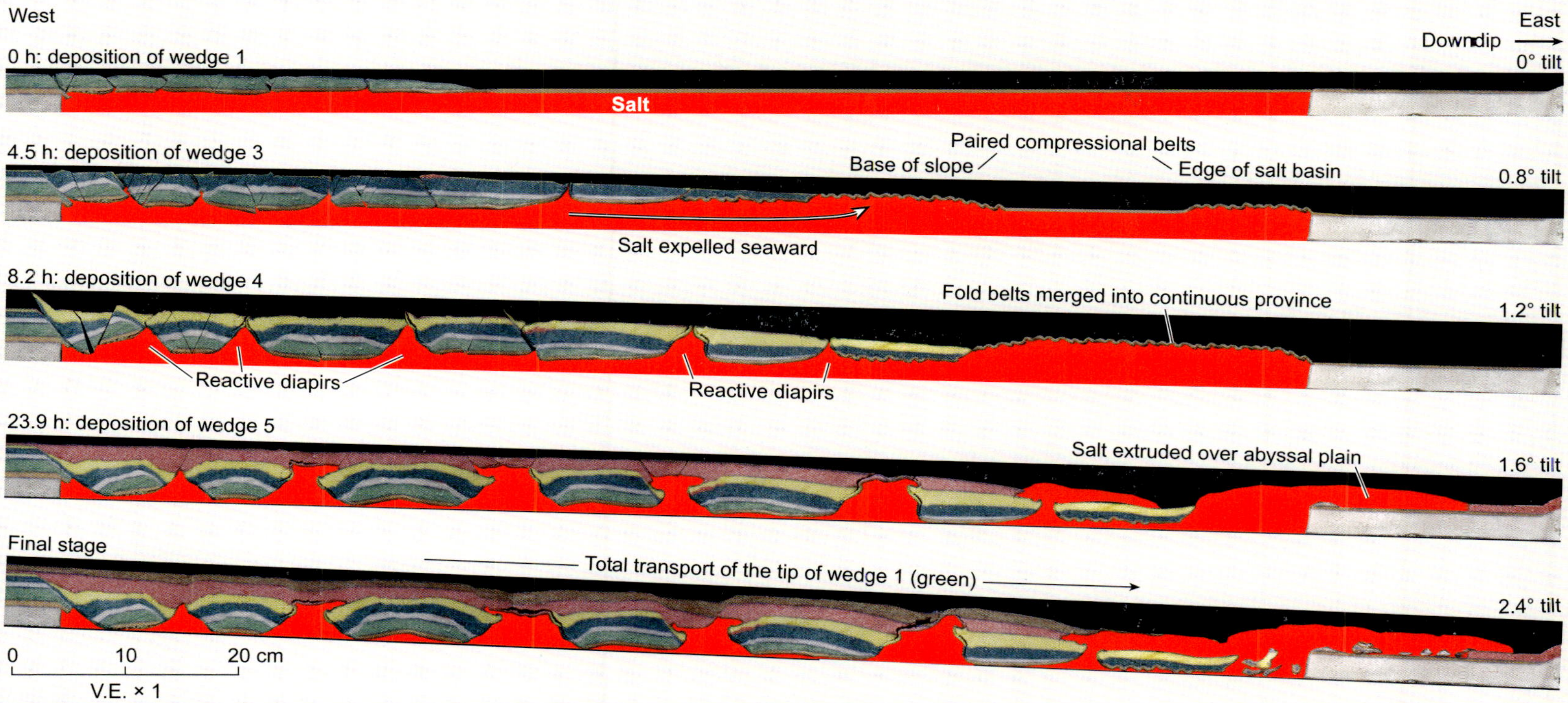

Figure 3.23. Extensional diapirism. Extension of a prograding wedge allowed triangular reactive diapirs to rise beneath the grabens. In advanced stages, diapirs extruded from some grabens (see overhead views in Figure 3.22). Extension of the wedge was balanced by shortening of the thin overburden at the downdip end. Initially two fold belts formed: at the base of the slope and at the edge of the salt basin. These fold belts later merged. Extension in the final stages of the model was accommodated by extrusion of salt and thrusting of the roof over the abyssal plain. Modified from Rowan and others (2004).

PHYSICAL MODEL—Effect of Aggradation Rate on Structural Style of Triangular Diapirs in Extension

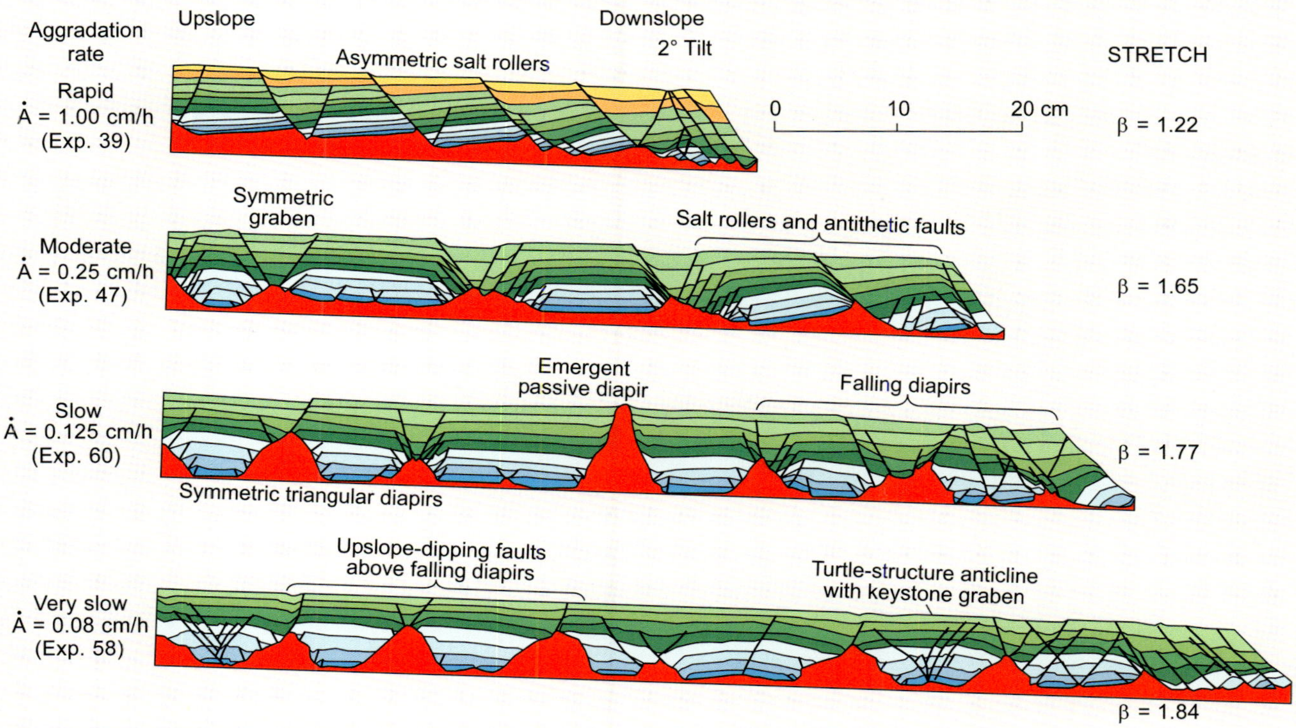

Figure 3.24. Extensional diapirism. Slow aggradation during regional extension tends to produce symmetric triangular diapirs. Also, the shallow burial of salt allows it to pierce to the surface until the source layer becomes depleted. Thus, some triangular diapirs were passive. Rapid aggradation, by contrast, produces asymmetric salt rollers during extension. All sediments were synkinematic. Diagrams are representative cross sections from separate models. Models by Shing-Tzong Lin and Bruno Vendeville.

PHYSICAL MODEL—Spreading of a Curved Passive Margin

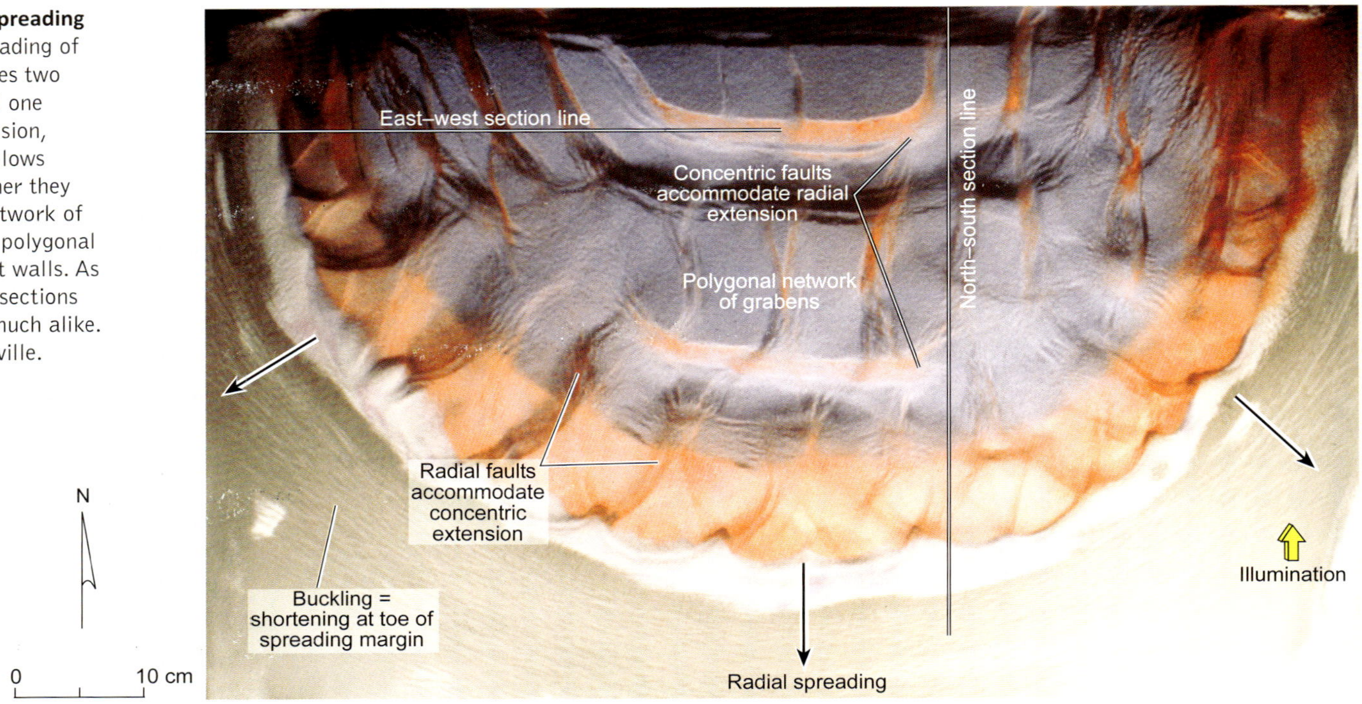

Figure 3.25. Radial spreading above salt. Radial spreading of a curved margin requires two fault systems—a radial one allows concentric extension, and a concentric one allows radial extension. Together they produce a polygonal network of grabens underlain by a polygonal network of reactive salt walls. As a result, strike and dip sections (Figure 3.26) appear much alike. Model by Bruno Vendeville.

PHYSICAL MODEL—Spreading of a Curved Passive Margin

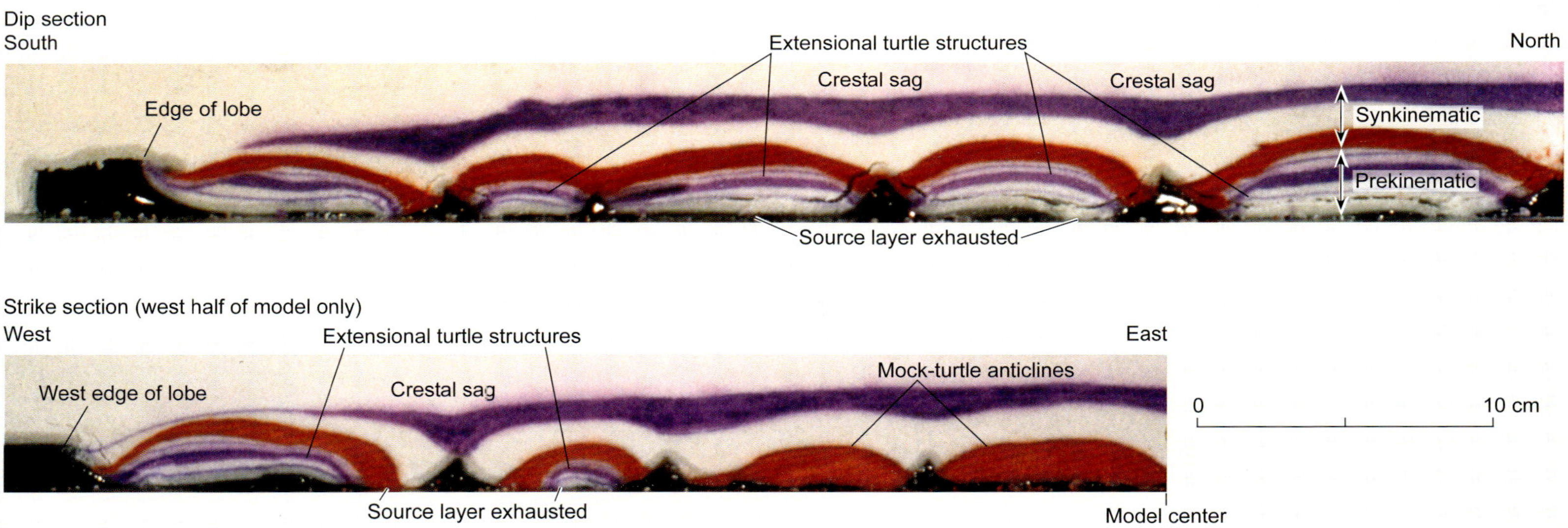

Figure 3.26. Radial spreading above salt. Where spreading is radial (see overhead view, Figure 3.25), strike and dip sections have similar structural styles. Triangular diapirs separate rafts of prekinematic strata whose flanks subsided to form extensional turtle structures. Crestal sags over many of these diapirs indicate that diapirs have fallen as a result of continuing extension and diapir widening after the source layer was greatly depleted. The strike section intersects two mock-turtle anticlines, which lack any prekinematic section. These formed when a preexisting diapir fell, forming a sag in its roof. Model by Bruno Vendeville.

PHYSICAL MODEL—Extensional Reactivation of a Long Triangular Salt Wall

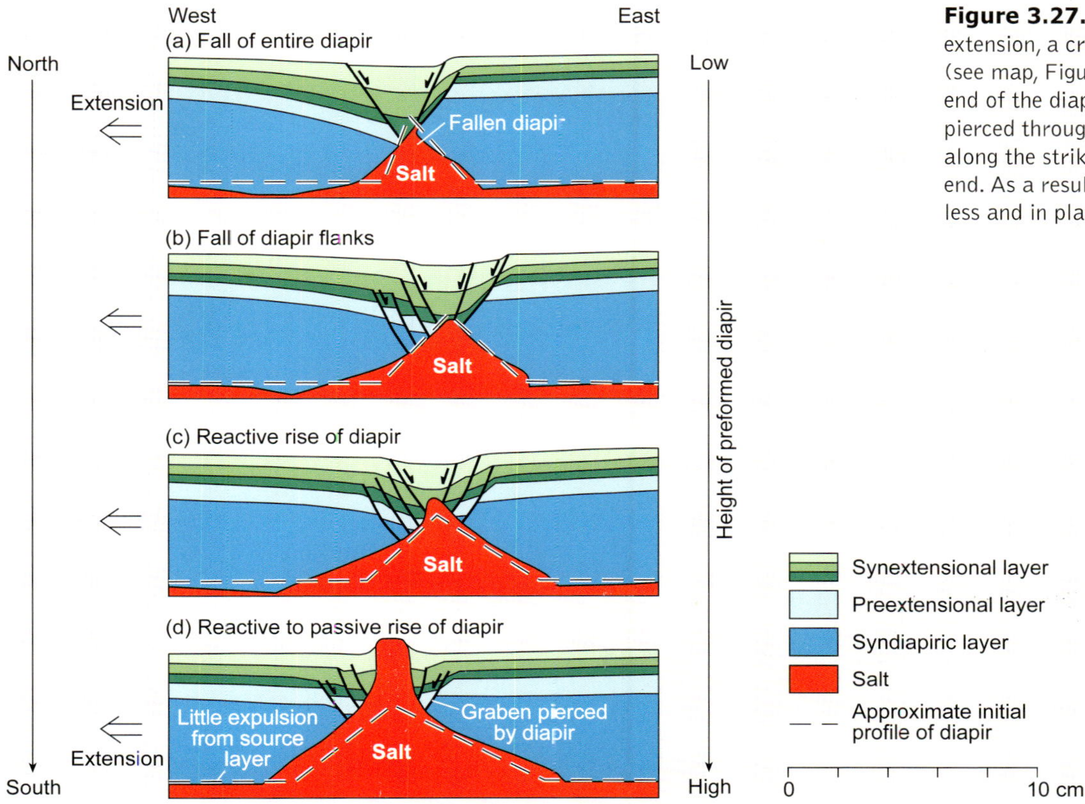

Figure 3.27. Extensional reactivation of a diapir. During extension, a crestal graben formed above a preexisting salt wall (see map, Figure 3.28). Extension was greatest above the south end of the diapir, where the roof was thinnest. Here the diapir pierced through the graben to emerge at the surface. Salt flowed along the strike of the wall to feed the diapir rising at its south end. As a result, more northern parts of the diapir rose much less and in places fell. Modified from Ge and others (1995).

PHYSICAL MODEL—Extensional Reactivation of a Long Triangular Salt Wall

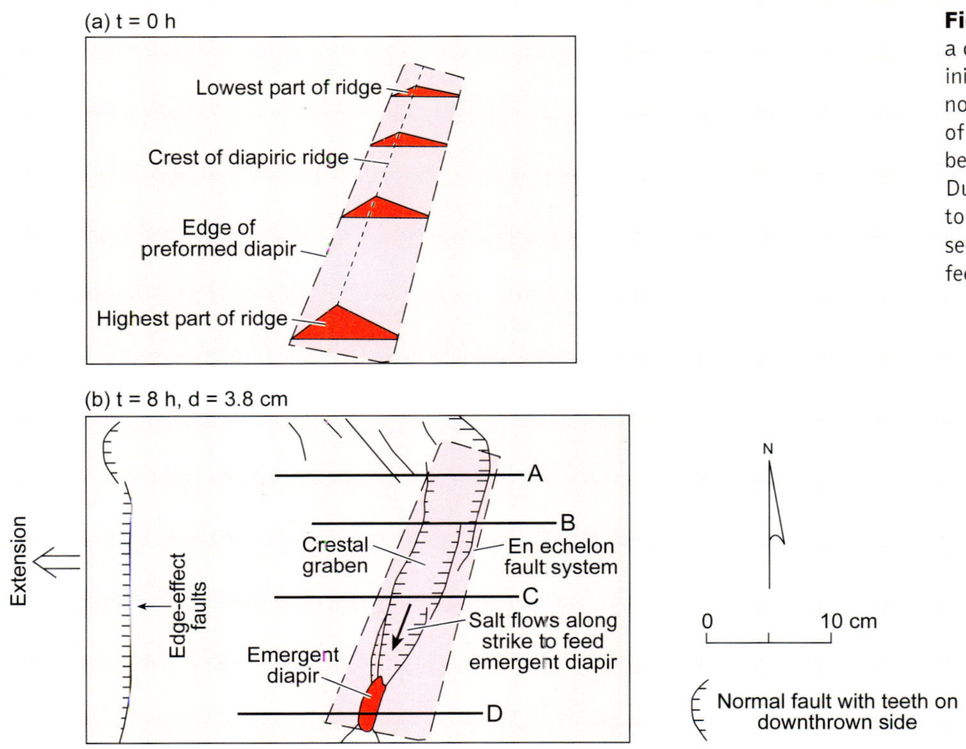

Figure 3.28. Extensional reactivation of a diapir. During extension, a crestal graben formed above a preexisting salt wall. The graben initiated in the south, where the wall was highest, and then propagated northward toward the lower end of the wall. The eastern boundary of this graben formed a right-stepping en echelon system perhaps because the preformed diapir was oblique to the extension direction. During extension, the tallest part of the preformed diapir pierced to the surface. Other parts of the diapir grew much less or fell (see sections, Figure 3.27) as salt flowed along the strike of the wall to feed its emergent south end. Modified from Ge and others (1995).

TRIANGULAR DIAPIRS

PHYSICAL MODEL—Salt Rollers above a Tilted Detachment

Figure 3.29. Reactive diapirism. Asymmetric reactive diapirs (salt rollers) are most common above tilted detachments. The tilt promotes seaward-dipping faults. Faulting results in differential structural thinning. Salt is expelled toward the extensionally thinned overburden and reduced load. This salt flow and the rotation of fault blocks form triangular diapirs in the fault footwalls. All units were prekinematic and tabular before deformation. Model by Bruno Vendeville.

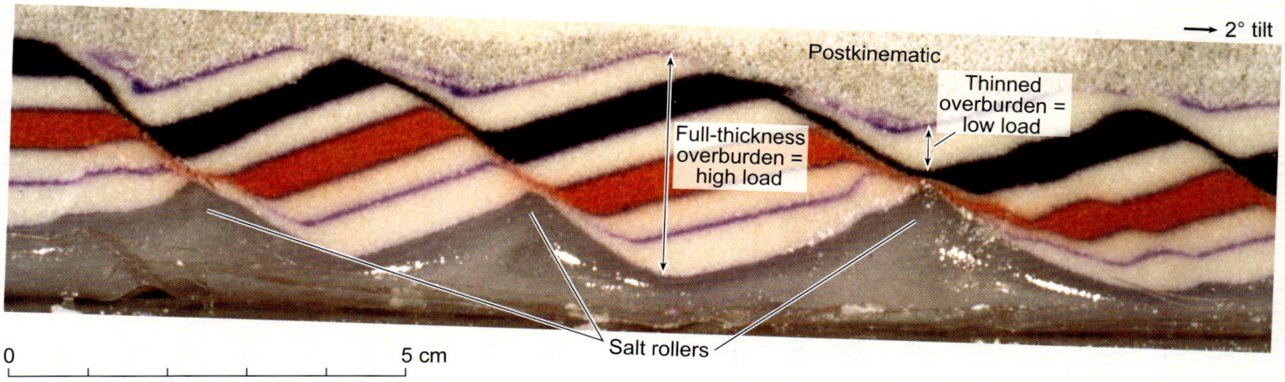

PHYSICAL MODEL—Raft Geometry above a Salt Detachment

Figure 3.30. Raft tectonics. Salt diapirs rose between separating raft blocks during extension. Salt diapirs pierced to the surface, where they extruded to form overhangs. Raft blocks are separated where faults have the most offset but connect with each other where fault throw and diapirism die out. Raft blocks thus relay like the fault systems. Model by Bruno Vendeville. Modified from Guglielmo and others (1997).

(a) Salt geometry with color contours

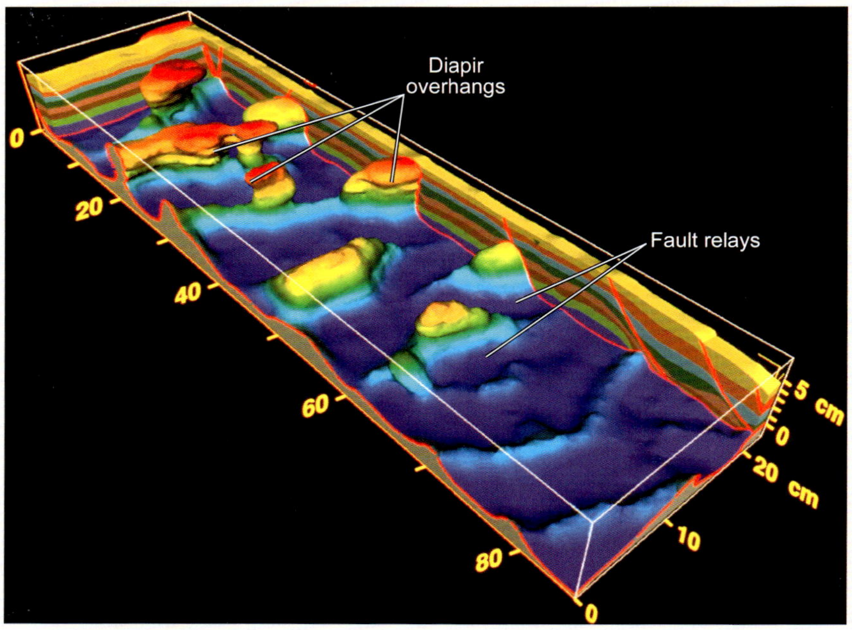

(b) Salt and overburden rafts

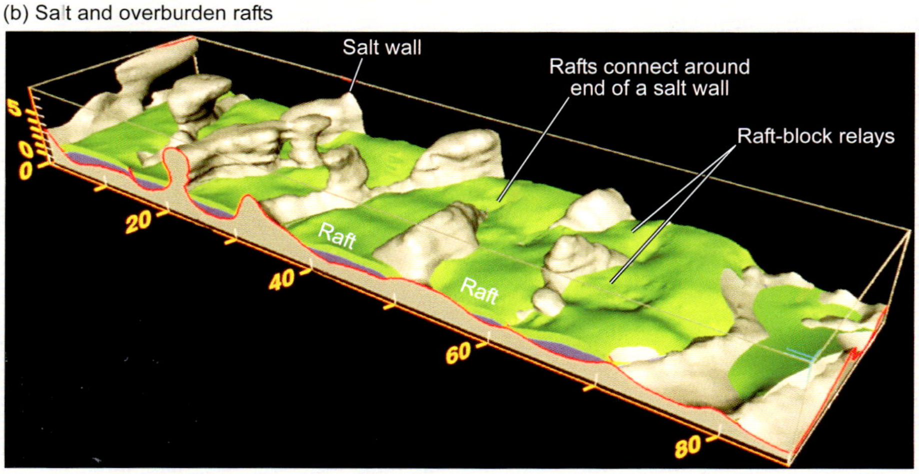

PHYSICAL MODEL—Weakly Inverted Triangular Diapir

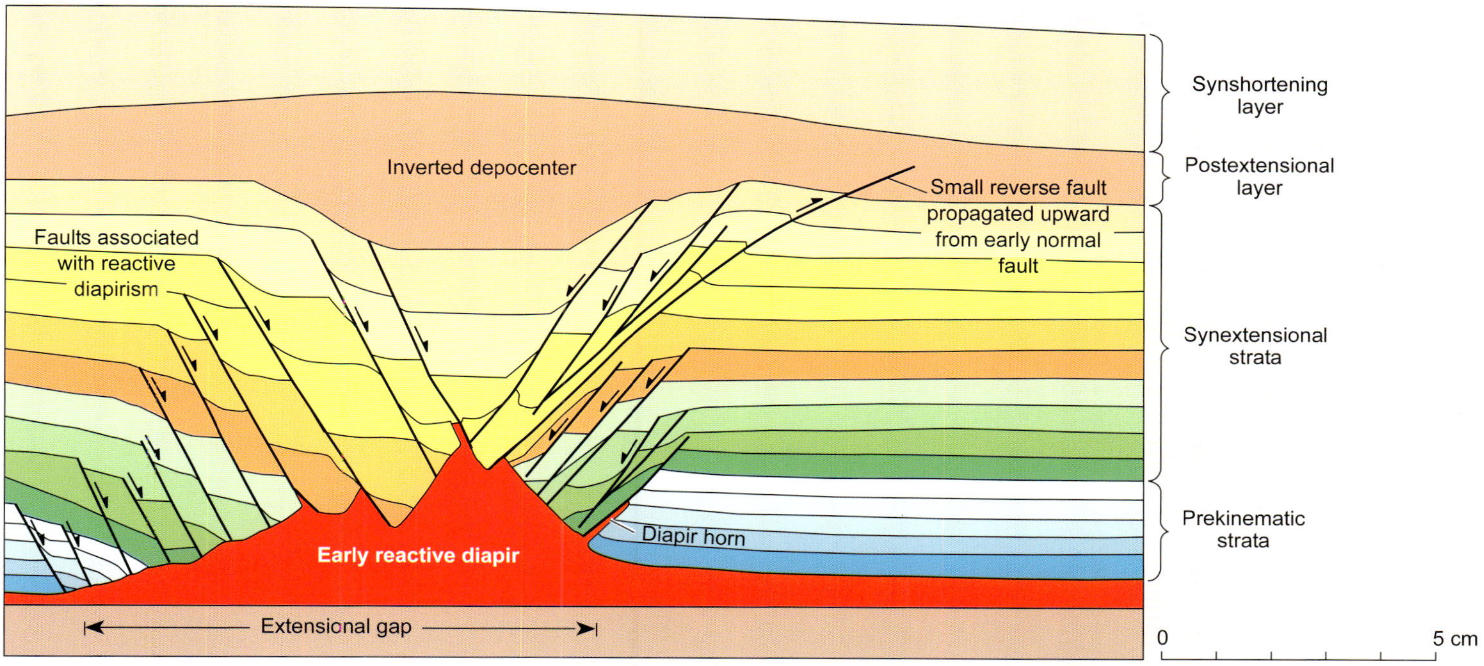

Figure 3.31. Shortening of a reactive diapir. The reactive diapir in the center of the model formed in extension, as indicated by the gap in prekinematic strata. The angular diapir margin is composed of several small cusps in the footwalls of individual faults. These cusps formed when the diapir began to fall because the rate of salt import from the source layer could not keep pace with diapiric widening. Mild shortening at the end of the experiment was localized over the crest of the diapir because the roof was thinnest and weakest there. The depocenter over the diapir crest was inverted to form a gentle anticline above the buried diapir. Most of the older normal faults did not reactivate as reverse faults during shortening. Instead, the lower segments of a few normal faults reactivated with reverse slip, which cut upward as new reverse faults beyond the graben margins. Model by Bruno Vendeville.

PHYSICAL MODEL—Triangular Diapir Triggered by Basement Extension

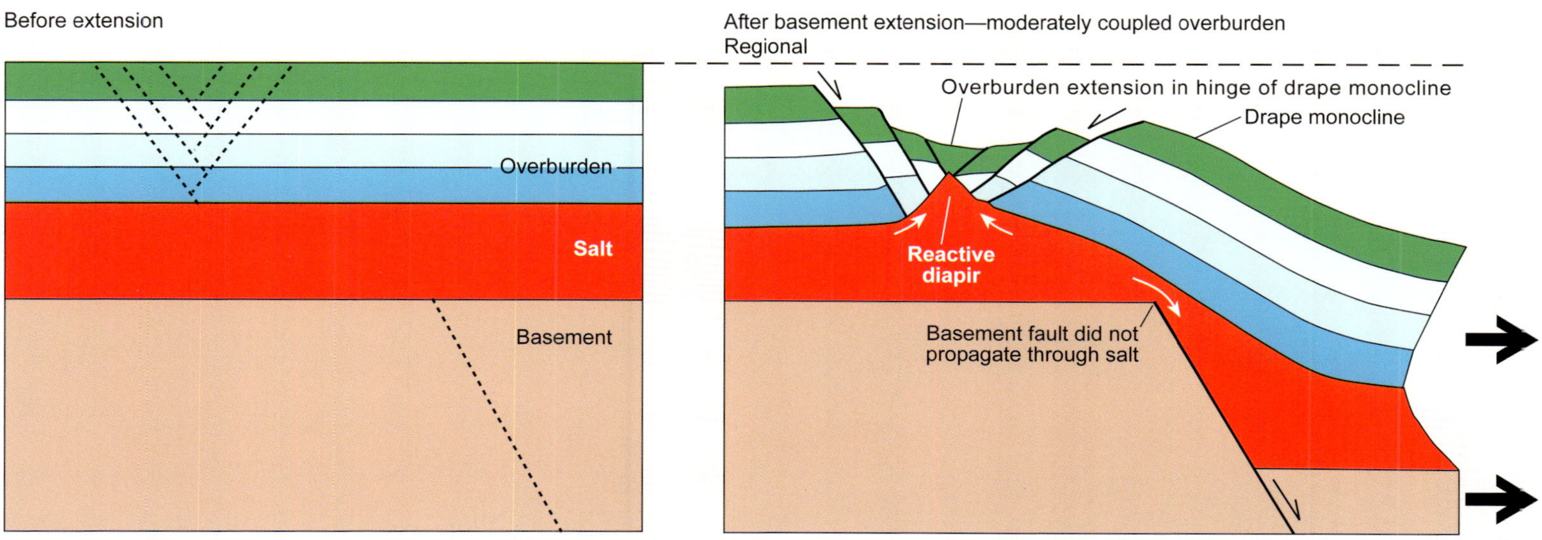

Figure 3.32. Basement extension below salt. Extension above and below salt was moderately coupled in this experiment. Space created by basement faulting was partly filled by flow of salt from above the upthrown block, but this was not enough to completely fill the graben. As a result, the overburden subsided to fill this extra space in the graben and formed a drape monocline over the edge of the basement fault block. Overburden extension was localized in the upper hinge of this monocline, nucleating a reactive diapir. Overburden faults were offset from the basement fault, and the basement fault did not propagate up through the salt. Modified from Jackson and Vendeville (1994).

NUMERICAL MODEL—Pressure Contours in a Reactive Diapir and Its Overburden

Figure 3.33. Growth of a reactive diapir. Reactive diapirs grow in response to load gradients. Here, the reduced load beneath the extensionally thinned graben in the center of the model formed a zone of low pressure in the salt. Salt flowed laterally along a pressure-head gradient into the zone of lower pressures, eventually forming a reactive diapir. Faulting caused abrupt lateral changes in pressure in the graben. GEOSIM finite-element model by J. P. Bobineau.

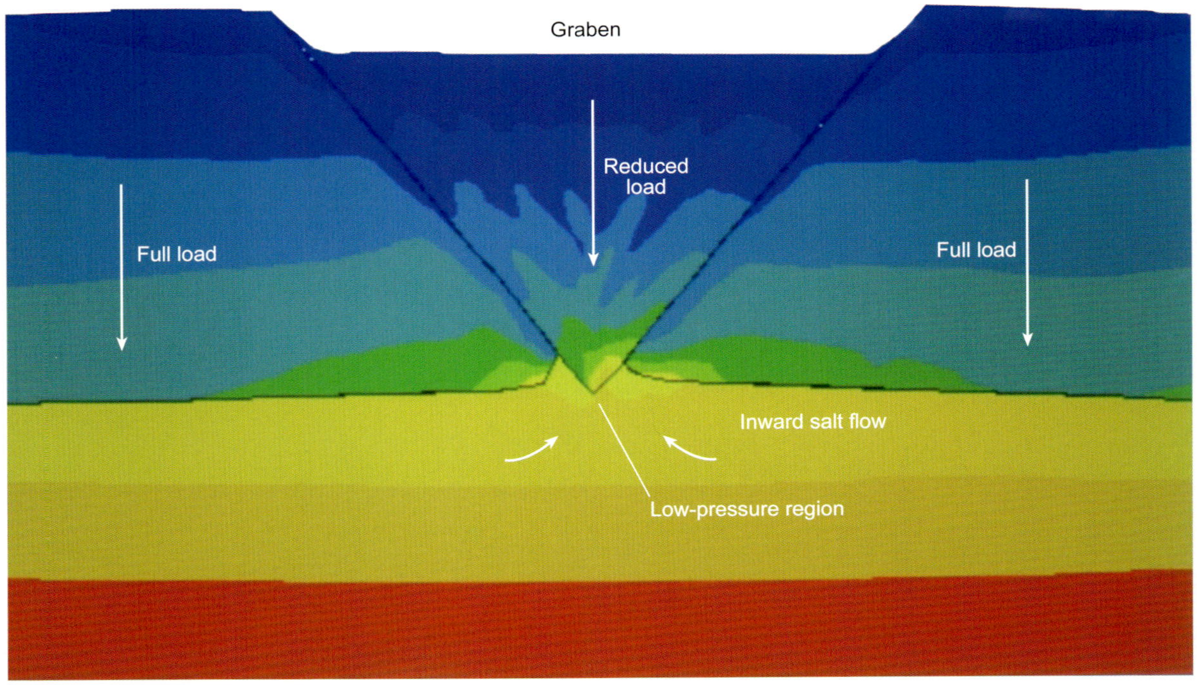

NUMERICAL MODEL—Reactive Diapirism Triggered by Extension above a Preexisting Salt Monocline

Figure 3.34. Growth of a reactive diapir. Extension at the upper hinge of a salt monocline is strongly asymmetric. The dominant fault propagated upward from the monocline crest and dipped in the direction of monocline dip. An antithetic fault eventually formed, but it remained subsidiary to the left-dipping structure. A reactive diapir rose where the two faults intersected. A more symmetric fault system forms when the preexisting structure is more symmetric. GEOSIM finite-element model by Dan Schultz-Ela.

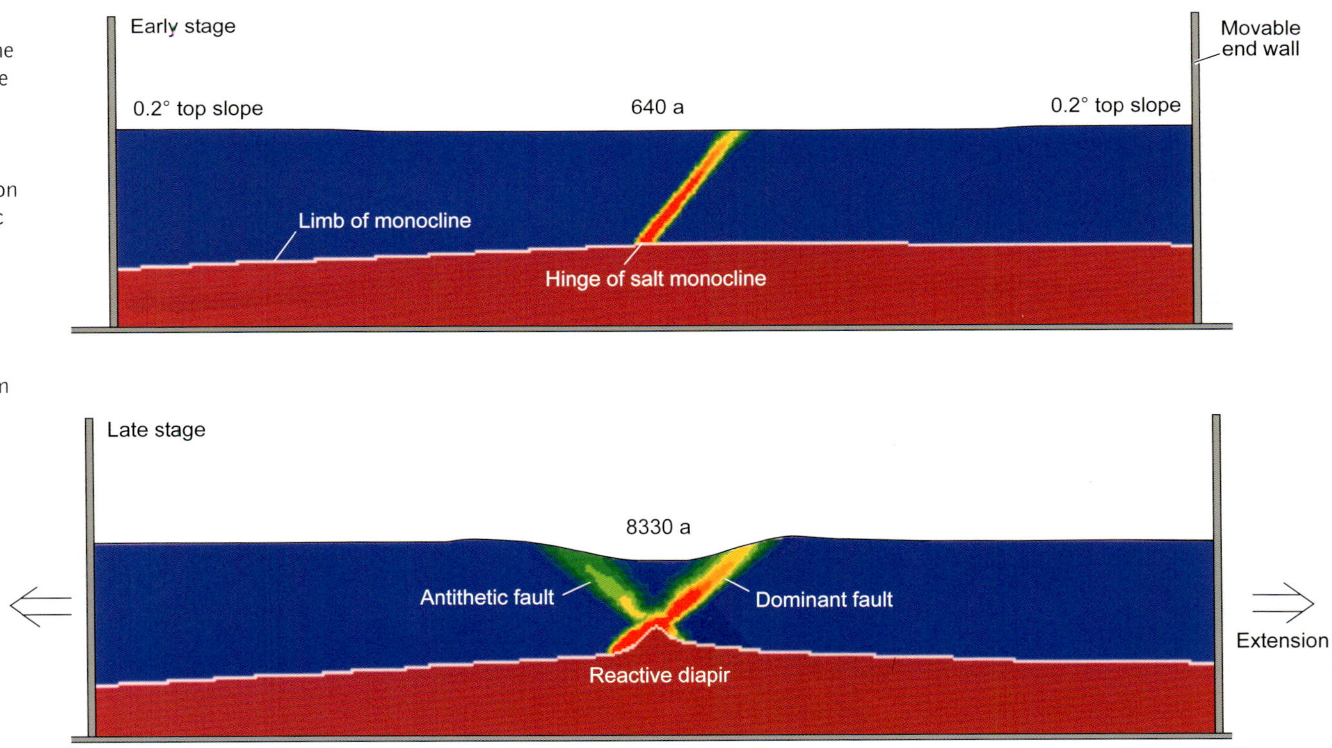

NUMERICAL MODEL—Asymmetric Reactive Diapirism above a Preexisting Salt Anticline

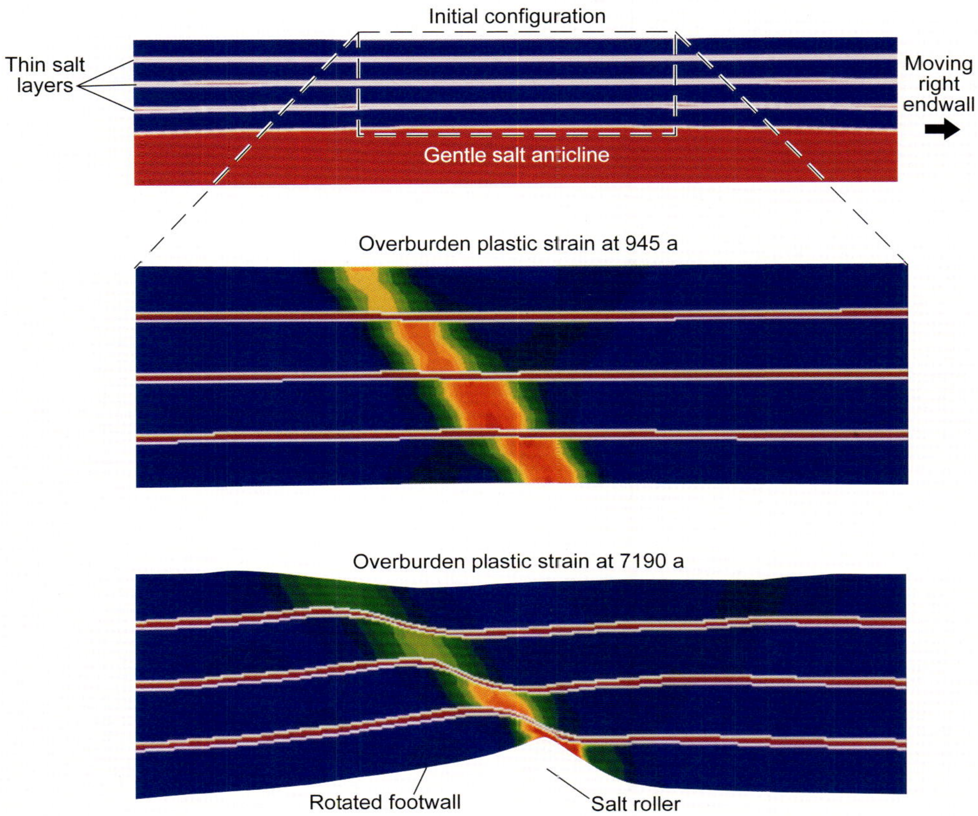

Figure 3.35. Growth of a salt roller. The initial fault pattern was asymmetric probably because only the right-hand endwall was moving. In models with isotropic overburden, the initial asymmetry quickly evolved into a more symmetric form. However, here where the overburden was anisotropic, asymmetry persisted because the decreased flexural strength caused by the presence of salt layers allowed the footwall to rotate. The resulting structure was an asymmetric reactive diapir, more commonly known as a salt roller. GEOSIM finite-element model by Dan Schultz-Ela.

NUMERICAL MODEL—Reactive Diapirism during Gravity Spreading

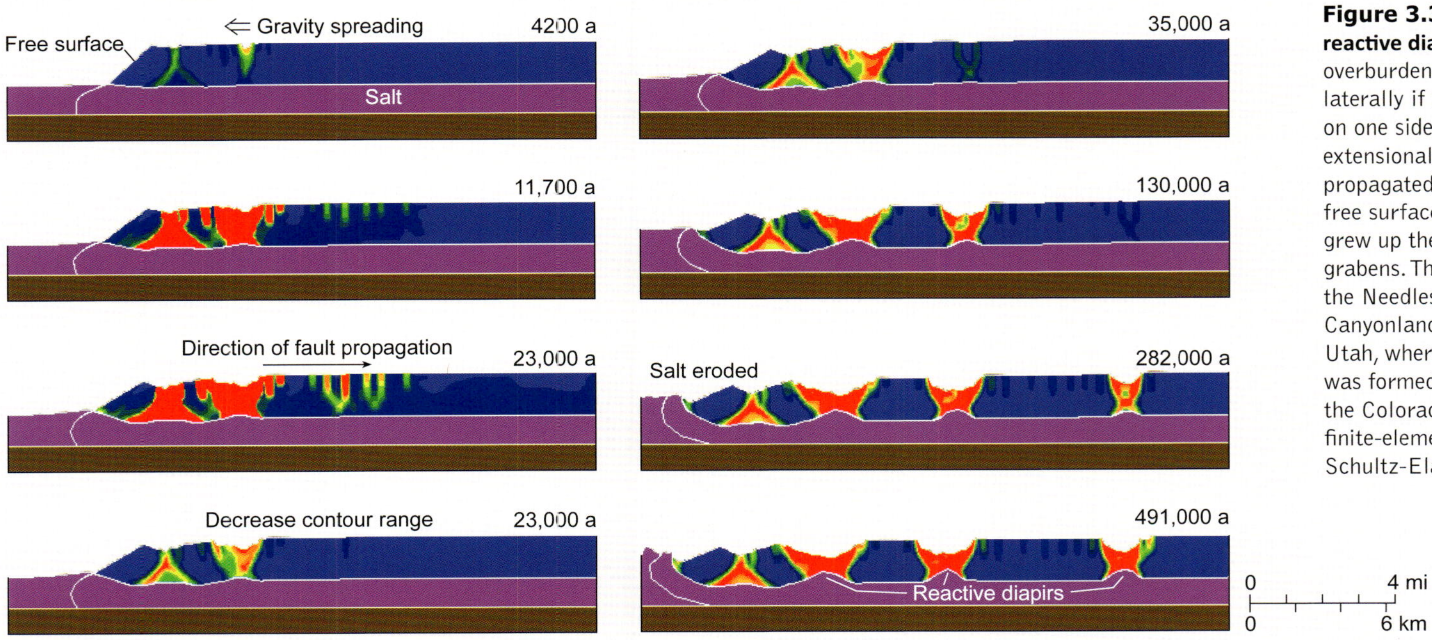

Figure 3.36. Growth of reactive diapirs. A block of overburden above salt spreads laterally if it has a free surface on one side. Spreading formed extensional grabens that propagated away from the free surface. Reactive diapirs grew up the axes of these grabens. This model simulates the Needles fault zone in Canyonlands National Park, Utah, where the free surface was formed by downcutting of the Colorado River. GEOSIM finite-element model by Dan Schultz-Ela and Patrick Walsh.

CONCEPTUAL MODEL—Diapir Growth Stages during Extension

Figure 3.37. Extensional diapirism. Without synkinematic fill, a graben locally thins the overburden above salt. During the reactive stage, thinning creates a gradient in the pressure head of salt, which causes salt to flow toward the axis of the graben. Faulting also weakens the overburden in the graben, making it easier for diapirs to rise. During the active stage, the roof has been thinned and weakened to the point where the pressure within the salt is sufficient to uplift and rotate the floor of the graben. Salt quickly breaks through the floor of the graben to be exposed at the sediment surface during the passive stage. Once the diapir is at the surface, diapir widening accommodates further extension. Modified from Vendeville and Jackson (1992a).

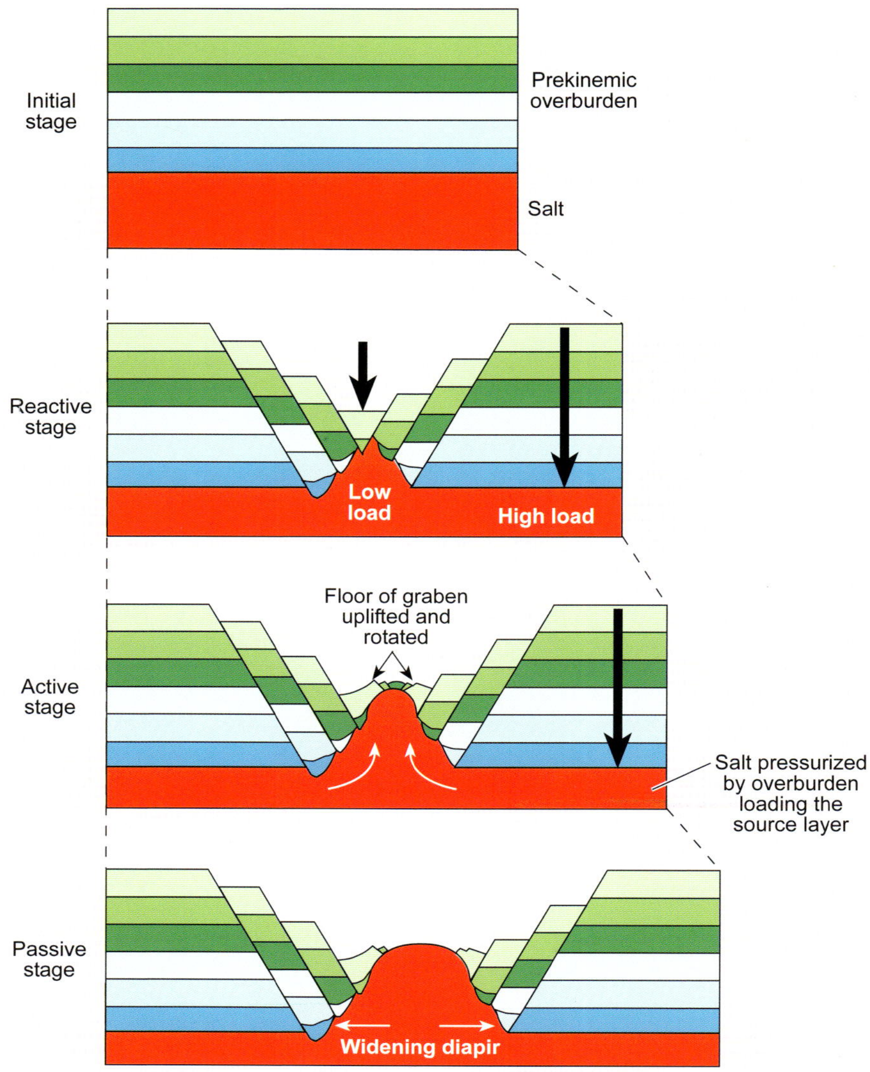

CONCEPTUAL MODEL—Comparison of Rising and Falling Diapirs

Figure 3.38. Extensional diapirism. The most unambiguously diagnostic feature of falling diapirs is an indenting graben, flanked by cusps of salt. Absent such a basin, it is often difficult to be certain whether a diapir is rising or falling because many features of a rising diapir can be found above some falling diapirs as well. In these instances, the best way to recognize whether a diapir has begun to fall is to construct a restoration. Modified from Vendeville and Jackson (1992b).

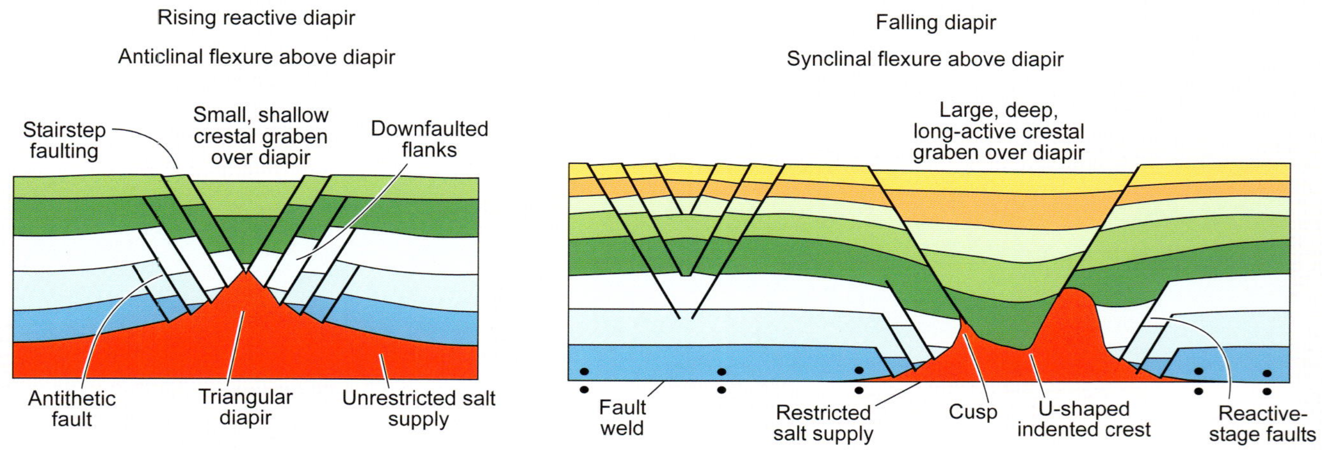

CONCEPTUAL MODEL—Factors Controlling Extensional Drape over Basement Faults

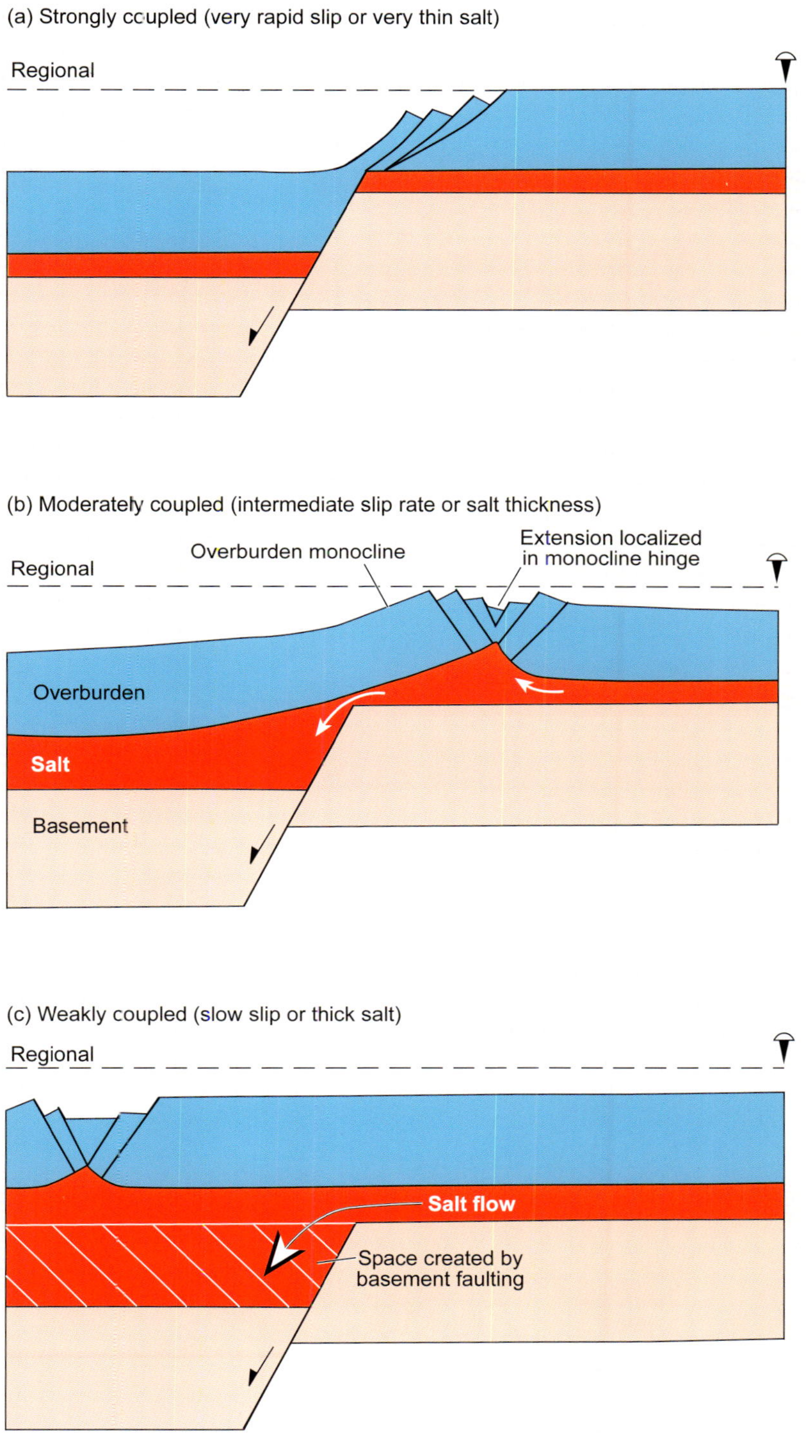

Figure 3.39. Basement extension below salt. Strong coupling between extension in the basement and the overburden is favored by thin salt, rapid slip, and high-viscosity evaporites. (a) When coupling is strong, basement faults propagate through the salt into the overburden. (c) When coupling is weak, space created by basement faulting is filled by flow of salt from above the upthrown block. The overburden stretches by the same amount as that of the basement, but there is no spatial correlation between reactive diapirs and basement faults. (b) In intermediate cases, salt flow fills some space created by basement faulting, but the rest is filled by monoclinal drape of the overburden over the fault-block edge. Overburden extension and reactive diapirism localize in the upper hinge of this monocline so are related to the position of the basement fault but laterally offset from the fault. Modified from Jackson and others (1994).

CONCEPTUAL MODEL—Influence of Synkinematic Sedimentation on Salt Flow above Basement Extension

Figure 3.40. Basement extension below salt. Whether salt flows toward or away from the upthrown block during subsalt normal faulting depends on the density of rock filling the downthrown basin. (a) If the fill density is less than salt density, which would include air or water fill, salt will flow downhill to the downthrown block, driven by an elevation head gradient. (b) If fill density is greater than salt density, salt will flow uphill from beneath the synkinematic basin, driven by the pressure-head gradient. The salt would lift the overburden on the upthrown block. If fill density equals salt density, no mass transfer of salt will occur. Modified from Jackson and others (1994).

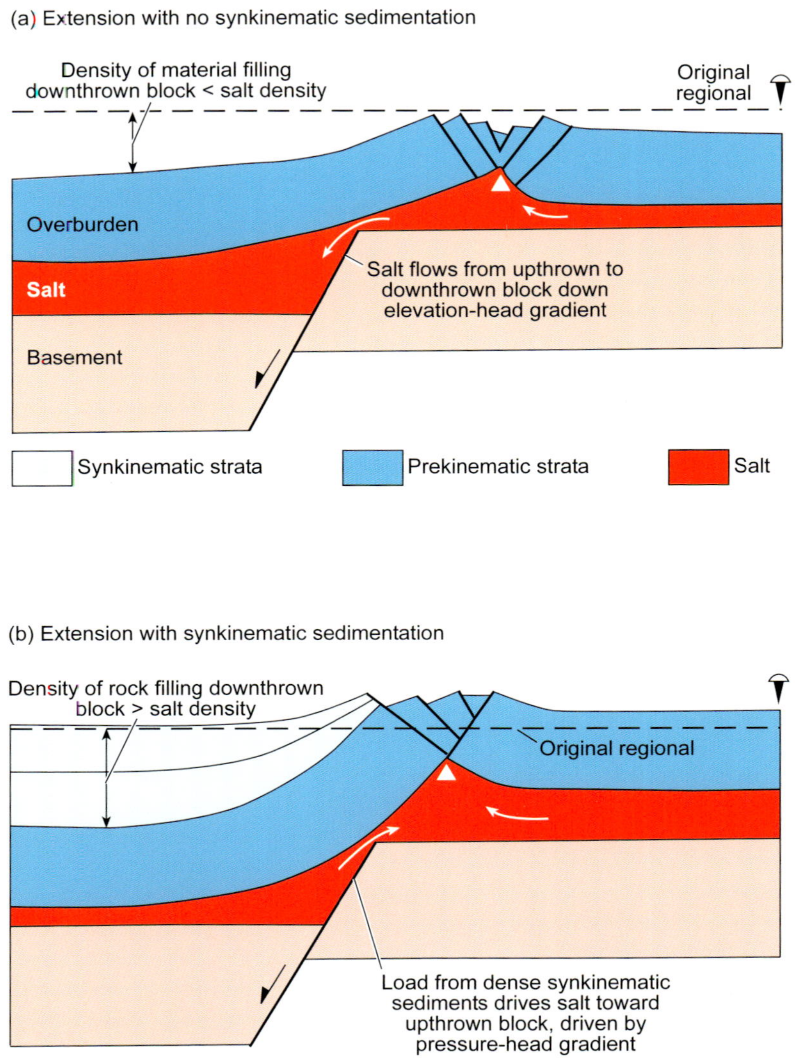

Chapter 4

Indented Diapirs

Sagging crests of some salt diapirs were once attributed to salt dissolution. However, this explanation was invalidated by the observation that sagging roofs also form above some shale diapirs. Physical modeling then suggested a solution: roofs subside because of regional extension, which causes the underlying diapir to widen and sag.

An indented diapir has an overlying graben that has subsided into its crest. The former crest of the diapir is preserved as a relic on either side of the indenting graben. If the graben extends all the way to the edge of the diapir, then the relic is a thin protrusion of salt that passes upward into the normal fault bounding the graben. These protrusions are remnants of the former salt crest and were not injected up the fault plane.

The more a diapir is laterally stretched, the deeper an indenting graben sinks. At moderate extension, the graben may form a small ornament on the diapir crest. Greater extension can cause the graben to subside until it welds onto the base of salt. This process splits the diapir in half, and the only evidence arguing for an origin as a fallen diapir is the anomalous youth of graben sediments resting on the basal weld.

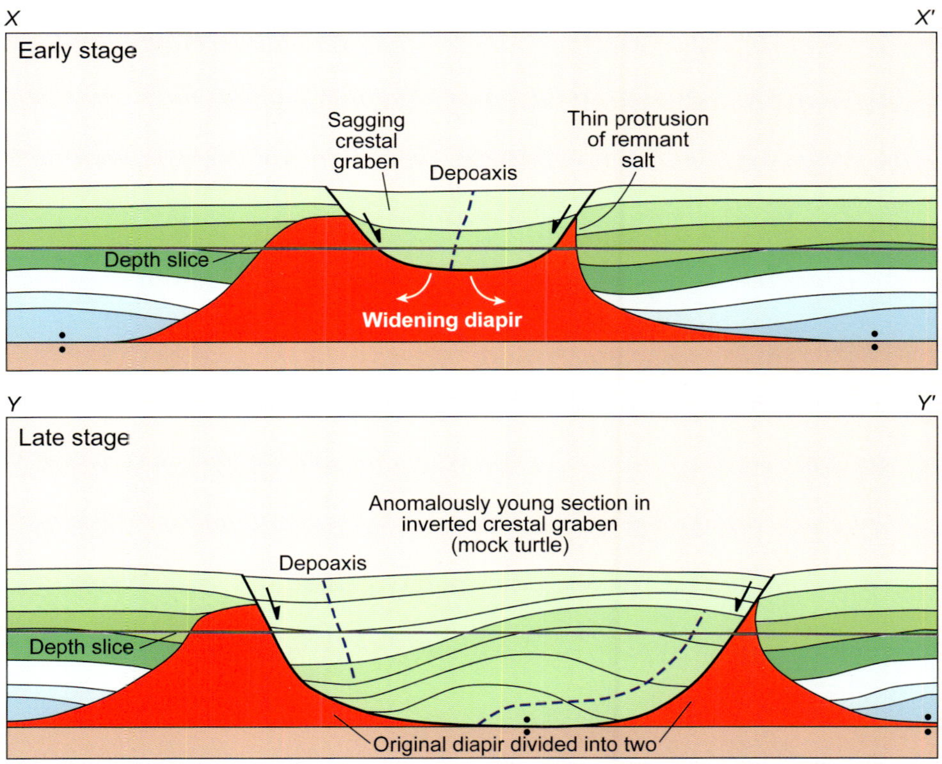

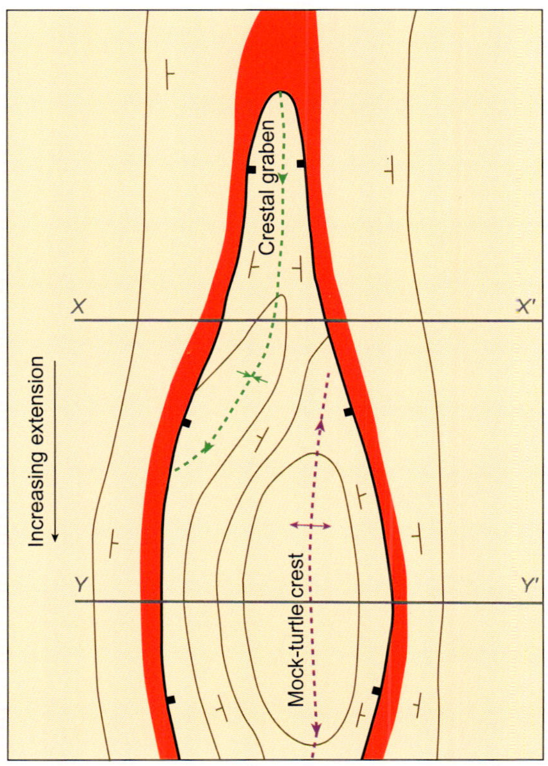

GULF OF MEXICO BASIN—Indenting Blocks at a Diapir Crest, Bay Marchand Field, Offshore Louisiana, USA

Figure 4.01. Extension above a diapir crest. Faults intersecting the top of the Bay March and Dome offset the top of salt, producing a jagged salt-sediment interface. Part of the roof has fallen, indenting the top of salt. Salt horns on either side of the indenting block are relics, indicating the elevation of the former top of the diapir. Modified from Abriel and others (1991).

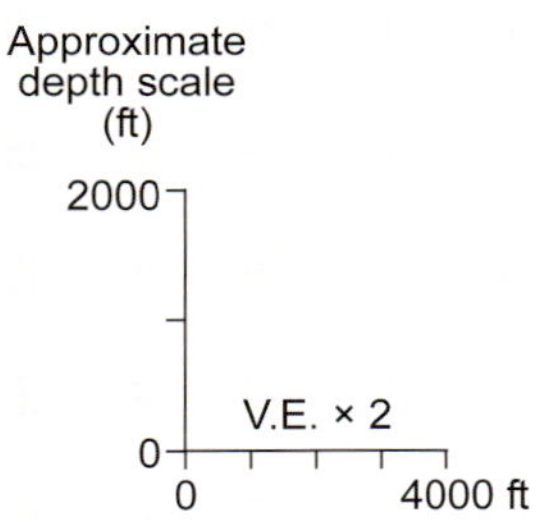

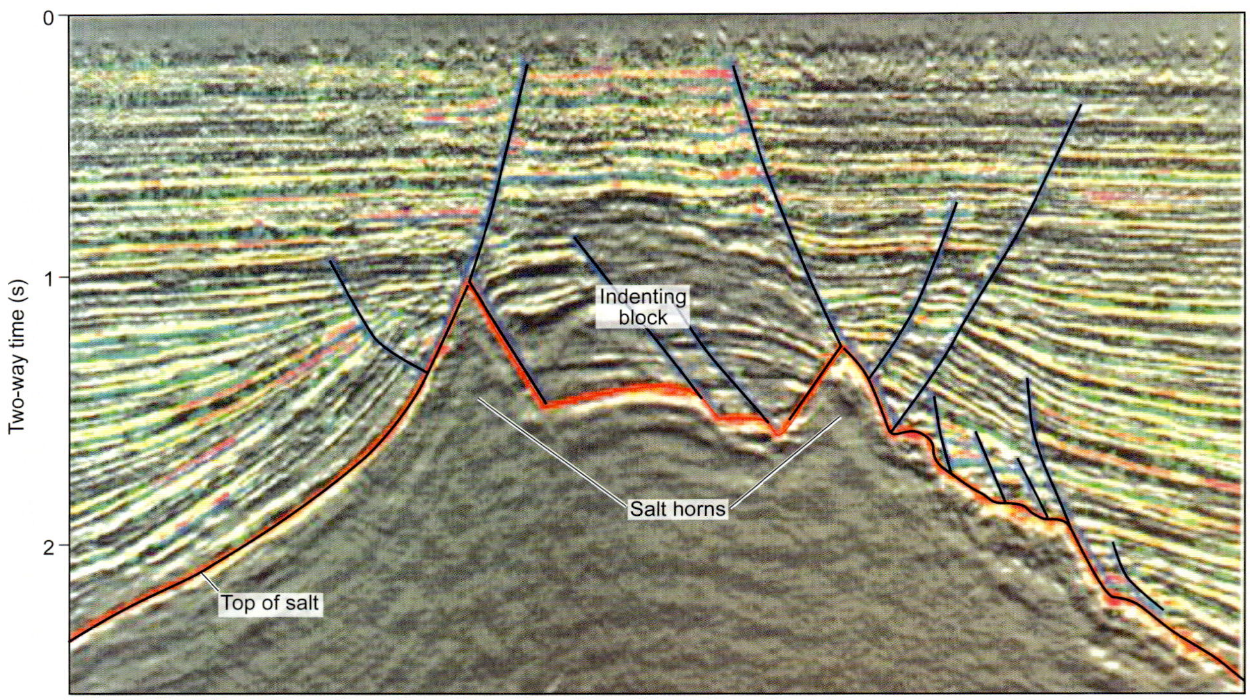

KWANZA BASIN—Sandstone Distribution in a Mock-Turtle Anticline, Quenguela, Angola

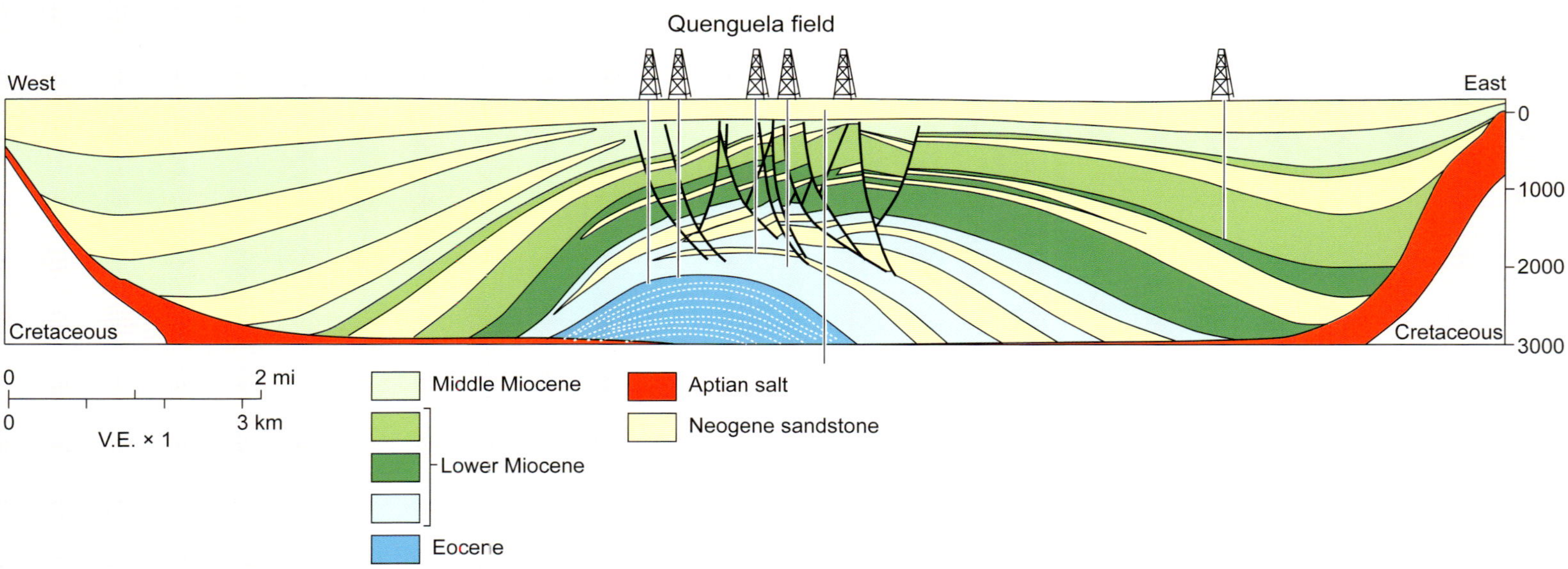

Figure 4.02. Growth of a mock-turtle anticline. A mock-turtle anticline forms through shifting salt expulsion during collapse of a salt wall (see restoration in Figure 4.03). Each sandstone reservoir thins from the expulsion axis onto the minibasin margins, and these geometries change with shifts in the locus of expulsion. The trap at Quenguela field is located where westward-thinning lower Miocene sands drape over the core of the mock-turtle anticline. The Eocene core of the anticline dates the onset of salt-wall collapse. Modified from Masson (1972).

KWANZA BASIN—Nonextensional Restoration of a Mock-Turtle Anticline, Quenguela, Angola

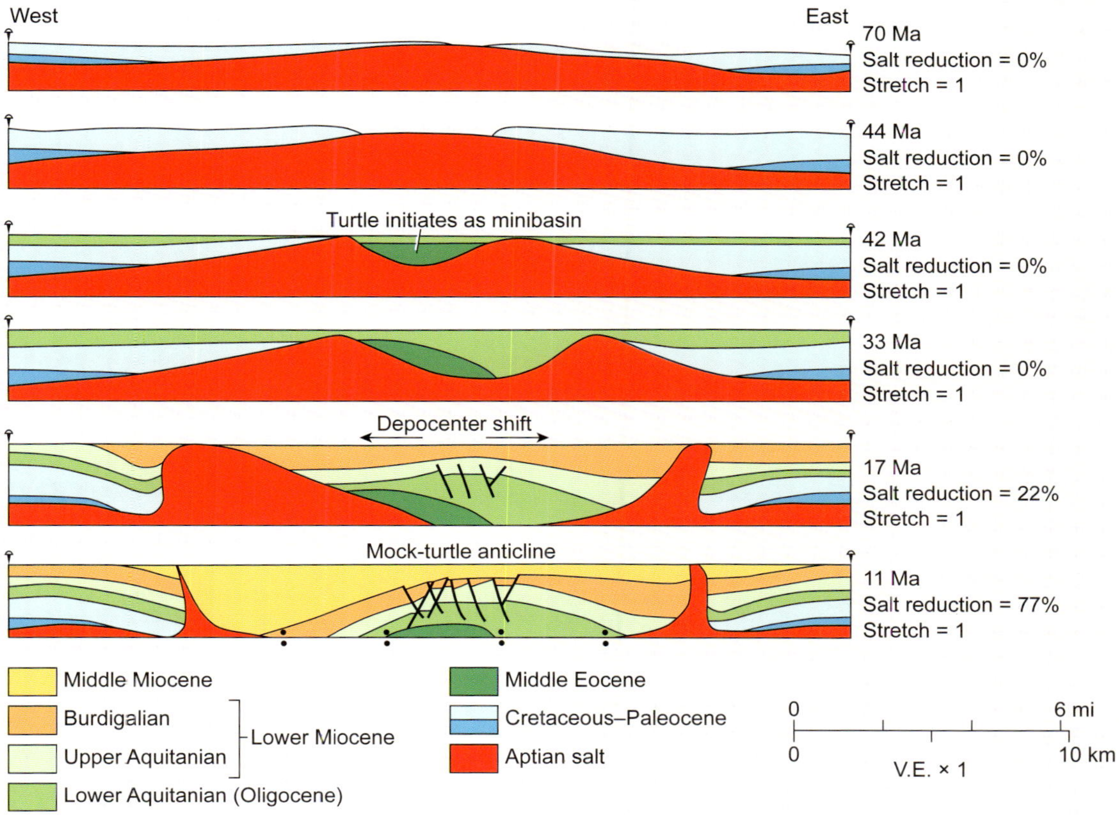

Figure 4.03. Growth of a mock-turtle anticline. A mock-turtle anticline may form by salt expulsion in the absence of extension (shown here), by extensional diapir fall (shown in Figure 4.04), or by a combination of the two processes. In this nonextensional restoration, the turtle initiated as a minibasin sinking into the crest of a broad salt high. Depocenters shifted east and west after this central minibasin welded at its base. Subsidence of the minibasin flanks left the early depocenter as an inverted structural high. This scenario requires reduction of more than 75% of the original salt area. Hudec and Jackson (2004) suggested, on the basis of the absence of compensating shortening downdip, that the Quenguela anticline formed without significant extension. The restoration is only qualitative because some subsurface units are shown truncating over time. See detailed Quenguela section in Figure 4.02. Restoration from Masson (1972).

KWANZA BASIN—Extensional Restoration of a Mock-Turtle Anticline, Quenguela, Angola

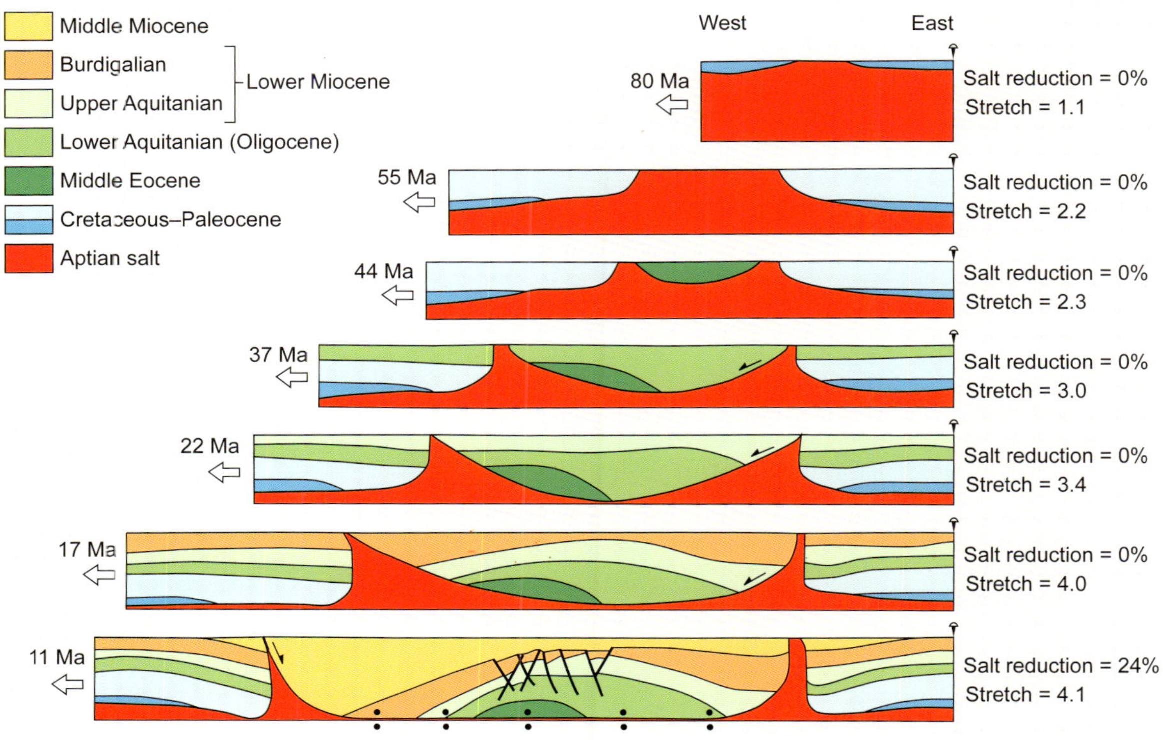

Figure 4.04. Growth of a mock-turtle anticline. A mock-turtle anticline may form by salt expulsion in the absence of extension (Figure 4.03), by extensional diapir fall (shown here), or by a combination of the two processes. Because the restoration was designed to accommodate minibasin subsidence by extension, much less salt dissolution was required than in the nonextensional model. However, to attain the stretch of 4.1 requires more than 20 km of extension. The absence of comparable shortening downdip led Hudec and Jackson (2004) to favor the nonextensional model. See detailed Quenguela section in Figure 4.02. Modified from Duval and others (1992).

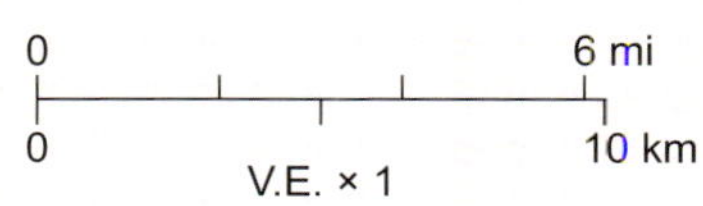

North Sea—Indented Diapir, Offshore UK

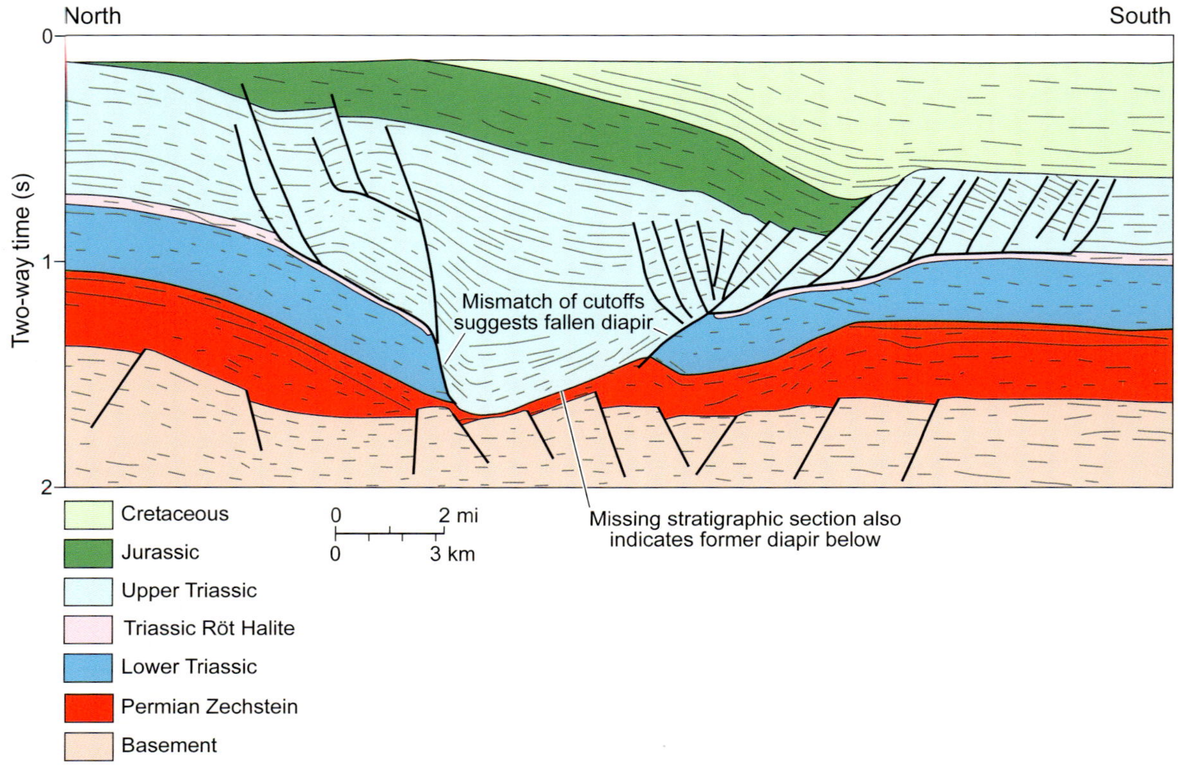

Figure 4.05. Extensional fall of a salt diapir. The mismatch of Lower Triassic cutoffs on opposite sides of the graben suggests that the two sides were originally separated by a diapir, which fell during the Late Triassic. If so, the lower part of the graben is bounded by salt welds rather than faults. Late Triassic extension was detached at both Zechstein and Röt salt levels. Shallow extension formed tilt blocks above Röt detachments. These detachments ramp down to Zechstein level in the area of the former diapir, where their slip contributed to extensional fall of the diapir. Modified from Stewart and Argent (2000).

Paradox Basin—Stages of Extensional Fall of the Gypsum Valley Diapir, Colorado, USA

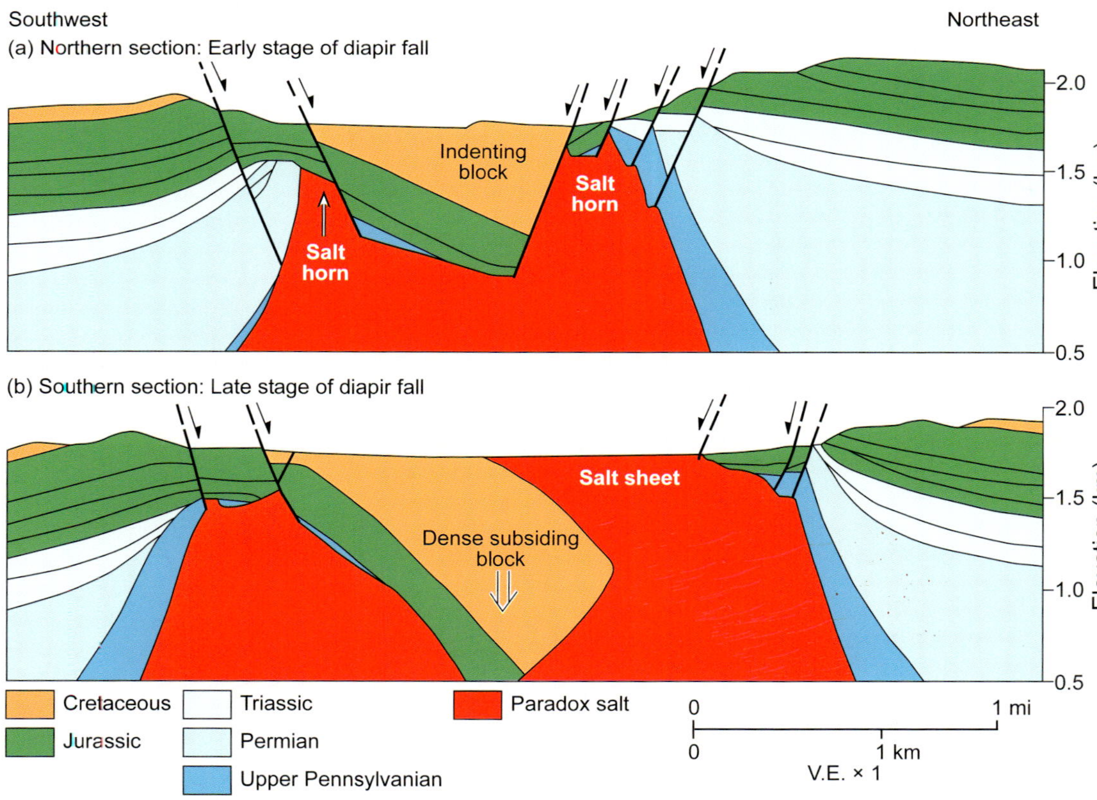

Figure 4.06. Extensional fall of a salt diapir. Serial sections of the Gypsum Valley diapir show stages of extensional diapir fall. (a) In the early stage, indenting roof blocks are downfaulted into the subsiding diapir, leaving relict horns of salt on either side. (b) At a more advanced stage, salt pierces the floor of the roof graben and extrudes over its former roof. These Mesozoic roof blocks are denser than the salt, adding gravitational subsidence to extensional subsidence. Modified from Ge and others (1995).

PARADOX BASIN—Variations in Diapir-Fall Geometry, Utah and Colorado, USA

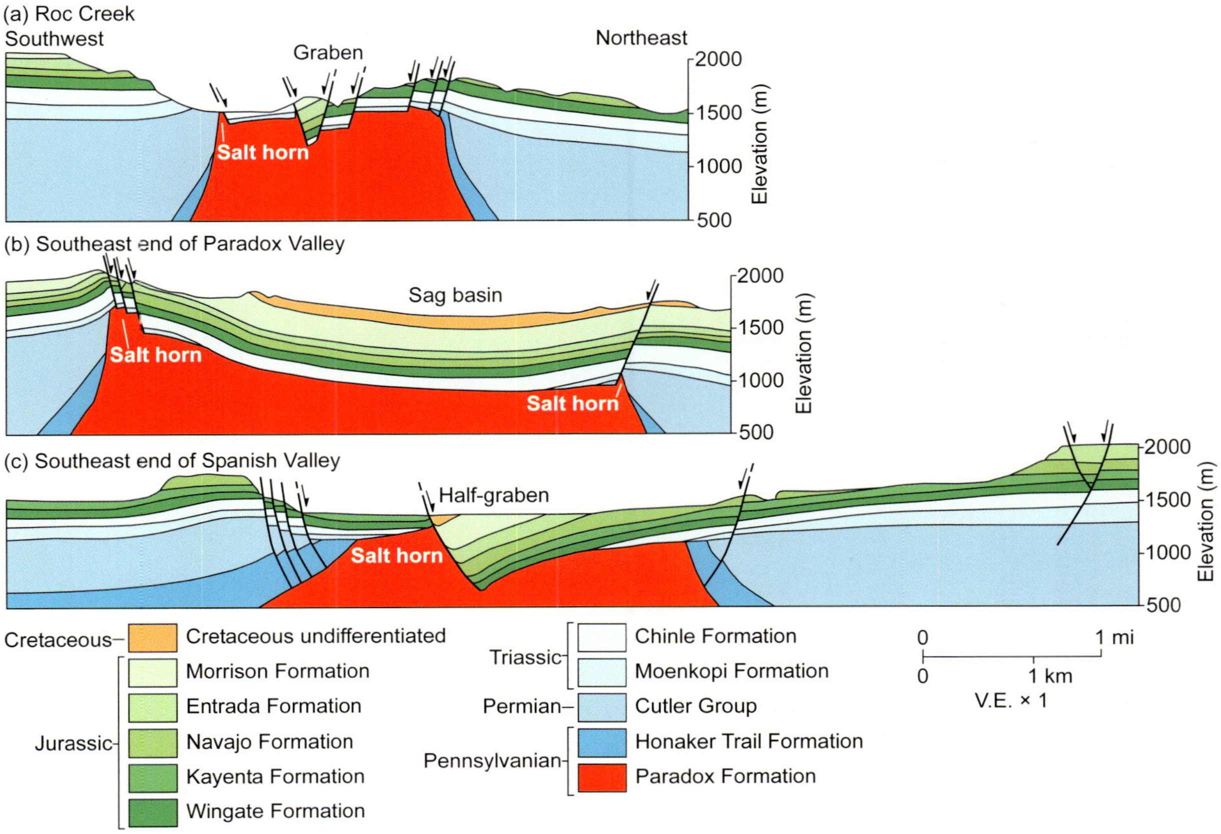

Figure 4.07. Extensional fall of salt diapirs. During extensional diapir fall, the diapir roof is downfaulted into the underlying salt, leaving relict salt horns on either side. Downfaulted roofs may form roughly symmetric graben systems (Roc Creek), broad grabens in which the central block is synclinal (Paradox Valley), or half-grabens (as in southeastern Spanish Valley). The extensional style is primarily controlled by the shape of the diapir and the magnitude of extension. Style can vary along the strike of a single salt wall. Modified from Ge (1996).

PARADOX BASIN—Indenting Blocks at the West End of the Onion Creek Diapir, Utah, USA

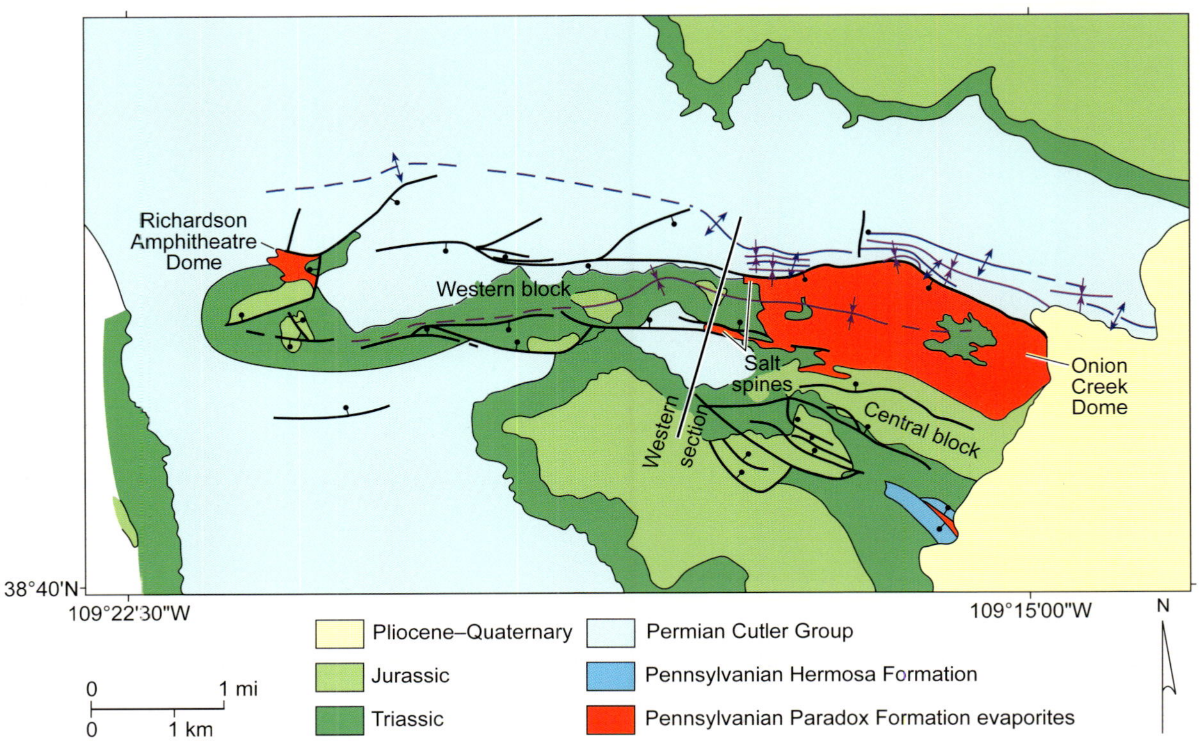

Figure 4.08. Extensional fall of salt diapirs. The Onion Creek and Richardson Amphitheatre Domes are culminations along the same west-trending salt wall. Cenozoic extension was focused in the weak roof of this salt high. Salt culminations rose to the surface as reactive diapirs. Lower parts of the salt wall fell to feed the rising culminations and were indented by subsiding roof blocks. A restoration (Figure 4.10) shows the reconstruction of the western indenting block. See also photo panorama in Figure 4.09. Modified from Doelling (1995).

INDENTED DIAPIRS

PARADOX BASIN—Indenting Blocks above the Onion Creek Diapir, Utah, USA

(a) Uninterpreted panorama

North South

(b) Interpreted panorama

North South

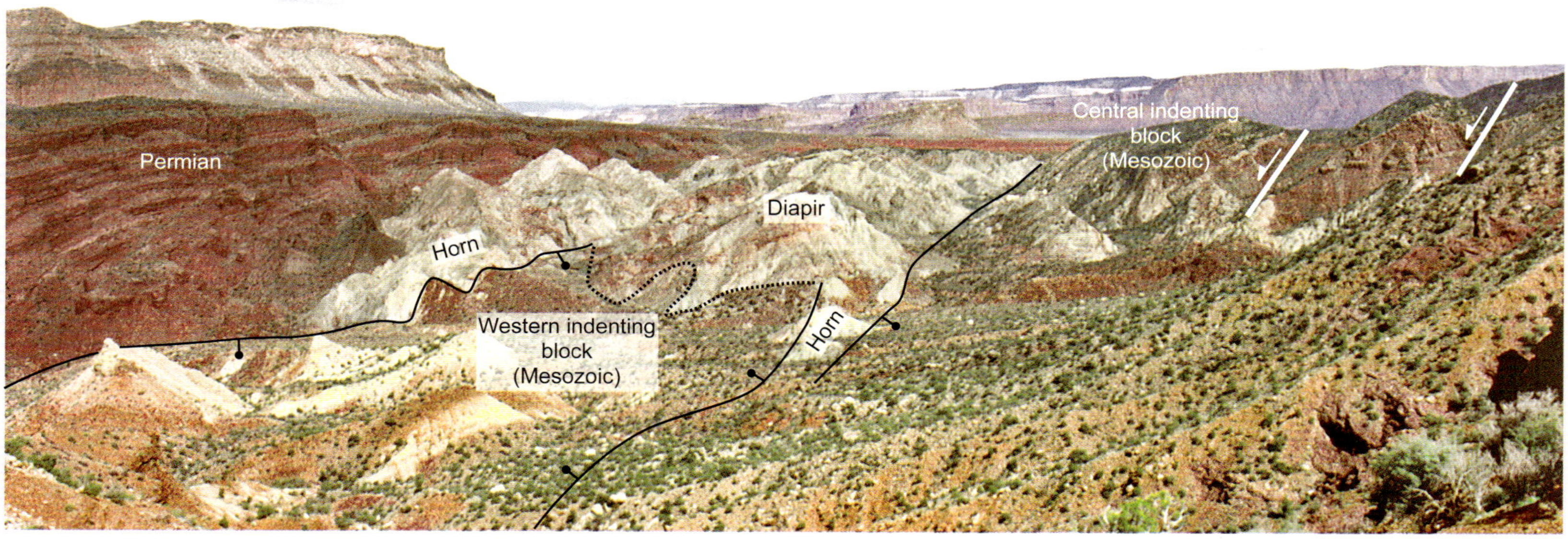

Figure 4.09. Extensional fall of a salt diapir. Cenozoic extension rejuvenated the crest of the Onion Creek Diapir and caused other parts of the roof to collapse, indenting the top of the diapir. Salt adjacent to the indented block formed thin horns, which are relics showing the former height of the diapir (see restoration of cross section of the west end of the diapir, Figure 4.10). This panorama shows two indenting blocks: a western block between two salt horns and a central block that dips steeply into the diapir face (see map, Figure 4.08). The exposed evaporites are only a small part of the diapir, which extends over a much wider area in the subsurface. View looking east. Photograph by Mike Hudec.

PARADOX BASIN—Indenting Blocks at the West End of the Onion Creek Diapir, Utah, USA

Figure 4.10. Extensional fall of a salt diapir. Cenozoic extension in the Paradox Basin stretched the Onion Creek diapir, causing it to fall as much as 2,500 ft along steep normal faults. These blocks indented the top of the diapir, leaving horns of salt in the fault footwalls (see map, Figure 4.08). Extension was much greater in the wider center of the diapir, suggesting that weak salt focused deformation there. Evaporites are exposed (see photograph, Figure 4.09), so salt dissolution also contributed to roof collapse. In the restoration, unfaulting the stratigraphy has offset the present-day topography. Currently eroded material shown as pale colors in the restoration. Modified from Boyers (2000).

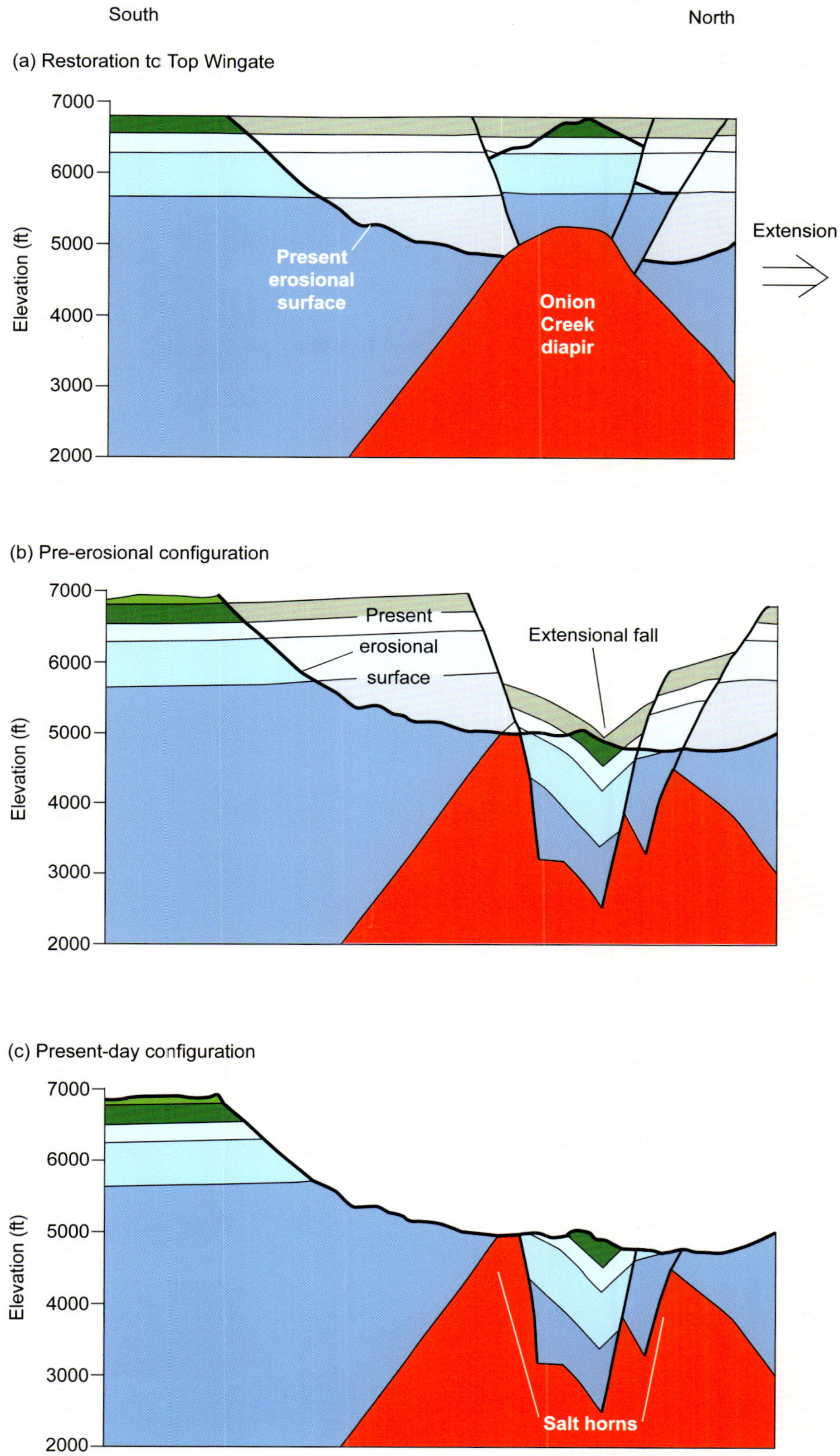

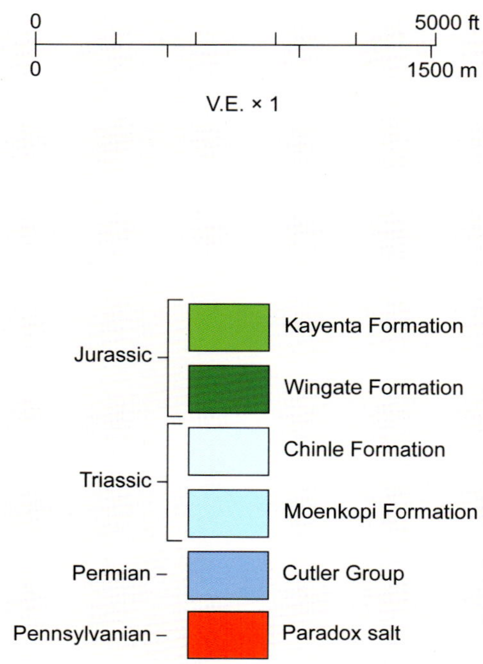

PHYSICAL MODEL—Indented Diapirs

Figure 4.11. Extensional fall of salt diapirs. Salt rose as reactive diapirs until the rate of salt flow into the diapir was restricted when the lower corners of extensionally tilted prekinematic overburden blocks impinged on the source layer. At that time the widening diapirs began to subside and were indented by crestal grabens (see restoration, Figure 4.12). Evidence that the salt was once higher is provided by relict horns and welds of salt. Modified from Vendeville and Jackson (1992b).

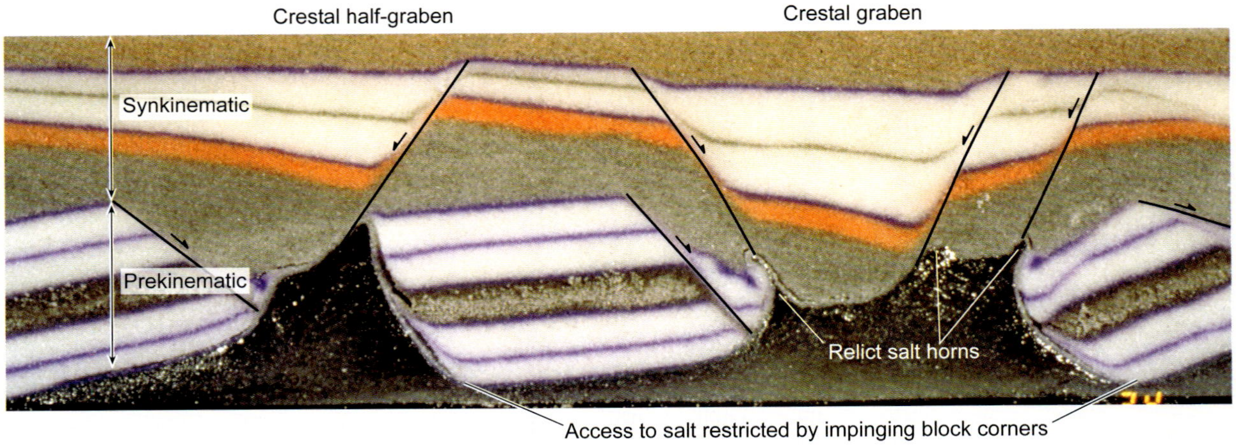

PHYSICAL MODEL—Restoration of Falling Diapirs

Figure 4.12. Extensional fall of salt diapirs. Diapirs originally grew during extension, rising as reactive, then passive diapirs in the gaps between separating fault blocks. Diapir rise continued until the corners of the subsiding fault blocks began to restrict the flow of salt from the source layer into the diapir. The widening diapirs could no longer be filled by new salt from below, so the diapir crests began to fall. Subsidence of basins above the diapirs left relict horns of salt in the footwalls of the salt. Note that these horns are not salt slivers intruded up the fault planes but, rather, relics of a once-taller diapir that were left behind as crestal grabens subsided and indented the diapir. See photograph of physical model in Figure 4.11. Modified from Schultz-Ela (1992).

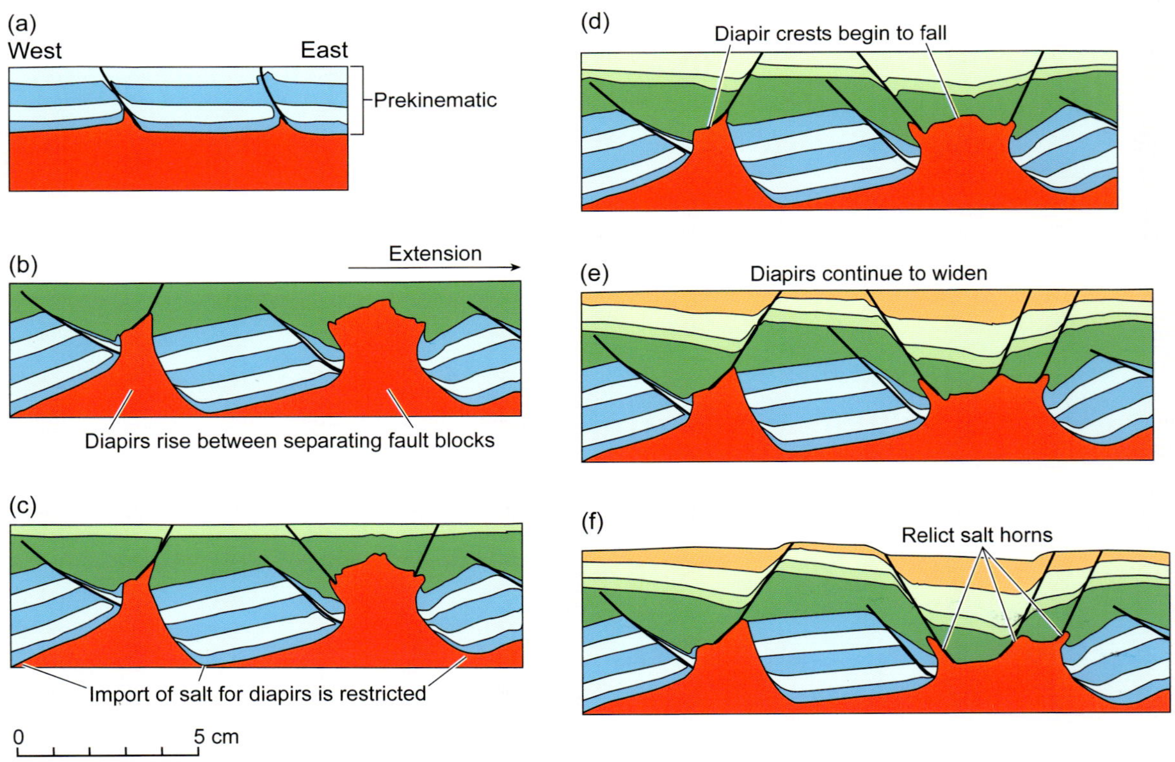

PHYSICAL MODEL—Extensional Fall of the Feeder of an Allochthonous Sheet

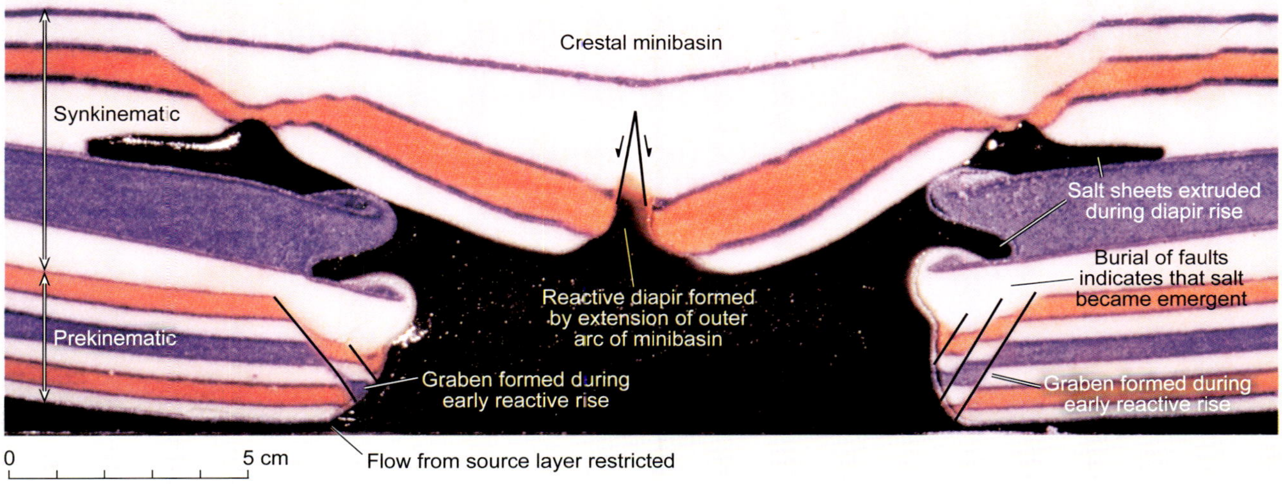

Figure 4.13. Extensional fall of a salt diapir. This diapir illustrates the three stages common to diapirs formed during extension. During the reactive rise phase, salt rose as a (presumably) triangular diapir up the axis of a graben in the overburden. Once the salt emerged at the surface, the faults became inactive as subsequent extension was accommodated by diapir widening. During the emergent phase, the diapir twice extruded salt sheets. The fall phase began once the subsiding blocks on either side of the diapir restricted the rate of salt flow from the source layer. Without inflow of new salt, the diapir continued widening, causing the roof to sag as a minibasin (see restoration, Figure 4.14). Model by Bruno Vendeville.

PHYSICAL MODEL—Restoration of Extensional Fall of the Feeder of an Allochthonous Sheet

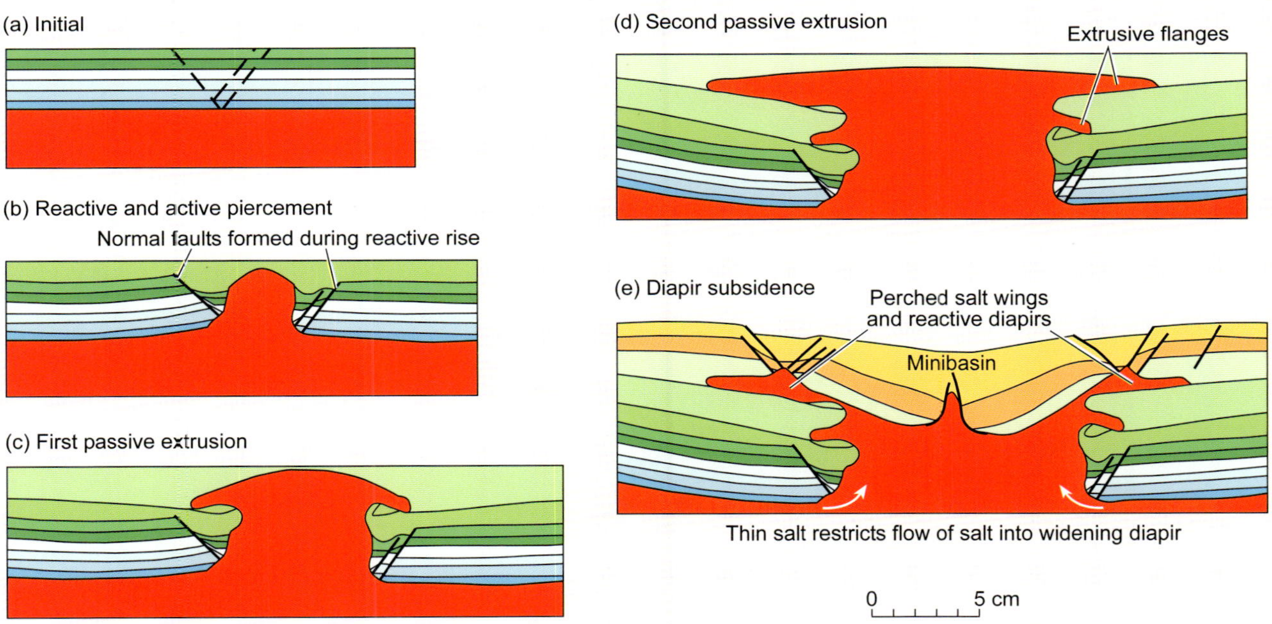

Figure 4.14. Extensional fall of a salt diapir. The diapir pierced to the surface by reactive and active diapirism. Thereafter, it repeatedly was buried, broke actively through to the surface, extruded, and was buried again. Repeated extrusion produced two sets of salt wings along the flanks of the diapir. Eventually, autochthonous salt thinned enough so that the slowed rate of flow into the diapir could no longer keep up with the rate of diapir widening, causing the diapir to fall. Subsidence of the crestal minibasin left the extrusive wings perched high on the flanks of the basin, connected to their source only by thin, soon-to-be welded stems. See photograph of physical model in Figure 4.13. Modified from Vendeville and Jackson (1992b).

PHYSICAL MODEL—Influence of Diapir Shape on Nonextensional Diapir Fall

Figure 4.15. Nonextensional diapir fall. If a diapir falls because of salt dissolution or withdrawal, extension in one part of the roof must be balanced by shortening elsewhere. The balance between extension and shortening is diagnostic of nonextensional diapir fall. Conversely, the absence of contractional structures is diagnostic of extensional fall. Collapsed roofs of these diapirs had extensional outer rims and contractional structures in the center. As shown here, the geometry of these systems depends largely on the shape of the original diapir crest. All units were prekinematic. Modified from Ge and Jackson (1998).

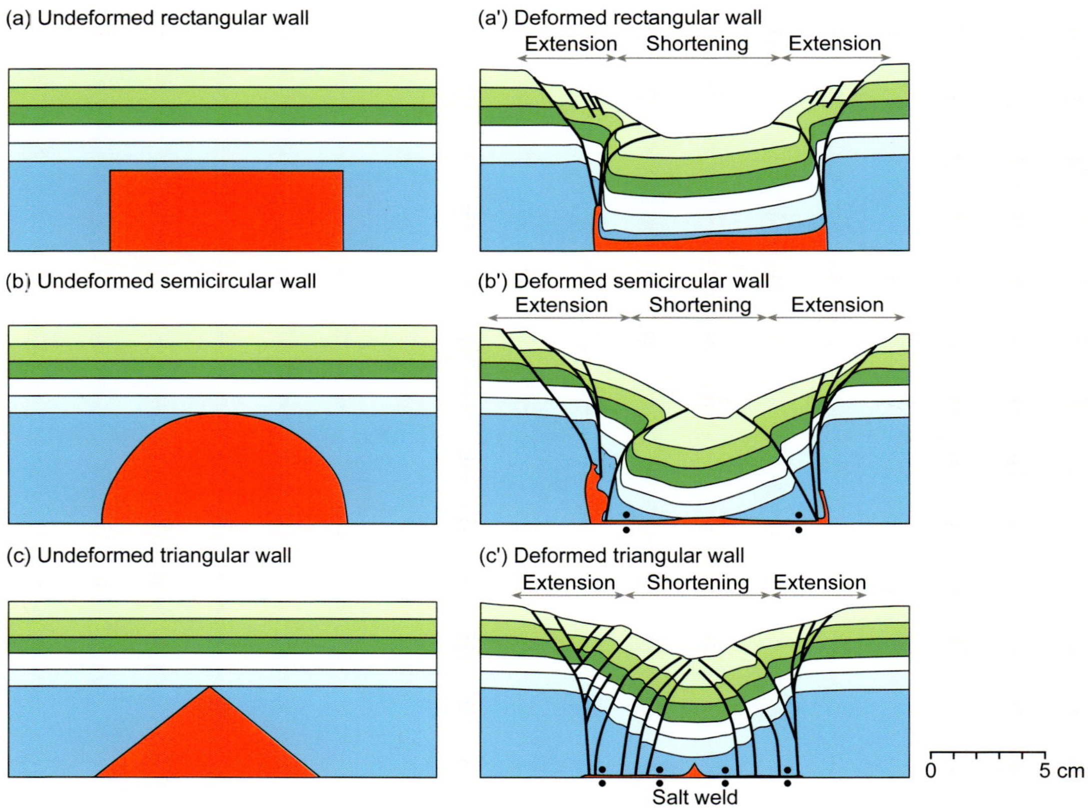

CONCEPTUAL MODEL—Formation of a Mock-Turtle Anticline

Figure 4.16. Extensional fall of salt diapirs. Stretching of the cover above salt initially leads to reactive diapirism, then passive diapirism as the salt reaches the surface. However, once the supply of salt from the source layer slows enough, further extension causes the diapirs to widen and fall. Depocenters form above the diapirs, and these subside as the diapirs are rifted apart. Eventually the depocenter may sink until its base welds against the base of autochthonous salt. Further subsidence of the depocenter flanks can cause the structure to invert, forming a mock-turtle anticline. Mock turtles are distinguished from standard turtles by their absence of oldest units in the core of the anticline. The age range of missing stratigraphy indicates when the original diapir was rising. Modified from Vendeville and Jackson (1992b).

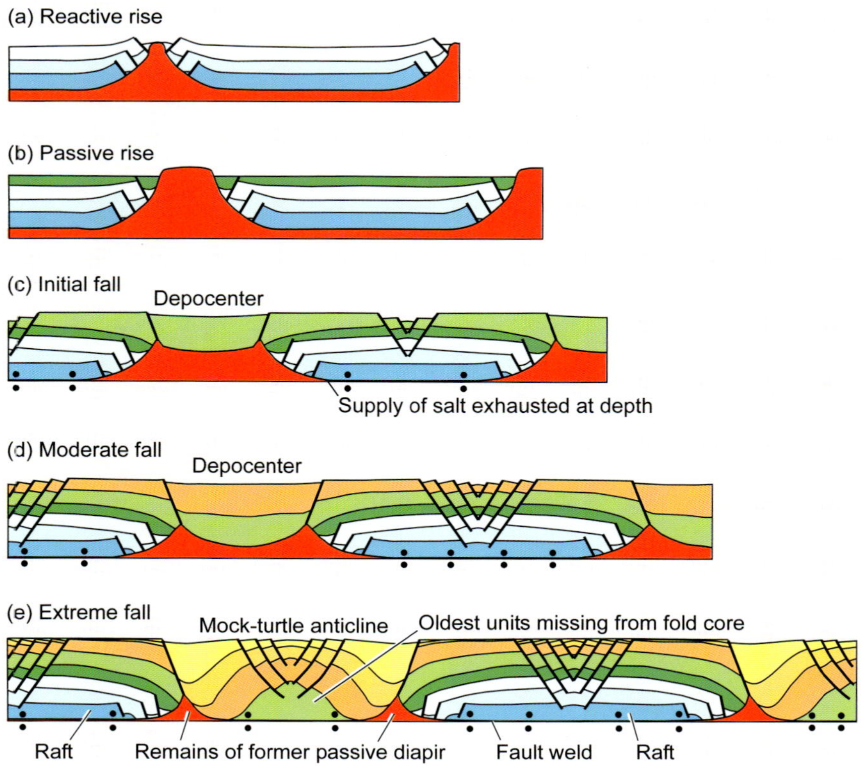

Conceptual Model—Rotation of Rafted Blocks during Extension

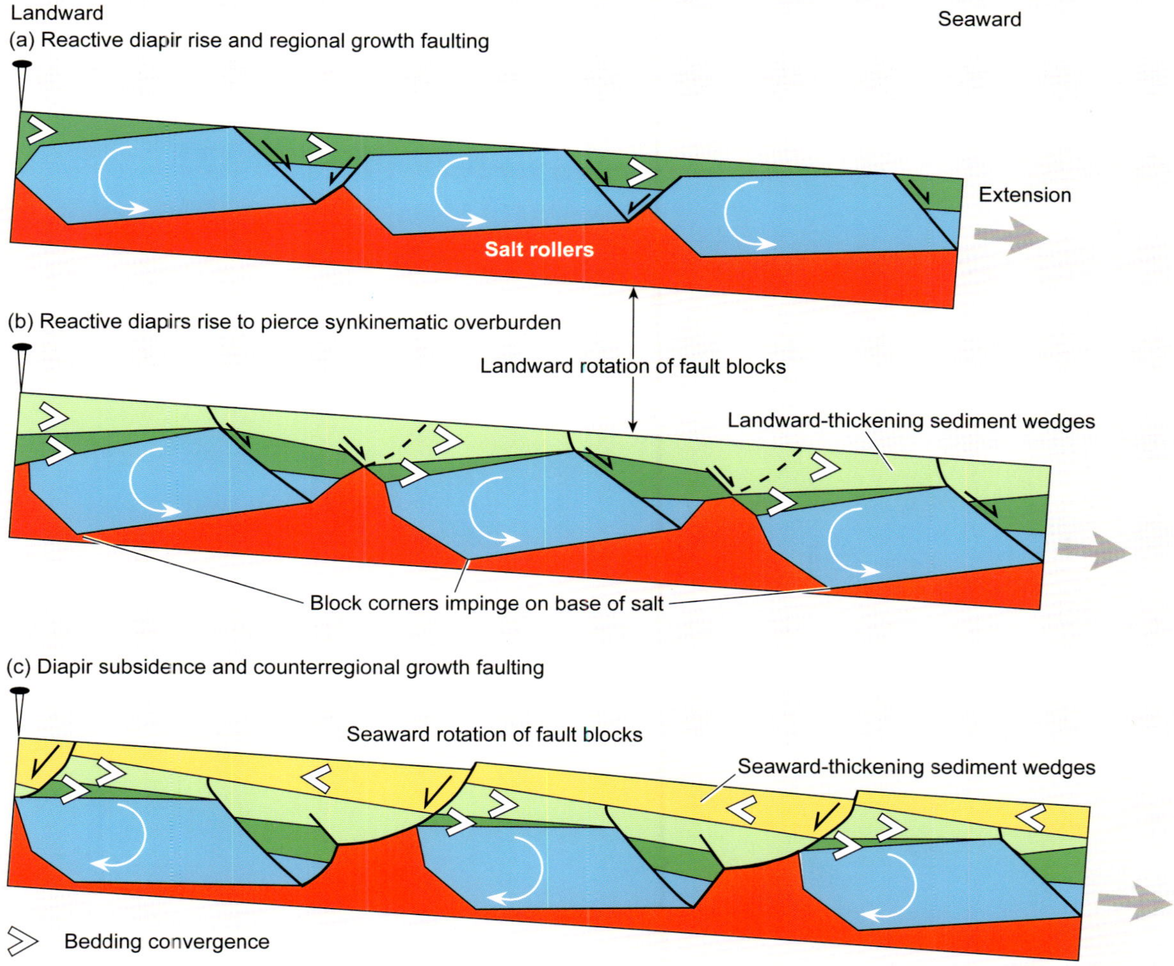

Figure 4.17. Raft tectonics. During slip of seaward-dipping normal faults, fault blocks rotate landward. This rotation continues until the lower corners of the fault blocks impinge on the base of salt, at which point they become grounded. After grounding, new landward-dipping (counterregional) faults may form, reversing the sense of rotation on the fault blocks. Each episode of rotation is recorded by wedges of sediment, tapering in opposite directions. Modified from Vendeville and Jackson (1992b).

Conceptual Model—Along-Strike Flow of Salt during Regional Extension

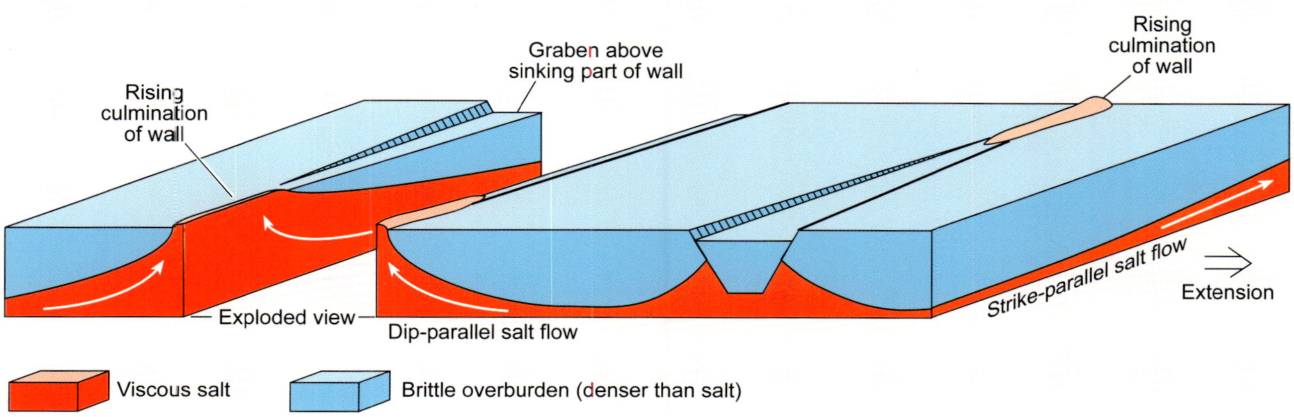

Figure 4.18. Extensional fall of a salt diapir. Once part of a salt wall begins to fall, subsidence may be enhanced by along-strike flow of salt from beneath the subsiding graben and toward parts of the wall that are still rising. This effect is increased if the overburden filling the graben has an average density greater than that of the salt. Along-strike flow may keep parts of the wall emergent even after the salt source layer is exhausted. Modified from Vendeville and Jackson (1992b).

Chapter 5

Extensional Salt Detachments

On most passive margins lacking salt, gravity-driven deformation is secondary. However, deformation intensifies greatly above salt because its weakness makes it an ideal detachment surface to lubricate the deformation of overlying sediments. As a result, many salt-involved passive margins have linked kinematic systems accommodating tens of kilometers of seaward translation.

Extensional salt detachments are present in nearly all salt-tectonic basins and are typically associated with extensional diapirs (Chapters 3 and 4). Structures presented in this chapter are those in which the diapirs are secondary to salt-detached normal faults.

Extensional salt detachments form where normal faults in the suprasalt section flatten into the salt layer. The absence of significant diapirs suggests an initially thin salt layer. However, the salt typically thins further by stretching and downdip flow during regional extension. Extensional salt detachments having large displacements are typically welded. Initially, extensional salt detachments consist of salt-detached normal faults, most of which typically dip basinward. Continued offset may eventually result in raft tectonics, in which fault hanging walls separate completely from their footwalls along the detachment as extension pulls the fault blocks apart.

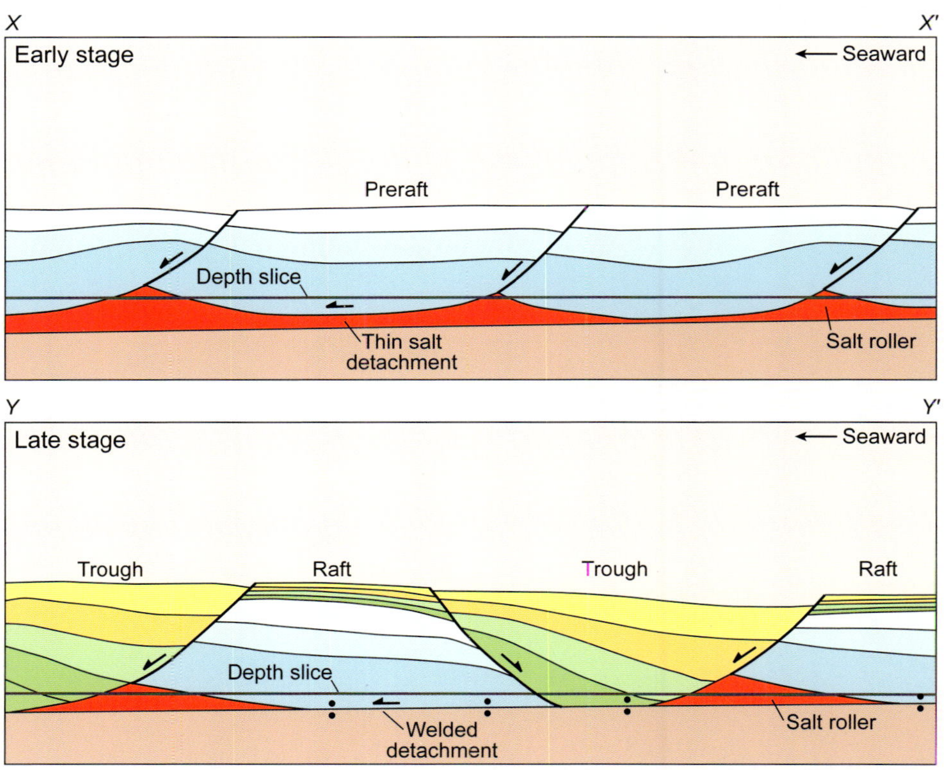

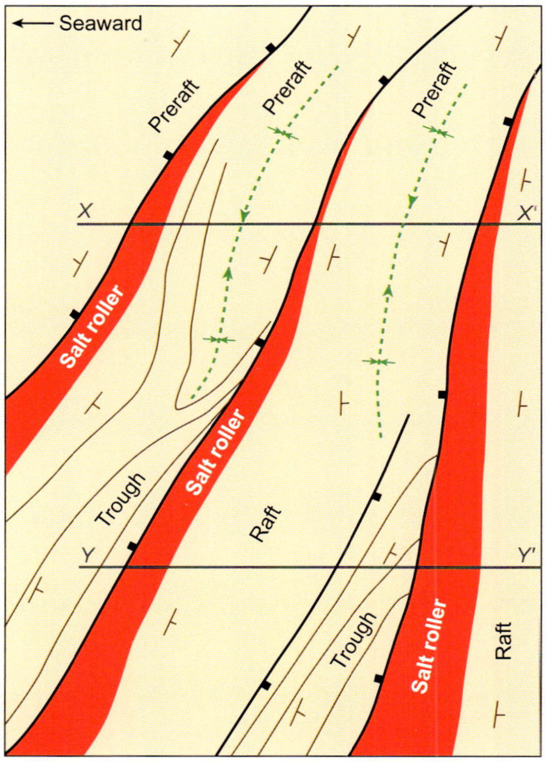

CAMPOS BASIN—Evidence of Radial Extension, Offshore Brazil

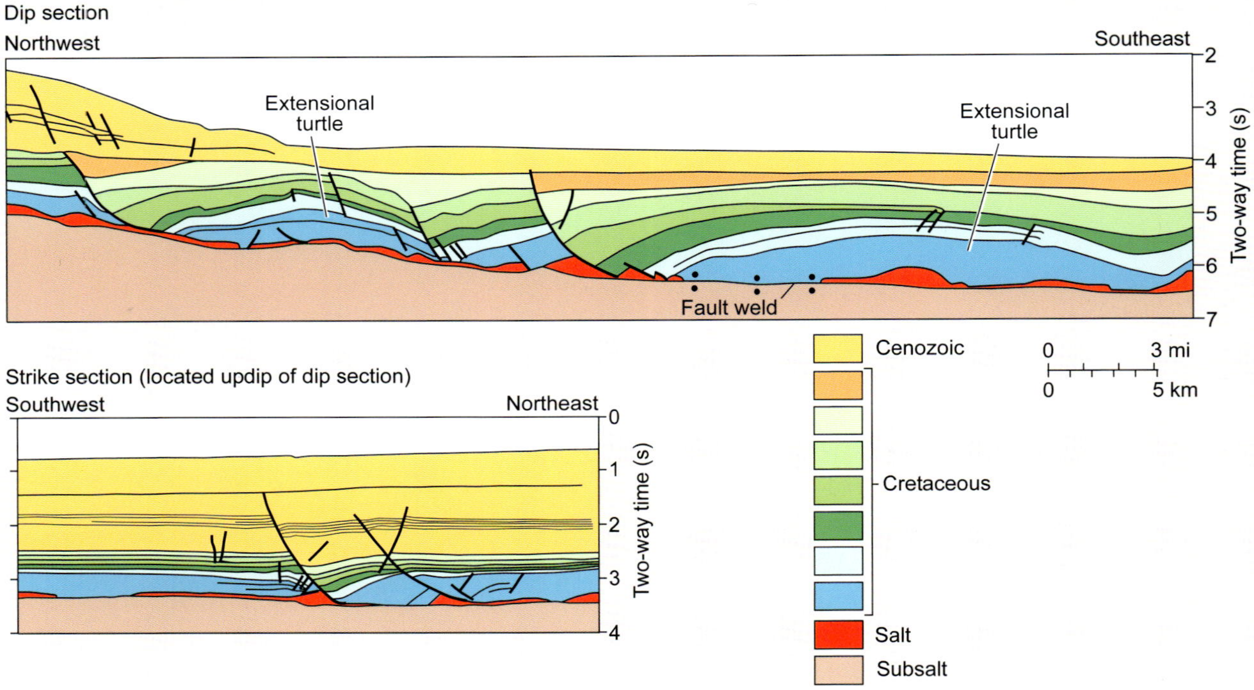

Figure 5.01. Extension above a thin-salt detachment. Bathymetric contours form a salient in the Campos Basin, leading to radial spreading. Because extension was three-dimensional, seismic sections show evidence of extension in both strike and dip directions. In the dip direction, spreading created several extensional turtle structures. Modified from Cobbold and Szatmari (1991).

EAST TEXAS BASIN—Extensional Breakaway along the Mexia-Talco Fault Zone, Texas, USA

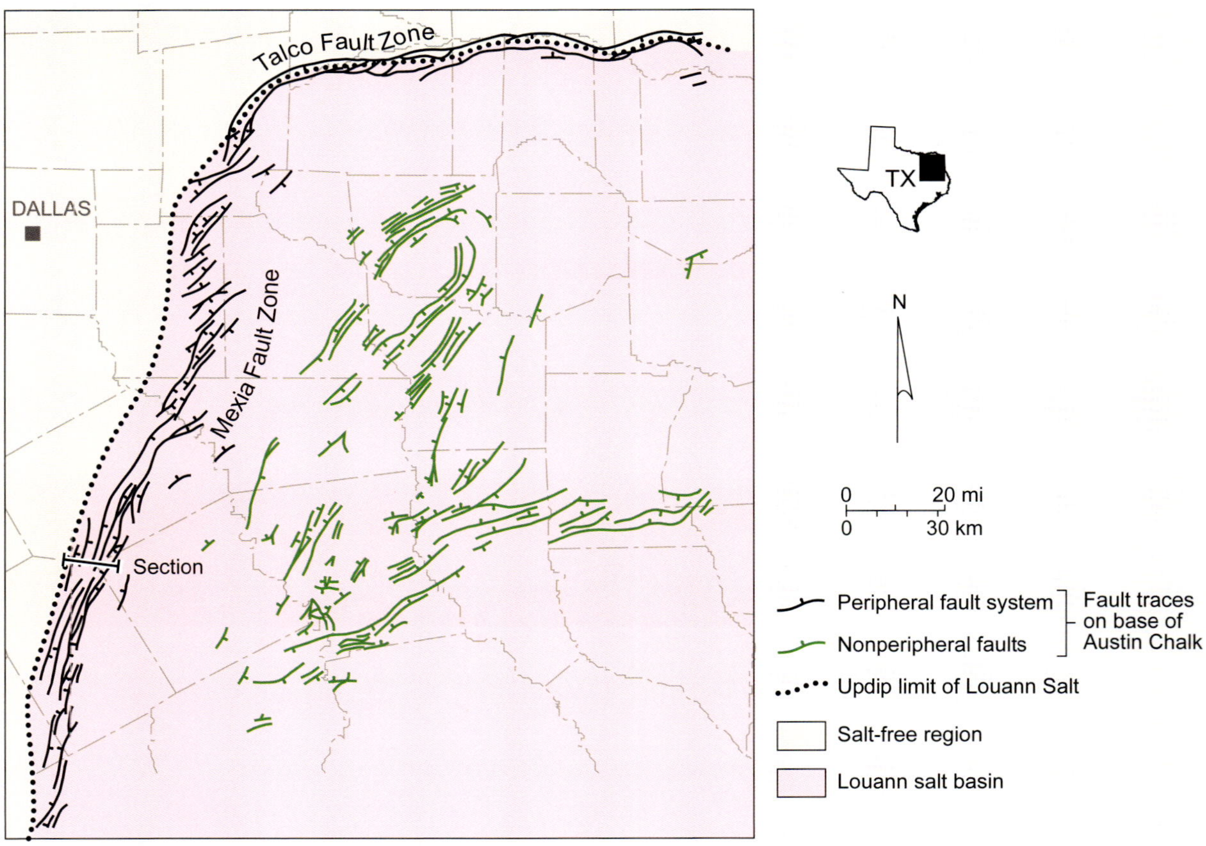

Figure 5.02. Extension above a thin-salt detachment. A peripheral fault system separates strata that moved along a salt detachment from stationary updip strata that did not overlie the detachment. In Texas, the Mexia-Talco Fault Zone lies along the edge of the Louann salt basin and forms the breakaway for the northern Gulf of Mexico gravity-spreading system (see section of the Mexia-Talco Fault Zone in Figure 5.03). Modified from Jackson (1982).

EAST TEXAS BASIN—Extensional Breakaway along the Mexia Fault Zone, Texas, USA

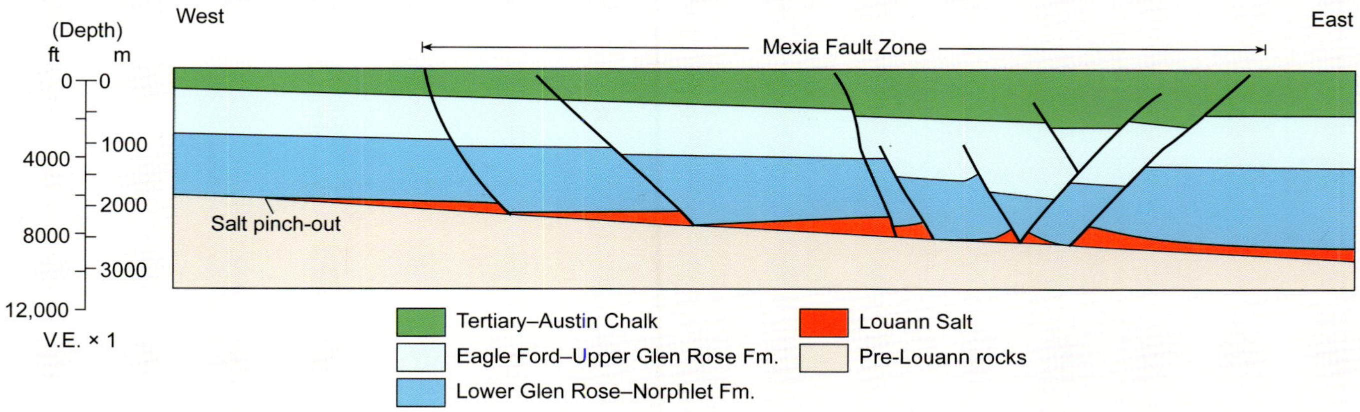

Figure 5.03. Extension above a thin-salt detachment. The Mexia Fault Zone is the extensional breakaway at the updip limit of the Louann salt basin (see map in Figure 5.02). Sediments seaward of the salt pinch-out were transported seaward above the salt detachment. Sediments landward of the salt pinch-out remained in place. Modified from Jackson (1982).

GULF OF MEXICO BASIN—Formation of the Wilcox and Queen City Depotroughs above an Extensional Detachment, Texas, USA

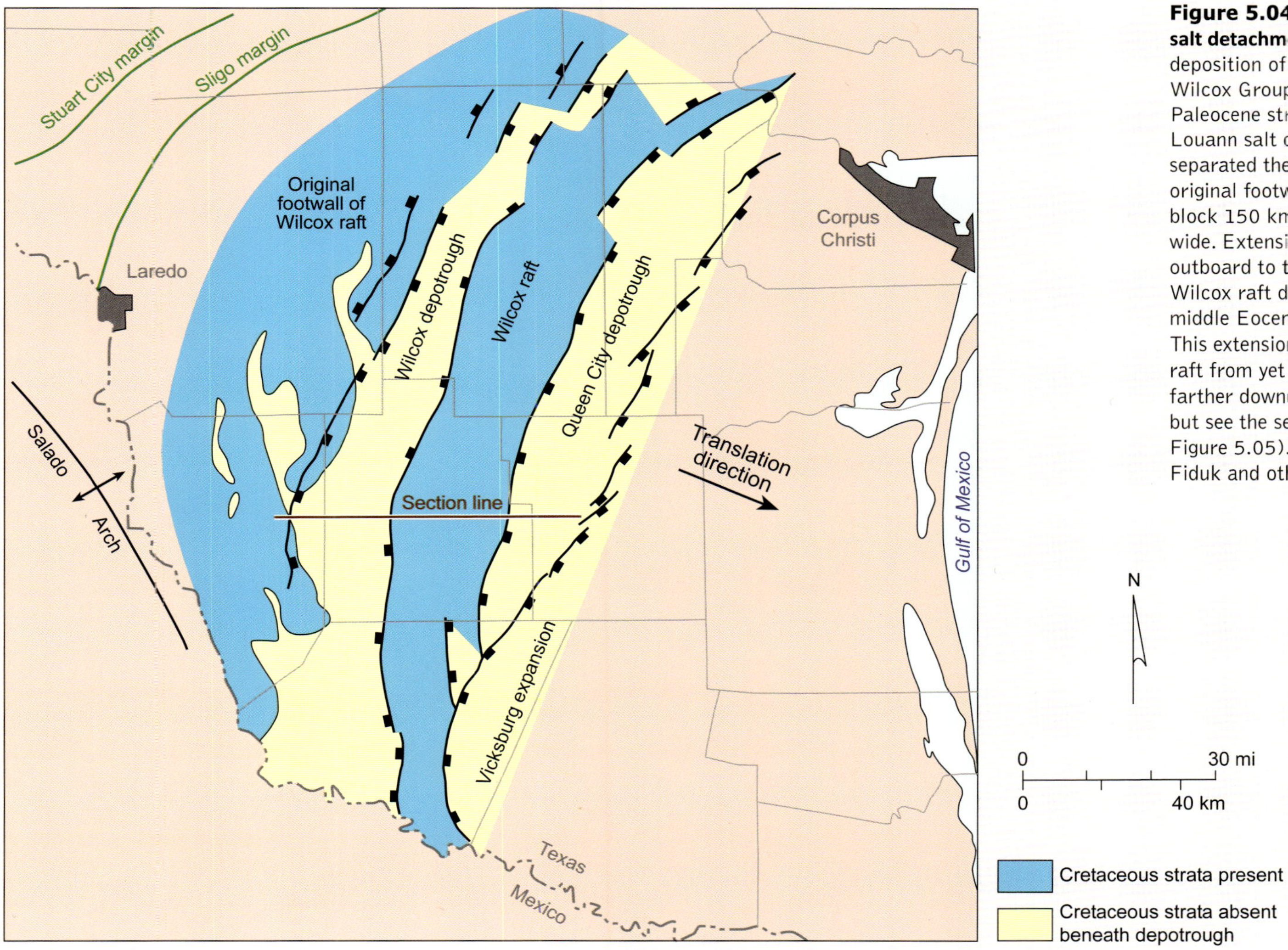

Figure 5.04. Extension above a salt detachment. Extension during deposition of the early Eocene Wilcox Group transported Jurassic–Paleocene strata seaward along the Louann salt detachment. Translation separated these units from their original footwall, forming a raft block 150 km long and 15 to 30 km wide. Extension then jumped outboard to the eastern flank of the Wilcox raft during deposition of the middle Eocene Queen City Group. This extension separated the Wilcox raft from yet another raft block farther downdip (not shown on map, but see the seismic section in Figure 5.05). Modified from Fiduk and others (2004).

Gulf of Mexico Basin—Wilcox and Queen City Depotroughs above an Extensional Detachment, Texas, USA

Figure 5.05. Translation above a salt detachment. A raft of Jurassic–Cretaceous sediments (the Wilcox raft) separates the Wilcox depotrough and Queen City depotroughs (see map in Figure 5.04). In one interpretation (Figure 5.06), the Wilcox depotrough formed as an extensional pull-apart. Alternatively, the structure may have formed dominantly through salt expulsion. The absence of thickness variations in the Wilcox raft suggests that the area was never underlain by much salt, supporting the pull-apart model. Modified from Fiduk and others (2004).

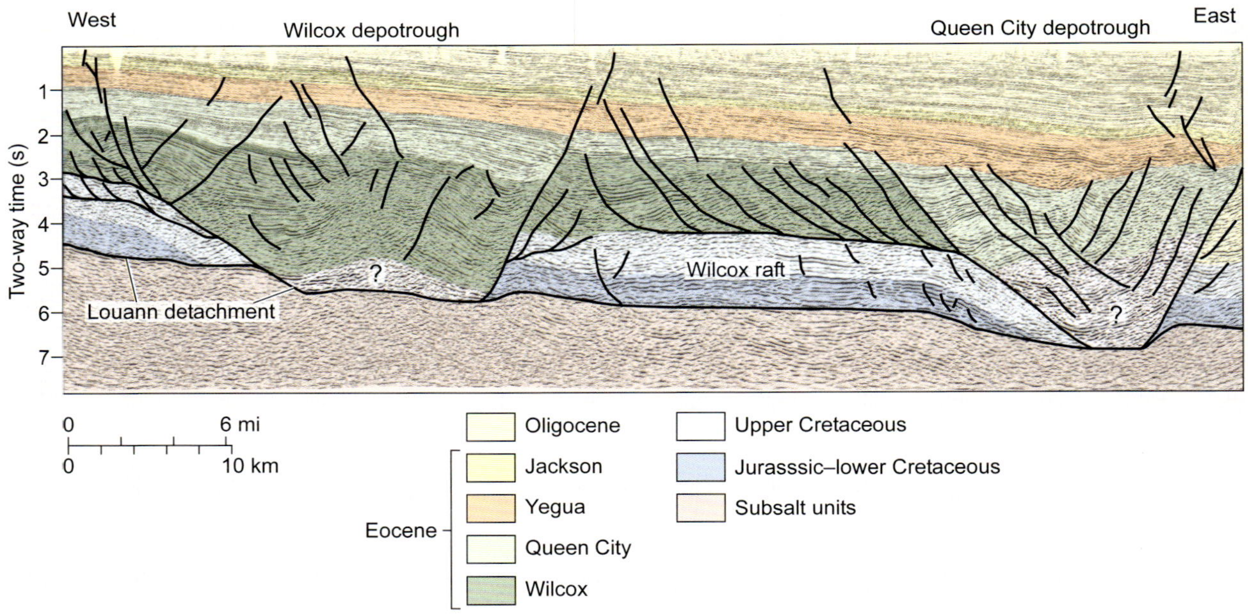

Gulf of Mexico Basin—Schematic Forward Model of Formation of Wilcox Depotrough, Texas, USA

Figure 5.06. Extension above a salt detachment. This model shows formation of the Wilcox depotrough without significant salt flow. Slip on a seaward-dipping listric fault initiated an asymmetric trough bounded on its eastern flank by a landward-dipping rollover monocline. A second, antithetic fault then formed as a bedding-plane detachment on the dipping limb of this monocline. Simultaneous slip on these two faults produced the final, roughly symmetric depotrough. A similar structure has been modeled along strike, where the dip directions of the initial fault and later bedding plane detachment are reversed. A transfer zone must lie between these two sections. See a nearby seismic section (Figure 5.05). Modified from Fiduk and others (2004).

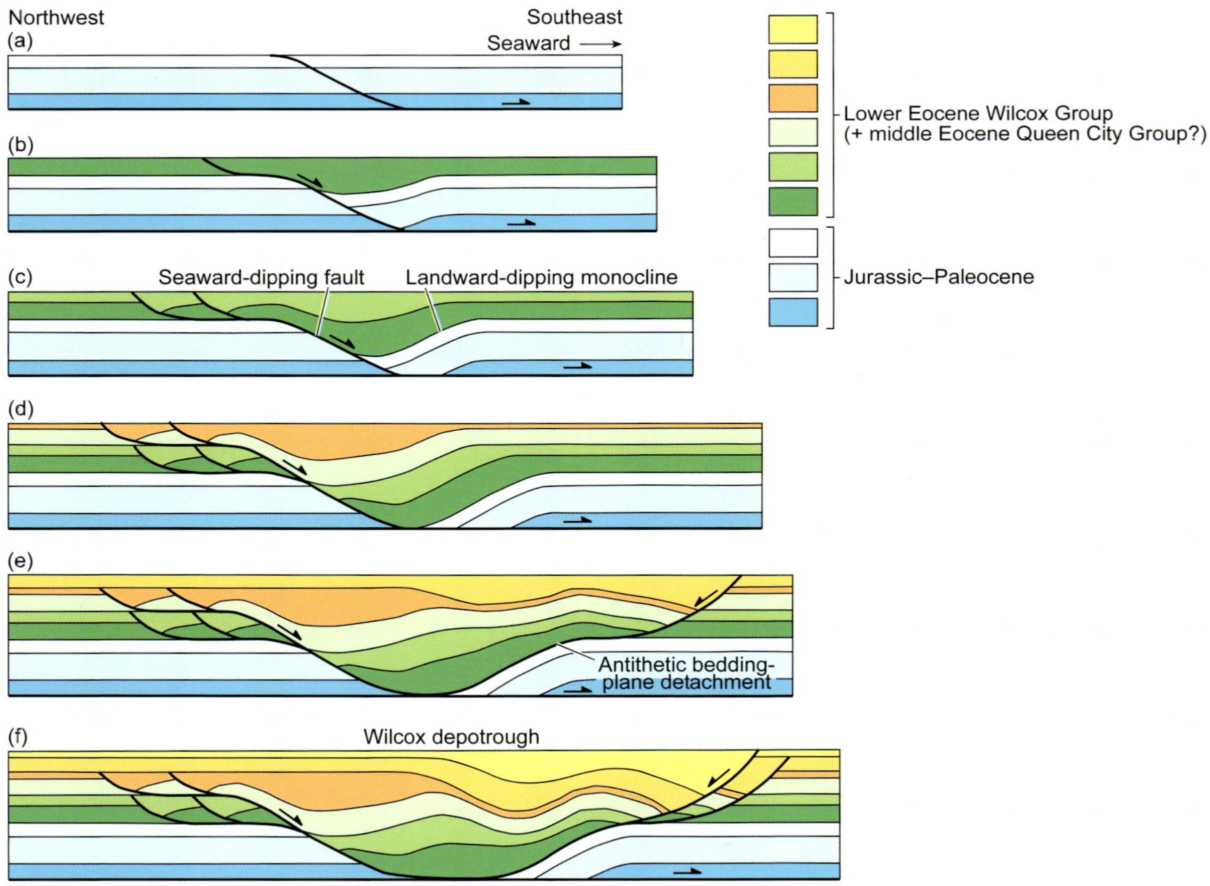

KWANZA BASIN—Stratigraphic Effects of Translation of Salt Diapirs down a Detachment Ramp, Offshore Angola

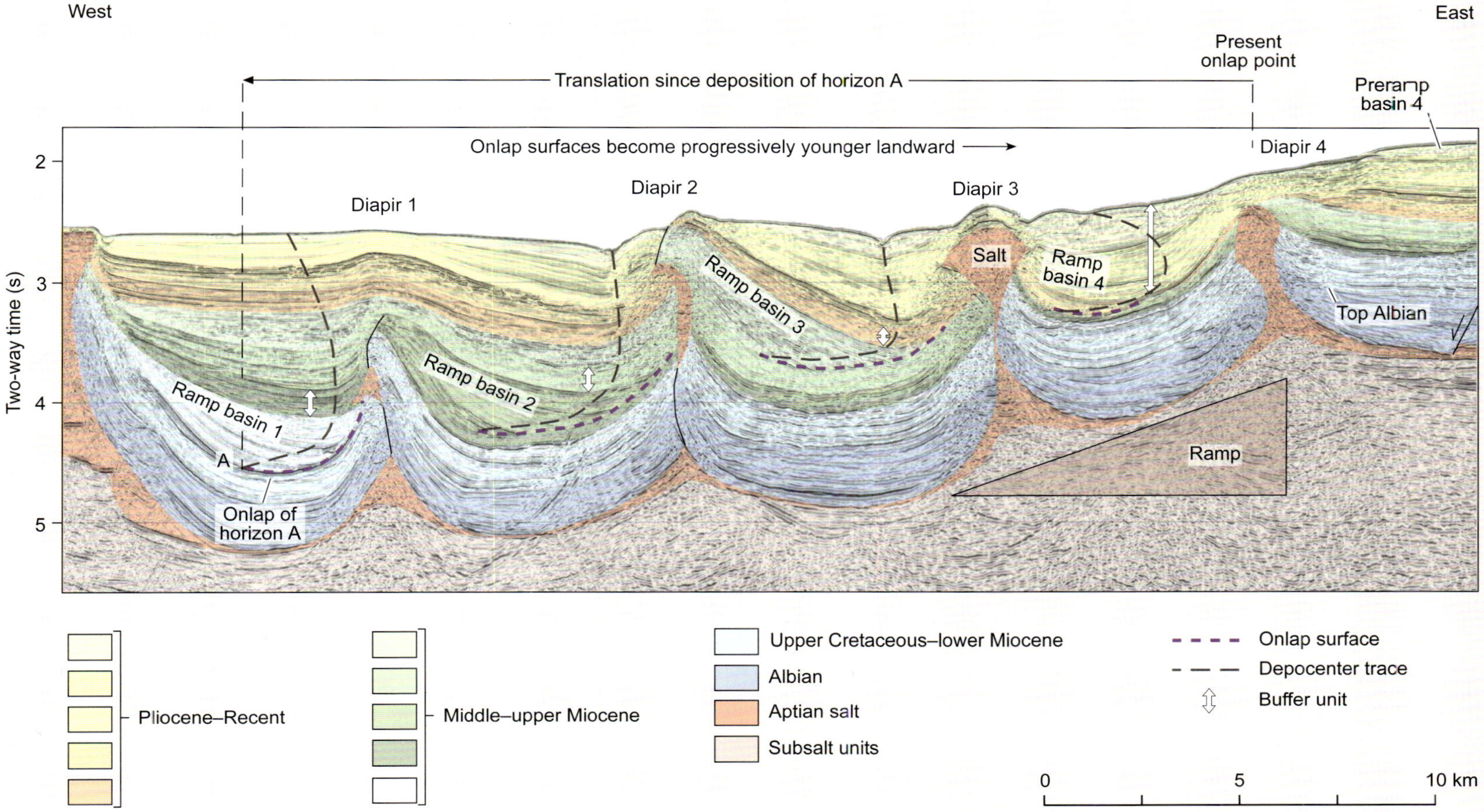

Figure 5.07. Translation above a thin-salt detachment. Diapirs resist being translated down detachment ramps because of the deep ramp basins on their downdip flanks. Thus, a diapir tends to stall and shorten once it reaches the top of a ramp (see schematic model, Figure 5.09). During this period, units updip of the ramp still translate seaward, but units downdip of the ramp stop moving. Translation resumes once the diapir is completely shortened and continues until another diapir reaches the ramp crest. In this example, three diapirs have translated down the ramp, and a fourth is currently shortening at the top of the ramp (see restoration in Figure 5.08). Each diapir has a ramp basin downdip, which formed while the diapir was approaching or at the ramp crest. The ramp basin is capped by a buffer unit, which records when the diapir was actively shortening at the top of the ramp. Modified from Jackson and Hudec (2005).

KWANZA BASIN—Stratigraphic Effects of Translation of Salt Diapirs down a Detachment Ramp with a Bathymetric Scarp

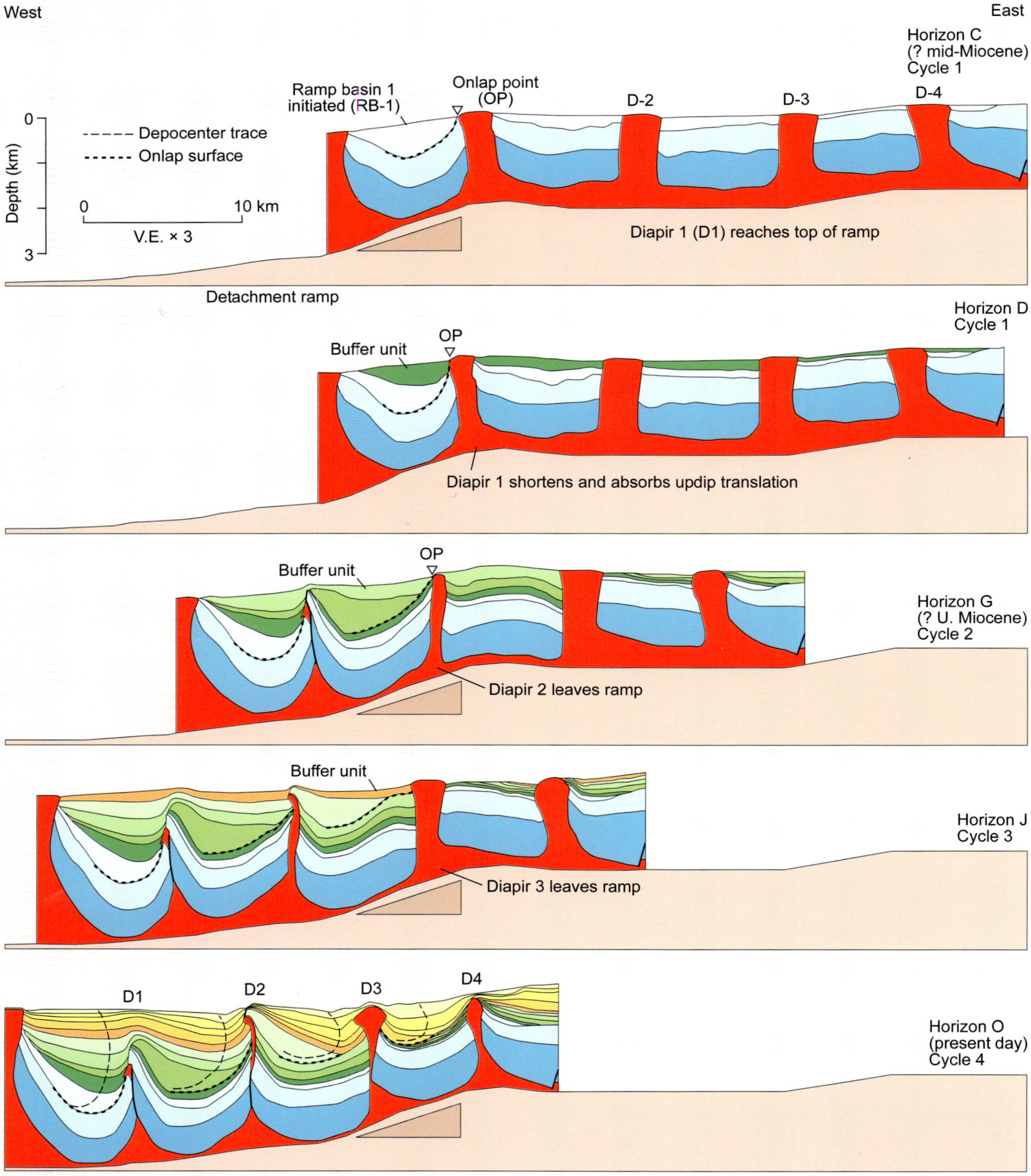

Figure 5.08. Translation above a thin-salt detachment. During seaward translation detached on salt, four diapirs were transported to the top of a detachment ramp. Once there, each diapir shortened, leading to deposition of a buffer unit that did not onlap in the ramp basin (see schematic model, Figure 5.09). Three diapirs were shunted down the ramp after they had been highly shortened. A ramp basin bounded by an onlap surface formed each time translation resumed. At the end of each translation phase, the next diapir reached the top of the ramp, and the cycle repeated. This cyclicity produced four consecutive ramp basins, becoming progressively younger landward. See seismic section in Figure 5.07. Modified from Jackson and Hudec (2005).

KWANZA BASIN—Stratigraphic Effects of Translation of Salt Diapirs down a Detachment Ramp, Offshore Angola

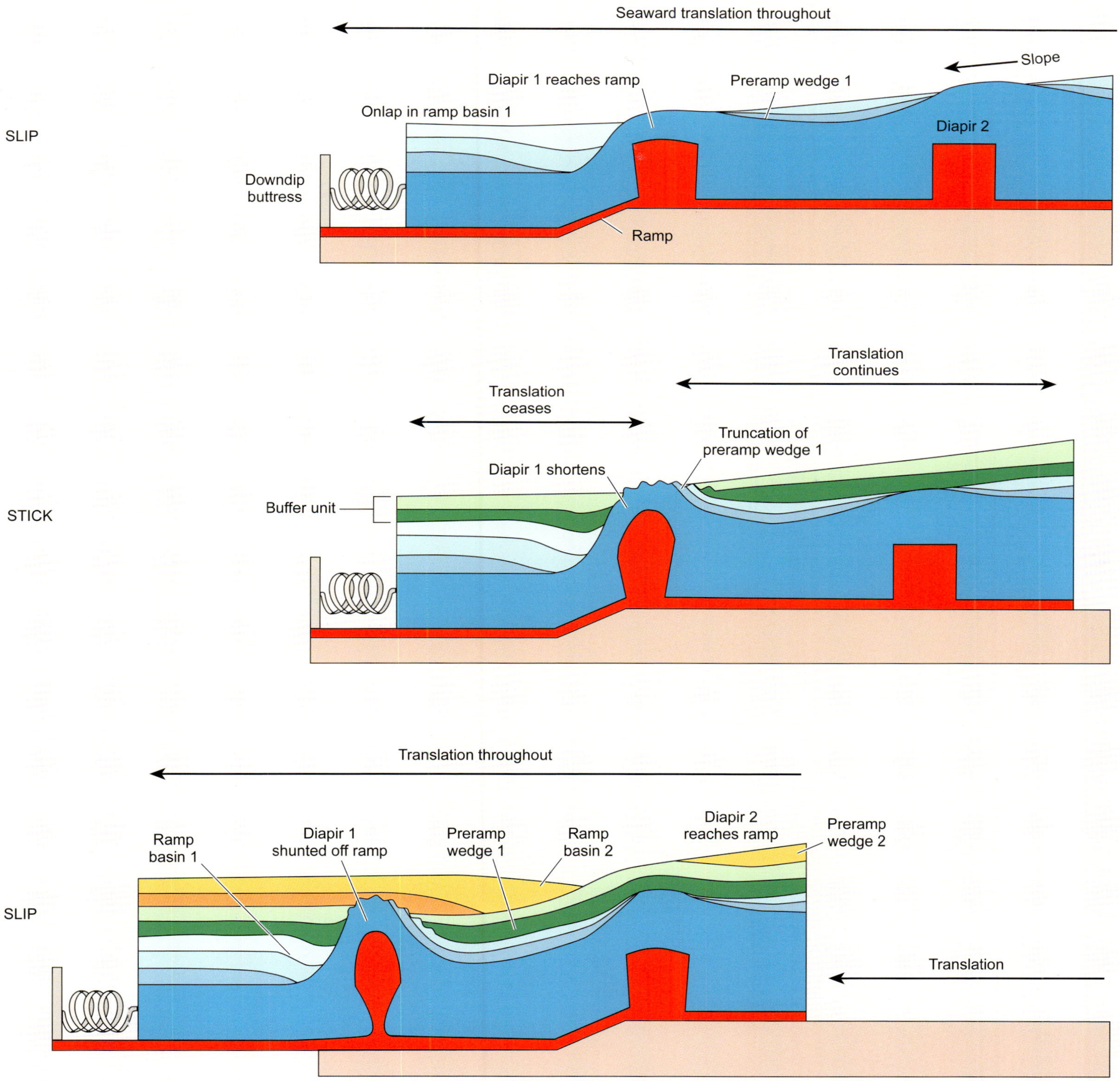

Figure 5.09. Translation above a thin-salt detachment. Translation of sediments down a detachment ramp creates a ramp basin. Continued translation forms a shingled stack of layers above the detachment as older layers move seaward. Diapirs resist being translated down ramps because of the thick ramp basin on their seaward side. Thus, diapirs tend to stall and shorten when they reach the top of a ramp. During this period, units updip of the ramp still translate seaward, but units downdip of the ramp stop moving. Translation resumes once the diapir is completely shortened and continues until another diapir reaches the ramp crest, after which the cycle repeats. These "stick-slip" cycles of diapirs at ramps produce a stratigraphic sequence that reflects the translation history. See a seismic example in Figure 5.07, and its restoration in Figure 5.08. Modified from Jackson and Hudec (2005).

Lower Congo Basin—Raft Tectonics, Basin Block 3, Offshore Angola

Figure 5.10. Extension above a thin-salt detachment. Extension of the Albian ramp margin translated Albian raft blocks down the Loeme salt detachment. Gaps between Albian fault blocks increase downdip, a result of progressively greater extension above a seaward-steepening detachment. This restoration suggests 7 to 10 km of extension in the eastern part of Block 3. Gaps between rafts in the restored map may indicate the presence of small diapirs, or they may be due to distortion of rafts during extension. Modified from Eichenseer and others (1999).

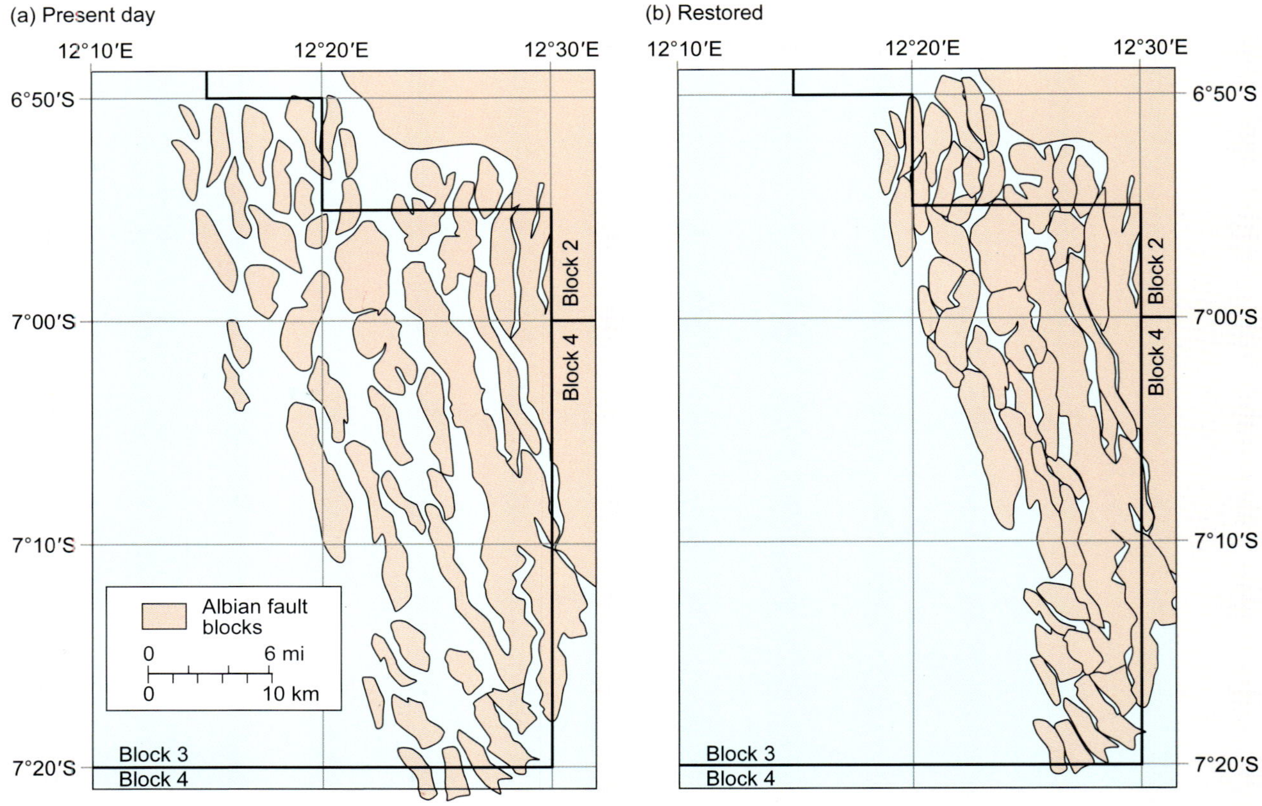

Lower Congo Basin—Detached Extensional Tectonics Southern Profile, Offshore Angola

Figure 5.11. Extension above a thin-salt detachment. Extensional faults become younger downdip, from Albian in the east to Paleocene–Eocene in the west. This progression is probably due to a combination of (a) westward progradation of the shelf edge, (b) thinner salt and greater shear resistance beneath eastern fault blocks, and (c) yoking of eastern fault blocks beneath thick overburden. Extension was reactivated by Oligo–Miocene progradation of the Congo fan into the area. The breakaway fault for this younger phase of extension overlies a flexure in the salt detachment above a presalt basement fault. Northward, this breakaway becomes the eastern edge of the Eastern Graben system. Modified from Anderson and others (2000).

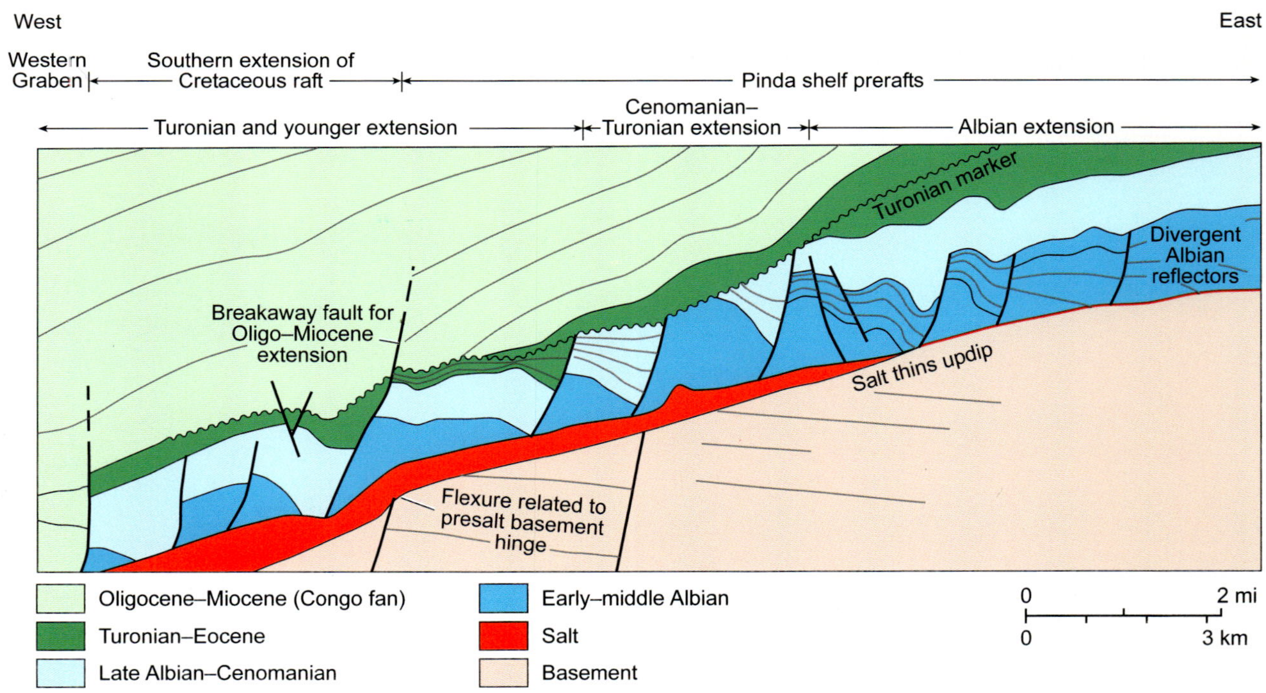

LOWER CONGO BASIN—Raft Tectonics, Offshore Angola

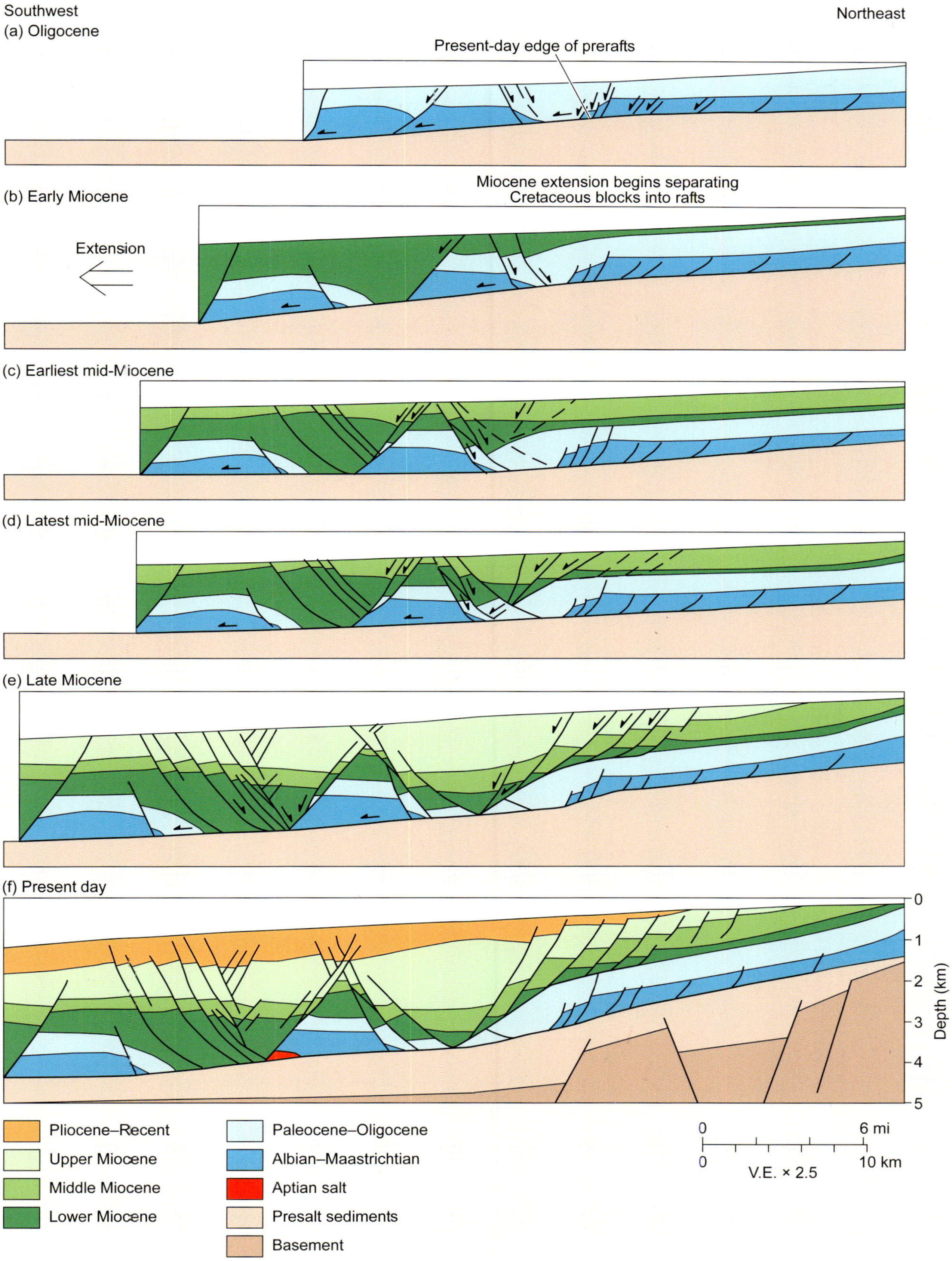

Figure 5.12. Extension above a thin-salt detachment. 17 km of Miocene extension separated Cretaceous units into rafts between Miocene troughs. This restoration suggests that there was only enough salt on the detachment to act as a lubricant. An alternative interpretation would place small diapirs between the rafts at the end of the Cretaceous. These diapirs would have localized Miocene extension and be preserved only as small triangles of salt in the footwalls of graben-bounding faults. Modified from Anderson and others (2000).

NORWEGIAN SEA—Extensional Drape above a Basement Fault, South End of the Nordland Ridge, Offshore Norway

Figure 5.13. Basement extension below salt. Offset on the Revfallet Fault during the Aptian–Cenomanian was partly absorbed by flow of Zechstein salt. Overburden strata were draped across the fault-block corner. The basement fault did not have enough throw to propagate upward into the overburden here, unlike along strike. The overburden was extended by a large detached fault system 15 km southeast of the Revfallet Fault. Modified from Pascoe and others (1999).

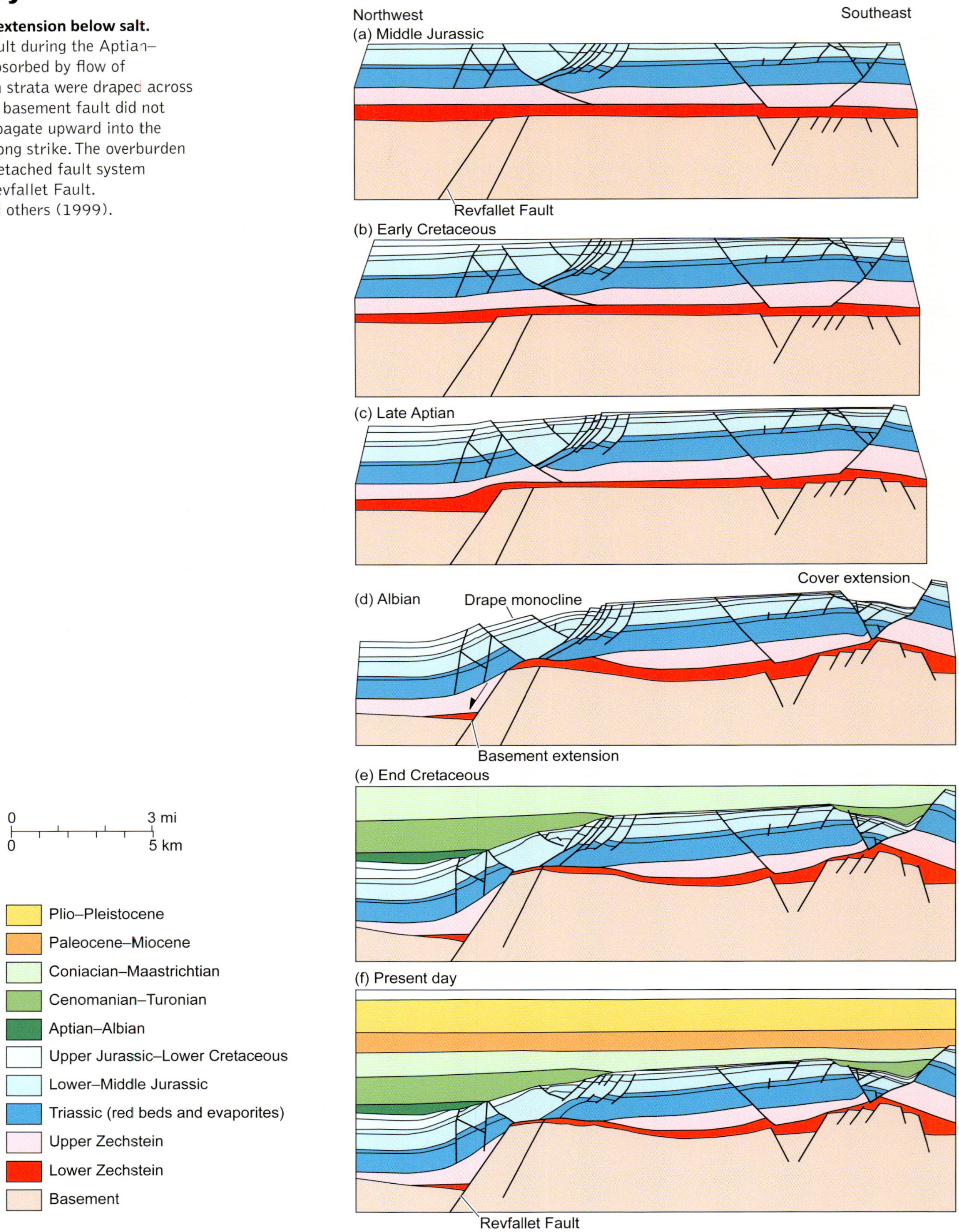

PYRENEES FOLD BELT—Inverted Extensional Structures, Spain

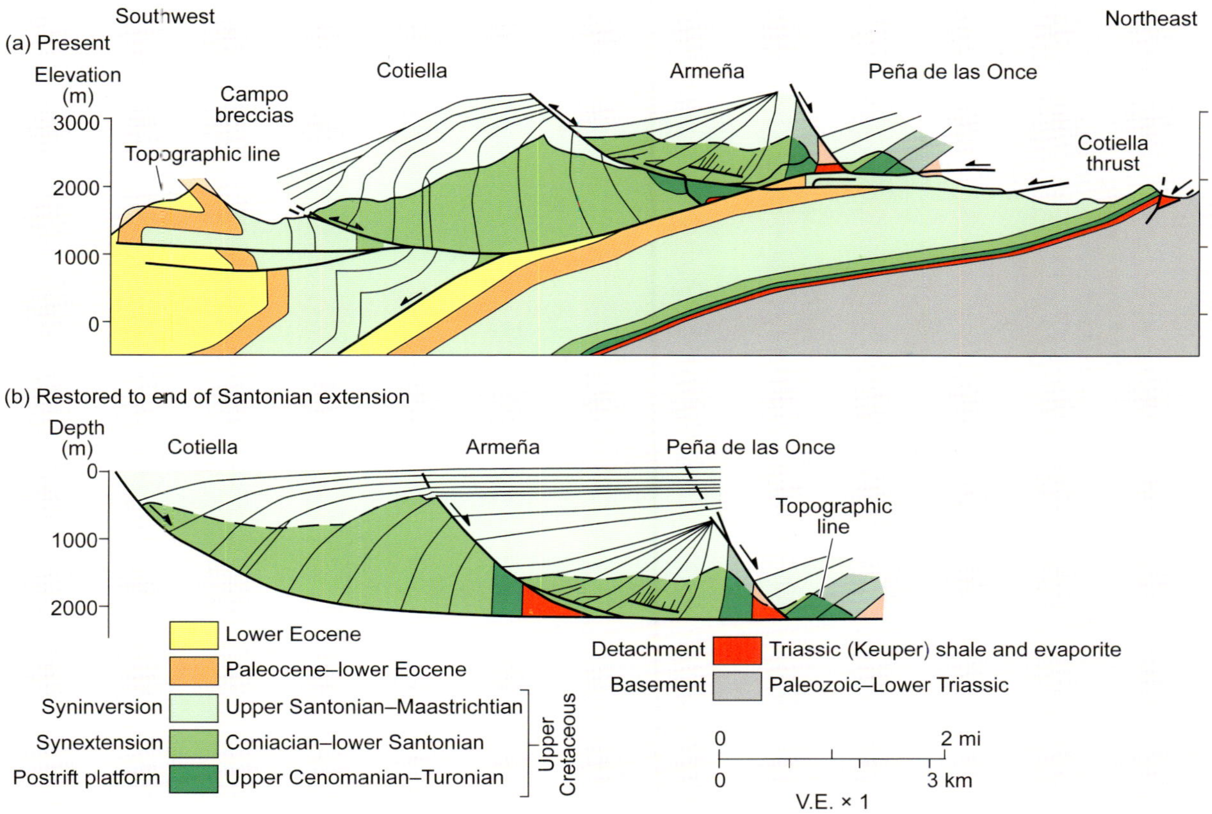

Figure 5.14. Extension above a thin-salt detachment. During Coniacian–early Santonian time, basinward tilting during thermal subsidence caused gravity-driven extension on the southern flank of the Pyrenean trough. At least 9 km of extension that detached on Keuper evaporites is recorded by large, rotated fault blocks with fanning internal stratigraphy. Later Pyrenean shortening reversed slip on some of these faults and thrust the entire complex 20 km southward over Eocene units. Modified from McClay and others (2004).

PHYSICAL MODEL—Relay Zone above an Extensional Detachment

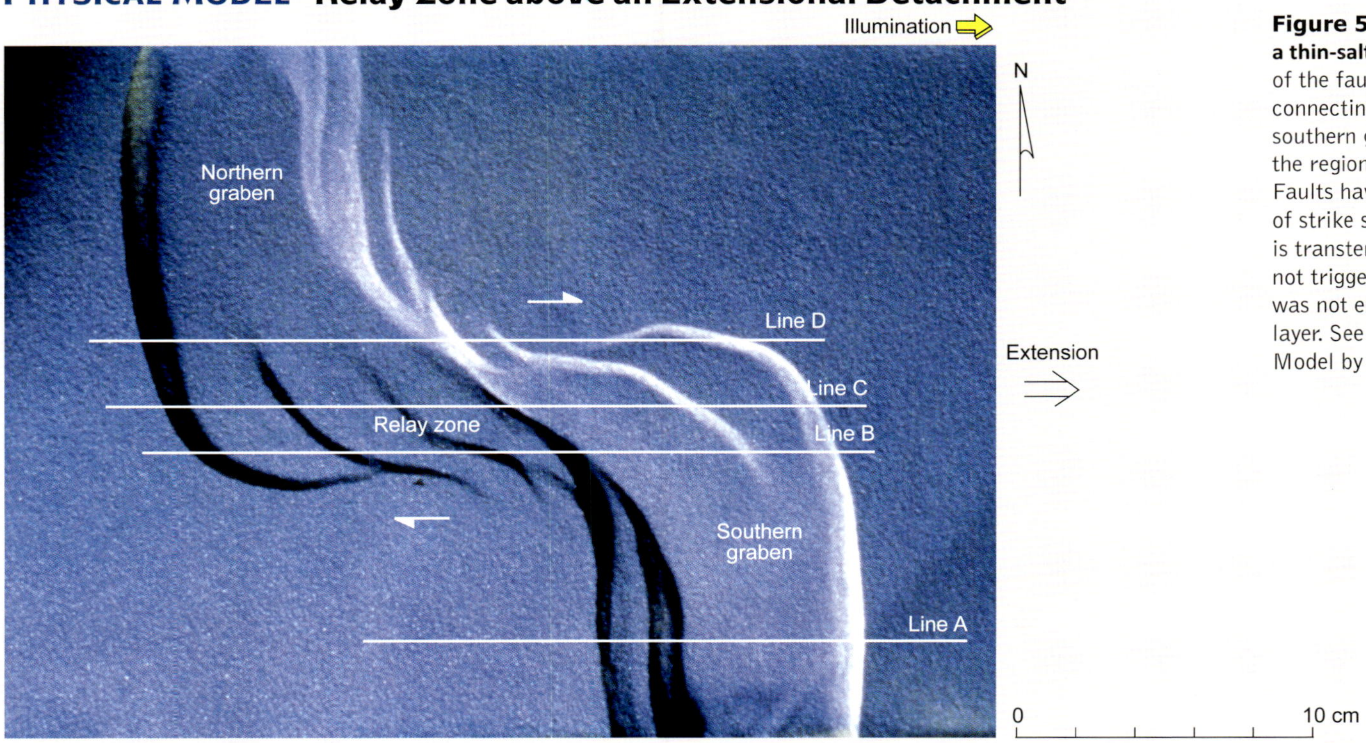

Figure 5.15. Extension above a thin-salt detachment. Most of the faults in the relay zone connecting the northern and southern grabens are oblique to the regional extension direction. Faults have a significant component of strike slip, and the relay zone is transtensional. Extension could not trigger diapirism because there was not enough salt in the source layer. See sections in Figure 5.16. Model by Bruno Vendeville.

PHYSICAL MODEL—Sections across a Relay Zone above an Extensional Detachment

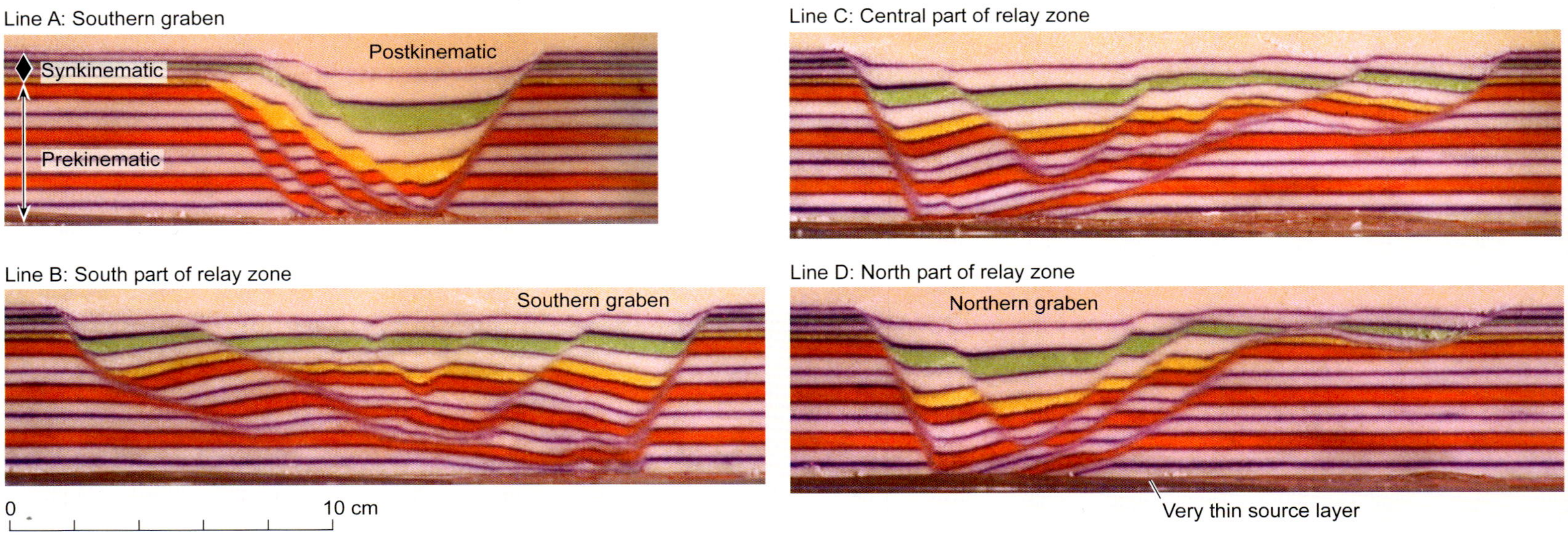

Figure 5.16. Extension above a thin-salt detachment. The northern and southern grabens both comprised a dominant fault matched by an array of smaller antithetic faults. Faults in the relay zone between the grabens have a lower apparent dip because they are cut obliquely to their strike (see map in Figure 5.15). The absence of reactive diapirs in the graben axes is due to insufficient salt in the source layer. Model by Bruno Vendeville.

PHYSICAL MODEL—Radial Spreading of a Curved Passive Margin

Figure 5.17. Extension above a thin-salt detachment. A convex passive margin increases in area as it spreads seaward. Two sets of normal faults accommodate the radial and circumferential extension during this spreading (see conceptual model, Figure 5.18). Expansion of the sedimentary wedge produced a compressional fold belt at the base of the slope. The arcuate trend of this belt mimics the shape of the margin. Modified from Gaullier and Vendeville (2005).

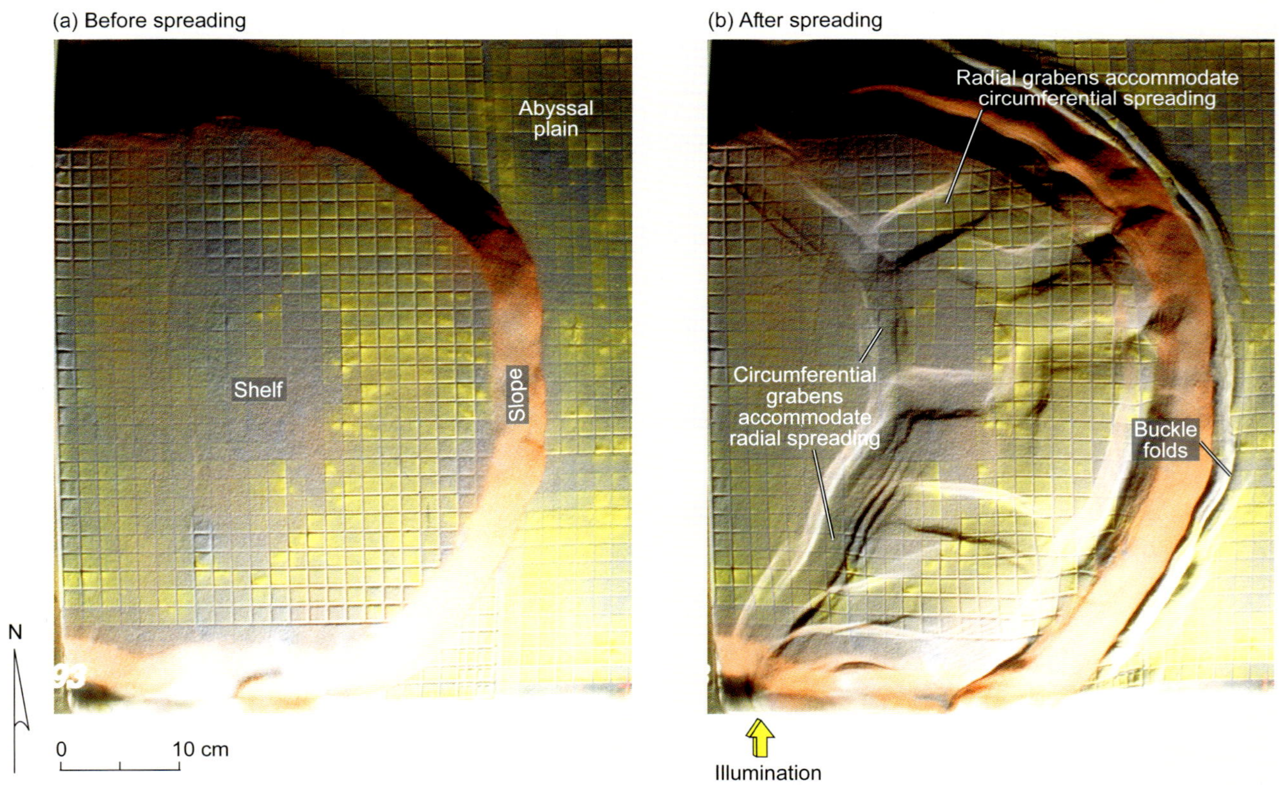

PHYSICAL MODEL—Radial Spreading of a Curved Passive Margin

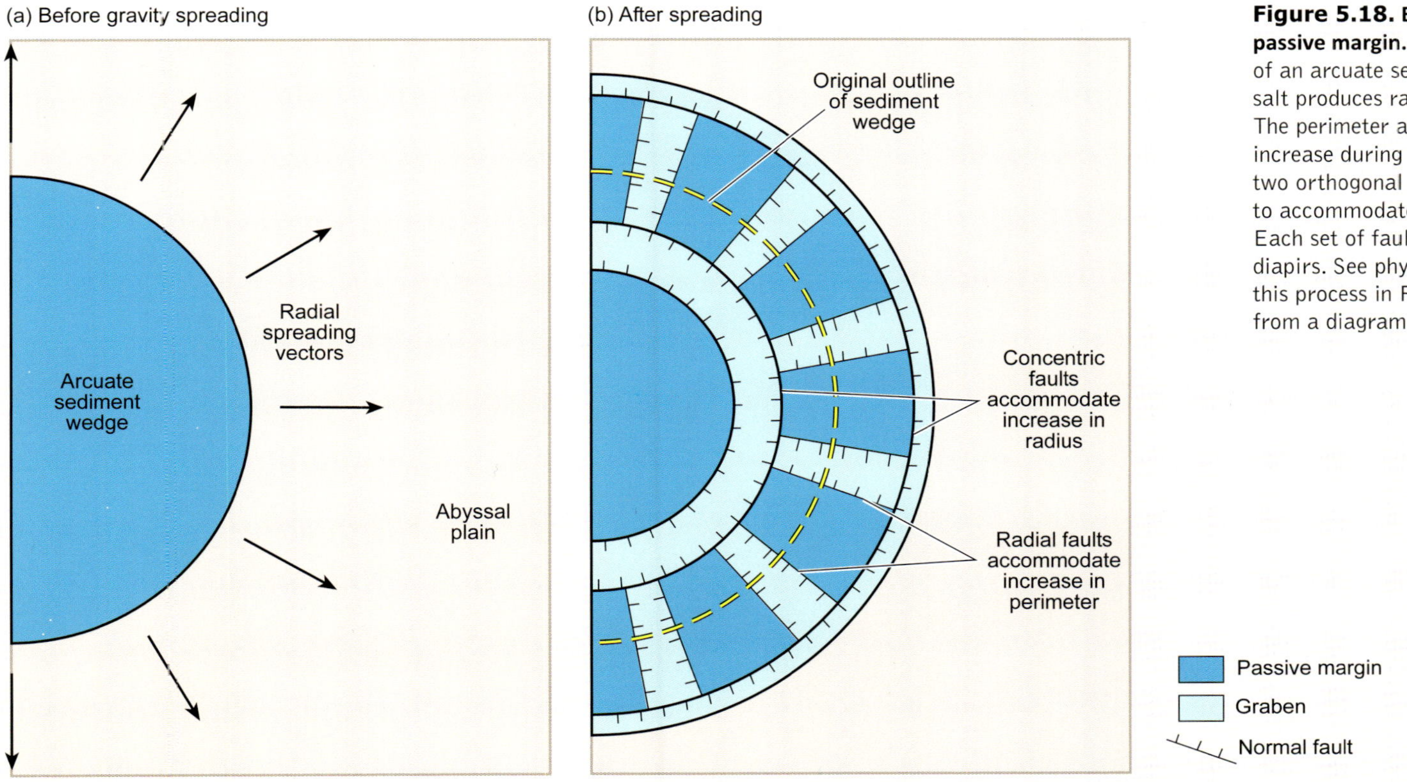

Figure 5.18. Extension on a curved passive margin. (a) Gravity spreading of an arcuate sediment wedge above salt produces radial movement vectors. The perimeter and radius of the wedge increase during spreading, requiring two orthogonal normal fault systems to accommodate the deformation. Each set of faults localizes reactive diapirs. See physical model illustrating this process in Figure 5.17. Modified from a diagram by Bruno Vendeville.

PHYSICAL MODEL—Influence of Basement-Fault Displacement on Extensional Drape

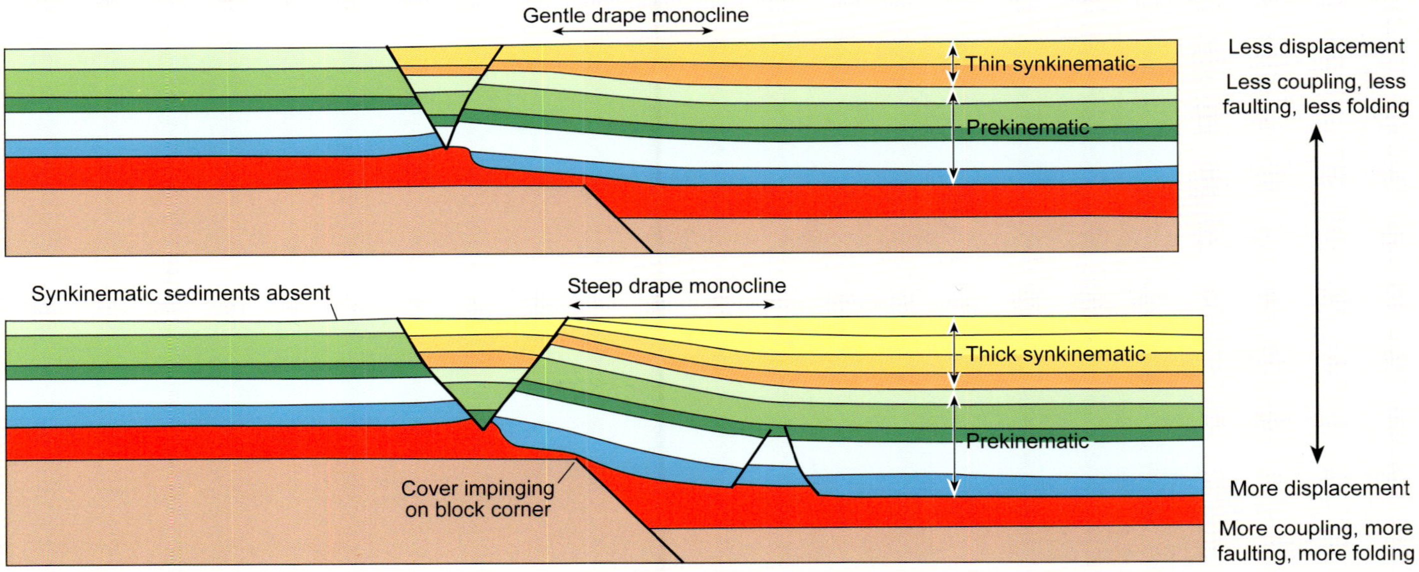

Figure 5.19. Basement extension below salt. Greater basement-fault displacement leads to greater coupling between basement and overburden deformation. Salt can decouple deformation because it flows down an elevation head gradient from the upthrown block to the downthrown block, detaching suprasalt from subsalt deformation and "smoothing out" the fault offset. Two processes hindered this salt flow. First, with greater fault throw (a) there may not have been enough salt to inflate the salt layer on the downthrown side and (b) the drape monocline began to impinge on the upthrown basement block corner, cutting off salt flow between the two blocks. Second, as thick synkinematic sediments accumulated on the downthrown side, the increased overburden thickness increased the pressure of the underlying salt. This increase created a pressure-head gradient toward the upthrown side. The pressure variation countered the elevation-head gradient and discouraged flow of salt into the graben, which would otherwise help to decouple deformation above and below salt. See also influence of salt thickness (Figure 5.20) and displacement rate (Figure 5.21). Modified from Withjack and Callaway (2000).

PHYSICAL MODEL—Influence of Salt Thickness on Extensional Drape

Figure 5.20. Basement extension below salt. Thicker salt layers promote decoupling between basement extension and overburden deformation. Where salt layers are thick, drape monoclines are broad. The salt detachment is efficient, and overburden faults may lie far from the causative basement faults. With decreasing salt thickness, monoclines are narrower and more curved in their hinges. Increasing hinge curvature tends to localize overburden extension nearer the basement fault. Finally, basement faults propagate directly into the overburden if salt is very thin or absent. See also influence of basement-fault displacement (Figure 5.19) and displacement rate (Figure 5.21). Modified from Withjack and Callaway (2000).

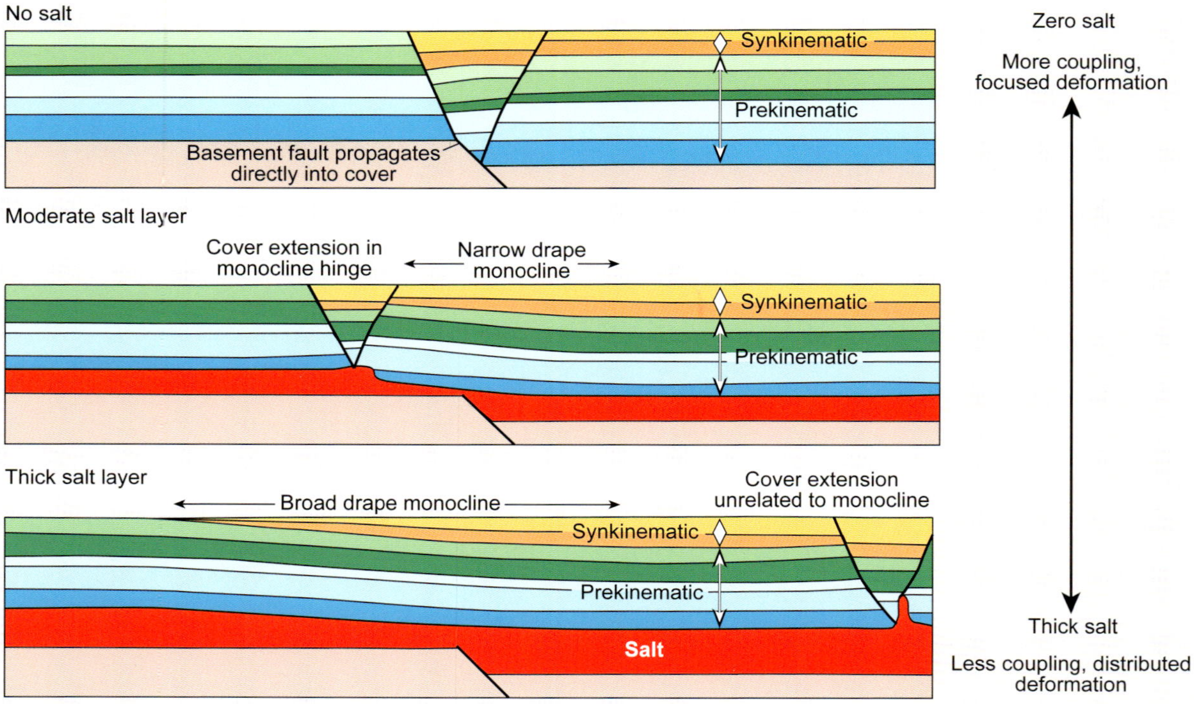

PHYSICAL MODEL—Influence of Displacement Rate on Extensional Drape

Figure 5.21. Basement extension below salt. Salt is able to decouple deformation because it can flow from the upthrown block to the downthrown block, enhancing the salt detachment and "smoothing out" the fault offset. However, if displacement rates are faster than salt flow rates, the salt has no time to flow across the fault plane. With no salt flow, basement and overburden deformation are strongly coupled. In extreme cases, a basement fault can propagate completely through the salt into the overburden. However, even the slowest fault slip (bottom section) lacks any thickening of salt on the downthrown side because thick synkinematic sediments there increased the pressure of the underlying salt. This increase created a pressure-head gradient toward the upthrown side. The pressure variation countered the elevation-head gradient and discouraged flow of salt into the graben, which would otherwise help to decouple deformation above and below salt. See also influence of salt thickness (Figure 3.20) and basement-fault displacement (Figure 3.19). Modified from Withjack and Callaway (2000).

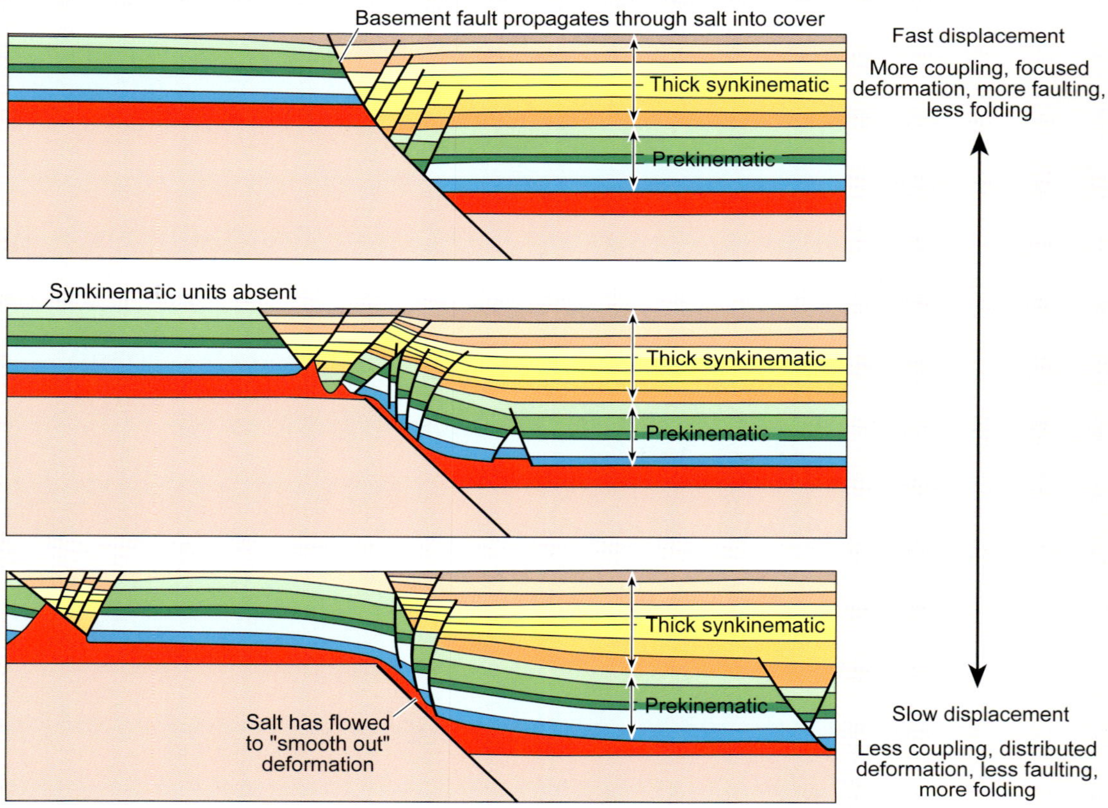

PHYSICAL MODEL—Extension above Transtensional Basement Step Separating Thick and Thin Salt

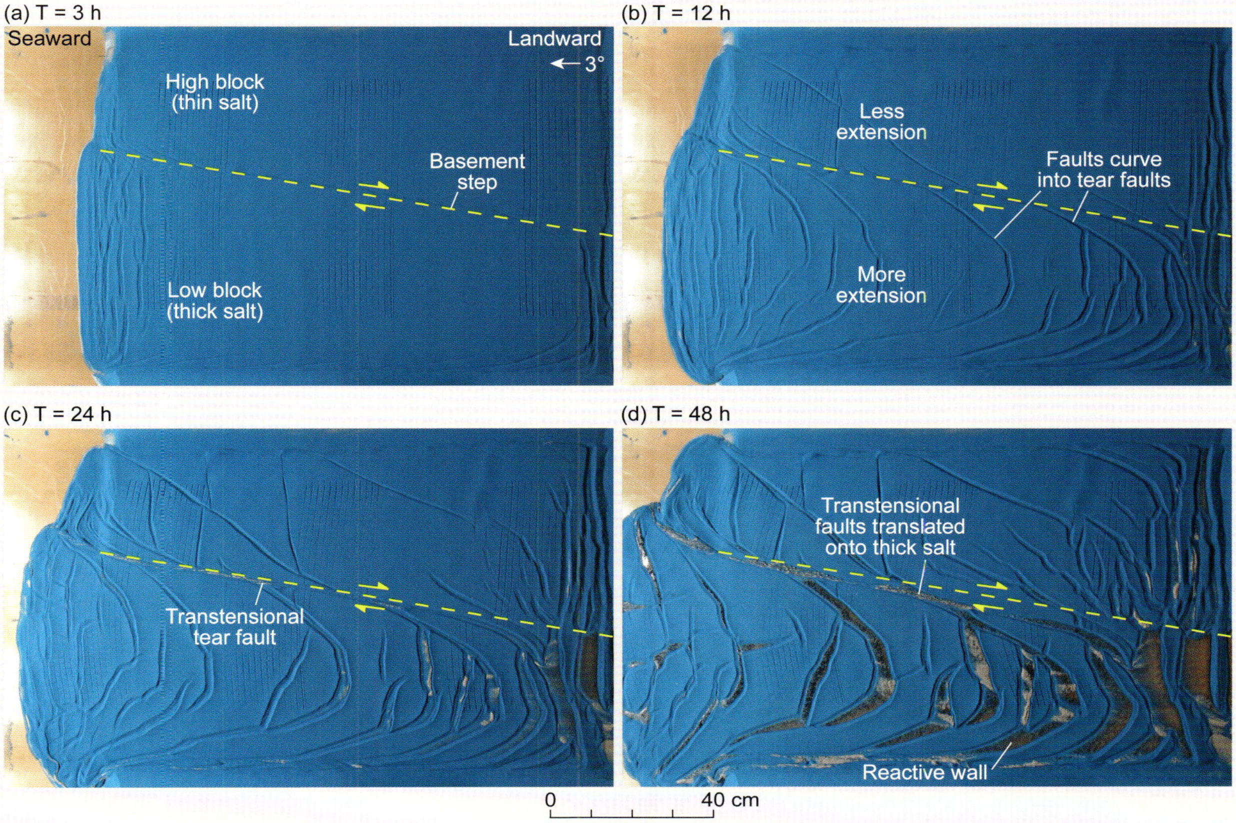

Figure 5.22. Extension above variable salt thickness. An oblique step in the base of salt separated thin-salt and thick-salt subbasins. These salt subbasins were buried beneath a 10-mm-thick, neutral-density roof and tilted seaward at 3° toward a free face downslope. Thick salt flowed downslope faster than thin salt because its boundary drag was proportionally less important. The roof above thick salt extended more than above thin salt, reflecting greater seaward flow. The two subbasins were separated by a transtensional dextral fault that was gradually transported downslope away from the boundary between the blocks. Compare with deformation above transpressional (Figure 5.23) and complex (Figure 5.24) basement steps. Model by Tim Dooley.

PHYSICAL MODEL—Extension above Transpressional Basement Step Separating Thick and Thin Salt

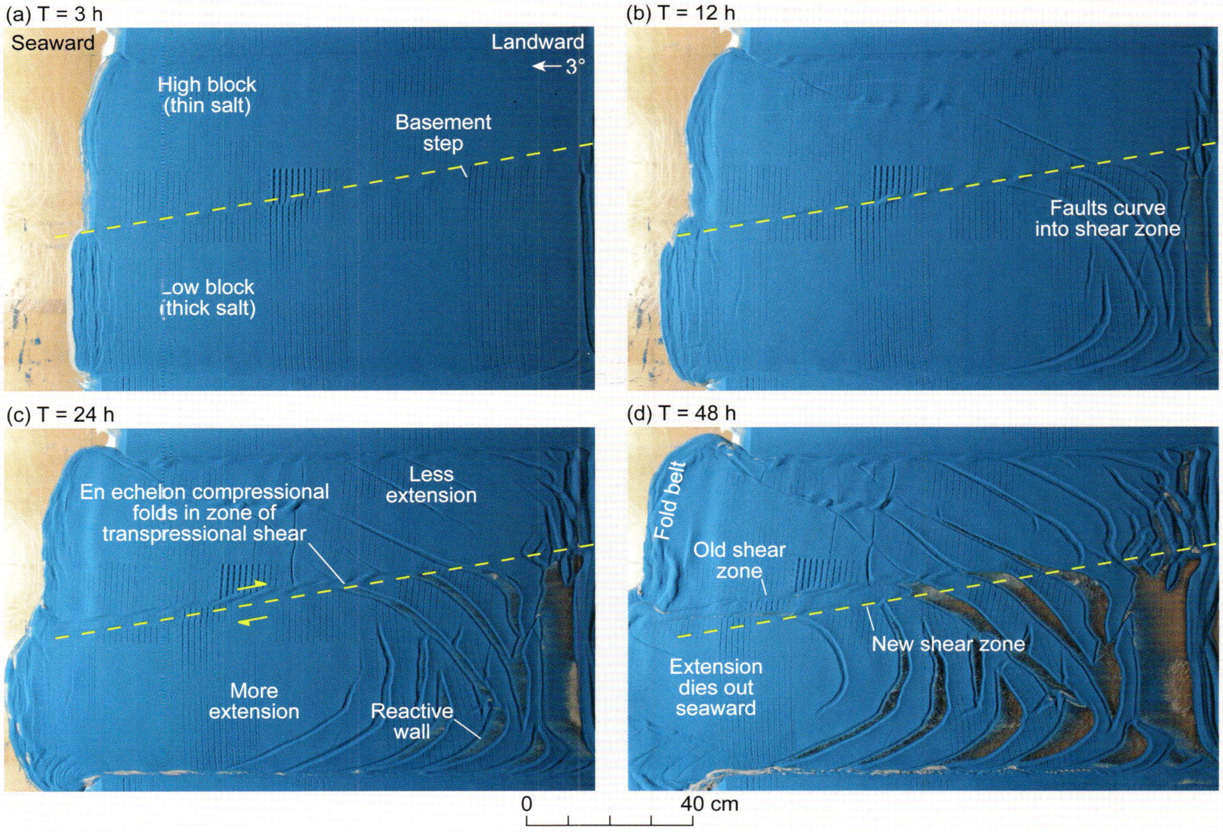

Figure 5.23. Extension above variable salt thickness. An oblique step in the base of salt separated thin-salt and thick-salt subbasins. These salt basins were buried beneath a 10-mm-thick, neutral-density roof and tilted seaward at 3° toward a free face downslope. Thick salt flowed downslope faster than thin salt because its boundary drag was proportionally less important. The roof above thick salt was extended more than above thin salt, reflecting greater seaward flow. Because the faster subbasin moved downslope into a narrowing gap, extension was inhibited in seaward parts of the block (compare with extension above transtensional step in Figure 5.22). The two provinces were separated by a zone of transpressional dextral shear, which was gradually transported into the thin-salt subbasin and was replaced by a new shear zone above the basement step. See also deformation above complex basement steps (Figure 5.24). Model by Tim Dooley.

PHYSICAL MODEL—Extension above Basement Block Separating Thick and Thin Salt

Figure 5.24. Extension above variable salt thickness. Restraining and releasing oblique steps in the base of salt separated thin-salt and thick-salt subbasins. These subbasins were buried beneath a 10-mm-thick, neutral-density roof and tilted seaward at 3° toward a free face downslope. Thick salt flowed downslope faster than thin salt because its boundary drag was proportionally less important. The roof above thick salt was extended more than above thin salt, reflecting greater seaward flow. Faster salt above the low block diverted around the high block, producing complex three-dimensional strain. Salt in the low block had to squeeze through a choke point where the thick salt was narrowest. Because salt slowed through the choke point, active extensional structures were rare there. Once through the choke point, salt accelerated as the thick-salt subbasin widened seaward, producing abundant extension. See animation. See also deformation above transtensional (Figure 5.22) and transpressional (Figure 5.23) basement steps. Model by Tim Dooley.

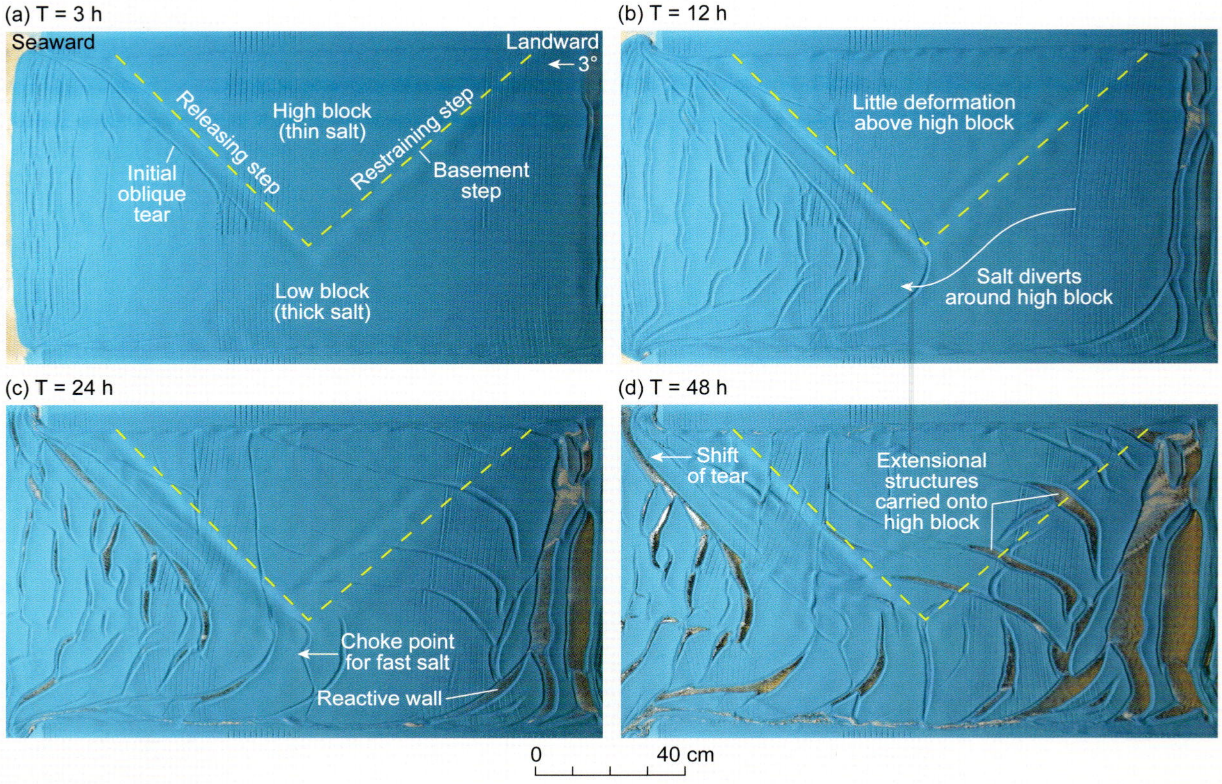

NUMERICAL MODEL—Evolution of Extensional Drape

Figure 5.25. Basement extension below salt. Salt acted as a detachment to decouple deformation above and below the salt. Prekinematic overburden draped over an active basement normal fault. Overburden extension localized in the upper hinge of this drape monocline, above the upthrown fault block. Continued extension formed a small reactive diapir below the overburden graben. An analogous fault zone formed at the lower hinge of the monocline but was inhibited by regional extension. Each fault zone at the upper and lower hinge of the monocline comprised paired extensional and contractional parts separated by a neutral surface of no strain. See also models with variable salt thickness (Figure 5.26), variable displacement rates (Figure 5.27), variable fault spacing (Figure 5.28), or variable displacement rates above multiple faults (Figure 5.29). GEOSIM finite-element model by Dan Schultz-Ela.

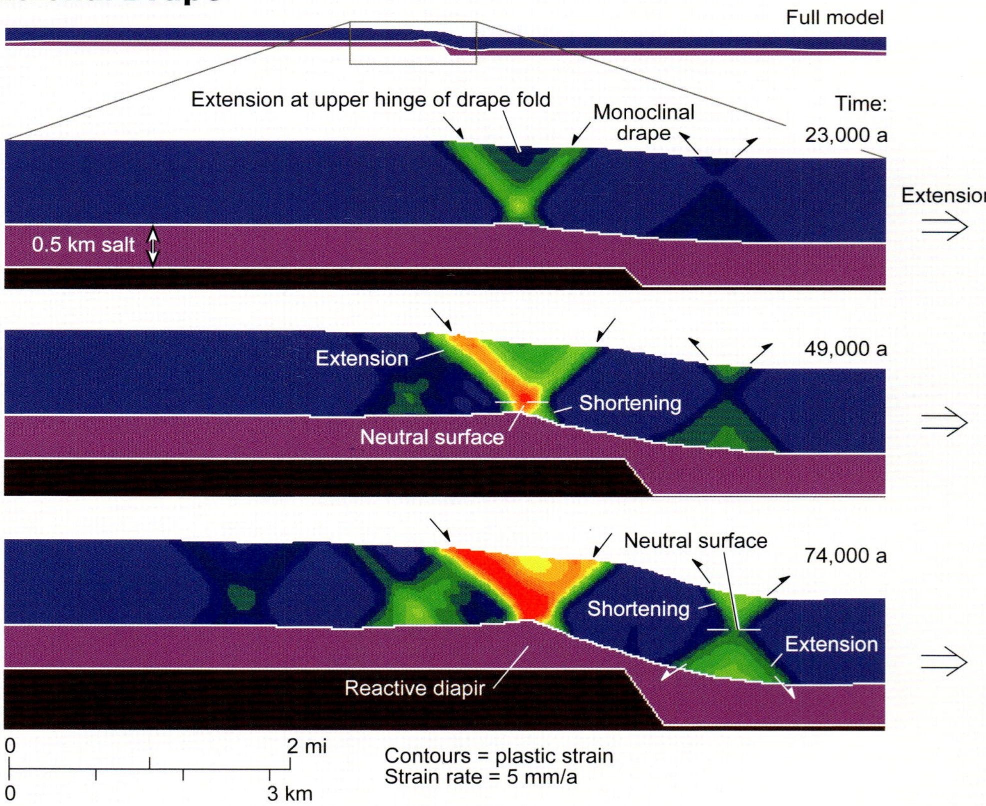

NUMERICAL MODEL—Effects of Variable Salt Thickness on Extensional Drape

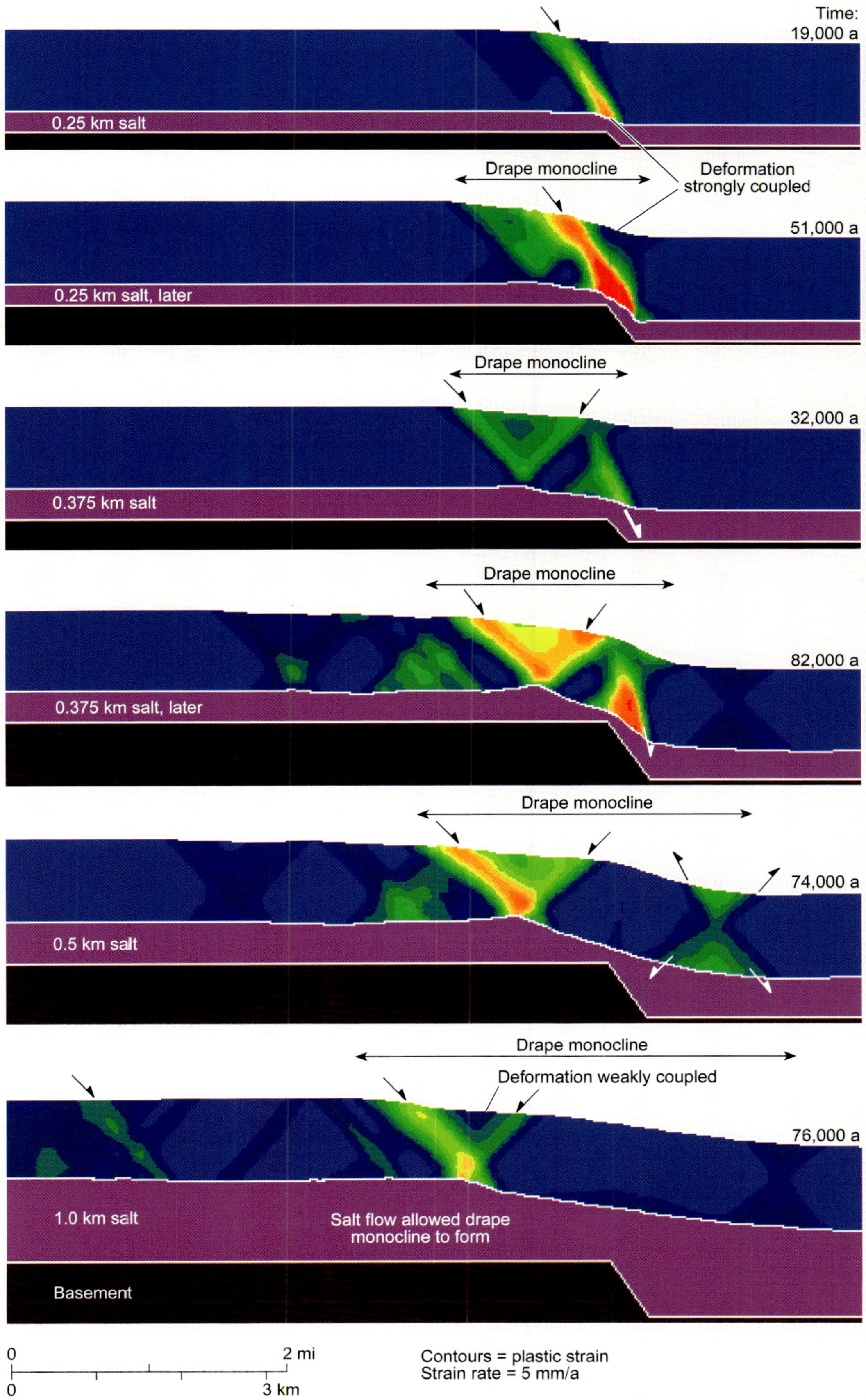

Figure 5.26. Basement extension below salt. Over thin salt, overburden deformation was strongly coupled to basement deformation. A much gentler, wider drape monocline formed above thick salt because salt flow was able to accommodate the drape, forming an efficient detachment. Extension above thick salt localized in the hinge of the drape monocline, laterally offset from the basement fault and well within the upthrown block. See also models with evolution of extensional drape (Figure 5.25), variable displacement rates (Figure 5.27), variable fault spacing (Figure 5.28), or variable displacement rates above multiple faults (Figure 5.29). GEOSIM finite-element model by Dan Schultz-Ela.

EXTENSIONAL DETACHMENTS

NUMERICAL MODEL—Effects of Variable Displacement Rate on Extensional Drape

Figure 5.27. Basement extension below salt. At high strain rates (top), there was no time for salt to flow in response to deformation. Stresses propagated through the salt, and overburden deformation was strongly coupled to basement structure. When deformation was slower, or salt thicker, the salt flow could partly accommodate overburden deformation. A broad drape monocline formed, and overburden extension localized in the upper hinge of the fold. The gentler and wider the drape monocline, the more offset there was between this hinge and the underlying basement fault. Slower strain rates and thicker salt thus both favor a more efficient salt detachment. See also models with evolution of extensional drape (Figure 5.25), variable salt thickness (Figure 5.26), variable fault spacing (Figure 5.28), or variable displacement rates above multiple faults (Figure 5.29). GEOSIM finite-element model by Dan Schultz-Ela.

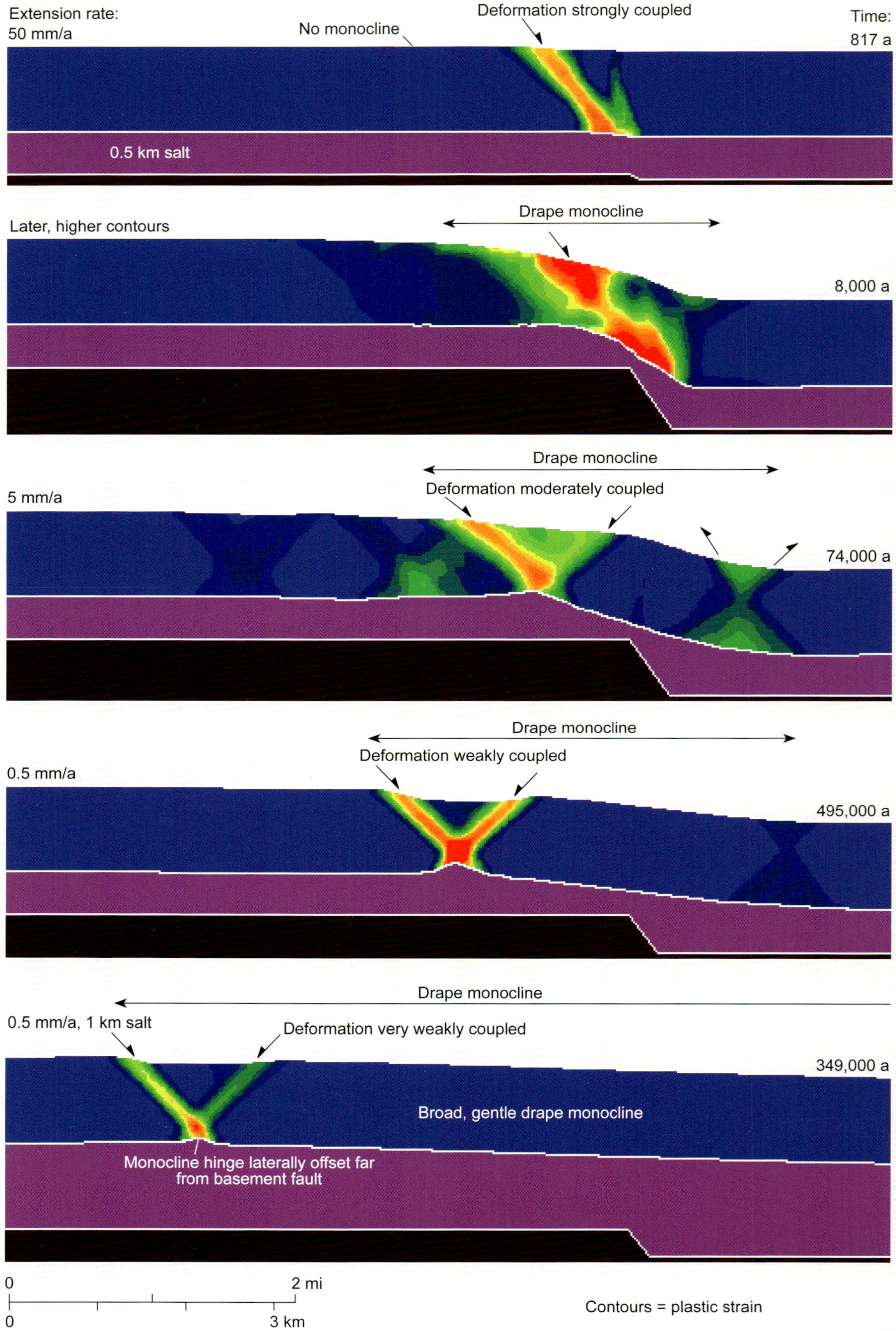

Numerical Model—Effects of Variable Fault Spacing on Extensional Drape

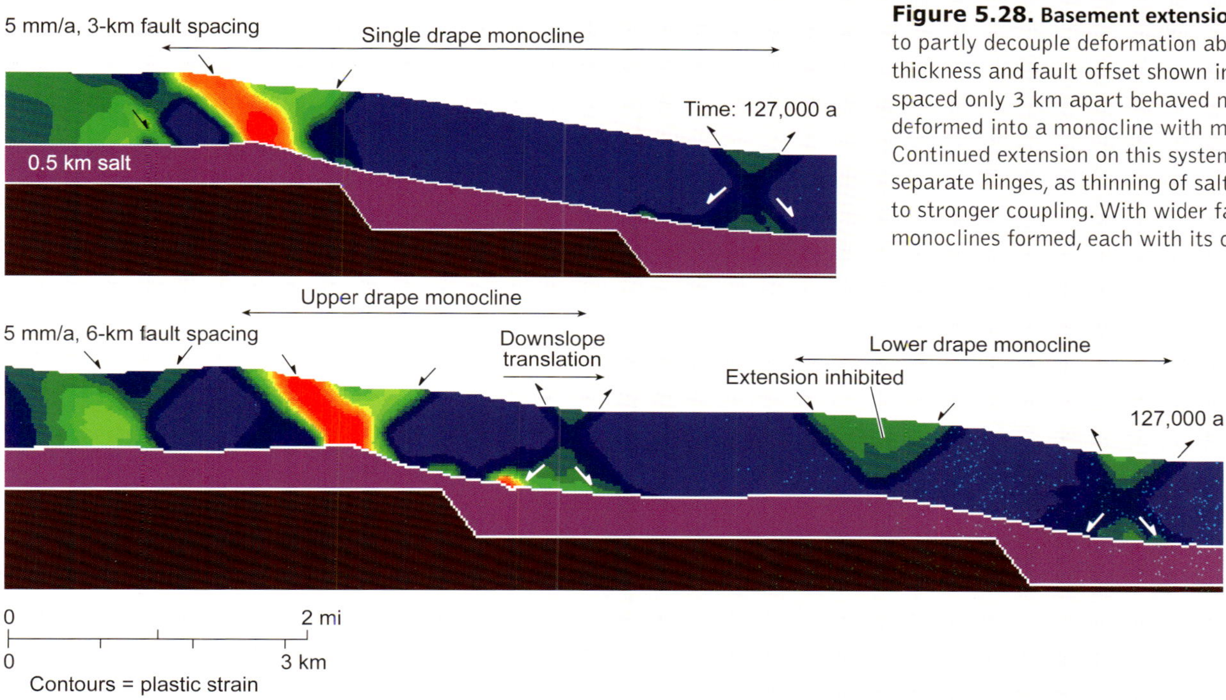

Figure 5.28. Basement extension below salt. Salt acted as a detachment to partly decouple deformation above and below the salt. With the salt thickness and fault offset shown in the upper section, two normal faults spaced only 3 km apart behaved much like a single fault. The overburden deformed into a monocline with most extension in its upper hinge zone. Continued extension on this system would have eventually produced two separate hinges, as thinning of salt over individual faults would lead to stronger coupling. With wider fault spacing (below), two separate monoclines formed, each with its own extensional zone in the upper hinge. Extension in the lower of these hinges was inhibited by compression from the higher block sliding down the topographic slope. See also models with evolution of extensional drape (Figure 5.25), variable salt thickness (Figure 5.26), variable displacement rates (Figure 5.27), or variable displacement rates above multiple faults (Figure 5.29). GEOSIM finite-element model by Dan Schultz-Ela.

Numerical Model—Effects of Variable Displacement Rate on Extensional Drape above Three Faults

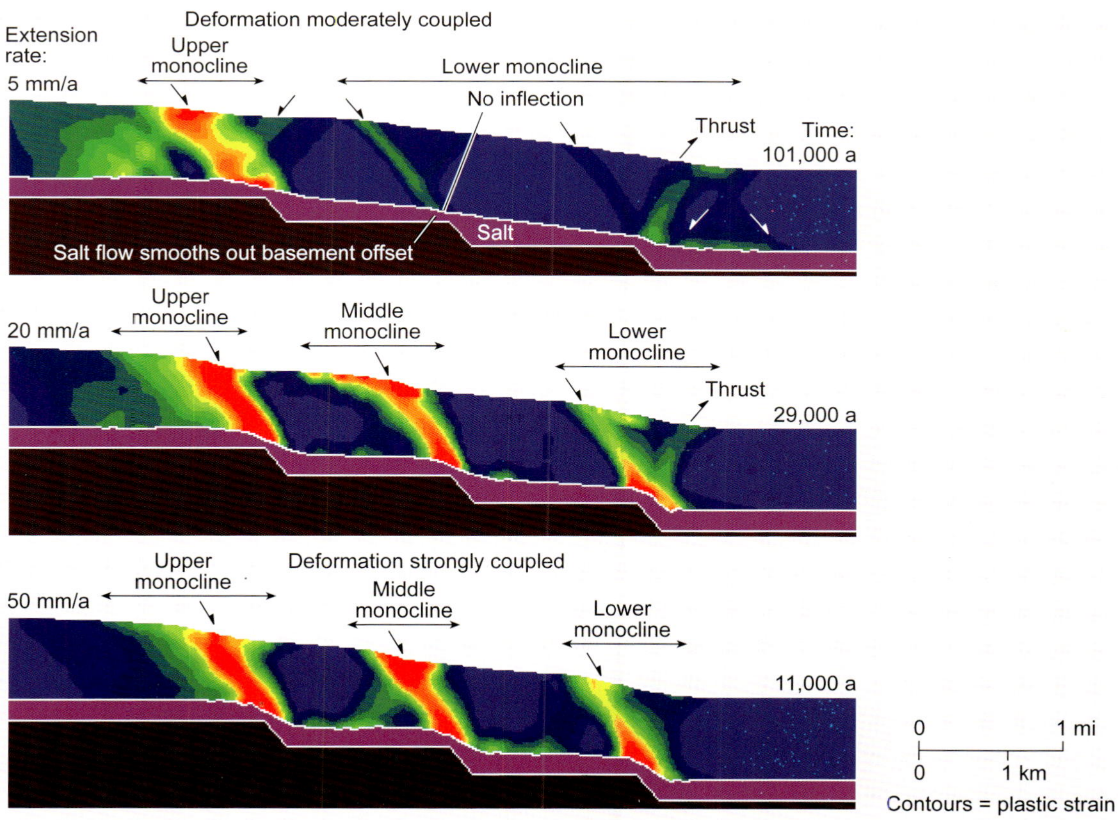

Figure 5.29. Basement extension below salt. Salt acted as a detachment to partly decouple deformation above and below the salt. With slow extension (upper section), salt flow had time to compensate offset on basement faults. There was virtually no inflection in the overburden above the central basement fault. The degree of coupling did, however, increase through time as salt thinned over the block corners. At higher strain rates (lower two sections), salt could not flow rapidly enough to fill the space created by basement offset, so the overburden draped across each subsalt fault, forming three narrow monoclines. Stresses propagated from the basement faults through the salt into the overburden. See also models with evolution of extensional drape (Figure 5.25), variable salt thickness (Figure 5.26), variable displacement rates (Figure 5.27), or variable fault spacing (Figure 5.28). GEOSIM finite-element model by Dan Schultz-Ela.

CONCEPTUAL MODEL—Influence of Salt Thickness on Decoupling during Basement Extension

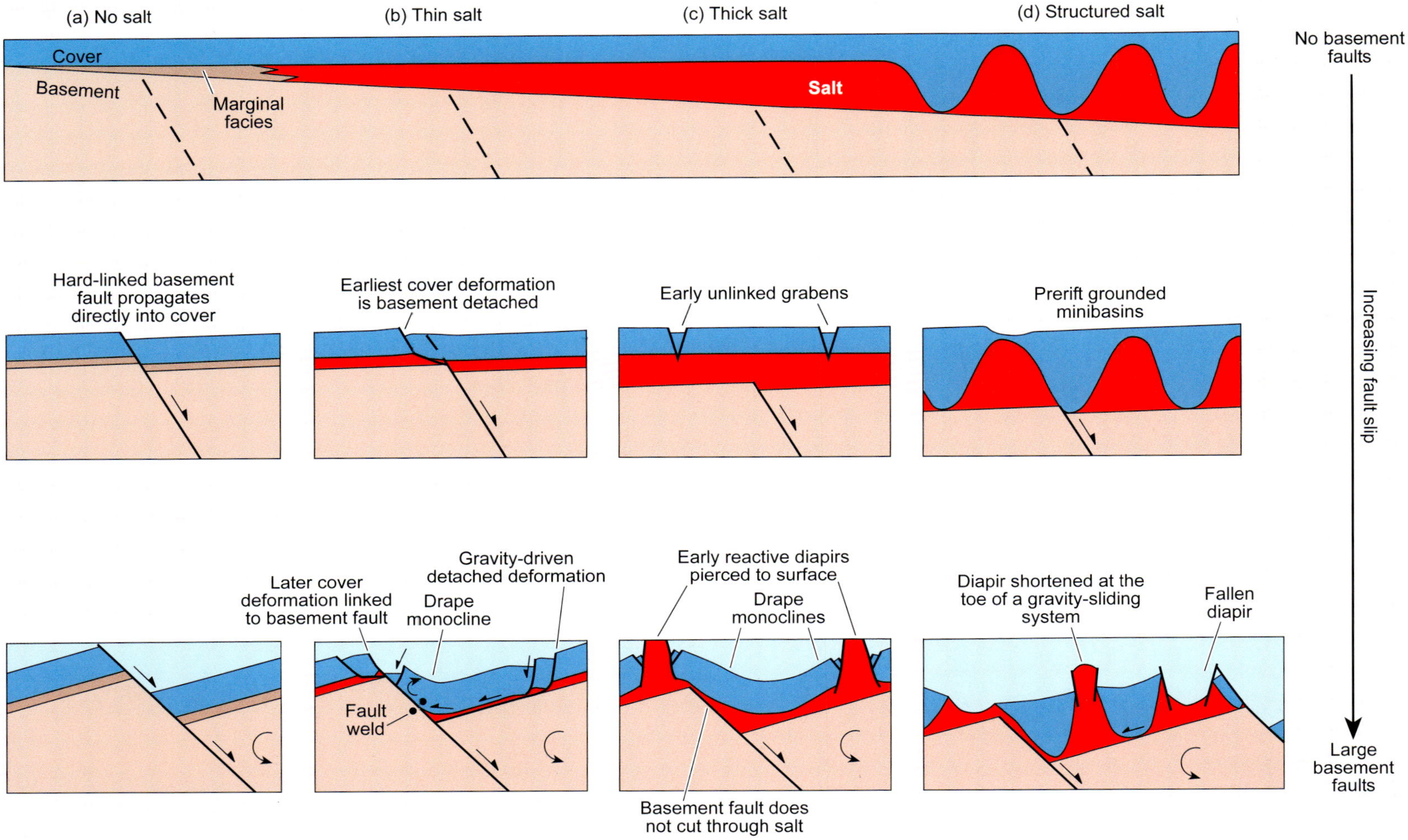

Figure 5.30. Basement extension below salt. Increasing salt thickness promotes decoupling of deformation above and below salt. Conversely, where salt is thin or absent, faults propagate directly from basement into the overburden. With thin salt, overburden deformation is initially detached from basement. Once the basement fault slips enough to offset the salt layer, however, the basement fault can propagate into the overburden. In areas of thick salt, overburden deformation may remain decoupled from basement structure. Detached extension in the overburden initiates reactive diapirs, which may then pierce to the surface. Finally, fault-block rotation locally tilts the salt layer, which may cause gravity-driven deformation. Salt structures at the downdip end of fault blocks (i.e., the toe of the local gravity-driven system) may shorten and continue to rise while their neighbors are falling extensionally. Modified from Stewart and Clark (1999).

Chapter 6

Contractional Salt Detachments

Mountain belts are among the most prominent landforms on Earth and can extend for thousands of kilometers. Most have large foreland fold-and-thrust systems, containing dozens of orogen-parallel thrust faults. Where fold-and-thrust belts overlie salt basins, as in the Jura Mountains and Sierra Madre Oriental and on the Potwar Plateau, thrusts extend much farther into the foreland, bulging outward as large salients in the mountain chain. This bulge occurs because mechanically weak salt forms an extremely efficient detachment. More recently, contractional salt detachments have also been discovered near the downdip ends of most salt basins on passive margins.

Like extensional salt detachments, contractional salt detachments are present in nearly all salt-tectonic basins. They are typically associated with contractional salt anticlines, salt-detached thrusts, and squeezed diapirs. Structures in this chapter are those in which salt structures are secondary to thrust faults.

In the absence of contractional anticlines and diapirs, dominant structures above contractional salt detachments are salt-detached thrust faults. Many thrusts carry thin salt at the base of their hanging walls, which have variable amounts of internal flow. In most respects, the structural style of salt-detached thrust belts is similar to that in thrust belts having no salt. However, thrust vergence tends to be more variable in salt-detached systems. Also, because the detachments are so weak, salt-involved thrust belts have a lower critical taper, so tend to be wider than nonsalt thrust belts if the salt basin is wide enough.

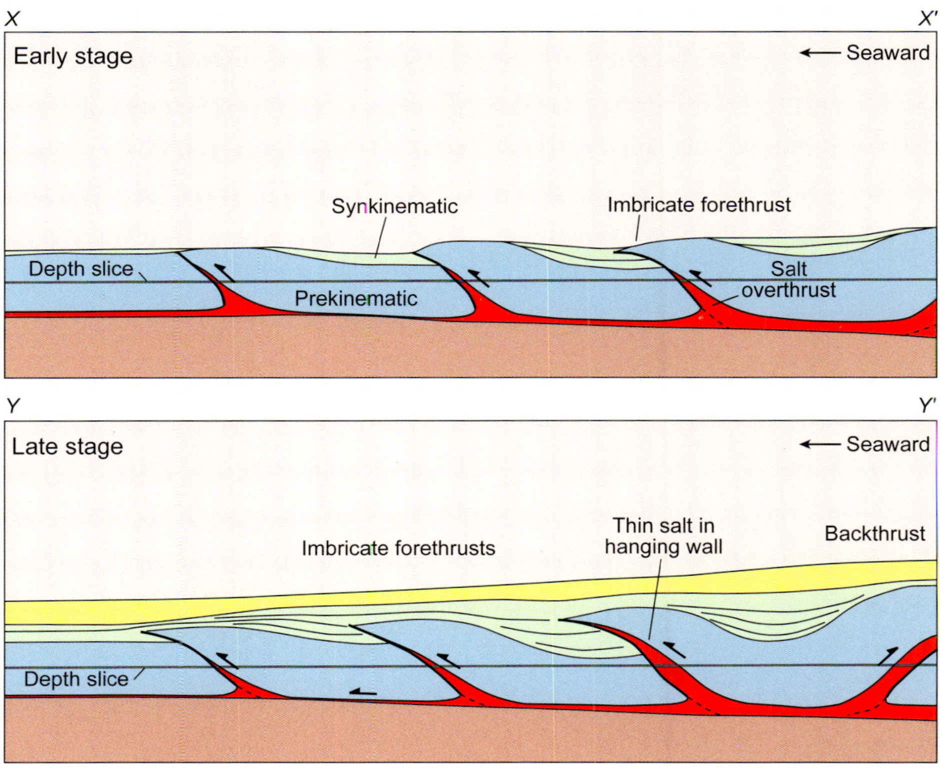

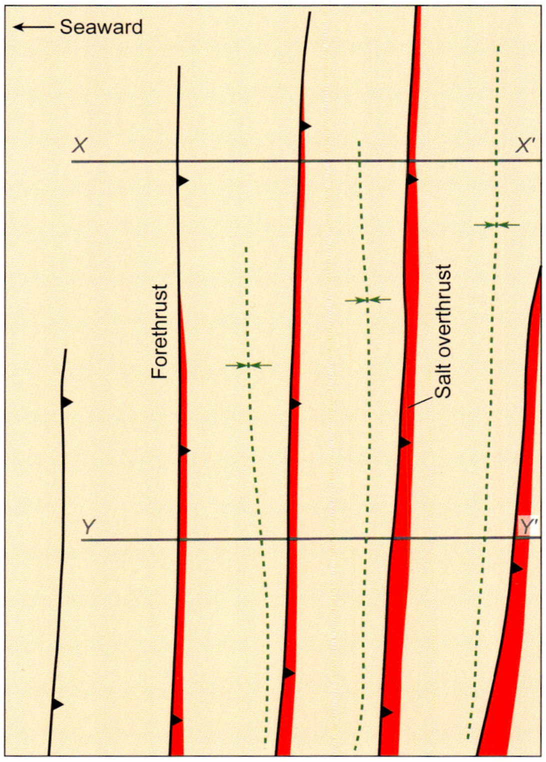

Aquitaine Basin—Salt Overthrust, Ossun Diapir, France

Figure 6.01. Alternating episodes of extension and shortening. The North Pyrenean Frontal Thrust is located in the position formerly occupied by the Ossun Salt Diapir. Stratigraphic patterns on either side of the diapir record alternating episodes of basement-involved extension and shortening. Extension produced subsidence; shortening caused buckling and uplift. The area was located along a transform plate boundary during the Mesozoic so that minor shifts in plate-boundary conditions could change the strain field from transpression to transtension and back again. After Canérot and others (2005).

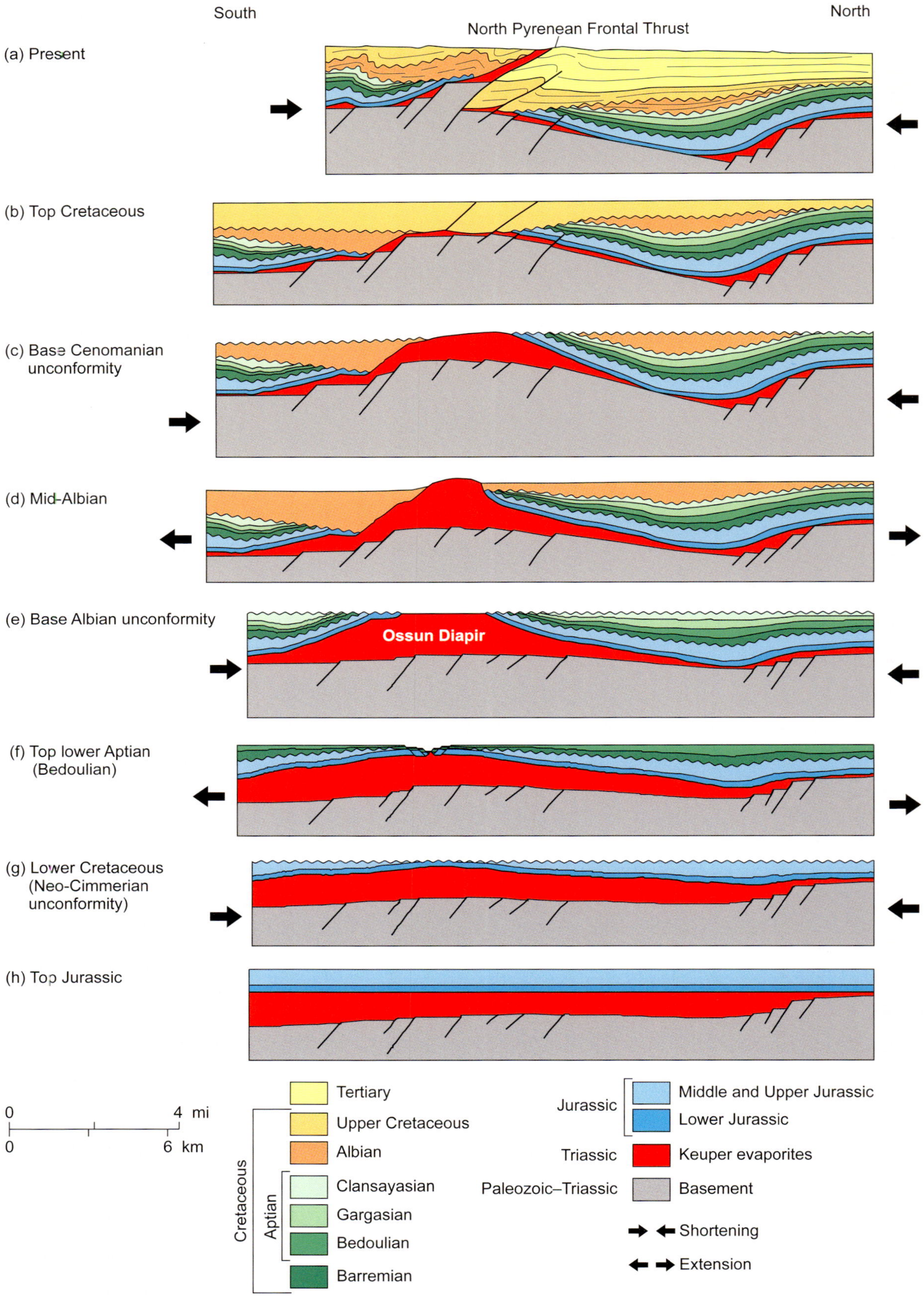

Gulf of Mexico Basin—Pop-Up Structure, Atwater Fold Belt, Atwater Valley Area, Offshore USA

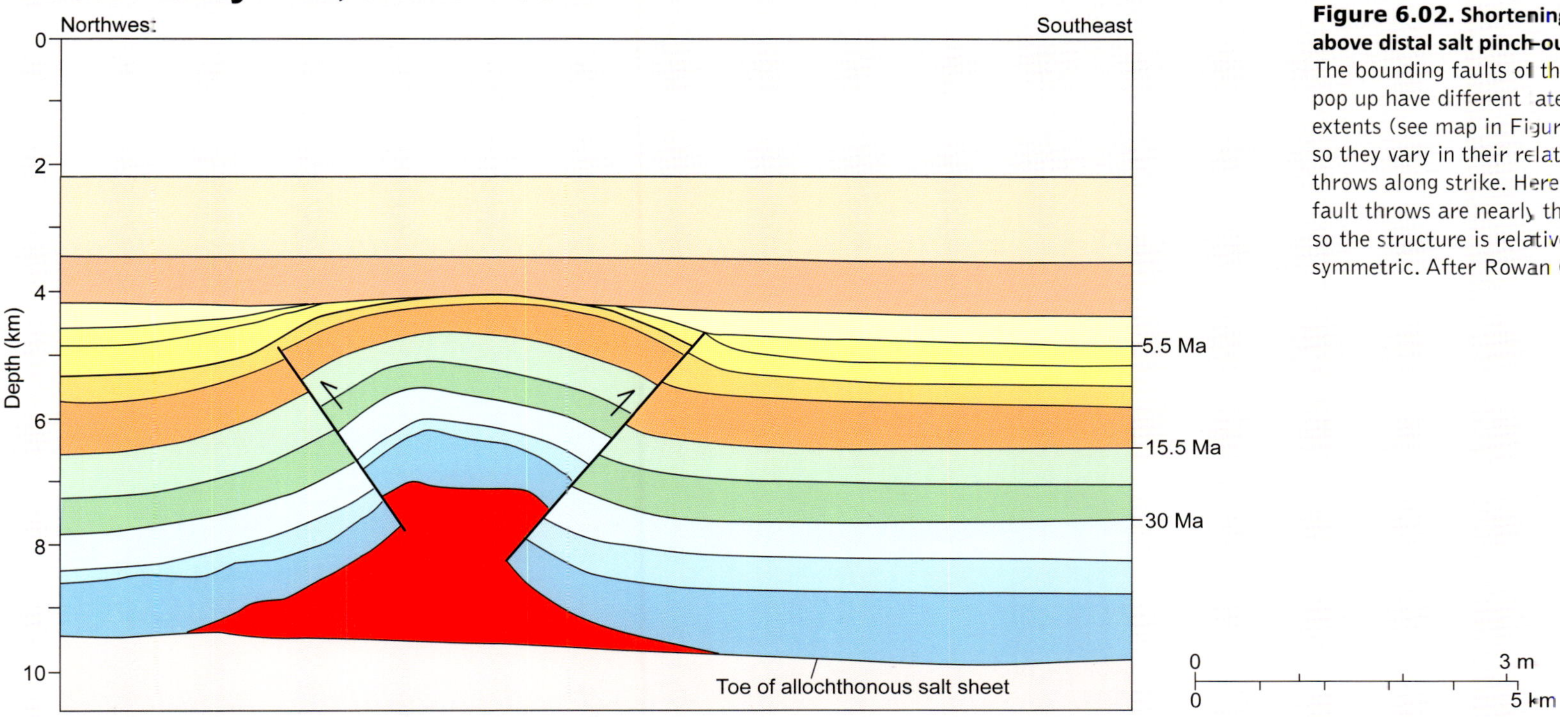

Figure 6.02. Shortening above distal salt pinch-out. The bounding faults of the pop up have different lateral extents (see map in Figure 6.03), so they vary in their relative throws along strike. Here the fault throws are nearly the same so the structure is relatively symmetric. After Rowan (1997).

Gulf of Mexico Basin—Map of Pop-Up Structure, Atwater Fold Belt, Atwater Valley Area, Offshore USA

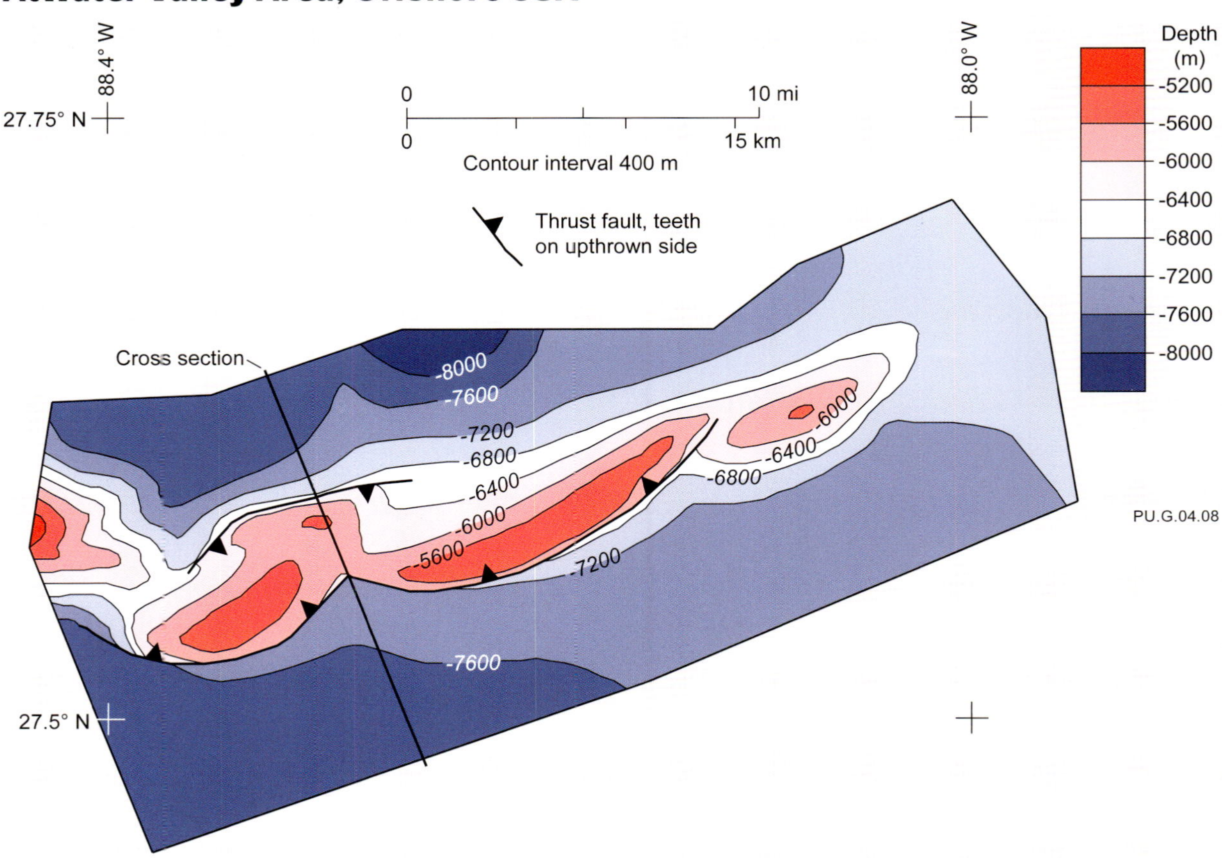

Figure 6.03. Shortening above distal salt pinch-out. The anticline is composed of three en echelon segments, with the pop-up structure lying to the west (see section, Figure 6.02). The geometry of forethrusts (south-verging) and backthrusts (north-verging) varies along strike. After Rowan (1997).

JURA FOLD BELT—Salt Overthrusts, Farisberg and Weissenstein Anticlines, Birtis, Switzerland

Figure 6.04. Shortening above thin salt. In this interpretation, both anticlines formed above major south-verging backthrusts. The Weissenstein Anticline is inferred to have deformed an older forethrust (middle section). Interpretations of these well-known anticlines are varied and controversial. The structures were originally interpreted as upright salt-cored anticlines by Buxtorf (1916) and Heim (1921). Conversely, Laubscher (1992) favored an upright anticline cored by a duplex antiformal stack. In all of these interpretations, the only role of salt is as a detachment surface. Because the salt is thin, fold and fault geometries are purely a result of thrusting. The scales of the restored sections are 65% of that in the present-day section at top. After Bitterli (1990).

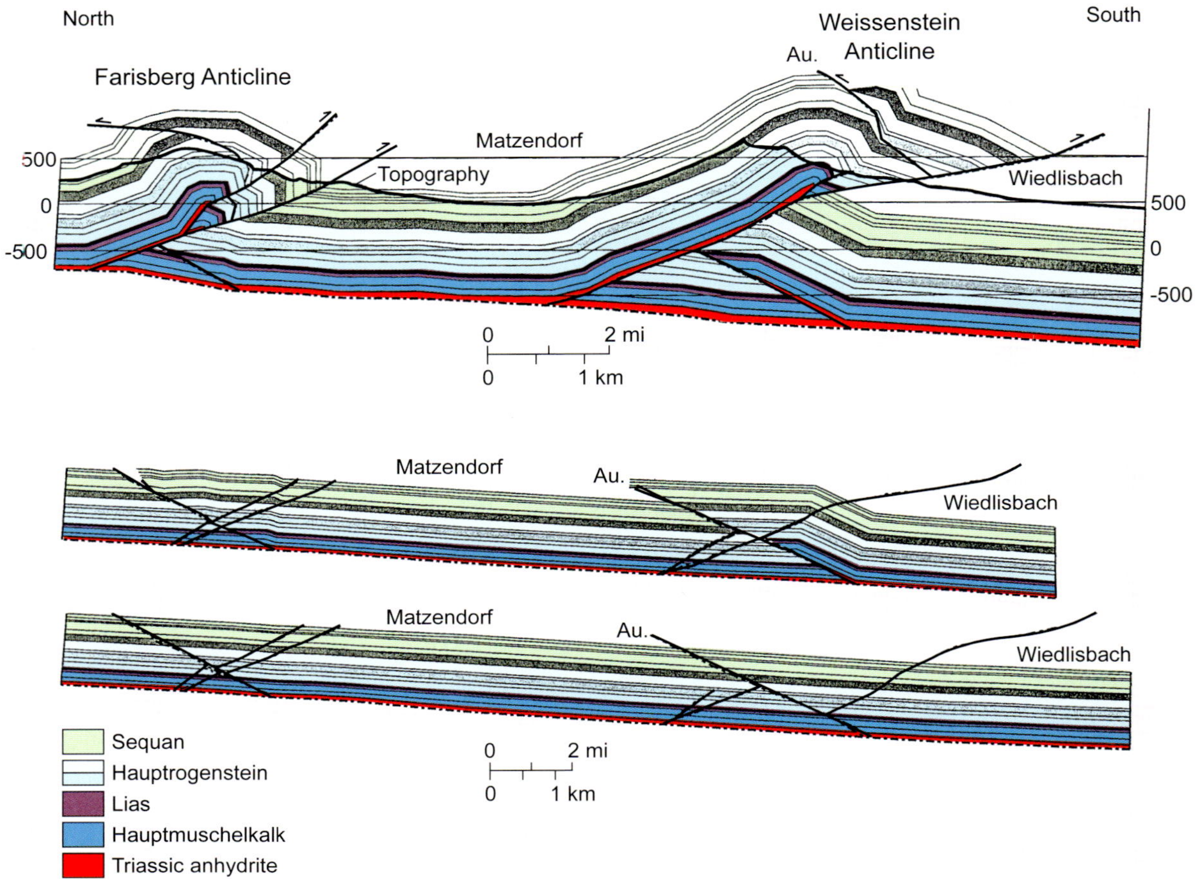

JURA FOLD BELT—Salt Overthrusts, Hauenstein Tunnel, Switzerland

Figure 6.05. Shortening above thin salt. The Hauenstein-Basis tunnel passes through two salt overthrusts. Both thrusts detach on a Permo-Triassic detachment below the Bundsandstein Formation. Because the salt here is very thin, its only role is as a detachment. Fold and fault geometries are purely a result of thrusting. The Upper Triassic Keuper, which elsewhere forms a major detachment in Alpine fold belts, plays a trivial role in the décollement here, perhaps as the result of a facies change. The Main Frontal Thrust separates the Tabular Jura from the Folded Jura. "Folded Jura" is somewhat misleading: most surface anticlines either are the result of ramp-flat thrust tectonics or are cored by thrust duplexes; most examples of the simple "Jura-style" folds are far from simple. After Buxtorf (1916) and Laubscher (1977, 1992).

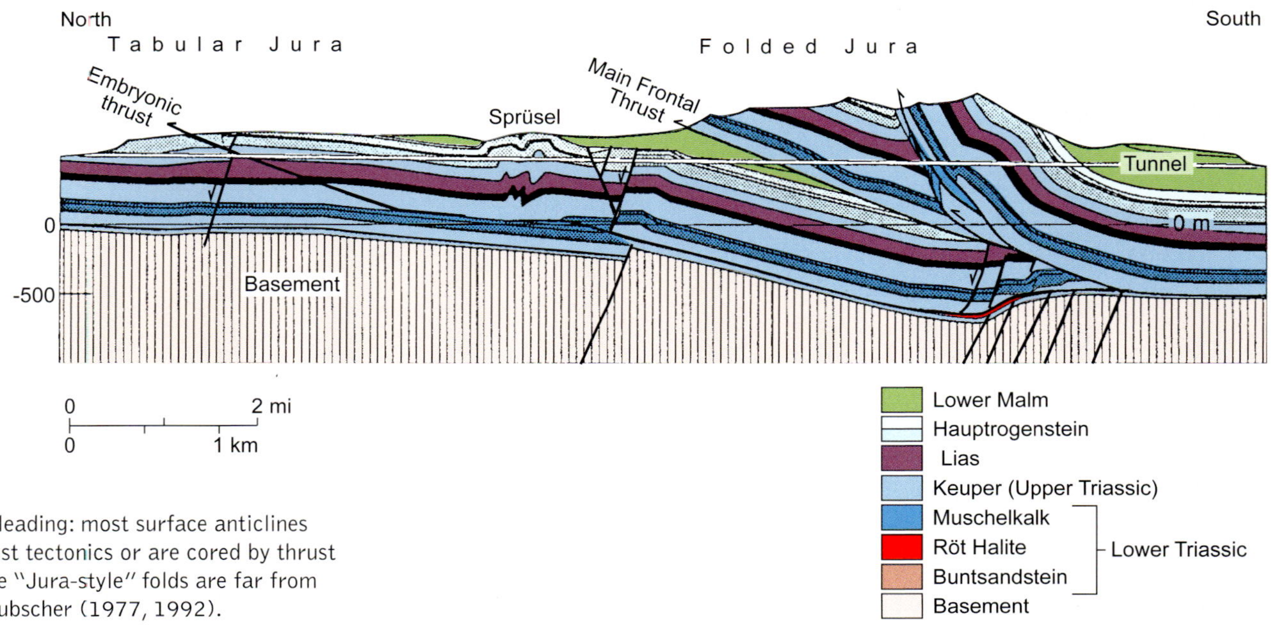

KWANZA BASIN—Salt Overthrusts, Offshore Angola

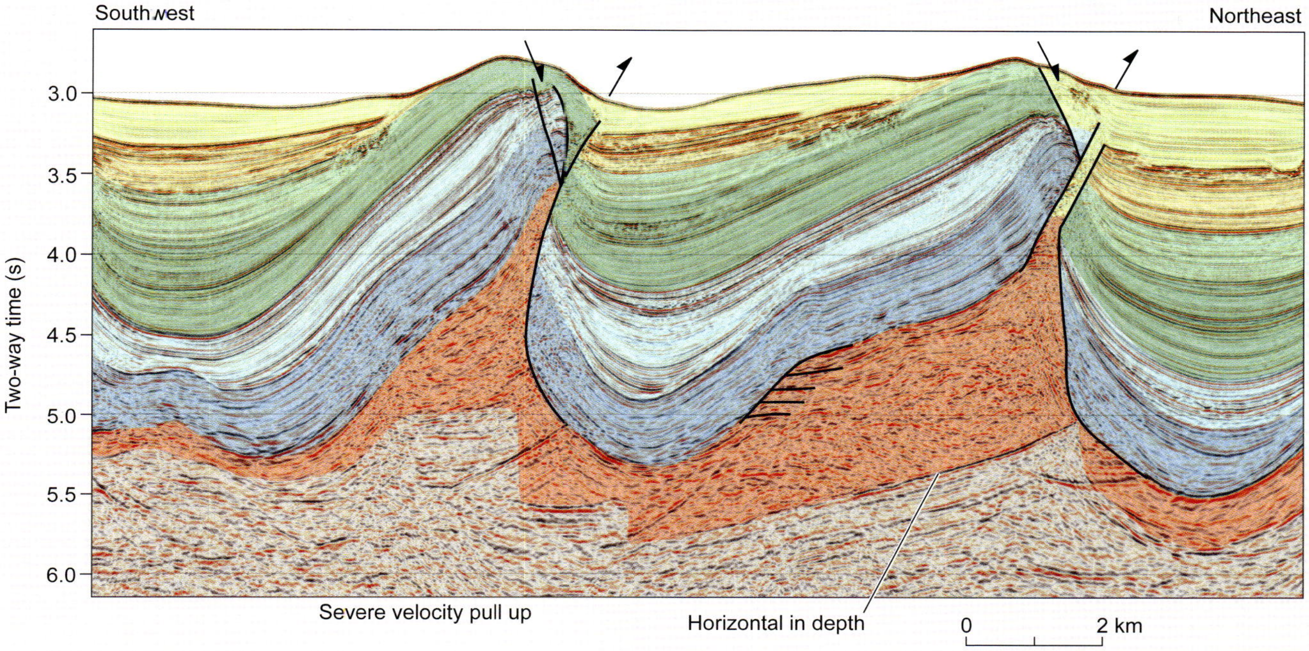

Figure 6.06. Shortening above thick salt. The deepest (medium blue) unit is roughly isopachous and thus prekinematic. Overlying synkinematic wedges in both the hanging wall and footwall thin toward the thrust faults. This thinning suggests that these landward-verging thrusts evolved from precursor anticlines whose crests were breached by small diapirs. Thrust formation is dated by the onset of a strongly asymmetric stratigraphic pattern during deposition of the pale green and yellow units. This thrust belt dies out southward across a major crustal boundary inferred to be the continental continuation of an oceanic fracture zone. Seismic data courtesy of WesternGeco. Interpretation by Martin Jackson.

LOWER CONGO BASIN—Salt Overthrust, Offshore Gabon

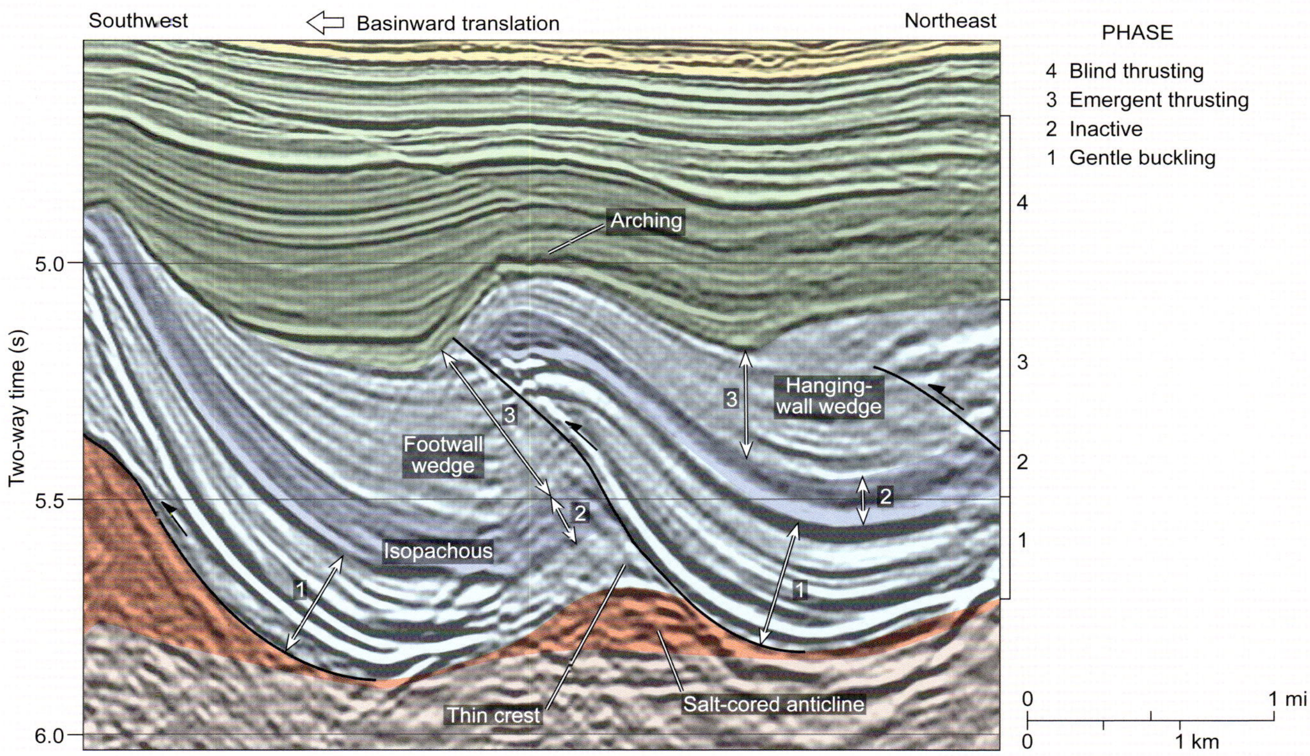

Figure 6.07. Shortening above thin salt. Growth stages of synkinematic thrusting. Gentle buckling formed long-wavelength, salt-cored anticlines. After a static tectonic period, emergent thrusting began; this was recorded by the asymmetry between a thick footwall wedge that thickens toward the overriding thrust and a thin hanging-wall wedge that thins toward the thrust. During the following stage of blind thrusting, the ramp anticline had seafloor expression (variably recorded by onlapping, thinning, and truncation), but the underlying thrust was buried. See restoration of a similar structure in Figure 6.08. Interpretation by Martin Jackson, Mike Hudec, and Dave Jennette. Seismic data courtesy of Total.

LOWER CONGO BASIN—Restoration of Salt Overthrust, Offshore Gabon

Figure 6.08. Shortening above thin salt. Mild buckling in the early Albian formed a precursor gentle anticline, which localized the later formation of a thrust. The onset of thrusting was recorded by the stratigraphic change from (1) symmetric wedges flanking the precursor anticline to (2) asymmetric wedges flanking the thrust. Salt expulsion from the deep anticline accommodated subsidence of a thick footwall wedge and folded the overlying thrust fault into its present-day sigmoidal shape. Translation up the thrust ramp rotated and thinned strata in the hanging wall. See a seismic example of a similar structure in Figure 6.07. Interpretation and restoration by Mike Hudec.

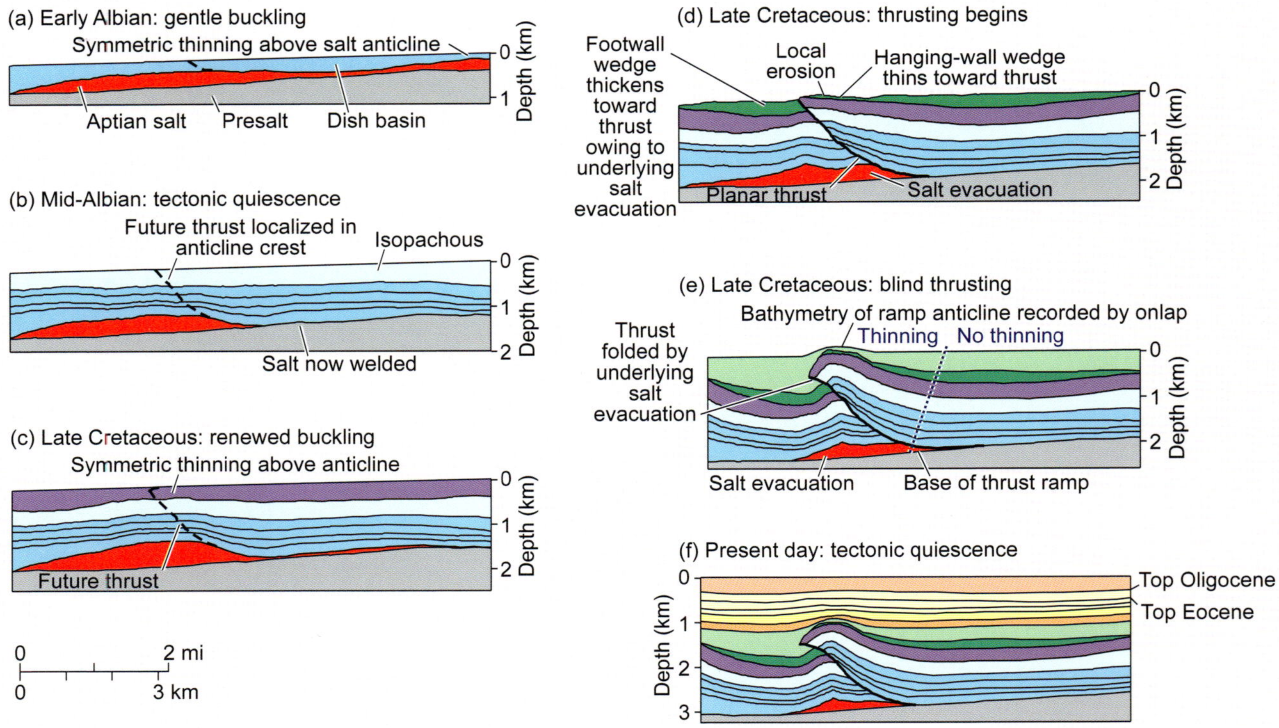

LOWER CONGO BASIN—Strike View of Salt Overthrusts, Offshore Gabon

Figure 6.09. Shortening above thin salt. In this strike view, these lateral ramps verge mostly toward the viewer, but they also have a component of vergence to the right. The lateral ramps nucleated along low salt ridges oriented obliquely down the regional dip. Differential sediment loading may have created the ridges. Interpretation by Martin Jackson, Mike Hudec, and Dave Jennette. Seismic data courtesy of Total.

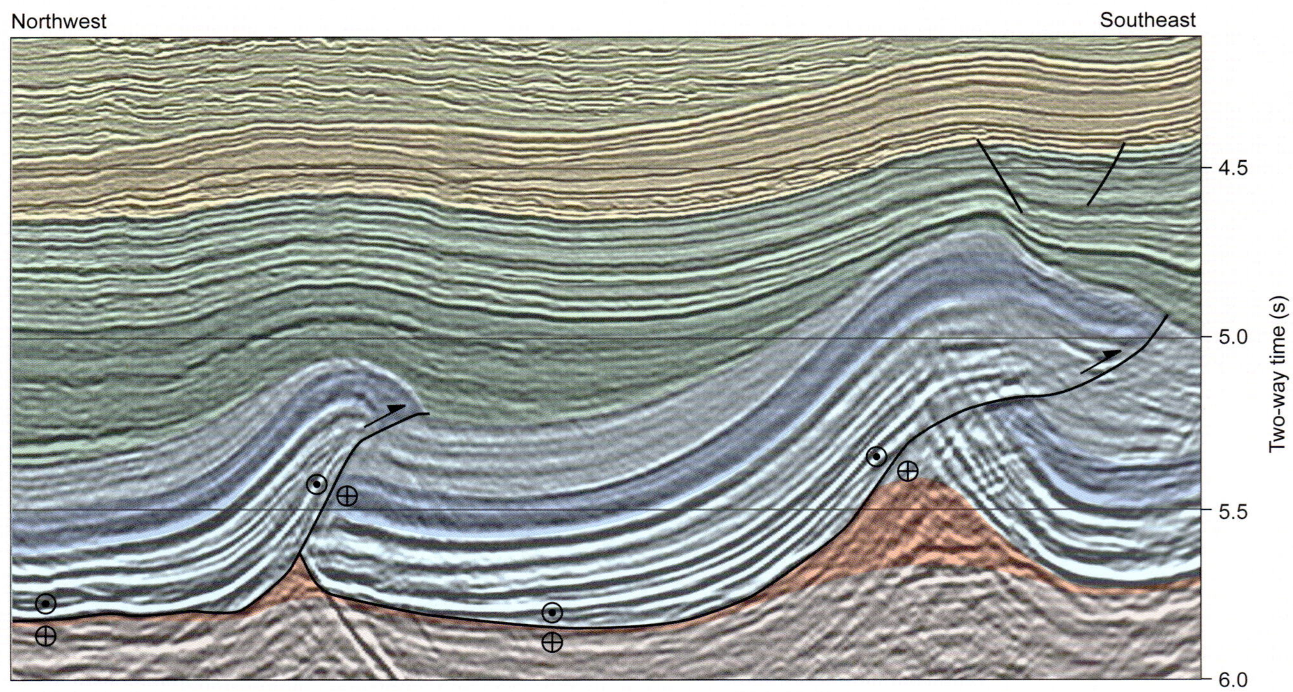

0 1 mi
0 1 km

LOWER CONGO BASIN—Embryonic Salt Overthrust, Offshore Gabon

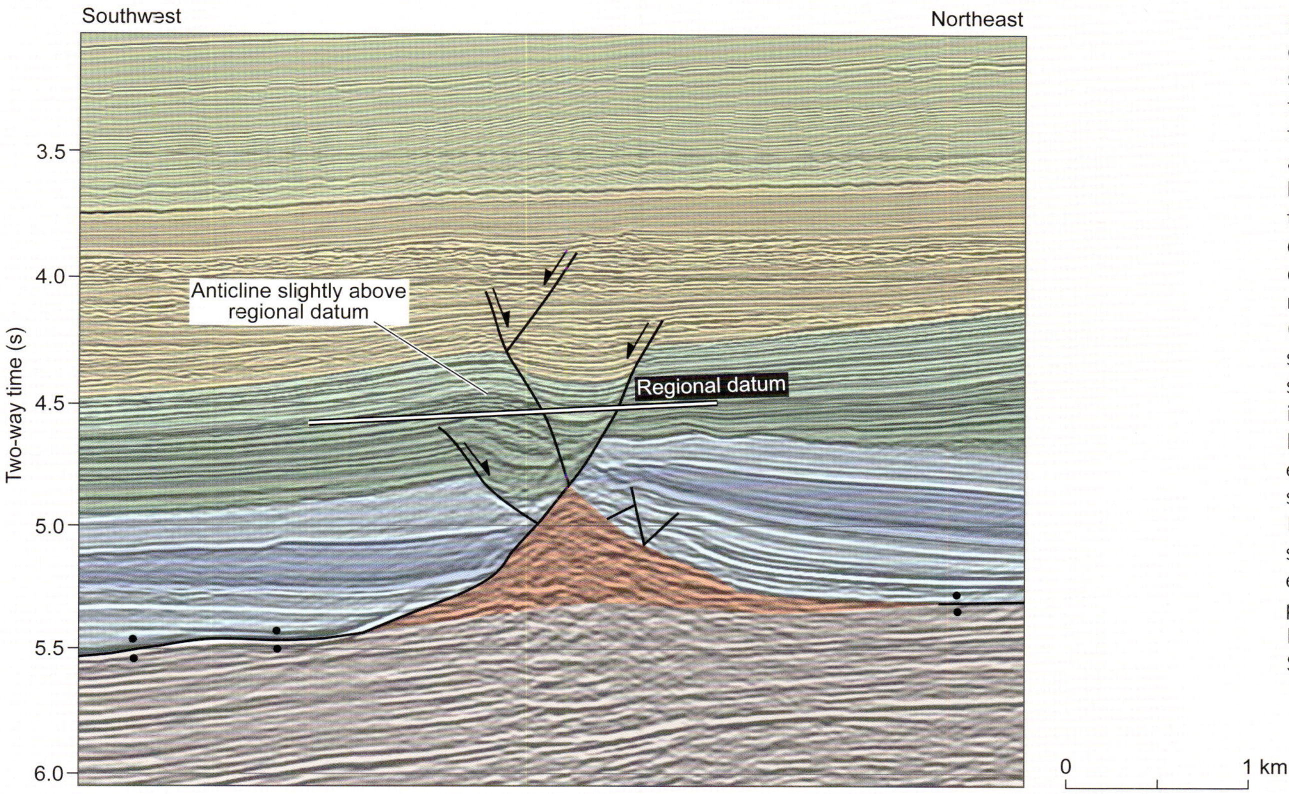

Figure 6.10. Subtle shortening of a diapir above thin salt. This structure lies along a salt ridge that extends outward from a tall salt dome. Shortening in the area is most intense at the dome because salt is much weaker than sediment and so localizes deformation. Deformation dies out gradually along the ridge moving away from the dome (compare this figure with strong shortening closer to the dome shown in Figure 6.11). Shortening in this section was too weak to leave definitive signs, apart from elevating the flanking anticline slightly above the regional datum. However, along-strike these subtle signs amplify into unmistakable evidence of shortening. Interpretation by Martin Jackson, Mike Hudec, and Dave Jennette. Seismic data courtesy of Total.

LOWER CONGO BASIN—Advanced Salt Overthrust, Offshore Gabon

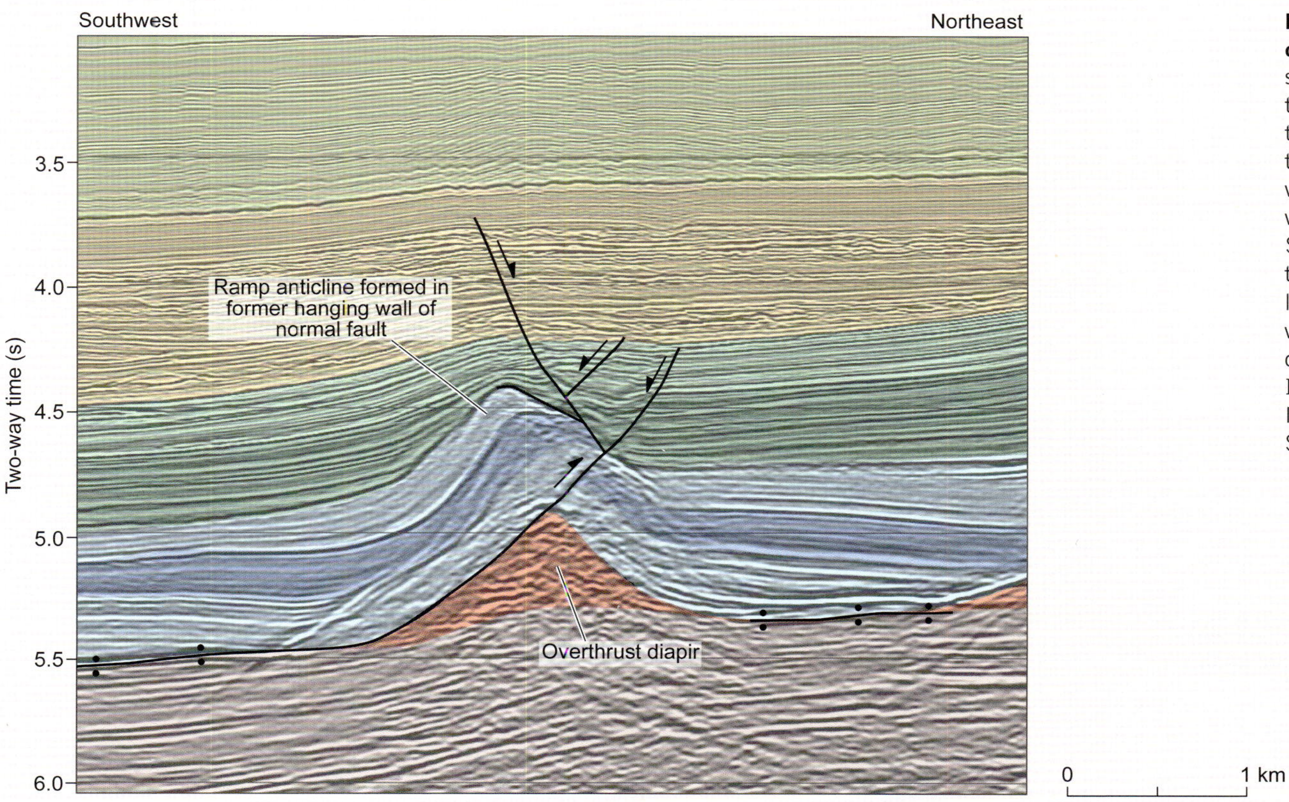

Figure 6.11. Strong shortening of a diapir above thin salt. The southwest flank of the diapir was thrust landward. The amount of thrusting gradually increases along the trend of the small diapiric wall and culminates where this wall joins a much taller diapir. Salt diapirs are much weaker than sediments, so they tend to localize deformation. Compare with subtle shortening of the same diapir along strike in Figure 6.10. Interpretation by Martin Jackson, Mike Hudec, and Dave Jennette. Seismic data courtesy of Total.

PARRY ISLANDS FOLD BELT—Eastern and Central Melville Island, Canada

Figure 6.12. Shortening in a multidetachment system. At the surface, the Parry Islands Fold Belt is expressed by long, east-west-trending folds, with few exposed faults. This is in contrast to the structural pattern at depth, where thrusts dominate in the cores of the folds (see Figures 6.13–6.17). The thrusts are kept from cropping out by a roof detachment. After Harrison (1995).

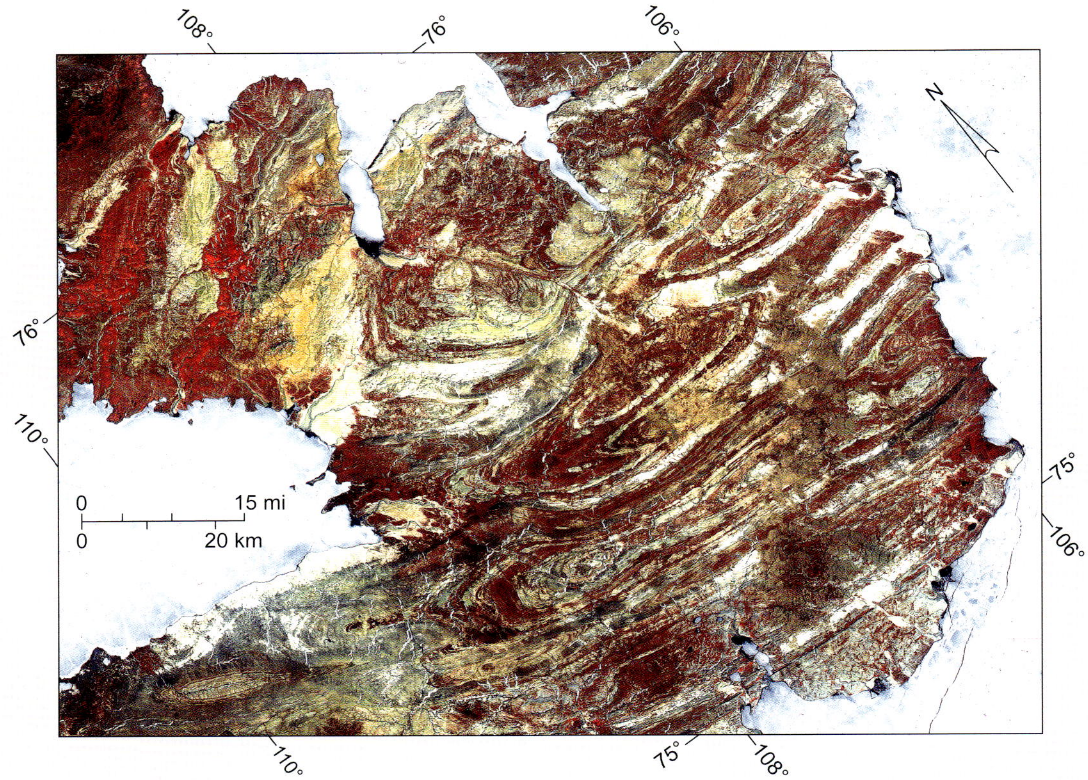

PARRY ISLANDS FOLD BELT—Richardson Point Pop Up 2, Melville Island, Canada

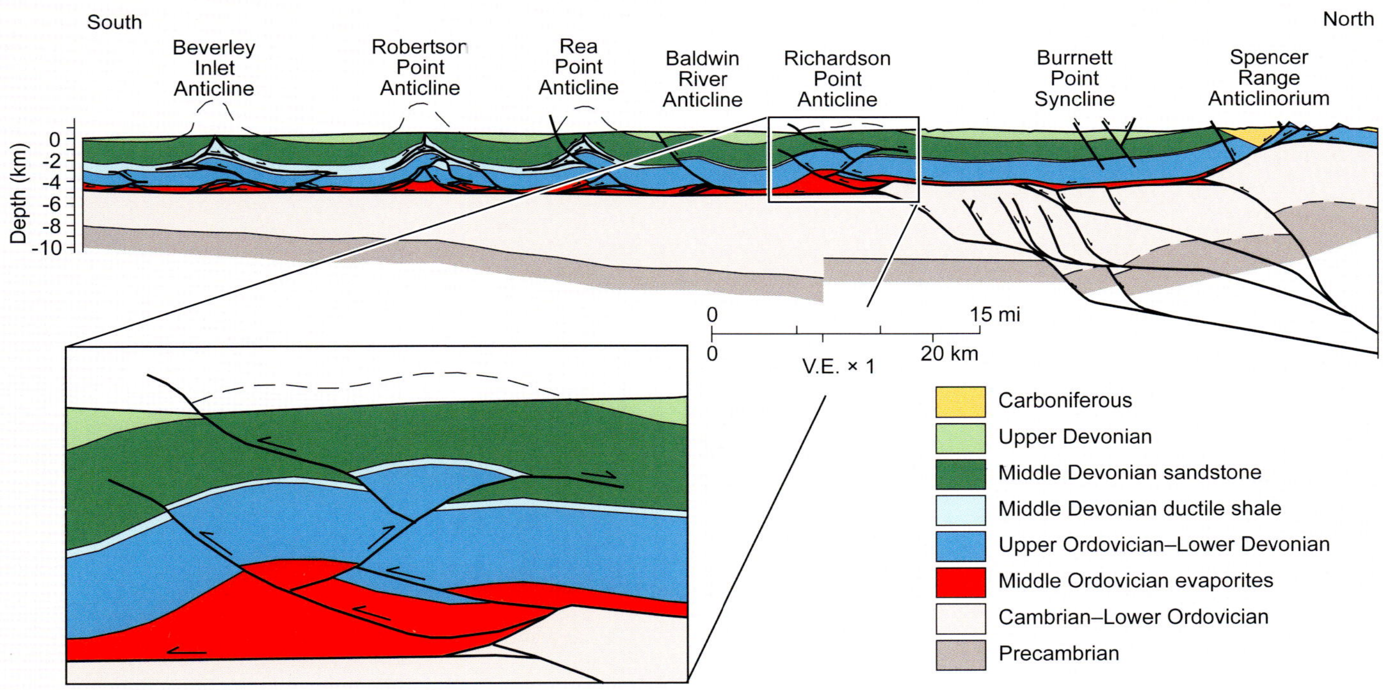

Figure 6.13. Shortening in a multidetachment system. The Richardson Point structure is here composed of two stacked pop-up structures. The shallow pop up formed along one of the flanking faults of the deeper pop up. The salt-detached Parry Islands Fold Belt is linked to basement-involved thrusts related to the Late Devonian–Early Carboniferous Ellesmerian Orogeny. After Harrison (1995).

PARRY ISLANDS FOLD BELT—Salt Overthrust, Rea Point Anticline, Melville Island, Canada

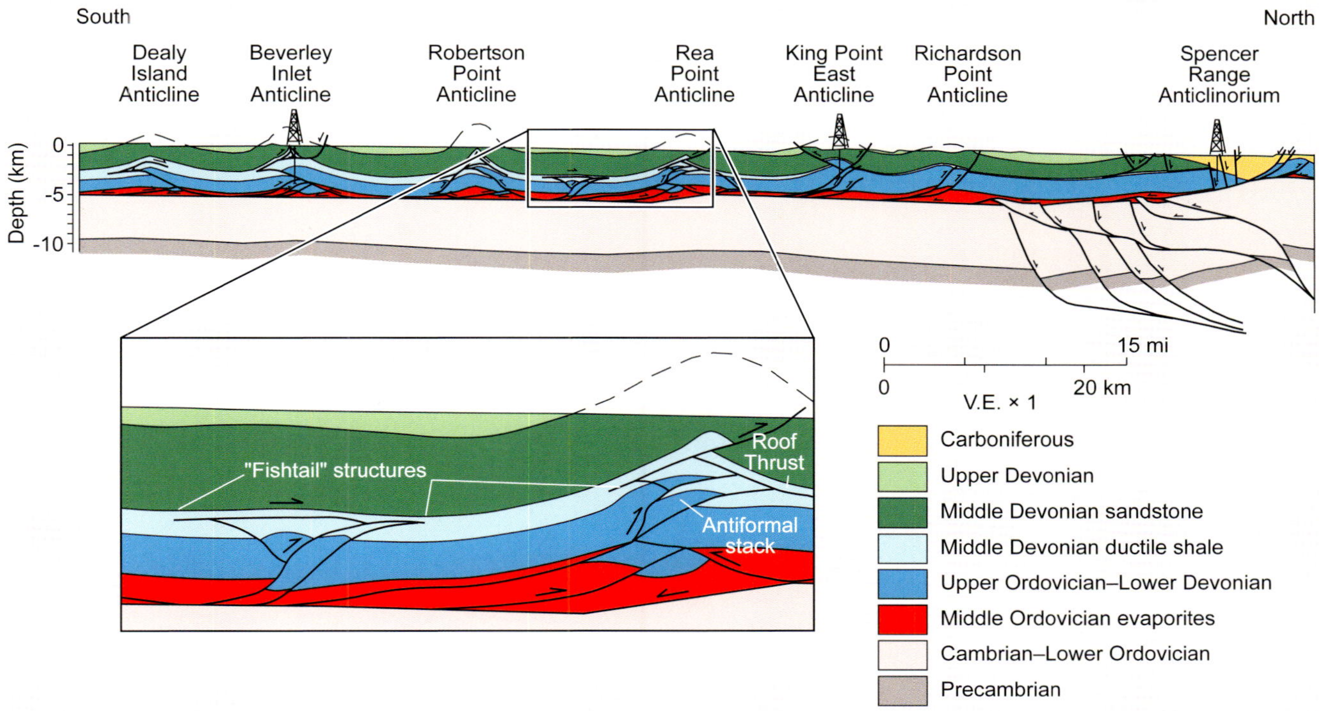

Figure 6.14. Shortening in a multidetachment system. The Rea Point Anticline is cored by an antiformal stack, in which numerous thrust sheets are stacked to arch the overlying roof thrust. Several "fishtail" structures reverse the direction of thrusting and tend to stack compressional structures. From Harrison (1995).

PARRY ISLANDS FOLD BELT—Salt Overthrust, Robertson Point Anticline, Melville Island, Canada

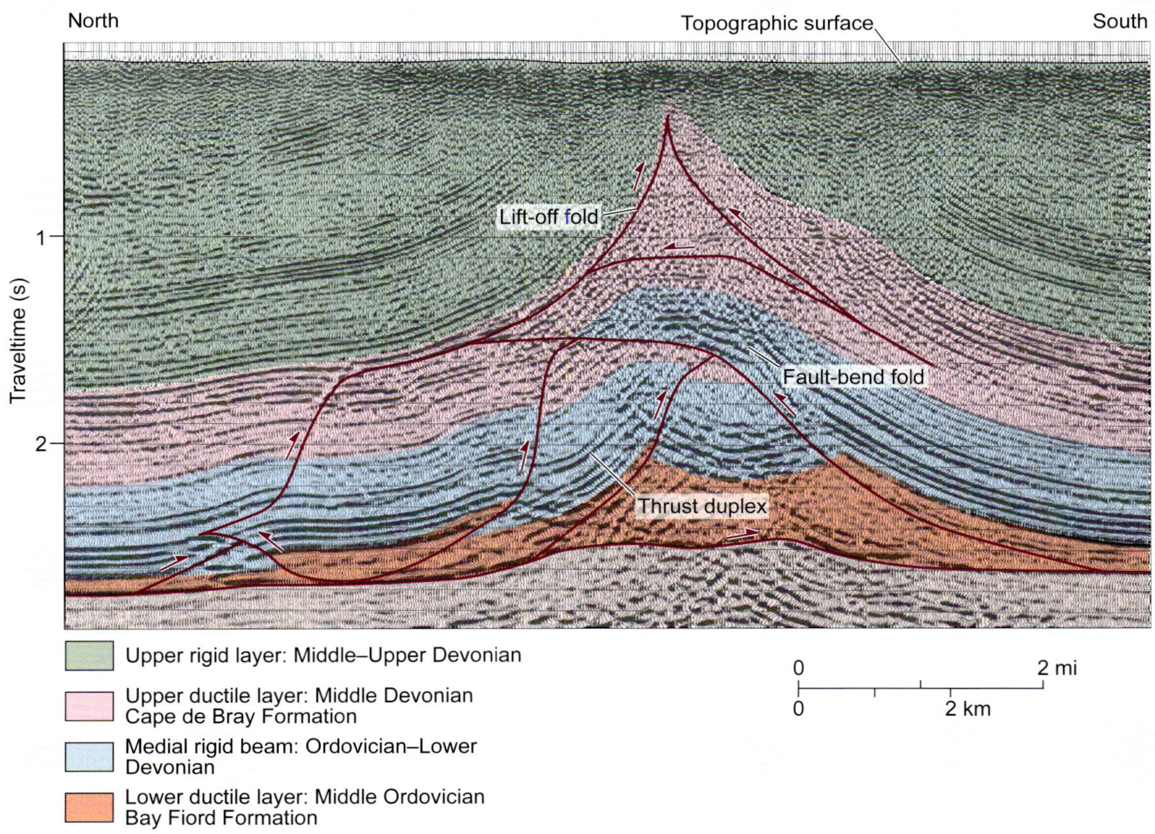

Figure 6.15. Shortening in a multidetachment system. Décollement surfaces exist in both the middle and top of the Cape de Bray Formation, producing a very complicated structure. The simple surface anticline is cored by a lift-off fold above the crest of a fault-bend fold that overlies a thrust duplex. From Harrison (1995).

PARRY ISLANDS FOLD BELT—Mecham River West Pop Up, Melville Island, Canada

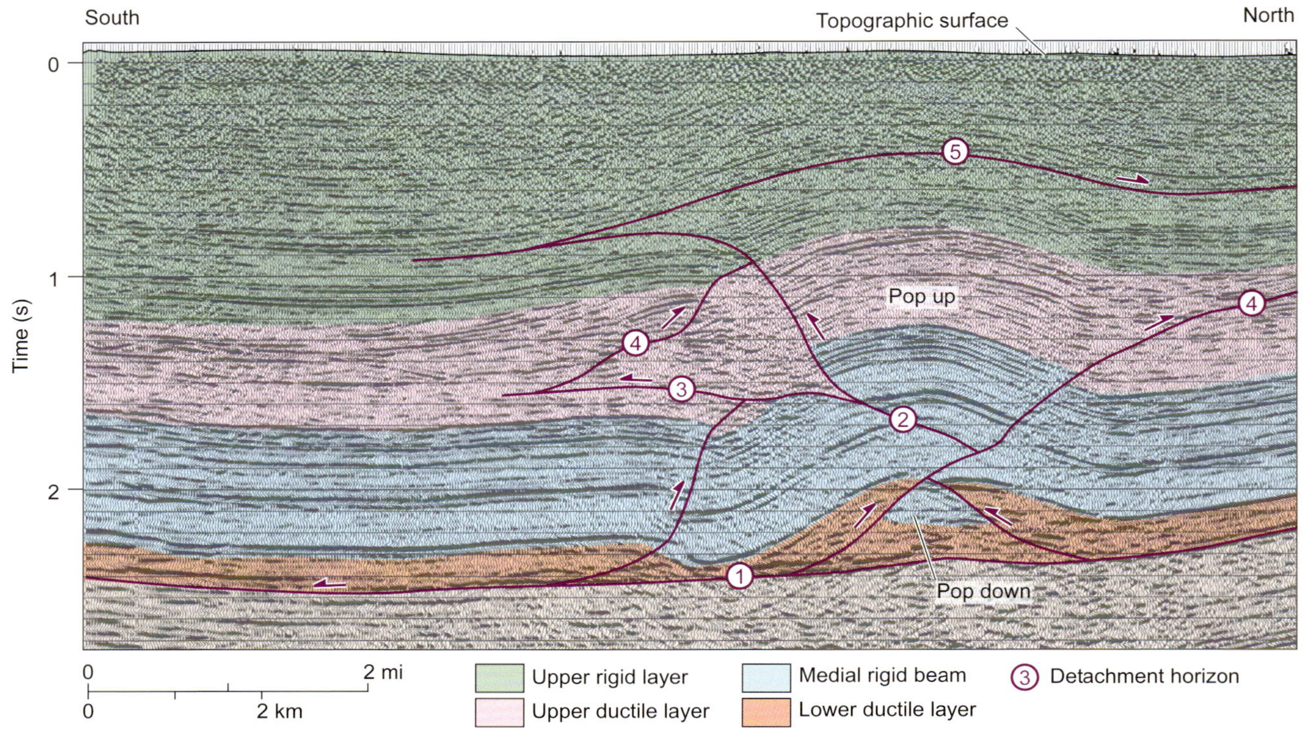

Figure 6.16. Shortening in a multidetachment system. The fault pattern is complicated by the presence of five detachment surfaces. As a result, the structure displays different geometries at different structural levels. This complexity is characteristic of fold belts having multiple detachments. After Harrison (1995).

PARRY ISLANDS FOLD BELT—Dealy Island Pop Up, Melville Island, Canada

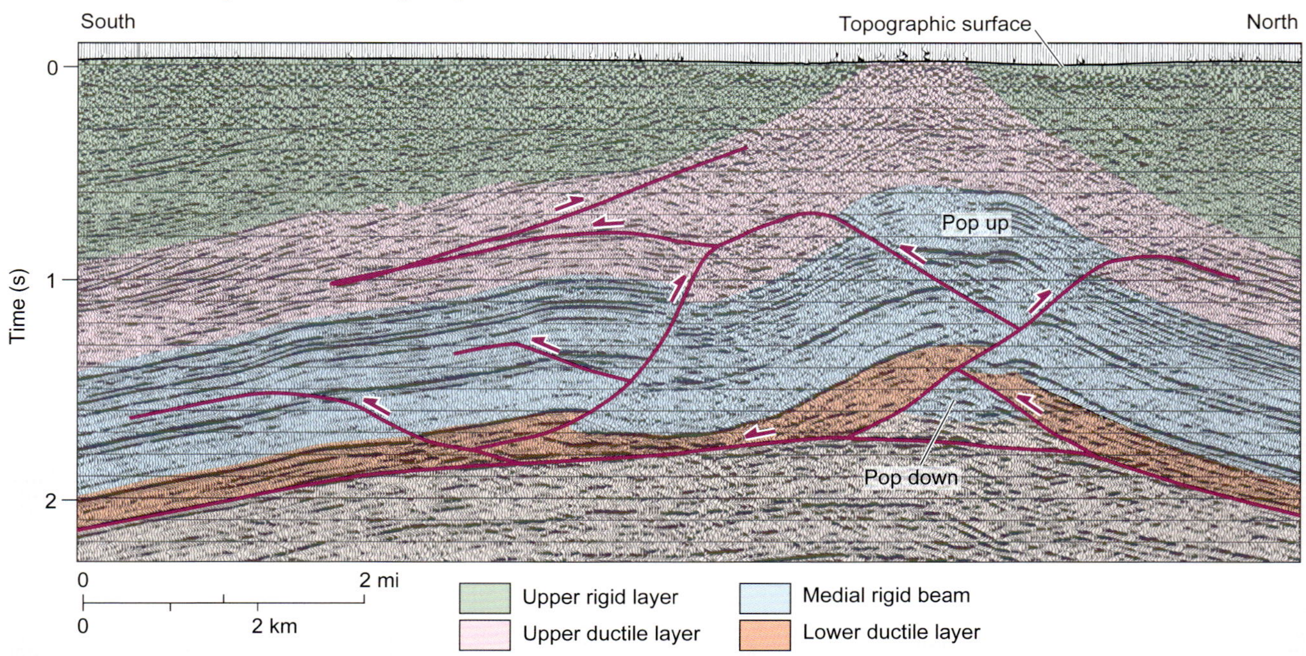

Figure 6.17. Shortening in a multidetachment system. The surface anticline overlies a pop up formed by a north-verging, salt-detached thrust and a single back thrust. The back thrust was offset by the north-verging thrust and so formed first. The crossing of the two thrusts forms an 'X' pattern, with a pop up at the top of the structure and a "pop down" at the base. After Harrison (1995).

POTWAR PLATEAU—Salt Overthrust, Pakistan

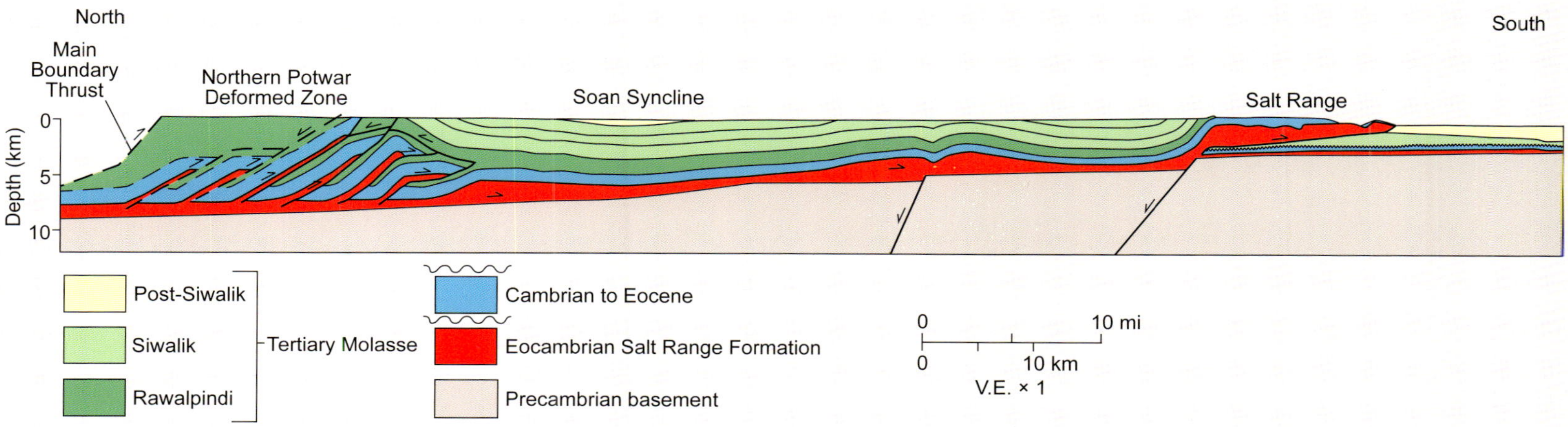

Figure 6.18. Shortening above salt of variable thickness. Two styles of contractional deformation are shown. Adjacent to the Main Boundary Thrust, the Northern Potwar Deformed Zone comprises closely spaced, overlapping thrust sheets, locally organized into duplexes. Salt thickens southward to a limit where the Salt Range forms a single allochthonous salt nappe. The intervening Soan Syncline was translated southward without being deformed because of the effective lubrication by the basal salt detachment. The Salt Range Nappe was apparently localized by a step in the subsalt basement. This step marks the southeastern limit of the original salt basin and may have deflected the southward translation of salt upward; 23 km of shortening is accommodated on this profile. From Baker and others (1988).

POTWAR PLATEAU—Salt Overthrust, Northern Potwar Deformed Zone, Pakistan

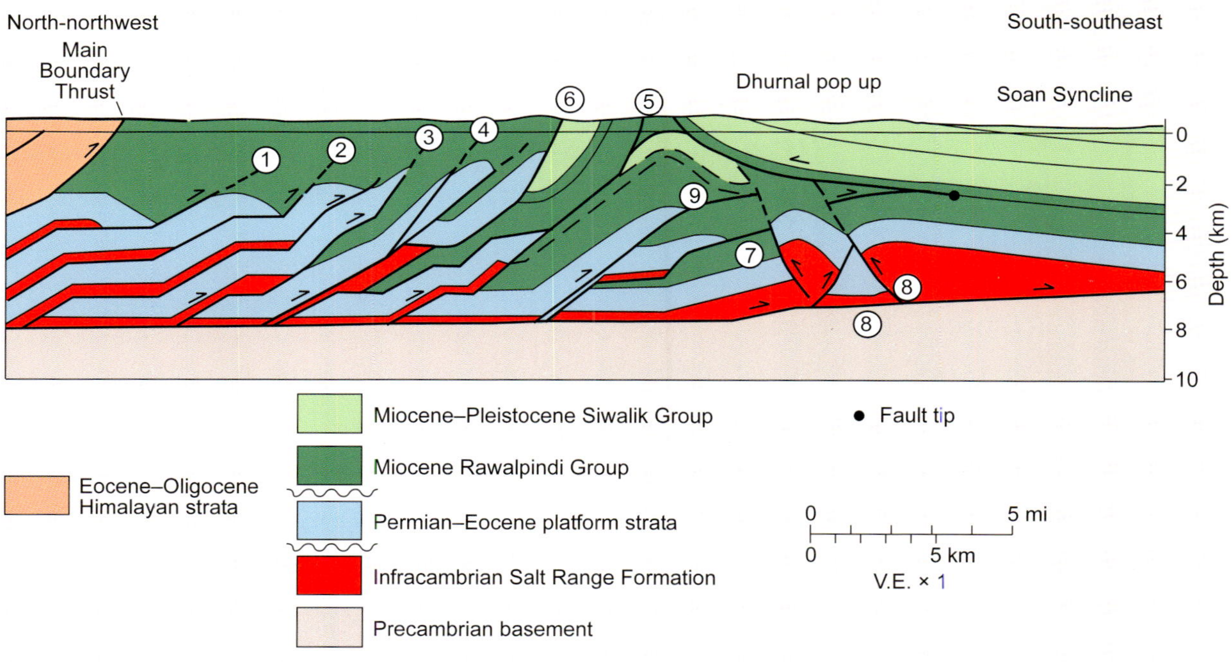

Figure 6.19. Shortening above salt of variable thickness. An imbricate thrust belt formed in the northern part of the section by detaching above thin salt. Southward, salt thickness increases, shortening magnitude decreases, and structural style becomes dominated by salt-cored folds and pop ups. Faults in this section are numbered from oldest (1) to youngest (9). Faults generally become younger toward the foreland, although faults 6 and 9 were both out of sequence. The authors estimate 55 km of shortening in this balanced section. From Jaswal and others (1997).

ZAGROS FOLD BELT—Kuh-e-Sabzu Pop Up, Iran

Figure 6.20. Shortening above thick salt. Pop-up structures are common in areas of thick salt for two reasons. First, thick salt reduces friction along the basal detachment so that bi-directional thrust vergence is more likely. Second, space problems at the core of the pop up can be accommodated by salt flow. After National Iranian Oil Company (1975a).

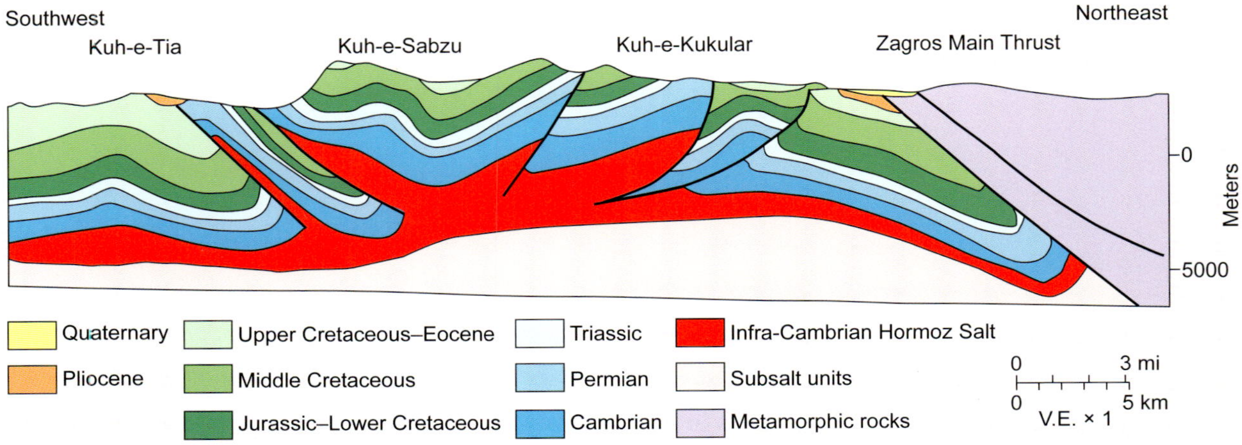

ZAGROS FOLD BELT—Kuh-e-Siah Pop Up, Iran

Figure 6.21. Shortening above thick salt. Pop-up structure forms the western termination of the Kuh-e-Siah anticline. The northern thrust separates the Kuh-e-Siah anticline from the en echelon Kuh-e-Khonj anticline. The map pattern suggests that the south-verging fault is primary and that the north-verging fault formed as a backthrust in the relay zone between the anticlines. Image courtesy of ExxonMobil. Interpretation by Mike Hudec.

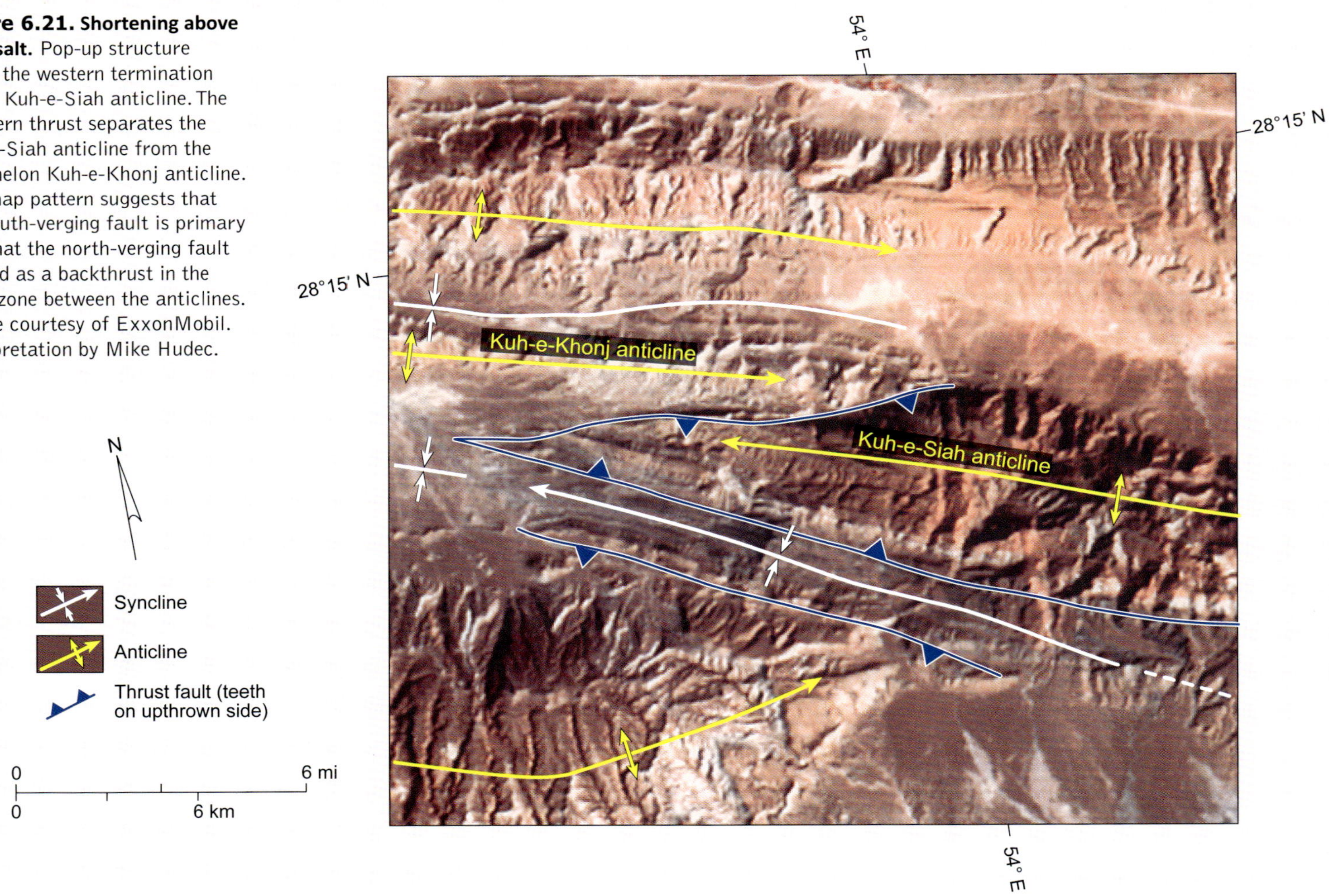

Physical Model—Salt Overthrusts at a Salt-Basin Edge

Figure 6.22. Shortening above thick salt. Forethrusts formed in prekinematic layers at the distal pinch-out of salt. Sediments beyond the pinch-out did not deform, illustrating how salt pinch-outs limit deformation. The ramp-anticline forelimbs steepened until they collapsed as slumps. Each forethrust overrode the slump debris. Model by Bruno Vendeville.

Physical Model—Salt Overthrusts Detached above a Frictional Detachment

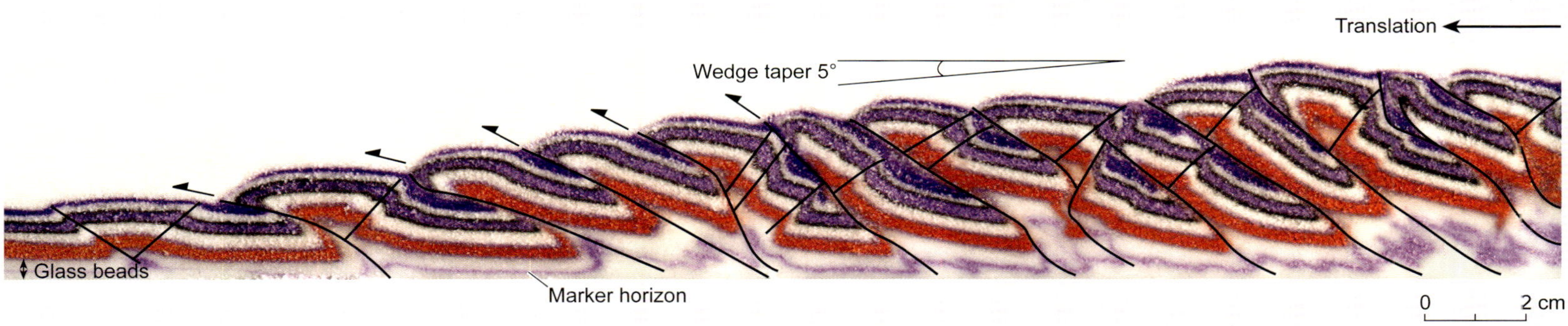

Figure 6.23. Shortening above thin shale. The dominant vergence toward the foreland is typical of thrusting above frictional detachments such as shale. Detachments above salt typically have less or no preferred vergence (compare with a salt detachment in Figure 6.24). Overriding forethrusts locked minor backthrusts. The moderate critical-wedge taper displayed here is typical for regions distant from the lateral boundaries of a shale-based thrust belt.
All layers were originally prekinematic and tabular, as seen in the foreland on the extreme left. The detachment layer contains a midlevel marker horizon. Modified from Costa and Vendeville (2002).

PHYSICAL MODEL–Salt Overthrusts above a Salt Detachment

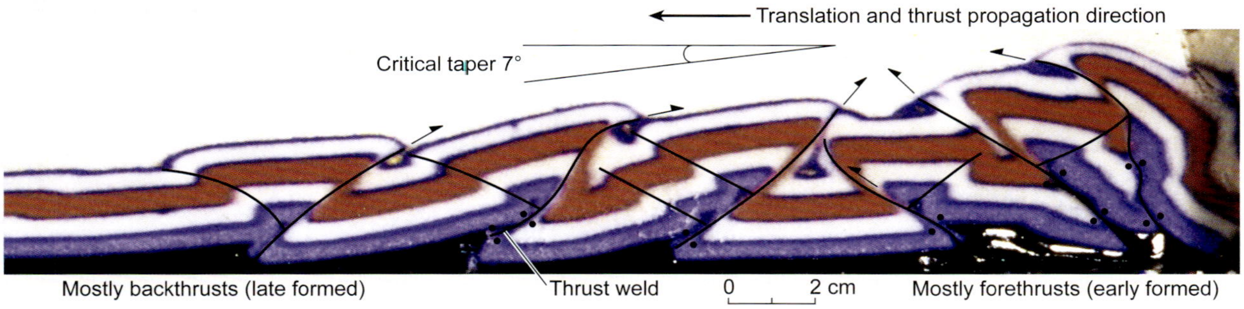

Figure 6.24. Shortening above thin salt. Thrust vergence was nonsystematic, which is typical of salt detachments (compare with a frictional detachment in Figure 6.23). Early thrusting was dominated by forethrusts, but later backthrusting became more common as the basal salt layer thickened as a result of shortening. Less typical are the observations that wedge taper was moderately high and salt overthrusts dominate rather than contractional anticlines. These features were caused by friction along sidewalls in a narrow experimental box. All layers were originally prekinematic and tabular, as seen in the foreland on the left. Model by Bruno Vendeville.

PHYSICAL MODEL–Salt Overthrusts, Pop Ups, and Contractional Anticlines

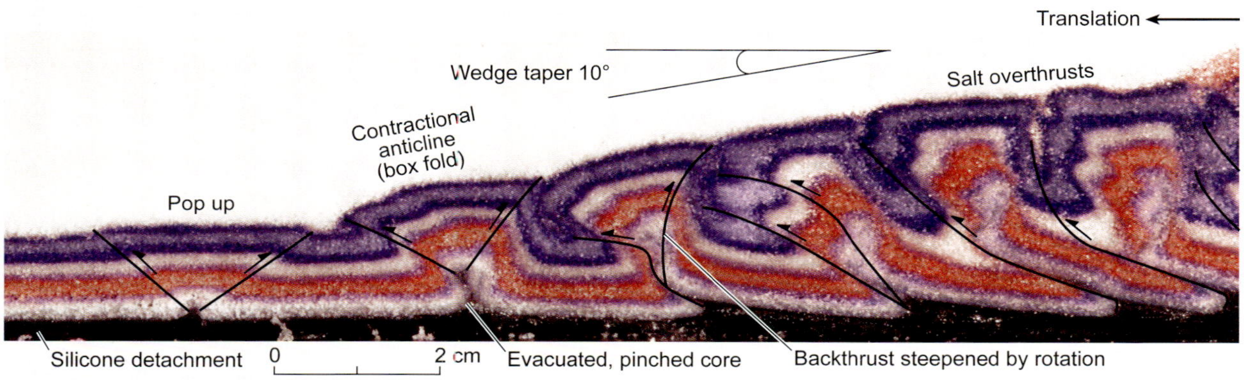

Figure 6.25. Shortening above thin salt. Thrust faults in the hinterland gave way to box folds and pop ups in the foreland. The steep critical-wedge taper and dominance of forethrusts are anomalous for a salt-based detachment. They were caused partly by the thinness of salt but mostly by friction along sidewalls because the experimental box was narrow. All layers were originally prekinematic and tabular, as seen in the foreland on the left. Model by Bruno Vendeville.

PHYSICAL MODEL—Variation of Pop-Up Style with Salt Thickness

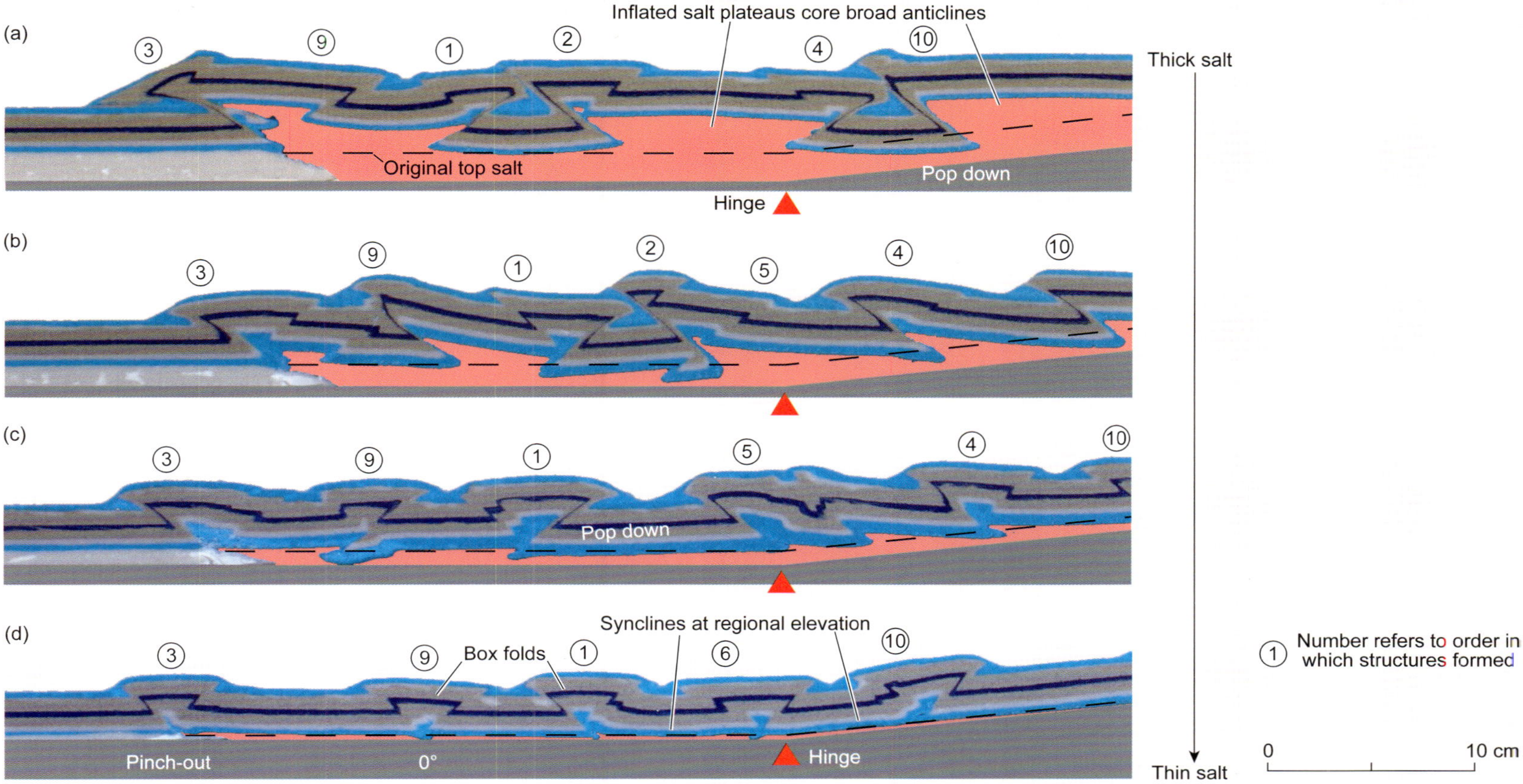

Figure 6.26. Shortening above variable salt thickness. Beds were shortened at the toe of a gravity-driven system above a salt layer that varied in thickness along strike, as shown in these serial sections. Pop-up structures formed in all areas, but their style varied with salt thickness. Above thin salt, broad, flat-bottomed synclines separated narrow pop ups. There was not enough salt available to flow into fold cores or to be expelled from beneath synclines, so the salt acted only as a detachment. Synclines therefore lie at their regional elevation, and anticlines form box folds without salt cores. With thicker salt, broad, flat-topped anticlines separate narrow pop downs, essentially a mirror image of the thin-salt structure. Inflated salt plateaus core salt anticlines. Shortening and thickening of the salt lifted almost all units above regional, although a few pop downs are below regional. Model by Tim Dooley.

PHYSICAL MODEL—Pop Ups in a Thrust Belt Detaching on Salt

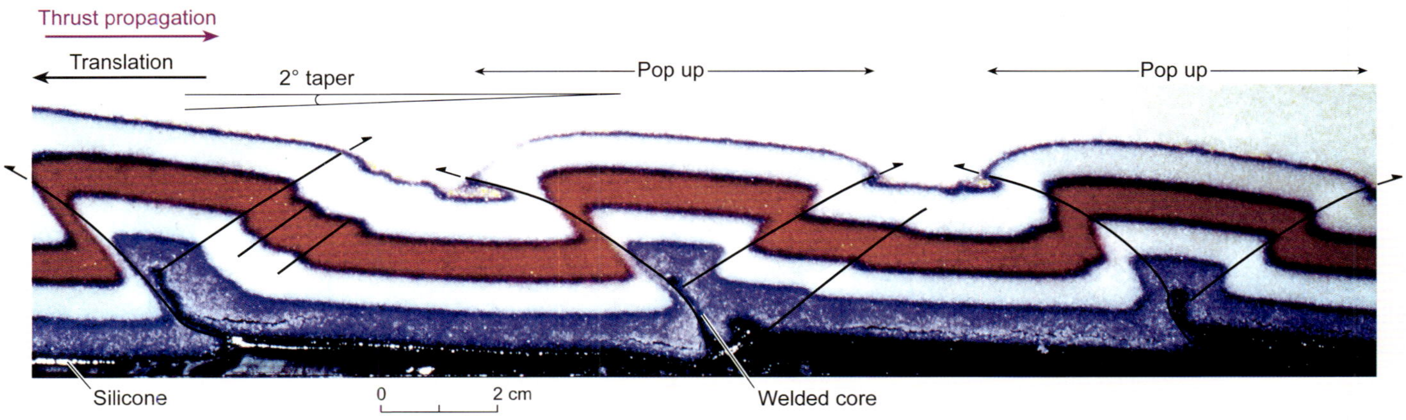

Figure 6.27. Shortening above thin salt. Backthrusts (verging to the left) are slightly more dominant than forethrusts. However, the preferred vergence is weak, as is typical above a weak salt detachment. Translation direction is opposite the direction of thrust propagation and the taper direction of the critical wedge. This unusual feature results from the lack of lateral friction because the sidewalls were also translating along with the right-hand endwall. The left-hand endwall was fixed and so acted as a buttress to translation. A similar mechanical response to lack of lateral friction might be expected in a particularly wide natural thrust belt. Model by Bruno Vendeville.

PHYSICAL MODEL—Pop Up in a Thrust Belt Detaching on Shale

Figure 6.28. Shortening above thin shale. The pop up was overridden by a forethrust from the right and buried by its slump deposits. This overriding would have retarded any further slip of the backthrust. Model by Bruno Vendeville.

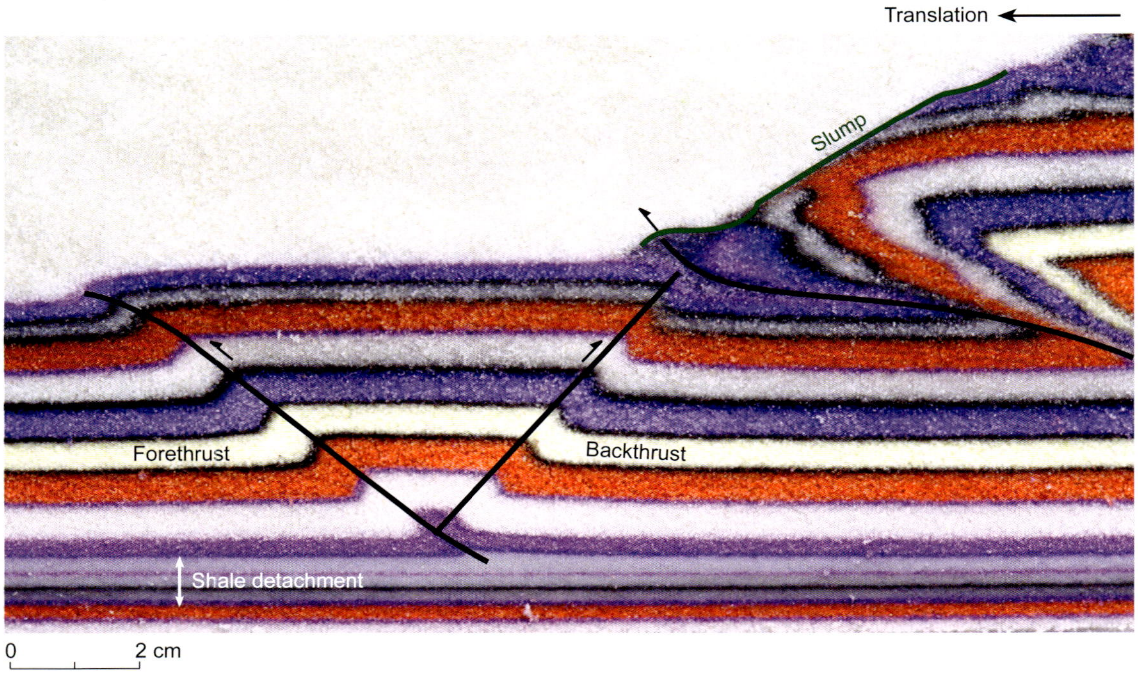

PHYSICAL MODEL—Pop Ups in Double-Detachment System

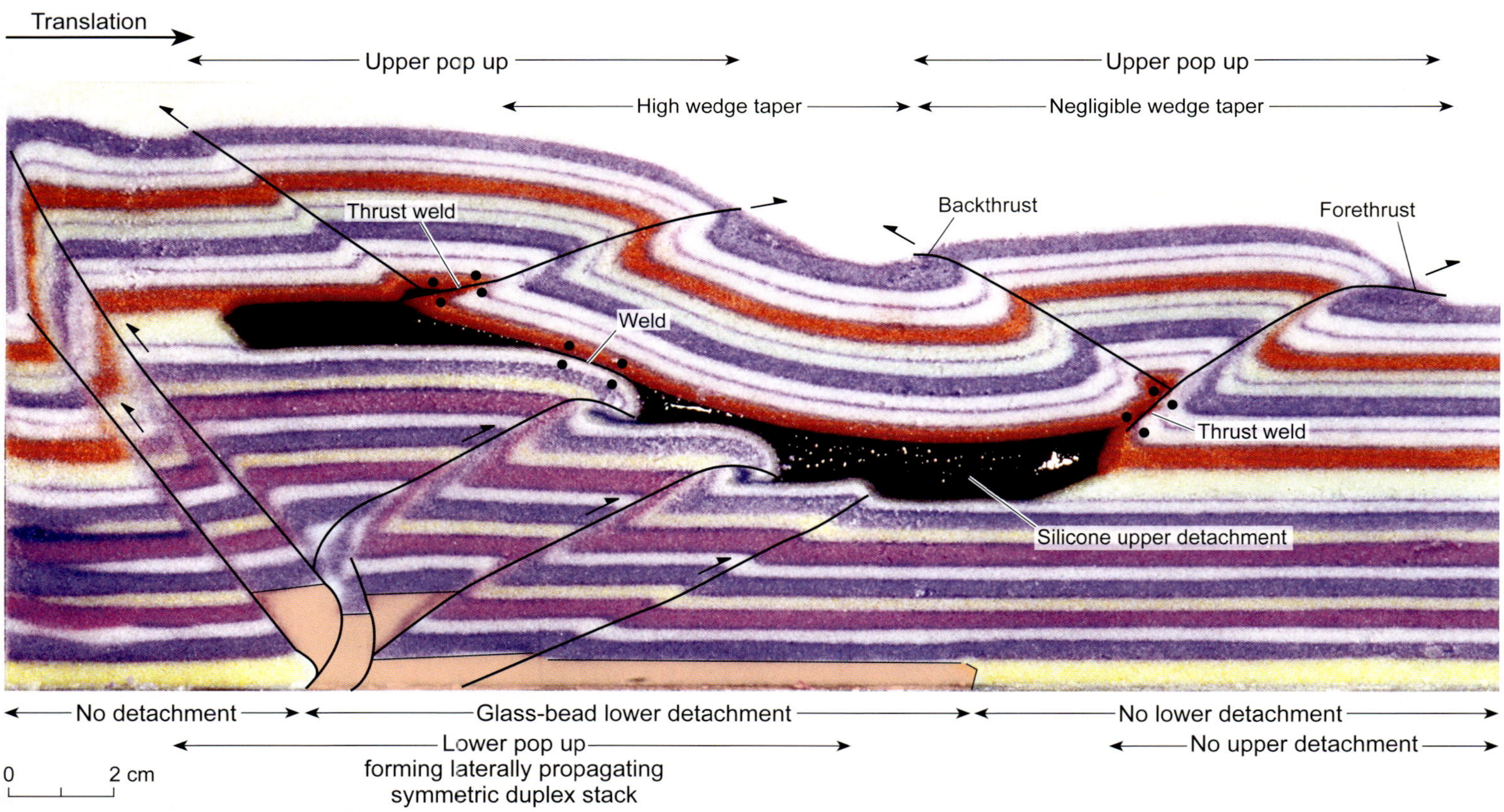

Figure 6.29. Shortening in a multidetachment system. Because both the weaker viscous detachment (silicone=salt) and the stronger brittle detachment (glass beads=shale) were narrow, deformation was strongly localized. Salt was expelled from cores of pop ups, creating thrust welds. Multiple thrusts in the lower pop up elevated the overlying strata and thrusts, thereby increasing the taper of the upper strata. All layers were initially tabular. Model by Bruno Vendeville.

Physical Model—Pop Up above an Active Salt Wall

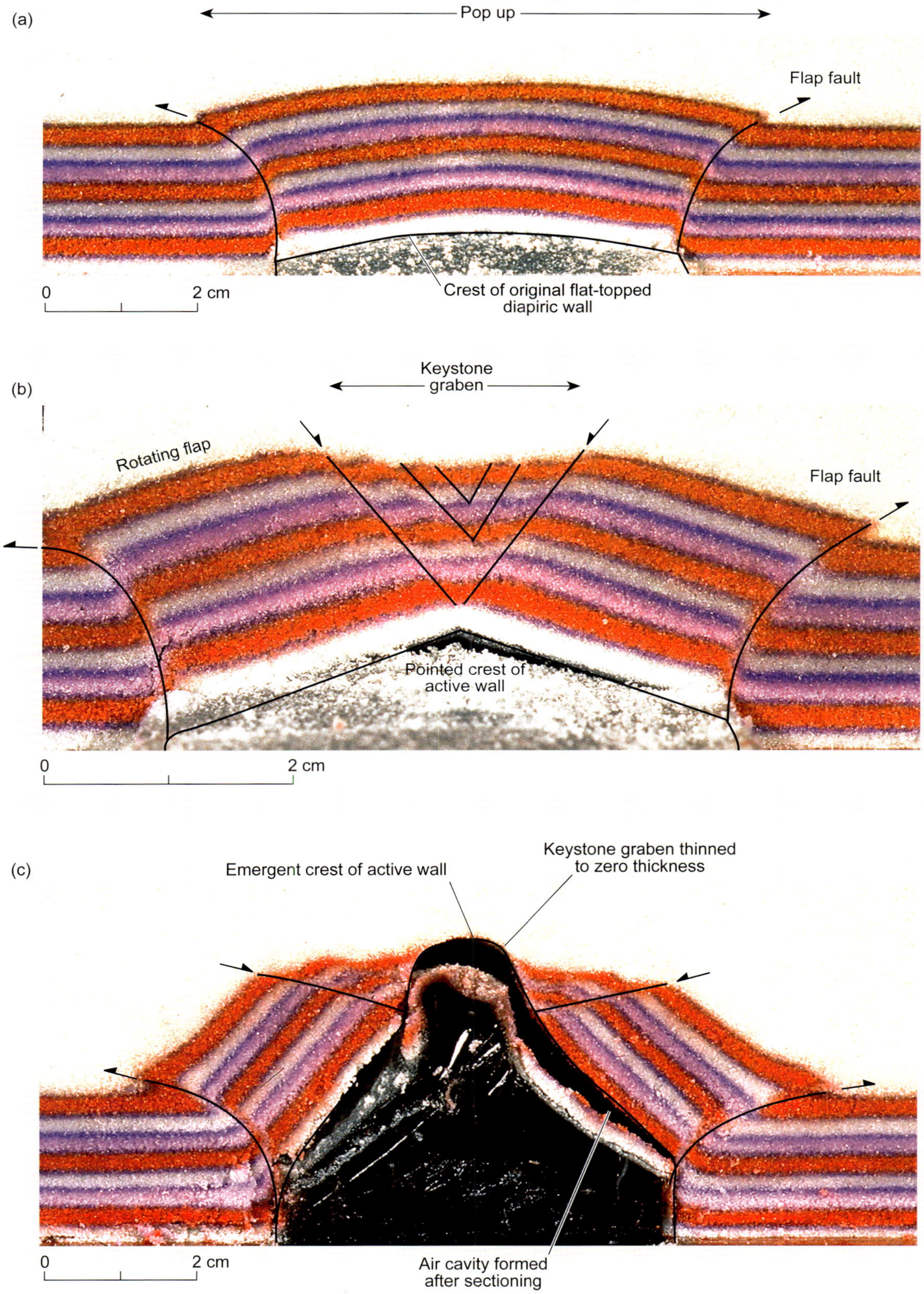

Figure 6.30. **Active diapirism.** Rotation and uplift of the flaps allowed the active diapir to push up and penetrate its roof. Three cross sections of differing structural maturity along strike illustrate the stages of emplacement. Lower parts of the diapir are not shown. (a) Outward-verging flap faults steepened downward from shallow thrust faults to vertical faults rooted along the vertical sides of the underlying diapiric wall. (b) Crestal graben formed over apex to accommodate flap rotation. (c) Active diapir broke through to the surface. Model by Bruno Vendeville.

PHYSICAL MODEL—Salt Overthrusts in a Multidetachment Thrust Belt

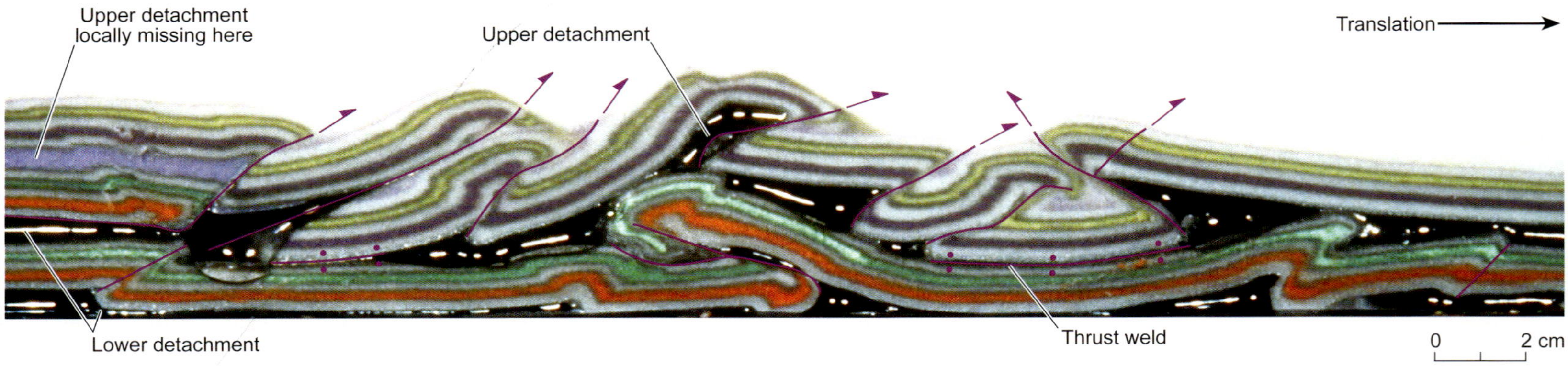

Figure 6.31. Shortening in a multidetachment thrust belt. Salt flow in the upper detachment was enough to largely decouple deformation. Deformation above the lower detachment created a few widely spaced structures, in contrast to the more closely spaced structures above the upper detachment. The two salt detachments were of equal strength and original thickness. Left-hand endwall moved to the right very slowly (0.5 cm/h), simulating very-low-viscosity detachments. On the left, the lower detachment was duplicated and stacked by thrusting. Here the upper detachment was never present, so shortening was entirely over the lower detachment. Thus, in the rest of the cross section, the upper detachment accommodated more shortening than did the lower detachment. Modified from Couzens-Schultz and others (2003).

PHYSICAL MODEL—Thick-Skinned Overthrusts

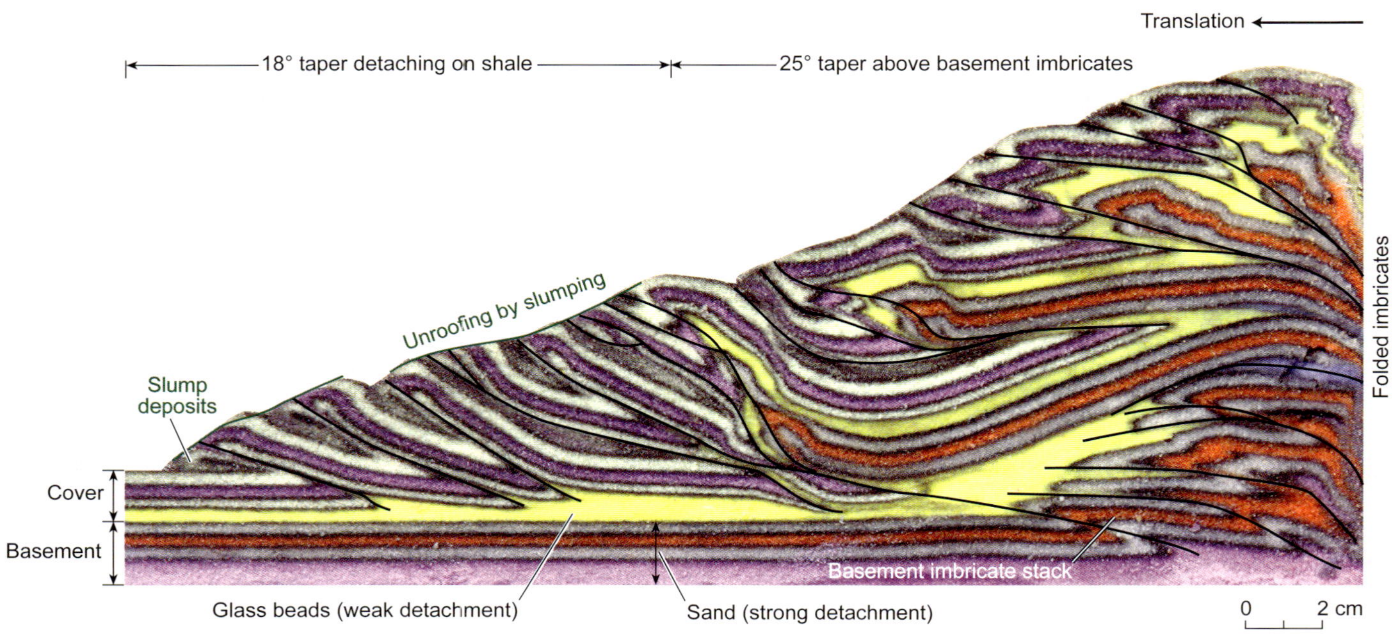

Figure 6.32. Shortening involving a shale detachment and the basement. The yellow shale detachment overlay a much stronger detachment within the shallow basement. Vergence was uniformly basinward, in contrast to the more variable vergence exhibited by salt-detached structures (Figure 6.24). On the right, thick-skinned shortening imbricated thrust sheets of basement, which stacked up to build an overall 25° dipping critical wedge, along with the cover units. Conversely, on the left, where shortening was thin skinned, the wedge taper above the shale was only 18°. The longer basement imbricates rode on the shale detachment while being folded. All layers were originally prekinematic and tabular, as seen in the foreland on the extreme left. Model by Bruno Vendeville.

NUMERICAL MODEL—Pop Up above Active Diapir at 152,000 a, Isotropic Roof with Shortening

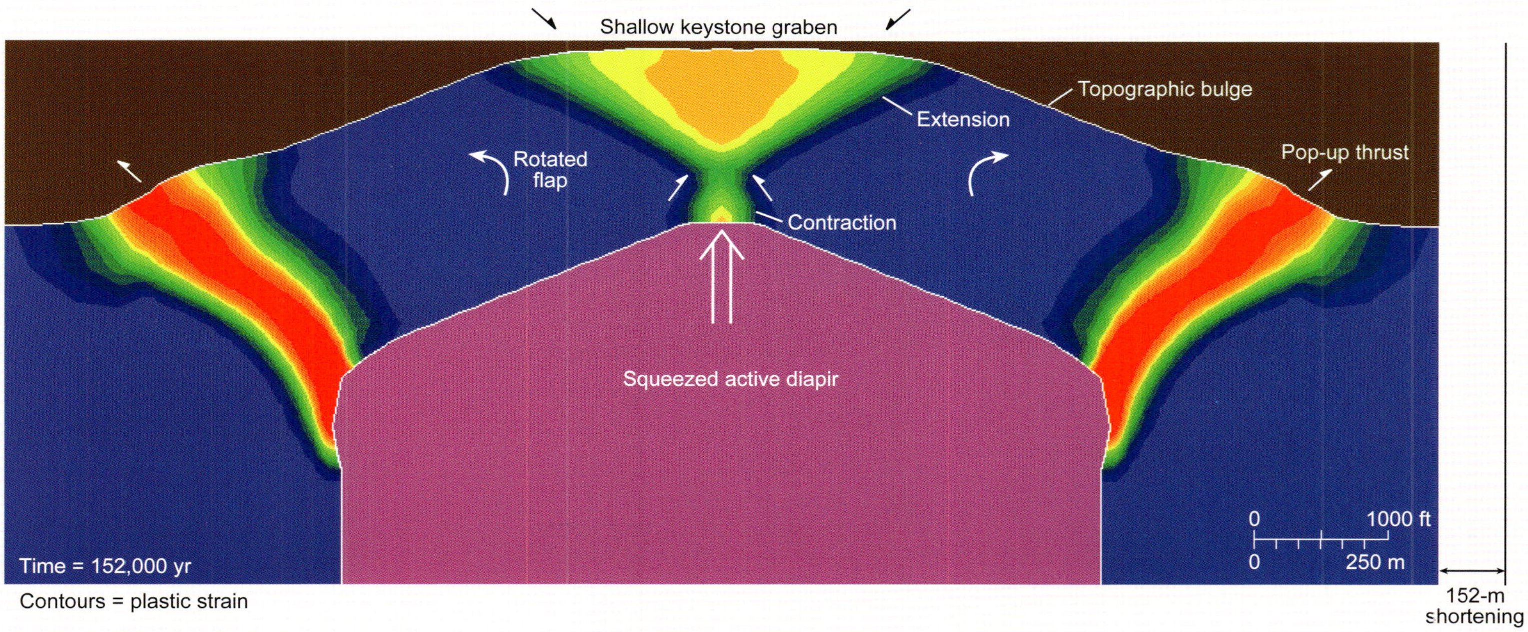

Figure 6.33. Halokinetic active diapirism during shortening. Outward rotation of the roof flaps caused the pop-up thrust faults to rotate to shallower dips through time. The zone of contraction immediately above the diapir crest inhibited the formation of a reactive diapir. Sediment:salt density ratio = 1.09. GeoSim finite-element model by Dan Schultz-Ela.

NUMERICAL MODEL—Pop Up, Isotropic Overburden

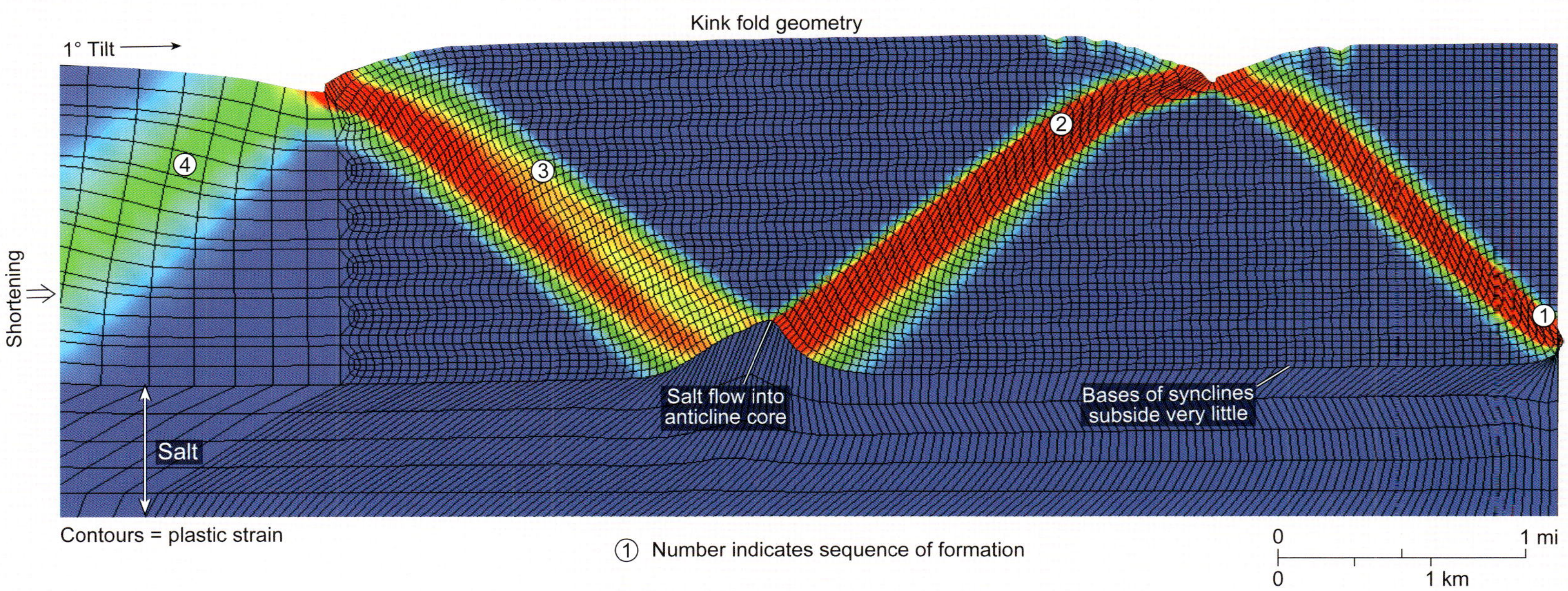

Figure 6.34. Shortening above thick salt. The model deformed along narrow shear bands (thrust faults). The resulting pop-up geometry was very different from the smoothly curved fold produced when salt interlayers were added to the same model setup. The shear bands formed sequentially from downdip to updip. The shear bands widened through time, eventually consuming the flat segment at the base of the anticline and creating a pointed fold core. Abaqus finite-element model by Dan Schultz-Ela.

CONCEPTUAL MODEL—Salt Overthrust, Vertical Transfer of Thrust Displacement

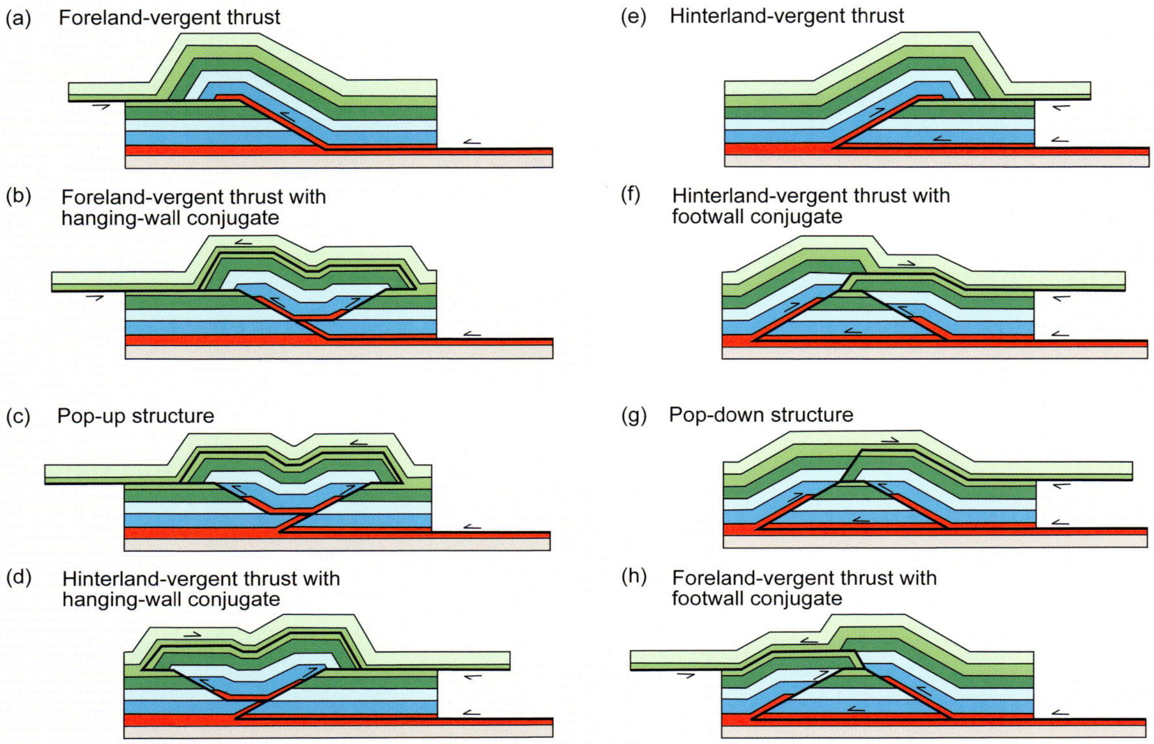

Figure 6.35. Shortening in a multidetachment system. Much structural variation can result from interaction between thrusts of opposing vergence above a salt detachment. All of these styles are based on actual structures in the Parry Islands Fold Belt, Canada (Figures 6.12–6.17). In map view, many of these styles could lie within a single fold system, where displacement is relayed between faults of opposing vergence. This style of deformation is most likely above thin salt. See also duplex styles above a salt detachment in Figure 6.36. Modified from Harrison (1995).

CONCEPTUAL MODEL—Salt Overthrust, Duplex Styles above a Salt Detachment

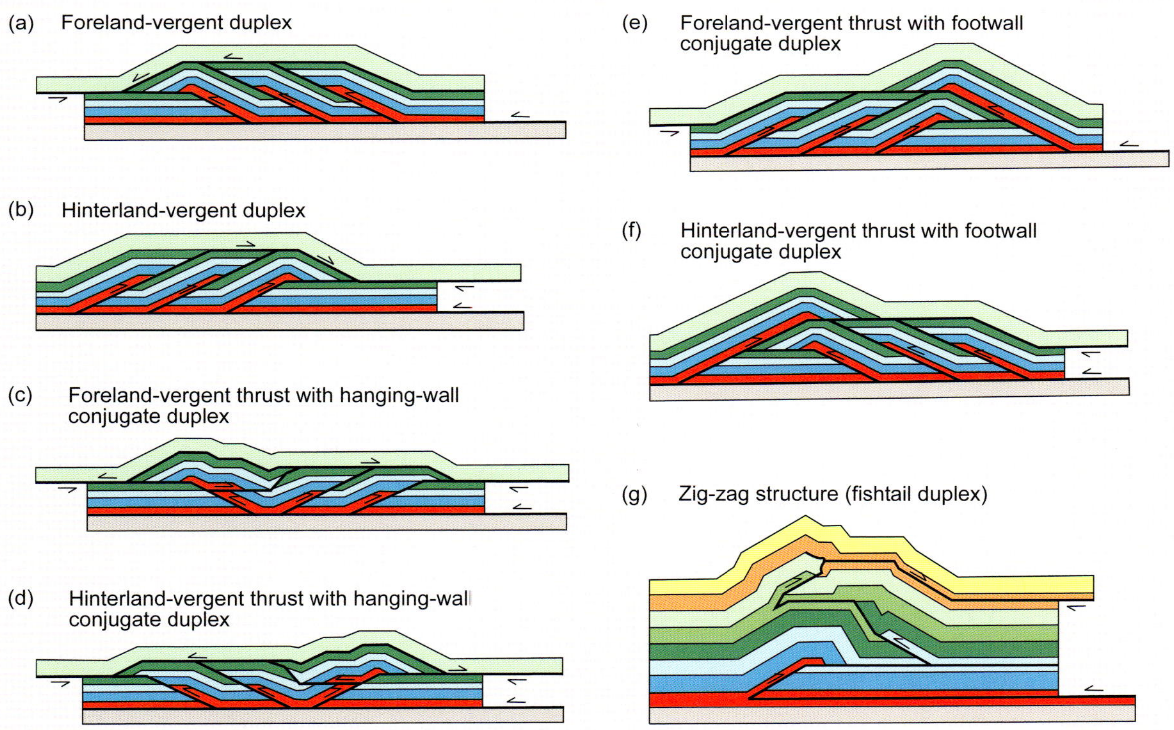

Figure 6.36. Shortening in a multidetachment system. Multiple detachment levels promote duplex structures in which broad anticlines are cored by complex thrust systems. The true complexity is hinted at by multiple dip panels on some fold limbs. This style of deformation is most likely above thin salt. All examples except c are based on actual structures in the Parry Islands Fold Belt, Canada (Figures 6.12–6.17). See also vertically linked thrust structures in Figure 6.35. Modified from Harrison (1995).

Chapter 7

Salt Sheets

When subaerial salt glaciers were discovered in Iran in the 1920s, they seemed strange, exotic, and rare. However, seismic surveys in recent decades reveal that salt sheets are common in the subsurface and form the largest of all salt structures. The Sigsbee salt canopy in the Gulf of Mexico covers 137,000 km^2, more than three times the area of Switzerland. Salt sheets have been intensely studied by hydrocarbon geologists because they offer challenges in drilling, seismic imaging, and geologic interpretation.

Salt sheets are subhorizontal or moderately dipping, sheetlike diapirs of allochthonous salt emplaced above sediments that are younger than the autochthonous source layer. Most salt sheets initiate either in the hanging walls of thrust faults or as extrusions from the crests of emergent diapirs. This chapter presents sheets fed from emergent diapirs; thrust-related sheets are described in Chapter 6. Furthermore, this chapter is restricted to salt sheets that have been little deformed after emplacement. More deformed salt sheets are presented in Chapters 8–10.

Diapir-fed salt sheets spread extrusively when salt rise is faster than sediment aggradation. Sheets gradually climb through progressively younger strata, producing an overall wedge shape. Variations in the ratio of salt-advance rate and sediment-aggradation rate produce ramps and flats in the base of the allochthonous salt. Adjoining salt sheets may coalesce to form salt canopies. Once salt sheets or canopies are buried, they can advance only by thrusting around their margins.

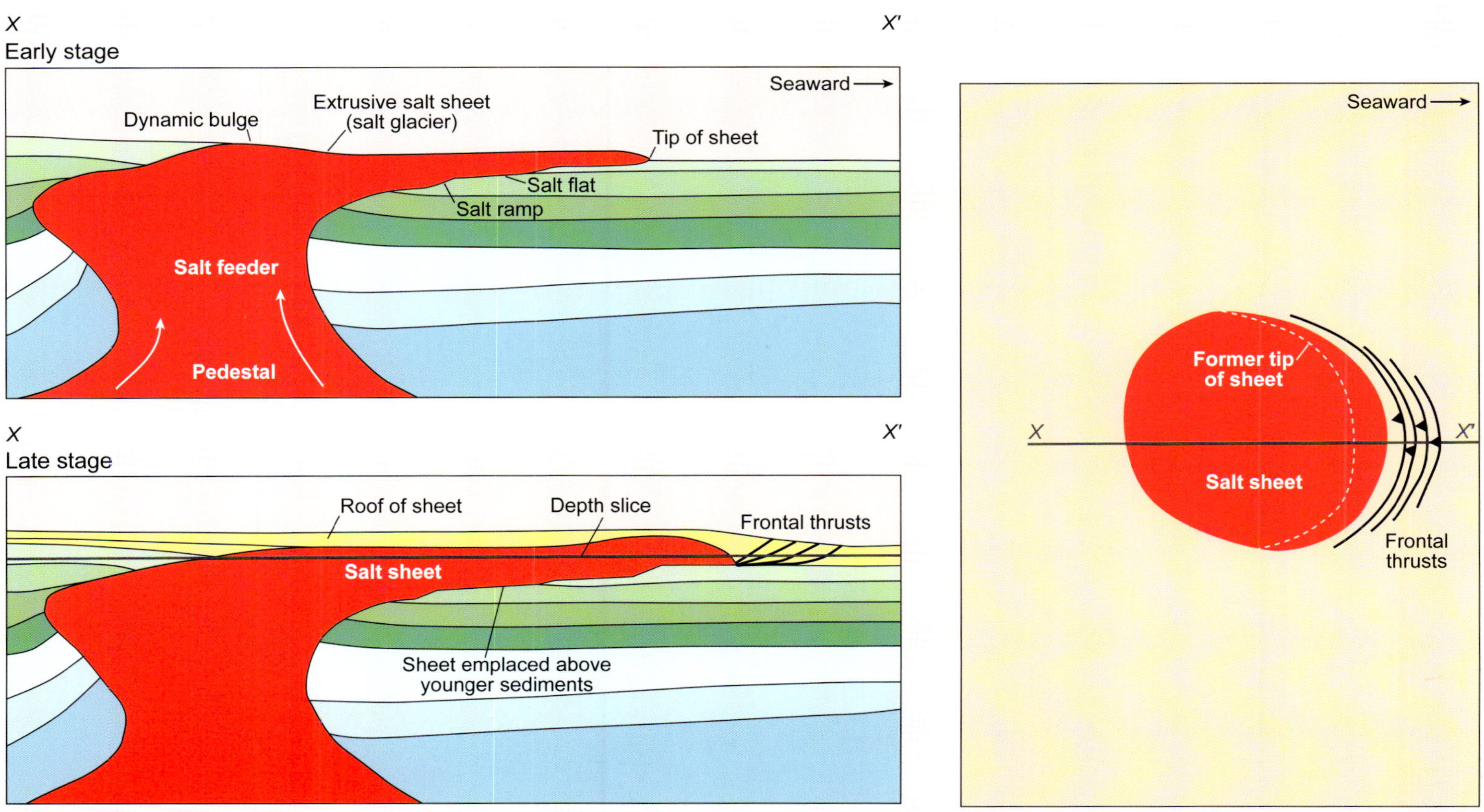

BETIC CORDILLERA—Guadalquivir Overthrust Salt Nappe, Spain

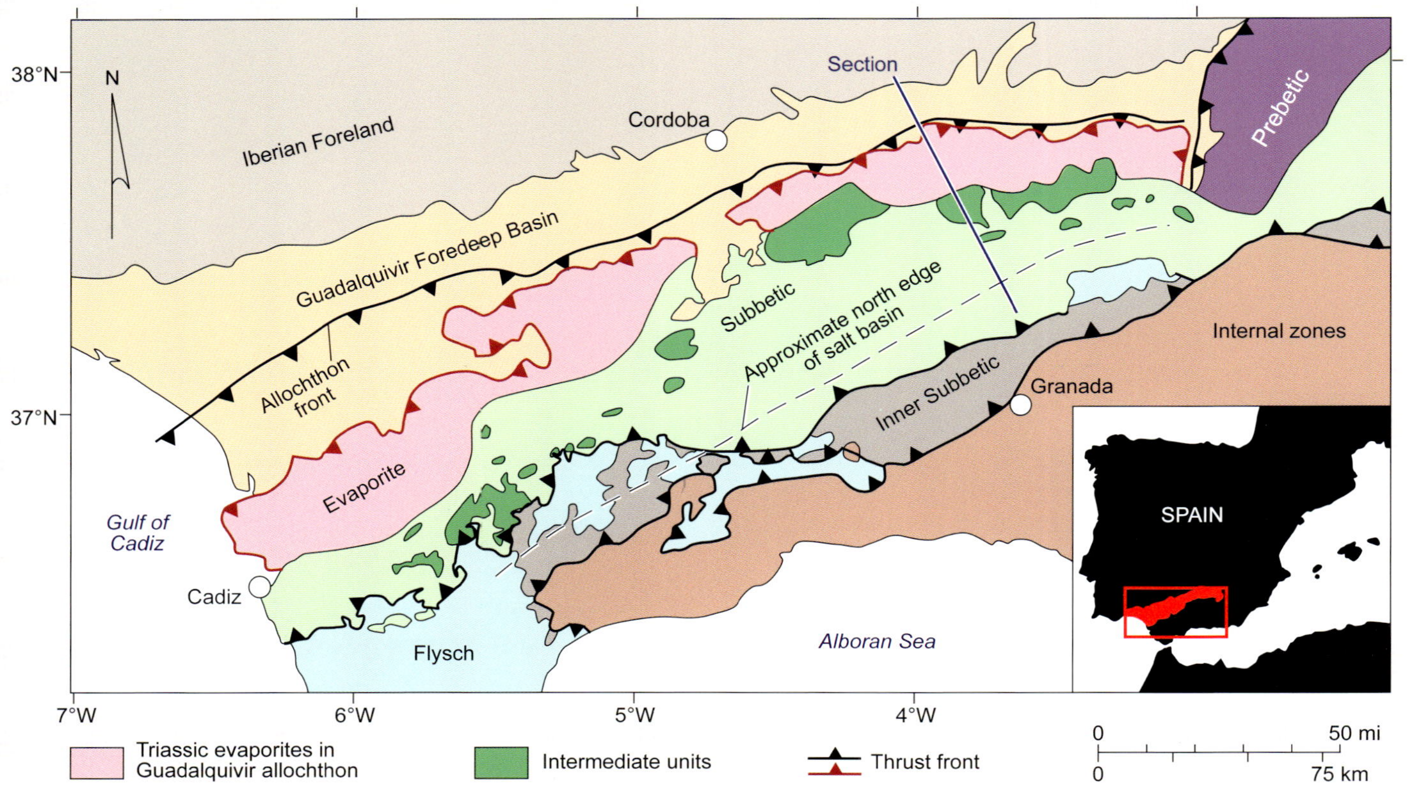

Figure 7.01. Salt sheet expelled by orogenic shortening. The broad salt nappe exposed in the Guadalquivir foreland basin was probably sourced from a basin roughly 40 km south of the salt front (see cross section, Figure 7.02). Salt was squeezed northward from beneath the Subbetic thrusts during Miocene shortening and is interpreted to be entirely detached from its source region. Modified from Berástegui and others (1998).

BETIC CORDILLERA—Guadalquivir Overthrust Salt Nappe, Spain

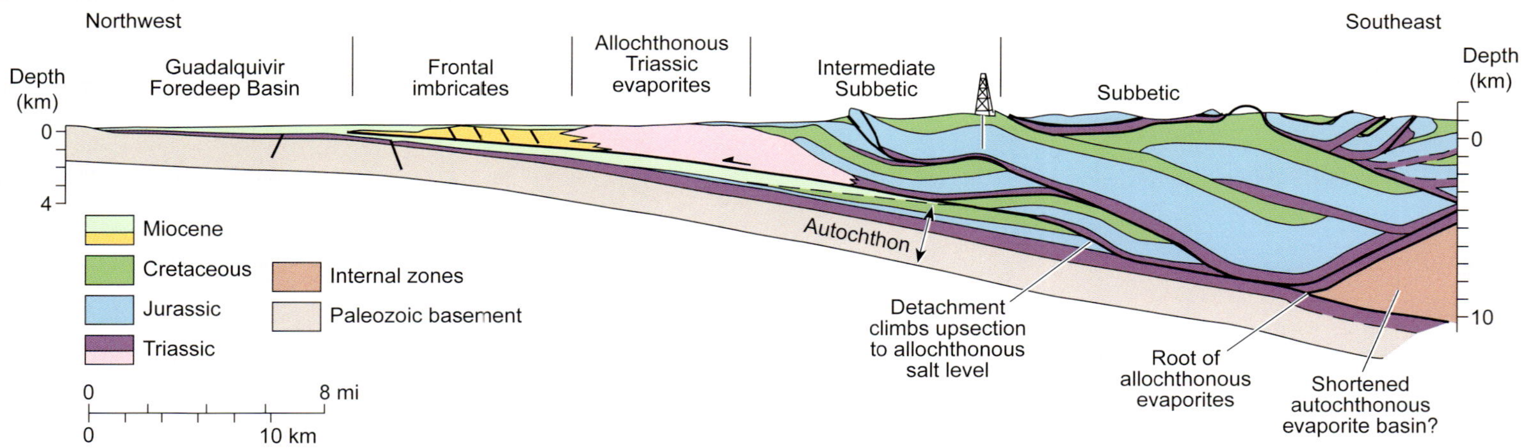

Figure 7.02. Salt sheet expelled by orogenic shortening. During the Tortonian (late Miocene), an allochthon of Triassic Keuper evaporites was squeezed northwestward 40 km over Paleozoic to upper Miocene sediments adjacent to the original salt basin (see geologic map, Figure 7.01). The allochthonous salt now forms a complexly deformed lozenge that is completely detached from its original salt basin. Advance of the allochthonous salt in front of the main thrust belt caused the detachment to climb upsection to the allochthonous salt level and also provided a lubricant for thrusting. Modified from Berástegui and others (1998).

GARMSAR BASIN—Extrusive Salt Sheet, Northern Iran

Figure 7.03. Evaporitic nappe escaping southward from the Alborz Orogen. The nappe of largely Eocene evaporites has spread southwestward from where it emerges in the Alborz mountain front. Flow of the nappe has dismembered and rotated kilometer-scale rafts of its Miocene (middle Upper Red Formation) roof. The salt nappe is actively spreading with a steep deformation front over the plains at its southwest end. East of Eivanekey, the feather edge of the nappe suggests that this part of it is wasting by dissolution. Locally, steep northward-dipping layering and fabrics at outcrop scale hint at the possibility of salt feeders hidden below the nappe. Salt stocks pierce the Alborz mountains northeast of the nappe root. Image courtesy of the Geological Survey of Iran.

GULF OF MEXICO BASIN—Growth of Mica Salt Sheet (Mitchell Dome), Mississippi Canyon Area, Offshore USA

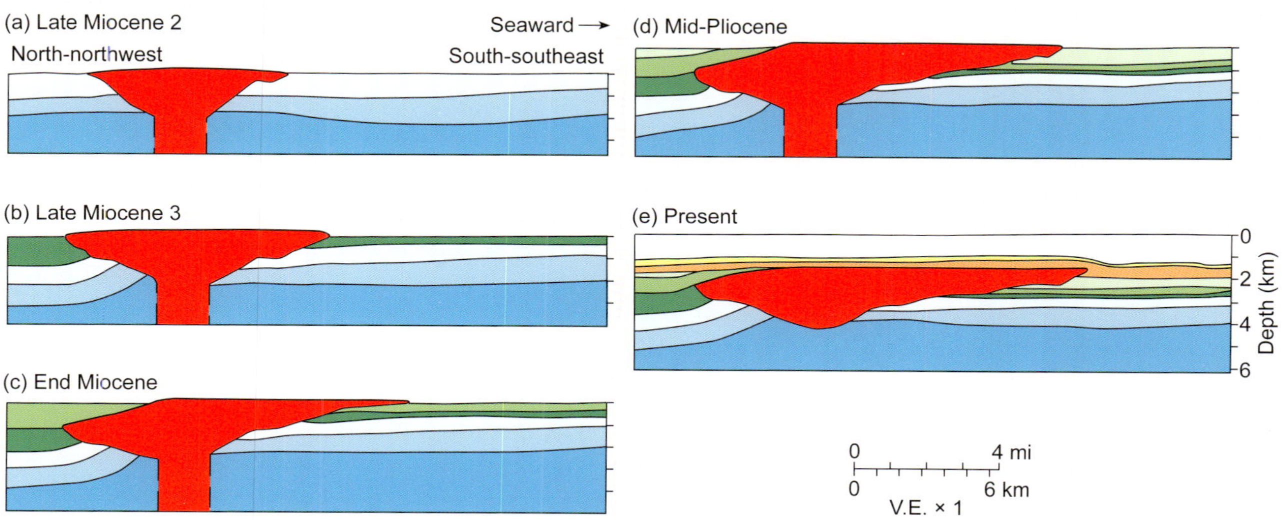

Figure 7.04. Passive growth of a salt sheet. The sheet advances glacially by extruding over adjoining sediments. The sheet spreads preferentially southward down the gentle regional dip in this direction. The landward side of the sheet was buried first, presumably because the sedimentation rate was higher here than on the distal side. In the Plio-Pleistocene, the entire sheet was buried and stopped advancing. Sediment retrodeformation was by vertical simple shear. The shape of the emergent salt at each stage was estimated from a 3-D geomechanical model. The feeder stem is generalized because it was invisible in the seismic section. Modified from Fletcher and others (1995).

GULF OF MEXICO BASIN—Sigsbee Salt Canopy Regional Section, Walker Ridge Area, Offshore USA

Figure 7.05. Transportation of a salt sheet roof. The salt-sheet has advanced 12 km in the Pleistocene, (shaded green), carrying its roof with it. A thrust fault separates the transported roof from underlying sediments. Although the roof is in the hanging wall of a major thrust system, all faults on top of the sheet in this profile are extensional. Ramps and flats in the base of salt may be related to changes in the relative rates of sheet advance and sedimentation during an extrusive phase. However, it is likely that the shallower steps are thrust geometries because we infer that the extrusive phase had ended by this time. Seismic data courtesy of CGGVeritas.

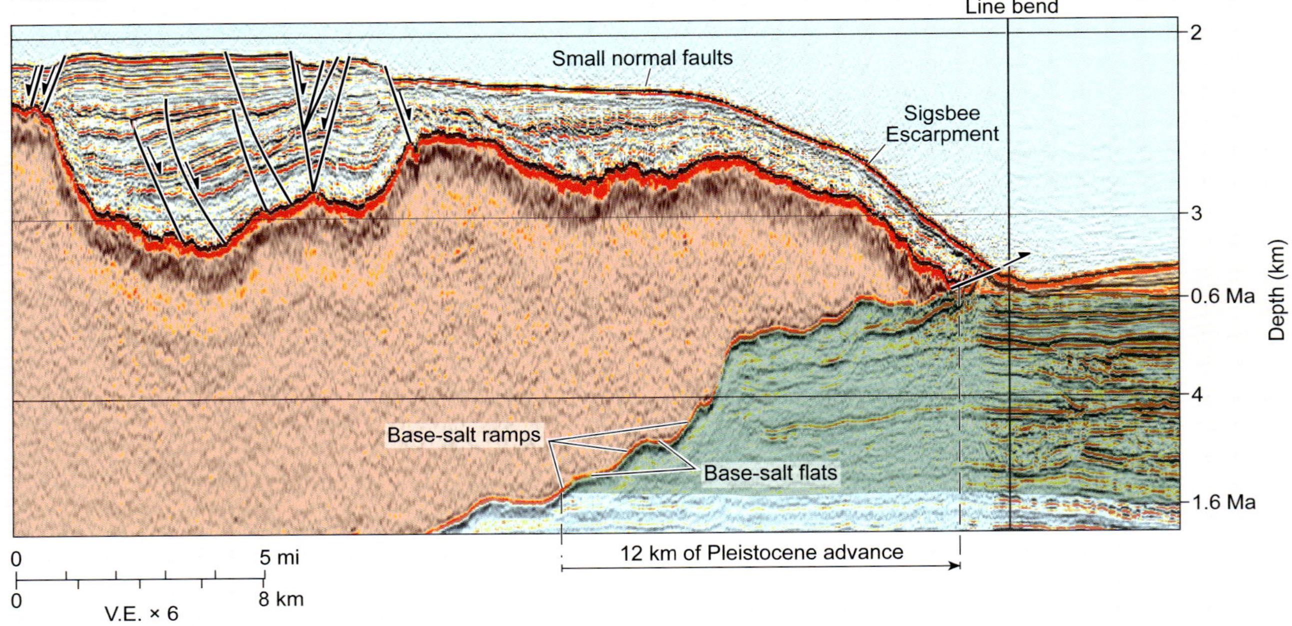

GULF OF MEXICO BASIN—Seismic Dip Map on Base of Salt at Mica Salt Sheet (Mitchell Dome) and Farnella Dome, Mississippi Canyon Area, Offshore USA

Figure 7.06. Passive growth of salt sheets. Extrusive advance of these two pancake-shaped salt sheets is recorded by alternating salt ramps (dark bands) and salt flats (pale gray–white), concentric around the diapir stems that fed sheet growth. Dark radial lines in both domes could be normal faults offsetting the base of salt. These two salt sheets coalesced along a suture to form part of a small canopy. Modified from Fletcher and others (1995).

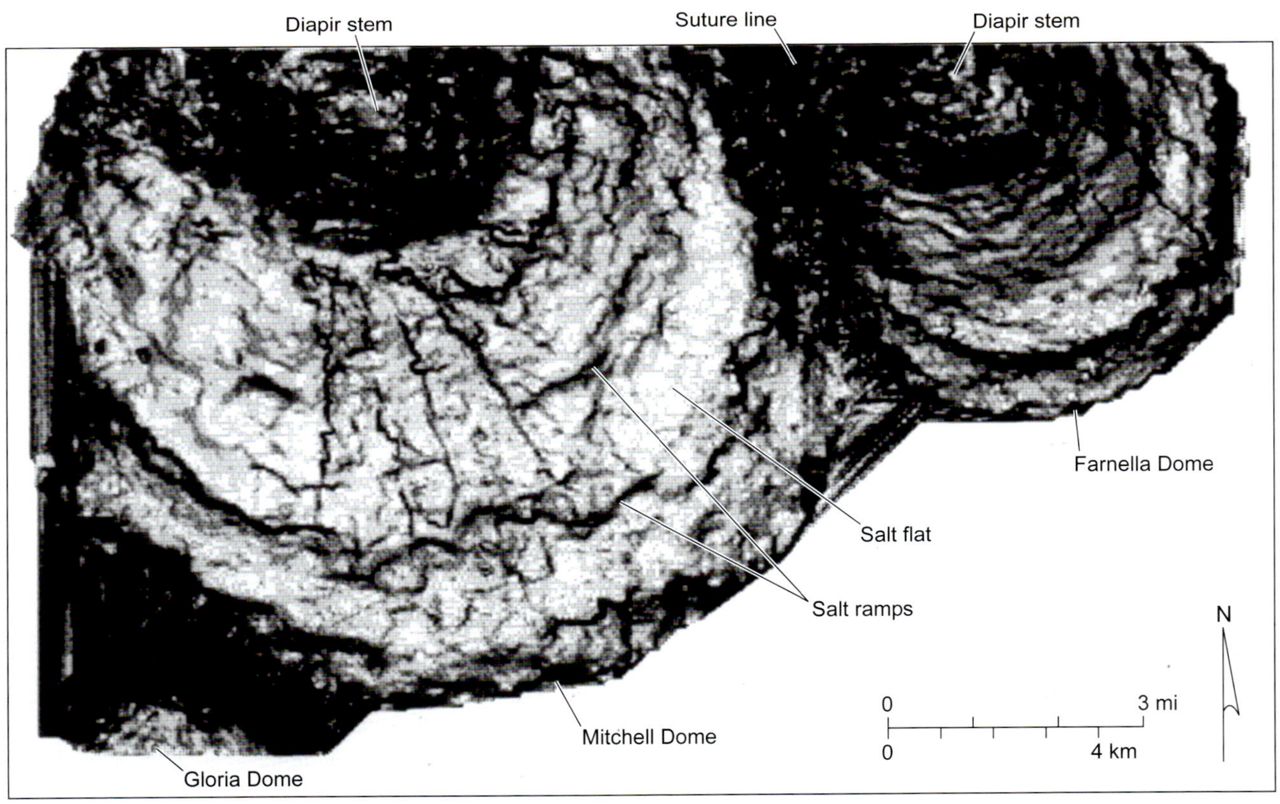

GULF OF MEXICO BASIN—Dip View of Exposed Toe of Sigsbee Salt Canopy, Walker Ridge Area, Offshore USA

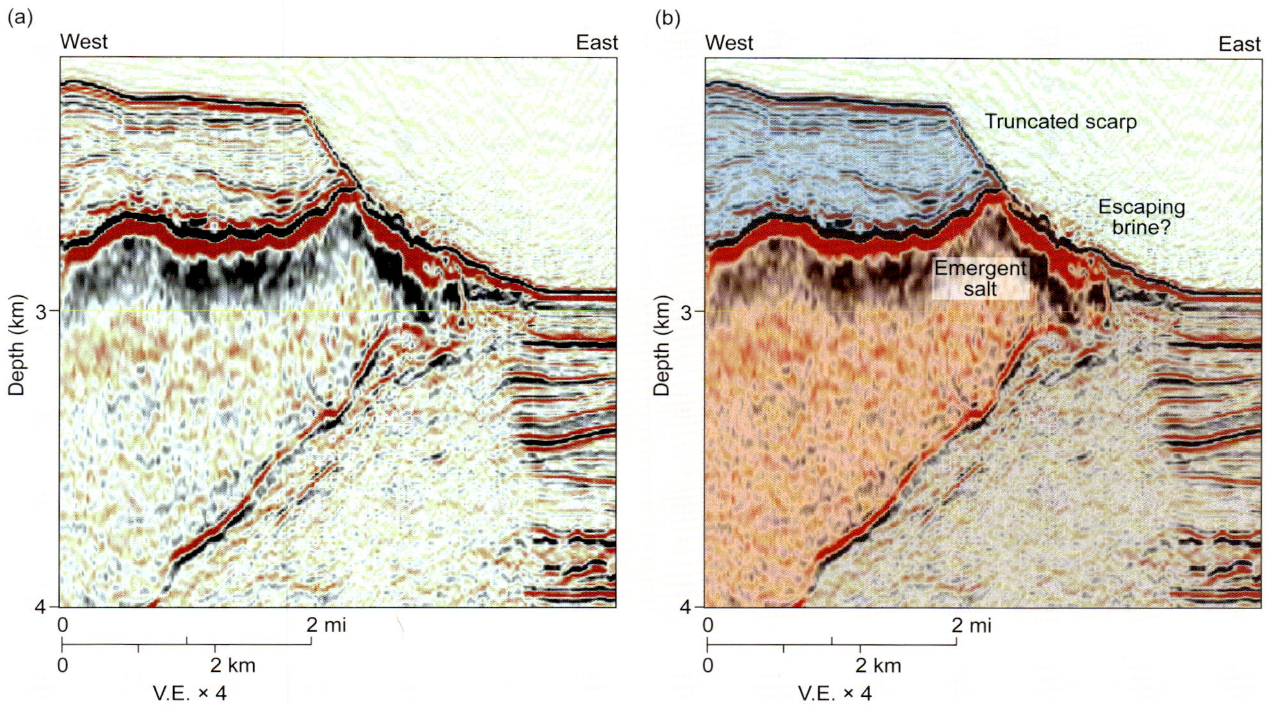

Figure 7.07. Breakout of a salt sheet. Erosion of the roof exposed the toe of the salt sheet. The exposed part of the sheet is not surging forward over the seafloor. On the contrary, it forms a small karst valley because dissolution rates are evidently faster than salt flow. Discontinuous seismic reflections in the water above the salt may represent brine flowing from the dissolving sheet. Seismic data courtesy of CGGVeritas. Modified from Hudec and Jackson (2006).

GULF OF MEXICO BASIN—Roof-Edge Thrust along the Sigsbee Escarpment, Green Canyon Area, Offshore USA

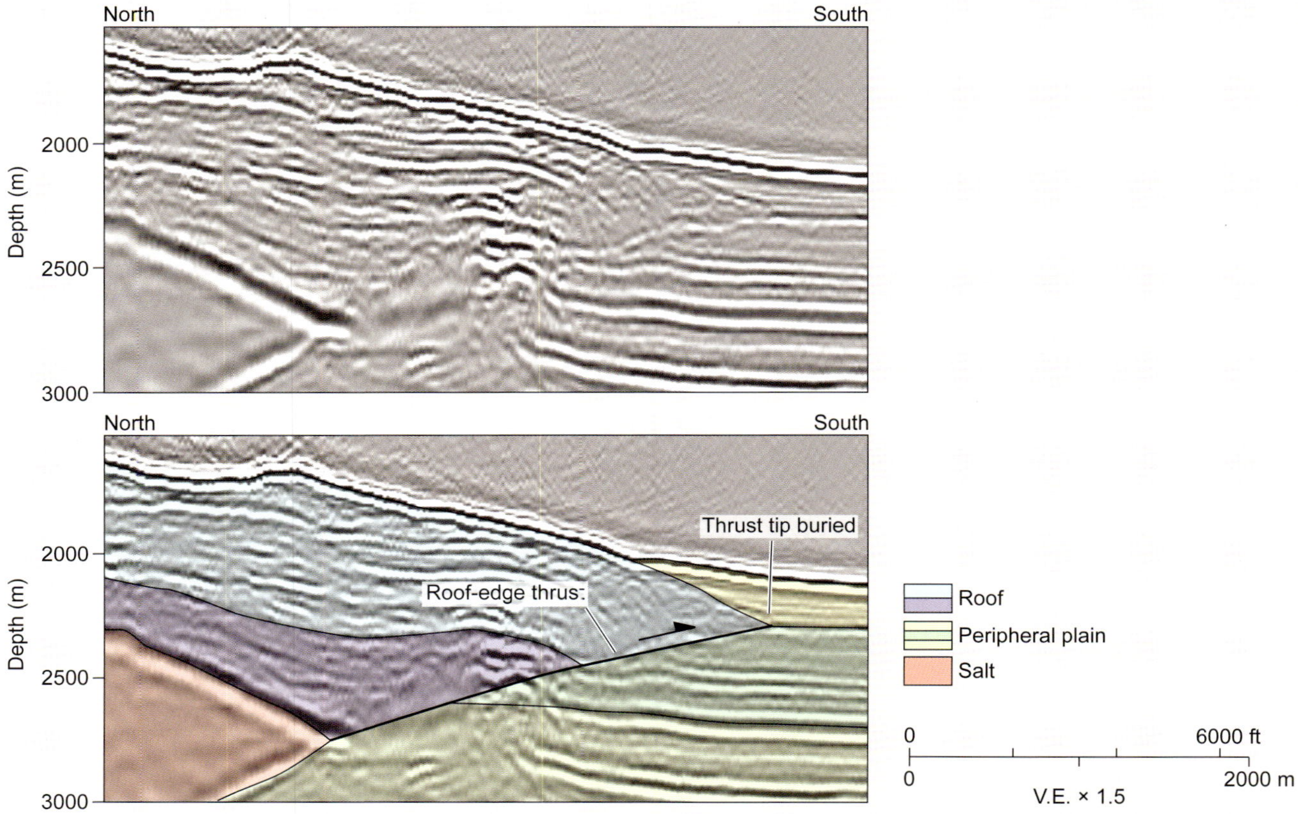

Figure 7.08. Thrust advance of a salt canopy. Here the buried Sigsbee salt canopy advanced along a roof-edge thrust that separated condensed stratigraphy in the sheet roof from thicker equivalents on the peripheral plain. Undeformed peripheral-plain strata (yellow) buried the salt tip, showing that this thrust is no longer active. A possible explanation for this inactivity is that the roof is now too thick (roughly 1 km) to be moved by the relatively weak forces driving gravity spreading of the canopy. Seismic data courtesy of BP, Unocal, and BHP Billiton. Modified from Hudec and Jackson (2009).

GULF OF MEXICO BASIN—Sigsbee Salt Canopy, Mad Dog Field, Green Canyon Area, Offshore USA

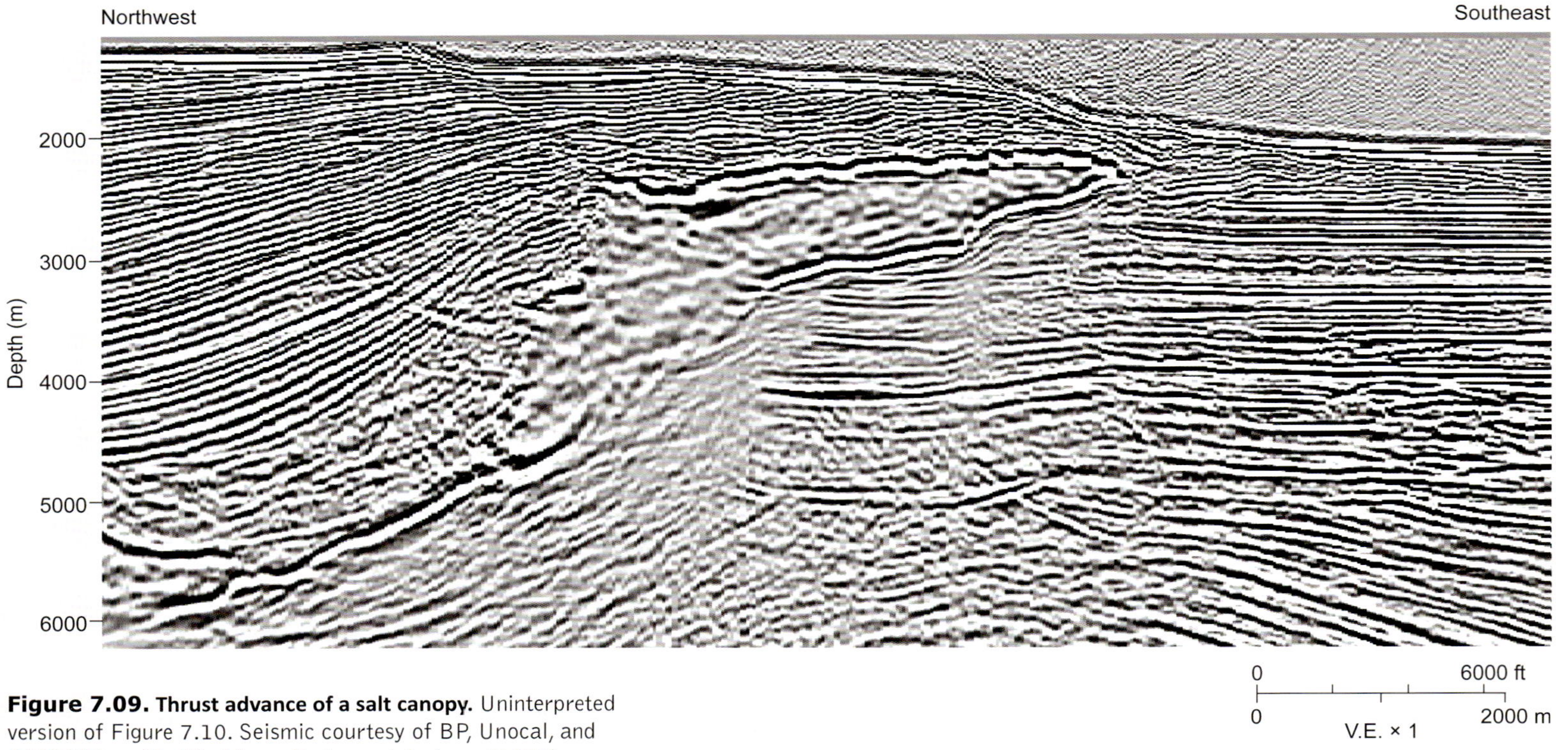

Figure 7.09. Thrust advance of a salt canopy. Uninterpreted version of Figure 7.10. Seismic courtesy of BP, Unocal, and BHPBilliton. Modified from Hudec and Jackson (2009).

GULF OF MEXICO BASIN—Sigsbee Salt Canopy, Mad Dog Field, Green Canyon Area, Offshore USA

Figure 7.10. Thrust advance of a salt canopy. Borehole data correlate units above and below the Sigsbee salt canopy near Mad Dog field. The condensed Pleistocene roof of the canopy has been thrust 7,500 m over the abyssal plain. Along most of its length, this thrust lies at or near the base of salt. Beyond the tip of the salt canopy, the thrust cuts up to the surface as a roof-edge thrust. Beneath the roof-edge thrust, an imbricate wedge formed when the advancing salt canopy telescoped the abyssal plain. Uninterpreted version of the seismic is shown in Figure 7.09, and a structural restoration of the section is presented in Figure 7.11. Seismic courtesy of BP, Unocal, and BHPBilliton. Modified from Hudec and Jackson (2009).

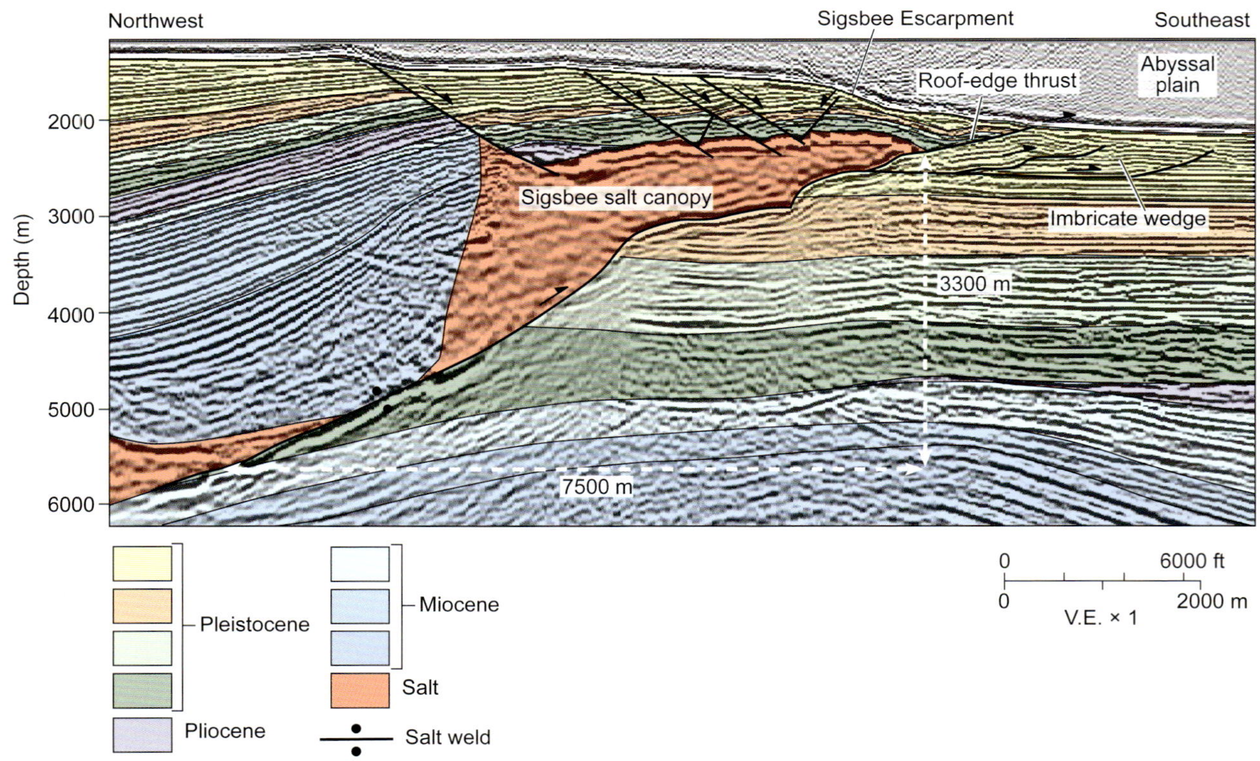

GULF OF MEXICO BASIN—Sigsbee Salt Canopy, Mad Dog Field, Green Canyon Area, Offshore USA

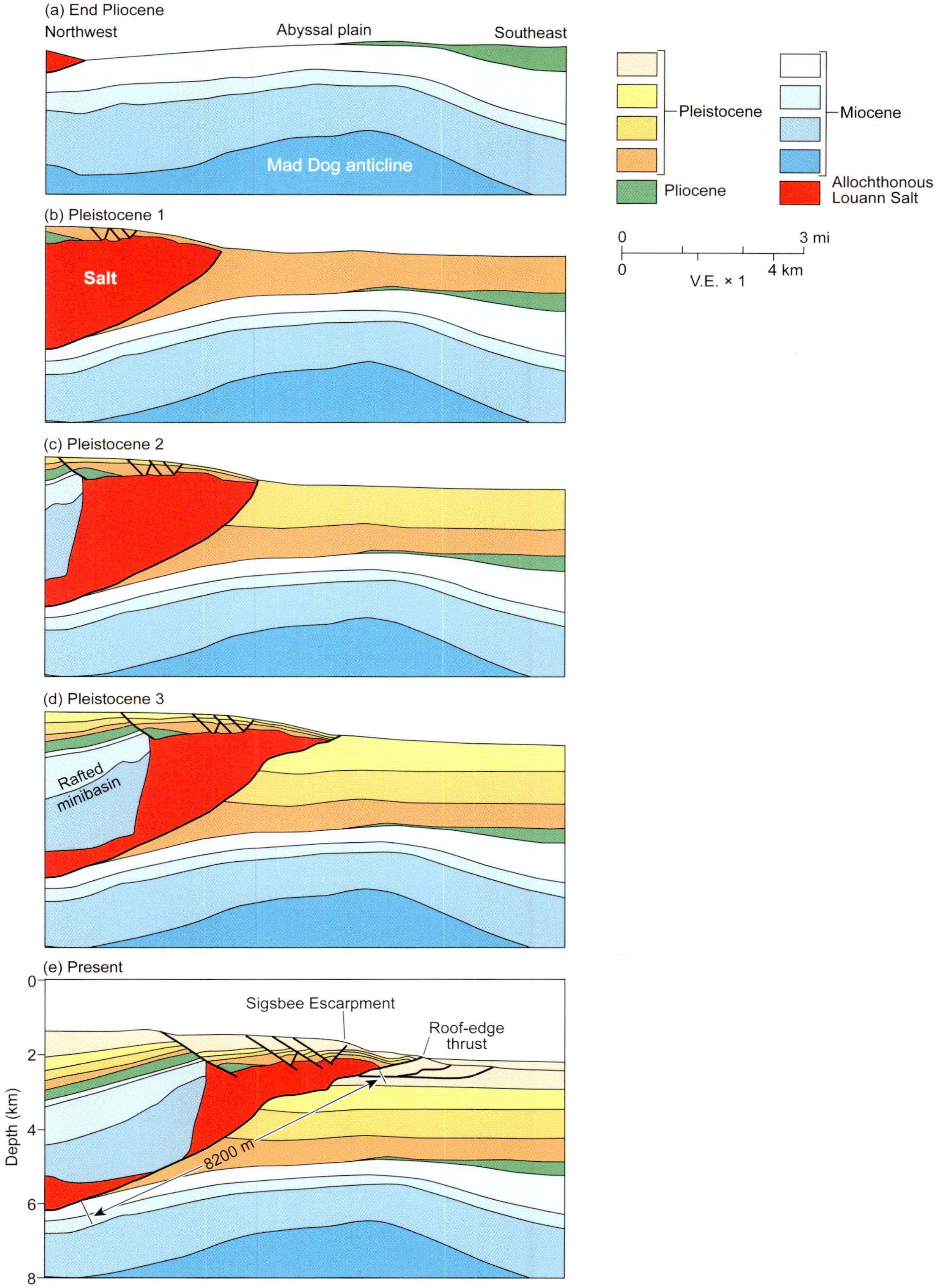

Figure 7.11. Thrust advance of a salt canopy. The Sigsbee salt canopy advanced across the top of the Mad Dog anticline during the Pleistocene. The canopy was topographically high during this interval, so sediments deposited in its roof were condensed relative to the abyssal plain. Roof sediments were transported along with the canopy, separated from their equivalents on the abyssal plain by a roof-edge thrust. Although the sediment-on-sediment part of this thrust is less than 2,000 m long, it accumulated more than 8,200 m of slip during the Pleistocene. In the Pliocene, the sheet may have advanced by extrusion because sediments of this age are not preserved on most of the roof. Restoration by Mike Hudec.

GULF OF MEXICO BASIN—Imbricate Wedge Involving a Thick Stratigraphic Sequence, Sigsbee Escarpment, Green Canyon Area, Offshore USA

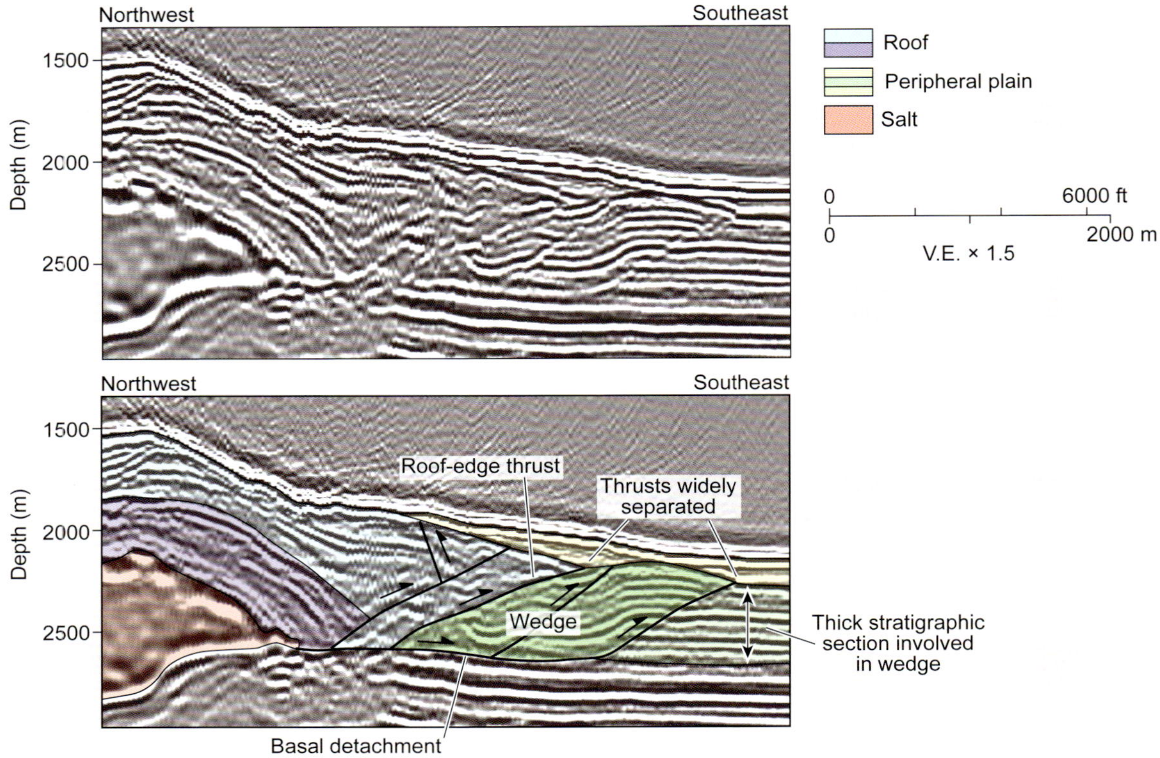

Figure 7.12. Thrust advance of a salt canopy. According to seismic character, the two southeastern thrusts imbricate peripheral-plain strata and so form an imbricate wedge. The section shortened in the wedge is about 400 m thick, so thrusts are widely separated and easily recognized. Furthermore, shortening in the wedge is small (< 500 m). Hudec and Jackson (2009) suggested that imbricate wedges involving thick sections have limited shortening because thicker sections are stronger and more able to resist compression. Seismic data courtesy of BP, Unocal, and BHP Billiton. Modified from Hudec and Jackson (2009).

GULF OF MEXICO BASIN—Imbricate Wedge Involving a Thin Stratigraphic Sequence, Sigsbee Escarpment, Alaminos Canyon Area, Offshore USA

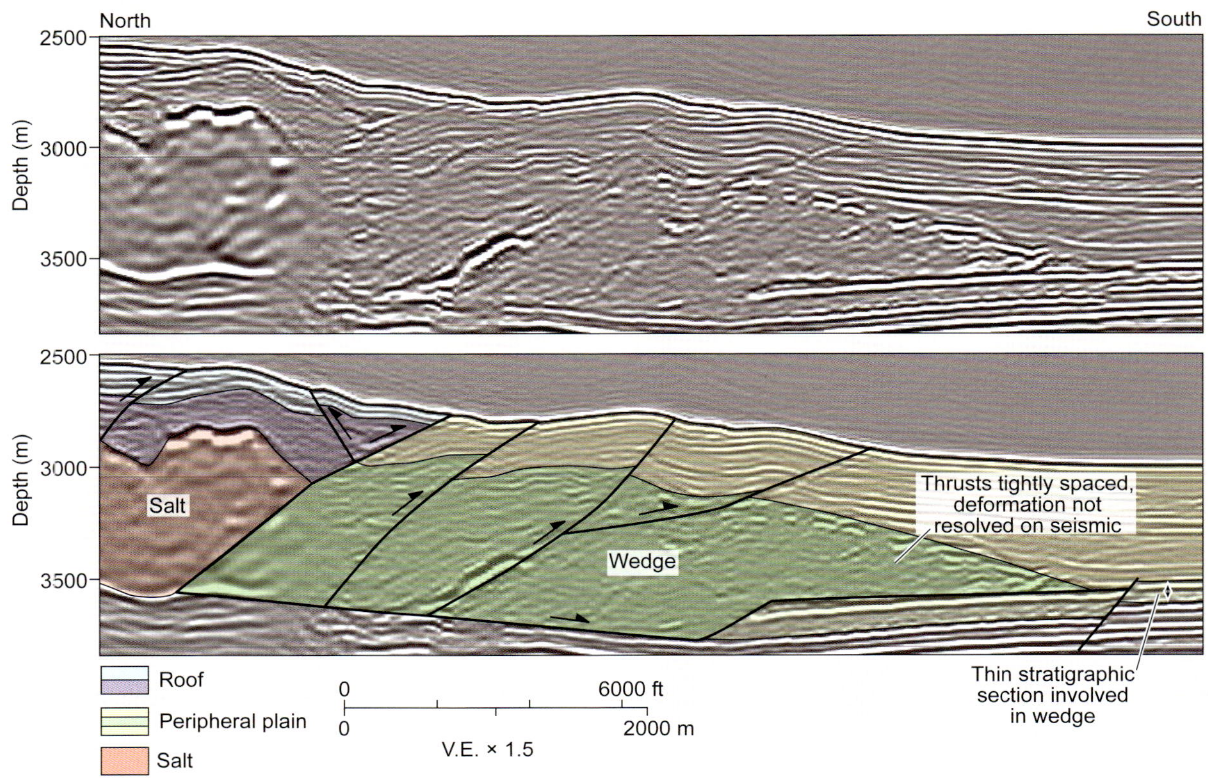

Figure 7.13. Thrust advance of a salt canopy. The thick wedge of largely incoherent sediment at the leading edge of the Sigsbee canopy is interpreted as an imbricate wedge, formed by shortening of peripheral-plain strata. The section shortened in the wedge was only 150 m thick, so most thrusts are tightly spaced and difficult to interpret on seismic data. Area balancing suggests that shortening here is around 22,000 m, many times greater than the less than 500 m of shortening calculated for the thick-section wedge (Figure 7.12). Hudec and Jackson (2009) suggested that imbricate wedges telescoping thick sections have less shortening because thicker sections are stronger and more able to resist compression. Seismic data courtesy of CGGVeritas. Modified from Hudec and Jackson (2009).

GULF OF MEXICO BASIN—Lateral Transition between Salt-Tip and Salt-Floor Imbricate Wedges, Sigsbee Escarpment, Green Canyon Area, Offshore USA

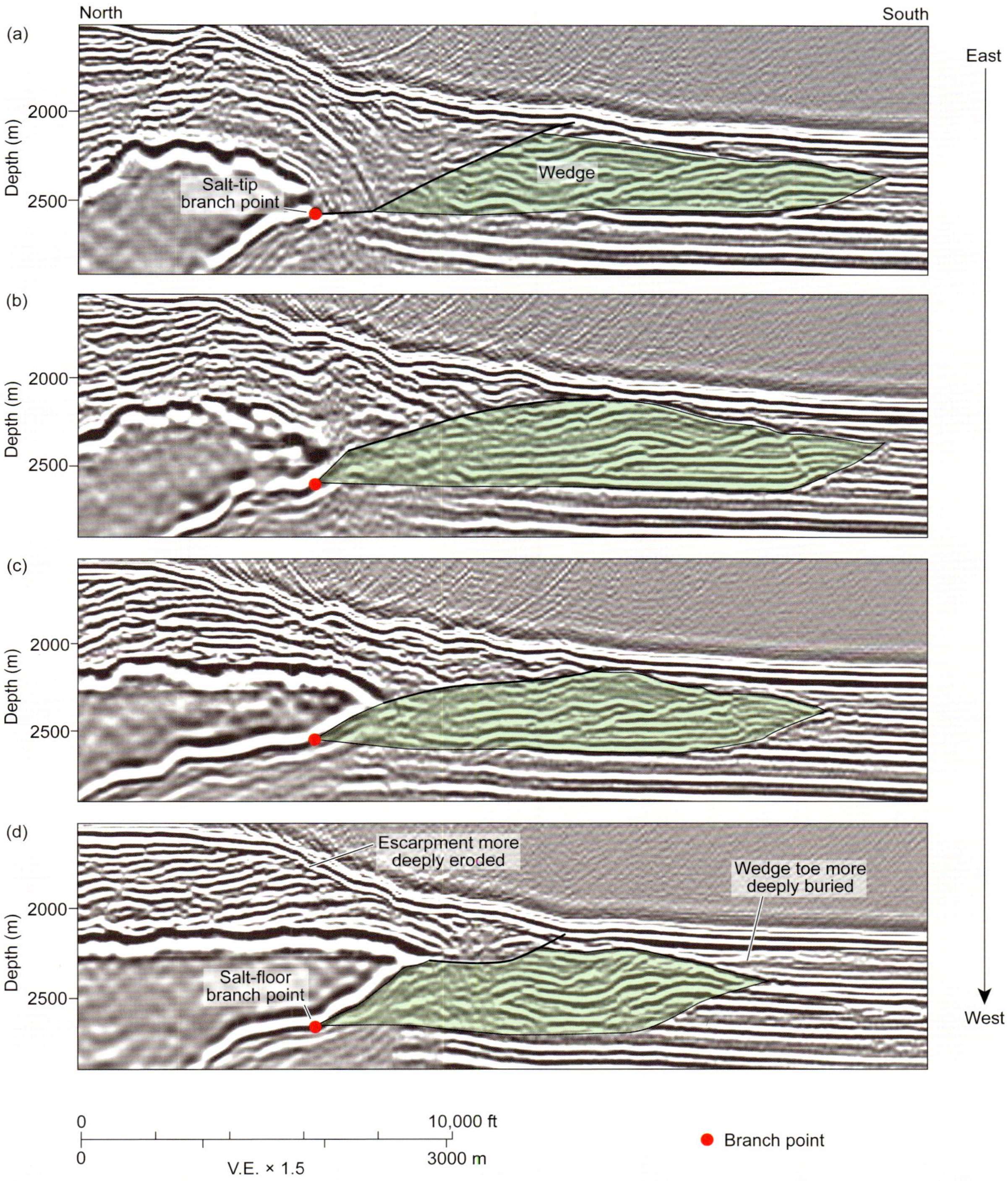

Figure 7.14. Thrust advance of a salt canopy. An imbricate wedge along the leading edge of the Sigsbee Escarpment is progressively overridden by salt along strike. This spatial series may mimic a time series because a wedge is overridden by a spreading salt canopy after the wedge becomes inactive. Consistent with this hypothesis, the most overridden wedge (d) appears to have been inactive the longest, judging by the greater thickness of its onlapping roof. Rise of the salt over the rear of the wedge uplifted the sheet roof, leading to progressively greater escarpment erosion to the west. Overthrusting of an imbricate wedge by salt suggests a detachment at or near the base of the salt canopy. Seismic data courtesy of BP, Unocal, and BHP Billiton. Modified from Hudec and Jackson (2009).

GULF OF MEXICO BASIN—Inactive Imbricate Wedges beneath a Roof-Edge Thrust, Sigsbee Escarpment, Alaminos Canyon Area, Offshore USA

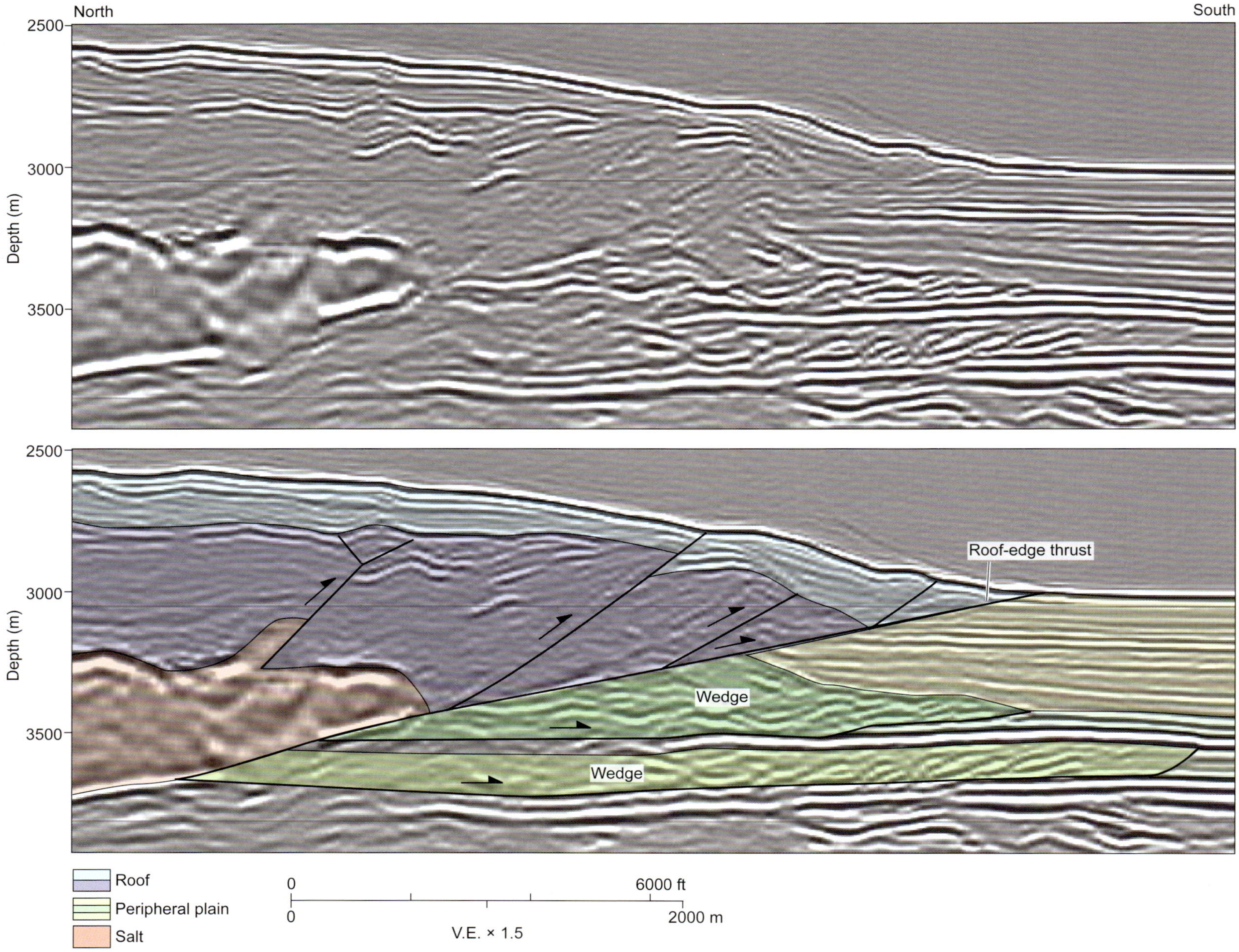

Figure 7.15. Thrust advance of a salt canopy. Two inactive imbricate wedges lie abandoned in the footwall of an active roof-edge thrust. Processes controlling formation and abandonment of imbricate wedges are poorly known. Factors favoring wedge formation could include (a) acceleration of sheet advance, which increases stress on peripheral-plain sediments; (b) increase in escarpment relief, which could also increase stress; (c) presence of a weak shale layer on the peripheral plain, which could provide a detachment for a wedge; and (d) an increase in peripheral-plain pore pressure, which would weaken peripheral sediments. Seismic data courtesy of CGGVeritas. Modified from Hudec and Jackson (2009).

GULF OF MEXICO BASIN—Salt-Roof Thrusts along the Sigsbee Escarpment, Walker Ridge Area, Offshore USA

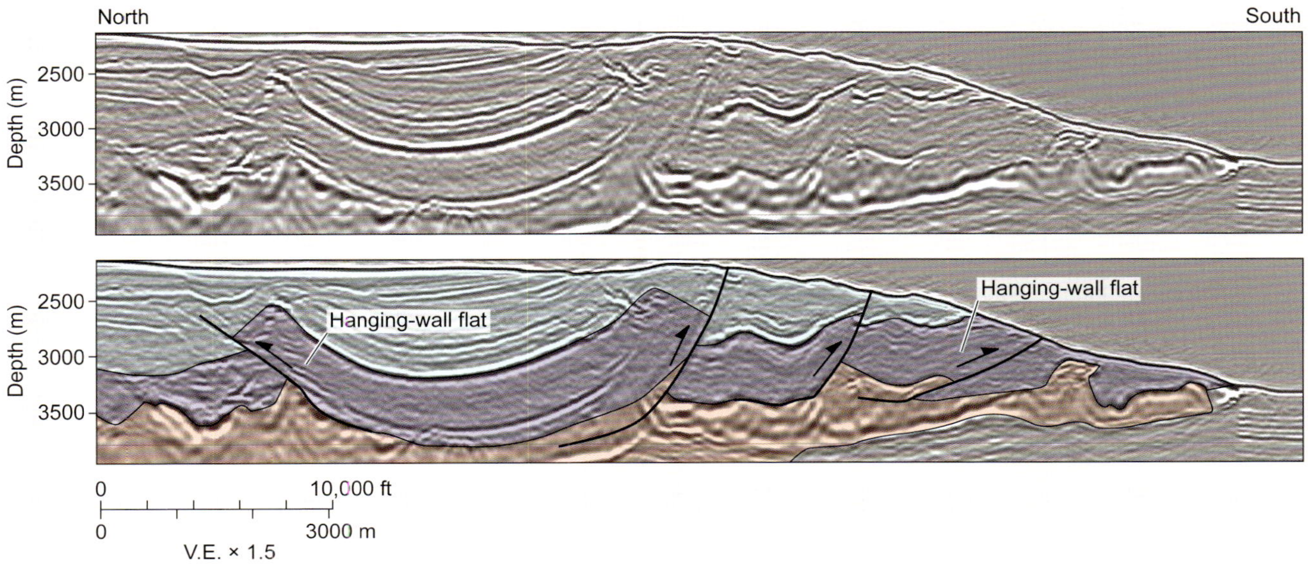

Figure 7.16. Thrust advance of a salt canopy. Salt-roof thrusts near the leading edge of a salt canopy form either because the leading edge of the canopy is the toe of a gravity-driven linked system or because the top of the sheet is advancing faster than its base (frontal rolling). The presence here of hanging-wall flats in the sediments above the thrusts suggests that the thrusts are detached at or just below the top of the salt. Seismic courtesy of CGGVeritas. Modified from Hudec and Jackson (2009).

GULF OF MEXICO BASIN—Water-bottom Dip Map along Part of Sigsbee Escarpment, Walker Ridge Area, Offshore USA

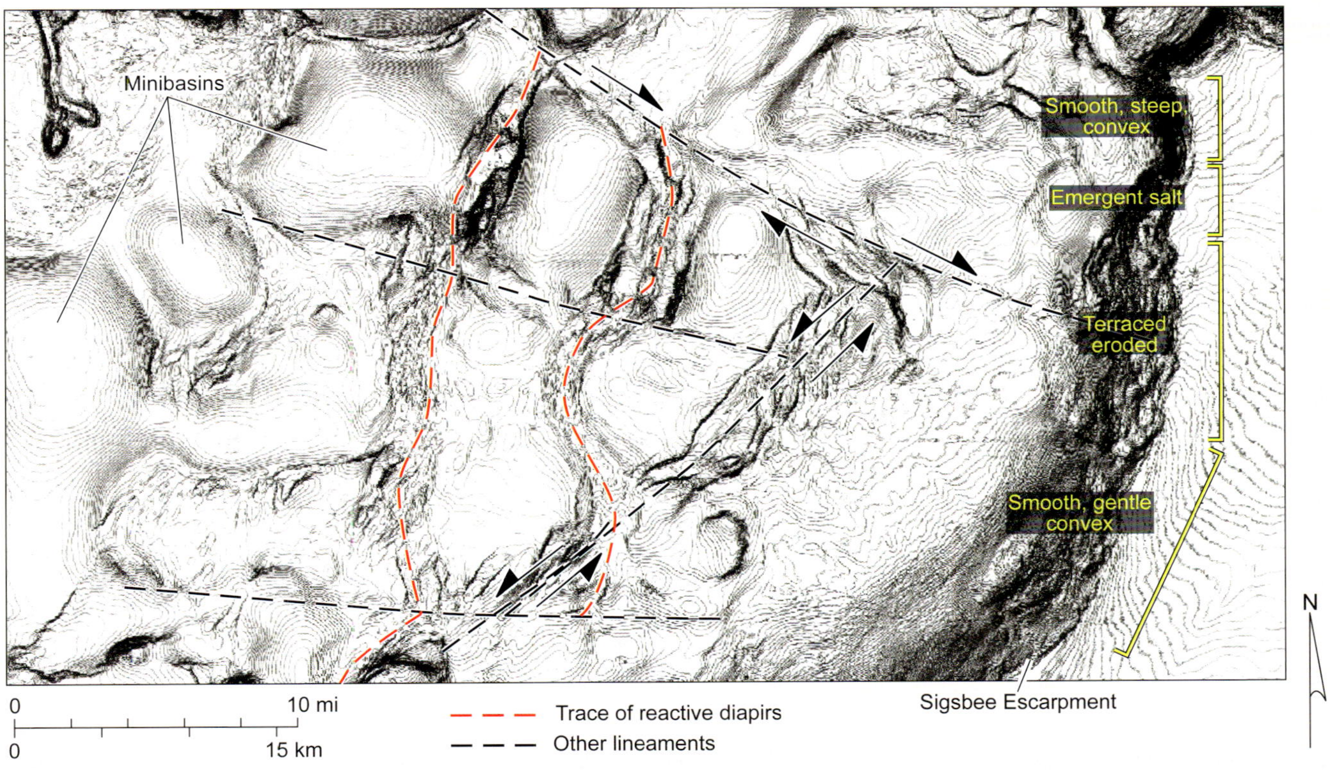

Figure 7.17. Segmentation of an advancing salt canopy. The frontal escarpment changes along the strike of the advancing salt canopy. Salt locally emerges at the seafloor but is buried elsewhere. In the south, the roof is intact, having a smooth, gently dipping convex-upward bathymetry. Farther north, the bathymetry is interrupted by slumps that form a terraced topography. In the far north, an intact roof has much steeper dips than in the southern area. The reasons for this variable bathymetry are unclear but may be related to shear zones whose pattern suggests differential flow within the salt sheet. En echelon faults and terminations of diapiric walls define strike-slip zones in the roof. Seismic data courtesy of CGGVeritas.

SALT SHEETS

GULF OF MEXICO BASIN—Early-Stage Suture of Salt Sheets, Green Canyon and Walker Ridge Areas, Offshore USA

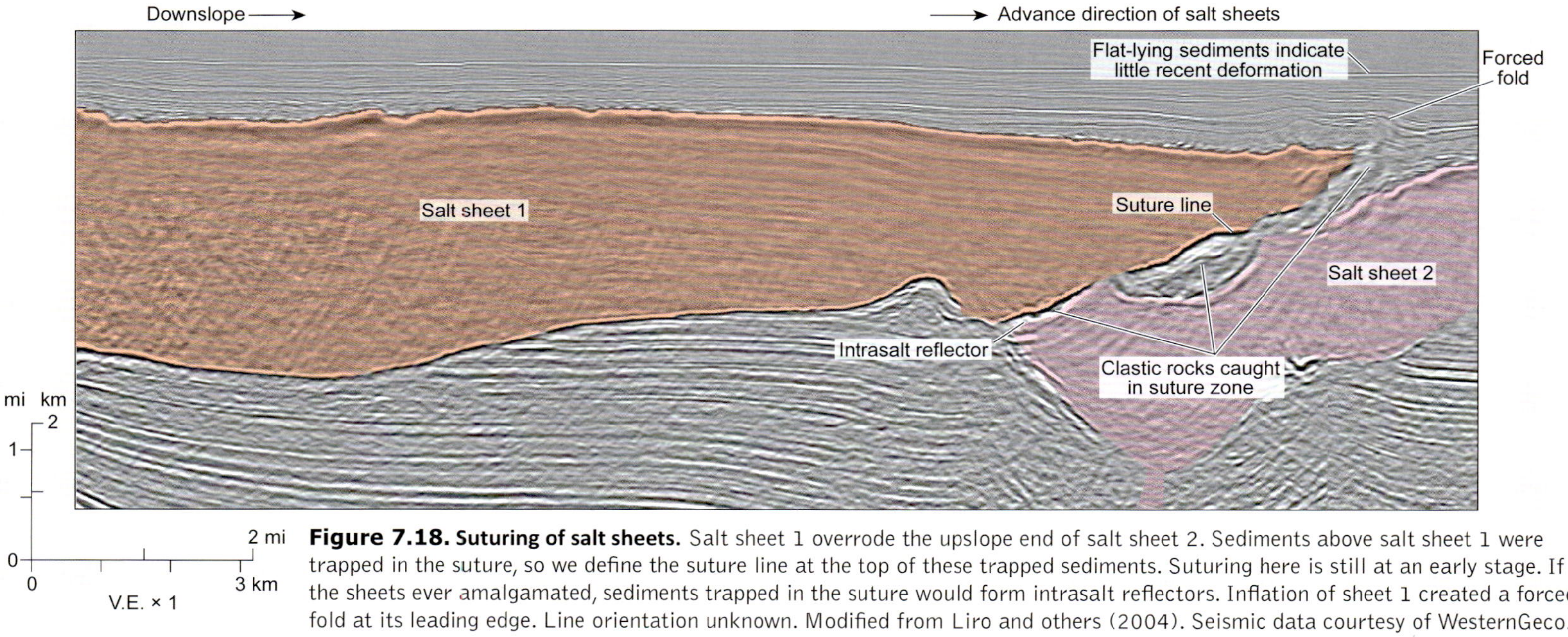

Figure 7.18. Suturing of salt sheets. Salt sheet 1 overrode the upslope end of salt sheet 2. Sediments above salt sheet 1 were trapped in the suture, so we define the suture line at the top of these trapped sediments. Suturing here is still at an early stage. If the sheets ever amalgamated, sediments trapped in the suture would form intrasalt reflectors. Inflation of sheet 1 created a forced fold at its leading edge. Line orientation unknown. Modified from Liro and others (2004). Seismic data courtesy of WesternGeco.

GULF OF MEXICO BASIN—Intermediate-Stage Suture of Salt Sheets, Green Canyon and Walker Ridge Areas, Offshore USA

Figure 7.19. Suturing of salt sheets. Salt sheet 1 overrode salt sheet 2. Sediments originally above salt sheet 2 were trapped in the suture, forming a chain of internal reflectors within the amalgamated canopy. The origin of the internal reflectors is clear on line B. Here the reflectors can be traced upward to the leading edge of the suture, where the sheets have not yet joined. Line orientations unknown. Modified from Liro and others (2004). Seismic data courtesy of WesternGeco.

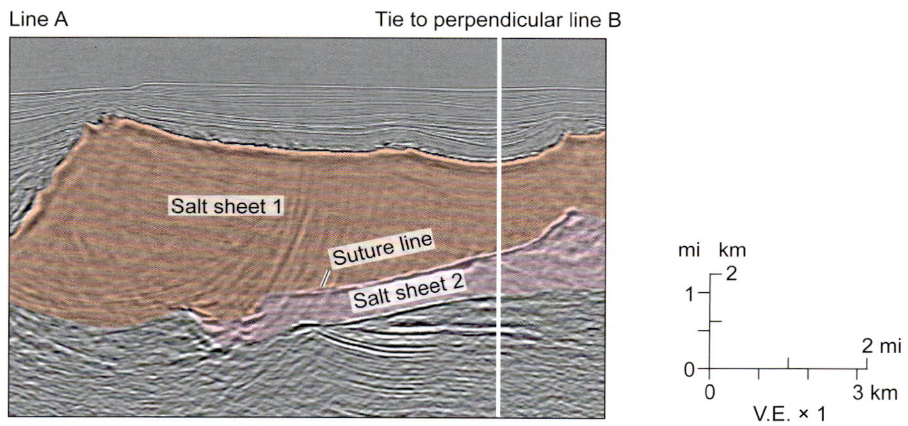

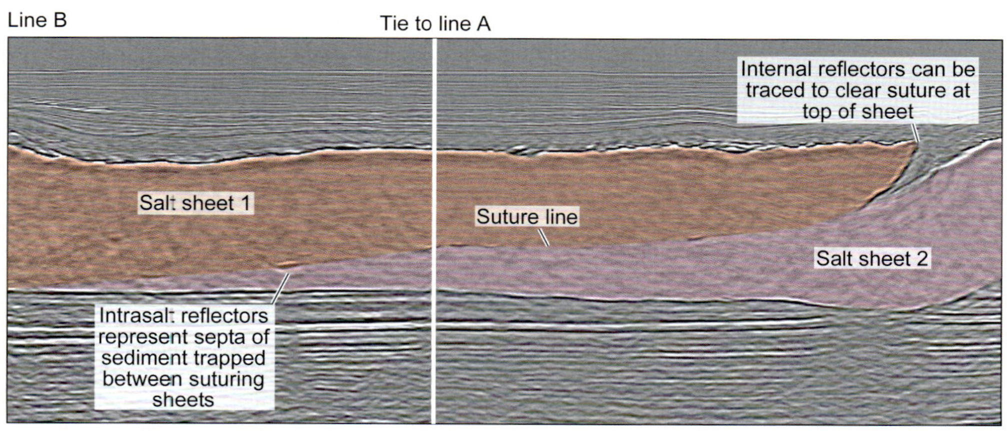

KATANGA FOLD BELT—Salt Sheet, West Lufilian Allochthon, Democratic Republic of Congo

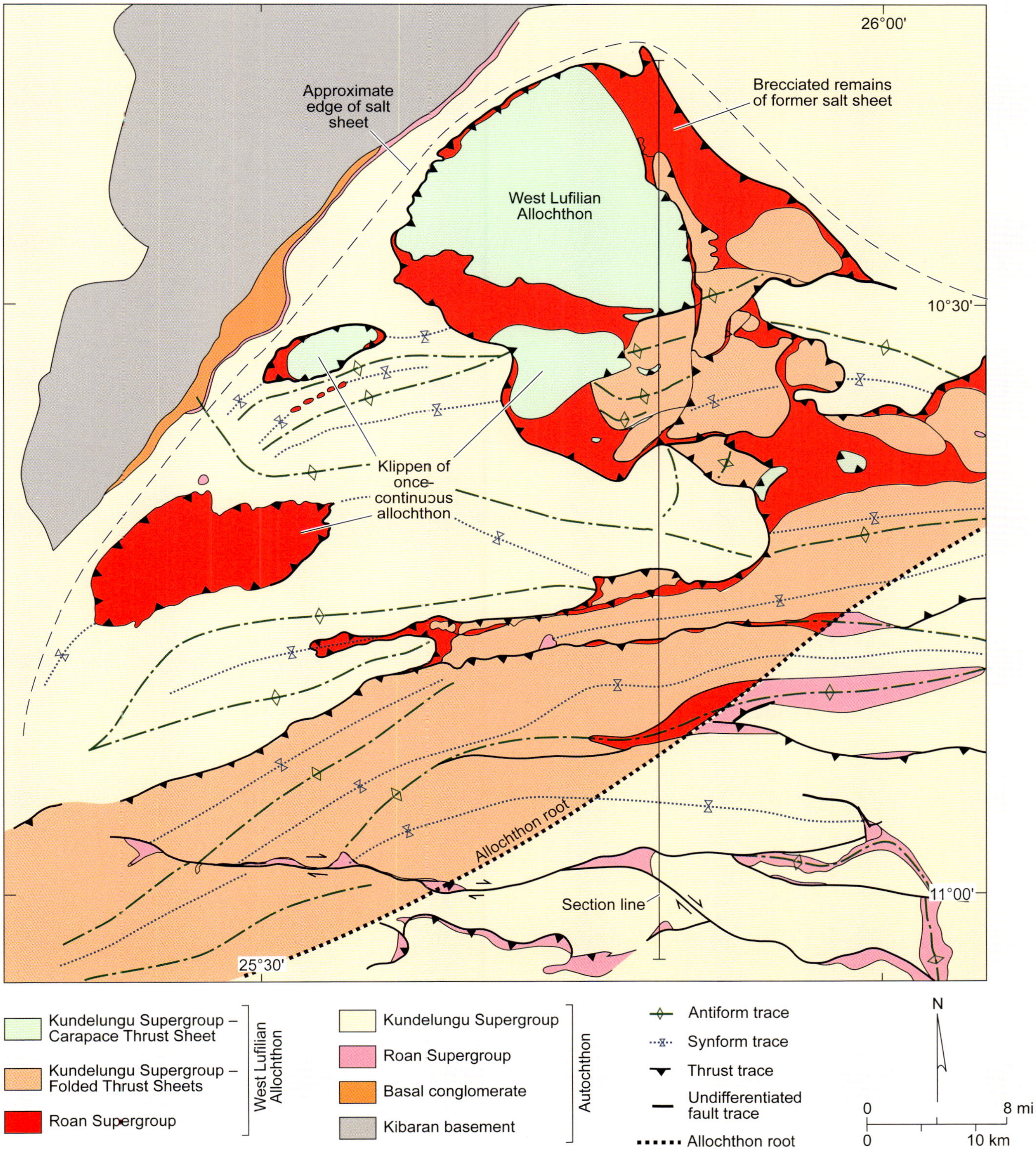

Figure 7.20. Orogenic shortening above a salt sheet. In the Katangan Copperbelt, the West Lufilian Allochthon was displaced more than 55 km northward over flat-lying sediments during the Neoproterozoic Lufilian Orogeny. Structural restoration (Figure 7.21) indicates that the allochthon advanced over a previously emplaced salt sheet containing Roan evaporites, which acted as a lubricant. The salt sheet thus controlled the location and extent of the West Lufilian Allochthon. The evaporites within the salt sheet are now dissolved, leaving a residual breccia sheet that covers more than 25,000 km² and is rich in cobalt and copper. The allochthon emerges at the surface 9 to 20 km northwest of its buried root. Modified from Jackson and others (2003).

KATANGA FOLD BELT—Salt Sheet, West Lufilian Allochthon, Democratic Republic of Congo

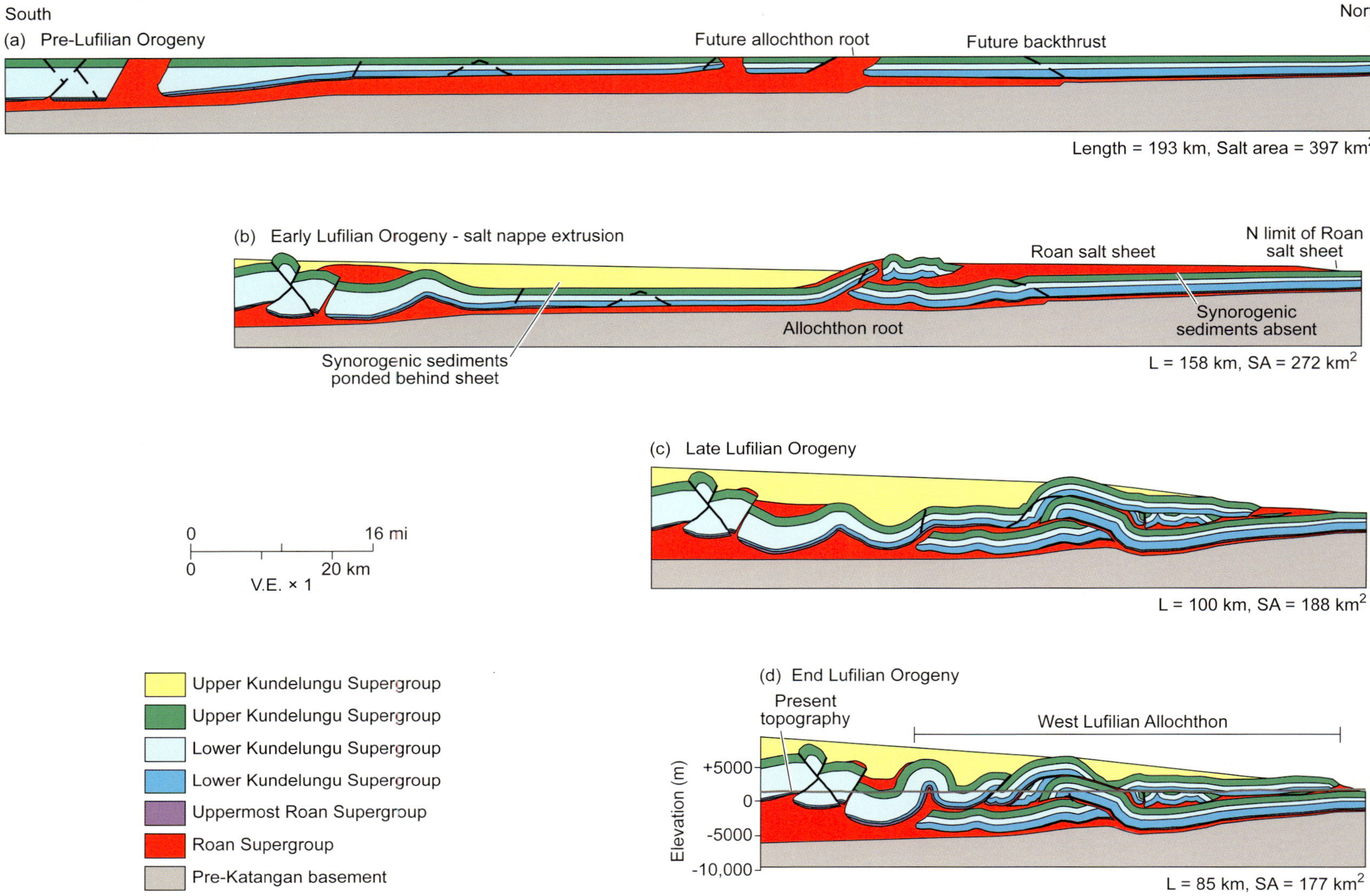

Figure 7.21. Orogenic shortening above a salt sheet. This restoration indicates that the West Lufilian Allochthon advanced more than 55 km over relatively flat-lying footwall strata along a single stratigraphic horizon, without overriding any synorogenic deposits, which would imply unreasonably rapid orogenic shortening. Alternatively, as shown here, before the allochthon was emplaced, a salt sheet extruded rapidly and blanketed the foreland with salt while synorogenic units were ponded in a basin to the south. The West Lufilian Allochthon then advanced more gradually, driven by orogenic advance across the lubricating salt sheet. Over time, the salt sheet lost volume by dissolution. See also the geologic map of the region (Figure 7.20). Modified from Jackson and others (2003).

Kwanza Basin—Angola Salt Nappe, Offshore Angola

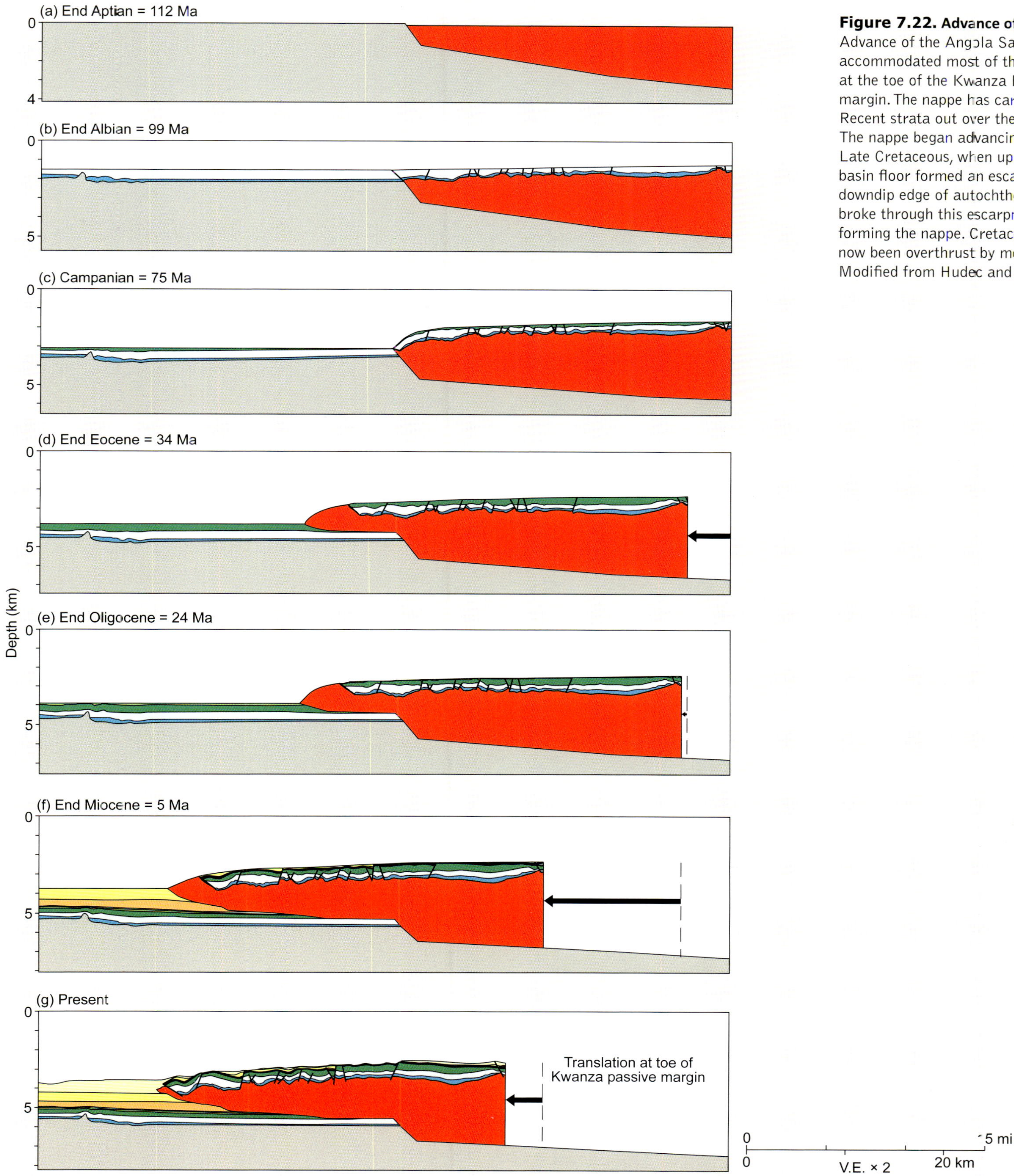

Figure 7.22. Advance of a salt nappe. Advance of the Angola Salt Nappe has accommodated most of the shortening at the toe of the Kwanza Basin passive margin. The nappe has carried Albian–Recent strata out over the abyssal plain. The nappe began advancing during the Late Cretaceous, when uplift of the basin floor formed an escarpment at the downdip edge of autochthonous salt. Salt broke through this escarpment to begin forming the nappe. Cretaceous strata have now been overthrust by more than 23 km. Modified from Hudec and Jackson (2004).

LOWER CONGO BASIN—Salt Sheet in Thrust Belt, Offshore Gabon

Figure 7.23. Salt sheet expelled during shortening. The salt sheet is within a toe-of-slope thrust belt. Salt sheets have formed here wherever thrusts intersect preexisting salt diapirs. The diapirs have all pinched shut. Most of the salt was displaced upward to form an allochthonous salt sheet. In many diapirs, including here, the pinched-off feeder was reactivated as a thrust weld during continued shortening. The onset of thrusting is recorded by the strongly onlapping and fanning wedge of sediment preserved in the thrust hanging wall. Interpretation by Martin Jackson and David Jennette. Seismic data courtesy of Total Astrid Marin Gabon.

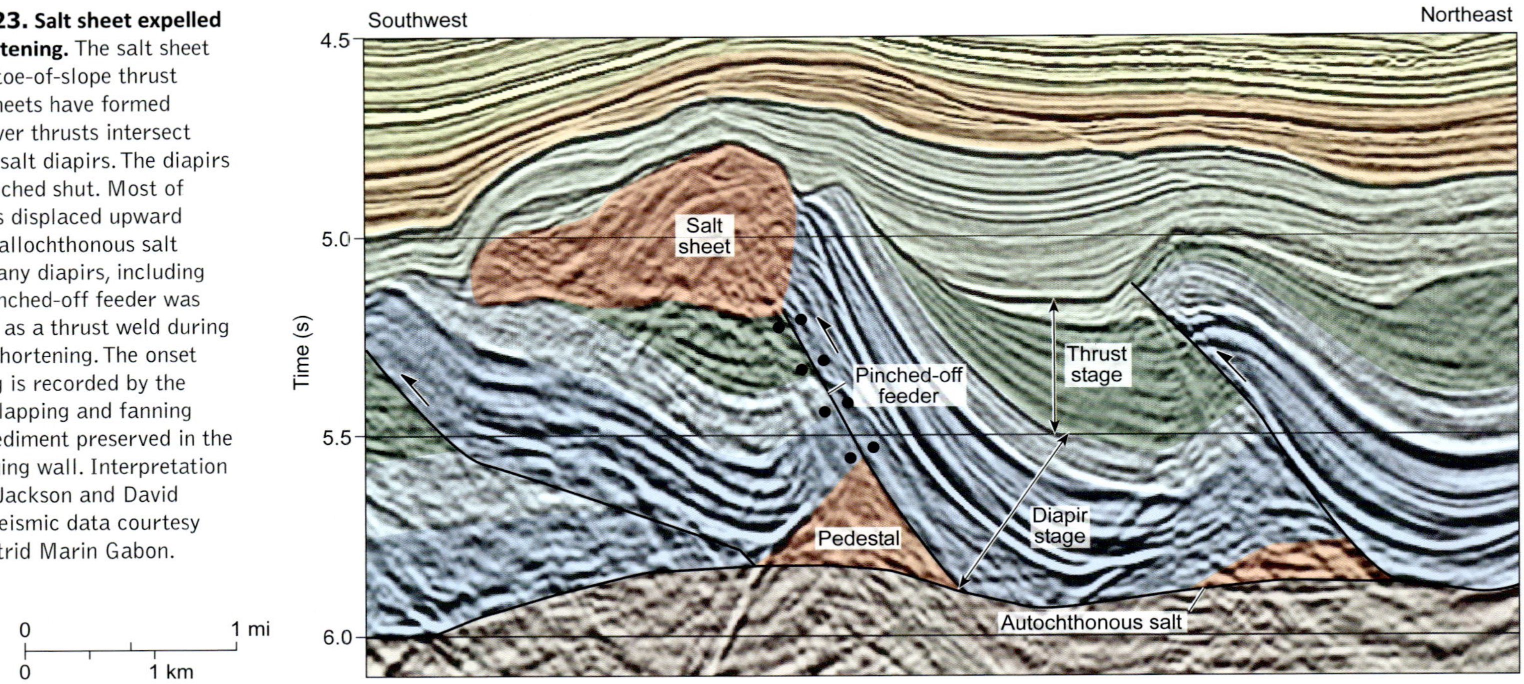

LOWER CONGO BASIN—Salt Sheet with Pinched-Off Feeder, Offshore Gabon

Figure 7.24. Salt sheet expelled during shortening. The salt sheet is within a toe-of-slope thrust belt. It formed when shortening pinched shut its feeder. After the sheet was extruded, the sheet was buried beneath a roof that became isopachous by the end of the Eocene, indicating that the sheet was quiescent. Then the sheet began to inflate and has been continually rising from the end of the Eocene to the present. Restoration (Figure 7.25) and regional relationships suggest that there was no shortening during this period. Thus, sheet inflation most likely results from importation of salt from out of the plane of section. Modified from Jackson and others (2008).

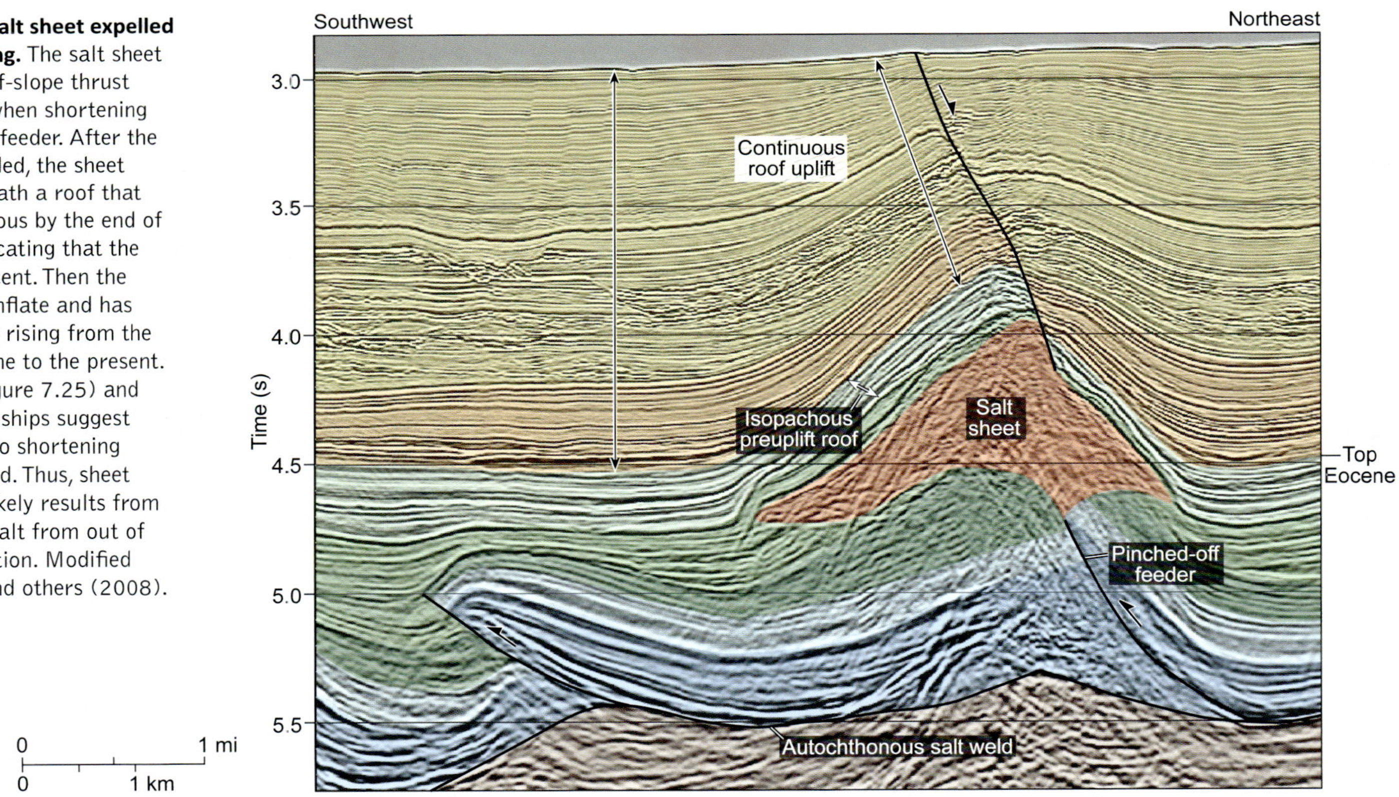

LOWER CONGO BASIN—Salt Sheet with Pinched-Off Feeder, Offshore Gabon

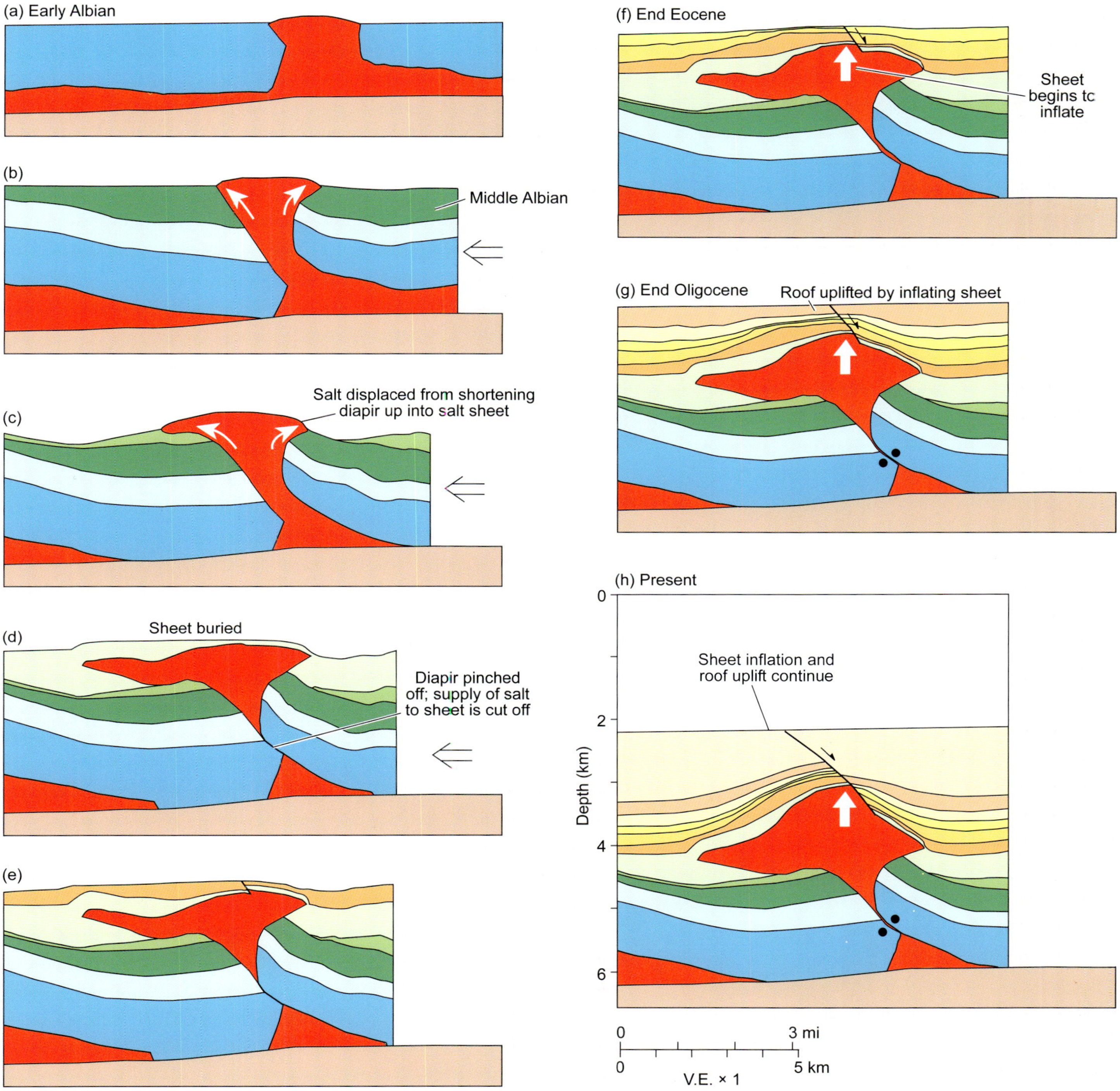

Figure 7.25. Salt sheet expelled during shortening. Shortening of the diapir displaced salt upward, where it extruded to form a salt sheet. Sheet extrusion continued until the feeder was pinched off and the sheet was buried. Near the end of the Eocene, however, the sheet began to inflate again, presumably because of salt supplied from out of the plane of section. Salt inflation from the Eocene to the present caused the roof to thin over the rising part of the sheet. The salt is evidently pressurized enough to lift a roof that is now more than 1 km thick. Interpreted seismic section is shown in Figure 7.24. Modified from Jackson and others (2008).

North German Basin—Oppershausen Salt Sheet, Germany

Figure 7.26. Salt-wing intrusion. There is no thinning above the salt wing on the southwest side of the diapir until the Late Cretaceous, which suggests that the salt was intruded beneath a roof that was at least 2 km thick. The thin Röt salt acted as a detachment during Late Cretaceous shortening. Units above the Röt shortened by detachment folding, whereas the lower and middle Buntsandstein shortened primarily by thrusting (see restoration, Figure 7.27). Zechstein salt flowed into the core of the detachment fold, filling the space between the two largely decoupled levels of deformation. Modified from Kockel (1990, 1998).

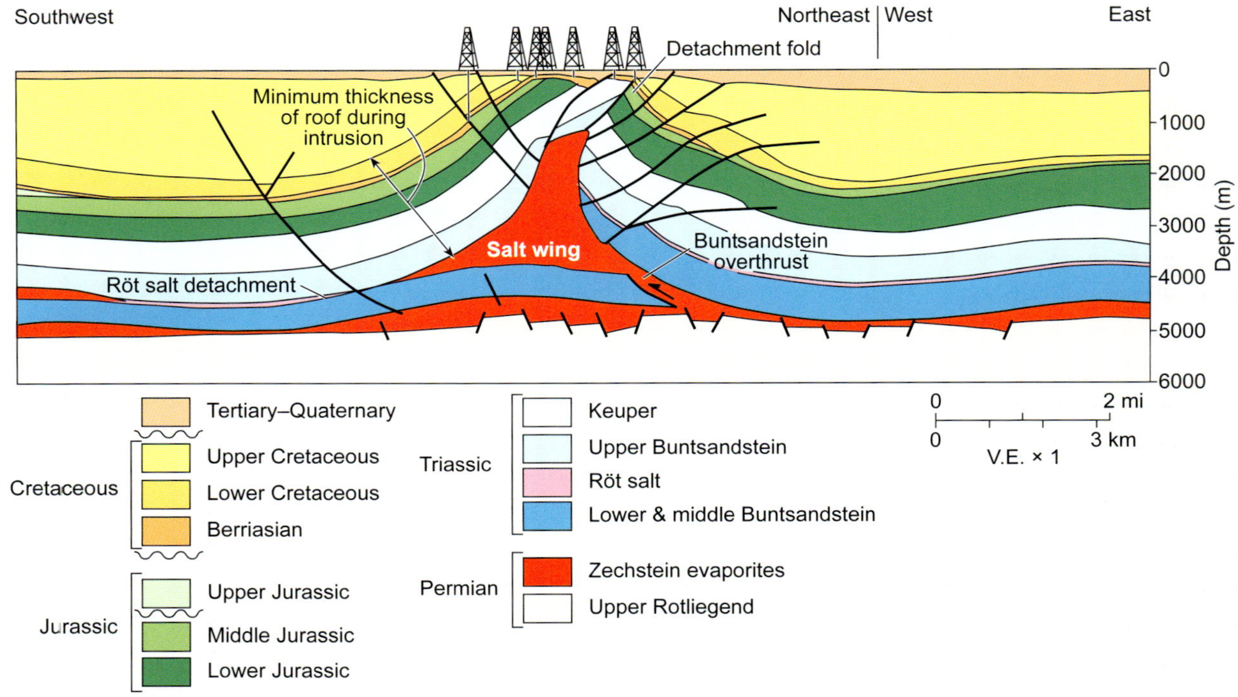

North German Basin—Oppershausen Salt Sheet, Germany

Figure 7.27. Salt-wing intrusion. Permian Zechstein salt intruded into the Triassic Röt salt during Late Cretaceous shortening. Before the Late Cretaceous, the structure was a gentle east-verging salt-cored anticline. During shortening, Zechstein salt rode up an east-verging thrust until it reached the weak Röt salt, where it filled the fold core. Mechanical arguments suggest that salt could not forcefully inject the Röt layer and lift a 2-km-thick roof. Rather, it is interpreted to have passively filled the core of the detachment fold that was rising as a result of regional shortening. The present-day section is a simplified version of the true cross section (Figure 7.26), modified from Kockel (1990). Restoration modified from Hudec (2004).

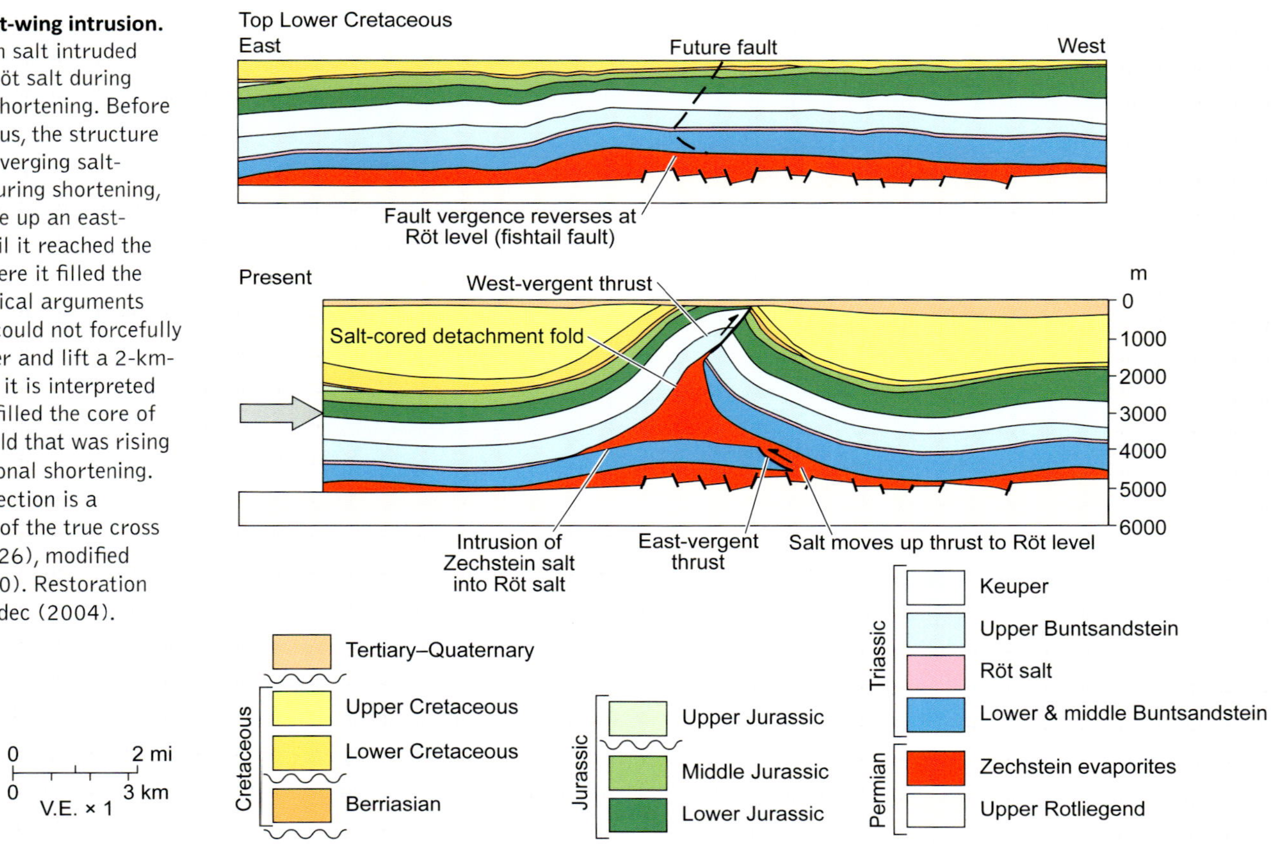

NORTH GERMAN BASIN—Wolthausen and Hustedt Sheets, Germany

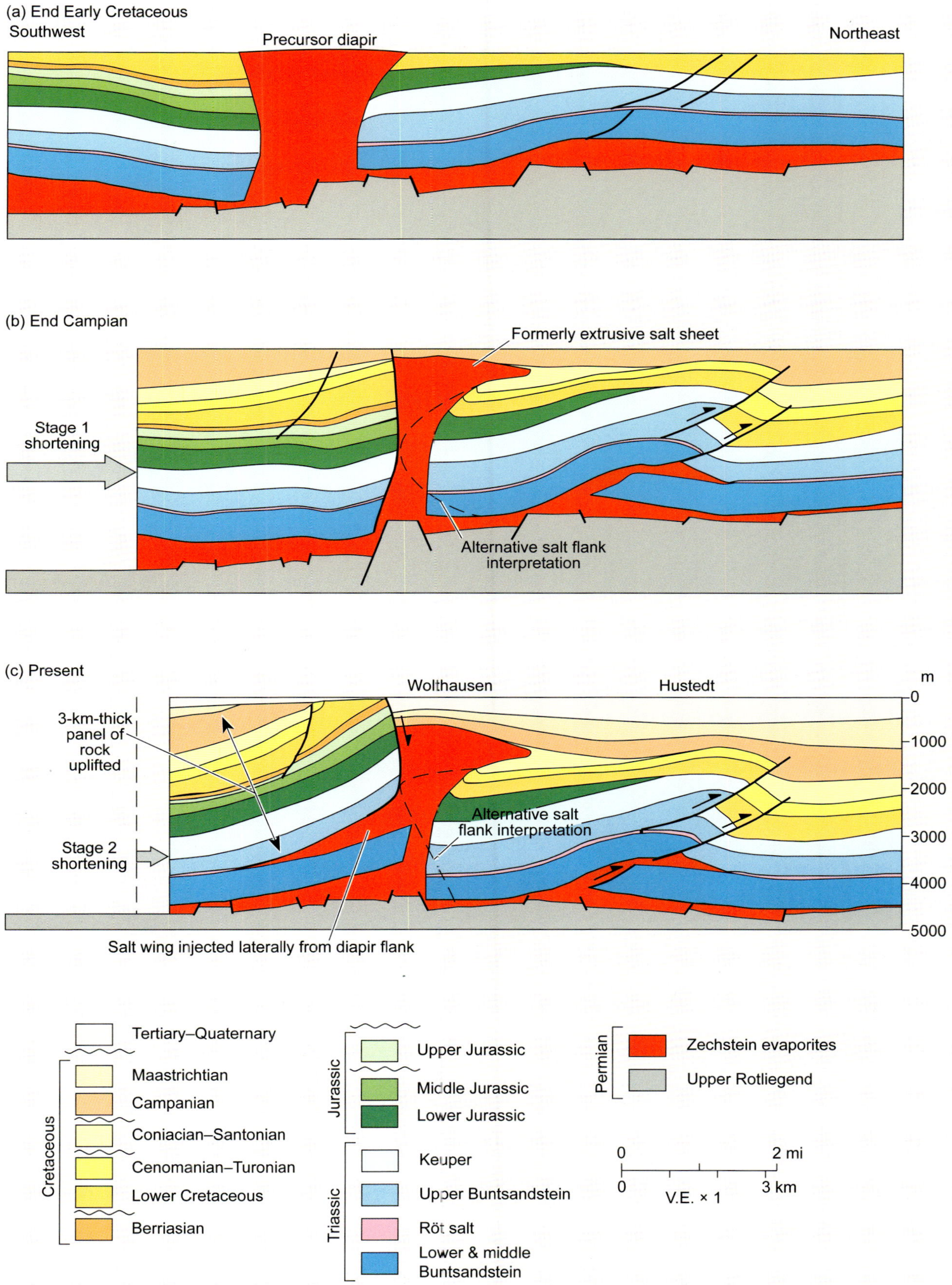

Figure 7.28. Salt-wing intrusion. The Wolthausen salt wing formed during the second of two phases of shortening of the Wolthausen salt wall. During the first phase (end Campanian restoration), salt displaced by diapir shortening was extruded from an emergent diapir. During the second stage, salt intruded laterally into the Röt salt layer, uplifting a 3-km-thick roof on the southwest side of the diapir. The presence of a diapir roof during the second stage of deformation helped to pressurize the salt enough to inject laterally. However, it is unclear why this roof was not itself uplifted and eroded during the second phase of shortening. The diapir may have actually pinched off during stage 1 shortening (as in the alternative flank interpretation), so that the high salt pressures causing injection were not transmitted to the top of the diapir. Section modified from Kockel (1990); restoration modified from Hudec (2004).

PRECASPIAN BASIN—Salt Sheet from Temir Block, Kazakhstan

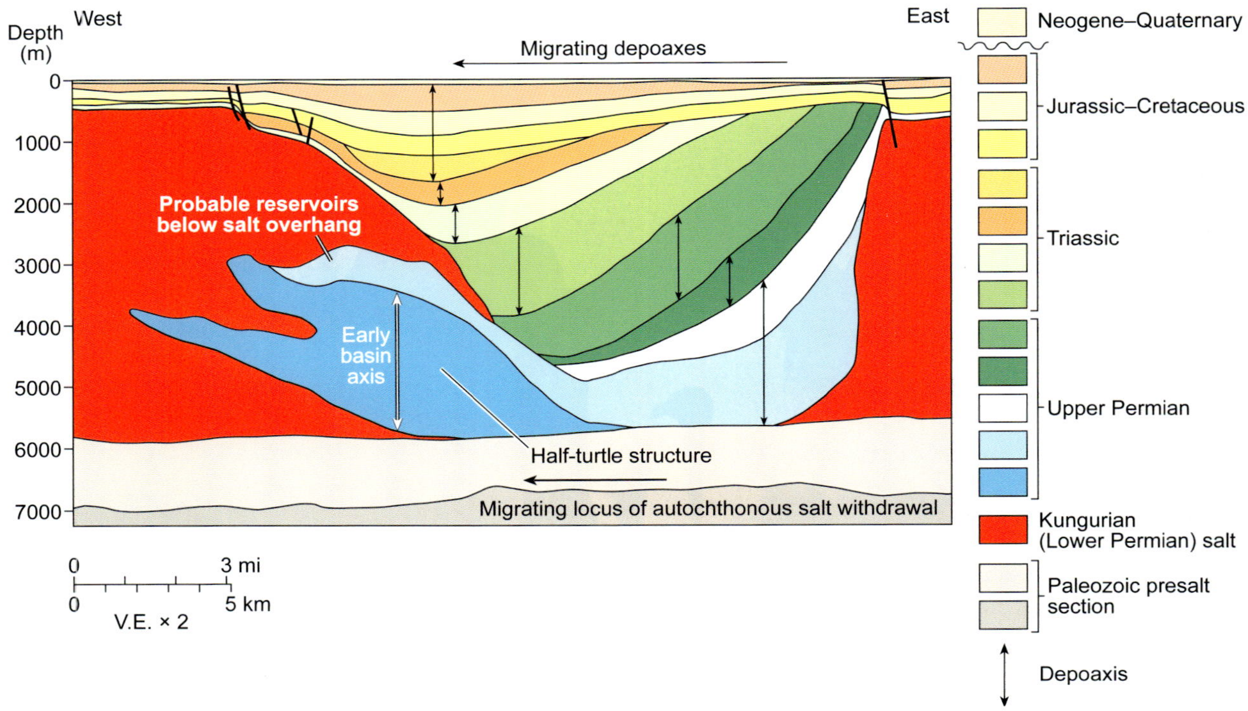

Figure 7.29. Folding of a salt sheet. A salt sheet was extruded over the west part of the minibasin during the Late Permian, when the main depoaxes lay well to the east of the sheet. Salt- withdrawal synclines then migrated westward, until autochthonous salt below the sheet became involved. Expulsion of autochthonous salt from beneath the sheet caused the suballochthonous salt section to subside. This subsidence formed a half-turtle structure and folded the formerly extrusive salt sheet, thus illustrating how preexisting salt sheets are deformed by all tectonic processes affecting the subsalt section. Modified from Barde and others (2002).

PRECASPIAN BASIN—Kum Salt Sheet, Kazakhstan

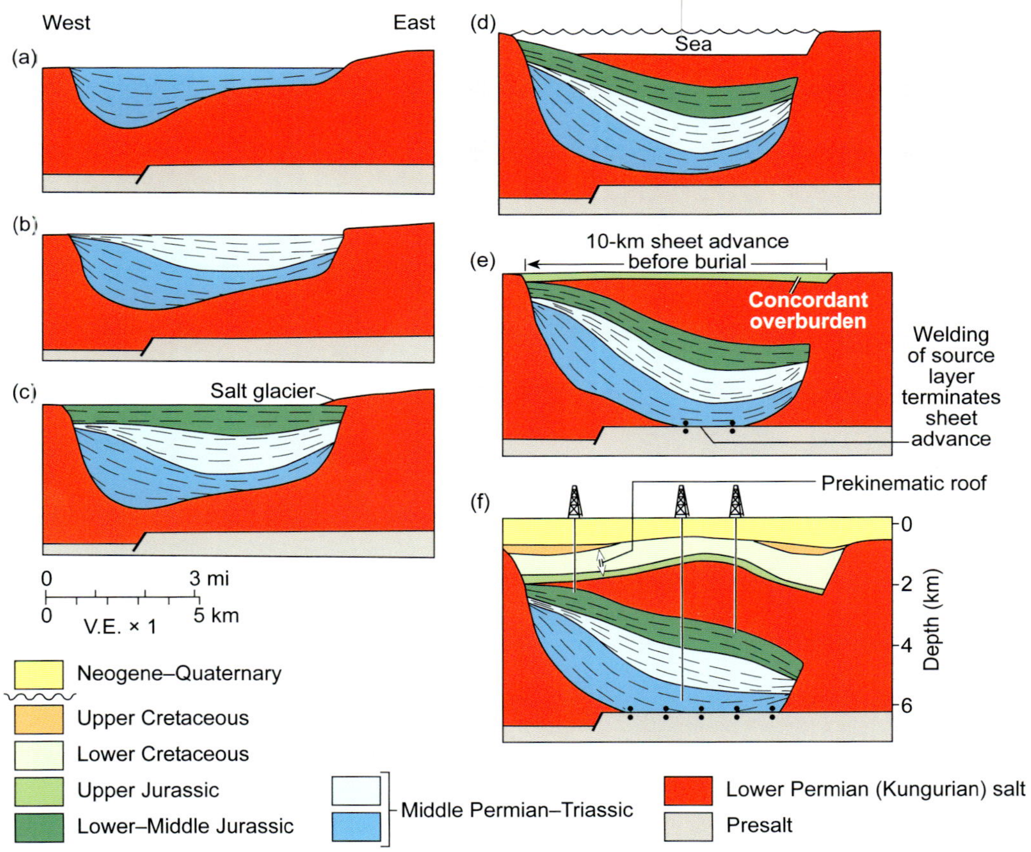

Figure 7.30. Rapid extrusion of a salt sheet. During Middle–Late Jurassic time, the Kum salt glacier spread across a single Middle Jurassic horizon a maximum of 14 km (10 km in this plane of section), impling that extrusion was much faster than sedimentation. The salt sheet stopped advancing once the underlying source layer was pinched off. The concordant isopachous Upper Jurassic–Lower Cretaceous roof shows that the salt then stagnated and was buried. Modified from Volozh and others (2003).

QOM BASIN—Extrusive Salt Sheet, Qom Kuh Diapir, Central Iran

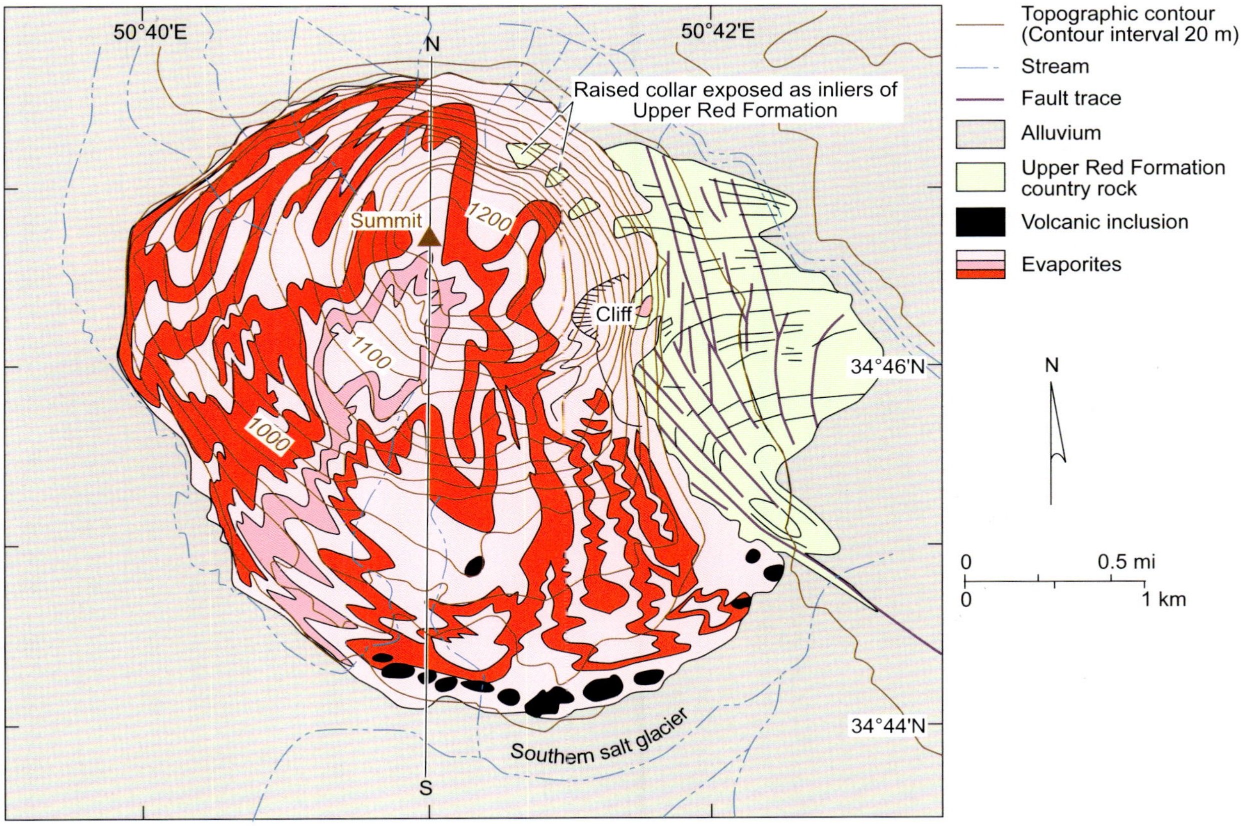

Figure 7.31. Extrusion of a salt glacier from an emergent diapir. Diapiric Eocene and Miocene evaporites form a 315-m-high dome about 2.5 km wide centered on the summit. From here, extrusive salt spreads radially, but mostly southward. The southern salt glacier is about 2 km long. The stratigraphy in the salt is arbitrary and intended to show the fold pattern. An upturned collar of country rock around the diapir's vent partly crops out as inliers in the northeast part of the diapir. This small diapir contains the world's finest example of a terminal moraine of stranded nonevaporite blocks (black) rimming the glacier (see Figure 7.33). After Talbot and Aftabi (2004).

QOM BASIN—Extrusive Salt Sheet Detail, Qom Kuh Diapir, Central Iran

Figure 7.32. Extrusive spreading of an emergent diapir. Diapiric Eocene and Miocene evaporites rise below the 315-m-high summit of the mountain and spread radially, but mostly southward. The dark-colored soil on the top of the sheet is a ferruginous residual soil left by dissolution of more soluble parts of the evaporite sequence. The eastern contact of the diapir, which follows the lower slopes of mountain, has been offset by faults shown on the geologic map (Figure 7.31). For further details, see Talbot and Aftabi (2004). Photograph by Martin Jackson.

QOM BASIN—Extrusive Salt-Sheet Igneous Blocks, Qom Kuh Diapir, Central Iran

Figure 7.33. Rafting and accumulation of igneous blocks in a salt glacier. The southern salt glacier of Qom Kuh, which is about 2 km long, is rimmed by the world's finest example of a terminal moraine of stranded nonevaporite blocks (see the geologic map in Figure 7.31). The igneous blocks (IB) in the moraine were probably originally interbedded in the Eocene salt. The diapir carried these blocks to the surface, then to the glacier terminus, where they were stranded by salt dissolution. The blocks have settled in a remarkably uniform chain along the glacier terminus. Reddish-brown residual gypsite soil covers most of the rock salt in this rugged terrain. White salt efflorescence coats the steepest slopes, from which soil has been washed off. For further details, see Talbot and Aftabi (2004). Photograph by Martin Jackson.

QOM BASIN—Extrusive Salt Sheet, Shurab Diapir, Central Iran

Figure 7.34. Emergence of a salt glacier from its vent. This quarried exposure captures a significant transition. On the right is a diapir stem, in which layering, axial surfaces, and fabrics are steep, in conformity with the lateral walls. Toward the left is a diapiric extrusion, in which these structures flop over to a subhorizontal attitude in conformity with the underlying bedrock and the inferred original topography. This original topography was destroyed as the salt glacier dissolved after emerging. Thus the present topography truncates the internal structures. Shurab diapir is about 60 km southeast of Qom Kuh. After Talbot and Aftabi (2004).

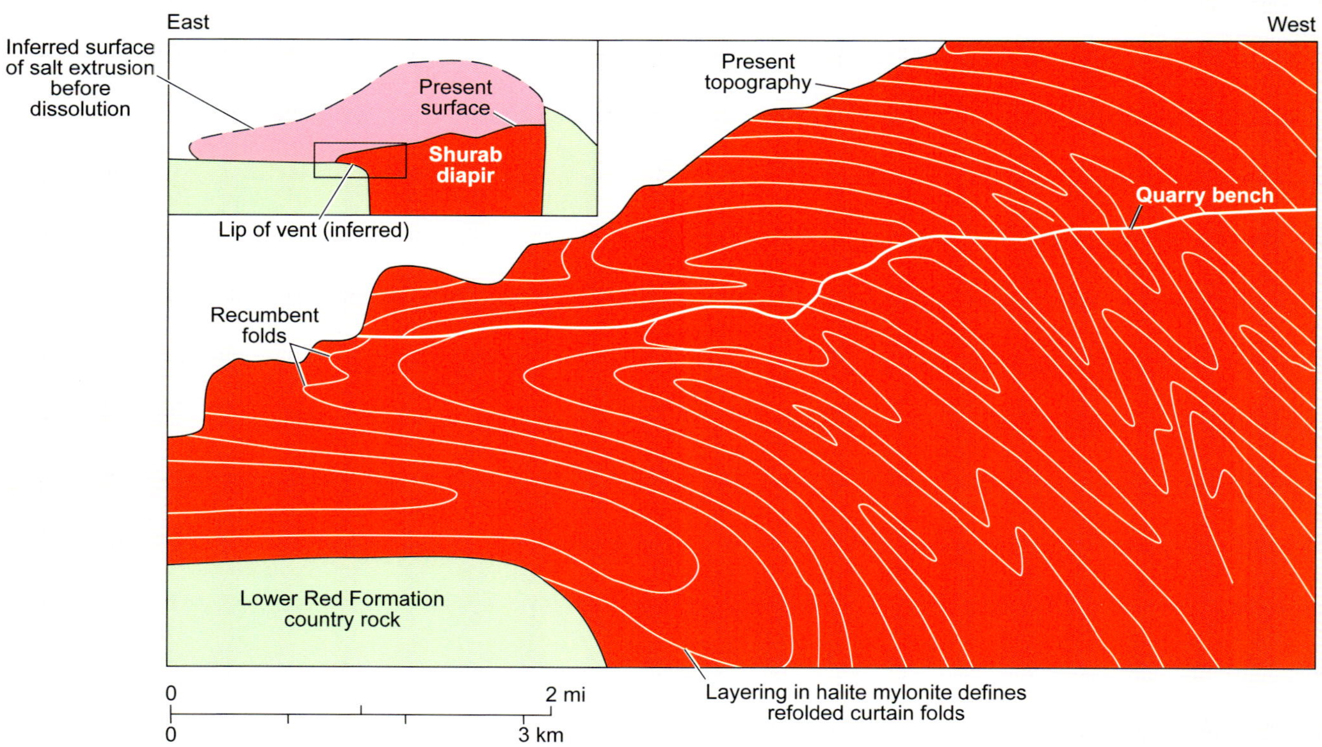

RIF BASIN—Overthrust Salt Nappe, Morocco

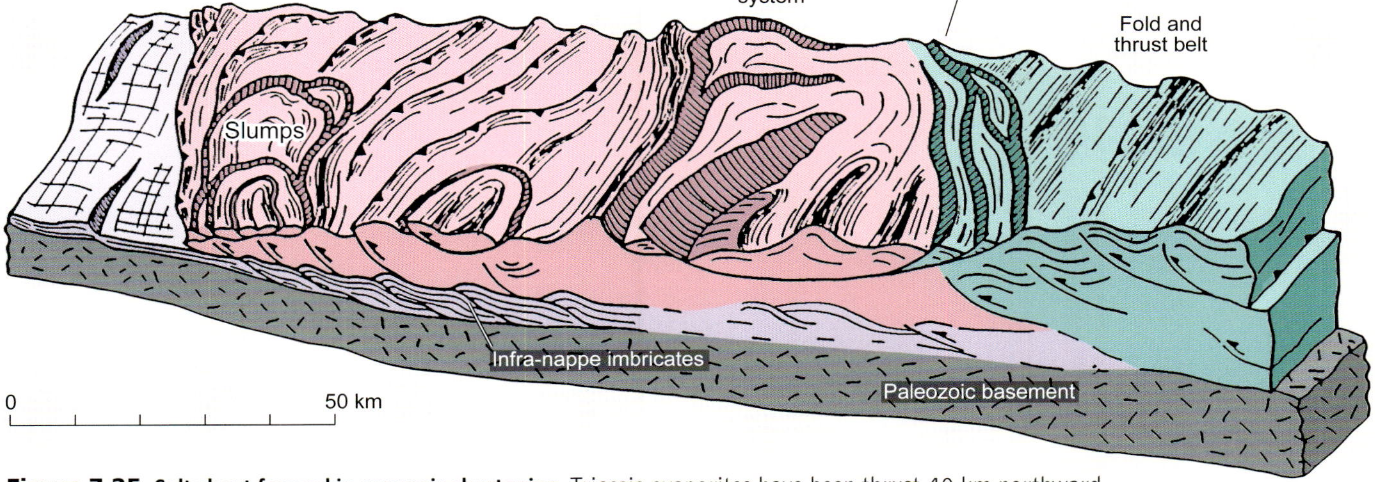

Figure 7.35. Salt sheet formed in orogenic shortening. Triassic evaporites have been thrust 40 km northward over rocks that were originally adjacent to the salt basin. The allochthonous salt has been shortened, with numerous internal imbricates. Upper levels of the salt are also undergoing gravitational collapse, forming slump deposits and normal faults. Some of these normal faults also affect roof strata at the south edge of the sheet. Modified from Flinch (1996).

ZAGROS FOLD BELT—Extrusive Salt Sheet, Anguru Diapir, Southern Iran

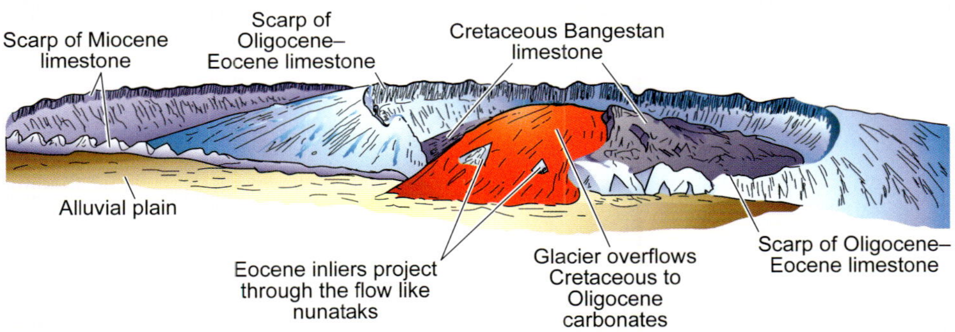

Figure 7.36. Extrusion of a salt sheet. The diapir emerges from a culmination of the Kuh-e-Gashu anticline. It is probably a preshortening diapir, whose present extrusion is caused by displacement of salt from the shortening feeder. The location of the diapir along the crest of the anticline is most likely not a coincidence; the diapir probably acted as a weak zone that localized later shortening. This sketch of Anguru by G. M. Lees, made in the early 1920s, is the first known illustration of an identified salt glacier (along with an accompanying photograph of Kuh-e-Namak in Bushehr Province). A modern photograph shows the characteristic smooth, parabolic profile of the summit dome. On the lower right of the summit dome, one of the Eocene inliers projects through salt, as sketched by Lees. After De Böckh and others (1929). Photograph by Martin Jackson.

Zagros Fold Belt—Extrusive Salt Sheets, Gach and Siah Tagh, Southwestern Iran

Figure 7.37. Extrusion of salt sheets. The domes are all near the crests of regional anticlines. Some domes probably existed before the Miocene Zagros shortening began. If so, they formed weak plugs that controlled the location of orogenic folds. Note, for example, the axial bend in the Kuh-e-Gach Anticline that links both the Aliabad and Gach salt plugs. This eastward view shows four emergent Zagros diapirs. Gach and Siah Tagh are two of the longest salt glaciers in Iran, with lengths of 7 km from the peak of their summits to the terminus. Both are salt fountains. The approximately 200-m-thick Gach salt glacier is markedly dissected and has no visible salt left. It comprises only insoluble blocks and gypsite soil. These block further flow of the main glacier. The approximately 500-m-high summit dome of Gach continues to extrude salt, but this flow has been diverted to a small flow off the northwest of the summit. The equally long salt glacier of Siah Tagh contains visible salt in the proximal two-thirds of its length. Dark fans collecting next to the large salt glaciers comprise dark rubble consisting mainly of black carbonates. Farther east, the 5-km-wide Chahal rises 2.2 km above sea level from the nose of a plunging anticline. This is the highest salt diapir in the Zagros foreland. Two small, young salt glaciers occupy ravines on the northern flank of the dome but are not visible at the scale of the image. This is a hemispherical dome evolving into the earliest stages of a salt fountain. This perspective view with 2× vertical exaggeration was created by draping a band 3-2-1 (RGB) image over an ASTER-derived digital elevation model and was acquired on August 10, 2001. Image from NASA/GSFC/METI/ERSDAC/JAROS and U.S./Japan ASTER Science Team.

Zagros Fold Belt—Extrusive Salt Sheet, Kuh-e-Namak and Southern Salt Glaciers, Panorama, Southwestern Iran

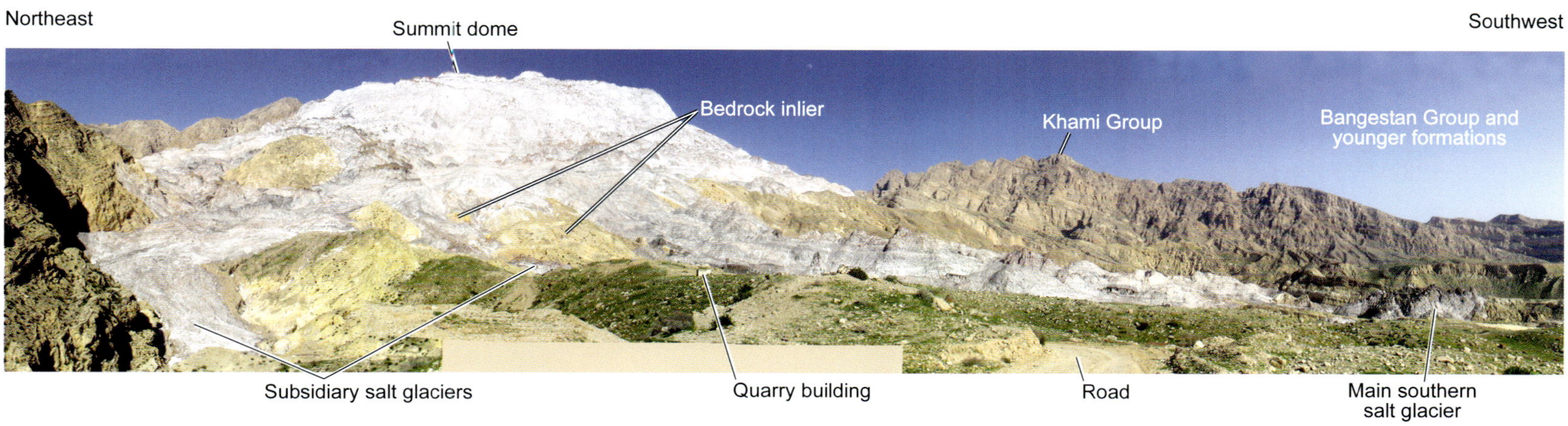

Figure 7.38. Salt-glacier extrusion. Salt flows from the summit dome above the diapir stem. The glacier spreads in several streams around the bedrock inliers. The leftmost small salt glacier is shown in close-up in Figure 7.39. Superb clean exposures are attributed to exposure of the southern face of the mountain to prevailing rain from the nearby gulf, which washes away the residual gypsite soil cap. Photograph by Martin Jackson.

ZAGROS FOLD BELT—Extrusive Salt Sheet, Kuh-e-Namak Summit and Southwestern Salt Glacier, Southwestern Iran

Figure 7.39. Dissolution of a salt sheet. This salt glacier was once much thicker, as shown by (a) the relict salt plastered high on the cliffs enclosing the valley, and (b) the feather edge of the downstream terminus, which suggests the glacier is wasting by dissolution faster than it is being supplied by salt. Thinning of the glacier has exposed several bedrock inliers. Salt flows from the summit dome above the diapir stem and spreads in several streams around the inliers. Salt is extruding over the southwestern limb of the Kuh-e-Namak Anticline. The position of this salt glacier on the overall structure is shown on the overview photograph in Figure 7.38. Modified from Hudec and Jackson (2006).

ZAGROS FOLD BELT—Thrust-Related Evaporite Sheets, Gachsaran Basin, Southwestern Iran

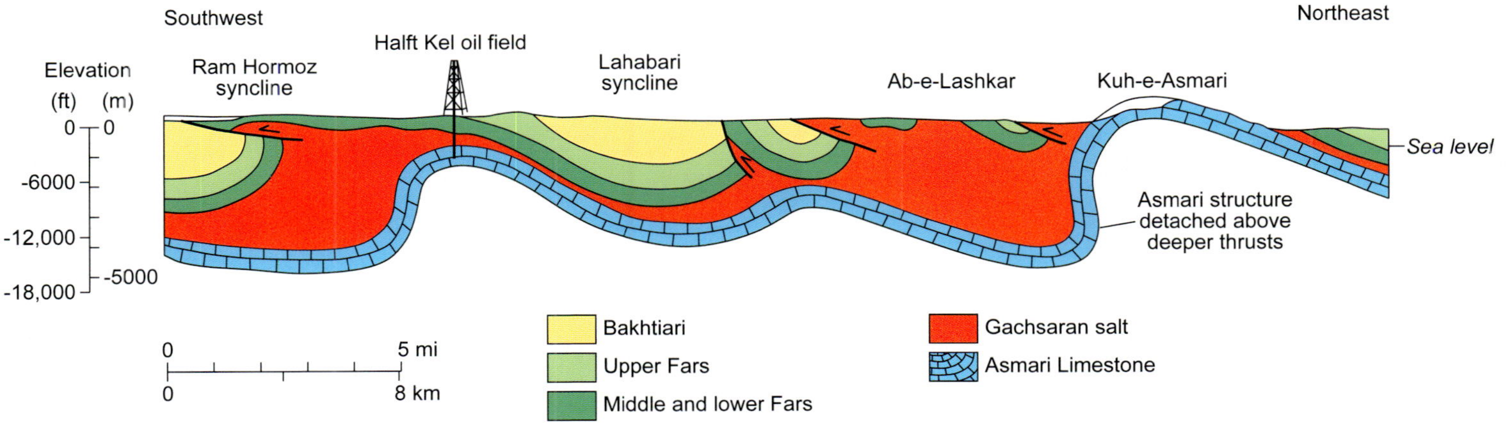

Figure 7.40. Expulsion of evaporite sheets from a fold belt. The ramp-fed extrusive sheets in the lowest part of thrust hanging walls advance over Pliocene and Pleistocene gravels. The lower Miocene Gachsaran unit comprises evaporites (anhydrite and subordinate halite), mudrock, marl, and sandstone. The mobility of the Gachsaran largely decouples shortening above and below. Thus, hydrocarbon-bearing anticlines in the Asmari carbonates (Oligocene to lowest Miocene) have no simple relationship with post-Gachsaran shortening structures. Here in the Dezful Embayment, the Neoproterozoic–Cambrian Hormoz evaporites are too thin to create emergent diapirs. The main detachment for the deep, thrust-related anticlines involving Asmari and older strata could be Cretaceous Garau shale or Upper Jurassic Hith/Gotnia evaporite. After O'Brien (1957).

ZAGROS FOLD BELT—Extrusive Salt Sheet, Ideal Profile, Southern Iran

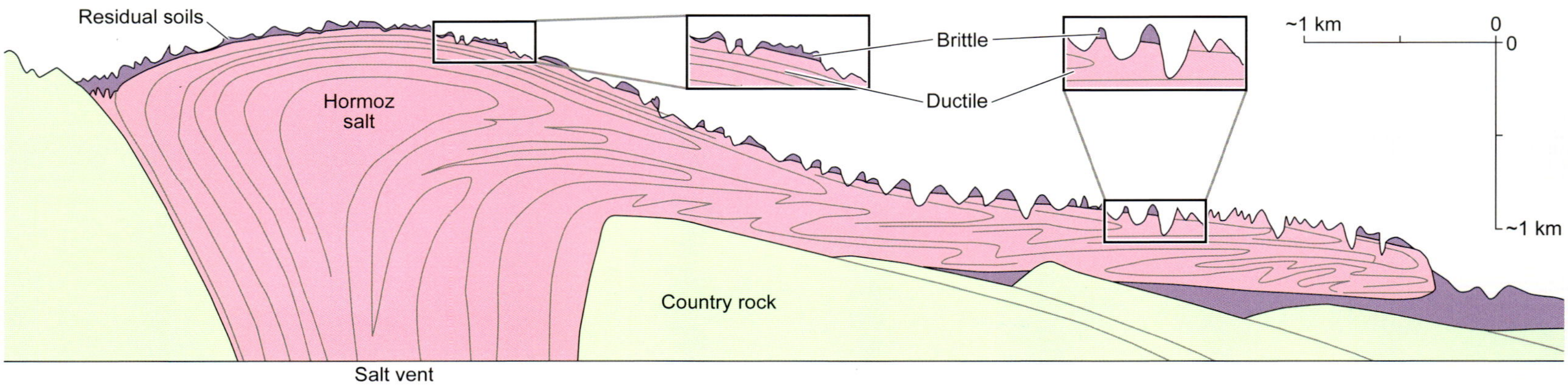

Figure 7.41. Flow of salt sheets. The diapir rises from a Zagros anticline and extrudes a salt glacier down the fold limb. Flow is driven by the surface slope from the summit above the salt vent down the glacier length. Internal structures (layering, fabric, and fold axial surfaces) conform to the steep margins of the diapir stem and to the subhorizontal rigid base and top free surface of the glacier. A brittle carapace of alternating sheets of salt and residual soil caps the extrusion. This cap slows the rate of salt dissolution on the sheet surface. After Talbot (1998).

ZAGROS FOLD BELT—Extrusive Salt Sheet, Ideal Internal Structure, Southern Iran

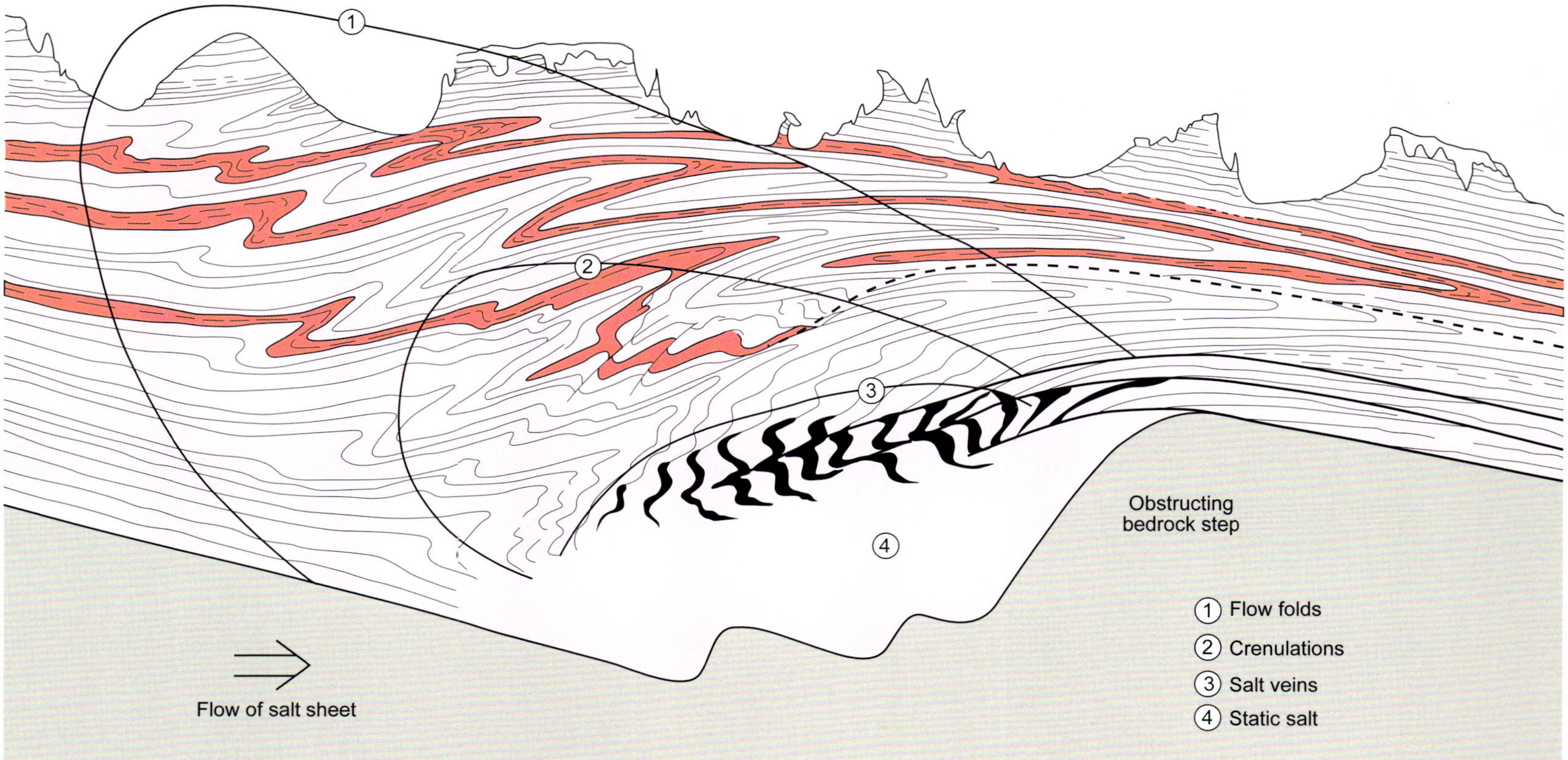

Figure 7.42. Flow of salt sheets. Salt sheets deform in response to flow over a step in the base of salt. Heavy curved lines separate four structural zones. The lowest zone is static salt, which is equivalent to dead ice in an ice glacier. Flow folds form wherever streamlines are oblique to the layering. The folds tighten downstream to the point where their hinges are barely recognizable. The folds become refolded multiple times downstream. The layering is inherited from bedding in the original evaporite. Internal structures are based on Kuh-e-Namak (Bushehr). After Talbot (1981).

ZAGROS FOLD BELT—Extrusive Salt Sheet, Kuh-e-Namak Southwestern Salt Glacier, Bushehr, Southwestern Iran

Figure 7.43. Folding in a salt glacier. These tight to isoclinal folds have axes parallel to the southward flow direction of the glacier. This orientation suggests that these folds are sheath folds, stretched in a downstream direction. Black salt layers are carbonate rich; red salt layers are ferruginous. In the background is Kuh-e-Darang, a buried but active Hormoz diapir. Photograph by Martin Jackson.

ZAGROS FOLD BELT—Extrusive Salt Sheet, Kuh-e-Namak Diapir, Bushehr, Southern Iran

Figure 7.44. Salt-glacier extrusion. Bedrock beneath the Kuh-e-Namak diapir was lithified before the salt extruded, so it was not deformed during sheet emplacement. The contact between the sheet and its basement is sharp, which contrasts with some of the deformation fabrics reported at the base of sheets emplaced over less lithified sediments. Prominent layering in the salt is obscured by a white veneer of reprecipitated salt weathering to a hackly texture, resembling white foam. Lower reaches of the glacier are visible in the distance. Beyond the major dry river bed is Kuh-e-Darang, a blind but active Hormoz diapir that has been growing since at least Permo-Triassic time and is overlain by a half graben of Jurassic–Cretaceous strata. Both Kuh-e-Namak and Kuh-e-Darang lie along the Kazerun fault, which is the northwestern boundary of the Hormoz diapir province. Photograph by Martin Jackson.

ZAGROS FOLD BELT—Extrusive Salt Sheet, Kuh-e-Namak Southern Salt Glacier Basal Inlier, Bushehr, Southwestern Iran

Figure 7.45. **Salt-glacier extrusion.** Bedrock beneath the Kuh-e-Namak diapir was fully lithified before the salt extruded, so the basal contact of the sheet is sharp. This sharpness can be seen where dissolution has exposed a window of limestone bedrock and the basal contact of the salt glacier. Layering in the basal red-black salt wraps over the basal swell of the bedrock. Thickness of the basal salt is a few tens of meters. Both this unit and the overlying pink-white salt have been deformed many times, mostly by isoclinal folding. Photograph by Martin Jackson.

ZAGROS FOLD BELT—Extrusive Salt Sheet, Jahani Diapir, Southern Iran

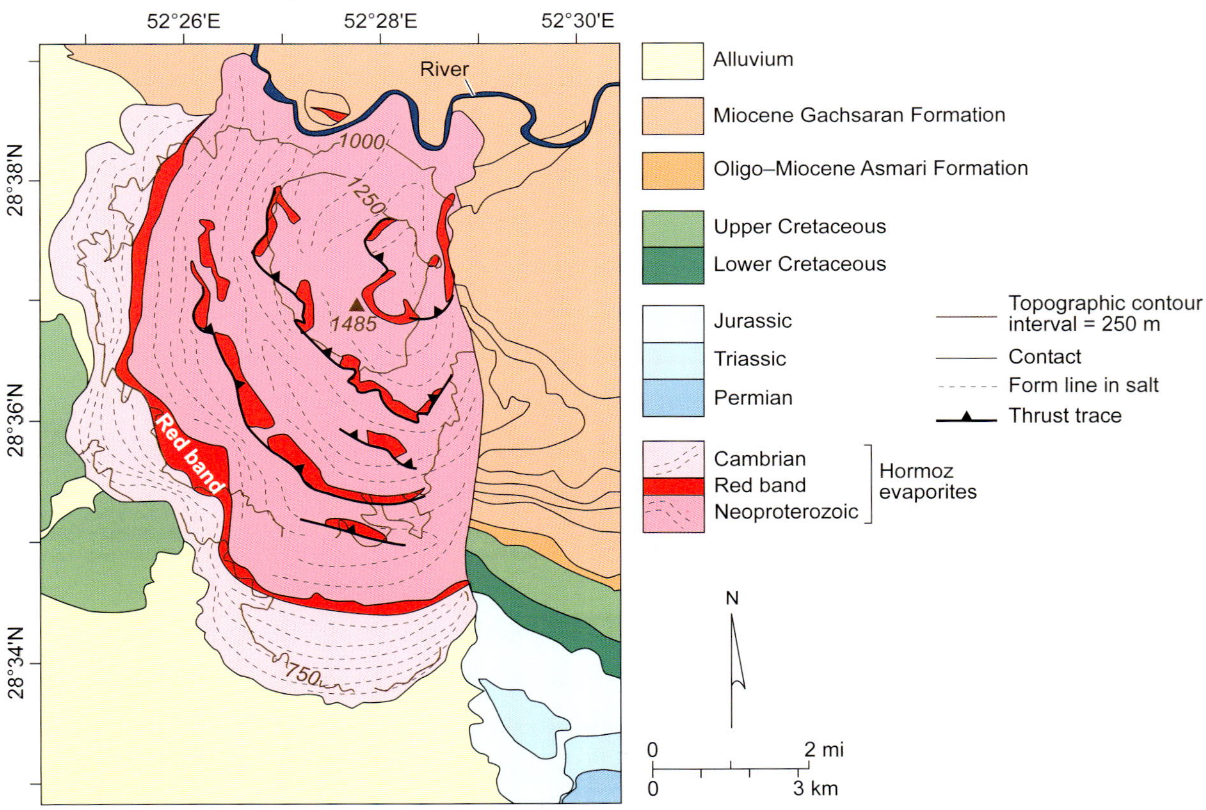

Figure 7.46. **Flow of salt sheets.** Neoproterozoic evaporites have overridden the originally overlying Cambrian salt. Both evaporites and the red band (sandstone, marl, and carbonate) separating them have been repeated by thrusts within the salt sheet. These thrusts form where rapidly spreading parts of the salt sheets override the slower frontal parts (see flow vectors in Figure 7.47). Jahani is one of the largest and most vigorously spreading salt fountains in the Zagros. Flow is largely southward. Jahani lies where a Zagros syncline is offset about 5 km by right-lateral oblique slip near the south end of the Mangarak transfer fault. After Talbot and others (2000).

ZAGROS FOLD BELT—Extrusive Salt Sheet, Jahani Diapir, Southern Iran

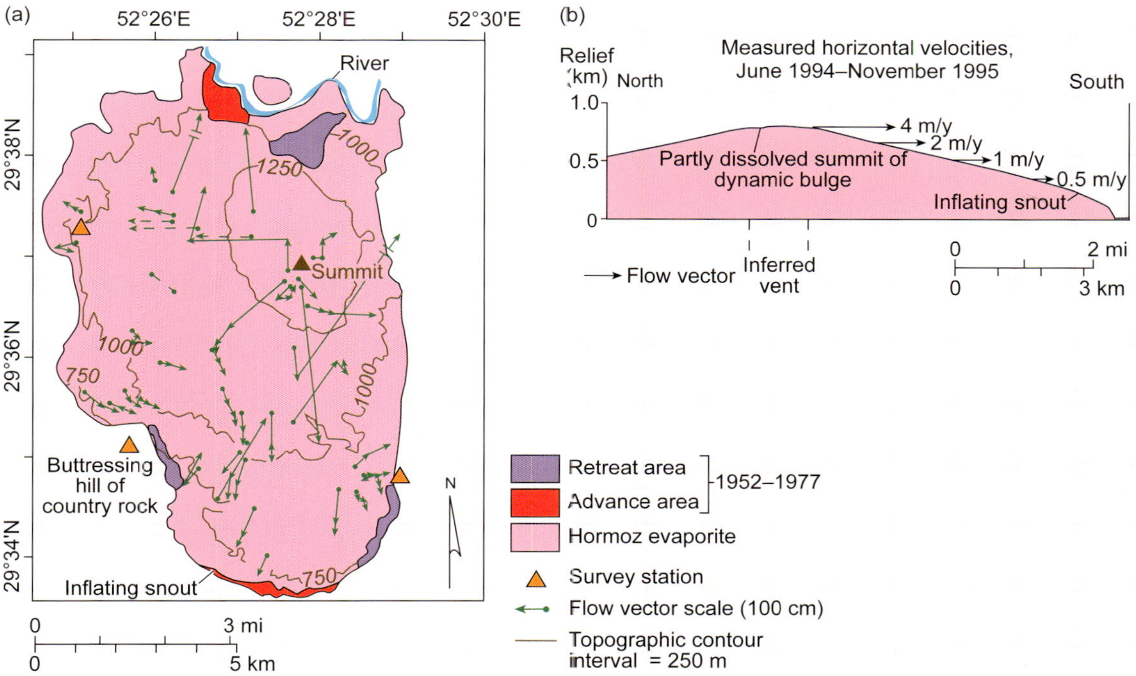

Figure 7.47. Spreading rates of a salt extrusion. Salt flow is most rapid near the crest of a salt sheet and slows toward the snout. This slowdown causes higher parts of the salt sheet to override the slower moving front, which can produce thrust repetitions in the salt (see geologic map in Figure 7.46). Calculations suggest that the rapid rate of salt flow may have nearly exhausted the estimated salt supply in the source layer in as few as 50,000 years. Jahani was surveyed over 4 years from three theodolite stations. Green arrows show velocity vectors over the time period shown. During this time, flow was highly unsteady. The surface dissolution rate is about 2–3 cm/year. The displacement vectors show 4 years of glacial flow. The scale of the vectors differs from that of the map. Areas of longer term advance and retreat were inferred by comparing aerial photographs and topographic maps of different ages. After Talbot and others (2000).

ZAGROS FOLD BELT—Extrusive Salt Sheet, Kuh-e-Namak Northern Salt Glacier Contact, Bushehr, Southwestern Iran

Figure 7.48. Spreading of a salt sheet. This lateral northwestern contact of the salt glacier forms a cliff for two reasons: (a) a brine stream is actively eroding it or (b) salt spreading rate appears to be high and increased by the locally steep slope of the glacier and the high extrusion rate of this entire northern salt glacier. Compare this cliff with the feather edge of the wasting southern glacier (Figure 7.39). The recumbent fold hints at a complex stack of sheath folds resulting from lateral salt extrusion. The salt glacier is extruding over the northeastern limb of the Kuh-e-Namak Anticline. Photograph by Martin Jackson.

PHYSICAL MODEL—Disruption of a Spreading Roof during Shortening, Part 1

Figure 7.49. Salt-sheet extrusion. Two buried stocks shortened at the downdip end of a salt-detached gravity-driven system. Diapir A shortened first because it lay updip of the hinge and because it was closer to the extensional fault (not pictured) that accommodated downdip translation. Shortening caused diapir A to inflate then extrude a salt sheet. Spreading of the salt sheet disrupted its roof into small rafts that were carried outward by the extrusive salt. The more-distal diapir B merely inflated at this point in the experiment, although later views (Figures 7.50 and 7.51) show that it eventually extruded its own salt sheet. Both diapirs shortened before any visible faulting in their wall rocks, but distortion of the marker grid shows that the overburden laterally compacted to compensate for diapiric strain. Thrust faults eventually propagated outward from both stocks, especially diapir A. See also oblique views of the experiment in Figure 7.52. Model by Tim Dooley.

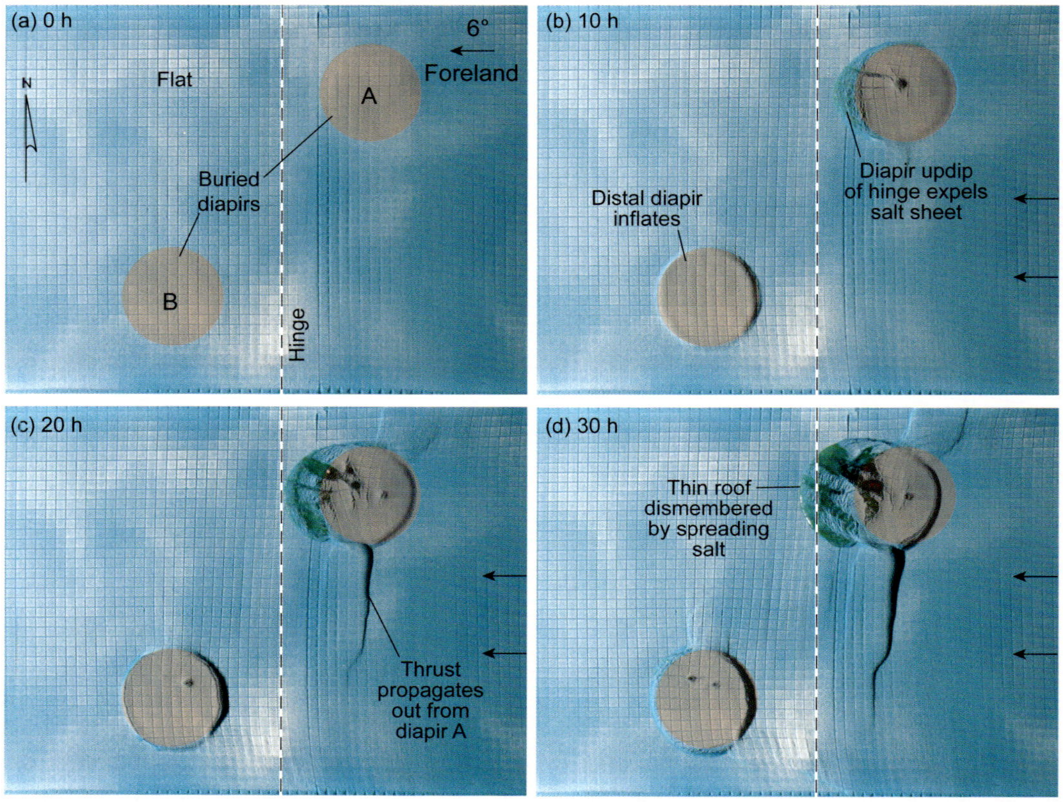

PHYSICAL MODEL—Disruption of a Spreading Roof during Shortening, Part 2

Figure 7.50. Salt-sheet extrusion. Two buried stocks shortened at the downdip end of a salt-detached gravity-driven system. Earlier views (Figure 7.49) showed that diapir A began shortening and extruding a salt sheet first, but after 40 hours diapir B also began to expel salt. As diapir B shortened, a thrust fault propagated northward into the transpressional zone between the two diapirs. This zone was bounded by strike-slip and oblique-slip faults. Salt-sheet spreading disrupted the thin roofs that originally covered the diapirs. Roof fragments were carried outward to the edge of the sheet, where they docked against the surrounding peripheral plain. These docked fragments locally blocked sheet advance, causing the sheet to break into sublobes that advanced between the stranded fragments. See later stages of the experiment in Figure 7.51 and oblique views in Figure 7.52. Model by Tim Dooley.

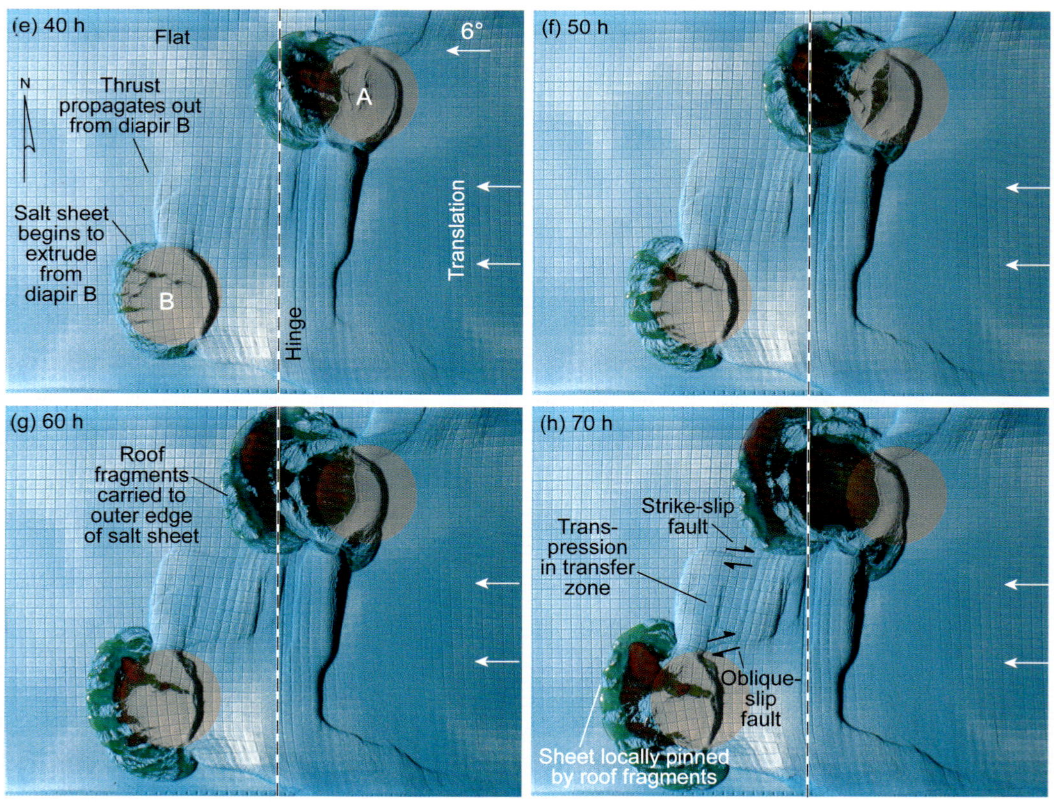

PHYSICAL MODEL—Disruption of a Spreading Roof during Shortening, Part 3

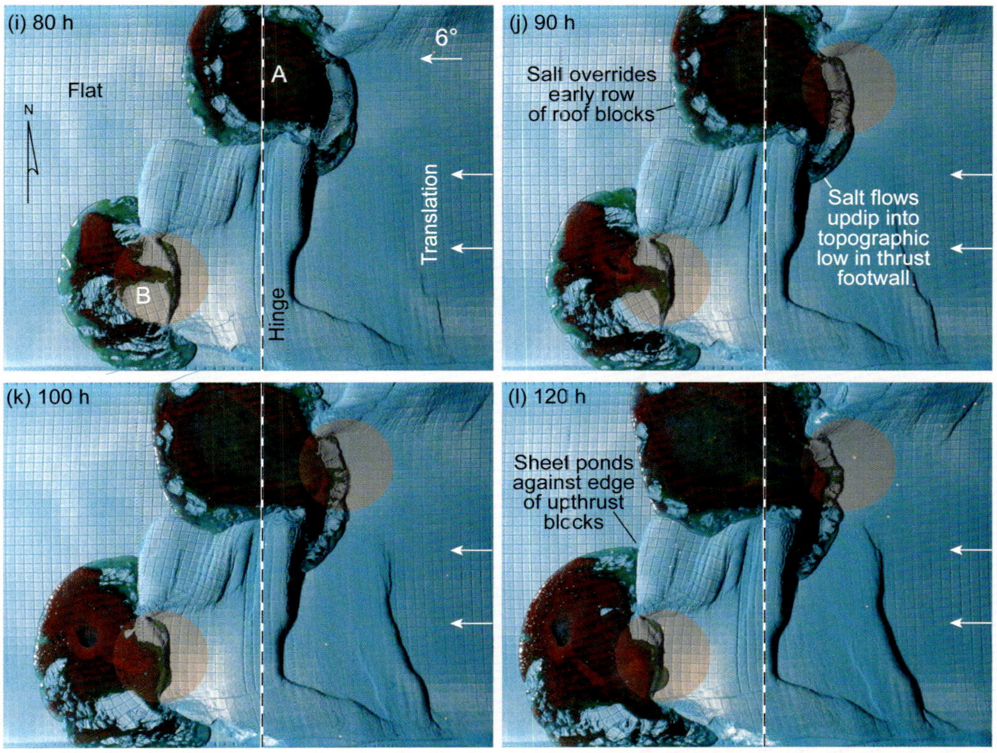

Figure 7.51. Salt-sheet extrusion. Two buried stocks shortened at the downdip end of a salt-detached, gravity-driven system. Because the salt sheets extruded in an active thrust belt, the sheets preferentially flowed into topographic lows in the footwalls of major thrusts. This flow produced a plan form very unlike the ellipse typical of sheets flowing freely down most continental slopes. Roof fragments carried to the edge of the sheets briefly obstructed sheet advance (see earlier stages of the experiment in Figures 7.49 and 7.50) but were ultimately overridden by the spreading salt. See also oblique views of the experiment in Figure 7.52. Model by Tim Dooley.

PHYSICAL MODEL—Oblique Views of Disruption of a Spreading Roof during Shortening

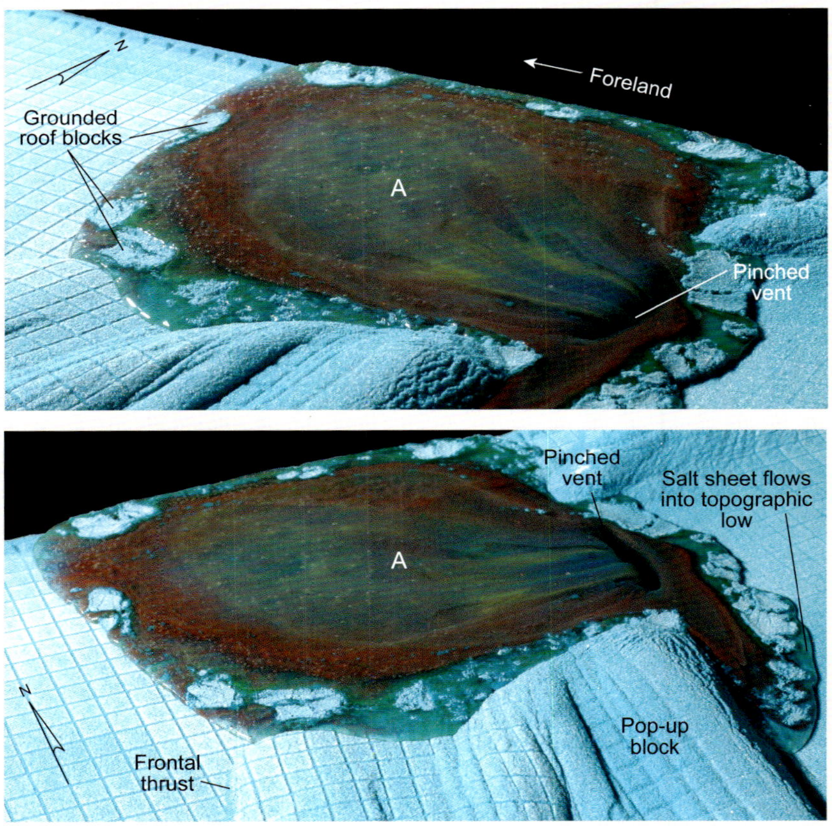

Figure 7.52. Salt-sheet extrusion. A buried stock shortened at the downdip end of a salt-detached gravity-driven system. Thrust faults associated with this east-west shortening propagated out from the diapir to form topographic highs north and south of the diapirs (see sequential overhead views of the experiment in Figures 7.49 through 7.51). Salt sheets expelled from the diapirs flowed around the topographic highs and ponded in low footwall blocks. Spreading salt also carried pieces of the disrupted diapir roof to the edges of the sheet, where they grounded against the surrounding peripheral plain. Some blocks transported early in sheet history were overridden by the salt and became entombed along the base of the sheet. Model by Tim Dooley.

Physical Model—Extrusion of Salt from an Inclined Stock above a Flat Source Layer

Figure 7.53. Salt-sheet extrusion. A foreland-inclined stock was shortened, expelling salt onto the sediment surface. Colors in the salt are passive markers in the diapir and the source layer. The markers track the source of the expelled salt. During early shortening, salt expulsion was as expected, with salt from deeper in the diapir expelled progressively later in the experiment. With greater shortening (shown here), plugs from the source layer began to be extruded. The underside view (d) shows that salt on all sides of the diapir flowed into the diapir, as expected in a flat source layer. However plugs 1 through 5, located in the thrust belt, flowed farther than plugs in the foreland because of (a) gravitational loading from the overlying thrust pile and (b) displacement loading of source layer from the moving endwall and the bottom of thrust sheets. Model by Tim Dooley.

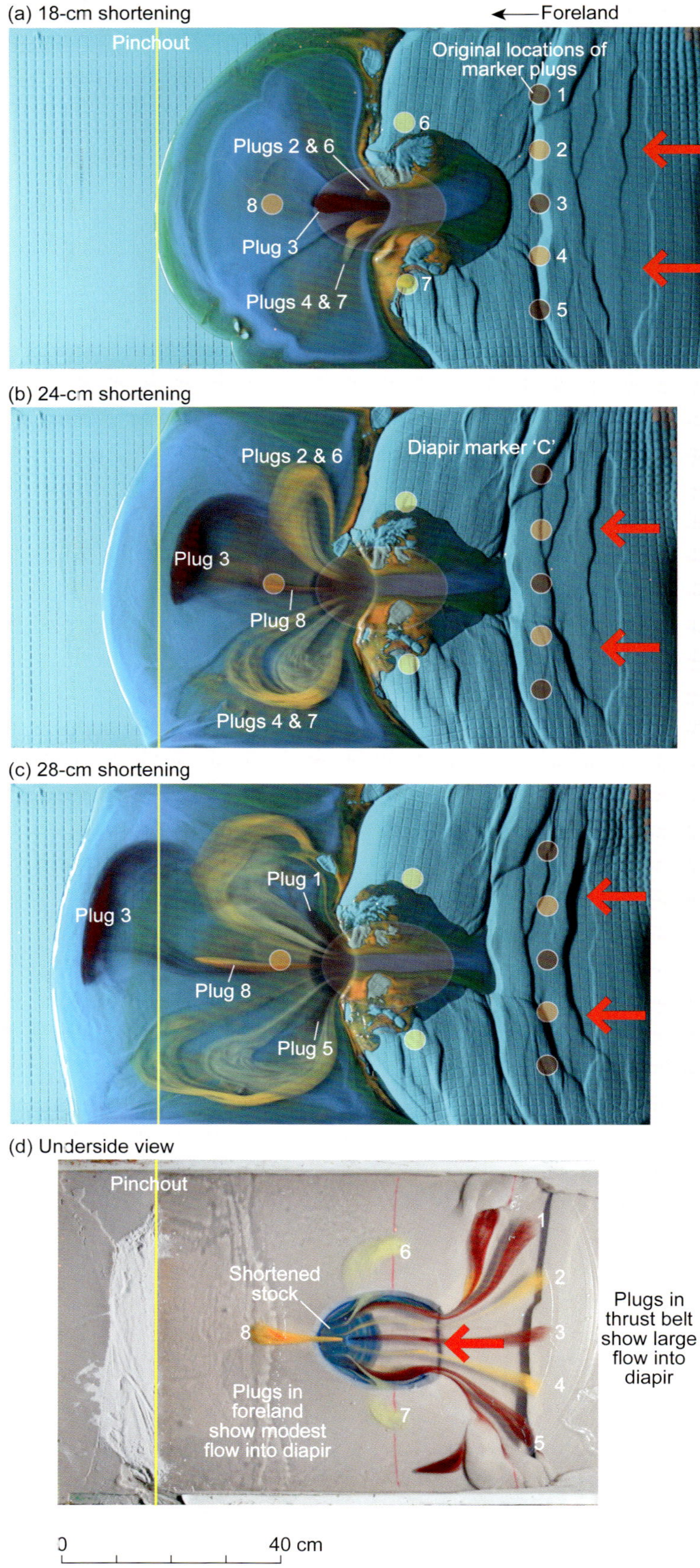

PHYSICAL MODEL–Salt Inflation prior to Salt-Sheet Breakout

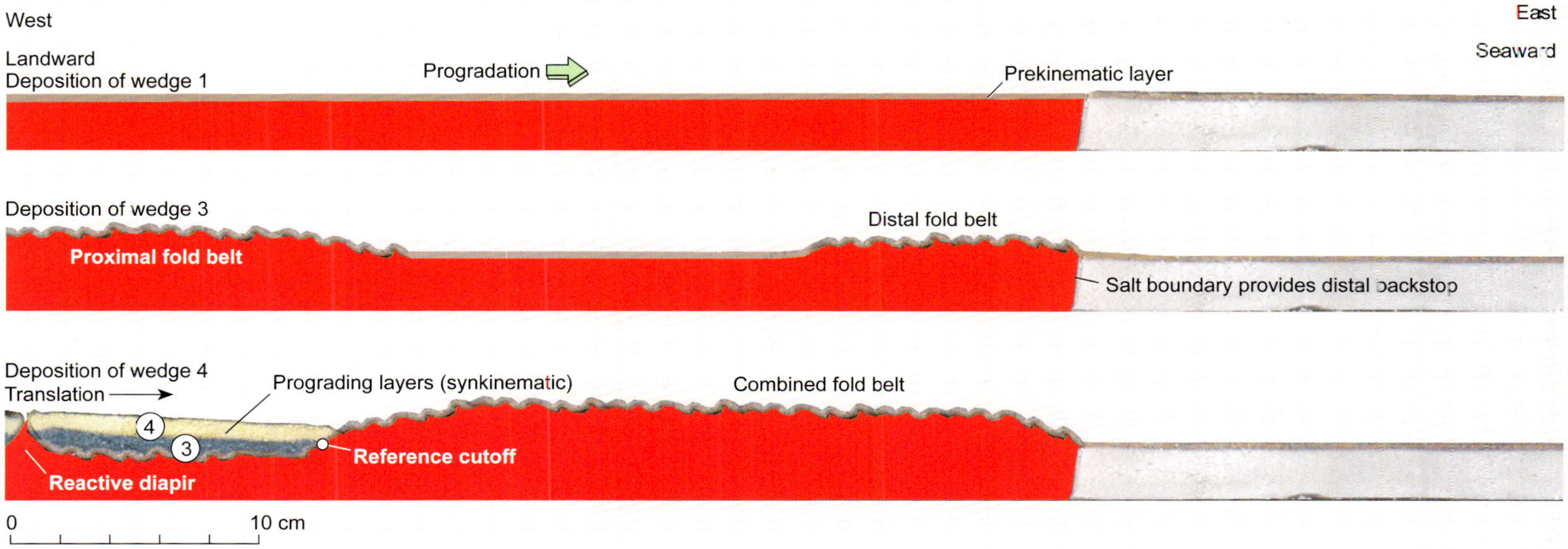

Figure 7.54. Inflation of fold belts on autochthonous salt before extrusion of salt sheets. A prograding sedimentary wedge drove salt seaward, inflating the distal end of a salt basin. Deformation of the wedge also created a linked kinematic system in the sedimentary roof, with updip extension linked to downdip shortening. Initially, two fold belts formed. The proximal fold belt formed at the break in slope just beyond the end of the sedimentary wedge. The distal fold belt formed at the seaward salt boundary. After deposition of wedge 4, the two fold belts merged, and progradation overran their proximal part. This thickening strengthened the overburden and prevented further folding. The reference cutoff point is tracked in the next stages of the restoration (Figure 7.55). Modified from Rowan and others (2004).

PHYSICAL MODEL–Salt-Sheet Breakout

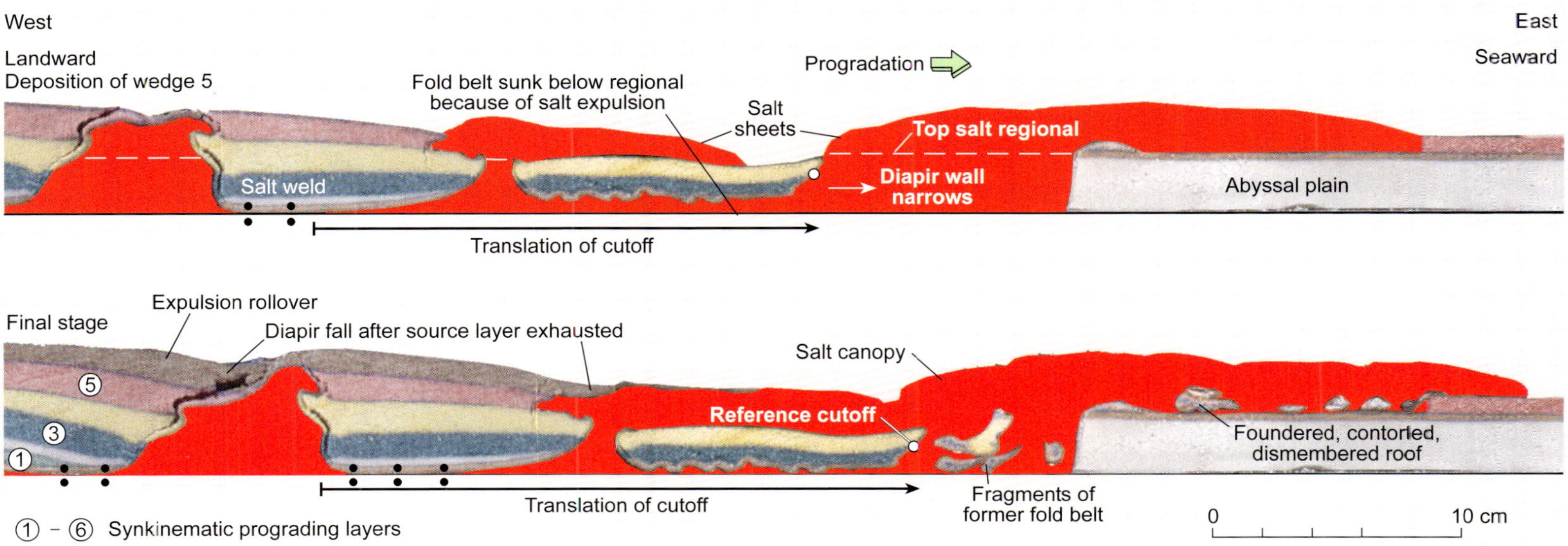

Figure 7.55. Breakout of a salt sheet and destruction of its roof. A prograding sedimentary wedge drove salt seaward, inflating salt at the distal end of the basin. Continuing the evolution shown in the previous stage (Figure 7.54), the inflated margin of the salt basin rose, extended, and pierced its thin roof. This vent released a large salt extrusion, which overrode the abyssal plain in the east. To the west, two diapirs and their encasing overburden moved eastward into the cross-section window because of regional translation. The proximal diapir fell because extension continued after the source layer welded. The distal diapir extruded ahead of the prograding wedge and eventually coalesced with the main salt sheet to the east, forming a canopy. Beneath this canopy, the old fold belt sank below regional into the less dense salt, along with contorted remnants of the same sedimentary roof. To show translation, the reference cutoff is tracked from the previous stage of the restoration. Experiment and restoration by Bruno Vendeville.

PHYSICAL MODEL—Salt-Sheet Advance during Sedimentary Loading

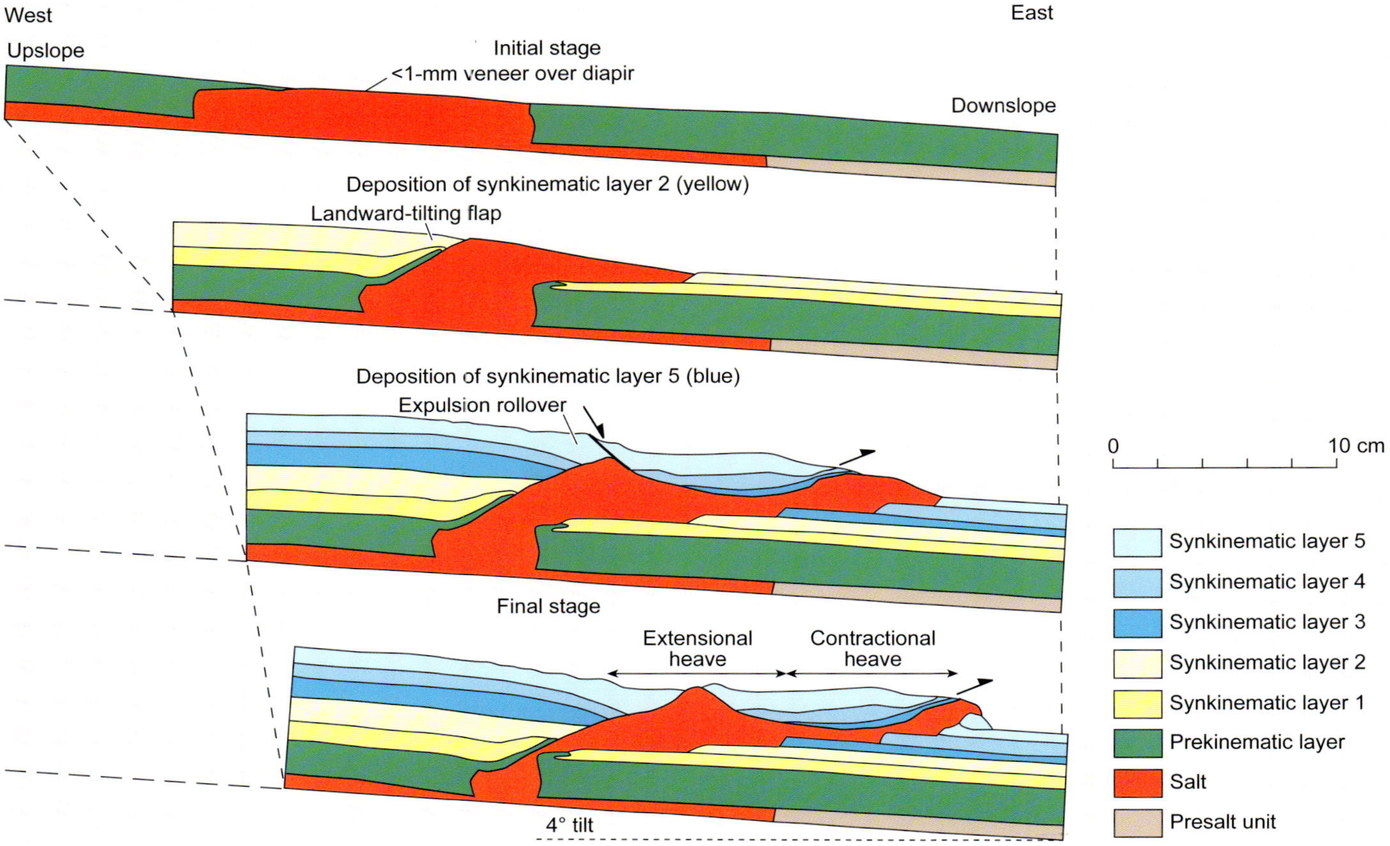

Figure 7.56. Overthrusting of the roof of a spreading salt sheet. Regional shortening greatly squeezed a squat passive salt wall, which reduced its width from 15 to 3 cm. Synkinematic layer 3 in the sheet roof was overthrust by 8 cm, which might seem to be caused by the regional shortening. However, the contractional heave in the east was balanced by 8 cm of extensional heave for the same unit in the center of the model, which suggests that overthrusting was directly related to rafting of the salt sheet's roof rather than being a direct consequence of regional shortening. Regional shortening is not evident in the prekinematic roof because it was initially so thin that it was effectively absent. The western flank of the diapir was a landward-dipping flap tilting away from the rising crest. As the feeder pinched off by regional shortening, this flap sagged onto the deflating salt sheet, forming an expulsion rollover. Model and restoration by Bruno Vendeville.

PHYSICAL MODEL—Multiple Breakouts of a Salt Sheet

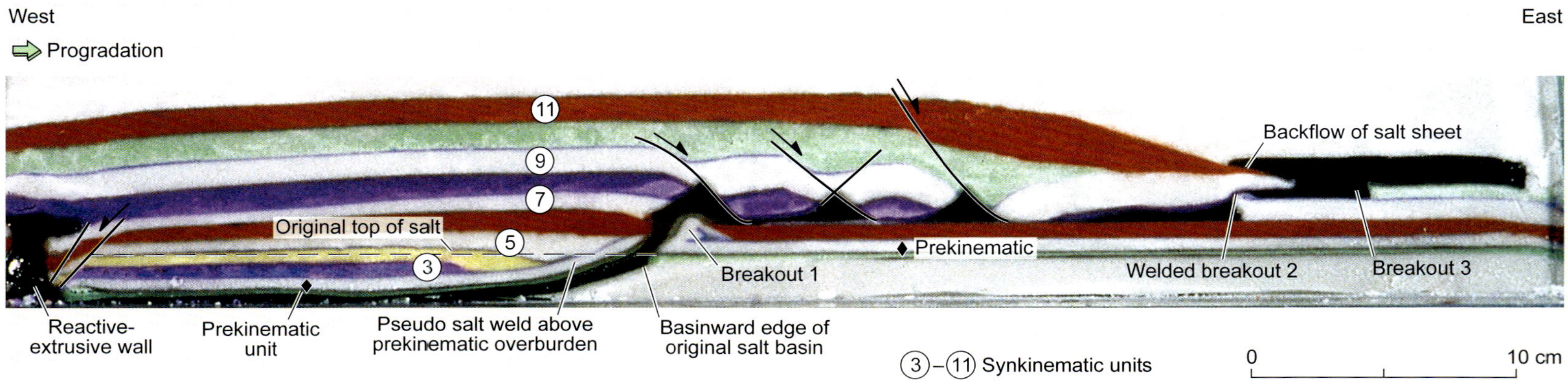

Figure 7.57. Roof deformation in a spreading salt sheet. Sediment progradation across an arcuate salt basin expelled salt seaward onto the abyssal plain. Salt extruded glacially from breakouts at three different stratigraphic levels. Each breakout records where the sedimentary roof ruptured and salt extruded ahead of the prograding load. Model by Bruno Vendeville.

PHYSICAL MODEL–Coalesced Salt Sheets at the Seaward Edge of a Salt Basin

Figure 7.58. Coalescence of salt sheets to form a canopy. Progradation of sediment across a salt basin squeezed the autochthonous salt seaward, forming a box fold at the seaward end of the salt basin. Three salt sheets then extruded across the abyssal plain from the forelimb of this fold. The three lower salt sheets merged as they spread radially. Their boundaries formed salt sutures. Their frontal margins were lobate, probably because stranded roof fragments created local barriers. Because of traction from the spreading salt sheets, their thin roofs wrinkled distally to form fold belts (shortening) and tore away proximally to expose the underlying salts (extension). An upper salt sheet overrode the northernmost lower salt sheet in piggyback style. Model by Bruno Vendeville.

PHYSICAL MODEL–Formation of Sedimentary Rafts above a Spreading Salt Sheet

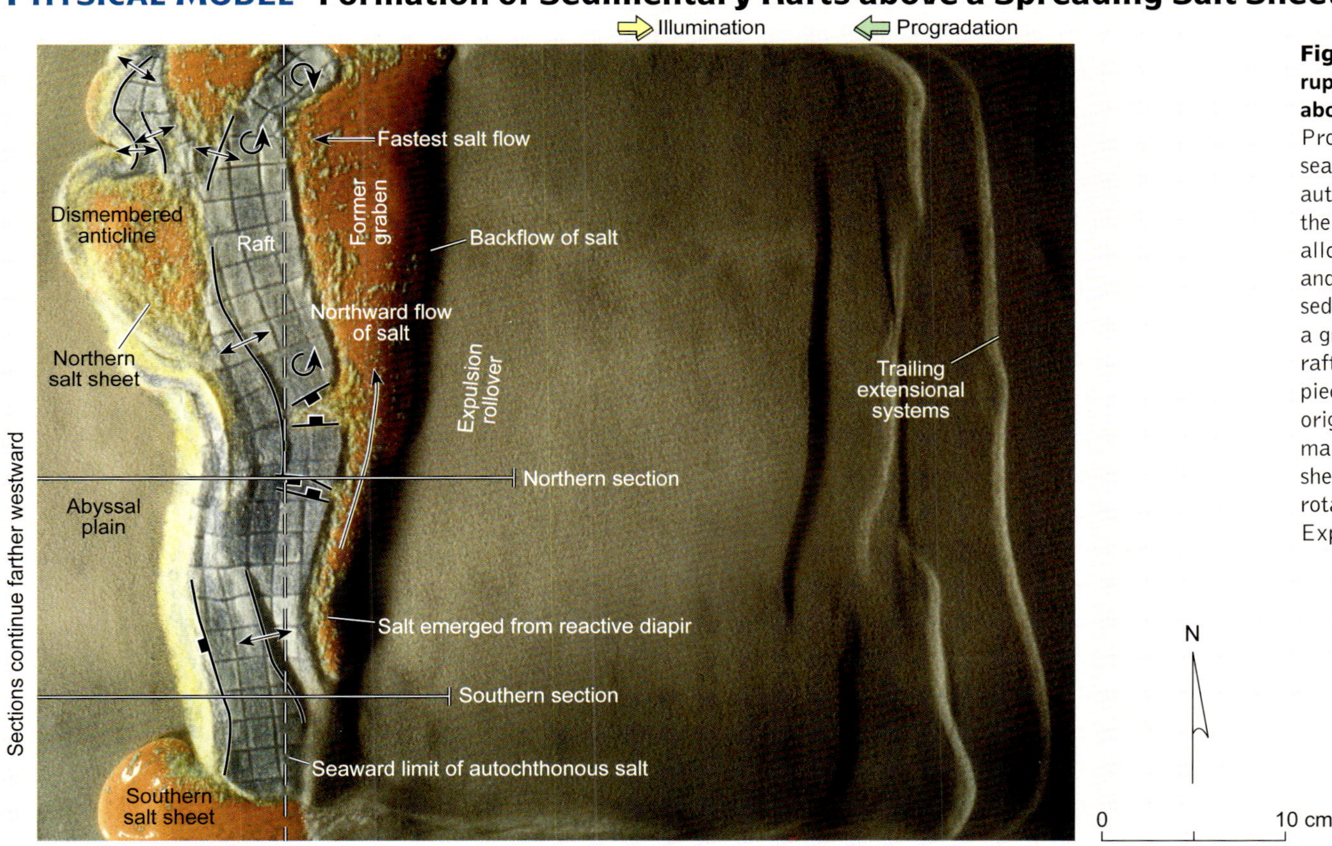

Figure 7.59. Extensional rupture of a sedimentary roof above spreading salt glaciers. Progradation expelled salt seaward from the edge of an autochthonous salt basin onto the abyssal plain. Flow of allochthonous salt stretched and ruptured the prekinematic sedimentary roof (marked by a grid) to form rafts. Distal rafts comprised dismembered pieces of the fold belt that originally formed the seaward margin of the spreading salt sheets. Rafts translated and rotated around vertical axes. Experiment by Bruno Vendeville.

SALT SHEETS

PHYSICAL MODEL—Multiple Salt-Sheet Levels in Front of a Prograding Sediment Wedge, Part 1

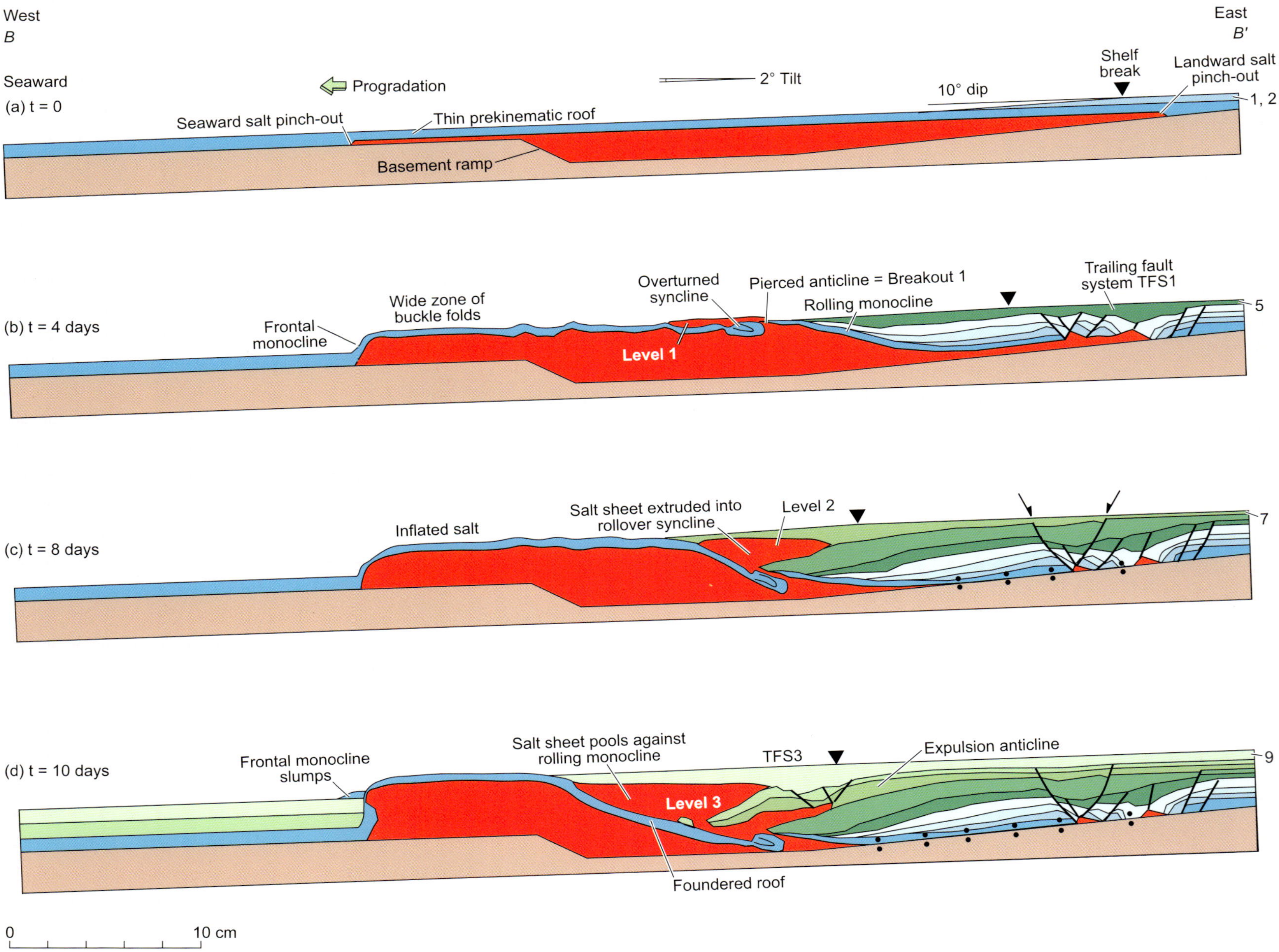

Figure 7.60. Imbricate stacking of salt sheets during progradation. Progradation of sediments expelled salt westward, inflating the distal part of the autochthonous salt basin. Salt first pierced a fold-belt anticline in the basin center to form an allochthonous salt sheet (level 1, after 4 days). This sheet continued to enlarge and climb against the foundering prekinematic roof and extrude again at two higher levels (levels 2 and 3, after 8 and 10 days, respectively) during sedimentation. Most extrusive salt flowed landward into the rollover syncline as progradation expelled underlying salt. The next series of stages in the restoration (Figure 7.61) shows a later and larger breakout along the frontal monoclinal scarp at the west end of the salt basin; this breakout was just beginning after 10 days. Model and restoration by Hongxing Ge.

Physical Model—Multiple Salt-Sheet Levels in Front of a Prograding Sediment Wedge, Part 2

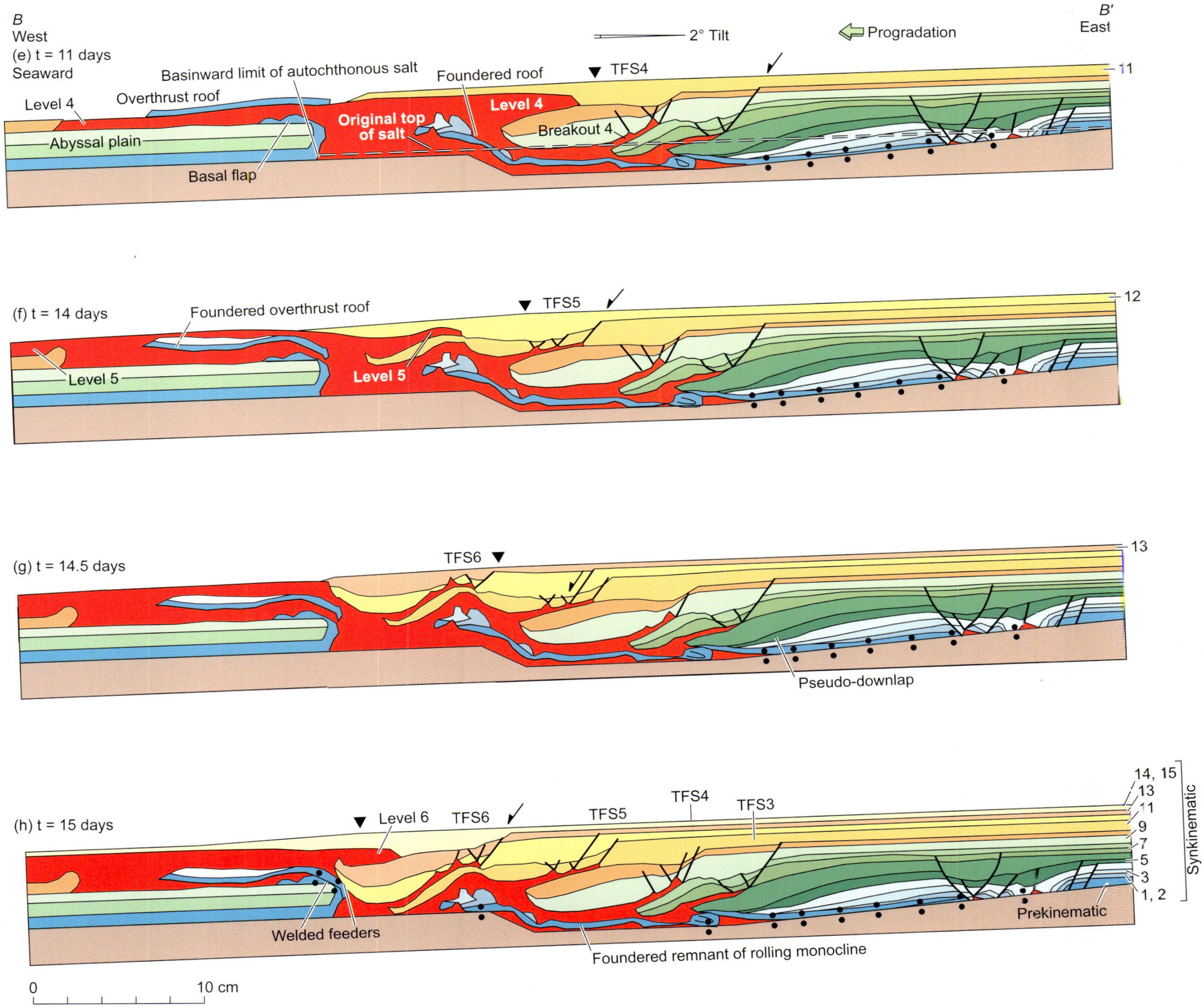

Figure 7.61. Imbricate stacking of salt sheets during sediment progradation. Progradation of sediments expelled salt westward, eventually leading to extrusion over the abyssal plain (see early stages of restoration in Figure 7.60). As progradation advanced, salt sheets continued to extrude at as many as six levels from vents successively farther seaward. Most extrusive salt flowed landward into the migrating rollover syncline as progradation expelled underlying salt. Later burial of each salt tongue expelled salt back into the main seaward-climbing salt body, which thinned or welded the salt sheets. The main seaward-climbing salt sheet rafted its thin prekinematic cover before this fragment foundered into the thick salt canopy. Model and restoration by Hongxing Ge.

PHYSICAL MODEL—Asymmetric Suture between Two Thinly Buried Salt Sheets during Oblique Collision

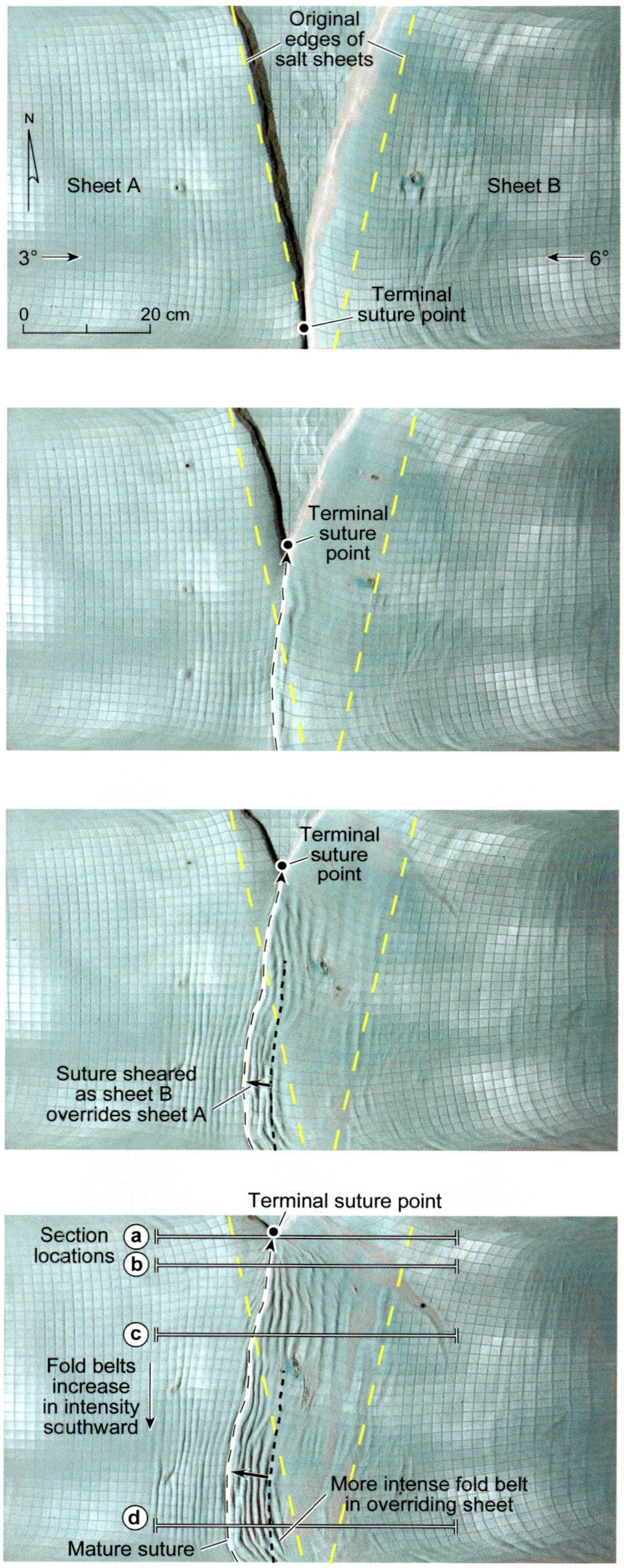

Figure 7.62. Suturing of salt sheets. Two buried salt sheets were tilted toward one another. Sheet A sloped east at 3°, and sheet B sloped west at 6°. Salt flowed downslope in both sheets, carrying their roofs with them, until the sheets collided to form a suture. The sheets first collided in the south, and then the terminal suture point propagated northward like a zipper closing. Serial sections (Figure 7.63) therefore show the evolution of an asymmetric suture from immature in the north to mature in the south. The suture was sheared westward as sheet B overrode sheet A. Fold belts formed above both salt sheets during collision but are more prominent in the overriding sheet, especially in the more-mature south. Model by Tim Dooley.

PHYSICAL MODEL—Asymmetric Suture between Two Thinly Buried Salt Sheets during Oblique Collision

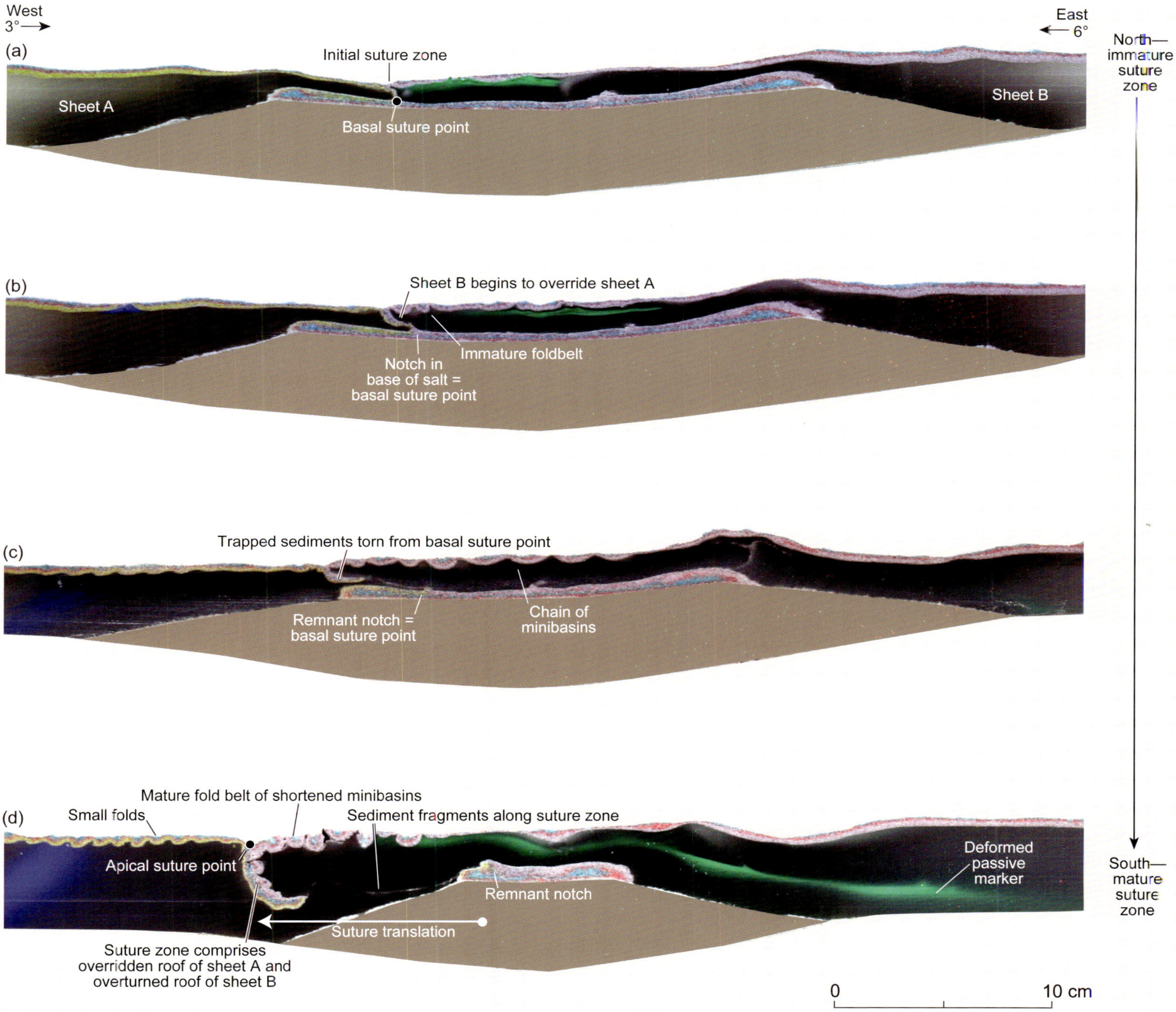

Figure 7.63. Suturing of salt sheets. Two salt sheets collided obliquely so that suturing began in the south and propagated northward (see overhead views in Figure 7.62). As a result, suturing reached a more advanced stage in the south than in the north, and serial sections show different stages in suturing. Sediments trapped between the salt sheets were bulldozed westward as sheet B overrode sheet A, eventually tearing away from the basal suture point that marked the original point of sheet collision. The only evidence of this transport is a trail of sedimentary fragments along the suture, connecting the bulldozed sediments back to the basal suture point. Fold belts formed above both salt sheets during collision but are more prominent in the overriding sheet. Model by Tim Dooley.

PHYSICAL MODEL—Salt Sheet Breaking through a Thick Roof

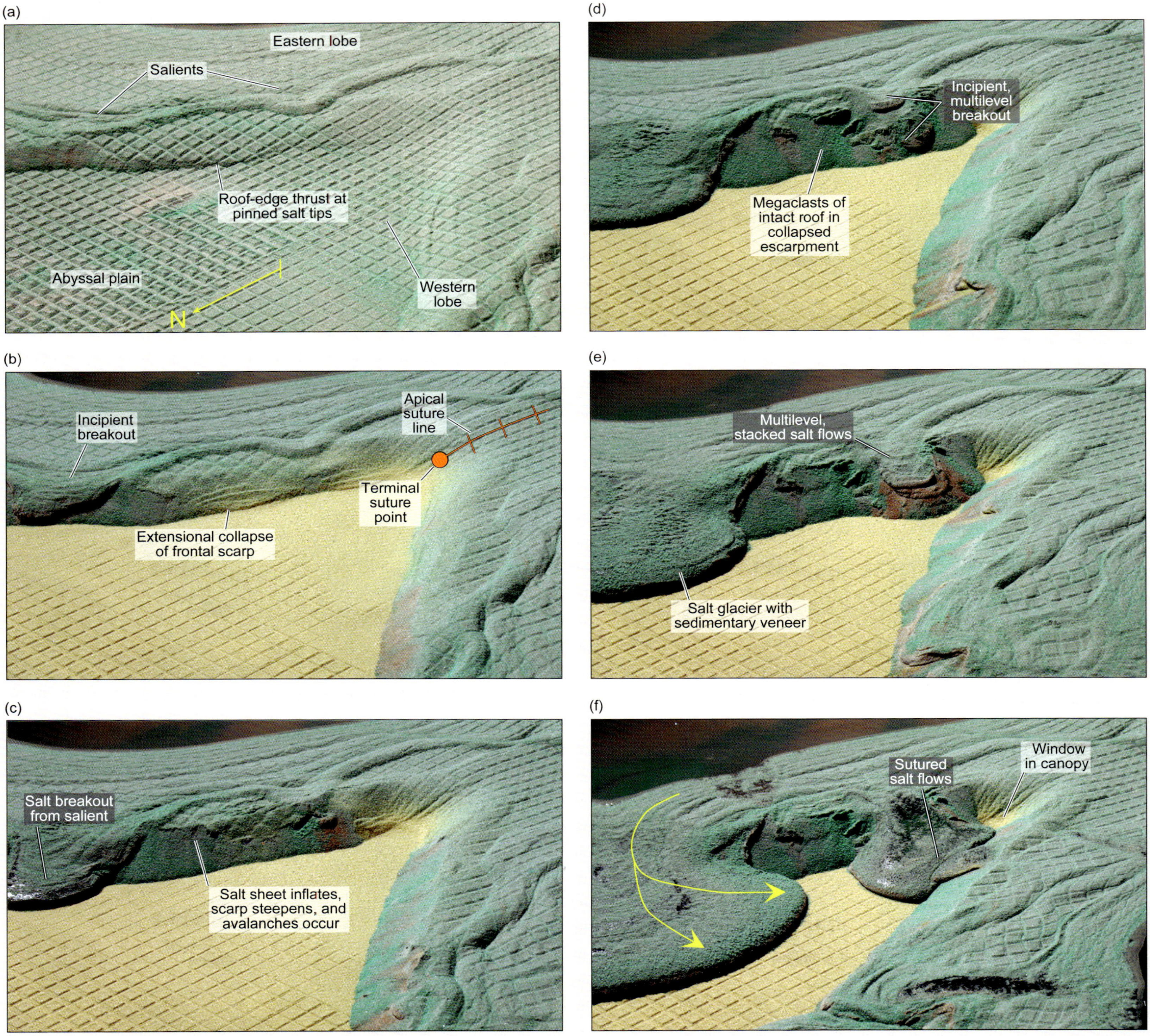

Figure 7.64. Breakout of salt sheets. A thick layer of sediment on the abyssal plain onlapped and pinned advancing salt sheets. The buried salt tips then inflated to form a steep frontal escarpment, which locally collapsed. Salt broke out over the buried salt tip from weak zones near the top of the escarpment and flowed down the escarpment onto the abyssal plain. Salt lobes from multiple breakouts eventually sutured with one another. Model by Tim Dooley.

PHYSICAL MODEL—Extrusion of Salt Sheets in Front of a Prograding Sediment Wedge

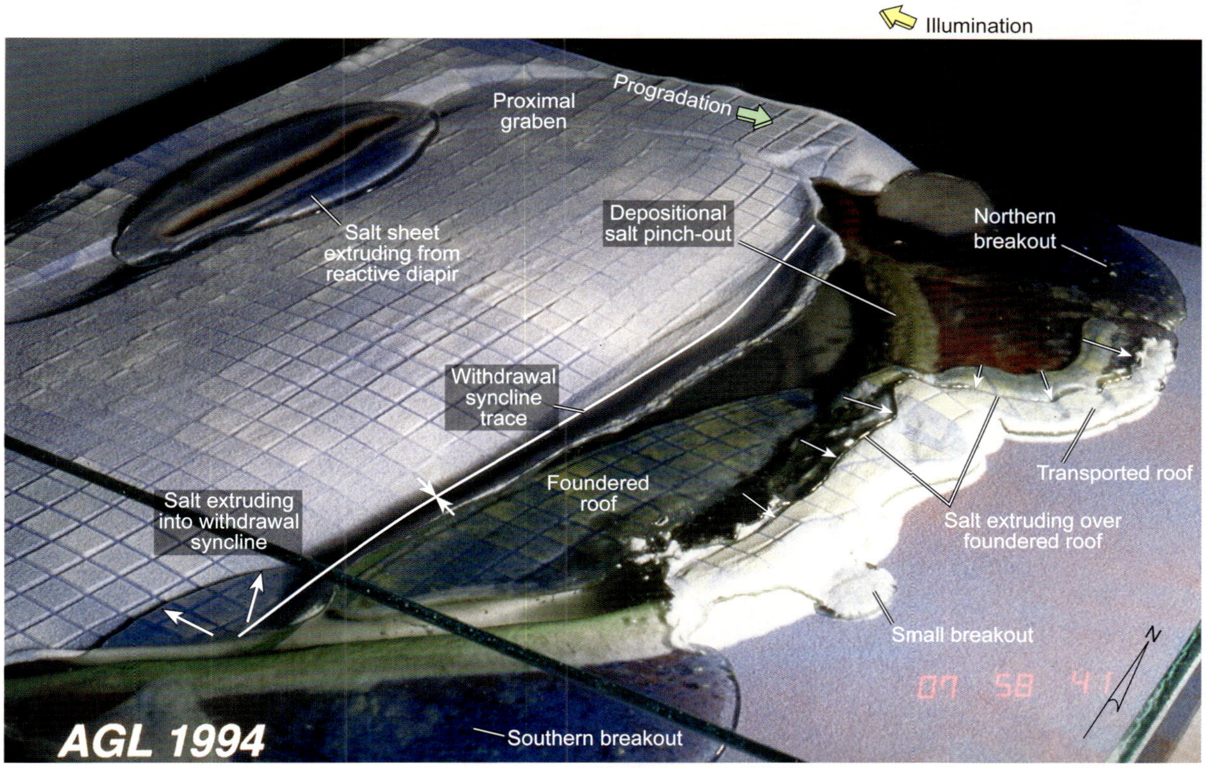

Figure 7.65. Growth of salt sheets during sediment progradation. Progradation across a salt basin having a convex pinch-out led to salt breakout at the north and south. Progradational bathymetry influenced deformation in several ways. First, the shelf was gravitationally unstable and spread to the east, causing salt to reactively pierce and extrude from a proximal graben. Second, withdrawal of salt just to the west of the vents feeding extrusion onto the abyssal plain produced a withdrawal syncline. Some of the extrusive salt flowed back updip into this synclinal low. For scale, model measures 59 cm north–south. Model by Bruno Vendeville.

PHYSICAL MODEL—Turtle Structure Formed during Loading of a Salt Sheet

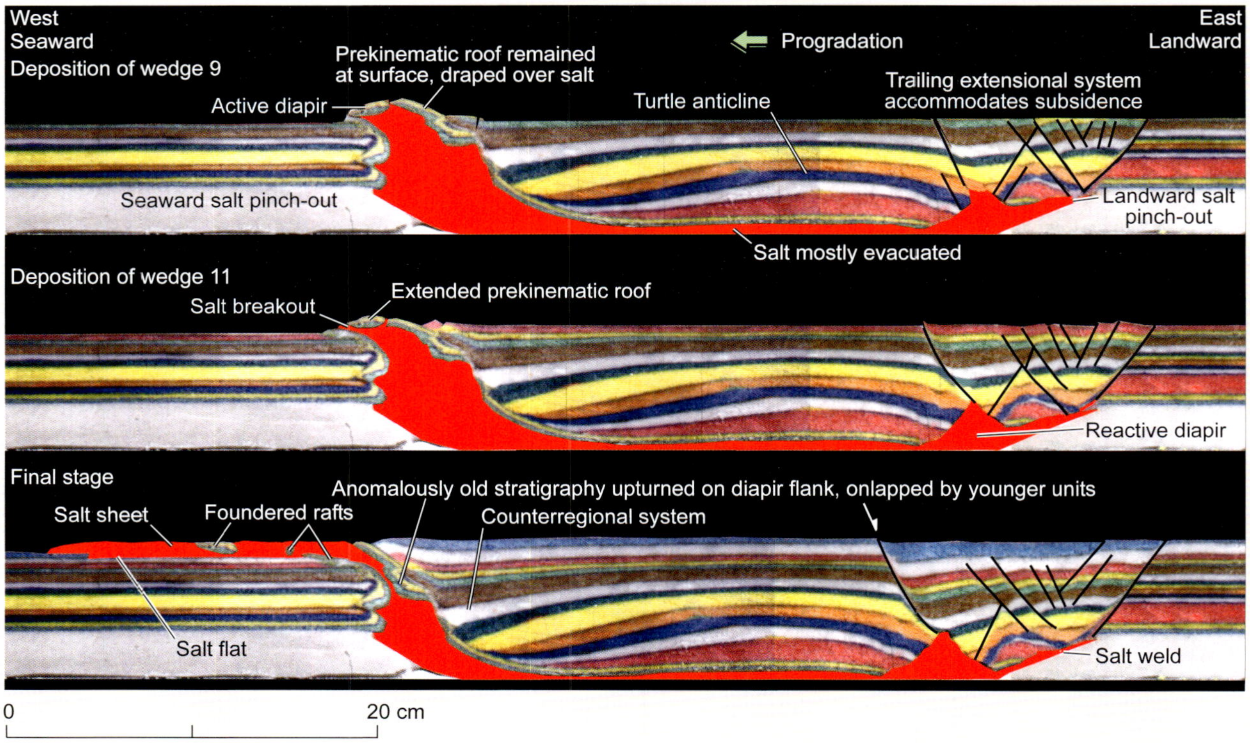

Figure 7.66. Extensional rupture of a prekinematic sedimentary roof above a spreading salt glacier. Progradation expelled salt seaward as part of a counterregional structure. Salt extrusion ruptured the sedimentary roof into rafts, which mostly foundered. The thin prekinematic roof forming the landward flank of the feeder remained near the surface throughout. This lack of significant burial shows how anomalously old stratigraphic units can be juxtaposed against much younger stratigraphy, even in areas not undergoing regional shortening. Experiment and restoration by Bruno Vendeville.

SALT SHEETS

PHYSICAL MODEL–Thrust Faults Formed during Advance of Buried Salt Sheets

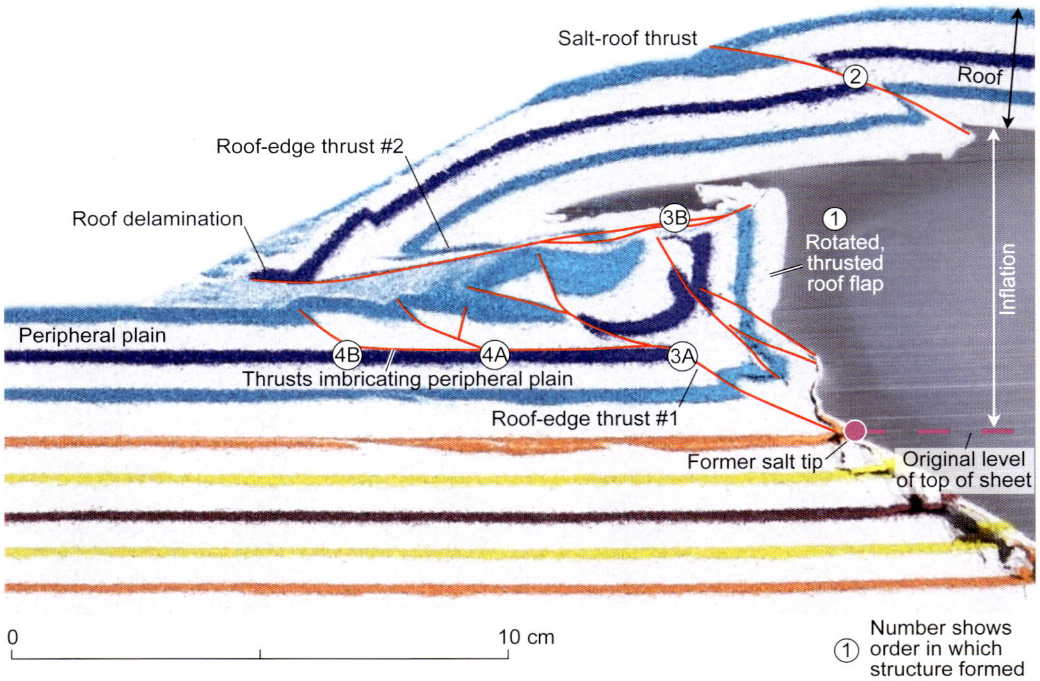

Figure 7.67. Overthrusting at the leading edge of a spreading salt sheet. (1) Inflation of the salt sheet rotated the front of the roof, producing a steeply dipping to overturned frontal flap. (2) Thrusting offset the advancing sheet roof. (3) Thrusts rooted into the salt tip (the edge of the roof) in two stages. An older roof-edge thrust cut upsection from the initial tip of the salt sheet (purple dot). A younger roof-edge thrust formed at the top of the rotated roof flap, where the roof was much thinner. (4) Thrusts imbricated the peripheral plain in front of the advancing allochthon. These frontal thrusts linked with the oldest roof-edge thrust. Model by Tim Dooley.

PHYSICAL MODEL–Compressional Rejuvenation of Buried Salt Sheets

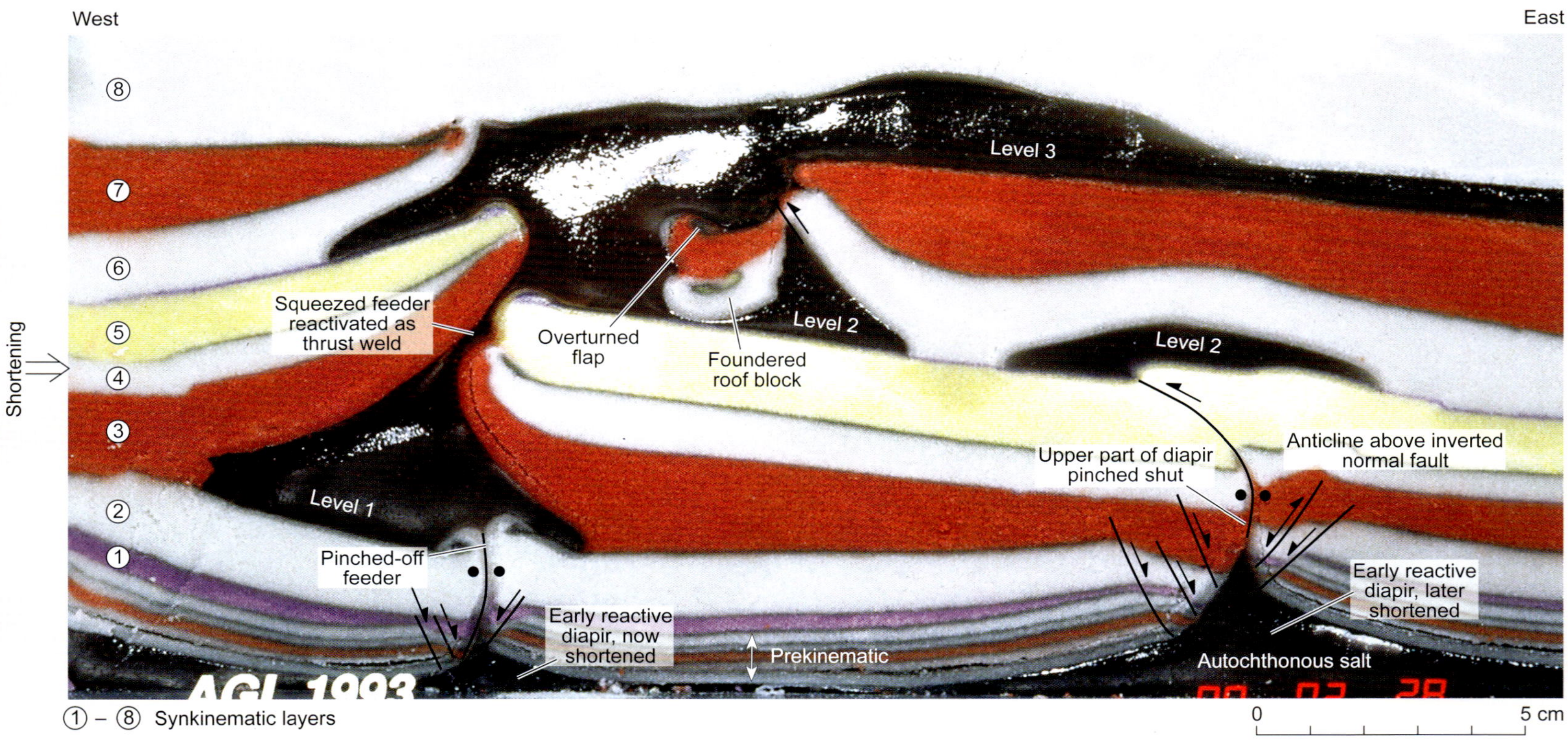

Figure 7.68. Shortening of salt sheets. Salt is the weakest part of most basins, so shortening tends to be localized in and above salt structures. In this model, each of the preshortening salt structures focused deformation in its roof. Feeders squeezed shut and were reactivated as thrust faults if they had at least one dipping limb. These thrusts propagated along strike into the sediments adjoining the diapir. Squeezing of feeders displaced salt upward to extrude on the surface. Model by Bruno Vendeville.

NUMERICAL MODEL—Flow in a Salt Sheet during Burial, No Feeder

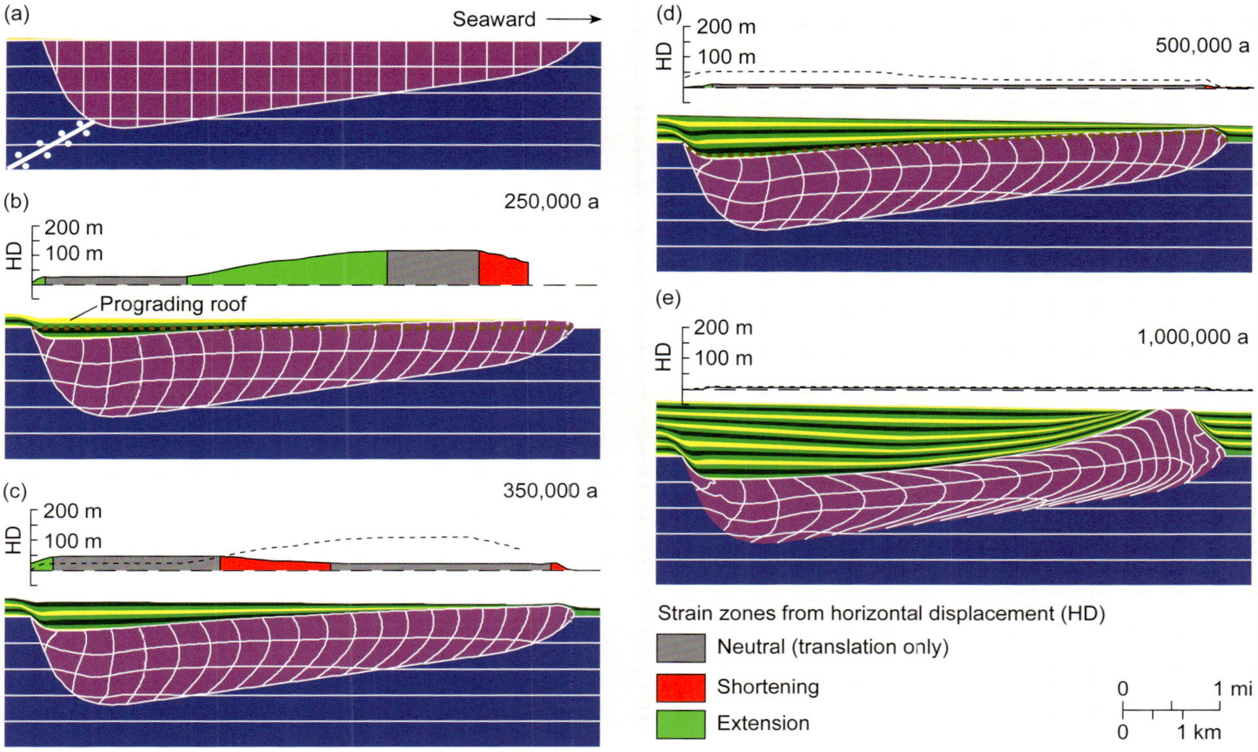

Figure 7.69. Sheet advance during sedimentation. Rapid burial of the salt sheet allowed sediment to prograde quickly across the sheet, forming a continuous roof that resisted further deformation. The thickening roof changed the flow of salt in two ways. First, frictional resistance at the base of the roof inhibited salt flow in the upper part of the sheet, which is indicated by landward curvature of the originally vertical white markers. Second, eventual burial of the sheet's toe kept it from advancing. Salt displaced by the subsiding minibasin inflated the toe and led to diapir piercement instead of sheet advance (stage e). Sediment aggradation rate = 1 mm/yr, surface slope of sedimentary wedge = 1°. Compare with the same model sourced from an inclined feeder in Figure 7.70. GEOSIM finite-element model by Dan Schultz-Ela.

NUMERICAL MODEL—Flow in a Salt Sheet during Rapid Burial, Inclined Feeder, Gentle Sediment Slope

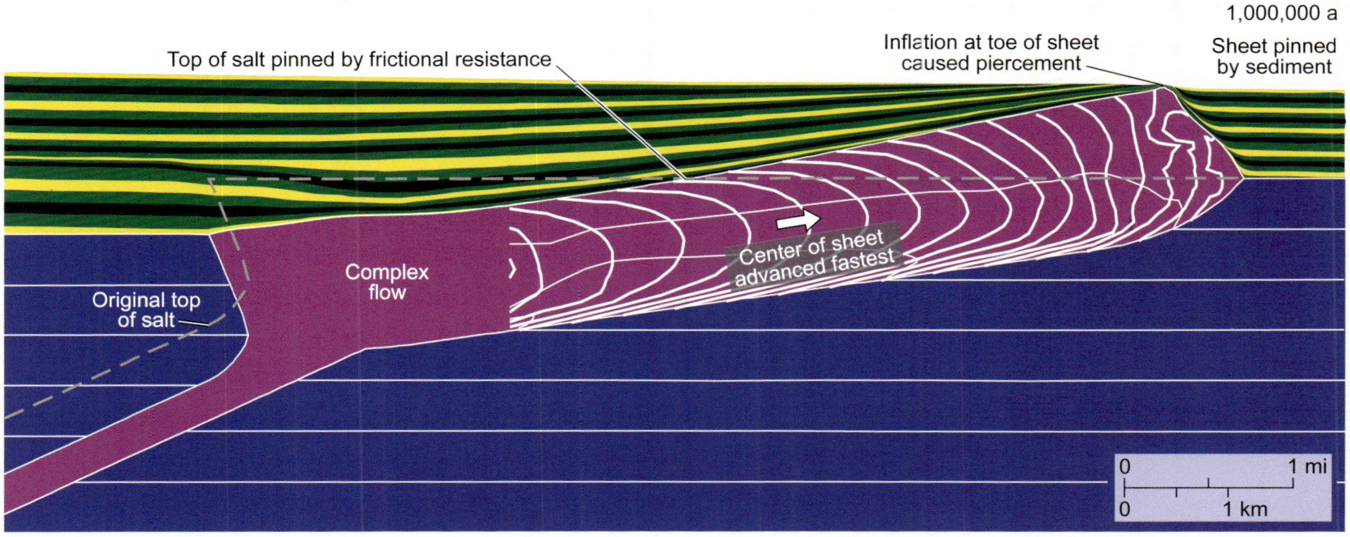

Figure 7.70. Sheet advance during sedimentation. Sediment prograded quickly across the salt sheet, forming a continuous roof that resisted deformation. Formation of the roof changed the flow of salt within the roof in two ways. First, frictional resistance at the base of the roof inhibited salt flow in the upper part of the sheet, which is indicated by landward-curving white marker lines. Second, onlapping sedimentation at the toe of the sheet blocked its advance. Salt displaced by the subsiding minibasin massively inflated the toe and led to diapir piercement of the sheet's shoulder instead of sheet advance. Sediment aggradation rate = 1 mm/yr, surface slope of sedimentary wedge = 1°. Compare with the same model without a feeder in Figure 7.69 and with progradation using a higher sediment slope in Figure 7.71. GEOSIM finite-element model by Ryan Mann; modified from Mann (1993).

NUMERICAL MODEL—Flow in a Salt Sheet during Rapid Burial, Inclined Feeder

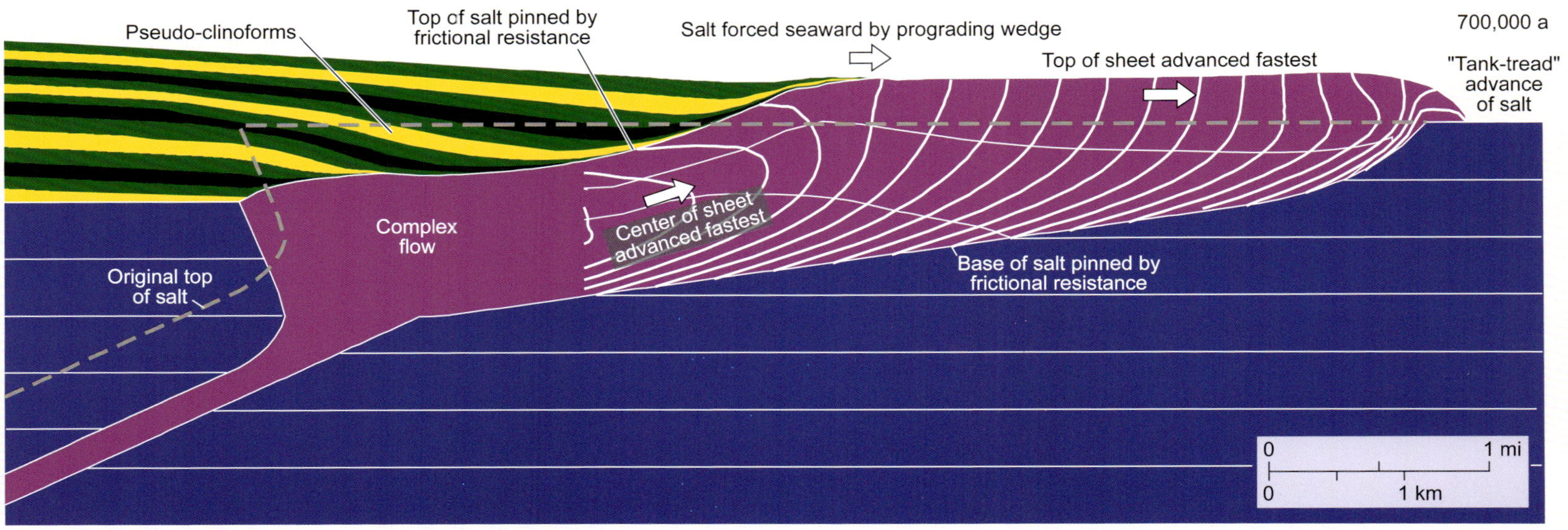

Figure 7.71. Sheet advance during sedimentation. Subsidence of a prograding minibasin above the updip end of the sheet displaced salt seaward, leading to advance of the salt sheet. Salt flow below the minibasin was restricted by friction at top and base of salt, so the middle of the sheet advanced fastest. The toe of the sheet was bare, allowing the top of salt to extrude in "tank-tread" style. The seaward-thickening and dipping sediment represents the early stage of a counterregional system. The apparent clinoforms were produced by deformation of onlaps during expulsion of underlying salt. Aggradation rate = 1.0 mm/yr, surface slope of sedimentary wedge = 3°. Compare with sheet advance produced by progradation with a lower sediment slope in Figure 7.70. GEOSIM finite-element model by Ryan Mann; modified from Mann (1998).

CONCEPTUAL MODEL—Models for Salt-Sheet Advance

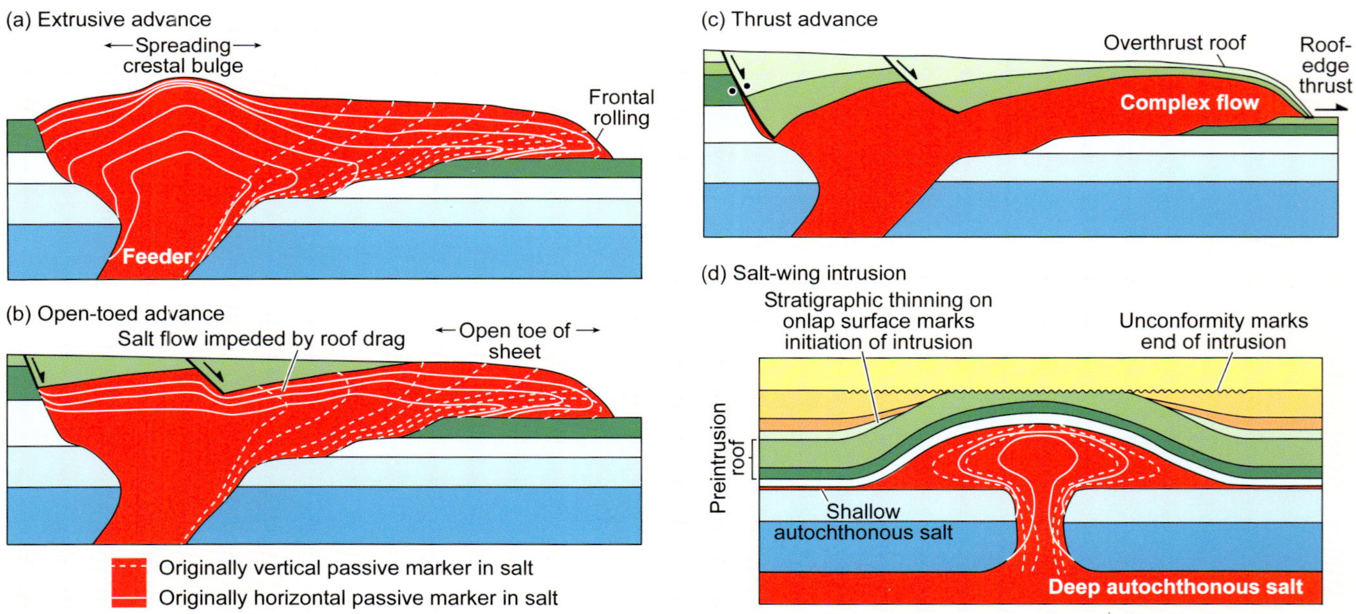

Figure 7.72. Emplacement of salt sheets. Salt sheets can advance in four ways. In (a) extrusive advance, an exposed salt sheet spreads at the surface without a roof. The salt flows outward by gravity, spreading from a summit formed by a dynamic bulge above the feeder. Deformation of passive marker lines here illustrates frontal rolling, which is common above lithified bedrock but less typical above unlithified subsalt strata. In (b) open-toed advance, the sheet is extrusive at its toe but is partly buried by a roof strong enough to modify the underlying salt flow. The roof resists salt flow. Traction exerted on the roof by flow of the underlying salt may extend a sufficiently weak roof. In (c) thrust advance, a buried sheet advances along with its roof in a thrust hanging wall. The roof is separated from peripheral sediments by a roof-edge thrust fault that projects from the tip of the salt to the surface. During (d) salt-wing intrusion, a wedge of salt is emplaced along a thin delaminating stratigraphic layer in the flanking country rock. The roof is typically uplifted as a detachment anticline during regional shortening. This arching typically leads to onlap or erosion of the roof strata. These processes can occur sequentially during growth of a sheet to produce salt-sheet lineages (Figures 7.75–7.77). Modified from Hudec and Jackson (2006).

CONCEPTUAL MODEL–Interpretation of Basal Cutoffs beneath an Extrusive Salt Sheet

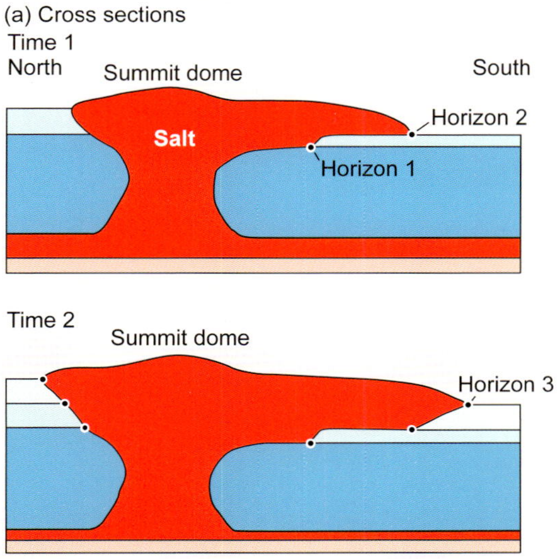

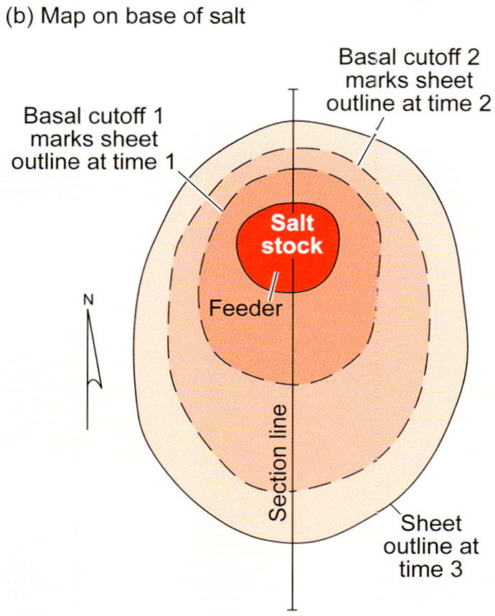

Figure 7.73. Emplacement of extrusive salt sheets. An extruding sheet climbs stratigraphically if it advances over aggrading sediments. The basal cutoff of a stratal horizon against the base of salt (a) thus marks the position of the sheet edge during deposition of that horizon, which can be tracked in map view (b). The map pattern of successive cutoffs is analogous to tree rings, in that they illustrate the growth of the sheet through time. Modified from Hudec and Jackson (2006).

CONCEPTUAL MODEL–Three-Dimensional Geometries of Plug-Fed Extrusions and Source-Fed Thrusts

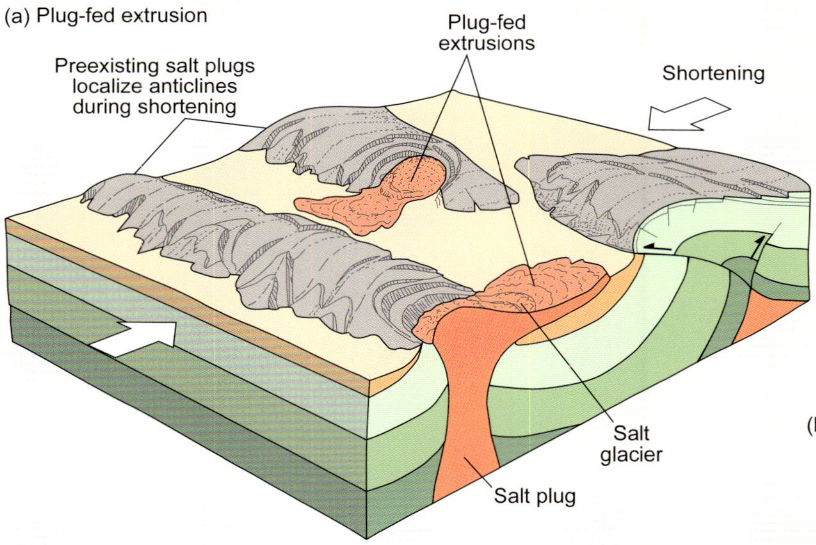

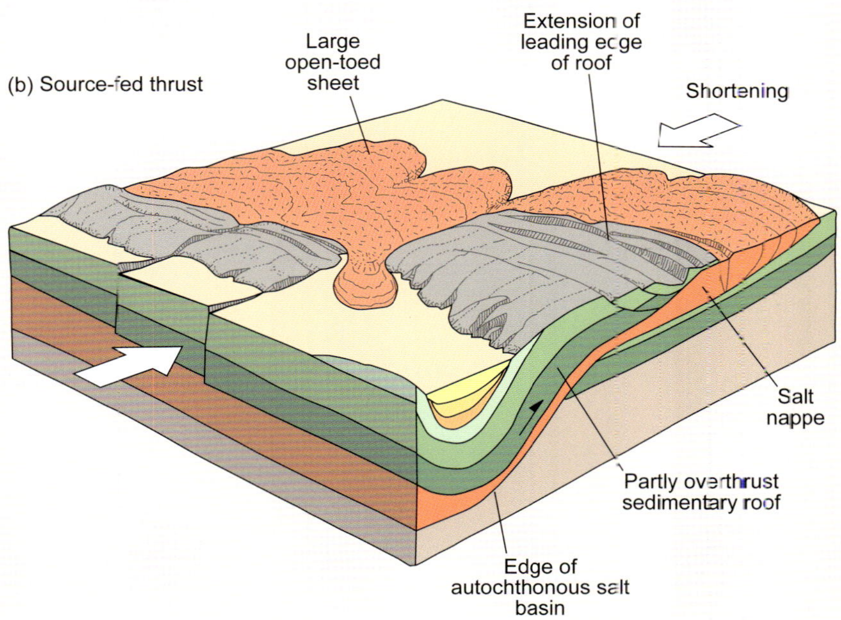

Figure 7.74. Emplacement of salt sheets. (a) Plug-fed extrusions form when salt extrudes from the crest of an upright stock or wall. Extrusion is particularly likely when the diapir is squeezed because salt is displaced out of the narrowing diapir and because the forced salt rise disrupts any roof barrier. Weak diapirs tend to localize compressional structures, so many plug-fed extrusions occur along regional thrust or anticline trends. (b) Source-fed thrusts are common along the edges of salt basins, where salt-detached translation ramps up from the edge of the salt layer. Such thrusting may occur at the downdip end of a passive margin salt basin or in an orogenic foreland (shown here). The uplifted hanging wall is prone to erosion. Salt exposed by erosion may advance as an open-toed sheet (see models for salt-sheet advance in Figure 7.72). See also restorations of plug-fed extrusion (Figure 7.75) and source-fed thrust (Figure 7.77). Modified from Hudec and Jackson (2006).

CONCEPTUAL MODEL—Lineages of Salt Sheets, Part 1: Plug-Fed Extrusions

Figure 7.75. Emplacement of salt sheets. Most salt sheets advance by extrusion, open-toed advance, or thrusting (see models for salt-sheet advance in Figure 7.72). These modes may combine in many ways over the life span of a salt sheet, but three evolutionary sequences (or lineages) are particularly common: plug-fed extrusions (shown here), plug-fed thrusts (Figure 7.76), and source-fed thrusts (Figure 7.77). A plug-fed extrusion flows from the top of an upright salt stock or salt wall. The sheet continues to advance while being partly and then completely buried. Eventually it is pinned by a roof too thick and strong to deform by gravity. Arrows refer to possible evolutionary paths between stages. The seaward edge of the Sigsbee salt canopy in the deep-water Gulf of Mexico consists primarily of amalgamated plug-fed extrusions that have reached the thrust-advance stage (d). See block diagram of plug-fed extrusions in Figure 7.74. Modified from Hudec and Jackson (2006).

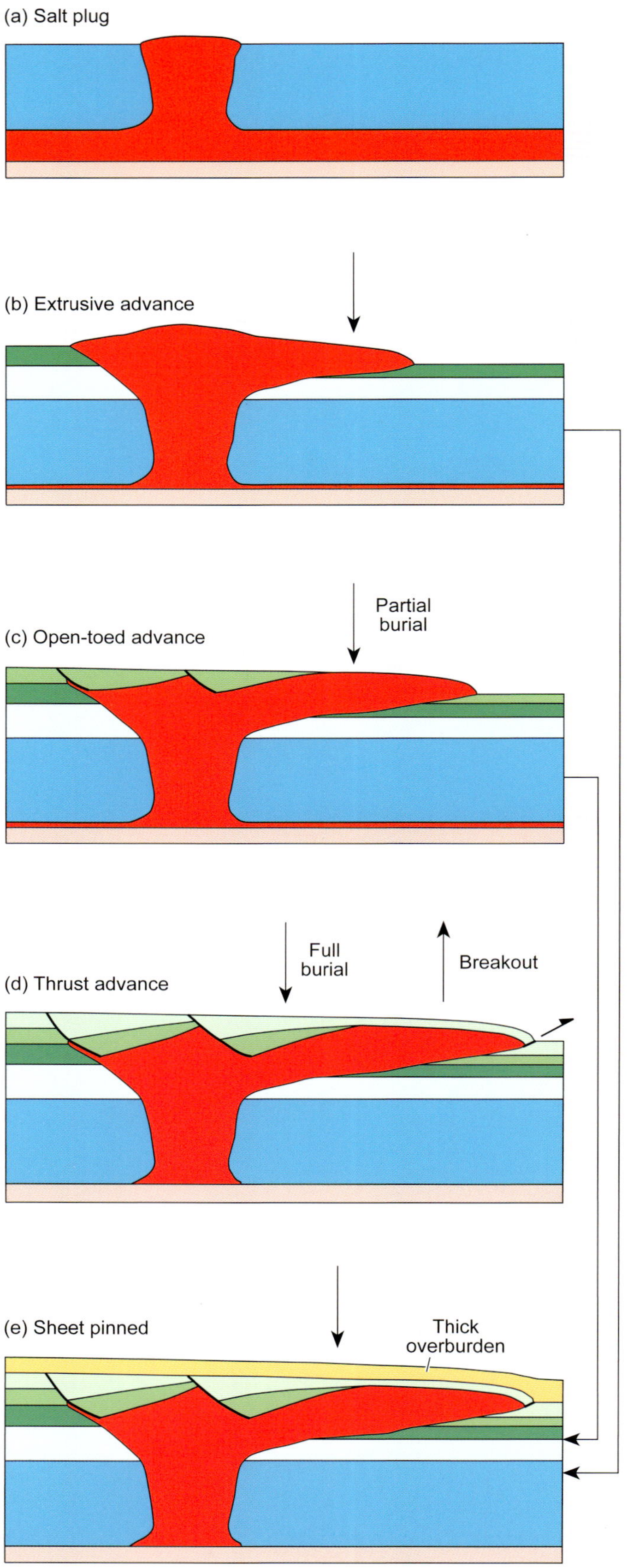

CONCEPTUAL MODEL—Lineages of Salt Sheets, Part 2: Plug-Fed Thrusts

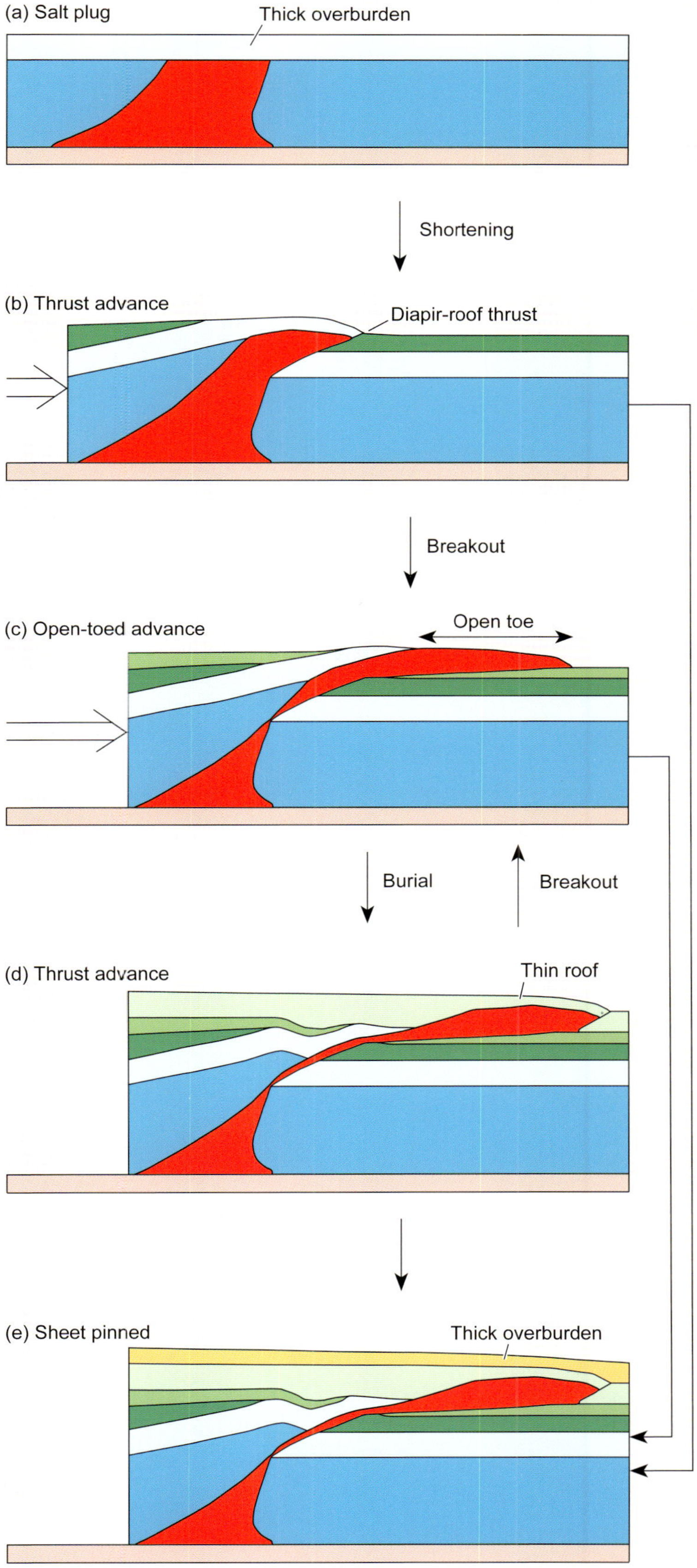

Figure 7.76. Emplacement of salt sheets. Most salt sheets advance by extrusion, open-toed advance, or thrusting. These modes may combine in many ways over the life span of a salt sheet, but three evolutionary sequences (or lineages) are particularly common: plug-fed extrusions (Figure 7.75), plug-fed thrusts (shown here), and source-fed thrusts (Figure 7.77). Plug-fed thrusts form when a diapir having a thick roof is shortened. The roof is too thick to be immediately disrupted, so the sheet initiates in the hanging wall of a diapir-roof thrust. If shortening continues, the sheet may break out at the surface and evolve into an open-toed sheet. Given sufficient sedimentation, this sheet is eventually buried. Compressional reactivation of buried diapirs is common in orogenic belts that involve salt basins and in some compressional provinces in deep-water passive margins. Modified from Hudec and Jackson (2006).

CONCEPTUAL MODEL—Lineages of Salt Sheets, Part 3: Source-Fed Thrusts

Figure 7.77. Emplacement of salt sheets. Most salt sheets advance by extrusion, open-toed advance, or thrusting. These modes may combine in many ways over the life span of a salt sheet, but three evolutionary sequences (or lineages) are particularly common: plug-fed extrusions (Figure 7.75), plug-fed thrusts (Figure 7.76), and source-fed thrusts (shown here). Source-fed thrusts form wherever translation ramps up from an autochthonous-salt detachment (the "source" in the term), carrying salt in its hanging wall. If shortening continues, the sheet may break through to the surface and evolve into an open-toed sheet. Given sufficient sedimentation, this sheet is eventually buried. Source-fed thrusts are common in orogenic belts that involve salt basins and at the downdip ends of passive-margin salt basins. See block diagram of a source-fed thrust in Figure 7.74. Modified from Hudec and Jackson (2006).

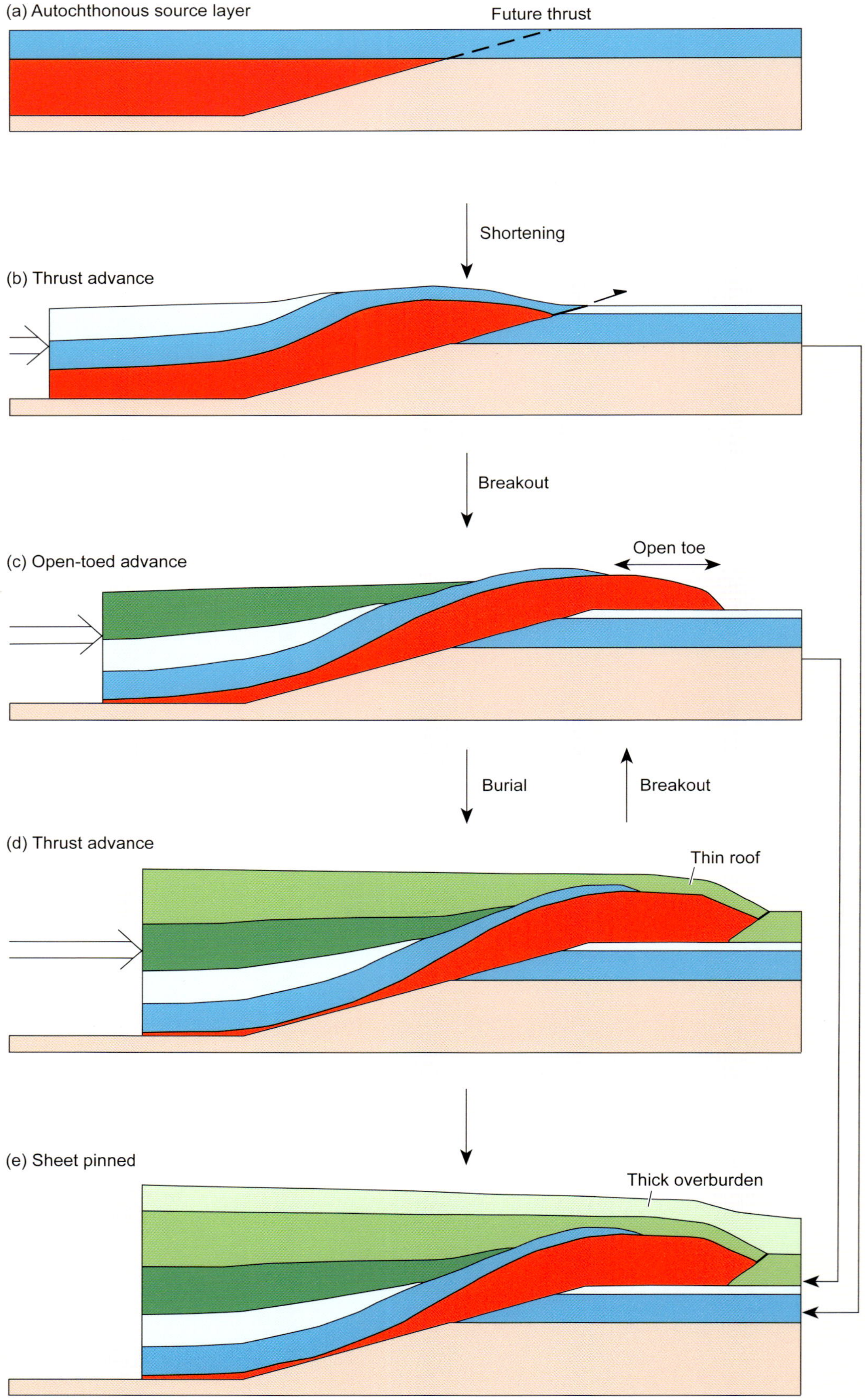

CONCEPTUAL MODEL—Disruption of Salt-Sheet Roof during Spreading

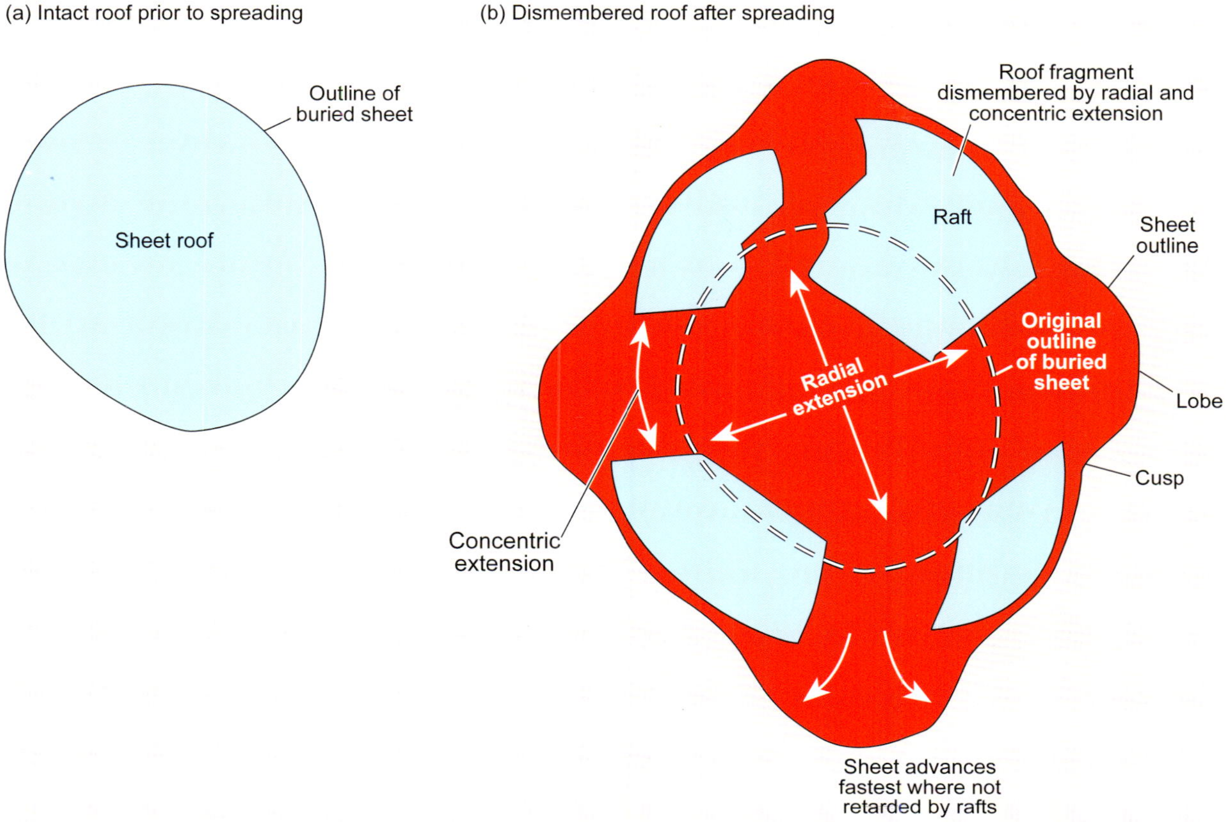

Figure 7.78.
Emplacement of salt sheets. A radially spreading sheet increases in surface area. Any roof above a sheet is dismembered into rafts by the combination of radial and concentric extension inherent in spreading. Large rafts locally inhibit spreading of the underlying salt, so the sheet advances fastest between rafts, creating a lobate outline of the sheet. Modified from Hudec and Jackson (2006).

CONCEPTUAL MODEL—Gravity Spreading of a Buried Salt Sheet

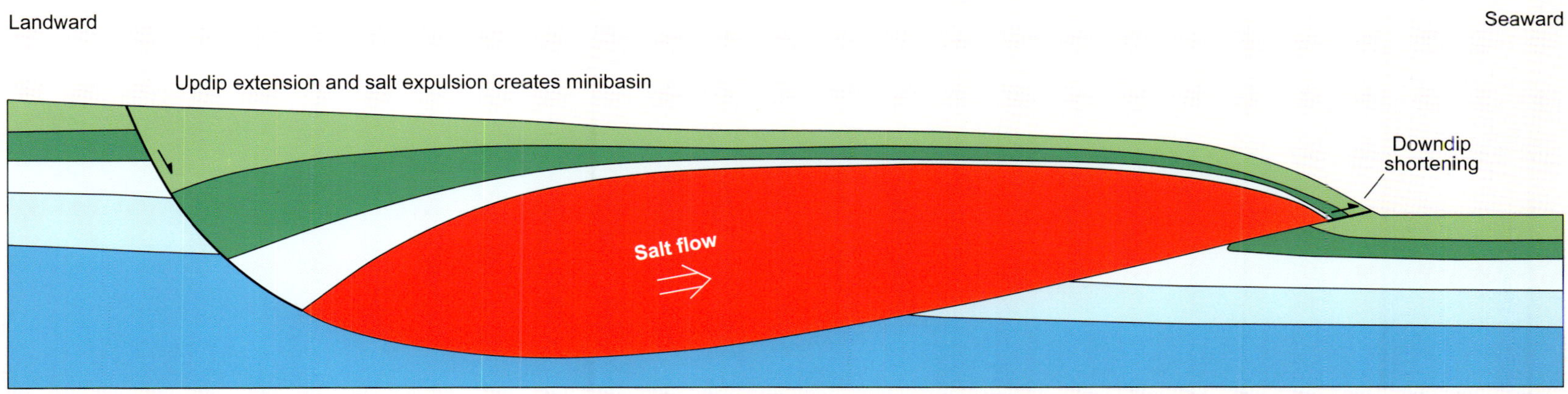

Figure 7.79. Advance of a buried salt sheet. A buried salt sheet continues to advance if its roof is thin enough to be deformed. This advance forms extensional basins near the updip end of the sheet and a compressional zone near the sheet toe. Modifed from Fletcher and others (1995).

CONCEPTUAL MODEL—Shear along the Base of a Salt Sheet

Figure 7.80. **Deformation of strata below a salt sheet.** This section is a composite of drilling results from three different hydrocarbon discoveries. The oldest roof strata comprise a condensed, laterally extended unit forming the base of a roho system. Deeper down, strata adjoining the base of salt are in places inverted, repeated, and sheared, which suggests that the sedimentary base of the sheet acted as a shear zone during or following salt-sheet advance. Modified from Harrison and Patton (1995).

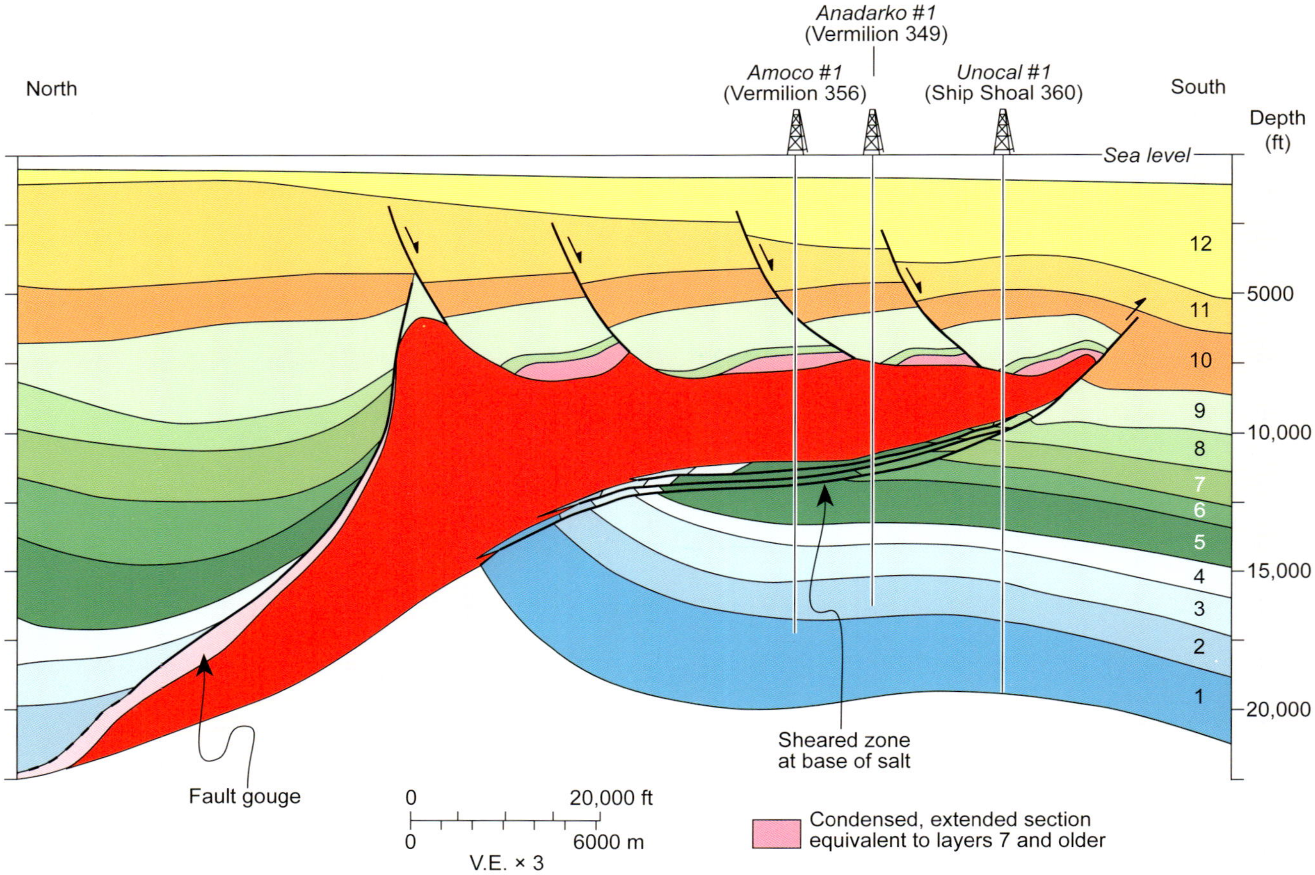

CONCEPTUAL MODEL—Classification of Sheet-Margin Thrust Systems

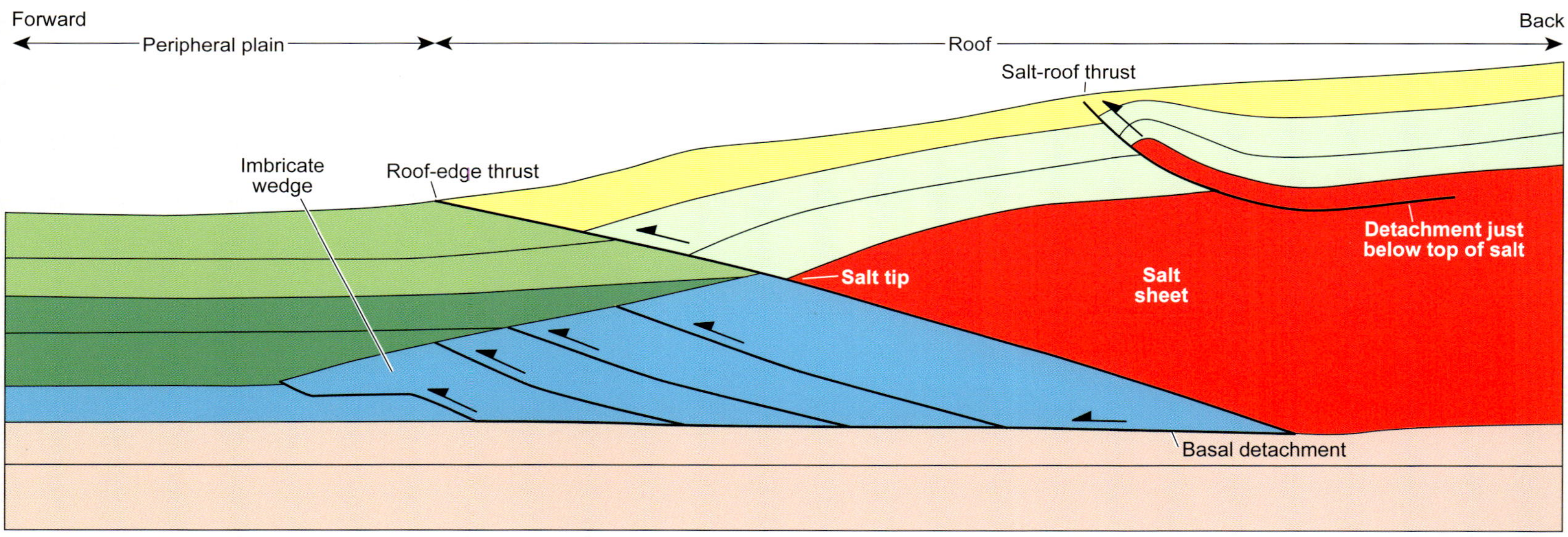

Figure 7.81. Thrust advance of a salt sheet. Three types of thrusts are recognized near the margins of salt sheets. Roof-edge thrusts connect to the salt tip and separate condensed roof strata from thicker equivalents on the peripheral plain. This is the dominant style of thrusting along the Sigsbee Escarpment in the deep-water Gulf of Mexico. Salt-roof thrusts shorten the salt roof and typically detach at or just below the top of salt. Imbricate wedges shorten sediments in the peripheral plain. Their basal detachments connect to the base of salt either at the salt tip or at the bottom of a ramp in the base of salt. Most imbricate wedges along the Sigsbee Escarpment are inactive and are buried by undeformed peripheral-plain sediments. Modified from Hudec and Jackson (2009).

CONCEPTUAL MODEL—Origin of Sheet-Margin Thrust Systems

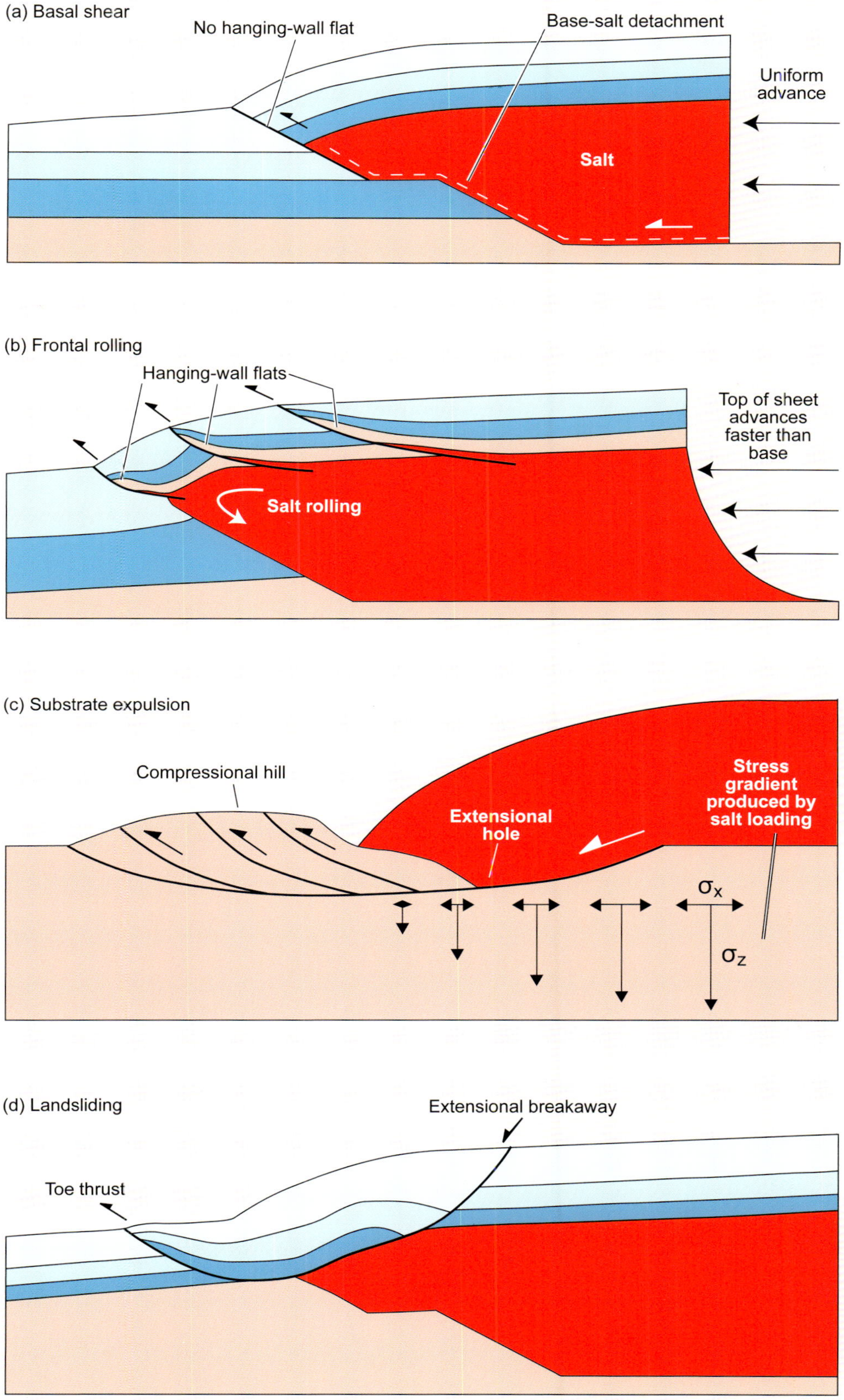

Figure 7.82. Thrust advance of a salt sheet. Many salt sheets and salt canopies have frontal thrust faults in the sediments around their margins. Four processes could produce these thrusts. (a) Slip along the base of a buried salt sheet would cut upsection at the edge of the sheet, producing a thrust fault in roof strata. (b) A sheet whose top moves faster than its base may advance by frontal rolling, where the top of salt rolls over at the front of the sheet to become the base of salt, akin to a tank tread. Roof strata cannot roll like ductile salt, so they shorten by thrusting. (c) Differential loading from the thinning salt sheet produces a stress gradient, which can expel weak sediment from beneath its leading edge. This process produces (in glacial terminology) a hill-hole pair, in which an extensional hole formed by sediment expulsion matches a compressional hill composed of thrust imbricates. (d) Landsliding off a frontal escarpment forms a linked system comprising an extensional breakaway and a contractional toe thrust. Observations along the Sigsbee Escarpment, deep-water Gulf of Mexico, suggest that basal shear a dominates there, although a canopy may produce thrusts by several processes (Figure 7.83). Modified from Hudec and Jackson (2009).

CONCEPTUAL MODEL—Evolution of a Sheet-Margin Thrust System

Figure 7.83. Thrust advance of a salt sheet. A hypothetical sequence shows how the style of salt-sheet advance might change through time (see origin of sheet-margin thrusts [Figure 7.82] and classification of sheet-margin thrusts [Figure 7.81] for more information on thrust styles). A single sheet may not experience all of these stages, but two or more stages are common. (a) An extrusive salt sheet differentially loads its substrate, expelling sediments ahead to form an imbricate wedge. (b) The salt sheet is buried but continues to advance along a roof-edge thrust. (c) The peripheral plain shortens to form an imbricate wedge, perhaps owing to the presence of a weak décollement. (d) The wedge becomes too thick and strong to shorten so becomes inactive and buried. Sheet advance continues on the roof-edge thrust. (e) Peripheral-plain sedimentation buries and pins the roof-edge thrust. The sheet inflates without advancing as salt is expelled seaward into the cross section. As a result, the frontal escarpment steepens and fails as a landslide. (f) Salt breaks through the roof thinned by the landslide, resuming advance along a new roof-edge thrust. Modified from Hudec and Jackson (2009).

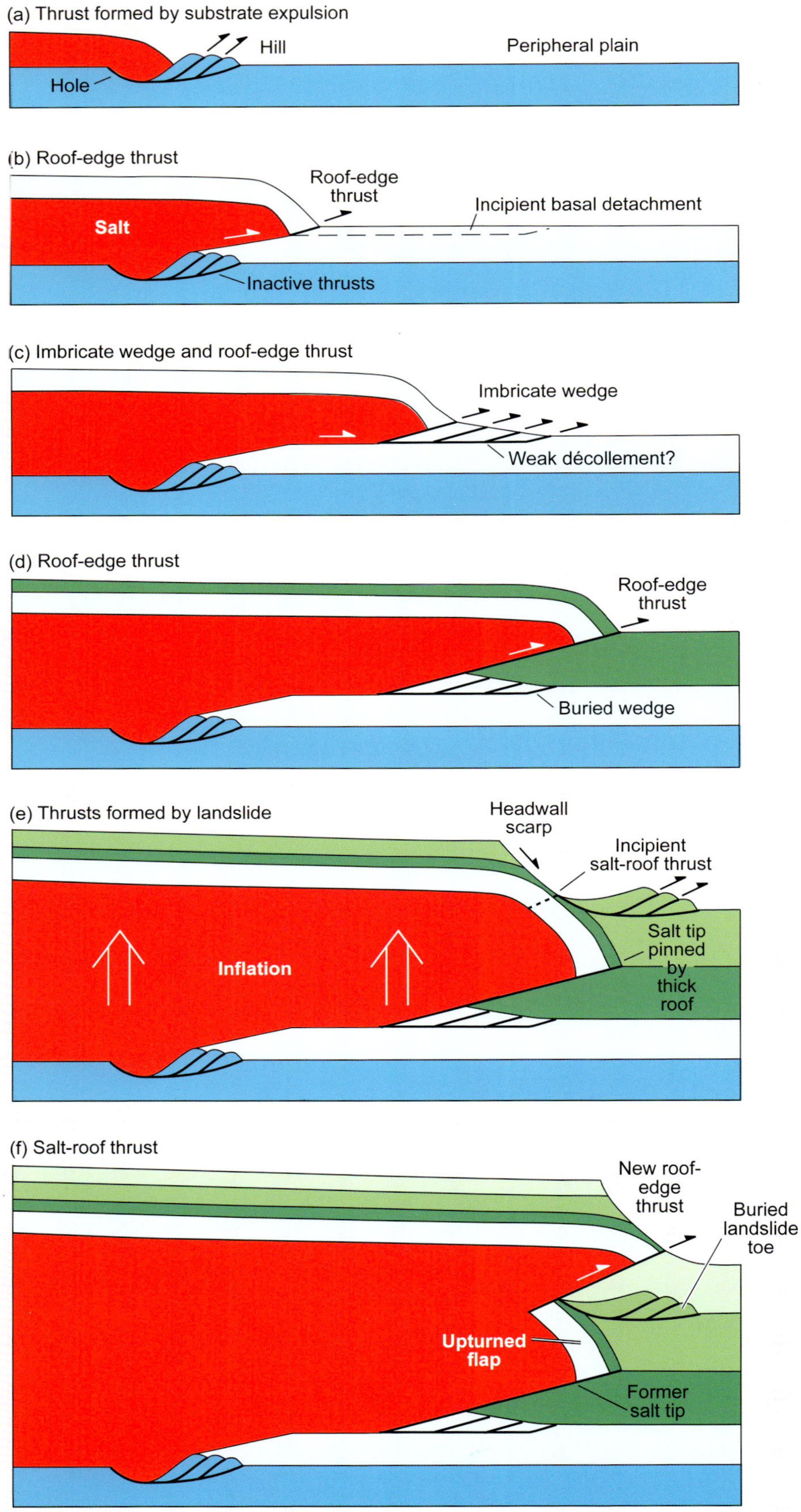

CONCEPTUAL MODEL–Modes of Salt-Sheet Breakout

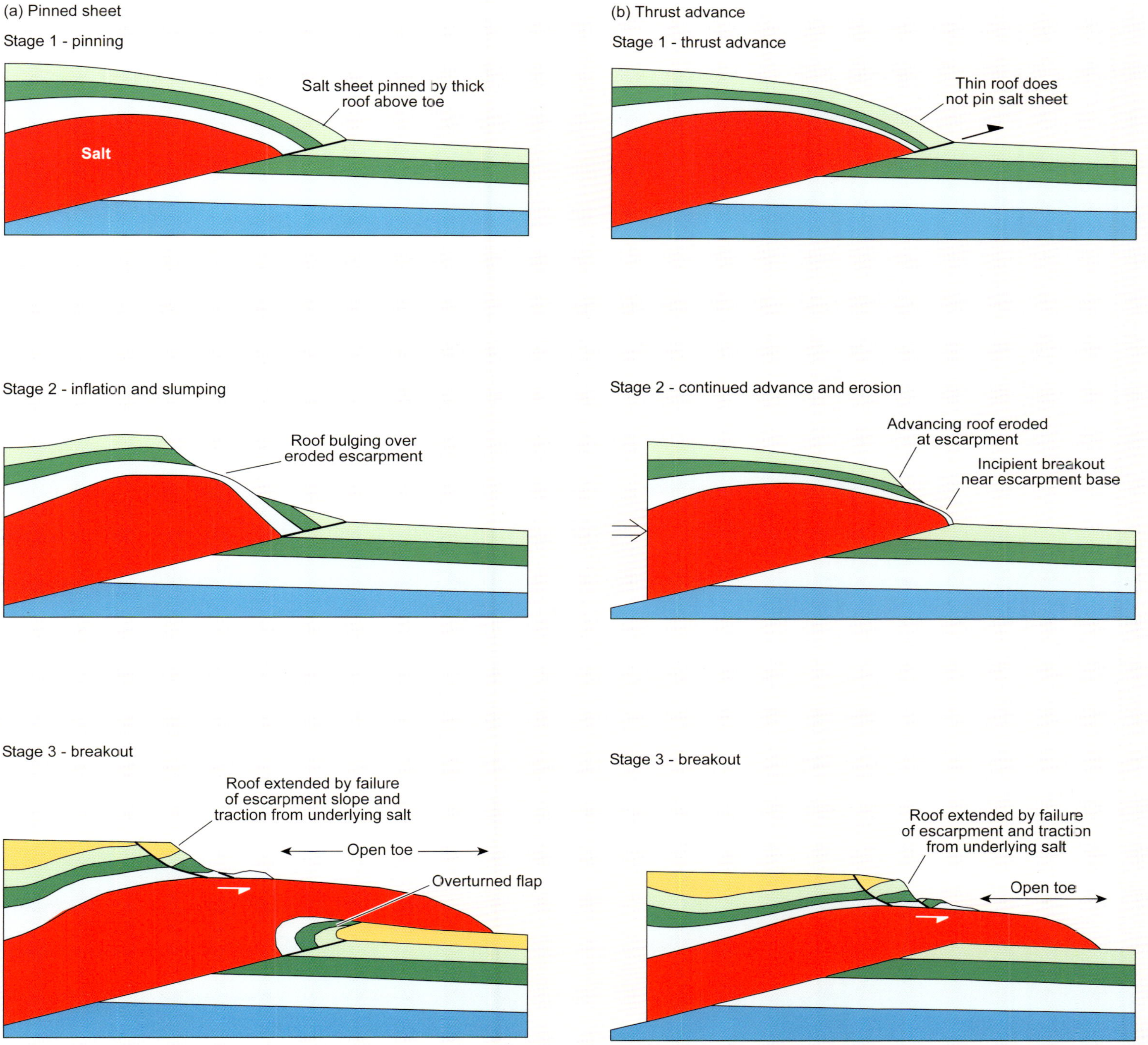

Figure 7.84. Breakout of salt sheets. The geometry of breakout at the toe of a salt sheet depends on whether the roof is thick and strong enough to pin the toe of the sheet. If so, (a) basins subsiding into the rear (mostly off section) expel underlying salt, which inflates the toe of the sheet. Inflation steepens the frontal escarpment, causing its upper part to fail. Salt breaks through the thinned roof near the escarpment crest. Roof strata at the escarpment below the breakout are inverted to form an overturned flap. If, however, the roof is too weak to contain the sheet, then (b) the roof and salt sheet advance together as a thrust hanging wall. If the leading edge of this thrust is sufficiently eroded, salt breaks out near the base of the escarpment, and open-toed advance can begin. See models for salt-sheet advance in Figure 7.72 and lineages of salt sheets in Figures 7.75 through 7.77. Modified from Hudec and Jackson (2006).

CONCEPTUAL MODEL—Criteria for Recognition of Salt Sutures in Section View

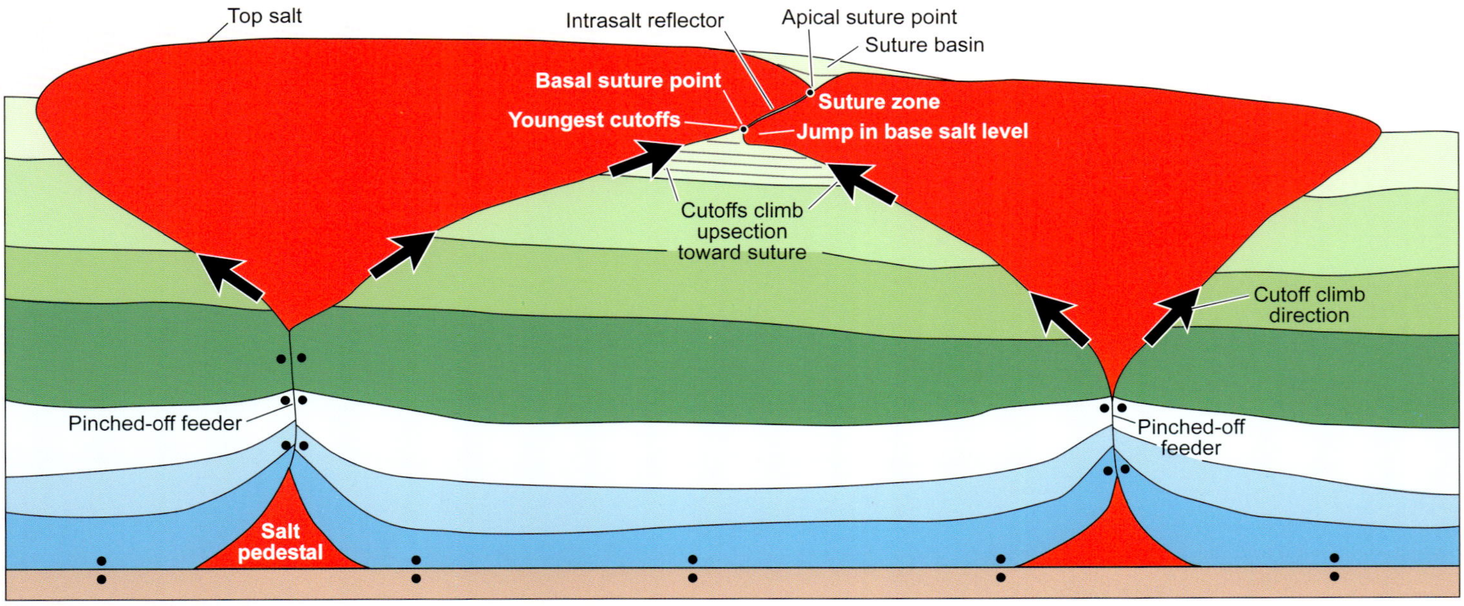

Figure 7.85. Growth of a salt canopy. Allosutures form when two salt sheets coalesce. Recognition criteria in cross section include (a) structural high in the base of salt at the basal suture point; (b) youngest subsalt cutoffs at the basal suture point; (c) cutoffs climbing up toward the basal suture point from both sides; (d) a suture basin above the apical suture point, with suprasalt sediments onlapping salt on either side; (e) plausible feeder locations identified for the sheets on either side of the suture; and (f) an intrasalt reflector or zone of reflectivity from sediments trapped in the suture between the two salt sheets when they merged. There may also be a jump in the level of the base or top of salt across the suture. See criteria for recognition of sutures in map view in Figure 7.86.

CONCEPTUAL MODEL—Criteria for Recognition of Salt Sutures in Map View

Figure 7.86. Growth of a salt canopy. Allosutures form when two salt sheets flow together. In map view, suture lines end at terminal suture points, which typically occur at cusps between lobate salt sheets. One or more suture lines also typically trend into windows through a salt canopy, where sheets have not quite coalesced. Finally, many sutures are marked by mappable zones of intrasalt reflectivity from sediments trapped between sheets when they merged. See criteria for recognition of sutures in section view in Figure 7.85.

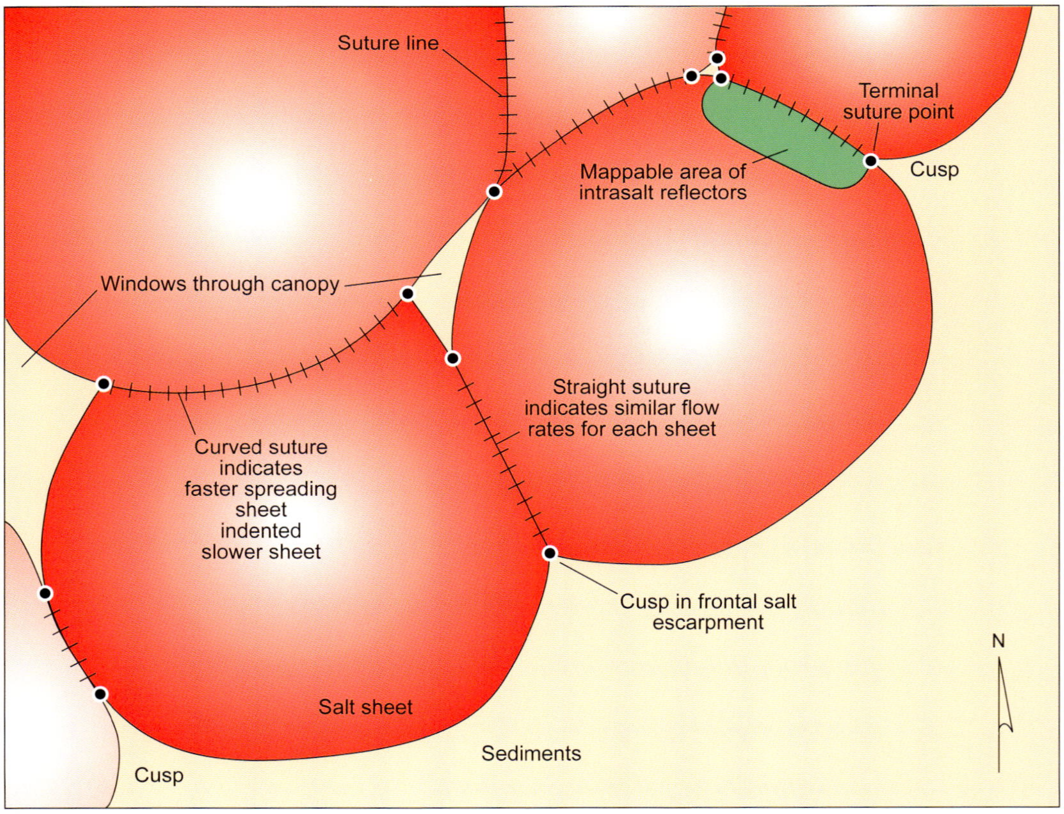

CONCEPTUAL MODEL—Terminology of Salt-Sheet Sutures

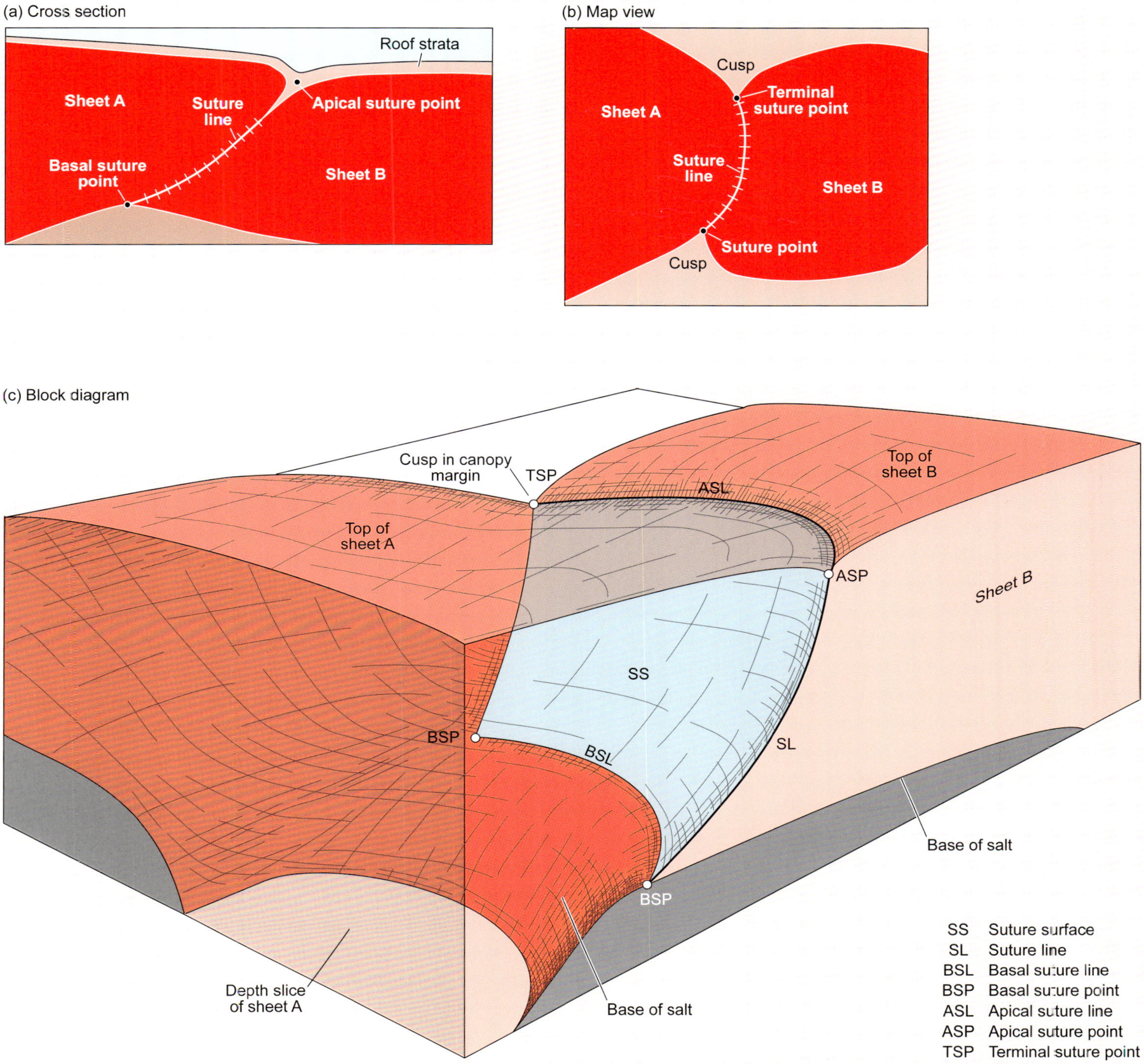

Figure 7.87. Suturing of salt sheets. We define a suture surface as the surface separating salt and any associated sediments derived from different allochthonous sheets or lobes fed by the same feeder. A suture line is the intersection of the suture surface with any two-dimensional surface such as a section or map. The apical and basal suture lines refer to the intersection of the suture surface with the top and base of salt, respectively. The two ends of a suture line on the plane of observation are suture points. Apical and basal suture points refer to the intersection of a suture line with the top and base of salt, respectively. A terminal suture point is the end of a suture line in map view.

Chapter 8

Minibasins

Crustal basins typically subside at rates ranging from 5 to 1000 m/m.y., limited by rates at which the mantle can flow. Because salt is much less viscous than the mantle, it responds much faster to sedimentary loading. Salt-floored minibasins thus subside at rates up to 10,000 m/m.y. Their small areas and extraordinary subsidence rates make minibasins unique laboratories of sedimentary processes.

A minibasin is a small basin subsiding into relatively thick, allochthonous or autochthonous salt. In the classic case, they are small depressions a few tens of kilometers in diameter sinking into a sea of salt. They are especially common above allochthonous salt sheets.

Minibasins sink fastest when they are denser than the underlying salt (density-driven subsidence). However, most minibasins do not become as dense as salt until relatively late in their histories. Prior to density inversion, minibasins can sink by a variety of mechanisms, depending on the bathymetry, tectonic regime, and sedimentary setting. Changes in these factors can change the subsidence pattern or even invert the entire minibasin. Minibasins ultimately stop sinking when they weld at their base, like a grounding ship. The geometry of the weld depends on whether the minibasin subsides onto a planar base of salt or into a funnel-shaped feeder for a salt sheet.

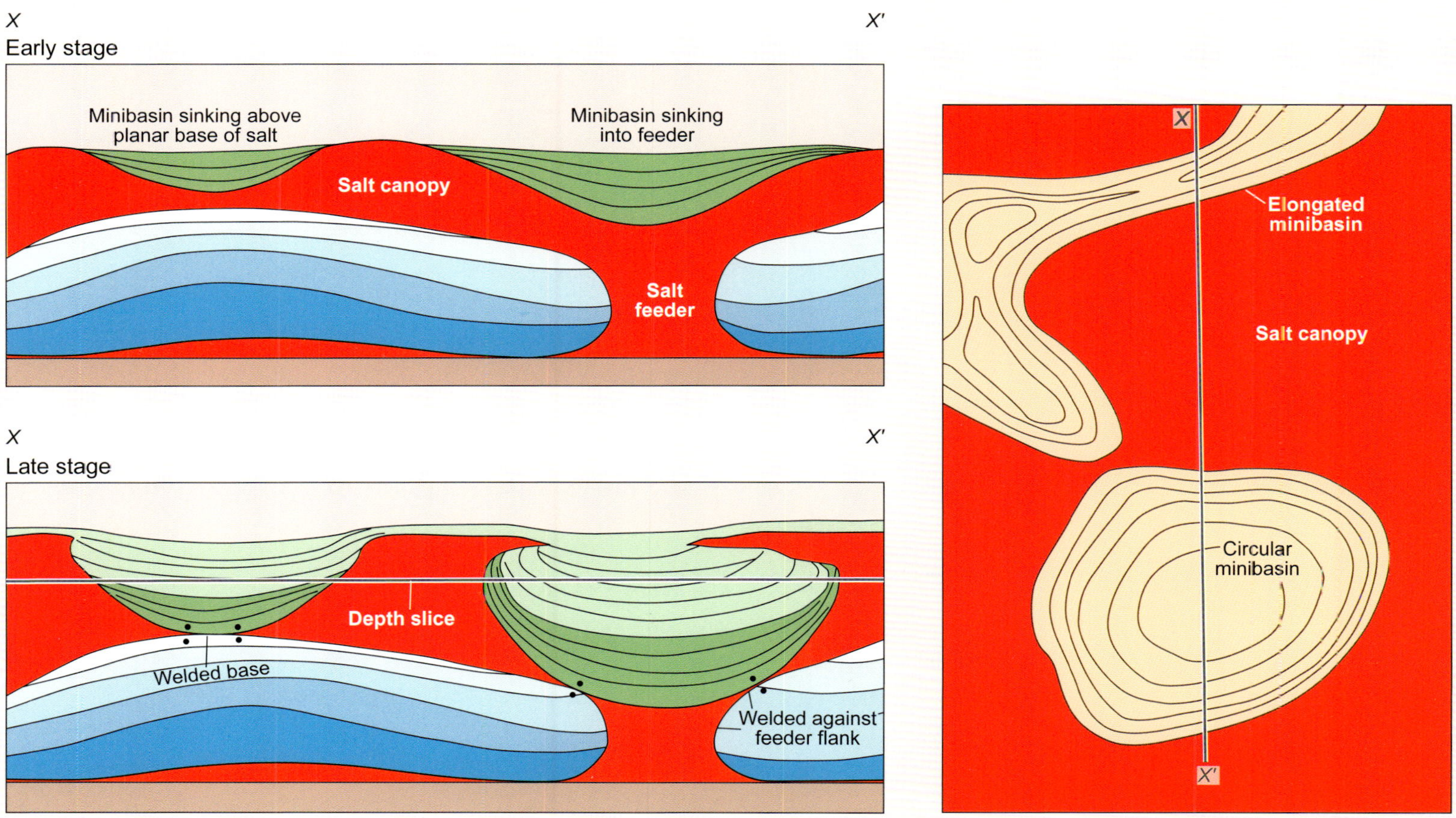

GULF OF MEXICO BASIN—Time Slice at 4.3 Seconds, Green Canyon Area, Offshore USA

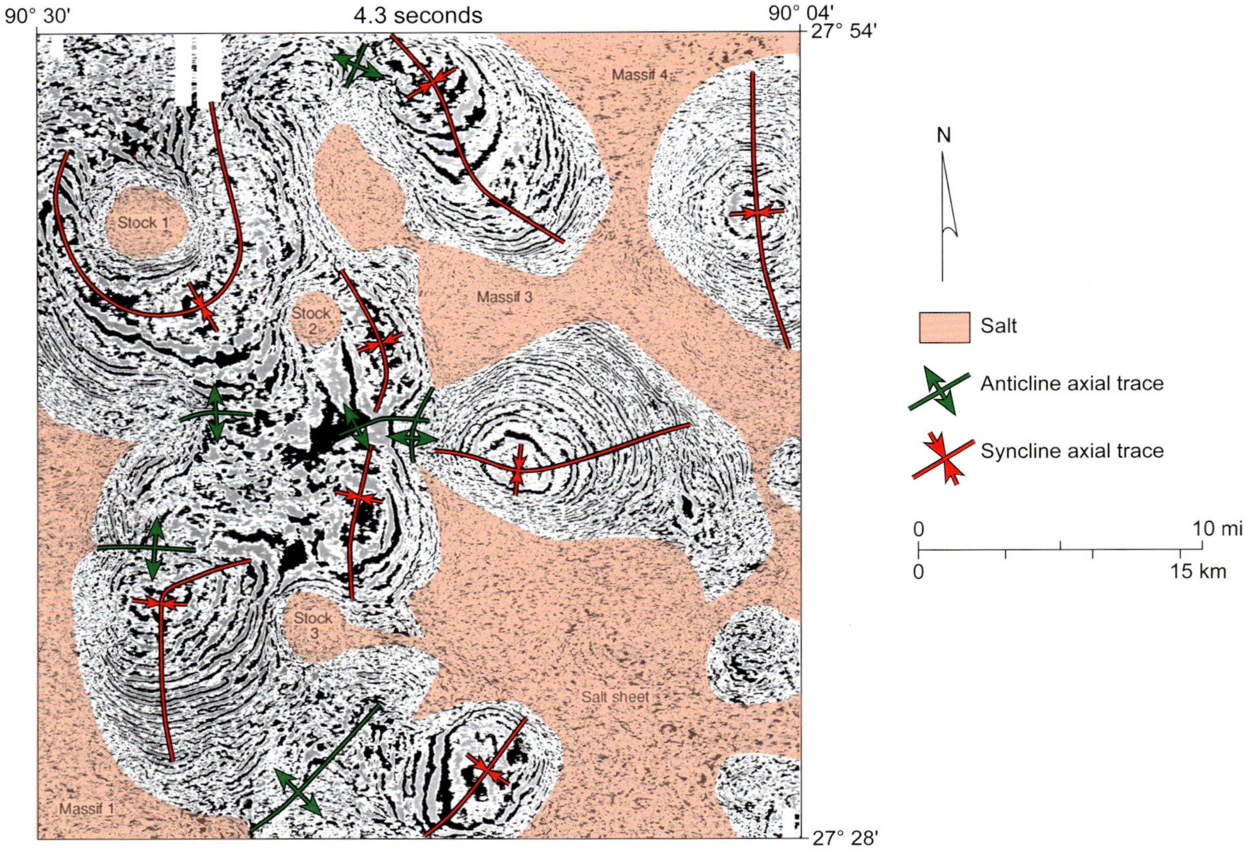

Figure 8.01. Minibasin subsidence. These minibasins are elliptical in map view. The west half of this slice shows small, oval minibasins separated by ridges and domes. These basins coalesce upward to form a single elongate trough. Modified from Seni (1994).

GULF OF MEXICO BASIN—Partly Evacuated Salt Sheet 1, Offshore USA

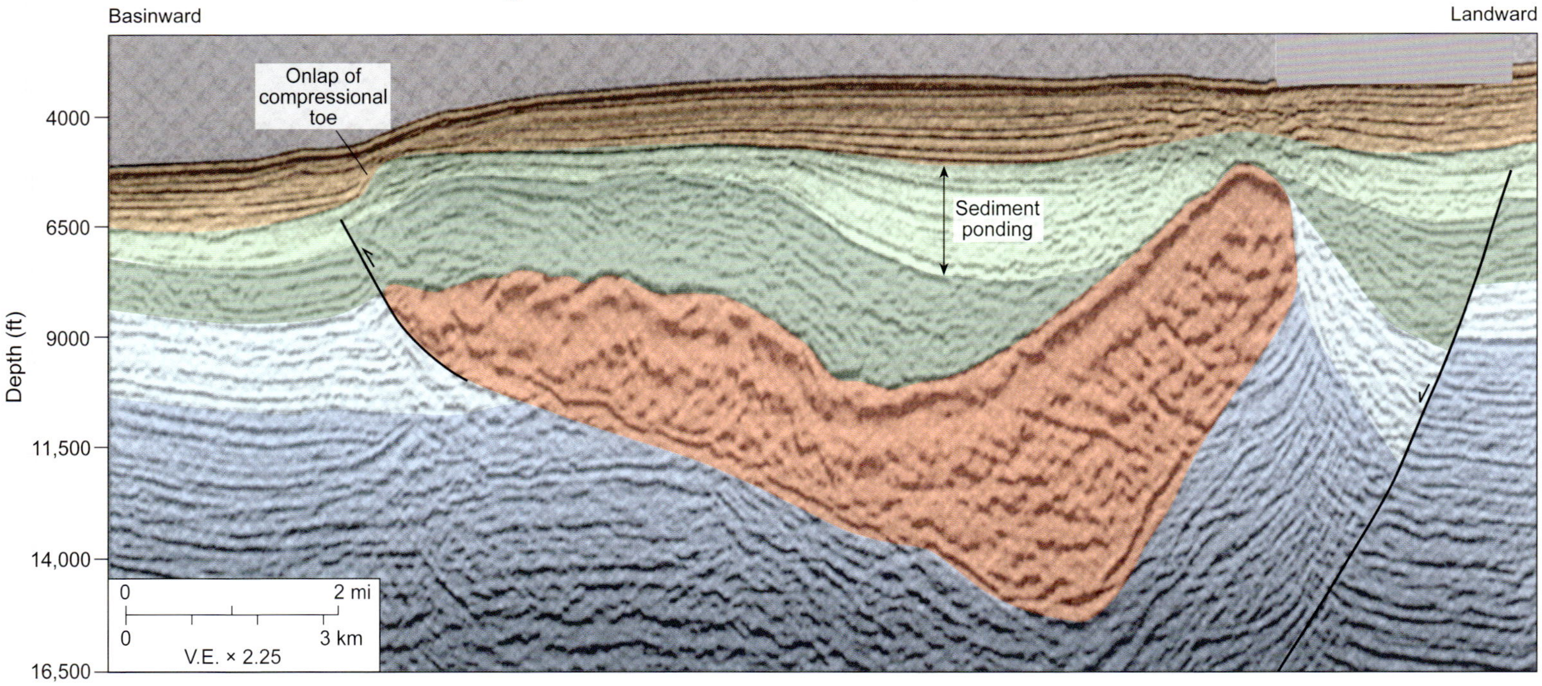

Figure 8.02. Minibasin subsidence. Subsidence into this minibasin and shortening at the toe of the sheet both slowed abruptly at the Top-Green unconformity. At that time, sedimentation switched from ponding in the minibasin to bypass across the top. The cessation of subsidence is curious because several thousand feet of salt remains beneath the minibasin. Evidently subsidence was being governed by some external tectonic control, which shut off at this time. Sediments in the minibasin were not yet compacted enough to be denser than salt, so subsidence stopped when the inferred external conditions changed. The inferred stem of the stock is not imaged in this section. See also a more evolved minibasin (Figure 8.03) and a welded minibasin (Figure 8.04). Data from Hodgkins and O'Brien (1994). Interpretation by Mike Hudec.

GULF OF MEXICO BASIN—Partly Evacuated Salt Sheet 2, Offshore USA

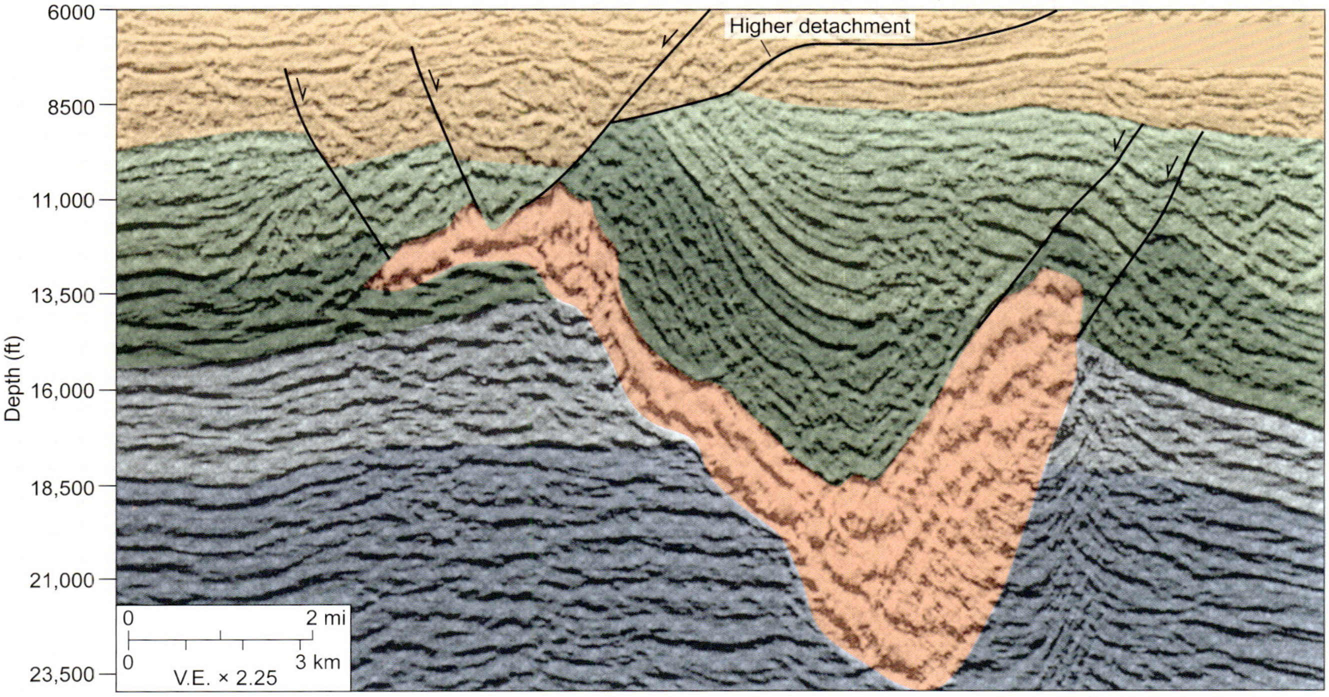

Figure 8.03. Minibasin subsidence. Most of the salt in this sheet has been expelled by minibasin subsidence. Sediments on the left edge of the minibasin are truncated against a dipping surface that may be a salt weld, marking the path that salt took as it escaped upward out of the more deeply buried sheet. See also a less evolved minibasin (Figure 8.02) and a welded minibasin (Figure 8.04). Data from Hodgkins and O'Brien (1994). Interpretation by Mike Hudec.

GULF OF MEXICO BASIN—Partly Evacuated Salt Sheet 3, Offshore USA

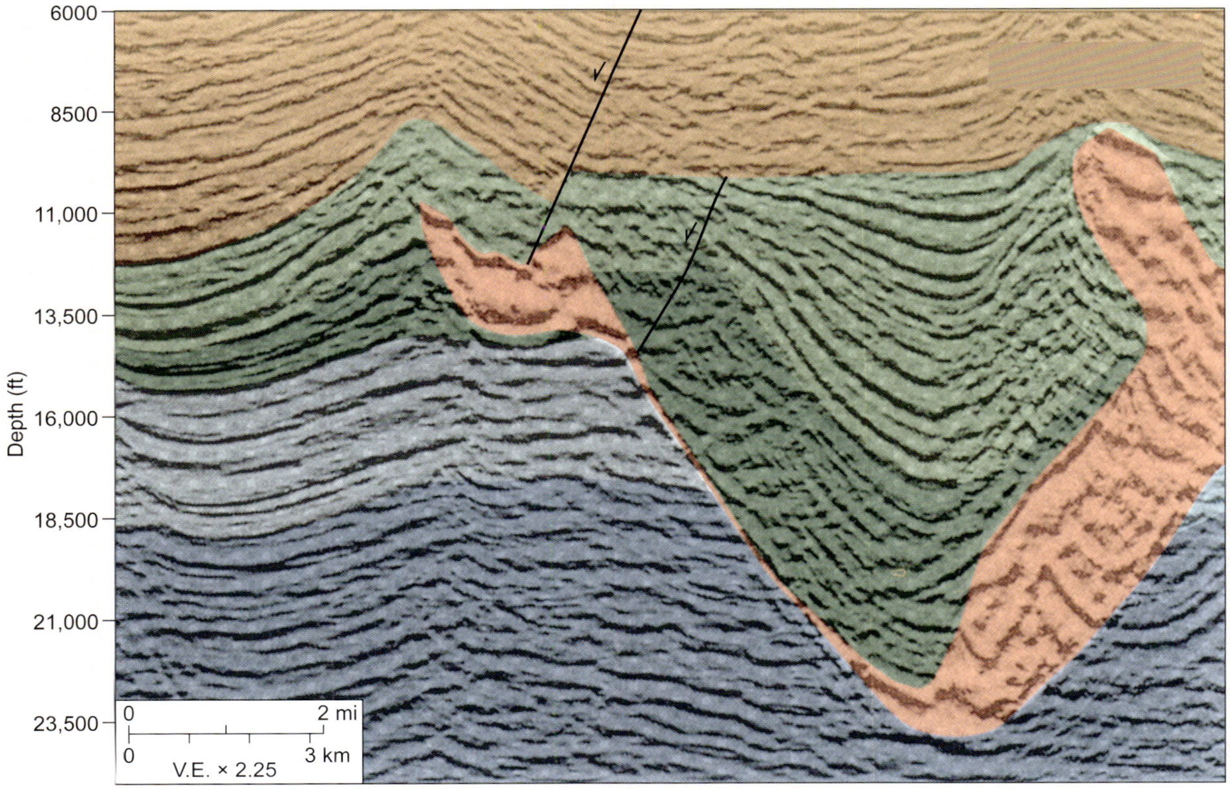

Figure 8.04. Minibasin subsidence. Advanced minibasin subsidence has created a salt weld along the left-hand base of the former salt sheet. The geometry of weld-parallel layers overlying truncated subweld sediments is diagnostic of an evacuated funnel-shaped stock. Salt-cored anticlines on both sides of the minibasin suggest that expulsion of salt from beneath the minibasin actively lifted the roofs of the flanking diapirs. See also a slightly subsided minibasin (Figure 8.02) and a moderately subsided minibasin (Figure 8.03). Data from Hodgkins and O'Brien (1994). Interpretation by Mike Hudec.

GULF OF MEXICO BASIN—Minibasin Sinking into Bowl Weld, Walker Ridge Area, Offshore USA

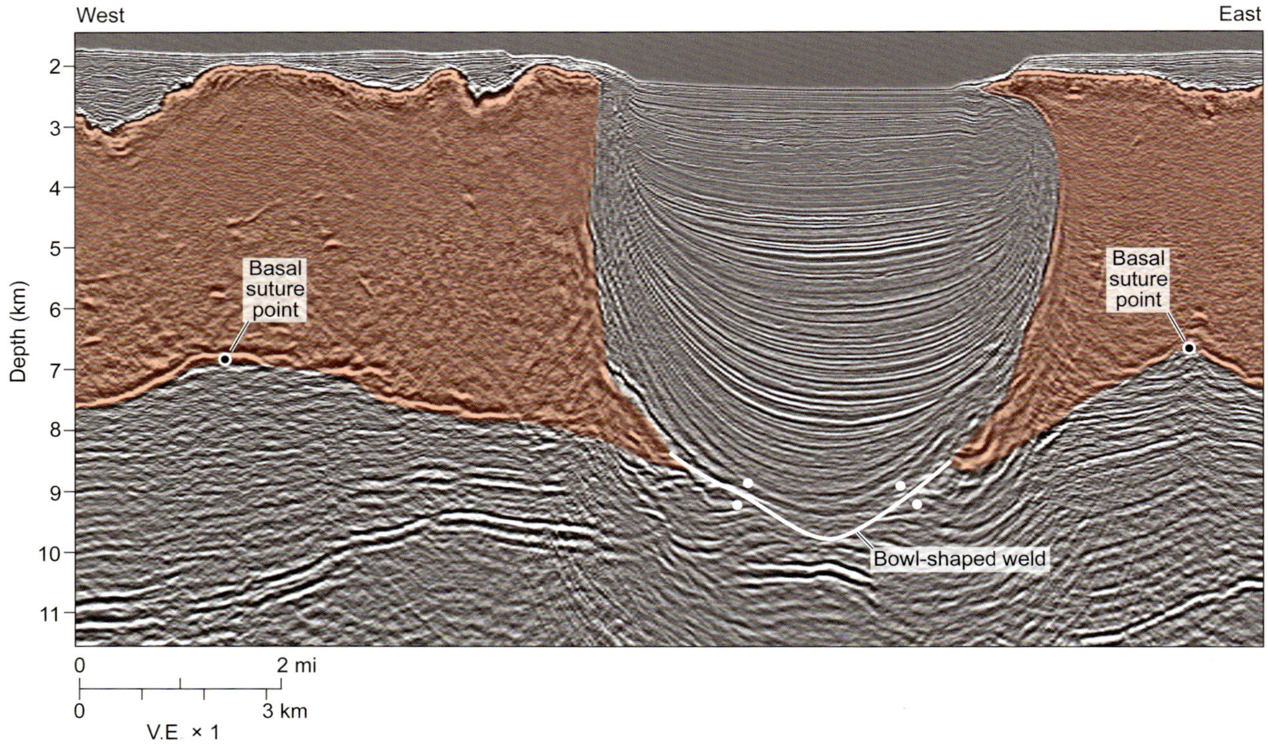

Figure 8.05. Minibasin subsidence. The minibasin sank into a bowl-shaped weld in the base of salt. This bowl-shaped low is probably derived from a feeder diapir that lies out of the plane of section. Isopach patterns in the bottom of the minibasin indicate that it began subsiding while sediments were thin. The spatial correlation of supracanopy minibasins with canopy feeders is common and may be due to either extensional diapir fall or decay of salt topography. Here the absence of extensional structures along strike with the feeder suggests that the minibasin probably initiated by decay of salt topography. Seismic courtesy of CGGVeritas. Modified from Hudec and others (2009).

GULF OF MEXICO BASIN—Seaward Migration of Depocenters in a Minibasin, Garden Banks and Keathley Canyon Areas, Offshore USA

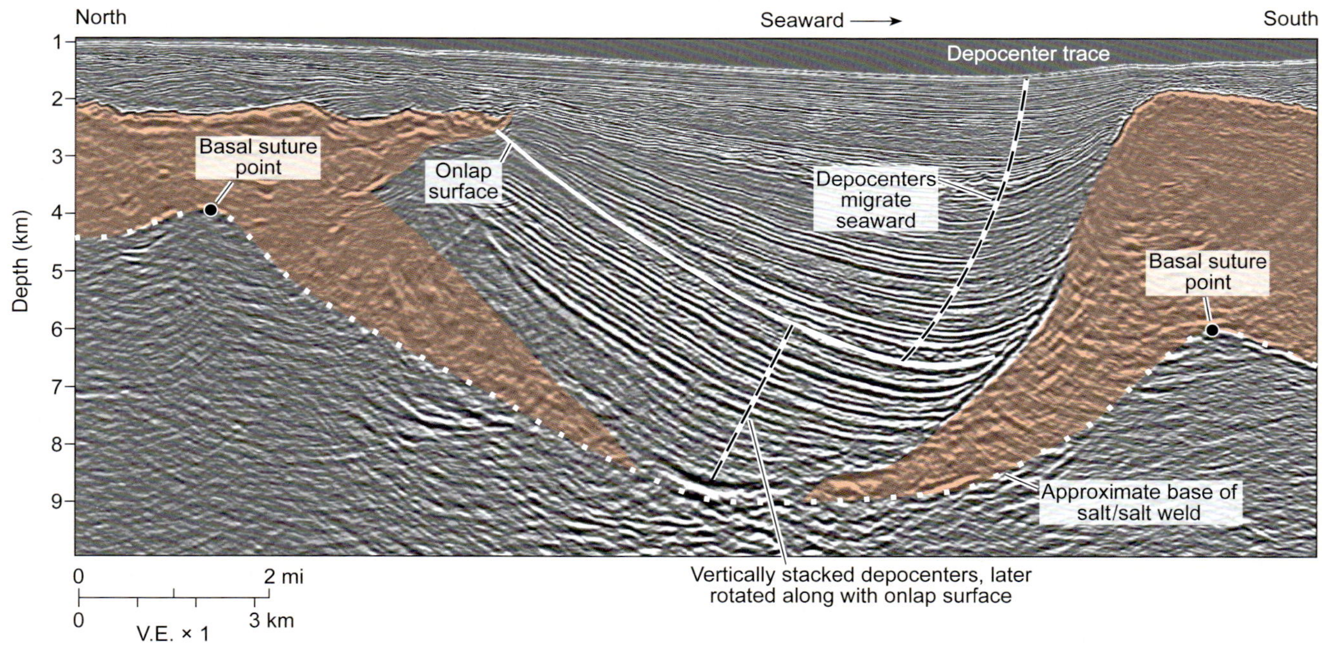

Figure 8.06. Minibasin subsidence. Depocenters in the minibasin migrated seaward through time during two stages of minibasin fill: an early phase in which depocenters stacked almost vertically in a symmetric minibasin and a later phase in which depocenters migrated seaward, making the minibasin asymmetric. Salt expulsion during the second phase rotated the older symmetric basin and its capping onlap surface. The phases are separated by a prominent onlap surface. We interpret the early phase to result from salt-topographic decay because the minibasin formed a syncline overlying a structural low in the base of the salt. The later seaward migration of depocenters was probably a result of sediment progradation that produced sedimentary topographic loading. Seismic courtesy of CGGVeritas. Modified from Hudec and others (2009).

GULF OF MEXICO BASIN—Sigsbee Salt Canopy Extensional Minibasin, Offshore USA

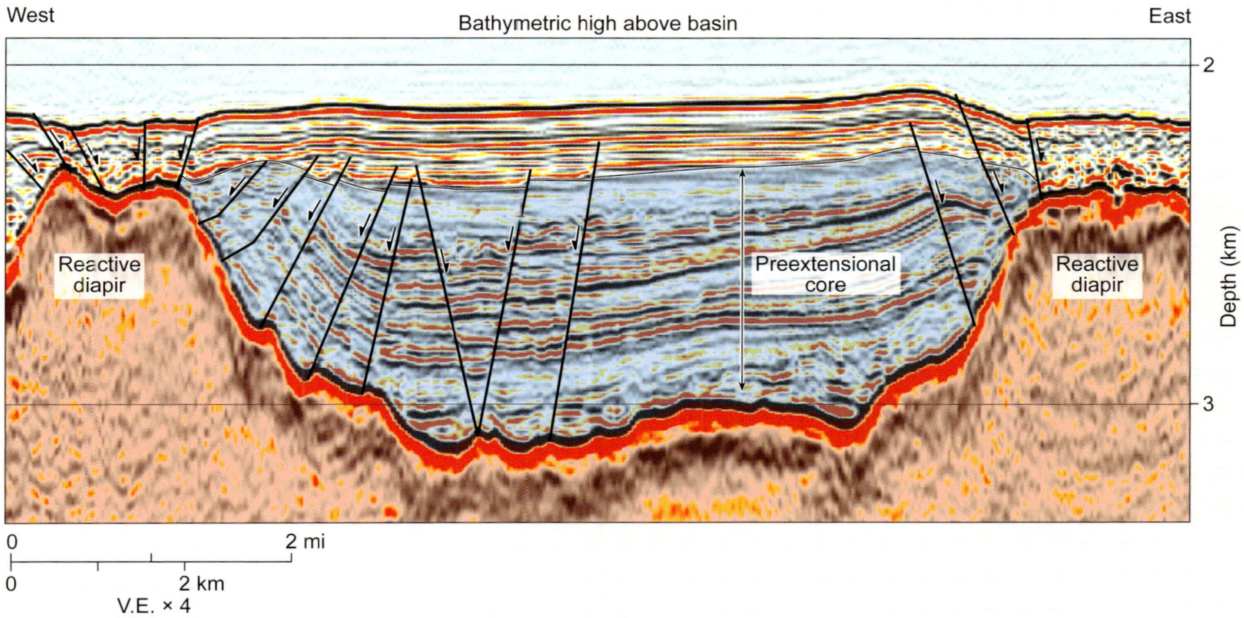

Figure 8.07. Extensional minibasin formation. The minibasin at the center of the section formed during extension, which caused reactive diapirs to rise on either side. Basin fill is now about 1 km thick, less than the estimated 1.5-2 km required to obtain a bulk density for clastic sediments greater than that of impure salt (2200 kg/cm³). The average density of the minibasin is less than that of salt, so it floats higher than the adjacent diapirs. As a result, sediment will be diverted away from this minibasin, so it is unlikely to evolve further as a depocenter. Thus, extension is not an effective way to initiate minibasins when the sediment is less dense than salt. Seismic data courtesy of CGGVeritas.

GULF OF MEXICO BASIN—Sigsbee Salt Canopy Complex Minibasin, Offshore USA

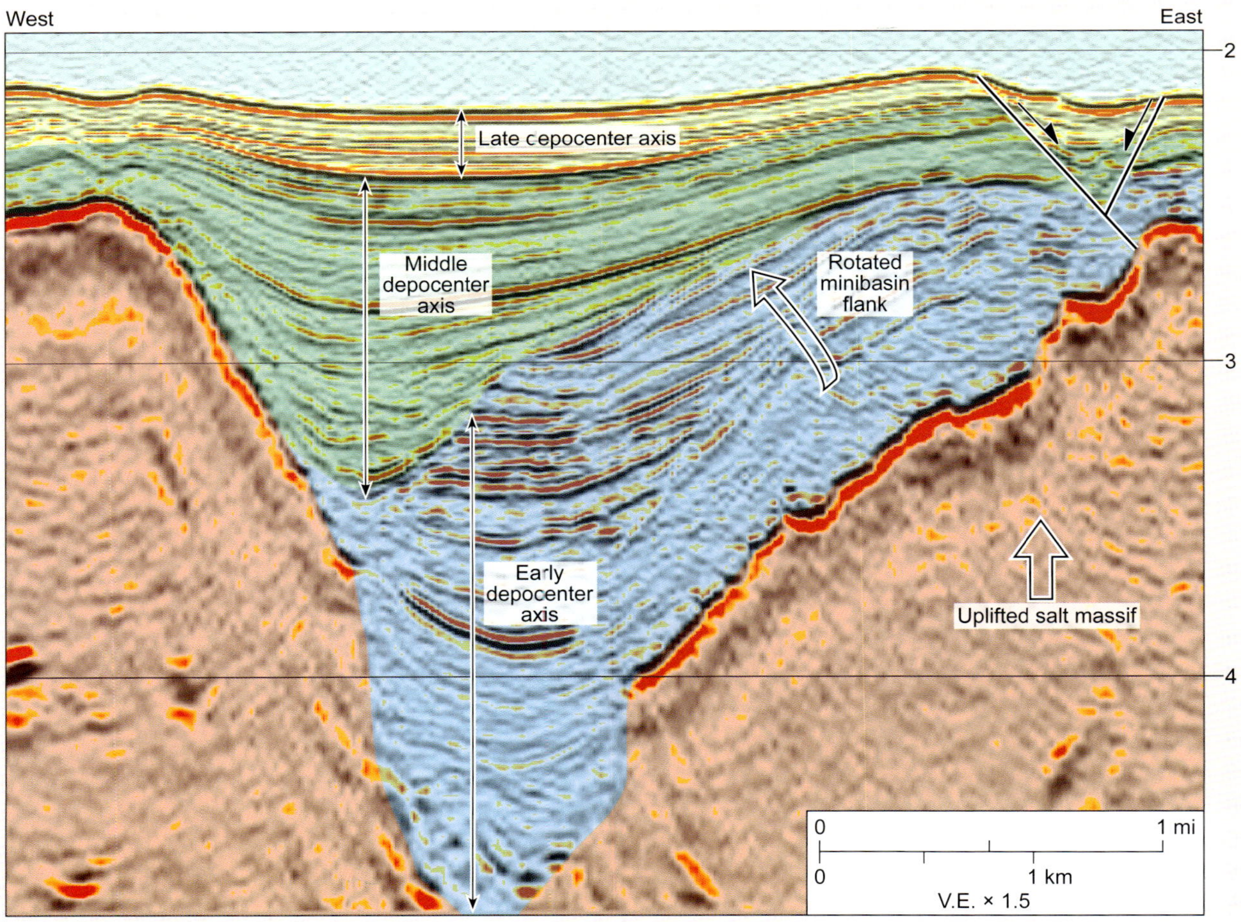

Figure 8.08. Shifting minibasin depocenters. The minibasin underwent three phases of subsidence, with each phase marked by a different subsidence axis. Depoaxes (vertical arrows) coincided with the thickest part of the minibasin in the early and late stages of subsidence, but during the middle stage the bathymetric low shifted to the west edge of the minibasin. This pattern cannot be explained by gravity-driven minibasin subsidence. Dip of beds within the minibasin suggests that the eastern salt massif was uplifted near the end of the early phase of deposition (blue interval), possibly as a result of shortening. This massif uplift rotated and uplifted strata on the east side of the minibasin. Accordingly, the depoaxis shifted westward away from this uplift during deposition of the green interval. Uplift of the eastern massif stopped by the end of the middle stage, allowing the depoaxis to shift back to the center of the basin, which was now dense enough to sink under its own weight. Modified from Hudec and others (2009).

GULF OF MEXICO BASIN—Sigsbee Salt Canopy Shortened Minibasin 1, Offshore USA

Figure 8.09. Minibasin shortening. This minibasin subsided in three stages, each of which was characterized by different patterns of reservoir deposition. The prekinematic stage is poorly resolved on the seismic profile. At that time, the minibasin had not yet formed, and sands were deposited without reference to the current basin configuration. During the succeeding stage of shortening, sands were deposited in small synclines between thrust anticlines. After shortening, the minibasin was thick enough to sink into the salt, so sands were deposited in a broad subsiding minibasin. Modified from Hudec and others (2009).

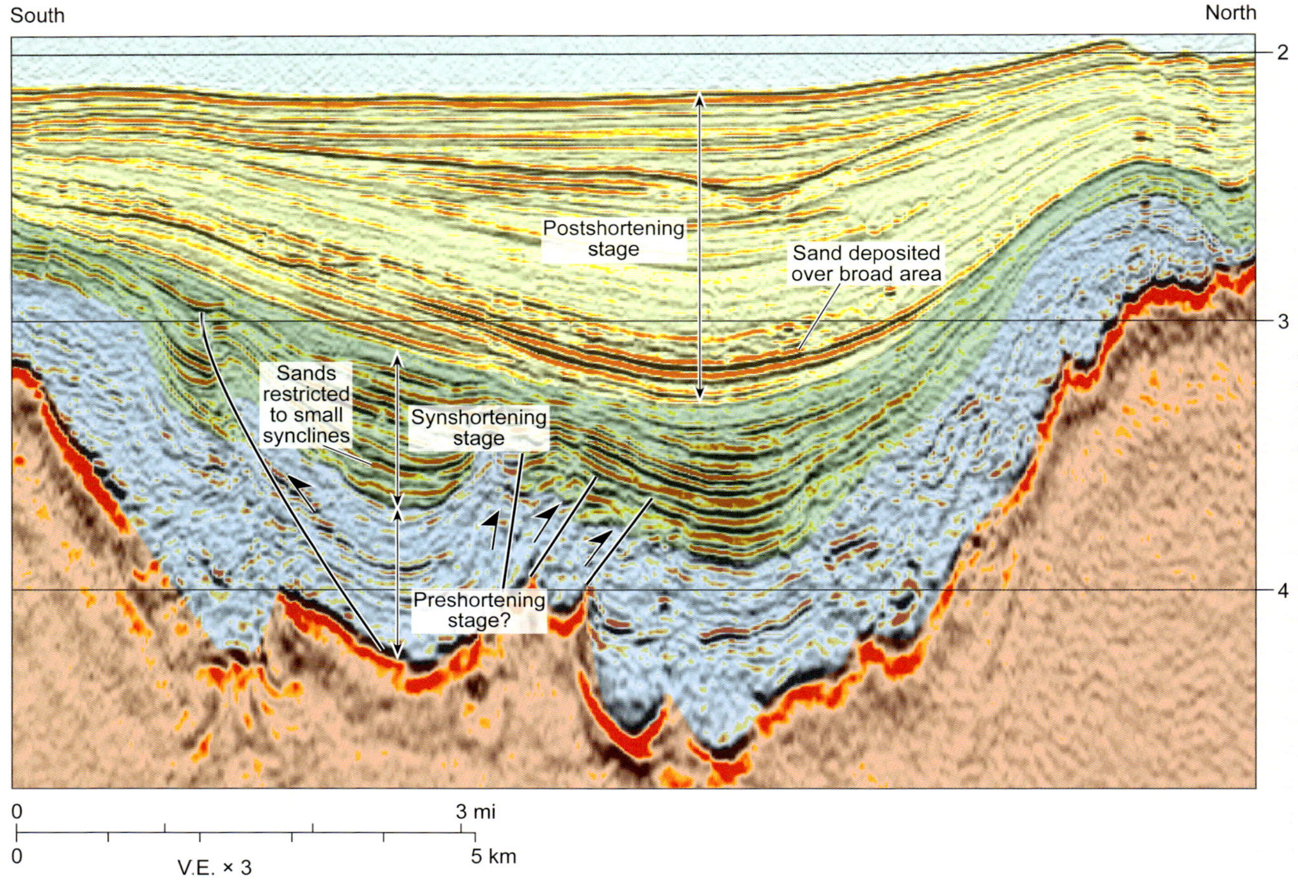

GULF OF MEXICO BASIN—Sigsbee Salt Canopy Shortened Minibasin 2, Offshore USA

Figure 8.10. Minibasin shortening. During shortening, the weak salt massifs on either side of the minibasin were preferentially shortened. They rose relative to the minibasin, leaving the minibasin as a depocenter. This is an effective way to make minibasins sink relative to their surroundings, even if their fill is less dense than salt. Seismic data courtesy of CGGVeritas. Modified from Hudec and others (2009).

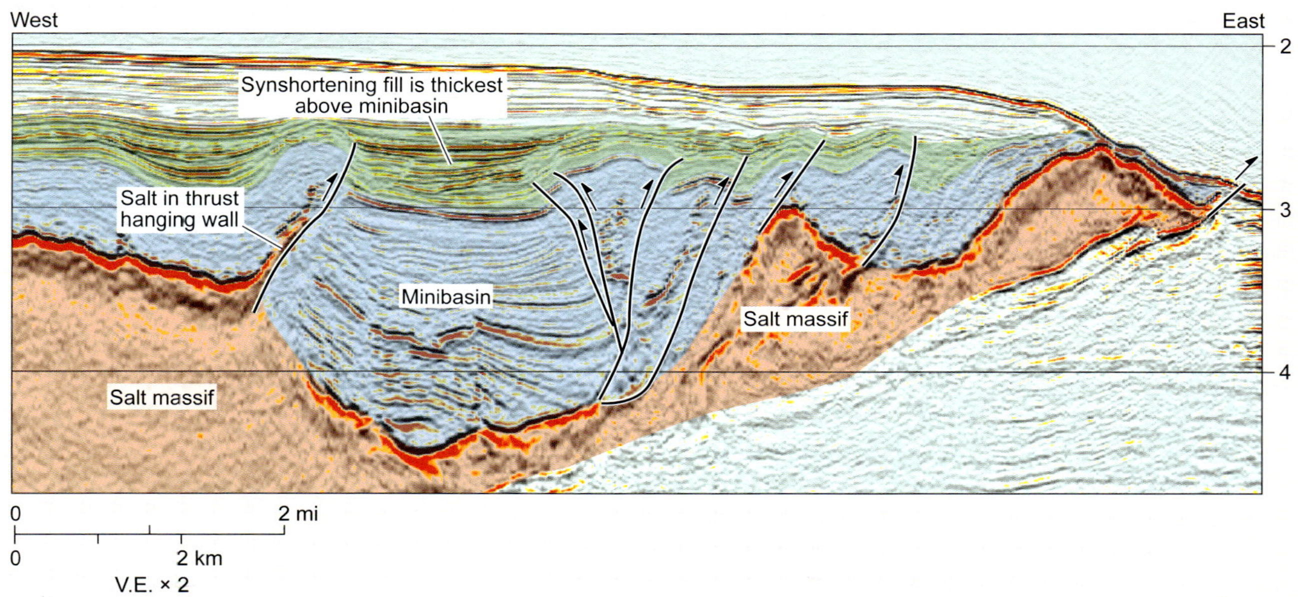

GULF OF MEXICO BASIN—Sigsbee Salt Canopy Minibasin Flank, Offshore USA

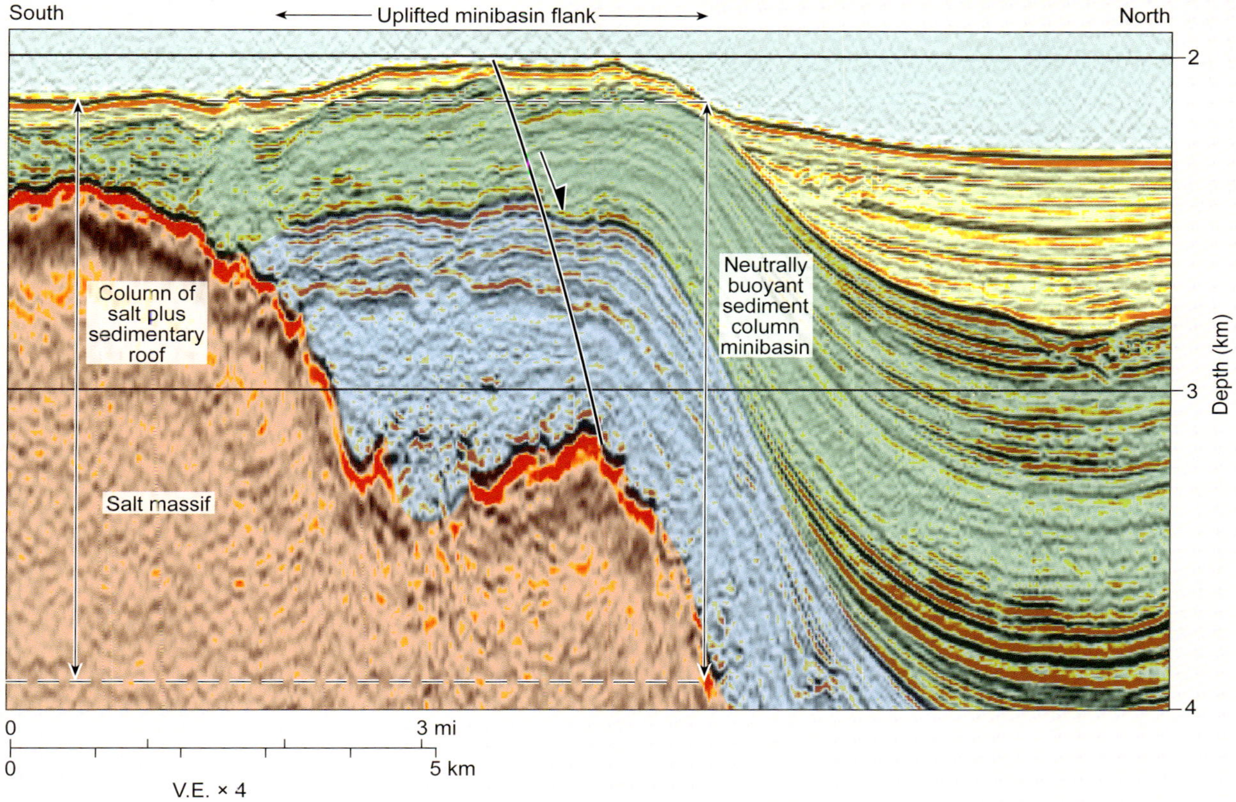

Figure 8.11. Uplift of minibasin flank. Fluid static forces cause the edge of this minibasin float higher than both the minibasin center and the adjacent salt massif. The minibasin center is thick enough to have a bulk density greater than that of salt, so the center forms a bathymetric low. The minibasin flank is thinner and less dense than salt, so it floats above the level of the massif. The part of the minibasin at an elevation identical to the salt massif is neutrally buoyant; it has an average density equal to the average density of the salt-sediment column at the massif. See conceptual model in Figure 8.31. Seismic data courtesy of CGG Veritas. Modified from Hudec and others (2009).

GULF OF MEXICO BASIN—Welded Salt Canopy, Grand Isle South and Ewing Bank Areas, Offshore USA

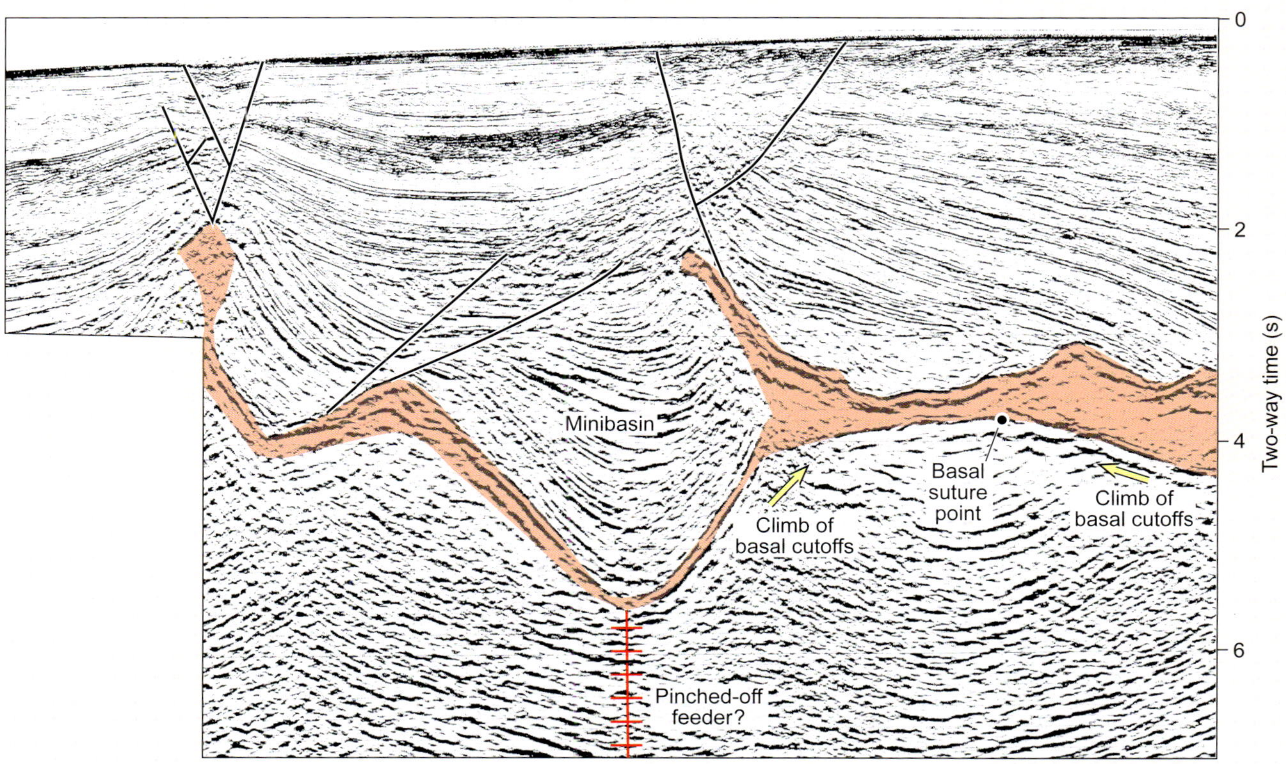

Figure 8.12. Coalescence of salt sheets and minibasin subsidence. A change in the direction of climb of cutoffs at the base of a salt sheet or equivalent salt weld suggests a suture, especially if the basal suture point is a structural high in the base of salt. The minibasin in the middle of the section is interpreted to have subsided into a winecup-shaped stock. Subsidence of the minibasin into this stock may have displaced the salt to form the extrusive salt sheets that flank the feeder. A feeder stock is not evident beneath the minibasin so may lie out of the plane of section. Modified from Diamond Geophysical (1997).

GULF OF MEXICO BASIN—Partly Evacuated Salt Canopy, South Timbalier South and Ewing Bank Areas, Offshore USA

Figure 8.13. Minibasin subsidence. Subsidence of the minibasin displaced salt onto the seafloor, triggering accelerated spreading of the salt sheet by 3.8 Ma. The bowl weld marks the original position of the base of salt. Younger sediments are juxtaposed against older strata across this weld. Modified from Rowan (1995).

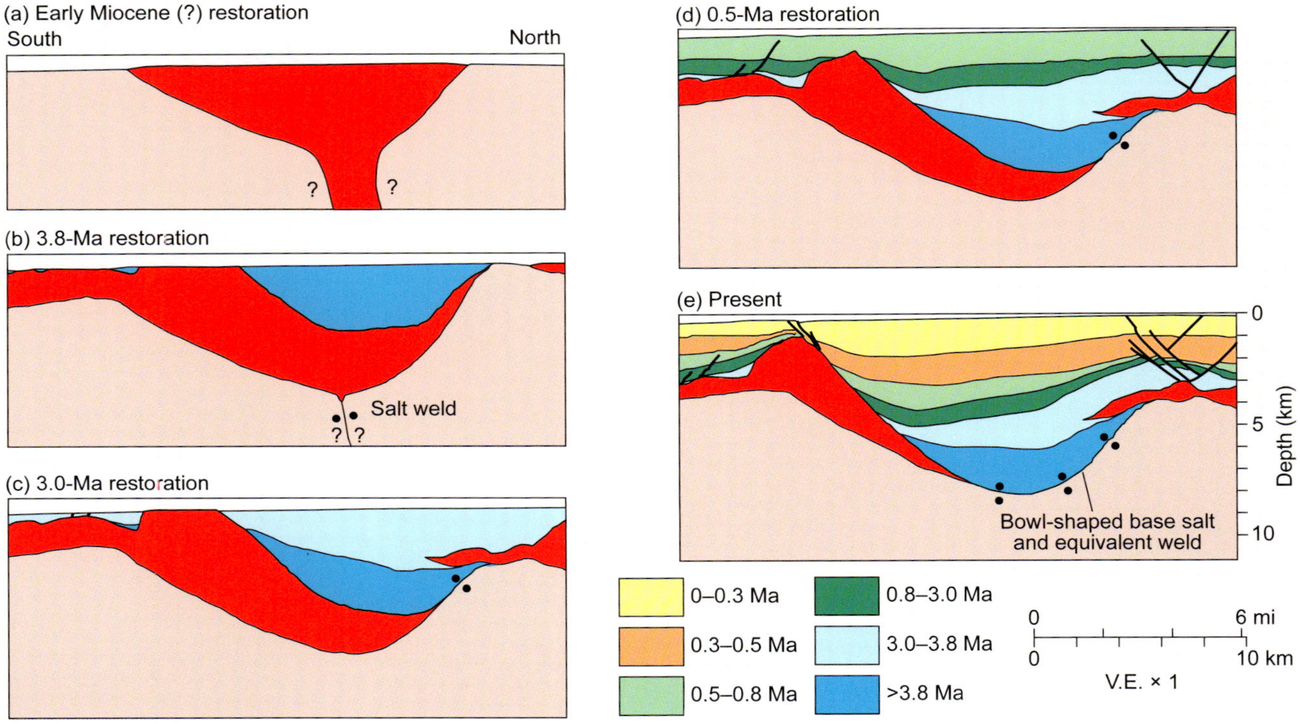

GULF OF MEXICO BASIN—Counterregional System, Salina Basin, Offshore Mexico

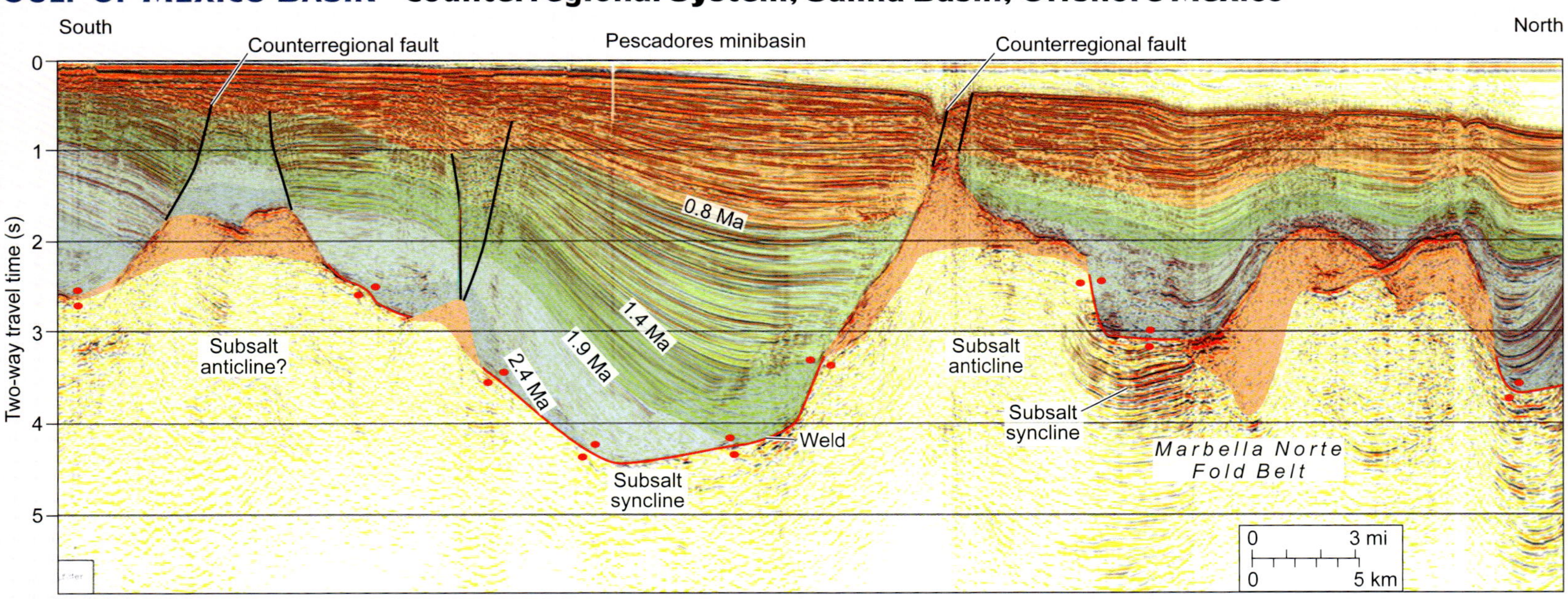

Figure 8.14. Minibasin subsidence. By the Pliocene, the salt sheet had spread over the top of the Marbella Norte Fold Belt. Thus, the undulose shape of the base salt is partly due to the topography of the underlying fold belt. Salt ponded and was thickest above the synclines and was thinnest over the anticline crests. The Pescadores minibasin subsided into thick salt that accumulated above a syncline. Basin subsidence was asymmetric, controlled by slip along a major counterregional surface that soles into the flank of a subsalt anticline. See restoration in Figure 8.15. Modified from Gómez-Cabrera (2003), with additional interpretation by Martin Jackson.

Gulf of Mexico Basin—Restoration of Counterregional System, Salina Basin, Offshore Mexico

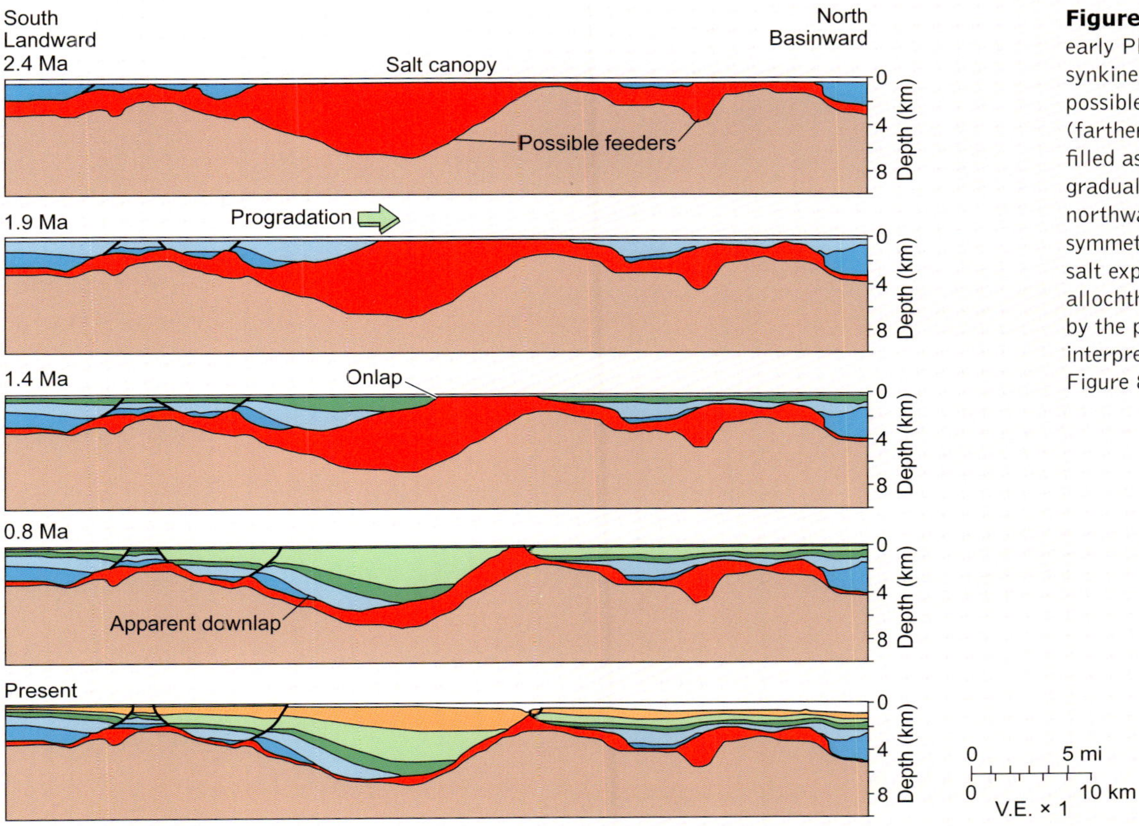

Figure 8.15. Minibasin subsidence. By the early Pleistocene, seaward-thickening wedges of synkinematic strata had begun to prograde over two possible feeders, one large (center) and one small (farther north). The basin above the larger feeder filled asymmetrically because of progradation, which gradually expelled the underlying allochthonous salt northward. The basin above the smaller feeder filled symmetrically. Onlaps were eventually deformed by salt expulsion into apparent downlaps. Feeders for the allochthonous salt canopy are inferred to be marked by the poorly imaged keels. Cross-section length was interpreted to remain constant. See seismic section in Figure 8.15. Modified from Gómez-Cabrera (2003).

Gulf of Mexico Basin—Partly Evacuated Salt Canopy 1, Ship Shoal South to Green Canyon Area, Offshore USA

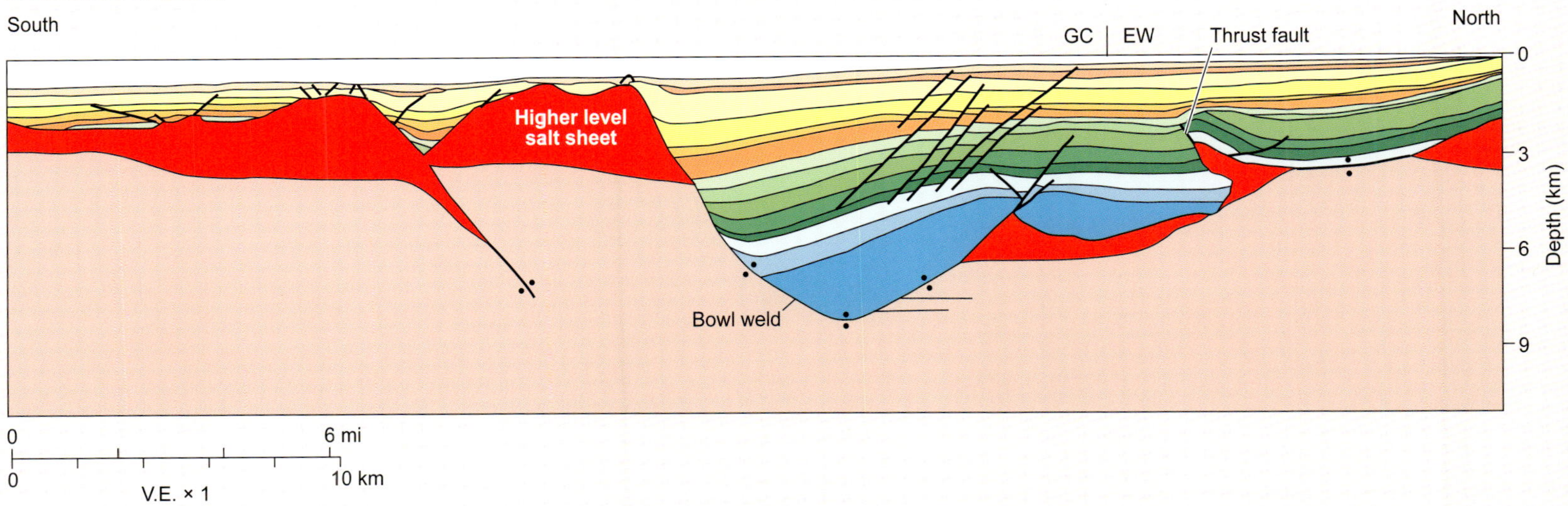

Figure 8.16. Minibasin subsidence. This section does not intersect the deepest part of the bowl weld, so the feeder is interpreted to be out of plane. Salt formerly in the feeder was displaced upward into a shallow salt sheet. The small thrust fault north of the bowl weld absorbed minor translation along the weld accommodated by uplift and extension farther north. Modified from McBride and others (1998).

GULF OF MEXICO BASIN—Partly Evacuated Salt Canopy 2, Ship Shoal South to Green Canyon Area, Offshore USA

Figure 8.17.
Subsidence into a funnel-shaped stock. Progradation of sediment southward across the salt sheet displaced extrusive salt basinward. Sediments were also transported southward, accommodated by slip along basinward-dipping normal faults. Salt evacuation was not completely efficient. Thus, several salt relics were trapped between subsiding minibasins. Modified from McBride and others (1998).

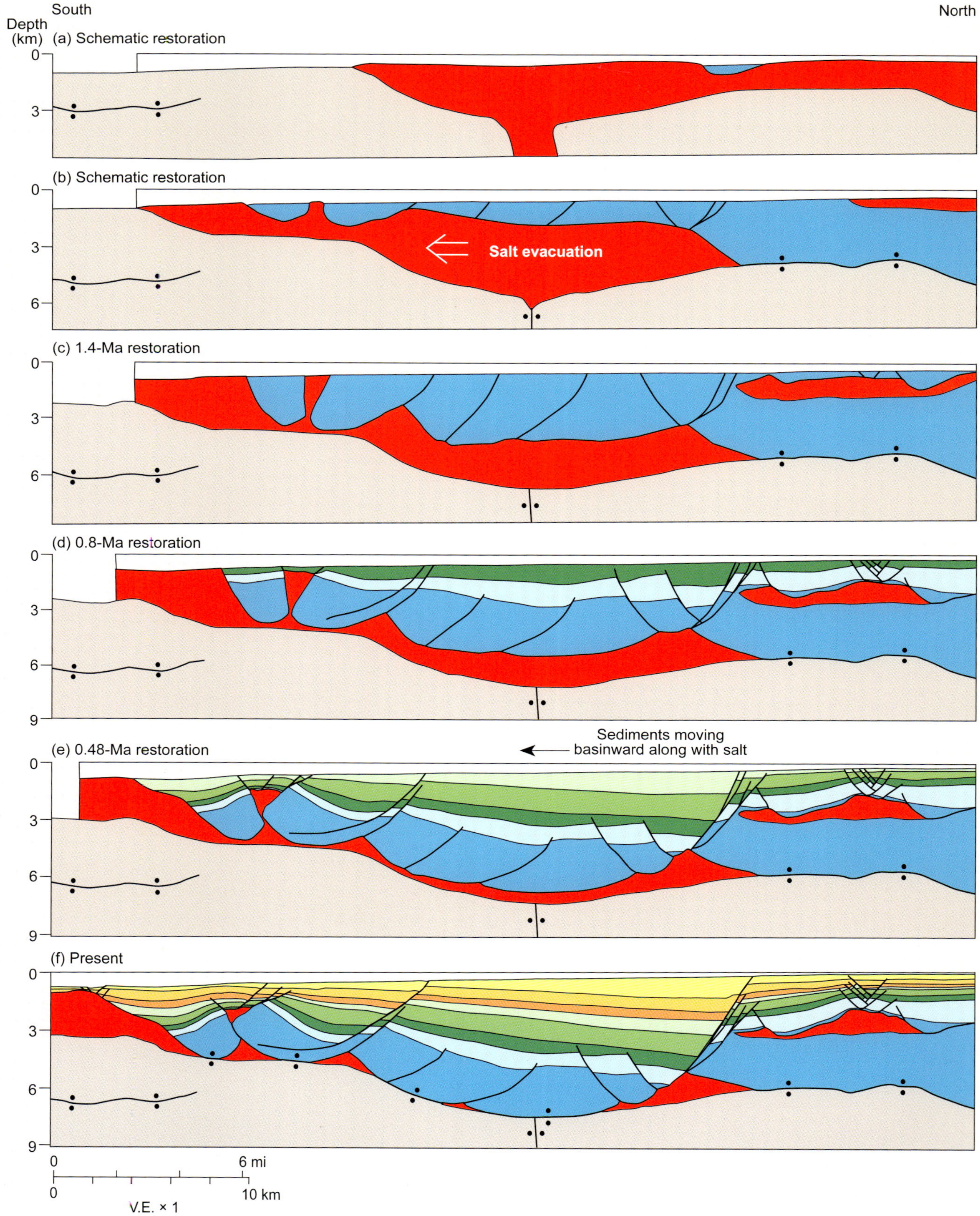

GULF OF MEXICO BASIN—Bowl Weld, Green Canyon Area, Offshore USA

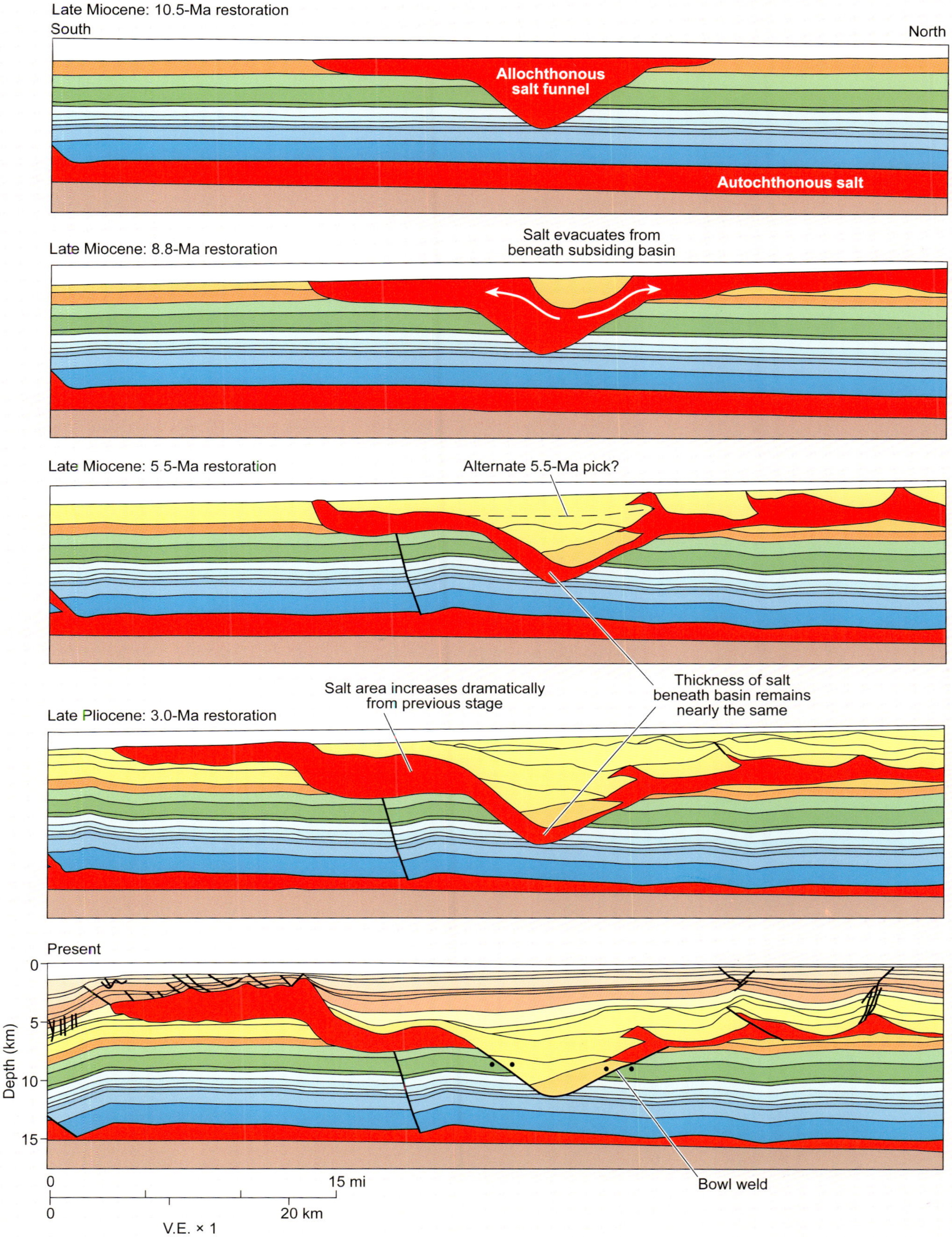

Figure 8.18. Minibasin subsidence. Sediments sinking into the base-salt structural low drove salt out into the salt sheets on either side. The restoration illustrates two anomalies in basin evolution, between 5.5 Ma and 3.0 Ma. First, the cross-sectional area of salt to the south of the base-salt structural low increases dramatically, with no obvious source for the salt. Second, the thickness of salt beneath the minibasin remains roughly constant, although the basin overhead is still subsiding. An alternative interpretation that might alleviate these problems would be to pick the 5.5-Ma surface lower in the basin. Doing so would make the basin fill thinner during the 5.5-Ma stage, leaving more salt beneath to be displaced by further subsidence in the Pliocene. Modified from McBride (1998).

PHYSICAL MODEL—CT Sections of Minibasin

Figure 8.19. Progradational infilling of a minibasin above a salt sheet. Subsidence of the minibasin above the northeast quadrant of the funnel expelled extrusive salt (white) southwestward. The proximal side of the salt funnel was welded and extended. All sedimentary layers were synkinematic except the uniform, darker, prekinematic layer encasing the original bowl of salt below the dashed line. Inset shows a horizontal slice in map view inscribed with section lines. Yellow layers are quartz sand; dark-red layers are garnet sand; medium-orange layers are glass microspheres. See deeper depth slices through the model in Figure 8.20. Experiment by Bruno Vendeville and Joel Le Calvez; CT scan by Rich Ketcham.

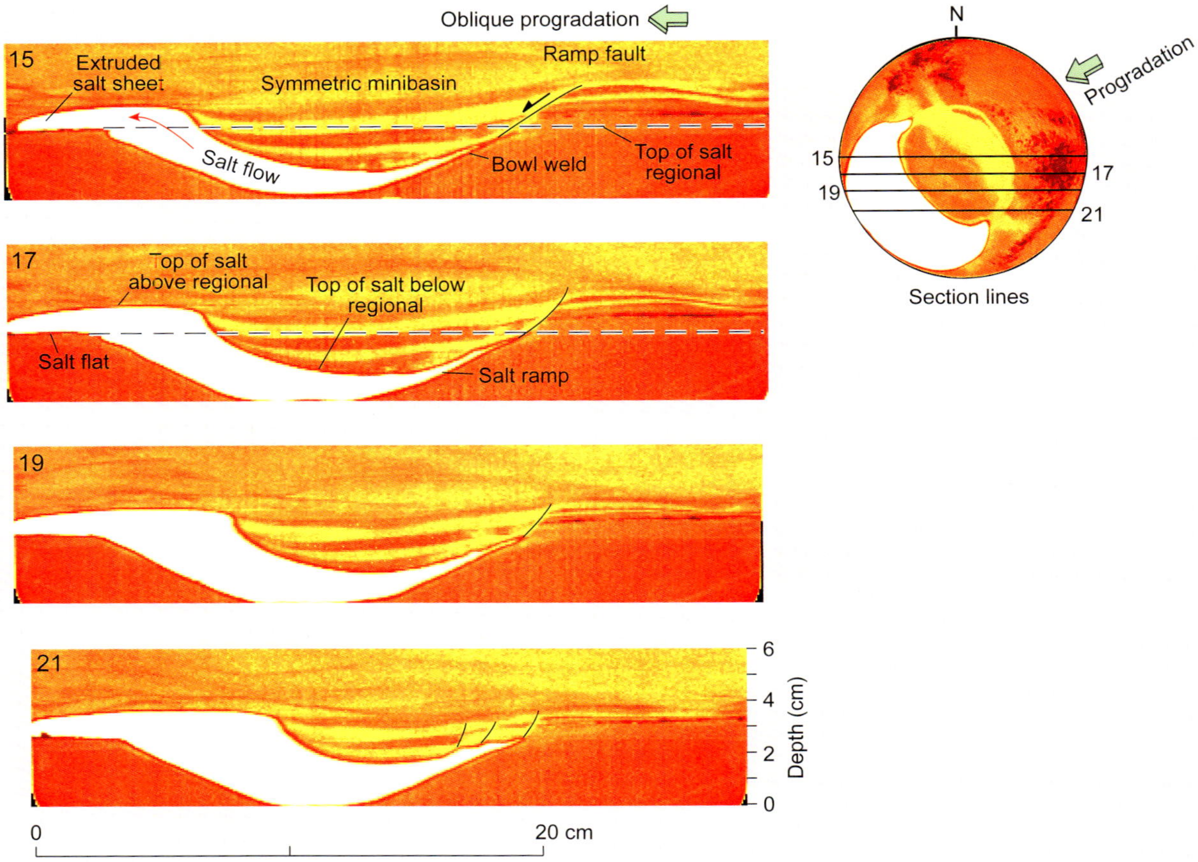

PHYSICAL MODEL—CT Depth Slices of Minibasin

Figure 8.20. Progradational infilling of a minibasin above a salt sheet. Subsidence of the minibasin expelled salt (white) toward the southwest as an extrusive flange. The position of the structural low in the minibasin is shifted slightly to the southwest in the higher slice in response to sedimentary progradation. All sedimentary layers in the minibasin were synkinematic. A prekinematic layer encased the original bowl of salt. Yellow layers are quartz sand; medium-orange layers are glass microspheres. See also sections through the model in Figure 8.19. Experiment by Bruno Vendeville and Joel Le Calvez; CT scan by Rich Ketcham.

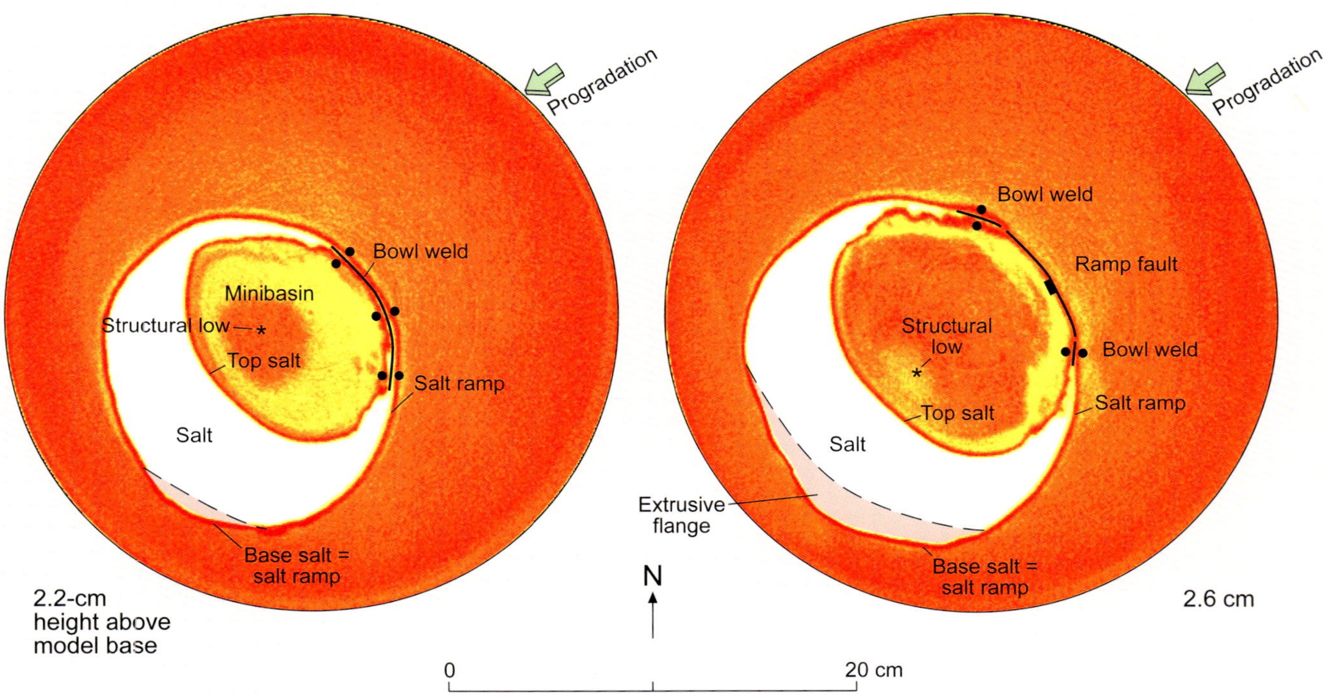

PHYSICAL MODEL—Subsidence of Aggrading Sediments into Salt Sheets

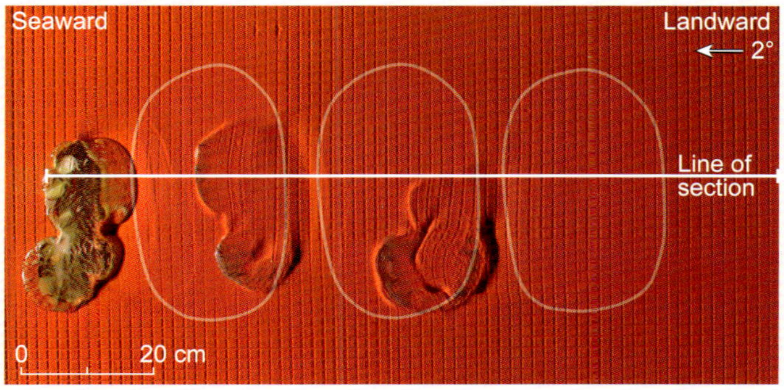

Figure 8.21. Minibasin subsidence. Three funnel-shaped salt sheets were aggradationally loaded on a 2° slope. Downdip flow of salt within each sheet inflated their seaward ends, uplifting the roof there. This seaward flow created accommodation at the updip ends of each sheet, which was filled with sediments to form U-shaped minibasins. Flow accelerated once sediments added were denser than salt. Continual seaward advance of salt in the vents at the end of the sheets created seaward-leaning diapirs over time. Once the U-shaped minibasins welded at their bases, the only salt available to feed higher extrusion was from deeper parts of the leaning feeders. Subsidence of overlying units into these feeders produced counterregional rotation of minibasin fill. Most salt extruding at the surface flowed downslope, but some flowed updip as salt wings into the bathymetric lows formed by the minibasins. When buried beneath dense sediments, these wings later evacuated back into the leaning feeders, as recorded by high-level evacuation grabens. Model by Tim Dooley.

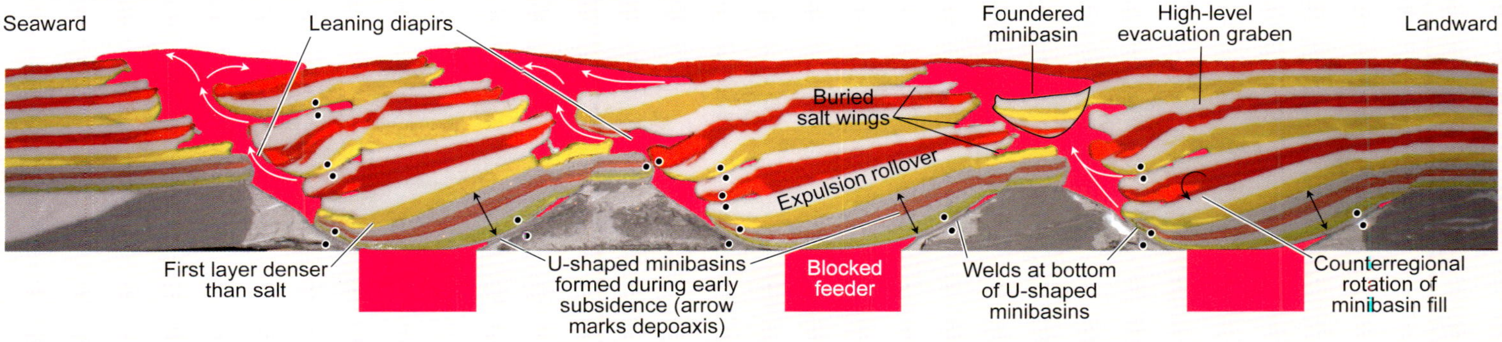

PHYSICAL MODEL—Subsidence of a Sediment Lobe into Salt Sheets

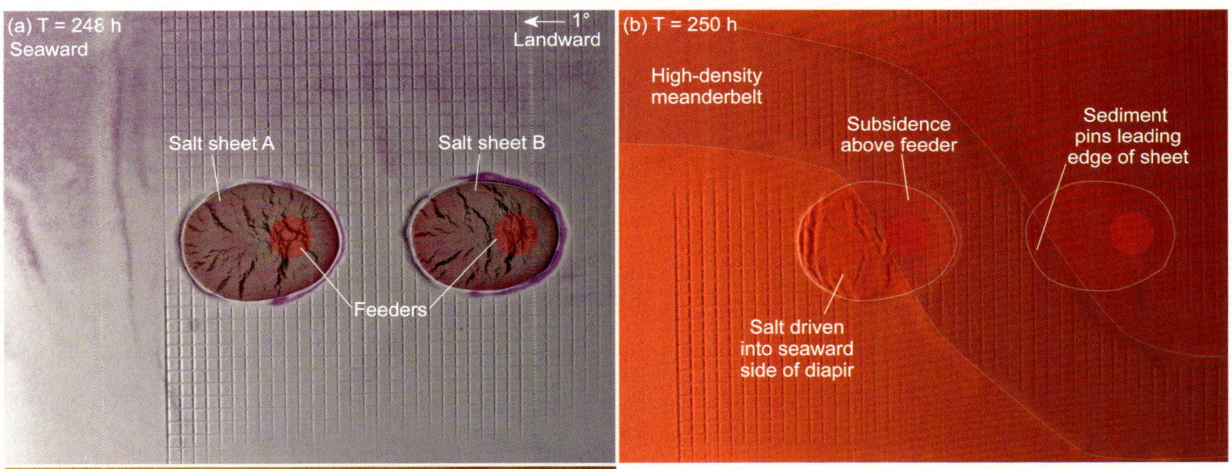

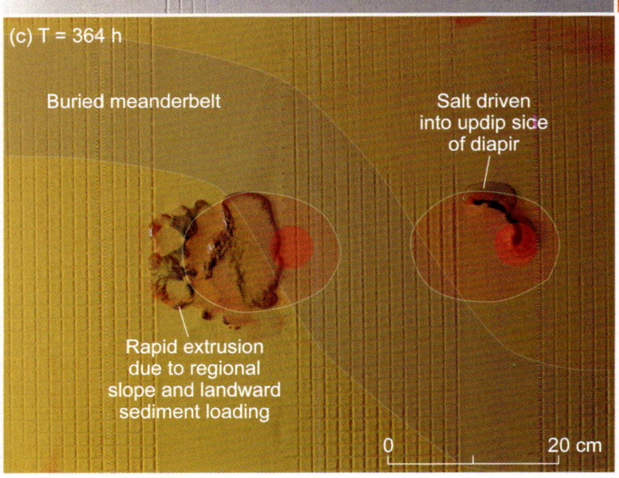

Figure 8.22. Subsidence into salt sheets. A meanderbelt of sediment denser than salt was deposited across an area containing two salt sheets. The belt crossed the landward end of salt sheet A and the seaward toe of salt sheet B. Subsidence of the belt into the updip end of salt sheet A accelerated seaward flow of salt within the sheet, forming a minibasin that sank into the landward end of the sheet above the feeder (see sections in Figure 8.23). Extrusion was rapid from the seaward end of the sheet because expulsion from beneath the minibasin and downslope salt flow both drove the salt seaward. In salt sheet B, the meanderbelt pinned the advancing toe of the sheet, forcing salt within the sheet to flow landward. Here the extrusion of salt was much less because minibasin expulsion and downslope flow drove salt in opposite directions. Model by Tim Dooley.

PHYSICAL MODEL—Subsidence of a Sediment Lobe into Salt Sheets

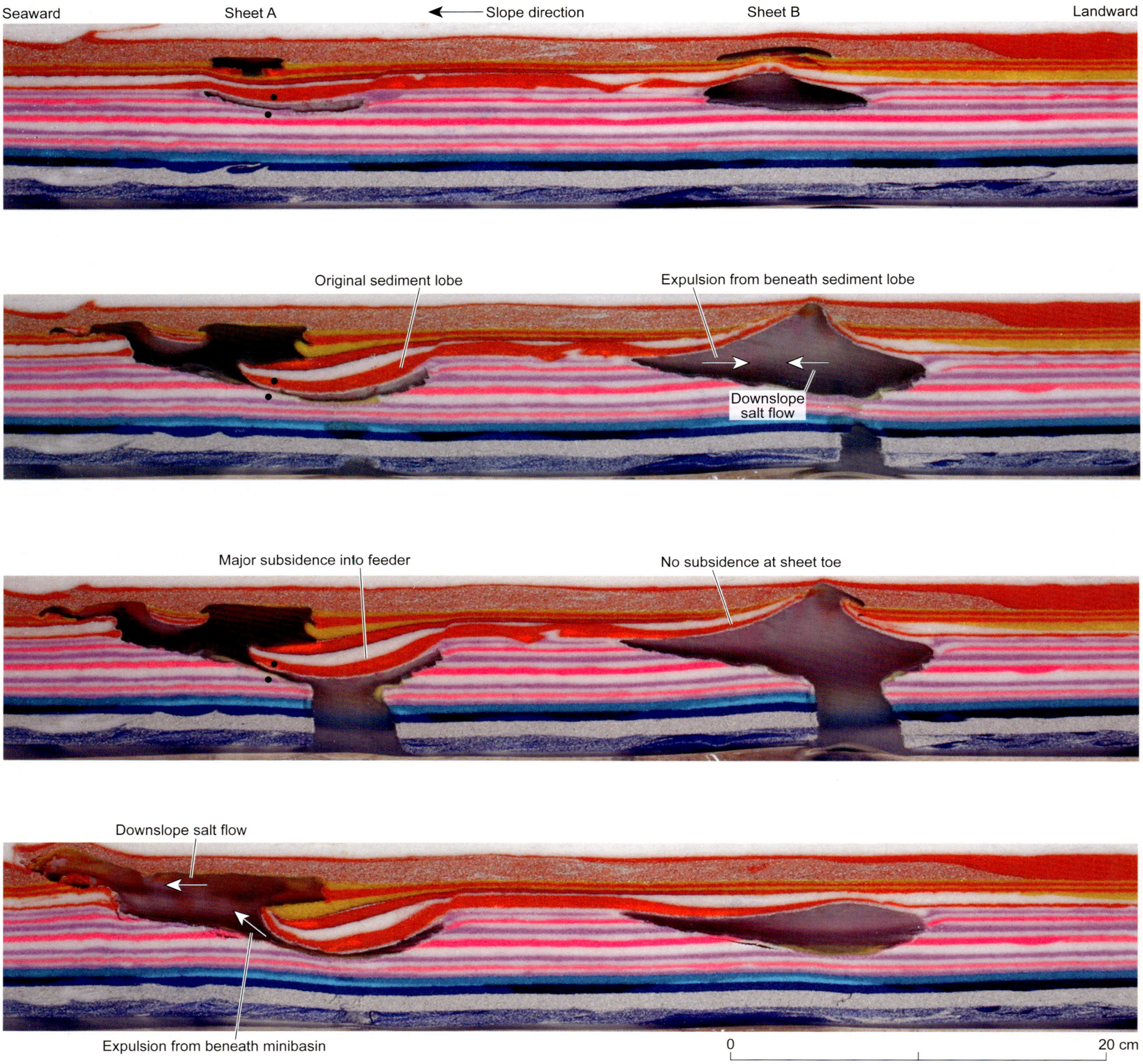

Figure 8.23. Subsidence into salt sheets. A meanderbelt denser than salt (deepest red layer) crossed two funnel-shaped salt sheets (see overhead views in Figure 8.22). The belt crossed the landward end of salt sheet A and the seaward toe of salt sheet B. In sheet A, salt was expelled efficiently from beneath the subsiding meanderbelt because minibasin subsidence and regional slope both drove salt seaward. As a result, the minibasin sank and welded against the base of the salt sheet. By contrast, in most areas the belt did not sink at all into sheet B because minibasin subsidence and regional slope drove salt in opposite directions. The meanderbelt blocked advance of sheet B but did not sink to form a minibasin, even though the belt was denser than the salt. Model by Tim Dooley.

NUMERICAL MODEL—Minibasin at Downdip End of Tilted Model

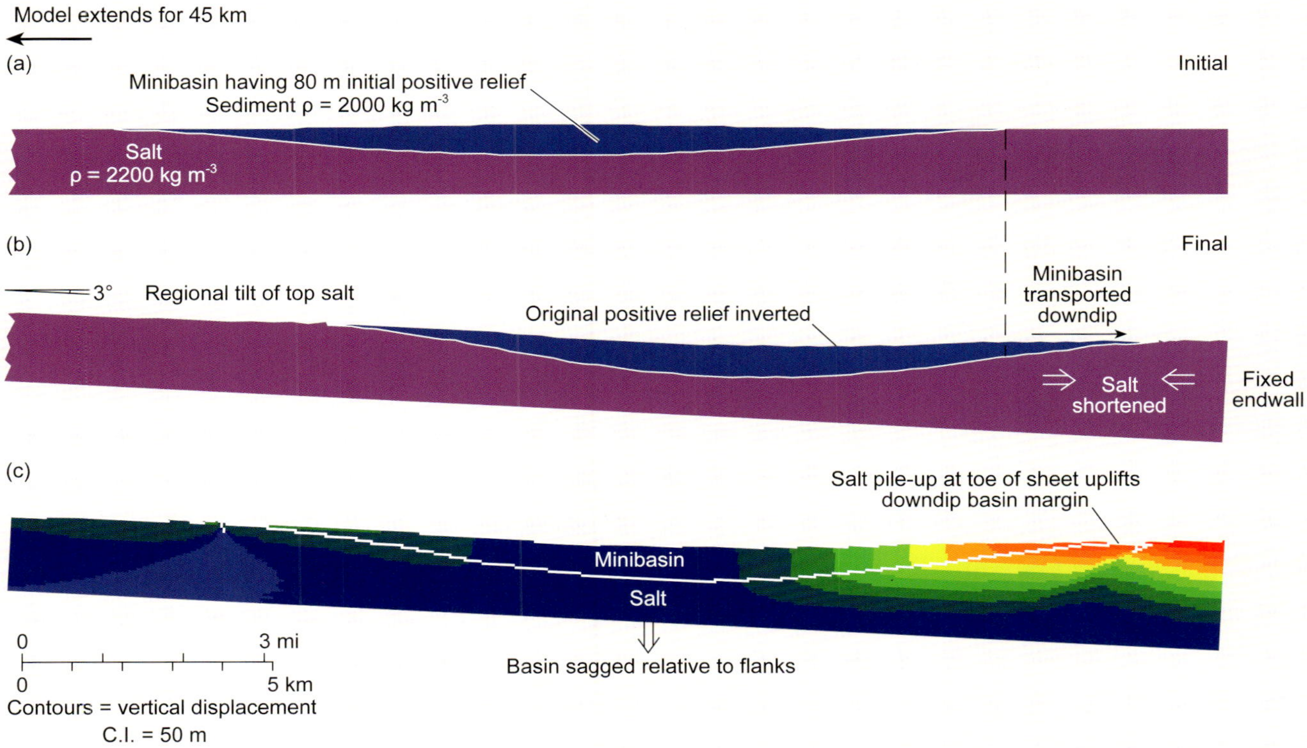

Figure 8.24. Minibasin shortening. Shortening caused subsidence of a minibasin that was less dense than salt. Tilting caused salt to flow downdip to the right, producing shortening at the downdip end of the model. Salt piled up beneath the downdip end of the minibasin, uplifting the margin and turning the minibasin into a bathymetric low. Compare with minibasin subsidence at the updip end of a salt sheet (Figure 8.25). Abaqus finite-element model modified from Hudec and others (2009).

NUMERICAL MODEL—Minibasin at Updip End of Tilted Model

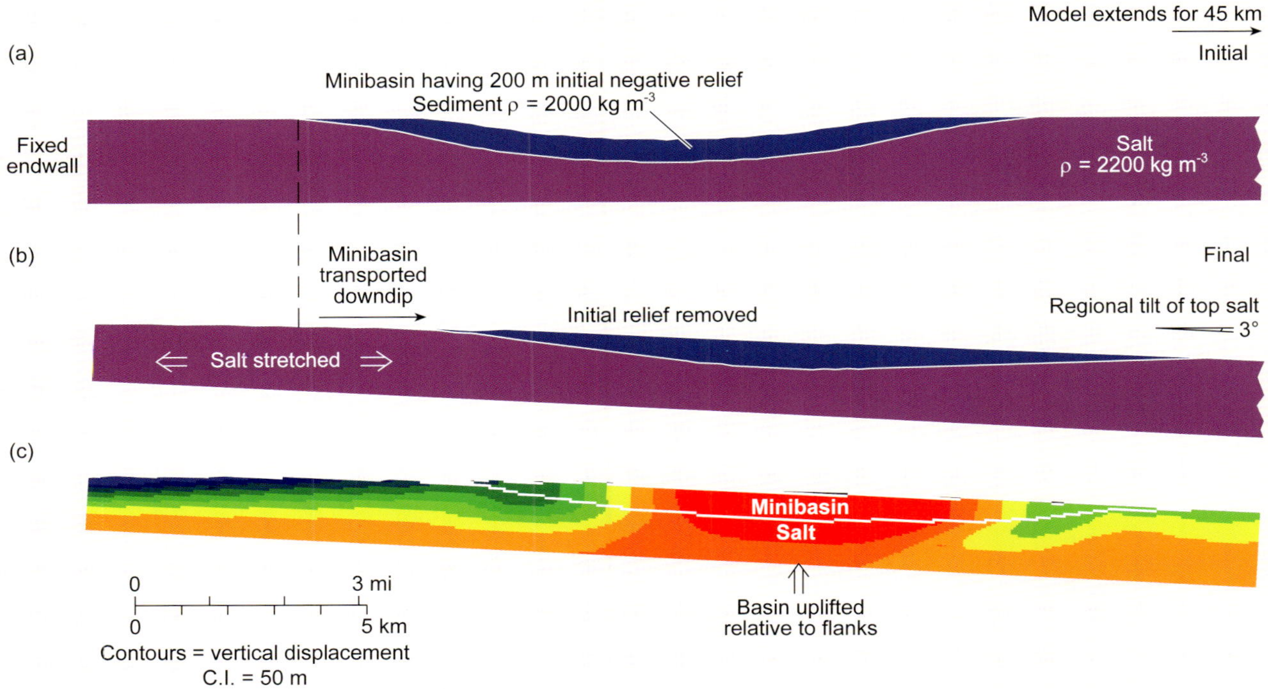

Figure 8.25. Minibasin extension. Extension caused uplift of a minibasin that was less dense than salt. Tilting caused salt to flow downdip to the right, producing extension at the updip end of the model. Extension thinned salt beneath the updip end of the minibasin, removing the initial basin sag and uplifting the minibasin center. Compare with minibasin subsidence at the downdip end of a salt sheet (Figure 8.24). Abaqus finite-element model modified from Hudec and others (2009).

CONCEPTUAL MODEL—Block Diagram of Minibasins above Allochthonous Salt

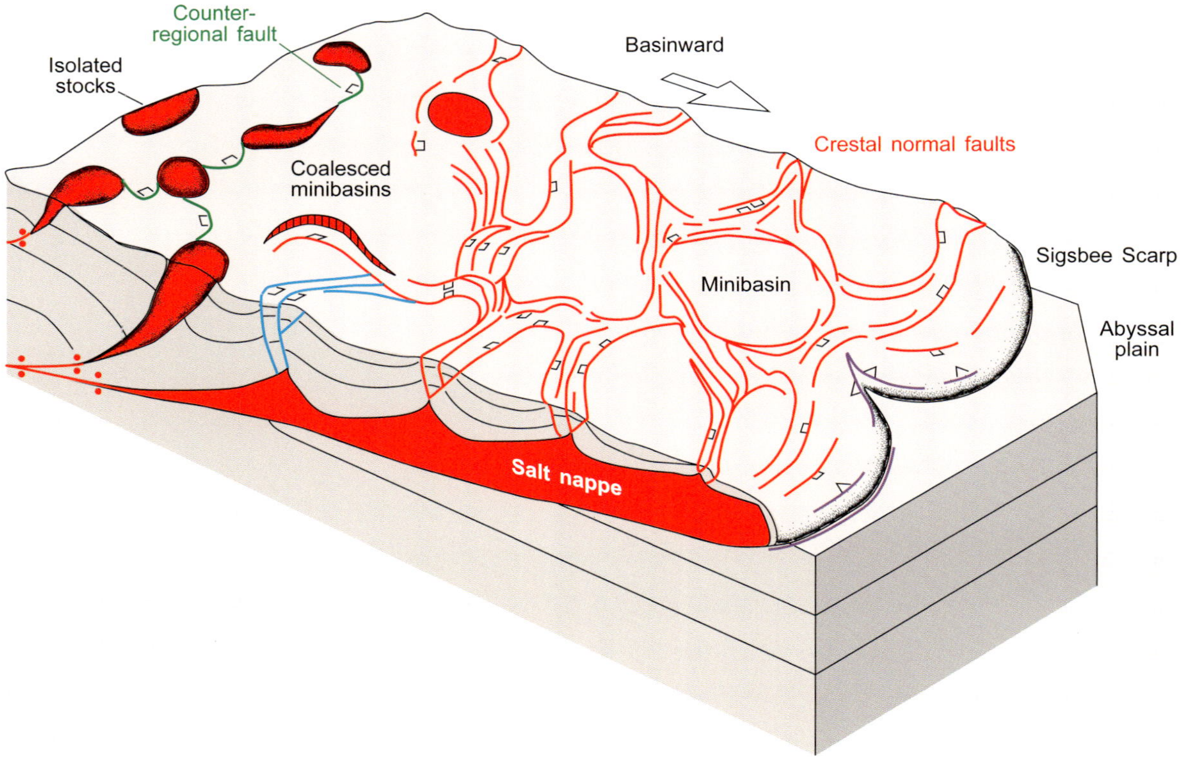

Figure 8.26. Subsidence into thick salt. Most minibasins above allochthonous salt are circular to elliptical in plan view. At early stages of development they are typically separated by a polygonal network of salt ridges. These ridges rise as reactive diapirs below extensional grabens, so the pattern of salt ridges is mimicked on the seafloor by swarms of normal faults. With additional sedimentation, this cellular pattern degenerates because mature minibasins tend to coalesce and merge into larger minibasin complexes. Modified from Rowan and others (1999).

CONCEPTUAL MODEL—Block Diagram of Salt-Stock Canopy

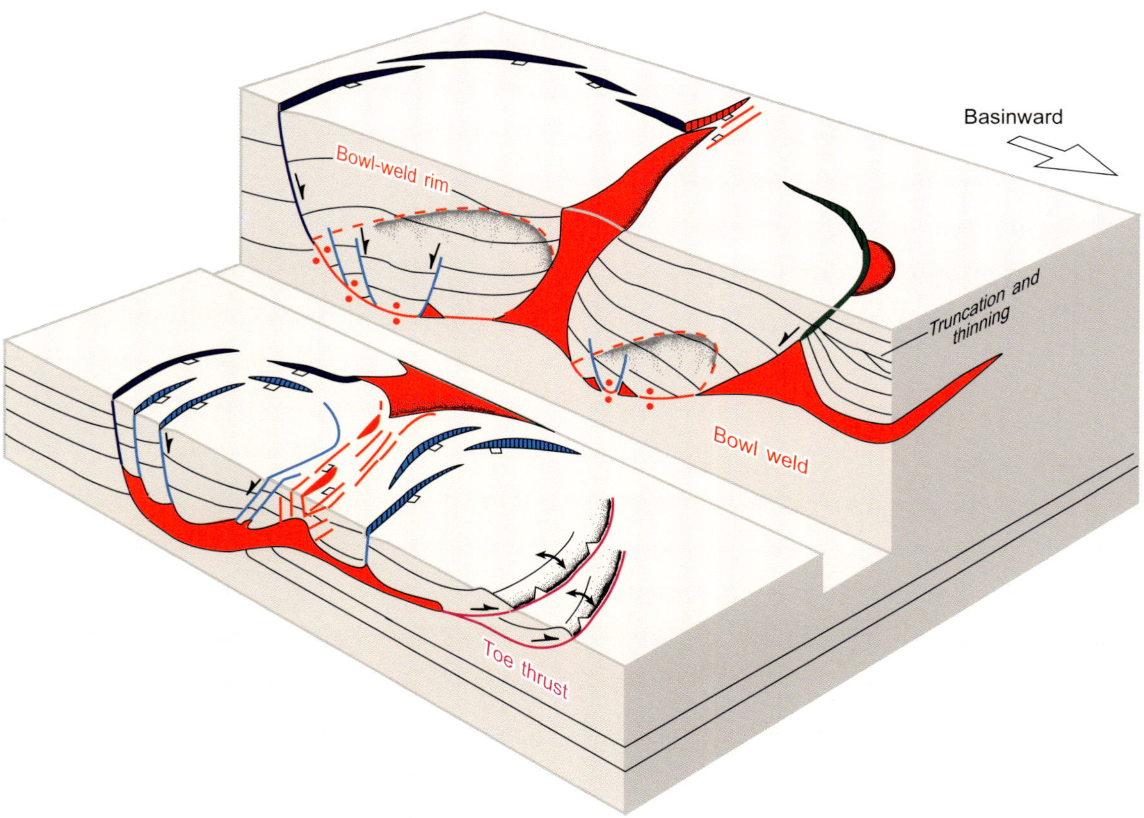

Figure 8.27. Geometry of a salt-stock canopy. A salt-stock canopy forms by coalescence of salt sheets fed by near-vertical salt stocks. The base of such a canopy is rugose, with large bowl-shaped depressions overlying the feeder stocks. As a salt-stock canopy is buried, minibasins subside into these depressions, displacing salt up onto base-salt ridges that separate the minibasins. The resulting pattern comprises elliptical basins separated by a polygonal network of salt diapirs. Modified from Rowan and others (1999).

CONCEPTUAL MODEL—Minibasins, Updip to Downdip Variations

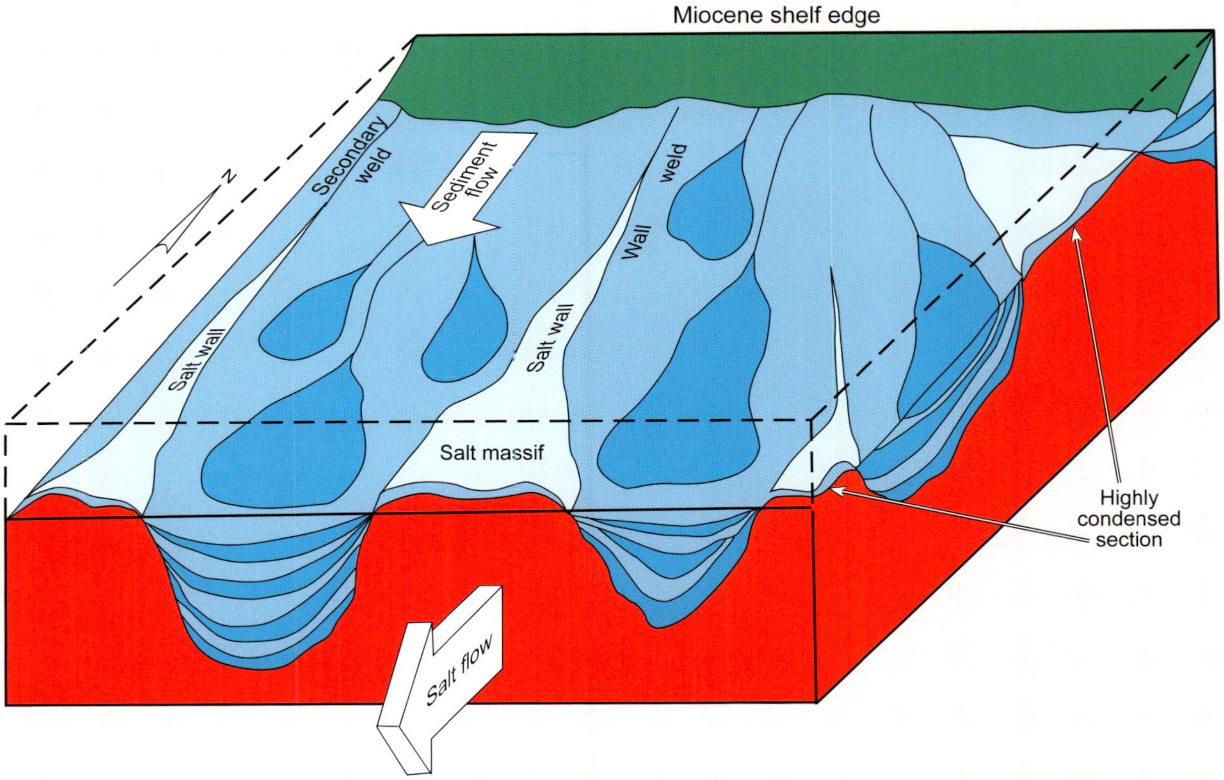

Figure 8.28. Progradation across thick salt. Immature minibasins are typically separated by broad salt massifs. In more mature areas, the diapirs separating minibasins are much narrower and may be welded. Hall and Thies (1995) suggested that in the Eugene Island/Ship Shoal/Green Canyon area of the Gulf of Mexico, subsiding minibasins have expelled salt from their bounding salt structures, narrowing them through time. Individual minibasins are separated by broad salt massifs near their distal, immature ends. As these basins are tracked north, the massifs separating them narrow into salt walls and eventually vertical secondary welds. They interpreted this sequence to indicate that salt is being expelled southward, along the strike of the diapirs. Modified from Hall and Thies (1995).

CONCEPTUAL MODEL—Minibasin Subsidence into a Feeder

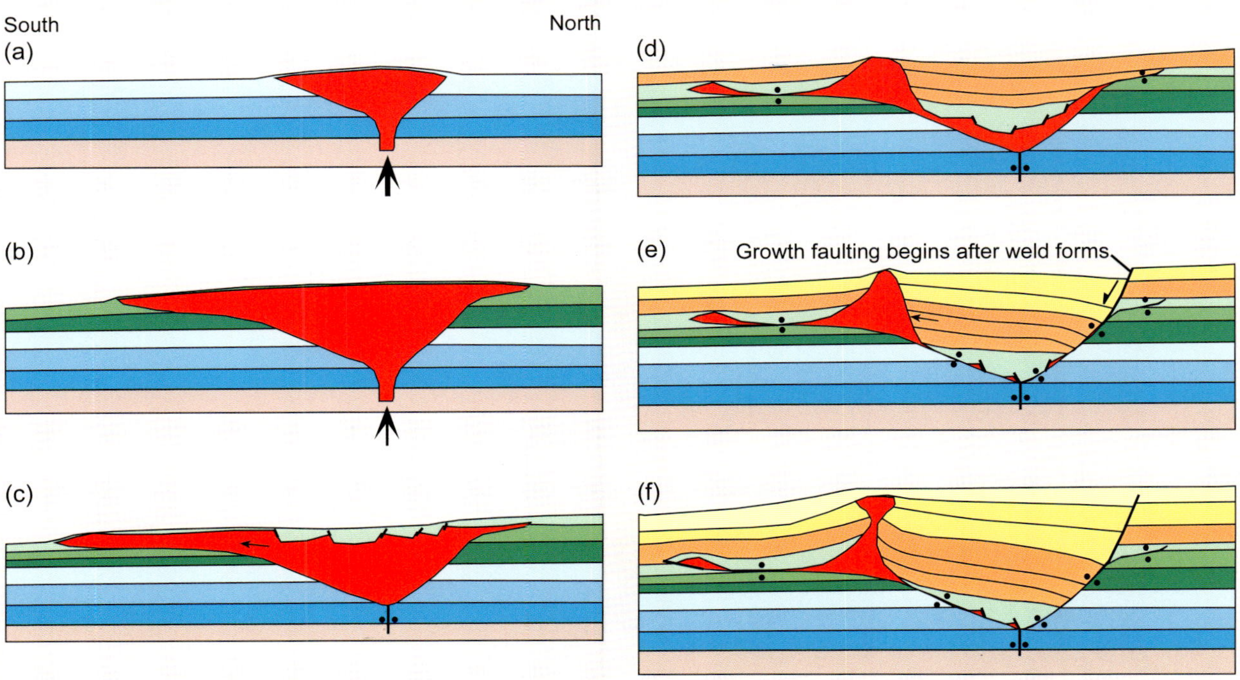

Figure 8.29. Minibasin subsidence. In this scenario, most of the salt sheet grows before the minibasin subsides. An alternative scenario would have the salt sheet growing during the main phase of minibasin subsidence, as salt displaced by the sinking minibasin extrudes over the seafloor. Modified from Rowan (1995).

CONCEPTUAL MODEL—Evolution of Sediment Pathways above Salt Sheets

Figure 8.30. Sedimentation above evacuating salt sheets. Extrusive salt forms bathymetric highs, tending to divert sediment fairways. Conversely, salt expulsion forms bathymetric lows and ponds sediment. Because salt sheets initiate extrusively and are then buried and evacuated, their effect on sedimentation patterns changes through time. Modified from Rowan and others (1995).

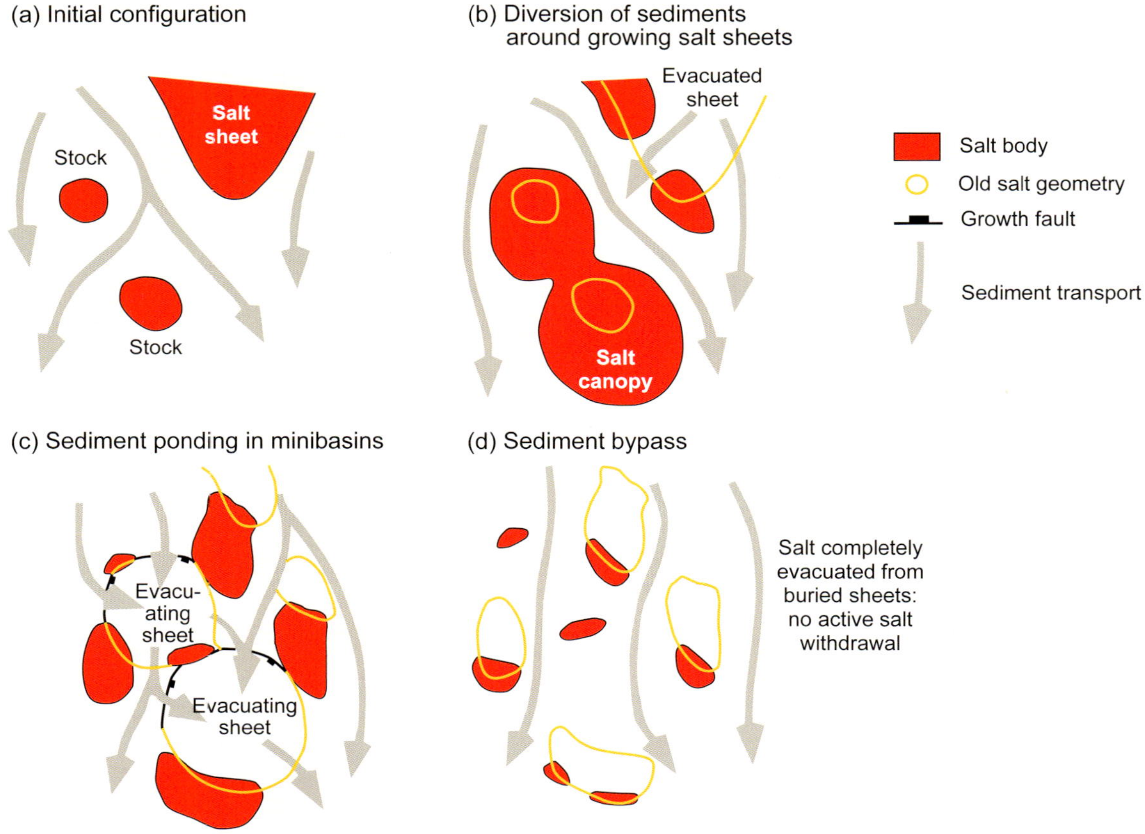

CONCEPTUAL MODEL—Minibasin Bathymetry during Density-Driven Subsidence

Figure 8.31. Uplift of minibasin flank. If a minibasin is in isostatic equilibrium, the flank will normally form a bathymetric high. This is because thinner parts of the minibasin have a lower average density and so float higher on the salt. If the average density of a portion of the minibasin is less than salt density, that portion of the minibasin will float above the level of the adjacent salt massif. See seismic example of this geometry in Figure 8.11. Modified from Hudec and others (2009).

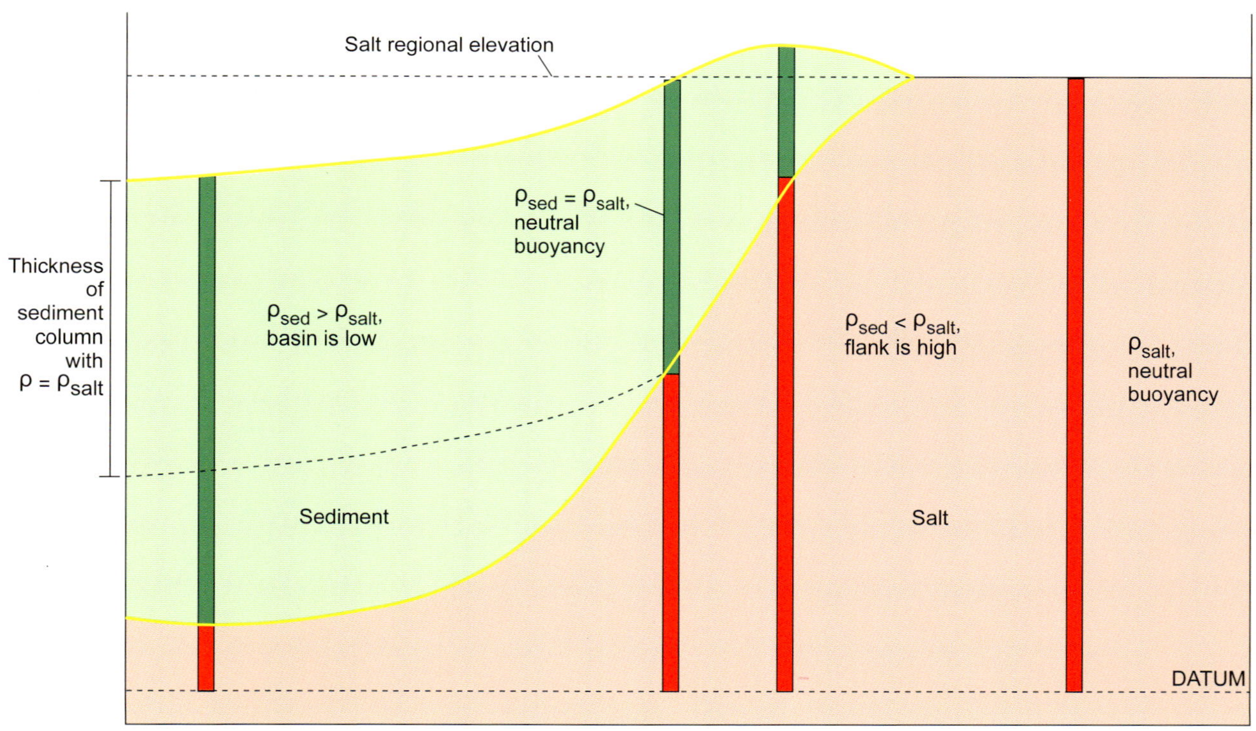

Chapter 9

Roho Systems

The classic linked kinematic system on a passive margin has an extensional zone on the outer shelf and upper slope, coupled with a contractional zone at the toe of the continental slope. On some salt-bearing passive margins, however, extension and contraction are more complexly partitioned. Salt canopies act as independent, high-level salt layers, each having its own zones of updip extension and downdip shortening, together known as a *roho system*. Where roho systems are present, the regional distribution of extension and contraction reflects stacked detachment systems, each having its own kinematics.

A roho system is a group of listric, basinward-dipping growth faults that sole out onto an allochthonous salt sheet or salt weld. Sediment wedges in the fault blocks dip and thicken landward but become younger seaward. Like minibasins and counterregional systems, they form as salt is evacuated from an allochthonous salt sheet.

Roho systems form as a roof extends above salt flowing seaward in an allochthonous salt sheet. This extension may be compensated by thrusting at the toe of a buried sheet. However, at an earlier stage the extension is absorbed by narrowing of the exposed toe of the sheet, leaving no record of shortening in roof strata despite large magnitudes of extension. Most normal faults in a roho system have salt rollers in their footwalls. Strike-slip fault systems form the lateral edges of rohos, separating seaward-translated strata in the roho from surrounding strata that have not moved.

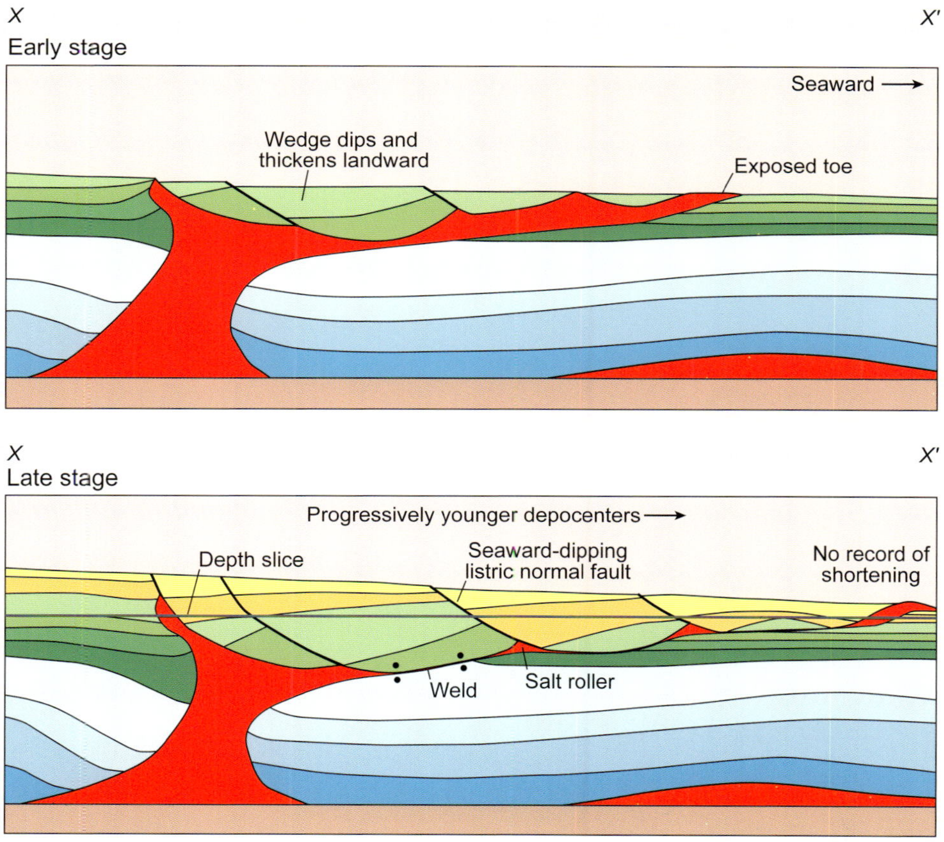

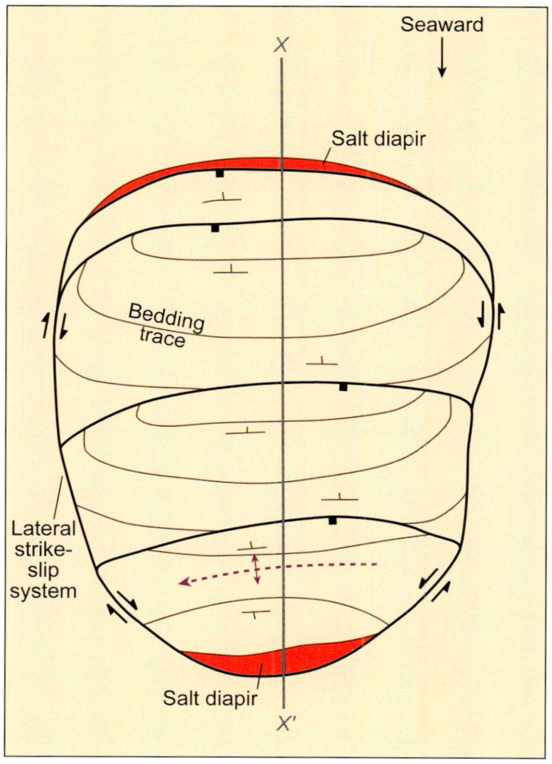

GULF OF MEXICO BASIN—Roho System, West Cameron, Offshore USA

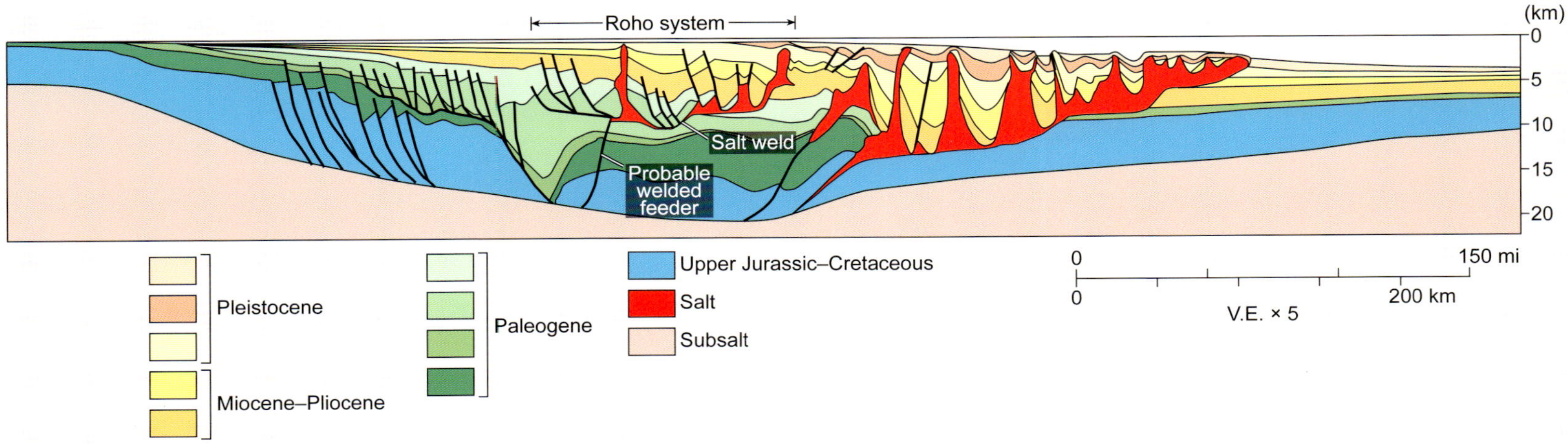

Figure 9.01. Extension above a salt sheet. The presence of salt structures along the subhorizontal detachment indicates that the detachment for the roho is actually a salt weld. Much of the salt originally in the sheet on which the roho system formed was expelled from the diapirs at the downdip end of the system and dissolved. The roho system was derived from a salt sheet emplaced during the Paleogene, probably via the steep salt weld connecting it to autochthonous salt at the base of the Upper Cretaceous. The sedimentary roof of the roho system extended from the late Paleogene to the Pleistocene. Modified from Peel and others (1995).

GULF OF MEXICO BASIN—Roho System, Whiting Dome, Offshore USA

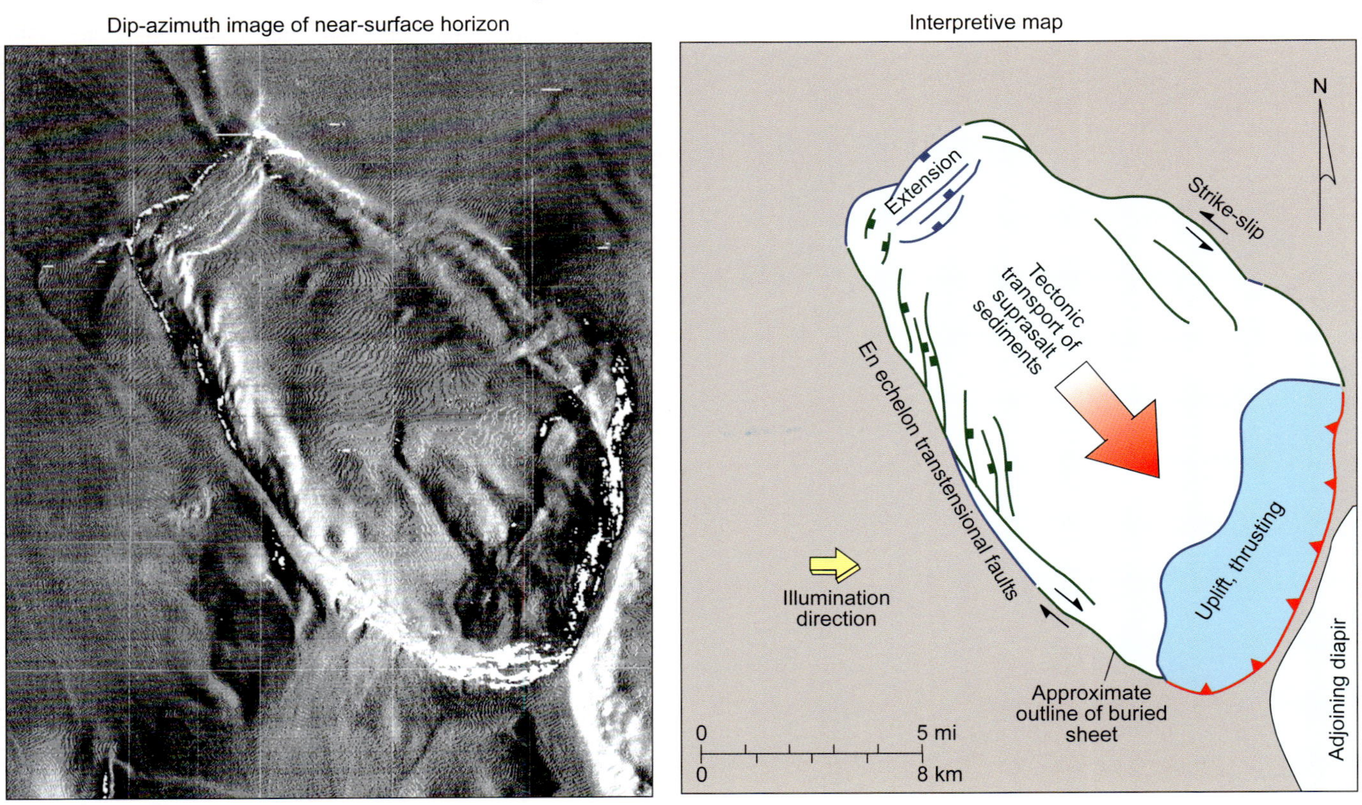

Figure 9.02. Extension and shortening above a salt sheet. Deformation of a shallowly buried salt sheet is imprinted on the structure of its sedimentary roof. Extension at the rear (northwest) of the sheet is balanced by uplift and shortening at the front (darker blue area). This translation of the roho roof relative to surrounding sediments is accommodated by strike-slip and transtensional lateral structures. Modified from Peel and others (1995).

GULF OF MEXICO BASIN—Rum Roho System, Offshore USA

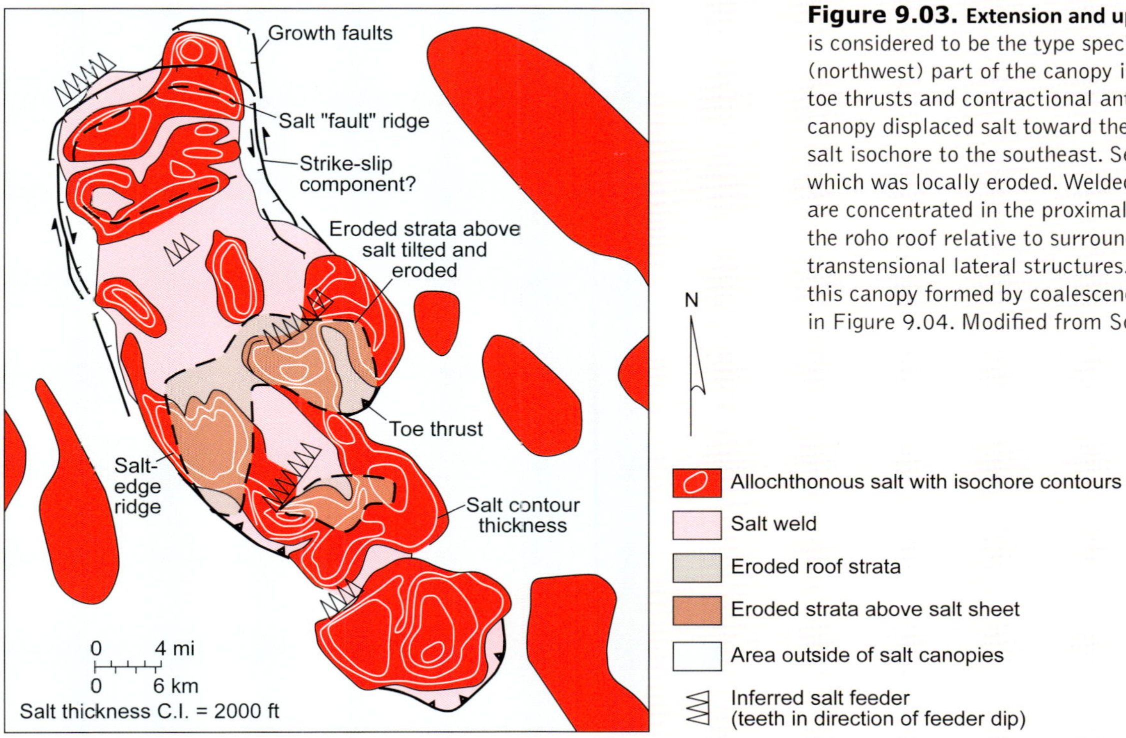

Figure 9.03. Extension and uplift above a salt canopy. This small salt canopy is considered to be the type specimen of a roho system. Extension in the landward (northwest) part of the canopy is balanced distally by shortening in the form of toe thrusts and contractional anticlines. Basin subsidence near the updip end of the canopy displaced salt toward the seaward end of the canopy, reflected in the thicker salt isochore to the southeast. Seaward transport of salt uplifted the canopy roof, which was locally eroded. Welded areas from which salt was largely expelled (pink) are concentrated in the proximal and lateral parts of the roho system. Translation of the roho roof relative to surrounding sediments is accommodated by strike-slip and transtensional lateral structures. The number of inferred salt feeders suggests that this canopy formed by coalescence of at least five diapirs. See section restoration in Figure 9.04. Modified from Schuster (1995).

GULF OF MEXICO BASIN—Rum Roho System, Offshore USA

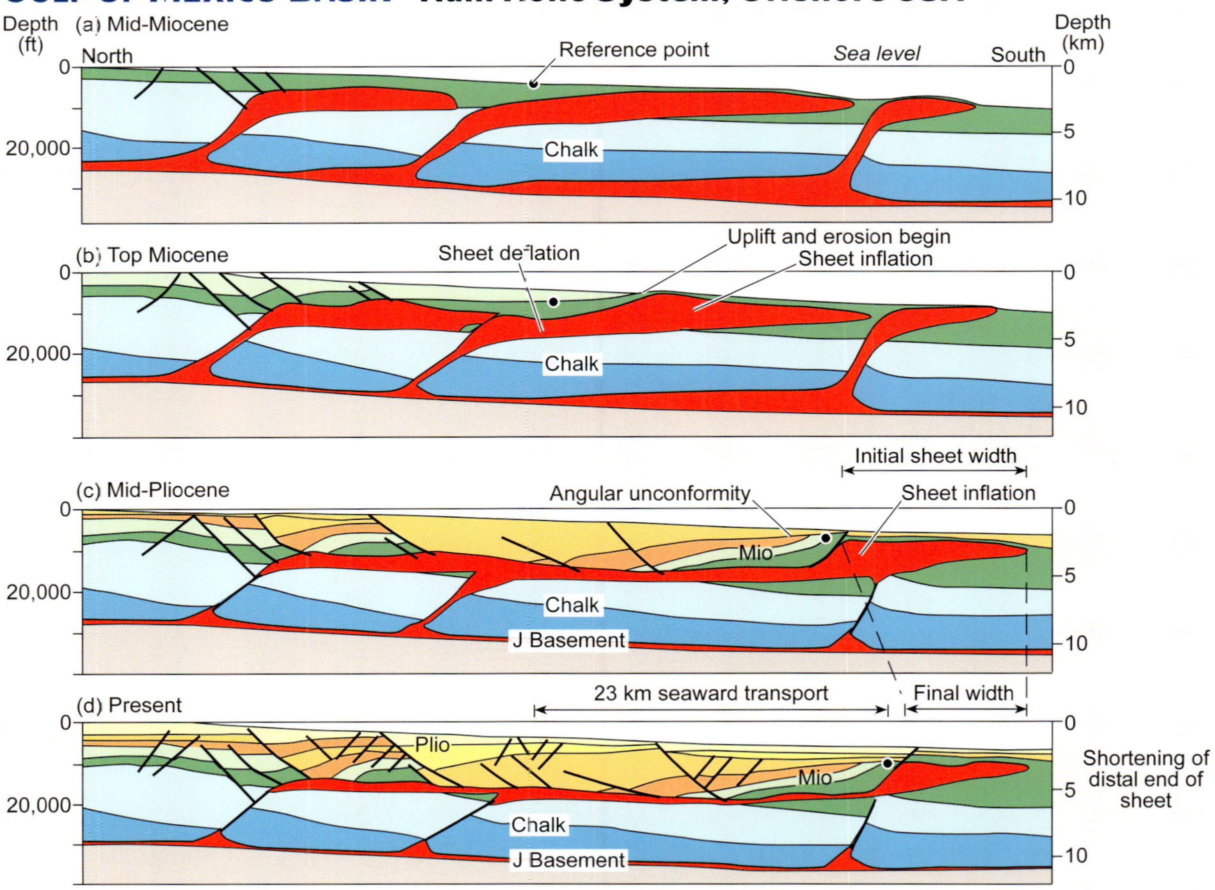

Figure 9.04. Extension and uplift above a salt canopy. This small salt canopy is considered to be the type specimen of a roho system. Extension at the landward (northwest) end of the sheet is balanced distally by shortening at the distal end of the sheet. The reference point tracks at least 23 km of extension and seaward translation. This large amount of extension, together with expulsion of underlying salt, created space for the anomalously thick Pliocene fill within the roho system. The salt was expelled from beneath the subsiding minibasins moving south, inflating the distal end of the sheet. Three of the inferred salt feeders for this canopy are shown in these sections. See map in Figure 9.03. Modified from Schuster (1995).

GULF OF MEXICO BASIN—Roho System, West Cameron South, Offshore USA

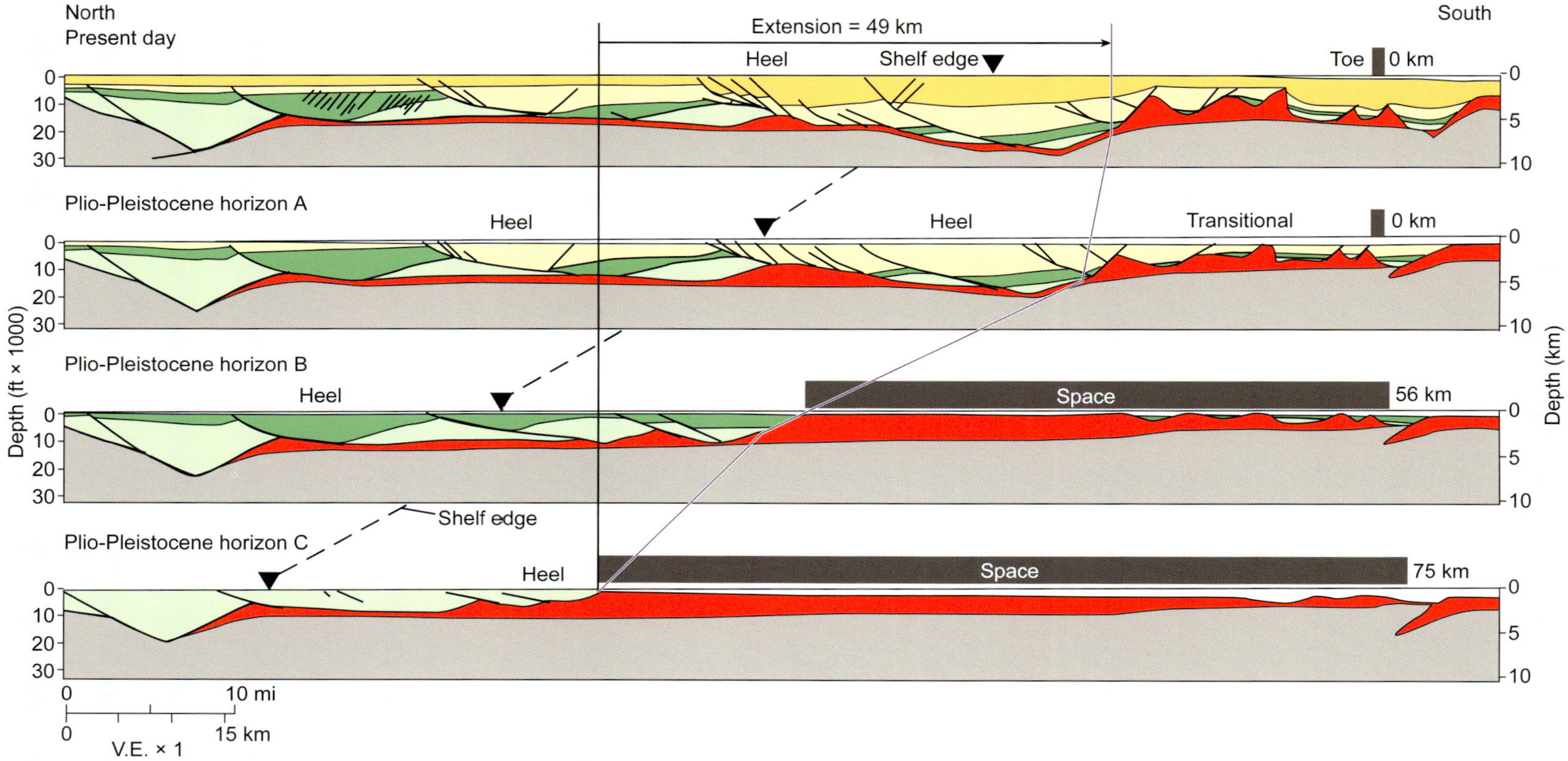

Figure 9.05. Extension above a salt canopy. The restoration shows huge extension (49 km) in the roof of this roho system. The purple line tracks the seaward translation of a point. Extension was accommodated by displacement of emergent salt (probably with a sedimentary veneer). "Space" refers to the width of this collapsible zone of emergent salt. "Heel" denotes landward-thickening sediment wedges typical of roho systems. Conversely, "toe" denotes seaward-thickening wedges typical of counterregional systems. Base of salt through time is schematic and is likely to have been much more irregular. Modified from Diegel and others (1995).

GULF OF MEXICO BASIN—Roho System, Mahogany, Offshore USA

Figure 9.06. Extension and shortening above a salt sheet. The map depicts the structure on top of salt; reddish colors are highest. The southward-dipping normal faults define a roho system in the north. These fault traces curve outward to align with the margins of the salt sheet. These margins are likely to have a strong component of strike slip, which is sinistral in the east and dextral in the west. Modified from Rowan and others (1999).

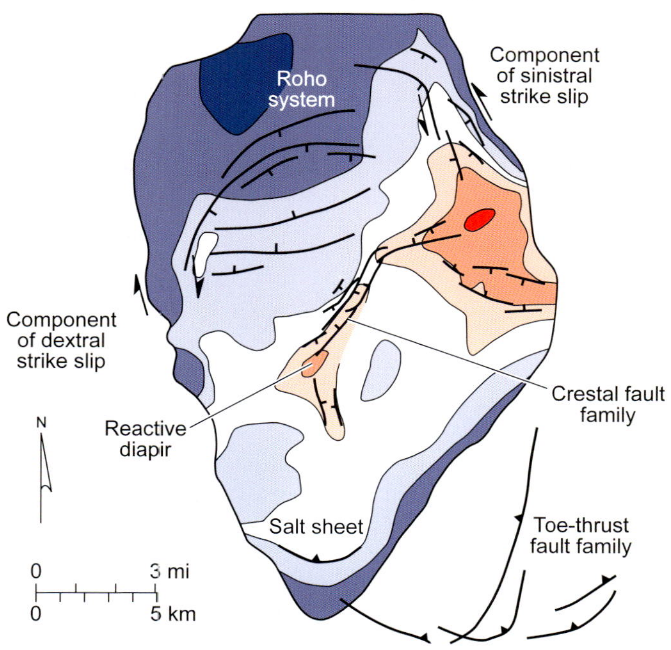

GULF OF MEXICO BASIN—Roho System, South Louisiana, USA

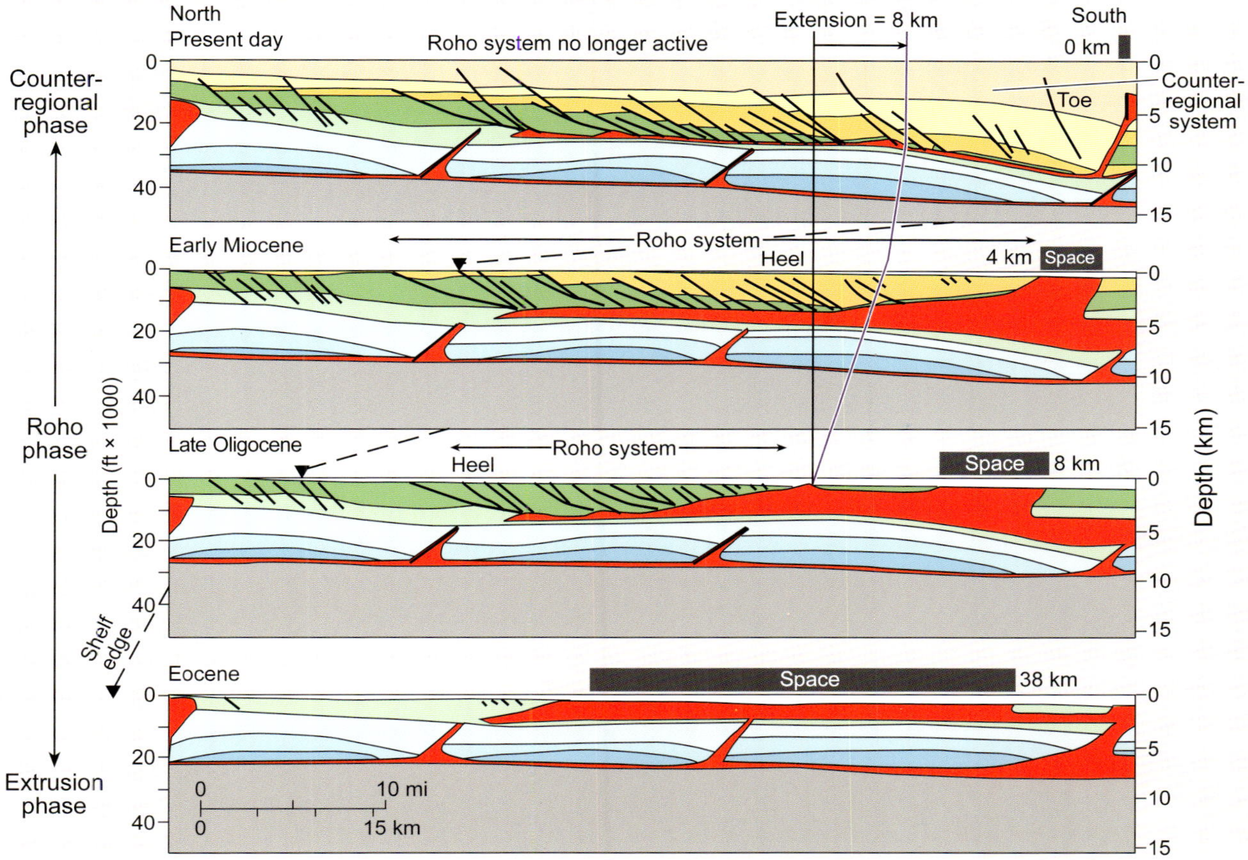

Figure 9.07. Extension above a salt canopy. The restoration shows a roho system that evolved into a counterregional system. During the roho phase, Oligocene and early Miocene strata extended, forming a series of landward-thickening sediment wedges. Extension was facilitated by emergent salt farther seaward, which acted as a collapsible "space." During the subsequent counterregional phase, this collapsible "space" shrank as the emergent salt was buried. Extension was hindered, so mid-Miocene and younger strata (pale-yellow, pale-brown) merely subsided by displacing underlying salt. This salt evacuation formed a counterregional system comprising seaward-thickening sediment wedges. The purple line tracks the seaward translation of a point through time. "Heel" denotes the landward-thickening sedimentary wedges typical of roho systems. Conversely, "toe" denotes seaward-thickening wedges typical of counterregional systems. Modified from Diegel and others (1995).

GULF OF MEXICO BASIN—Roho System, Terrebonne Trough, South Louisiana, USA

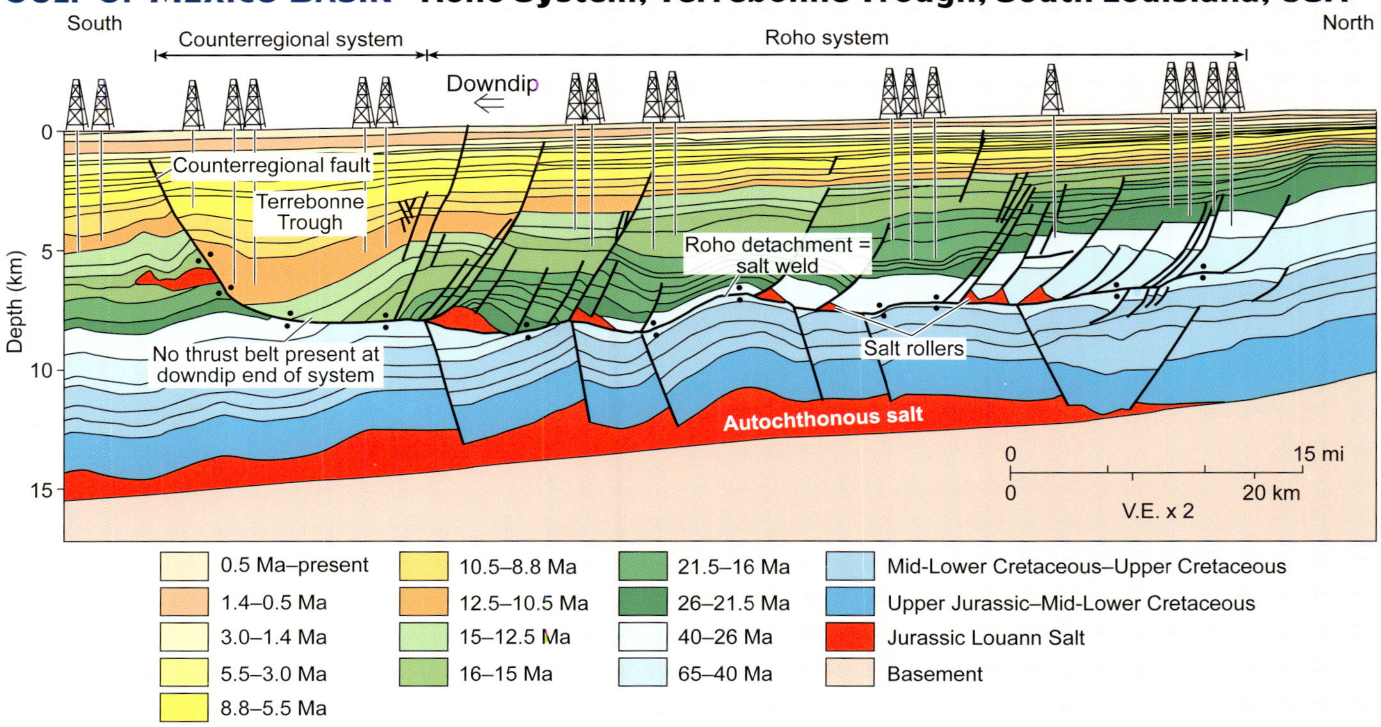

Figure 9.08. Extension and subsidence above a salt canopy. The section shows a transition from an updip (northern) roho system to a downdip (southern) counterregional system, forming the Terrebonne Trough. Most of the precursor salt canopy was expelled along the roho base, leaving only isolated salt rollers. No signs of shortening are visible in strata of the Terrebonne Trough. Thus, extension in the roho systems was accommodated by lateral closure of a counterregional diapir, which is now represented by a counterregional weld. See restoration in Figure 9.09. Modified from McBride (1998).

GULF OF MEXICO BASIN—Roho System, Terrebonne Trough, Offshore USA

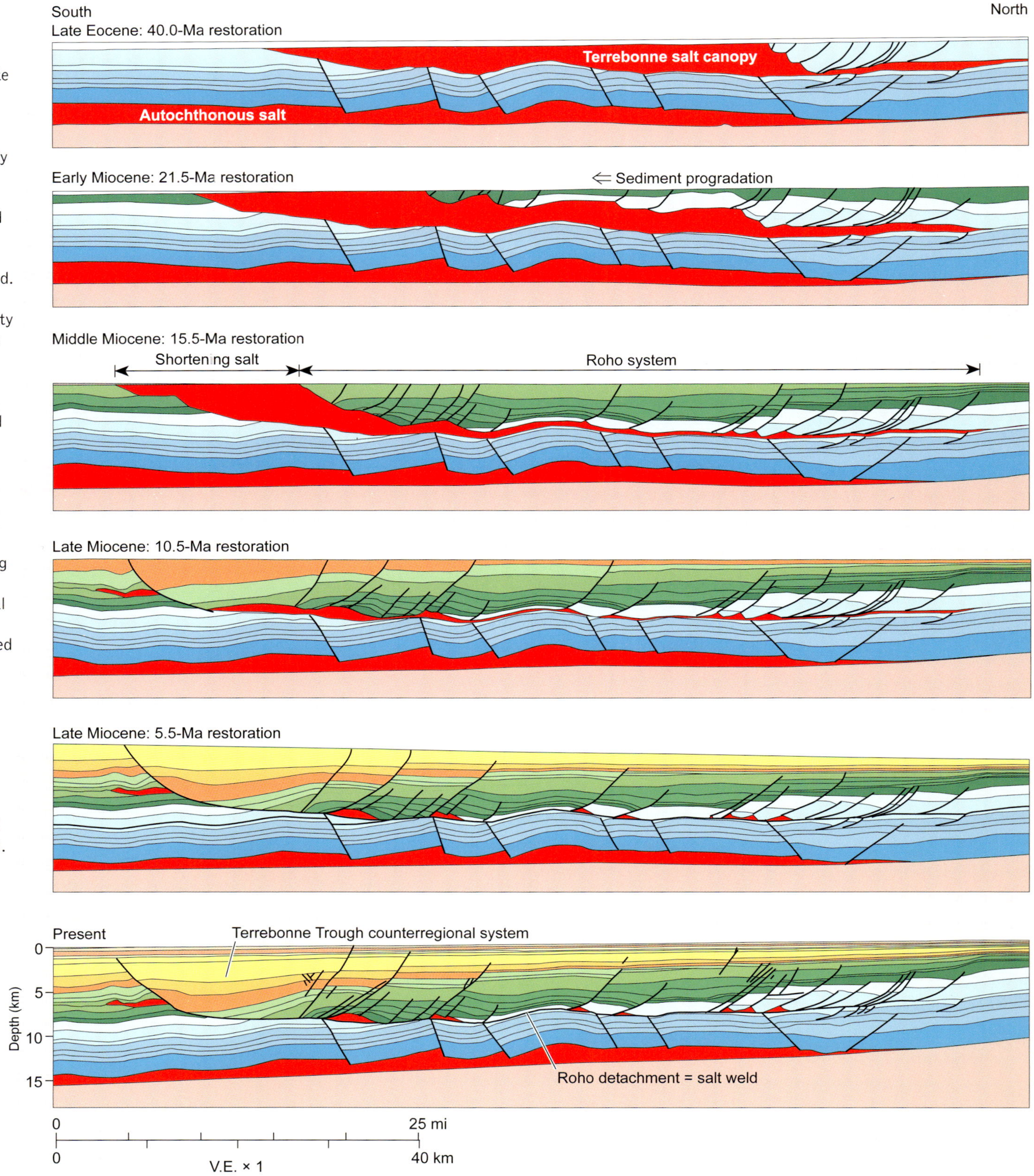

Figure 9.09. Extension and subsidence above a salt canopy. By the late Eocene, a salt canopy was emplaced above stretched Mesozoic strata. During the early to middle Miocene, progradation expelled salt southward toward the toe of the canopy. Sediments deposited above the salt extended. This stretching was driven mostly by gravity spreading but perhaps also by traction-imposed southward flow of the underlying salt. Sediments spread seaward into the adjacent salt massif. In the late Miocene, the residual emergent salt sheet in the south was finally buried as salt was expelled along strike or dissolved to form a counterregional system. Removal of this salt accommodated the Terrebonne Trough. Most of the precursor salt canopy was expelled along the base of the roho system, leaving only isolated salt rollers. See section in Figure 9.08. Modified from McBride (1998).

PHYSICAL MODEL—Roho System Section 5

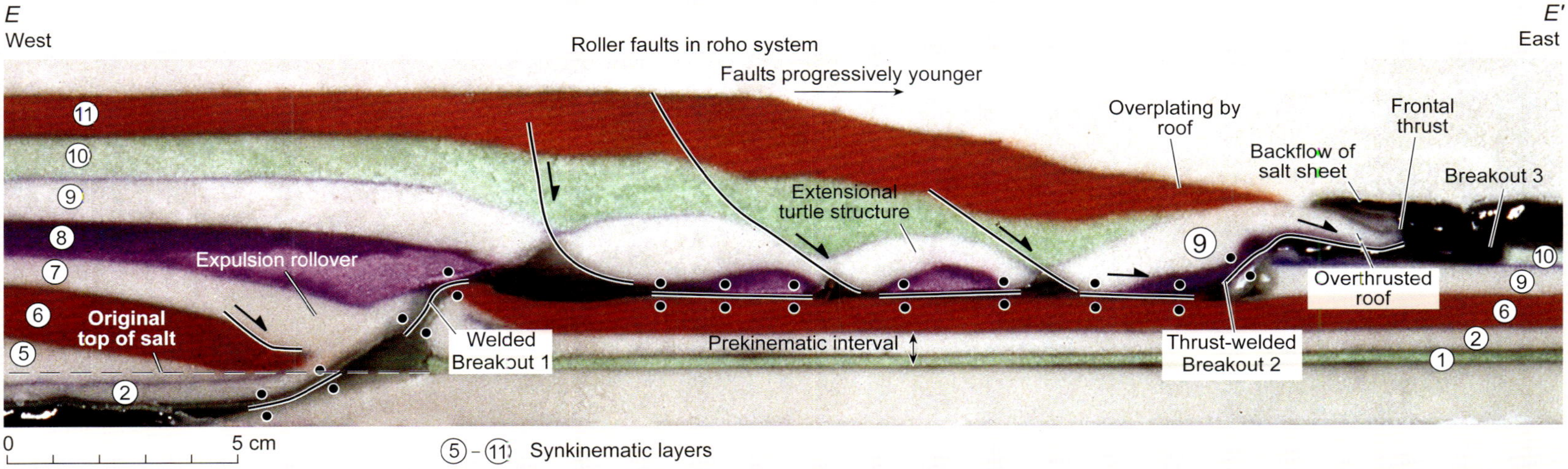

Figure 9.10. Extension above a salt sheet. The section shows the full extent of a roho system detached on a salt sheet extruded through Breakout 1. Several features characteristic of roho systems are well displayed: (1) listric normal faults soling out on a subhorizontal salt weld; (2) landward-thickening, seaward-younging sediment wedges; and (3) overthrusted roof of white unit 9 in the roho toe. The distal toe of the roho system climbed through Breakout 2 and Breakout 3, which mark the progressive escape of salt expelled from beneath the prograding load. See also sections 1 and 2 from the same experiment (Figures 9.11 and 9.12). Model by Bruno Vendeville.

PHYSICAL MODEL—Roho System Section 2

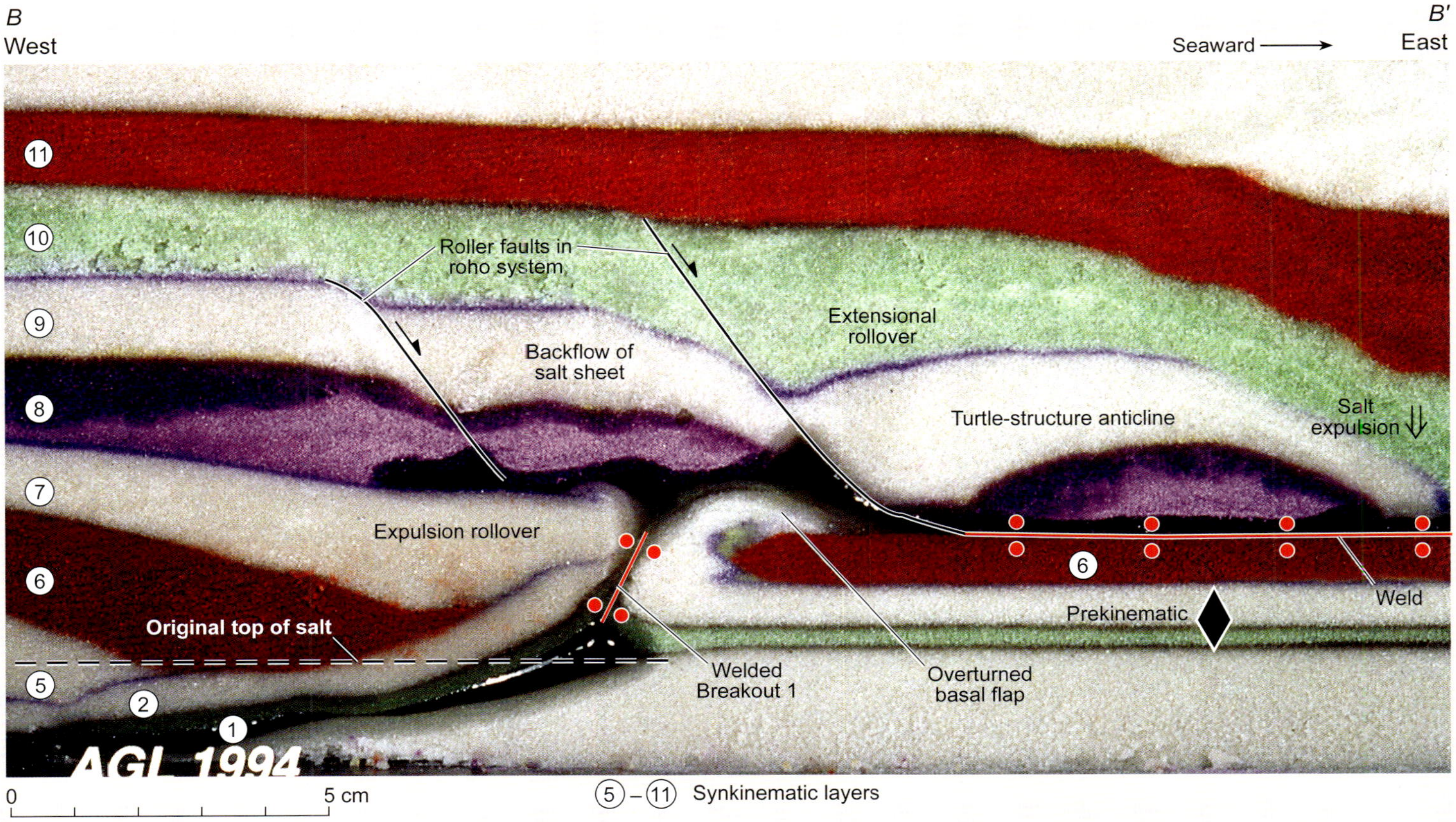

Figure 9.11. Extension above a salt sheet. The section shows details of a roho system that formed above a salt sheet extruded through Breakout 1, which is here welded. The distal (eastern) block of the roho contains a turtle-structure anticline cored by the purple layer 8. The turtle formed by a combination of extensional rollover (west flank) and salt expulsion (east flank). The proximal (western) block of the roho detached on allochthonous salt that extruded landward as a backflow into an expulsion rollover. See also sections 1 and 5 from the same experiment (Figures 9.10 and 9.12). Model by Bruno Vendeville.

PHYSICAL MODEL—Roho System Section 1

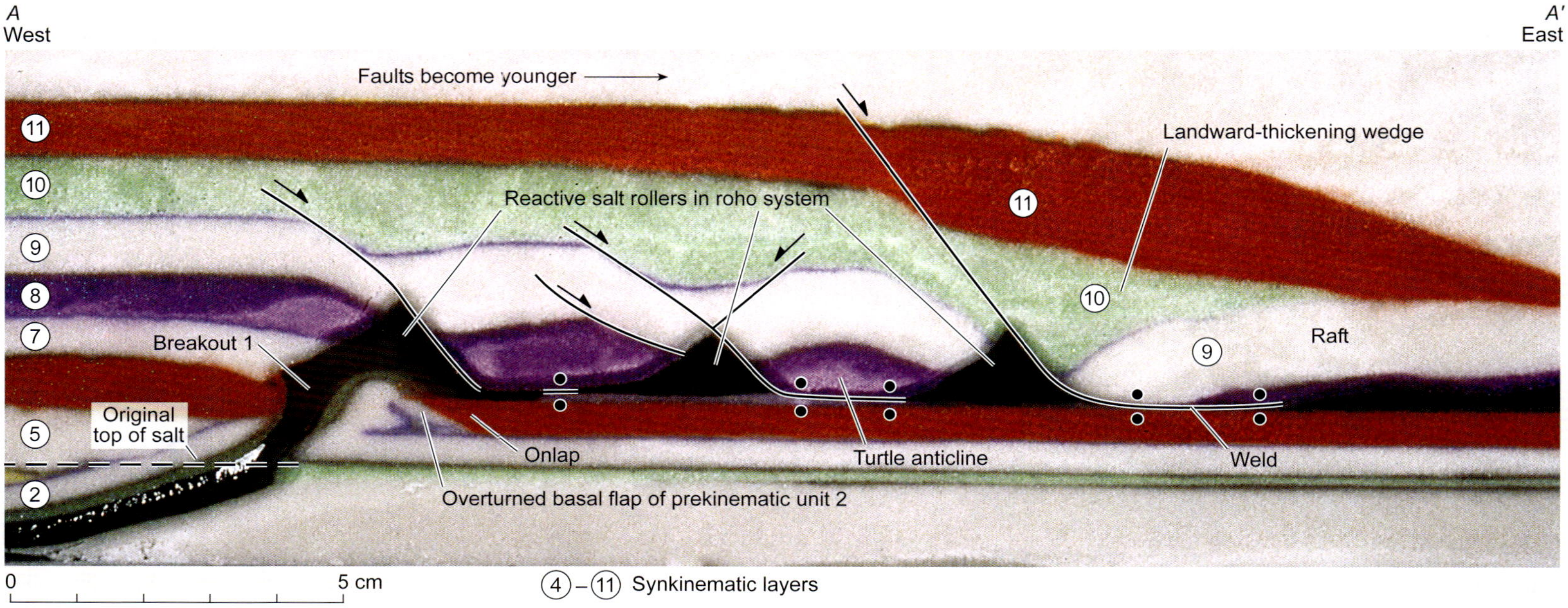

Figure 9.12. Extension above a salt sheet. The section shows details of a roho system that formed above a salt sheet extruded through Breakout 1. As typical of roho systems, roller faults became younger seaward because extension migrated seaward, along with progradation. The salt detachment was mostly welded, having remnant salt preserved as salt rollers in the footwall of each fault. Full section is longer than shown. See also sections 2 and 5 from the same experiment (Figures 9.11 and 9.10). Model by Bruno Vendeville.

CONCEPTUAL MODEL—Allochthonous Salt Geometries and their Resulting Allochthonous Salt Systems

Figure 9.13. Evacuation of allochthonous salt canopies. Evacuation of salt from buried canopies dominates most deformation in the northern Gulf of Mexico. Burial of asymmetric tongues or salt-tongue canopies can produce either stepped counterregional or roho systems. Formation of roho systems is favored by slope on the sediment surface, which drives seaward translation, and by long salt canopies, which provide space for a linked kinematic system to form. Burial of coalesced salt stocks produces a salt-stock canopy. Modified from Rowan and others (2001).

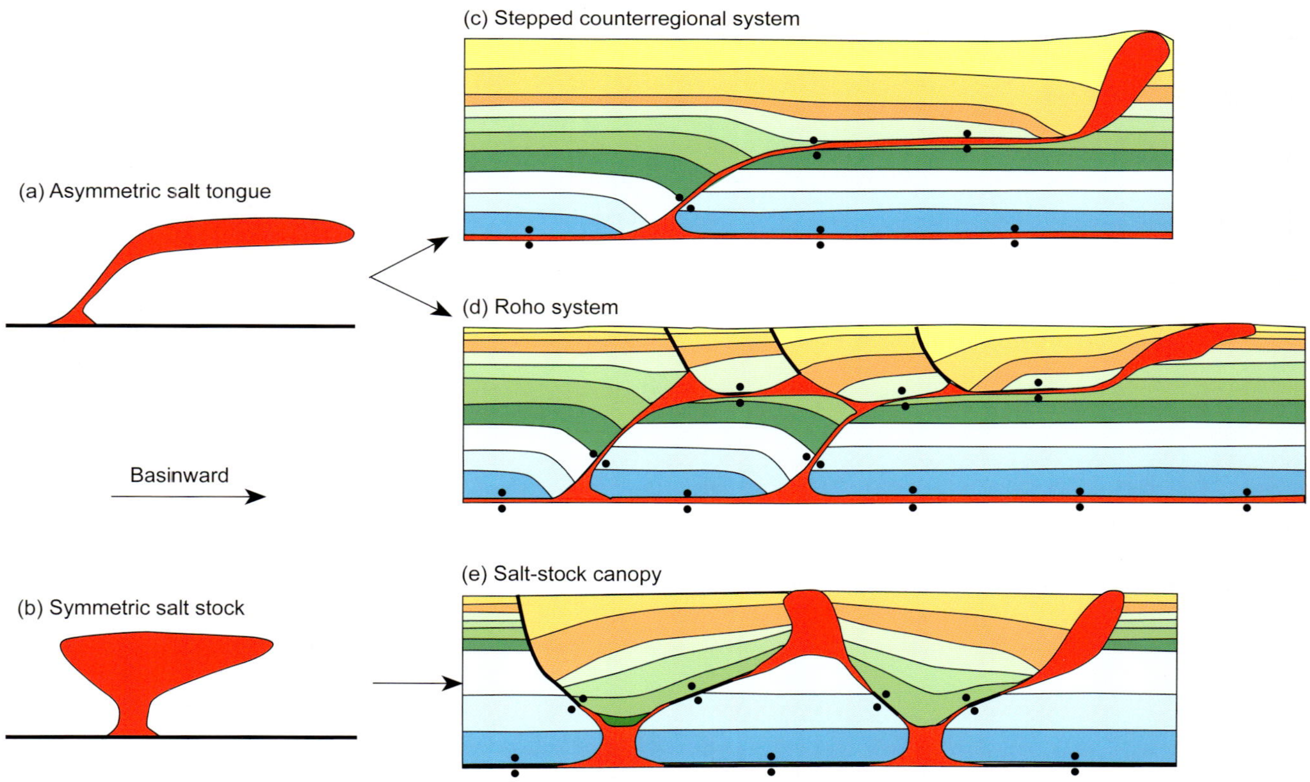

CONCEPTUAL MODEL—Location of Allochthonous Salt Systems on a Passive Margin

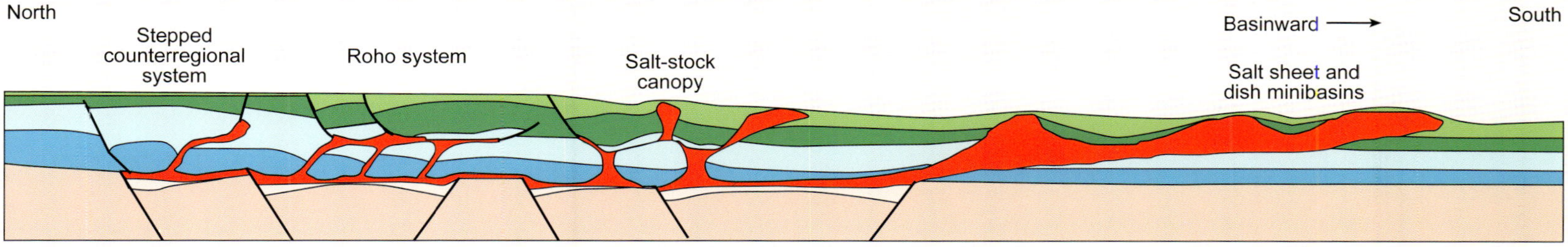

Figure 9.14. Allochthonous salt geometries. Landward allochthonous salt systems tend to be older and to have less preserved salt. Most salt sheets are welded, so there is little active subsidence and deformation. Farther seaward, salt systems are still actively deforming. Many salt sheets in deep water still have thick salt, and active deformation results in significant bathymetric relief. Stepped counterregional and roho systems are typical of the shelf and upper slope. Salt-stock canopies and thick salt sheets are typical of the continental slope. Modified from Rowan and others (1999).

CONCEPTUAL MODEL—Components of a Roho System

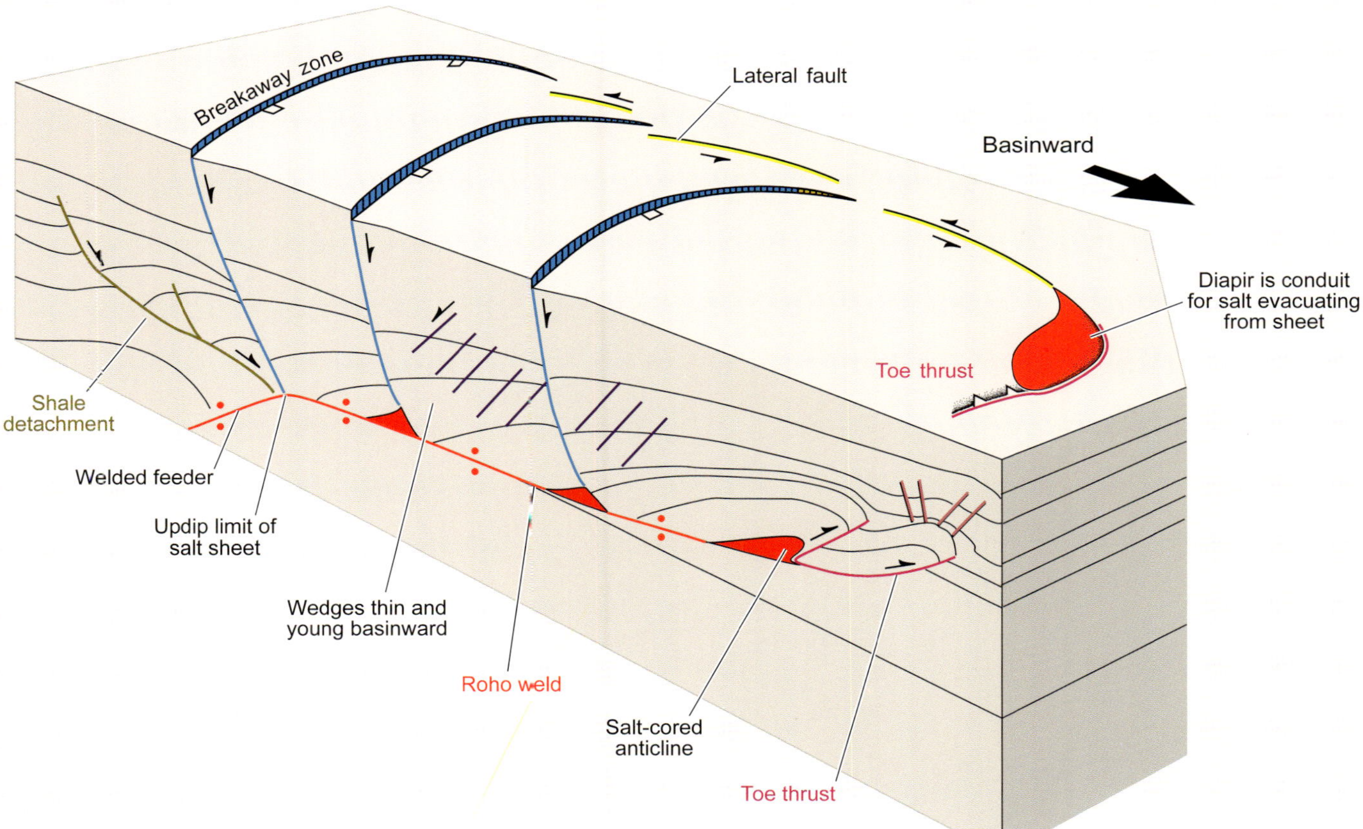

Figure 9.15. Allochthonous salt geometries. A roho system comprises basinward-dipping growth faults detached on an allochthonous salt sheet or canopy. Salt evacuates from the sheet during faulting, eventually producing a fault weld. The basal detachment can extend beyond the updip limit of the original salt sheet as a shale detachment. The basinward end of a roho system is characterized by salt diapirs, which serve as conduits for salt evacuated from the sheet, and by contractional structures. The contractional structures are kinematically linked to the extension in the updip part of the sheet and record the amount of translation that occurred after the toe of the sheet was buried. Modified from Rowan and others (1999).

Chapter 10

Counterregional Systems

Early workers on the U.S. Gulf of Mexico shelf identified what appeared to be large, landward-dipping growth faults. However, efforts to build structural restorations of these inferred normal faults failed because the hanging wall stubbornly refused to fit against the footwall when retrodeformed. The reason for this poor fit became clear when workers realized that despite their appearance, these counterregional systems had little extension, but instead formed mainly by salt evacuation. Their name derives from the fact that they dip counter to the regional dip of the basin.

A counterregional system is a landward-dipping salt system whose upper part is a weld, remnant diapir, or normal fault. The lower part of the system is largely welded and may ramp continuously down to the source layer or may have flat-lying segments. If the system has more than one ramp, it is a *stepped counterregional system*. Sediment wedges in the system dip and thicken seaward, unlike those in a roho system.

Counterregional systems form as sediment progrades over a salt sheet or salt canopy. Salt is expelled progressively seaward, allowing overlying sediment to subside and eventually weld against the base of salt. The rotation of sediment wedges onto the basal weld is an *expulsion rollover*. It resembles an extensional rollover, but scattered relict salt along the weld reveals its true origin. In map view, the seaward end of a counterregional system can include seaward-leaning diapirs spaced along the weld like pearls on a string.

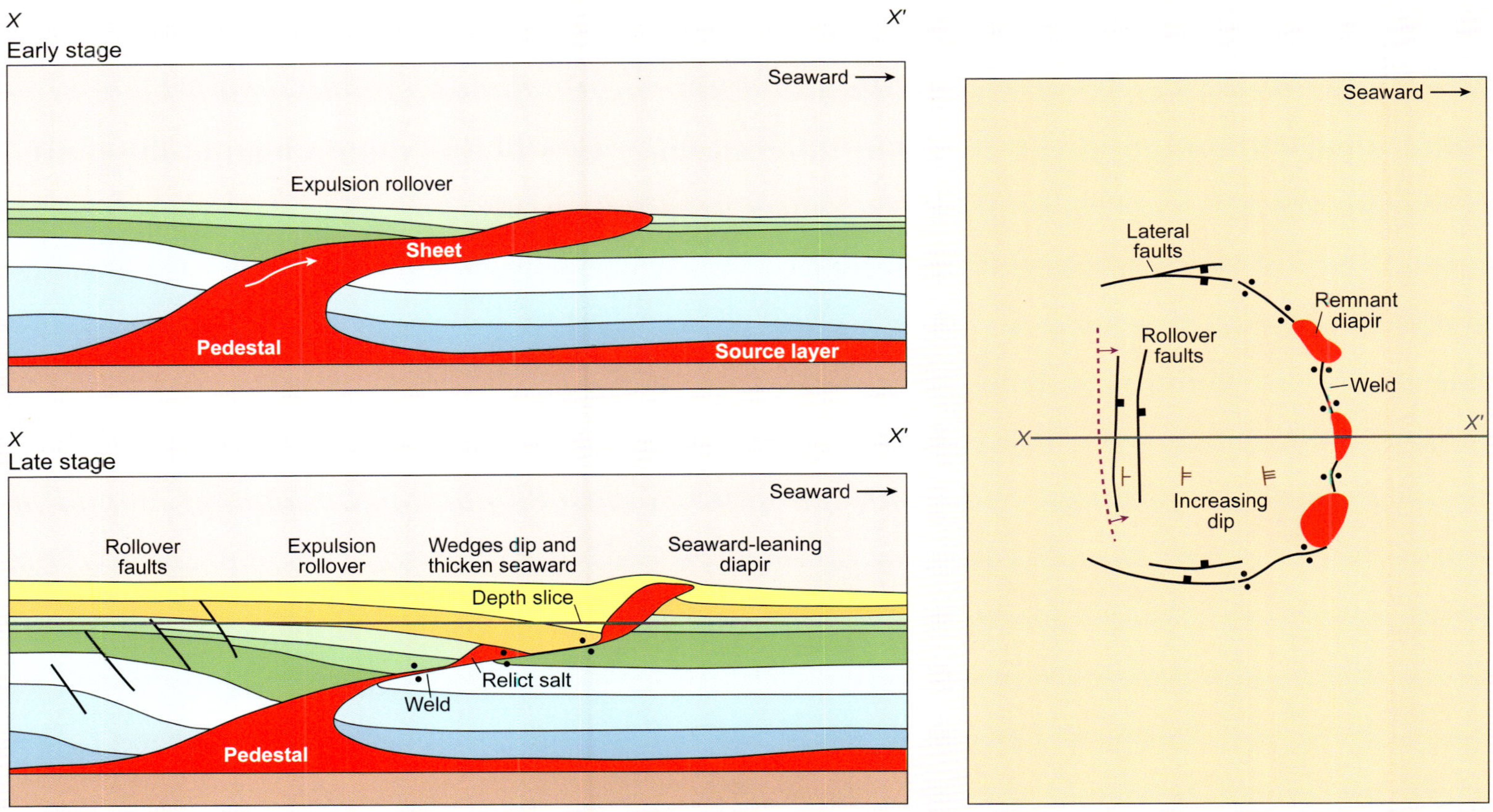

GULF OF MEXICO BASIN—Salt Evacuation Systems, Onshore and Offshore Southeastern Louisiana, USA

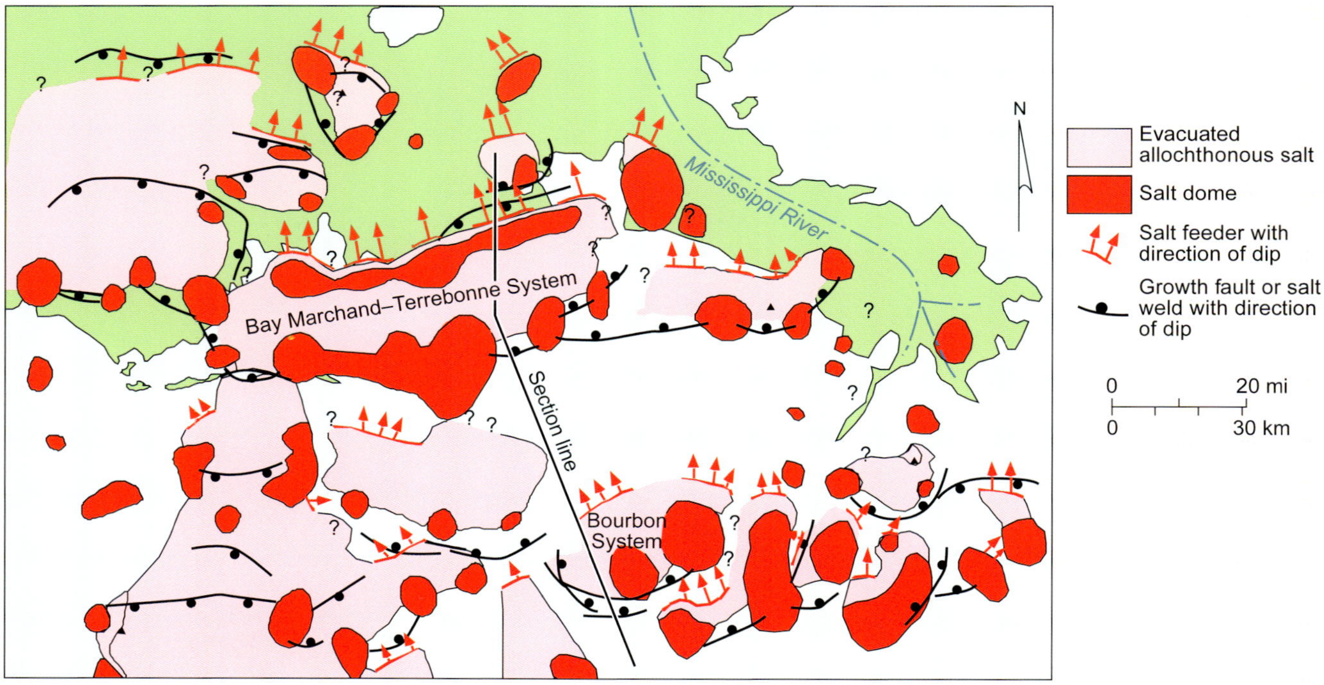

Figure 10.01. Progradation across salt canopies. Stepped counterregional salt systems consist of three parts, from landward to seaward: one or more landward-dipping feeders, an evacuated allochthonous salt canopy, and another series of landward-dipping structures, comprising a variable mixture of normal faults, salt welds, and domes sourced from the now-evacuated sheet. The Bay Marchand–Terrebonne and Bourbon systems are the best-known examples, but other salt systems in this area show a similar pattern. The size of the counterregional system depends on the number of salt stems feeding the original salt canopy. Examples shown here range from 170 to 3000 km^2. See restoration of section line in Figure 10.02. Modified from Schuster (1995).

GULF OF MEXICO BASIN—Bay Marchand-Terrebonne and Bourbon Stepped Counterregional Systems, Louisiana Shelf, Offshore USA

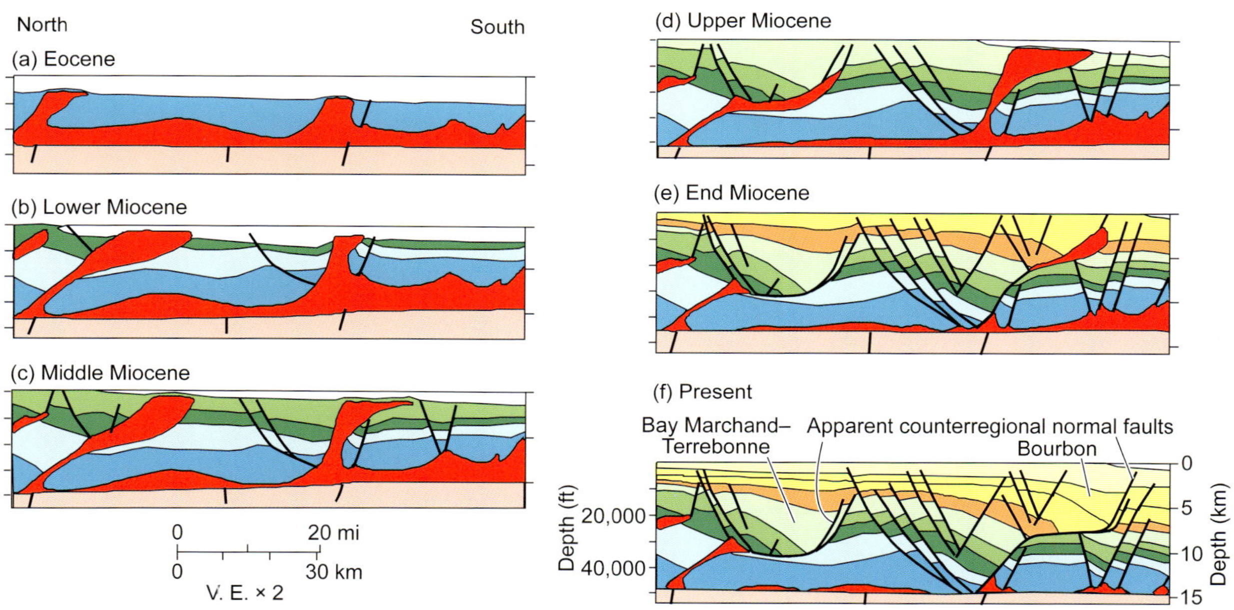

Figure 10.02. Progradation across salt sheets. Evacuation of salt from a buried salt sheet can produce a geometry resembling a large landward-dipping (counterregional) extensional fault. However, formation of these structures is largely a result of salt expulsion and involves no net extension. This example shows two stepped counterregional systems. Because they are produced by sediment progradation, the more landward one is older. The precursor salt sheets emerged on the continental slope and were evacuated mostly when buried beneath the continental shelf. See map in Figure 10.01. Modified from Schuster (1995).

Gulf of Mexico Basin—West Delta/South Pass Counterregional System, Louisiana Shelf, Offshore USA

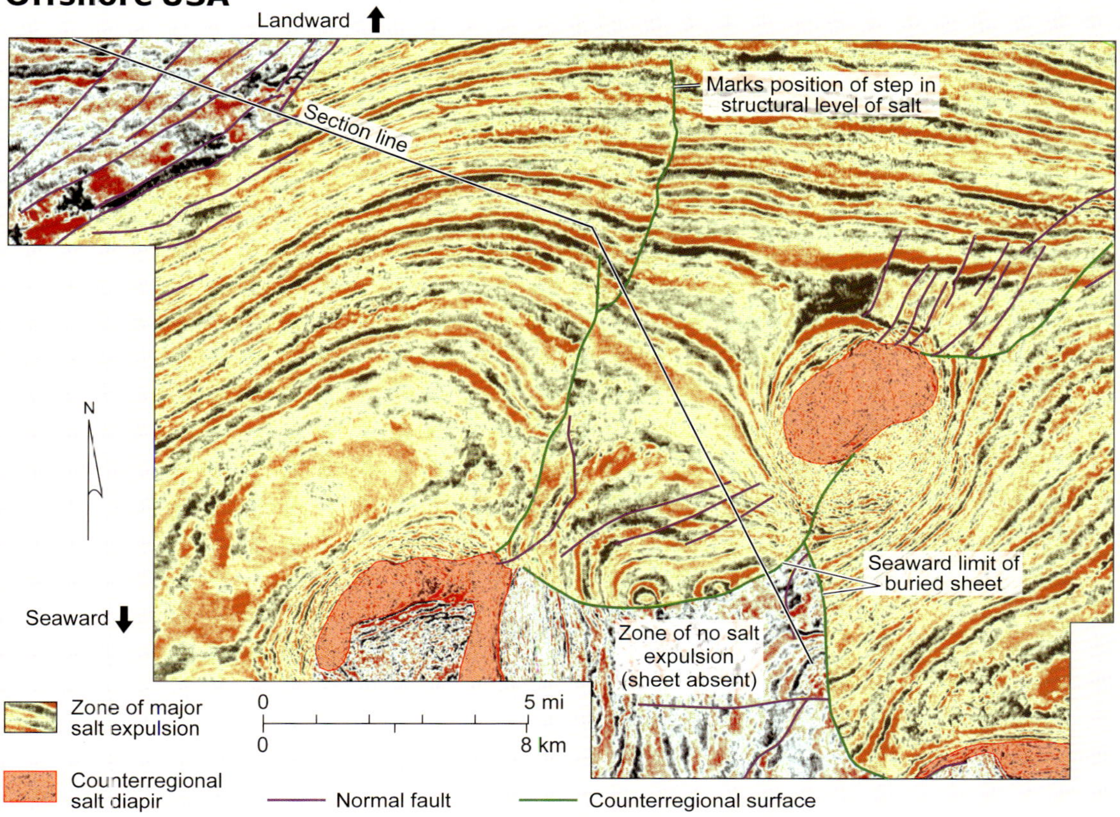

Figure 10.03. Progradation across a salt sheet. Counterregional structures separate areas that have undergone different amounts of salt expulsion. Common locations for these structures are the downdip edge of an evacuated salt sheet and steps in the structural level of salt (see seismic section in Figure 10.04). At higher structural level, many counterregional surfaces are normal faults. At deeper levels closer to the evacuated salt sheet, these surfaces are fault welds. In map view, counterregional surfaces commonly connect stocks that were used to evacuate salt from a buried sheet. If these diapirs are not colinear, the counterregional structure may contain major changes in trend. Modified from Rowan and others (1999).

Gulf of Mexico Basin—West Delta/South Pass Counterregional System, Louisiana Shelf, Offshore USA

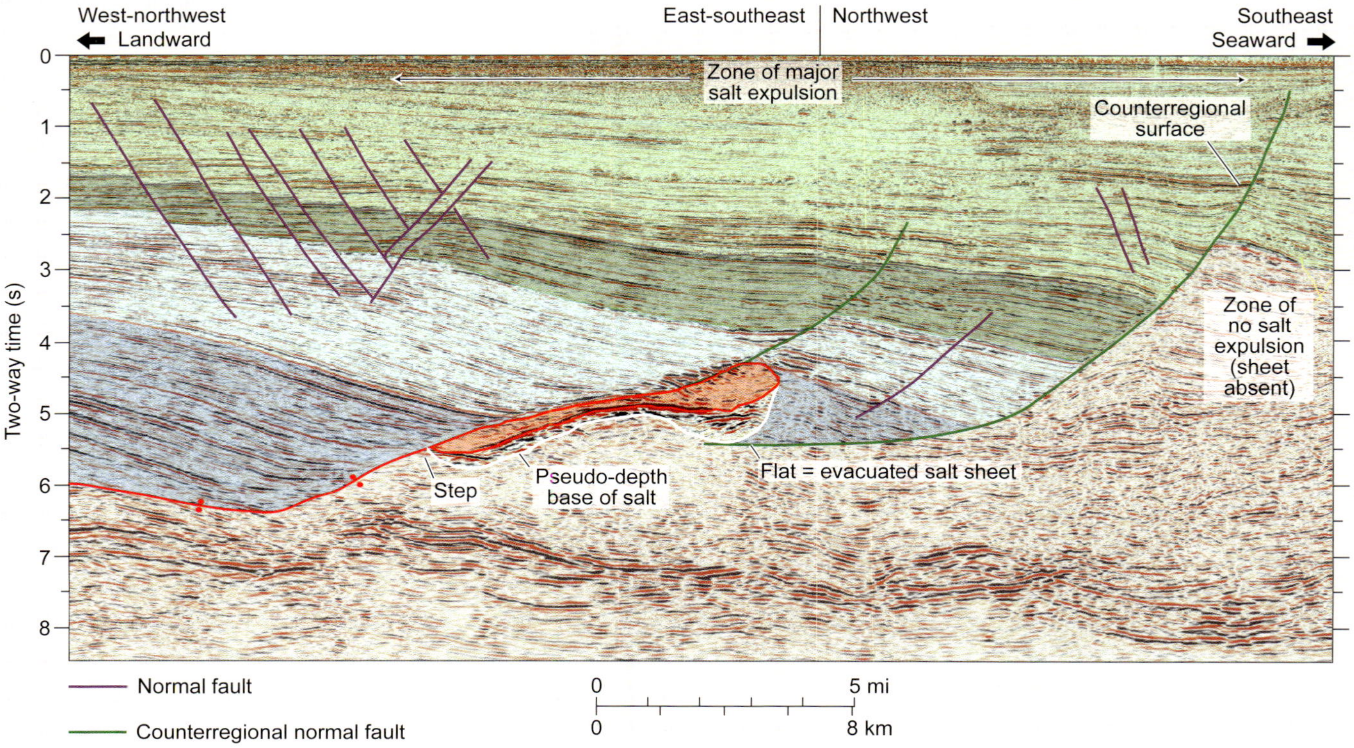

Figure 10.04. Progradation across a salt sheet. Counterregional surfaces separate areas that have undergone different amounts of salt expulsion. Here the landward green fault separates strata above the step from those above the flat, and the seaward fault marks the downdip limit of the system. The geometry resembles a basinward-dipping (counterregional) extensional fault but is actually mostly a product of salt expulsion. See seismic time slice at 2800 ms in Figure 10.03. Modified from Rowan and others (1999).

GULF OF MEXICO BASIN—Counterregional System, Salina Basin, Offshore Mexico

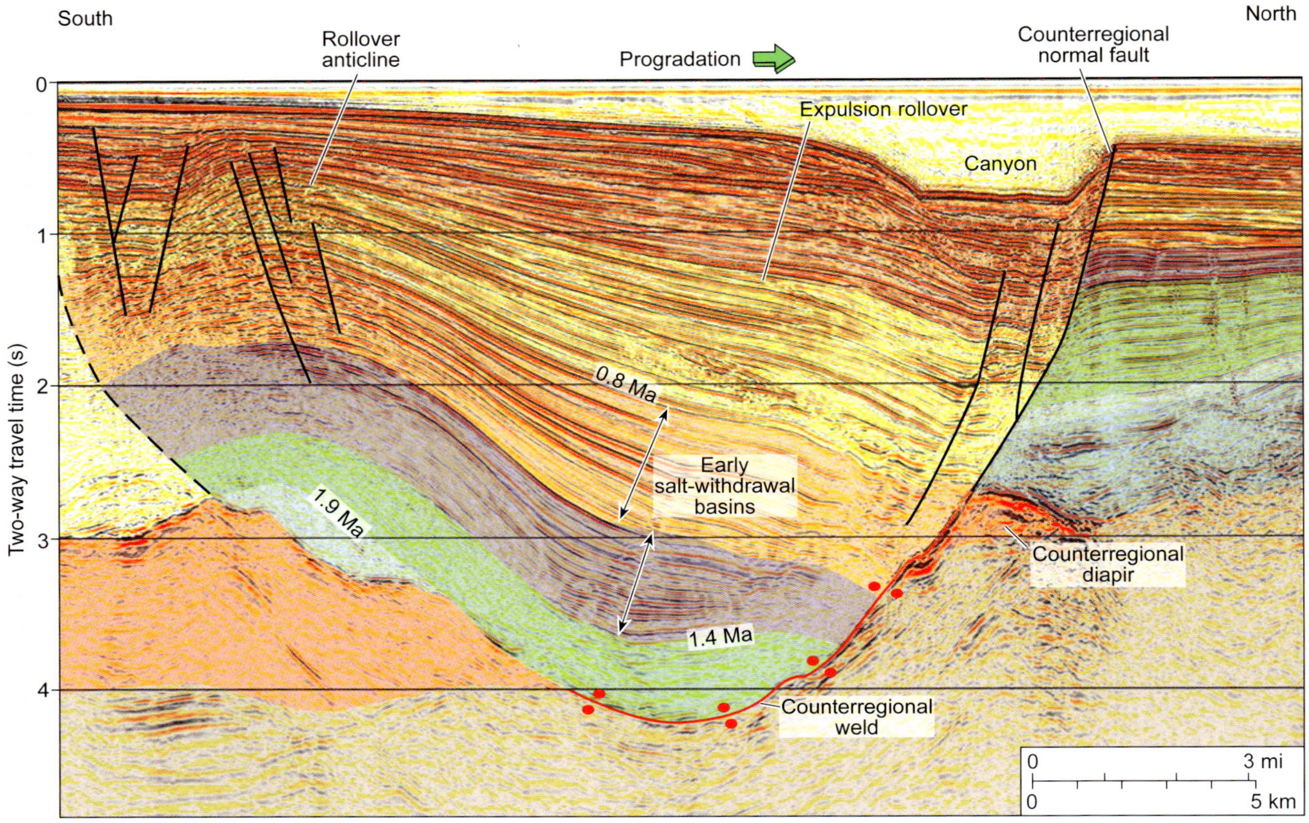

Figure 10.05. Progradation across thick salt. Shallow units thicken into the counterregional surface. The surface resembles both an extensional rollover and an expulsion rollover. However, deeper units have depocenters away from (not next to) the counterregional surface, indicating that salt expulsion was important during these phases. The normal faults at the crest of the anticline accommodate bending required for rollover. A submarine canyon is localized at the main scarp of the counterregional fault. From Gómez-Cabrera and Jackson (2008).

PHYSICAL MODEL—Counterregional System

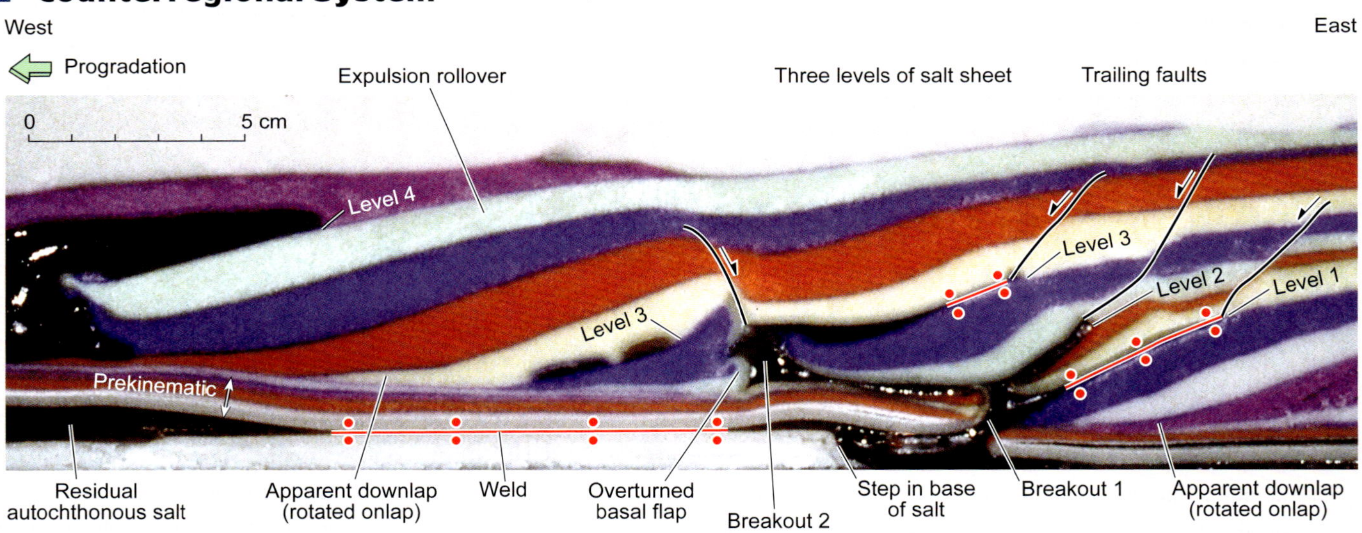

Figure 10.06. Formation of expulsion rollover. Slow westward progradation of synkinematic strata across an autochthonous salt basin expelled salt to the west. This activity progressively inflated the west part of the salt basin, which created a monocline that rolled seaward. Synkinematic strata that originally onlapped the rolling monocline subsided as salt was expelled seaward. This subsidence deformed the onlaps into apparent downlaps and created an expulsion rollover. Model by Hongxing Ge.

PHYSICAL MODEL—Evolution of Counterregional System during Progradation

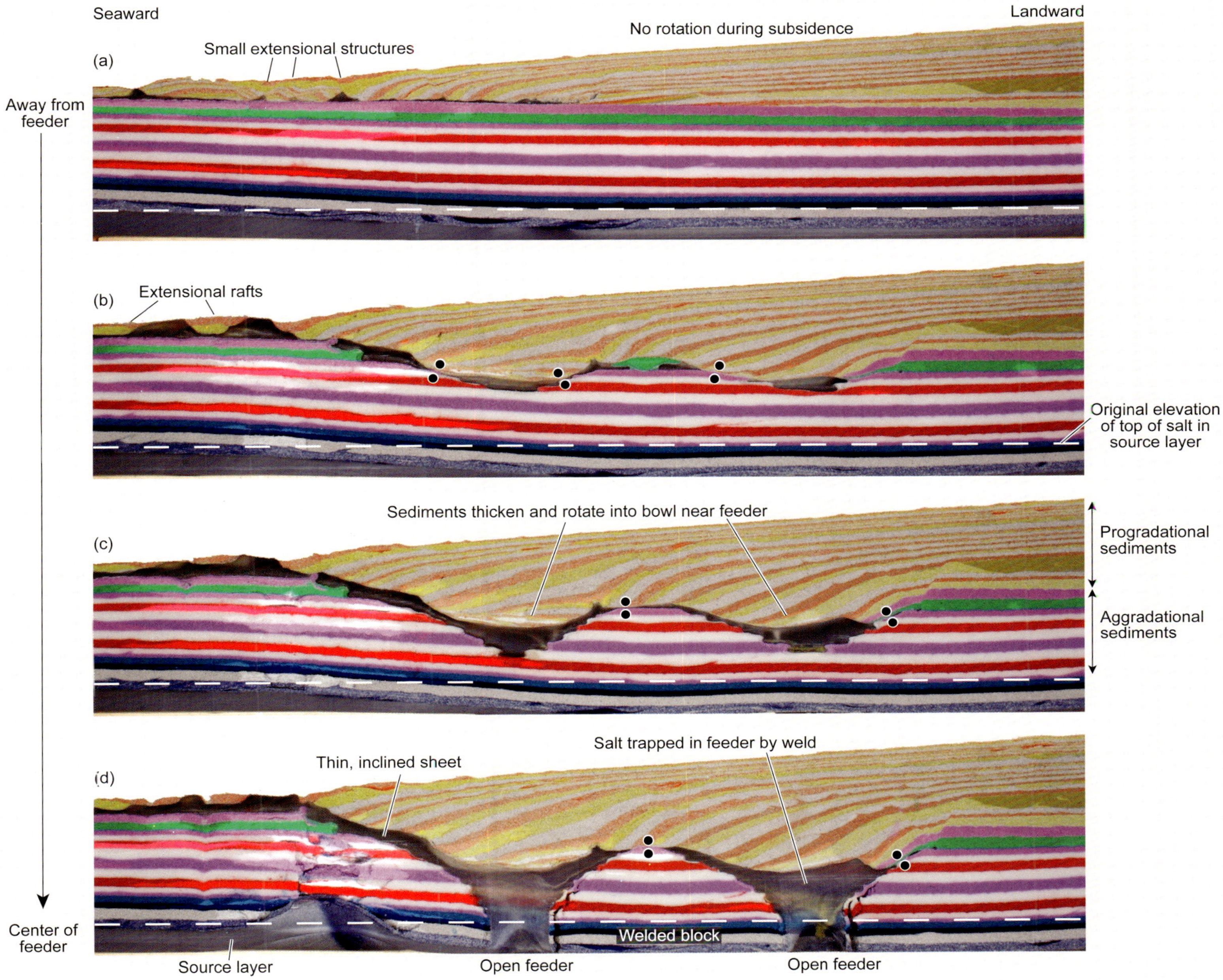

Figure 10.07. Progradation above a salt canopy. A salt canopy grew by passive diapirism during aggradation above a source layer. Sediments then prograded across the top of this canopy. Shown here are serial sections across the model, with section d crossing the canopy feeders. Progradation loaded the canopy and expelled salt seaward. Seaward flow of salt carried its roof with it at the thin tip of the prograding wedge. The seaward end of most wedges were thus disrupted by small normal faults or broken into extensional rafts that were carried away by flowing salt. Farther updip, the roofs were too thick and strong to break, so they subsided into the underlying canopy without faulting. The effect of this subsidence depended on the geometry of the underlying canopy. Away from the feeder, the sediment wedge subsided into the thin, tabular salt canopy without much rotation. Nearer the feeder, thicker salt was present in the bowl adjoining the feeder. The roof thickened and rotated into this bowl. Salt was trapped in the landward feeder by a weld separating it from the seaward feeder. Salt continued to escape from the seaward feeder via a thin, inclined sheet to the surface. Model by Tim Dooley.

Physical Model—Restoration of Counterregional System

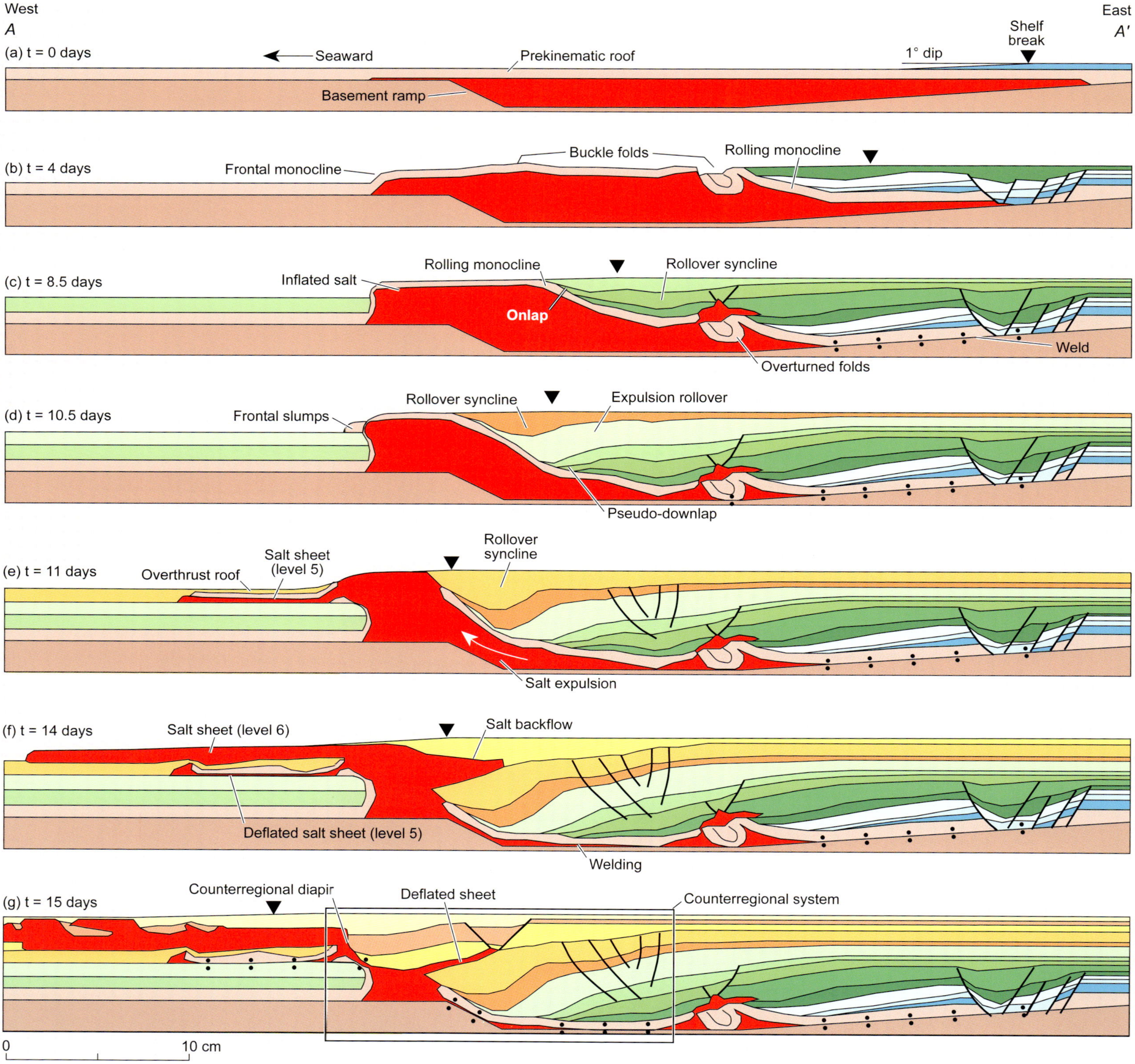

Figure 10.08. Formation of expulsion rollover. Progradation of sediments expelled salt westward, inflating the distal part of the autochthonous salt. Salt flow also created an expulsion rollover, which was separated landward by a rollover syncline from a rolling monocline near the shelf break. The monocline separated a landward region of expelled salt from a seaward region of inflated salt. Synkinematic strata that originally onlapped the rolling monocline became deformed by underlying salt expulsion to form a pseudo-downlap above the region of expelled salt. At the downdip end of the system, inflated salt eventually escaped up a counterregional diapir, which periodically extruded salt sheets. Model and restoration by Hongxing Ge.

CONCEPTUAL MODEL—Characteristics of a Stepped Counterregional System

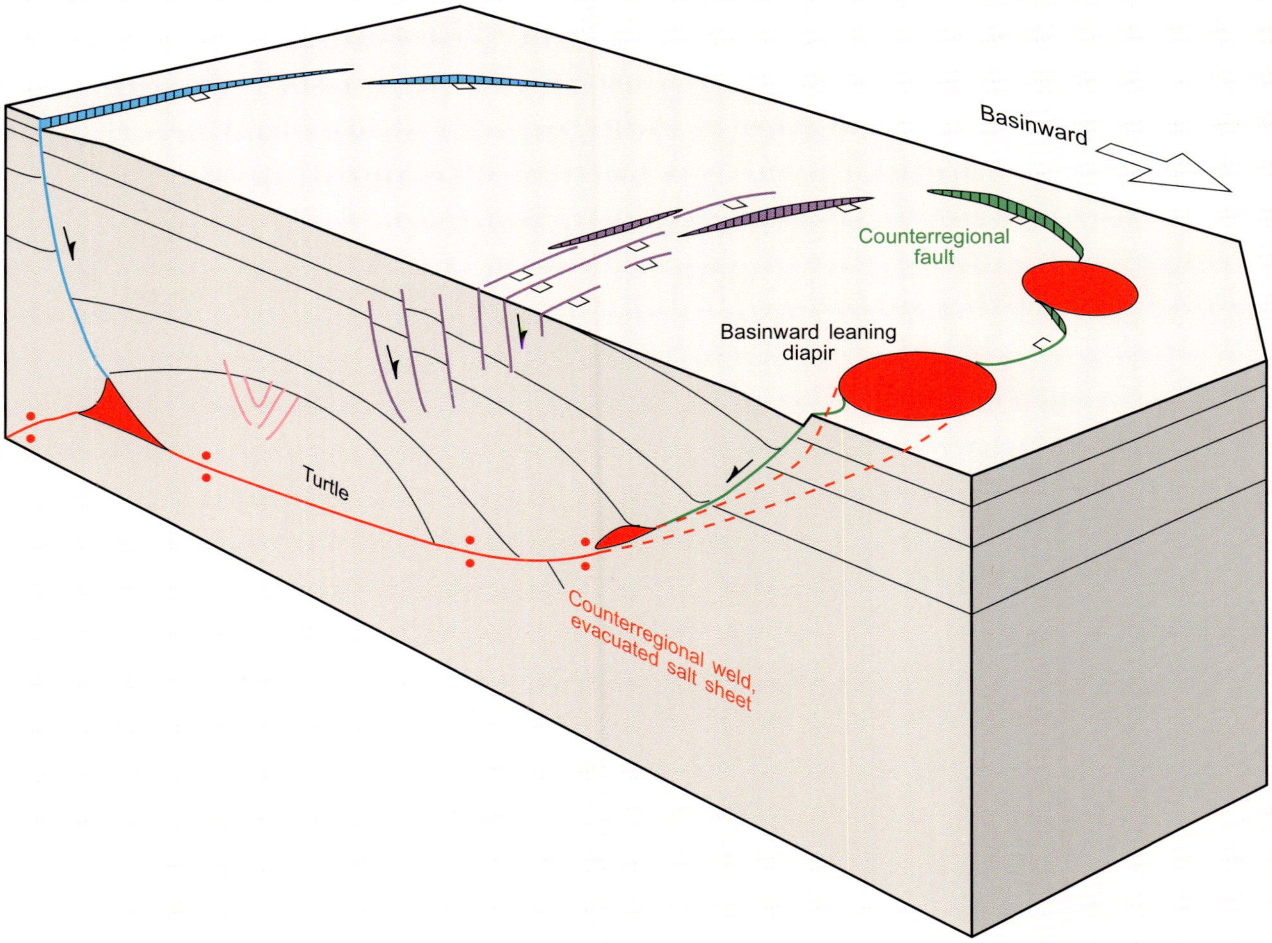

Figure 10.09. Progradation across a salt sheet. A counterregional salt system forms through evacuation of salt from a buried salt sheet. Strata above the sheet subside during evacuation more than surrounding areas, so the outline of the buried sheet is commonly traced by a ring of inward-dipping, arcuate normal faults and fault welds (blue and green fault systems). Basinward-leaning diapirs at the basinward end of the system represent pathways by which salt escapes from the buried sheet. These diapirs are typically connected by counterregional surfaces. Modified from Rowan and others (1999).

CONCEPTUAL MODEL—Formation of a Stepped Counterregional System

Figure 10.10. Progradation across a salt sheet. The shape of a stepped counterregional system is produced by progressive basinward expulsion of salt up alternating basinward-leaning diapirs (steps) and salt sheets (flats). Progradation of sediments across the allochthonous salt acts like a rolling pin to force salt basinward. The resulting salt weld resembles a basinward-dipping (counterregional) extensional fault but is actually a weld produced by salt expulsion. The system accommodates negligible net extension. Modified from Rowan (1995).

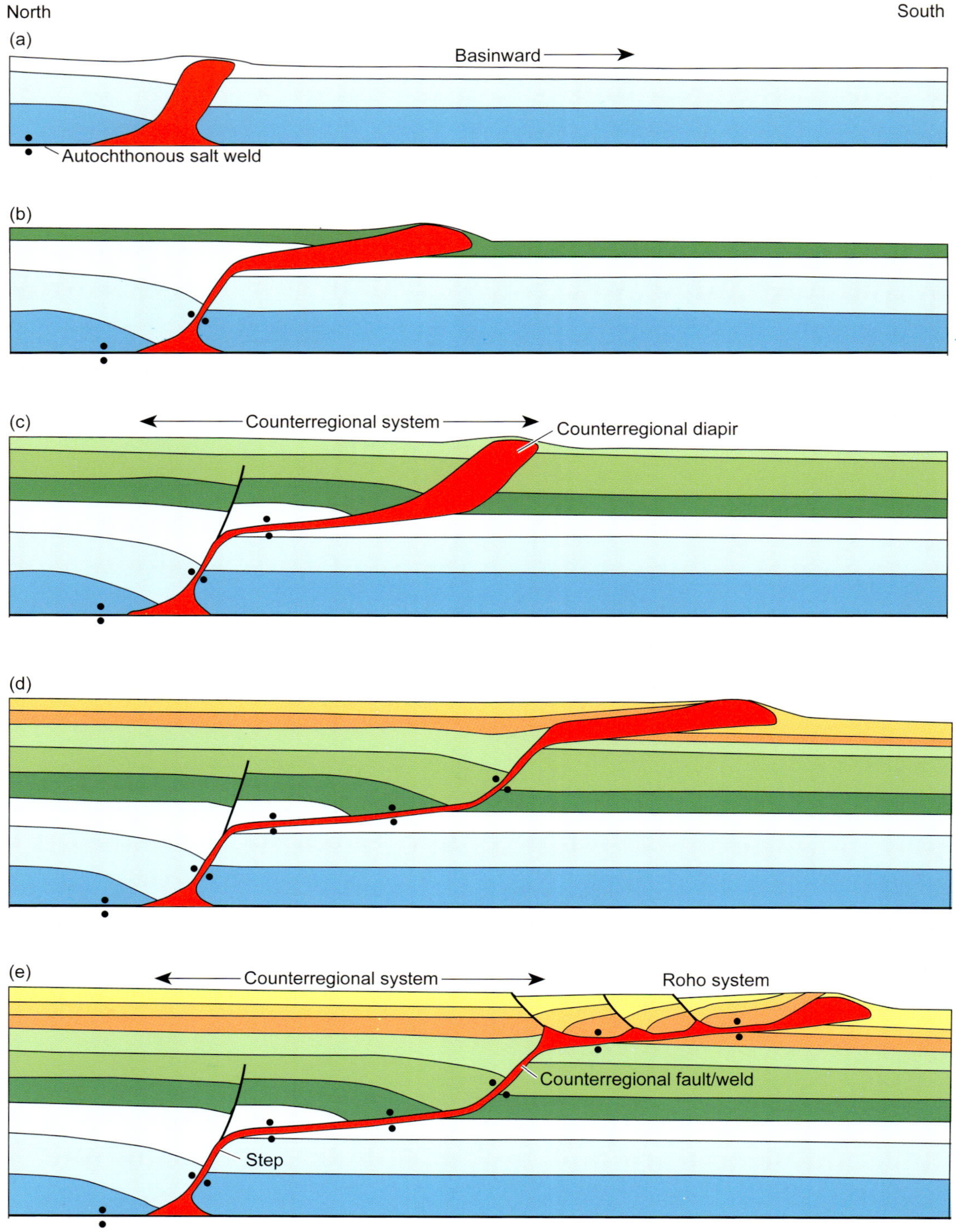

Chapter 11

Location Maps

This chapter contains location maps for images in the book version of *The Salt Mine*. Location maps for figures in the digital version of *The Salt Mine* are included on the DVD in the pocket in the back of the book.

These maps place images in their geologic and geographic contexts, allowing geoscientists to associate structures with other data from the same basin and to use the images to compare different basins.

The index map below shows figure numbers of more detailed maps, each of which displays all available locations of figures from that area and provides a figure number for each location. A few figures have no map locations typically because (a) source material published by others does not provide an exact location or (b) terms of data release do not permit us to show the location. Despite these limitations, we hope that the maps provide a useful and attractive supplement to the other chapters in the book.

Global index map

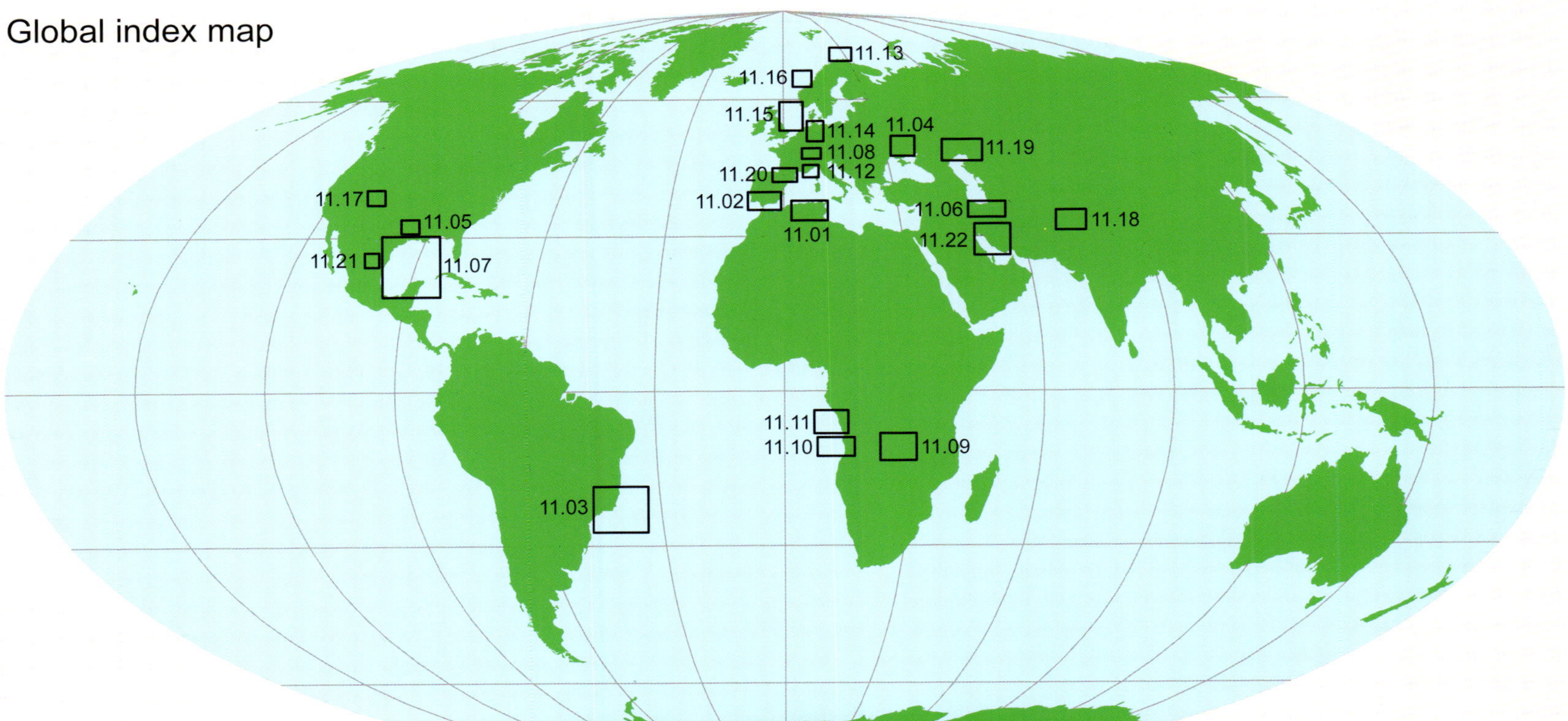

Mollweide Projection
Central Meridian: 0.00

FIGURE 11.01—Atlas Fold Belt, Algeria and Tunisia

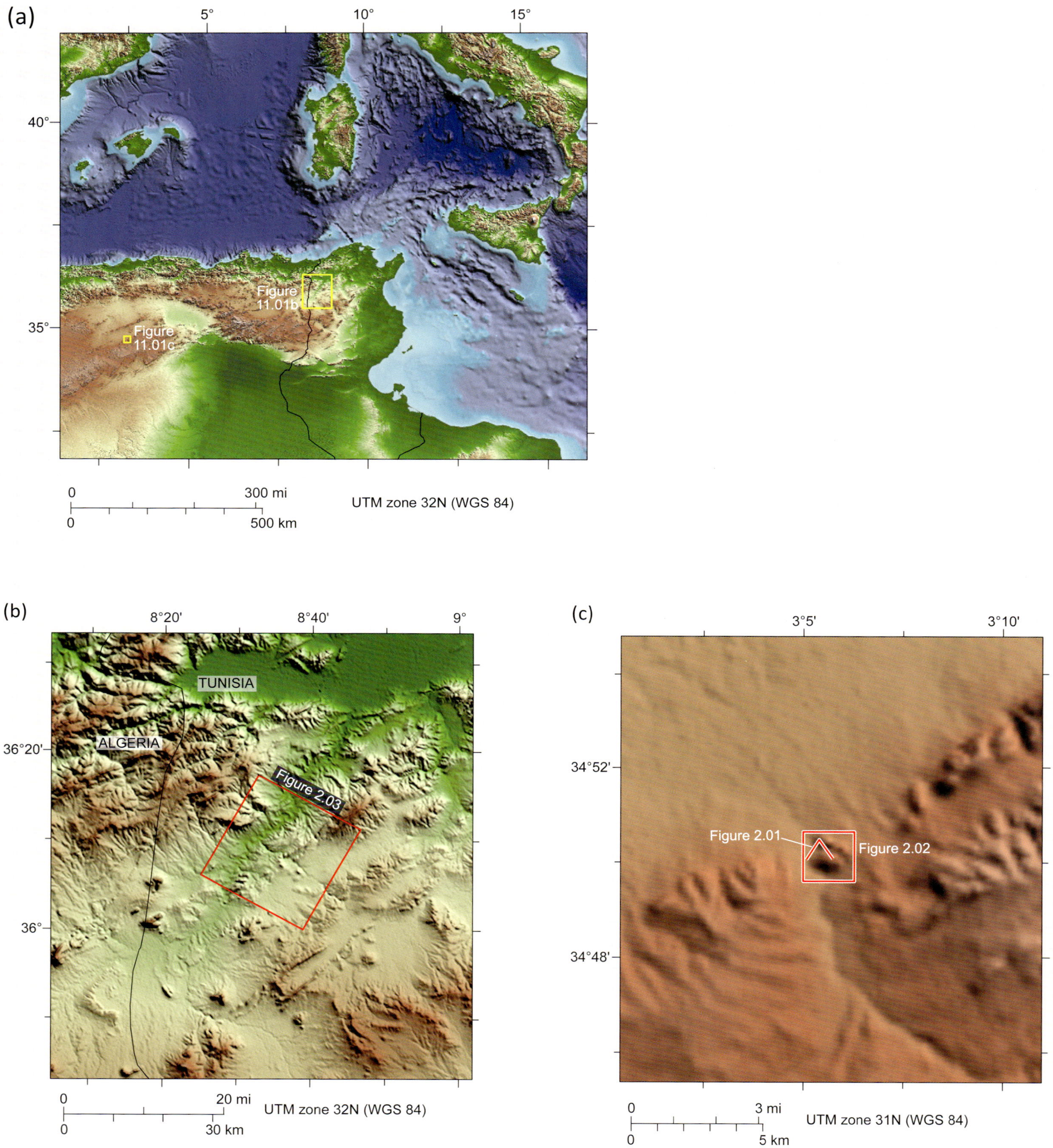

FIGURE 11.02—Betic Cordillera, Spain

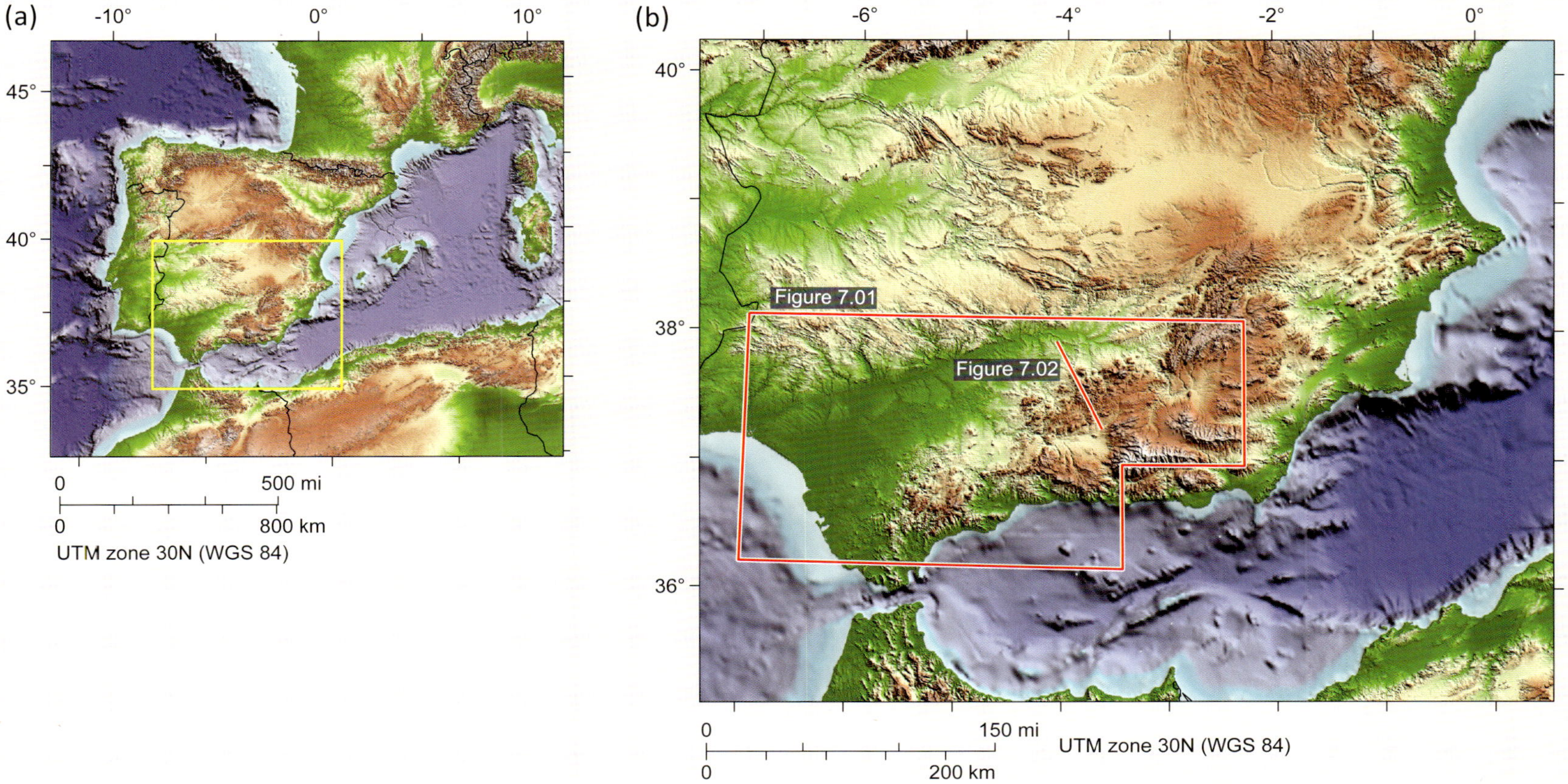

FIGURE 11.03—Campos Basin, Offshore Brazil

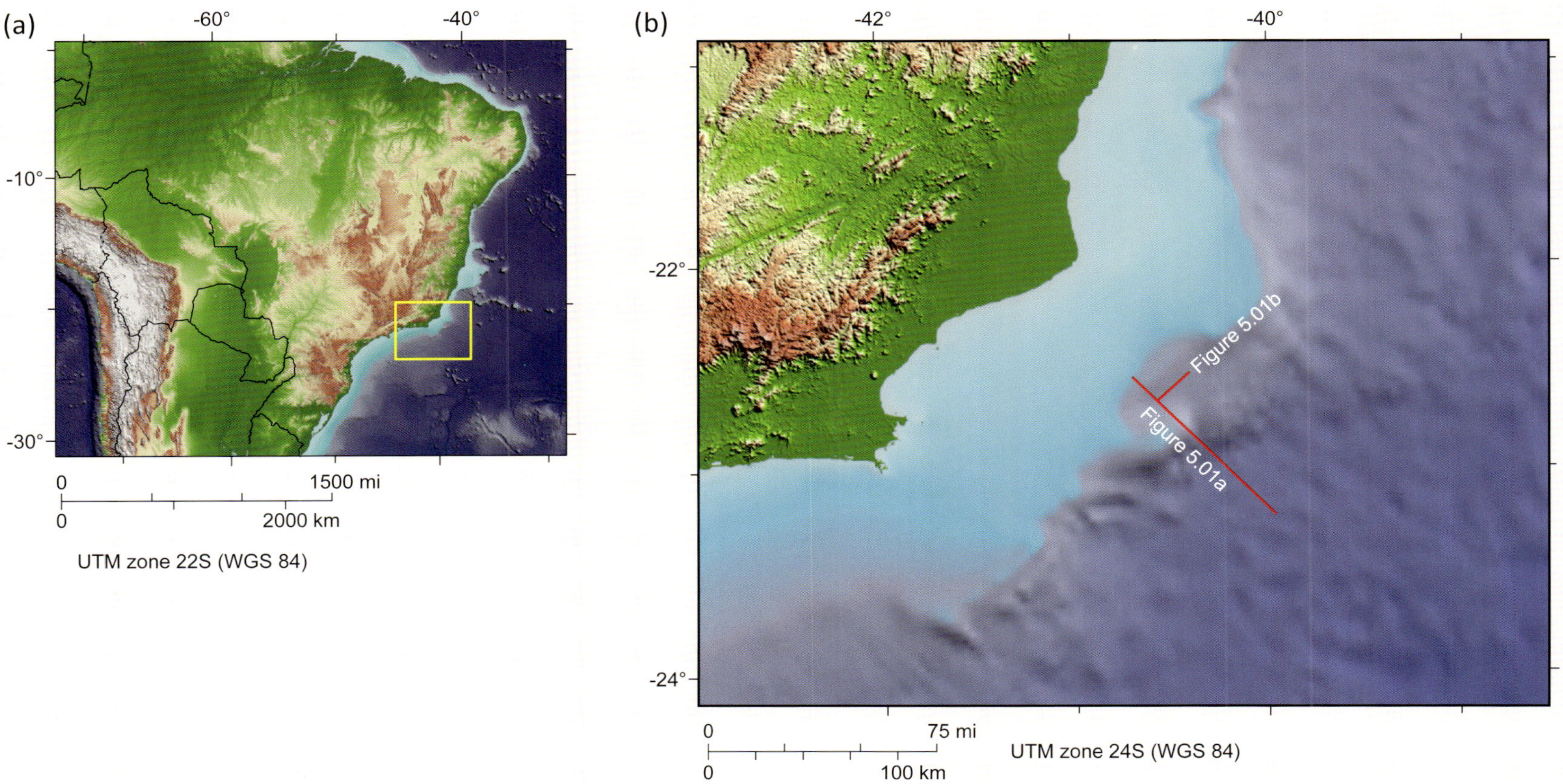

FIGURE 11.04—Dnieper-Donets Basin, Ukraine

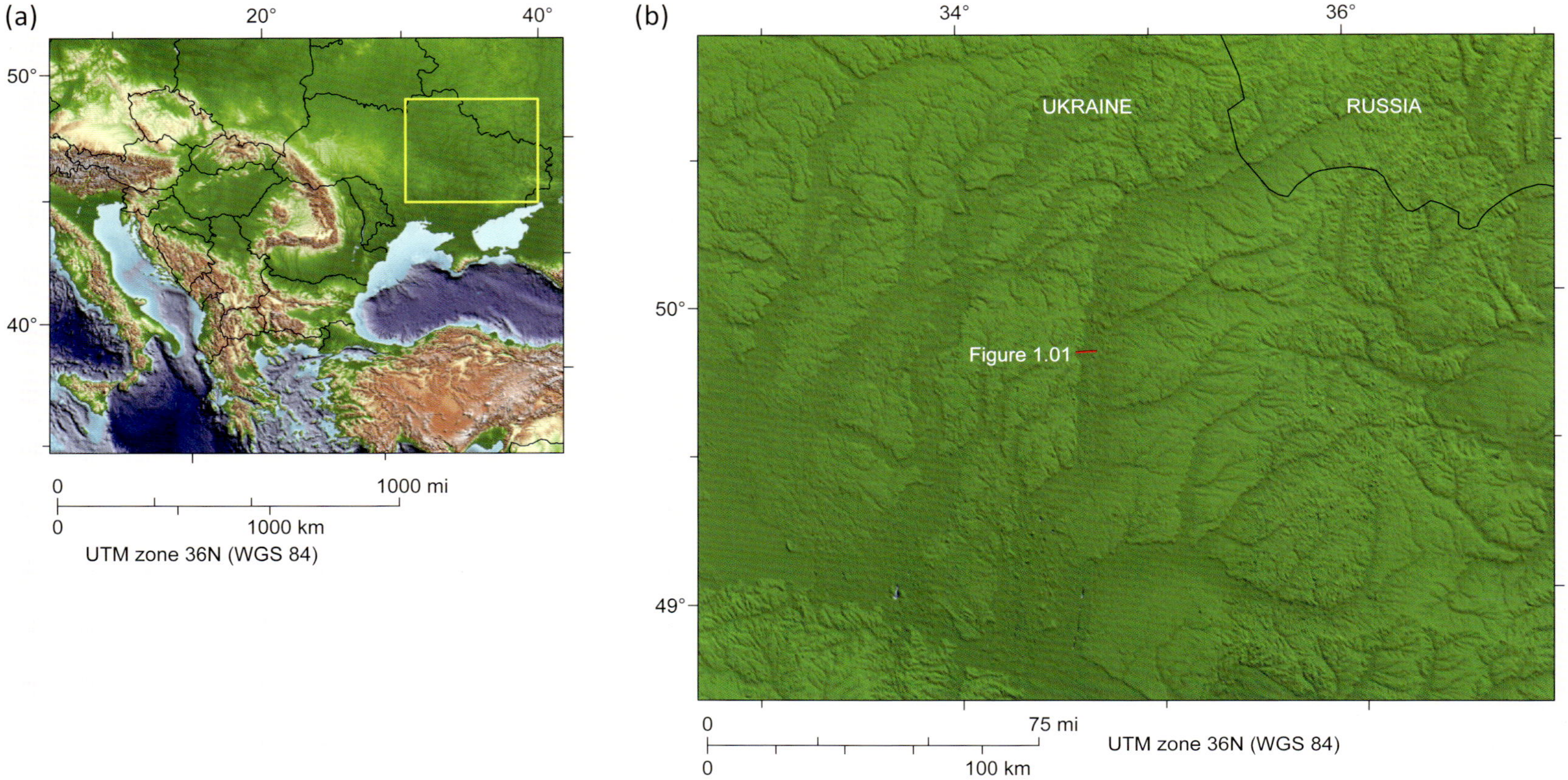

FIGURE 11.05—East Texas Basin, USA

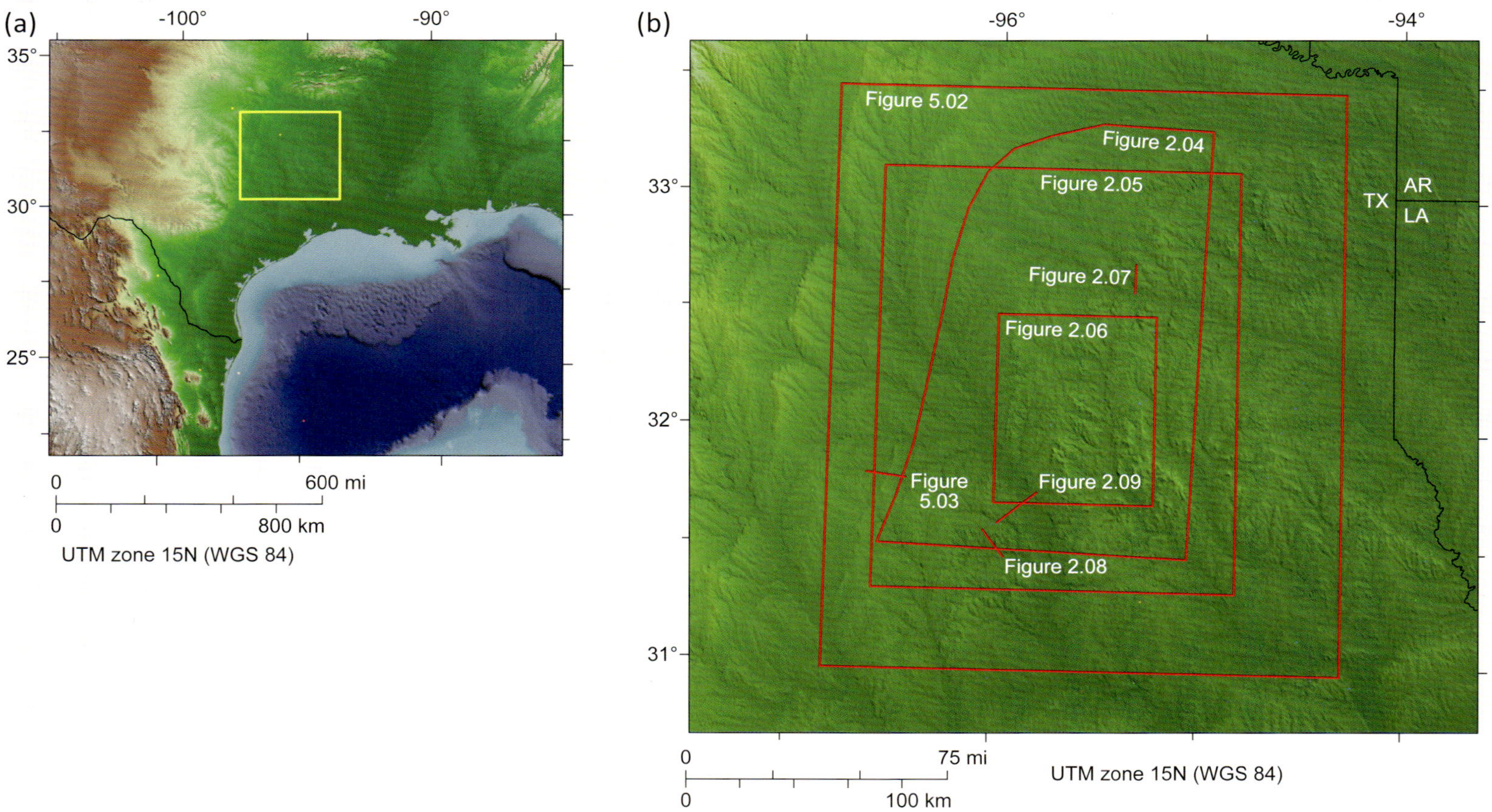

FIGURE 11.06—Garmsar, Great Kavir, and Qom Basins, Iran

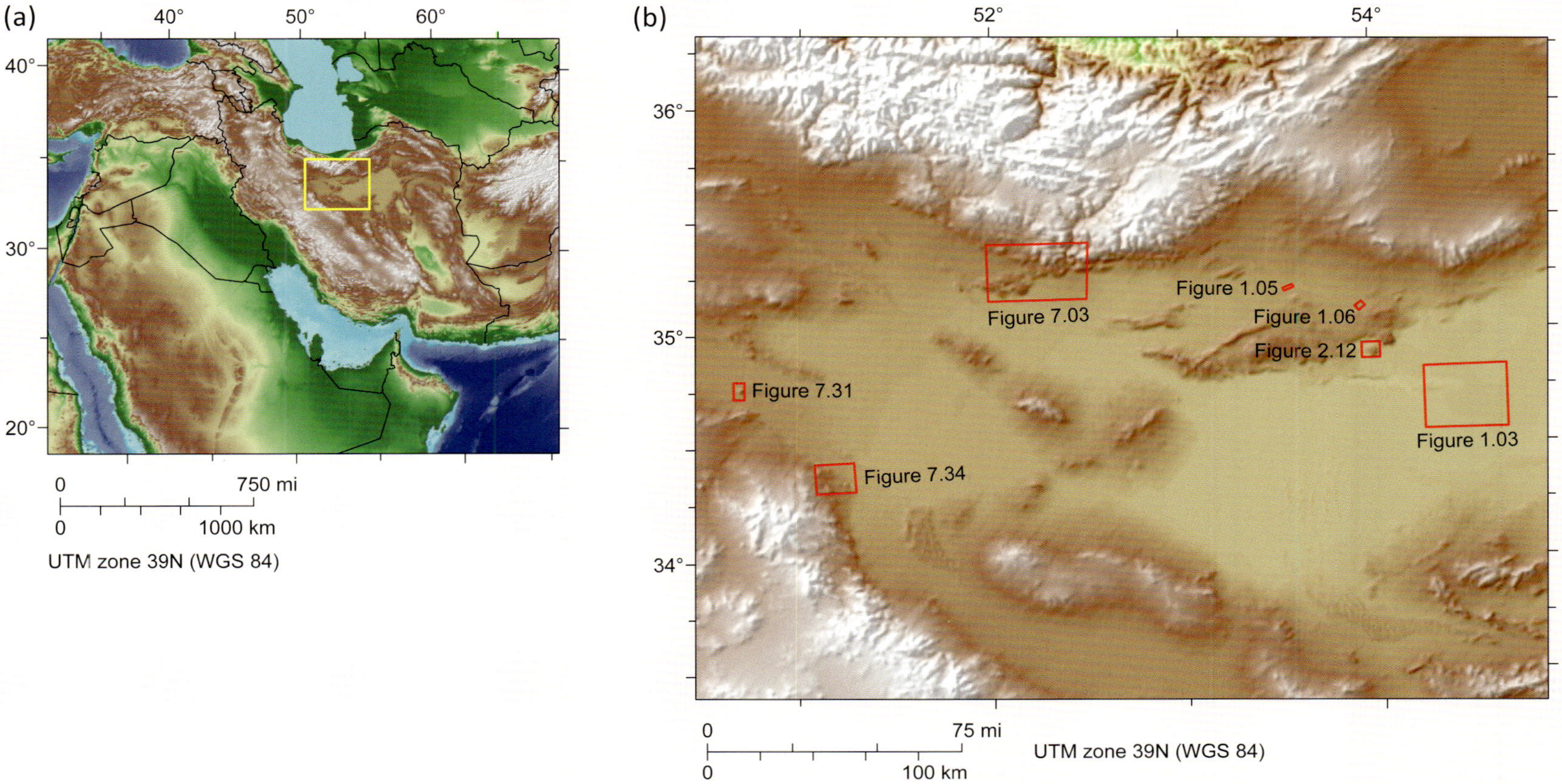

FIGURE 11.07—Gulf of Mexico Basin, Offshore Mexico and Onshore and Offshore USA

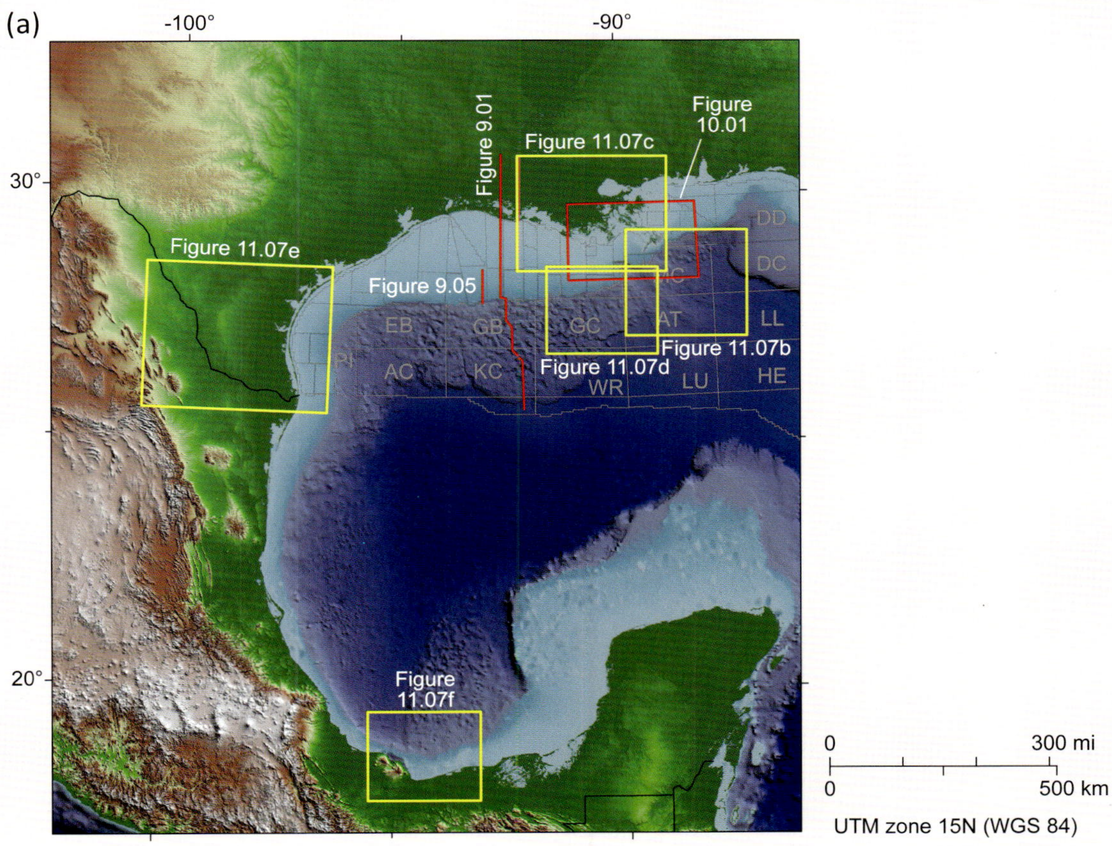

FIGURE 11.07 (CONTINUED)—Gulf of Mexico Basin, Offshore Mexico and Onshore and Offshore USA

(b)

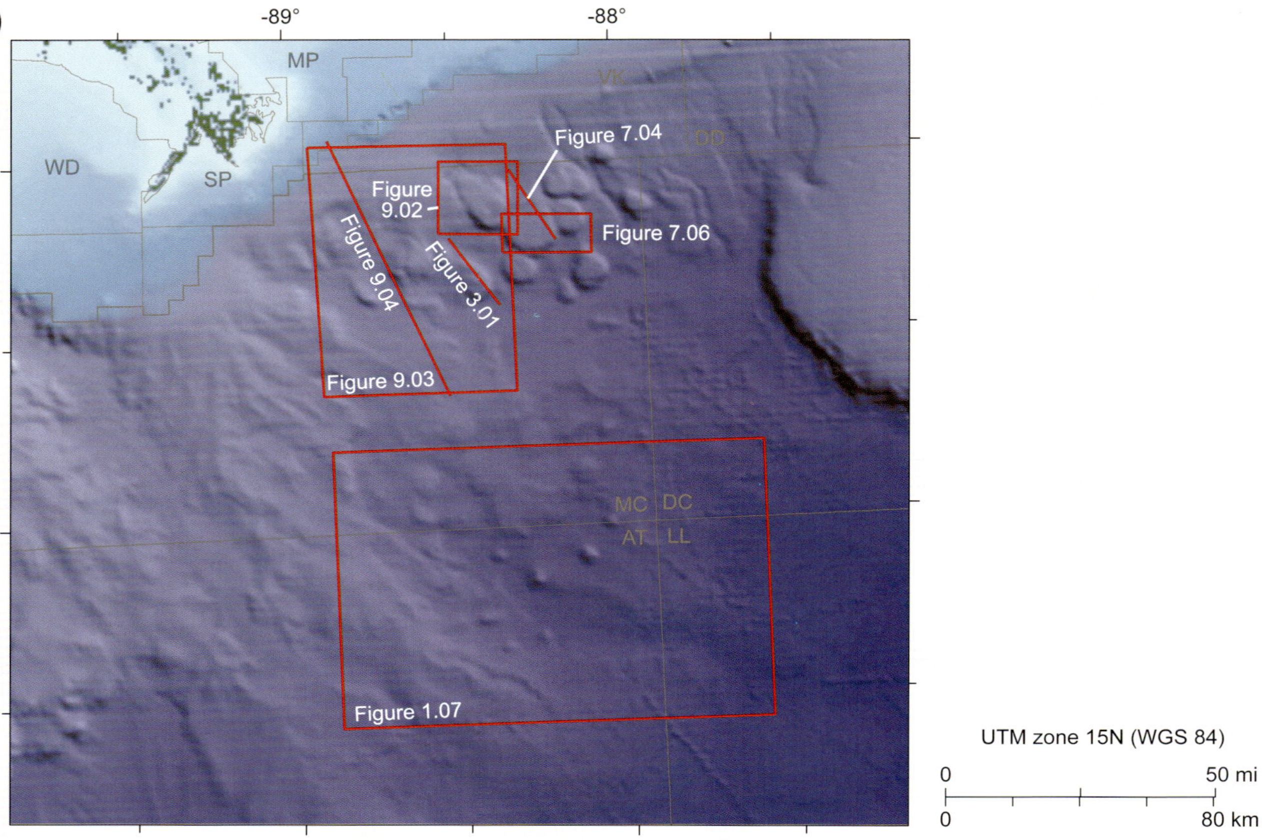

(c)

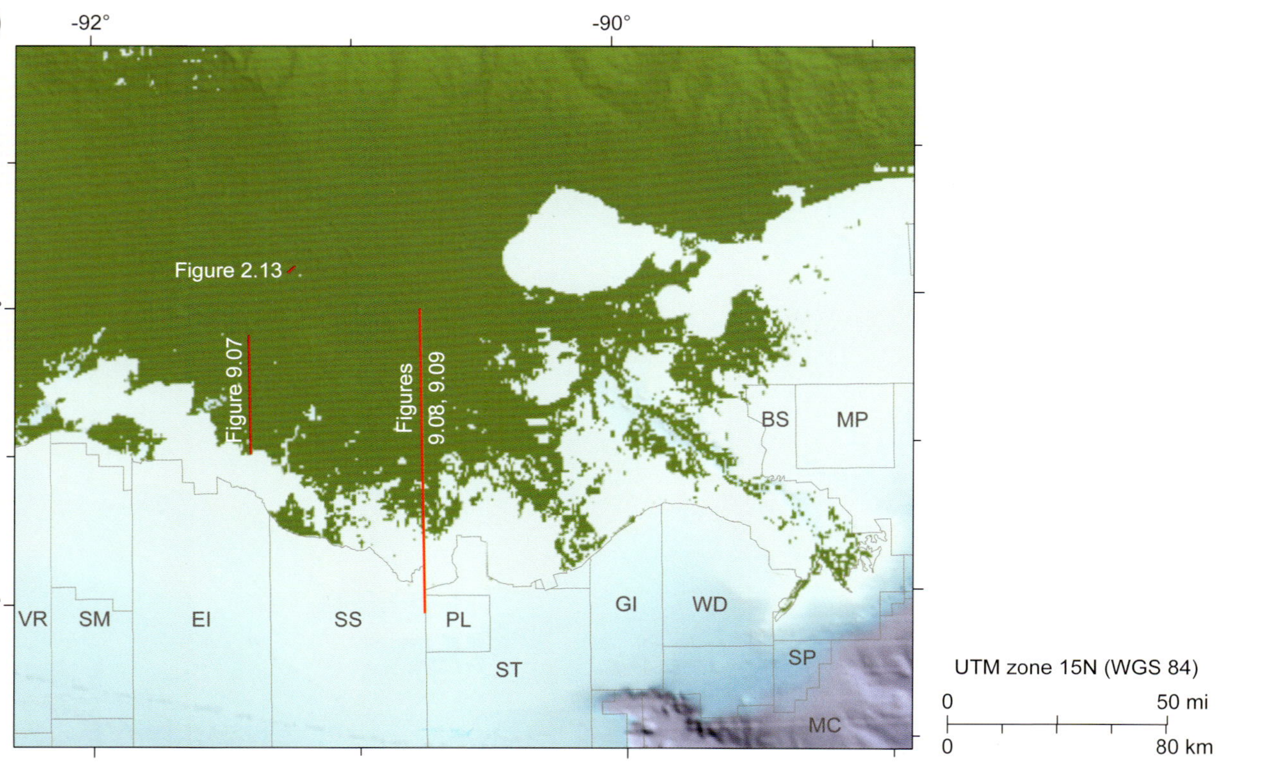

FIGURE 11.07 (CONTINUED)—Gulf of Mexico Basin, Offshore Mexico and Onshore and Offshore USA

(d)

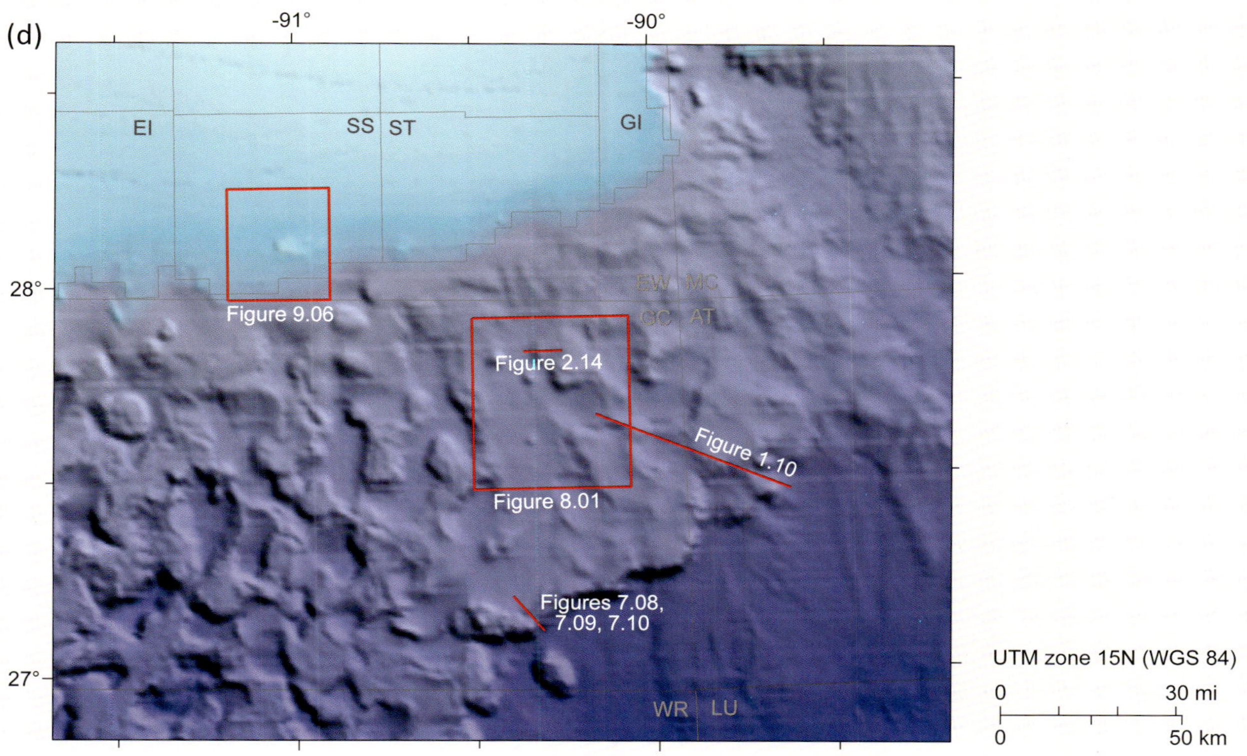

(e)

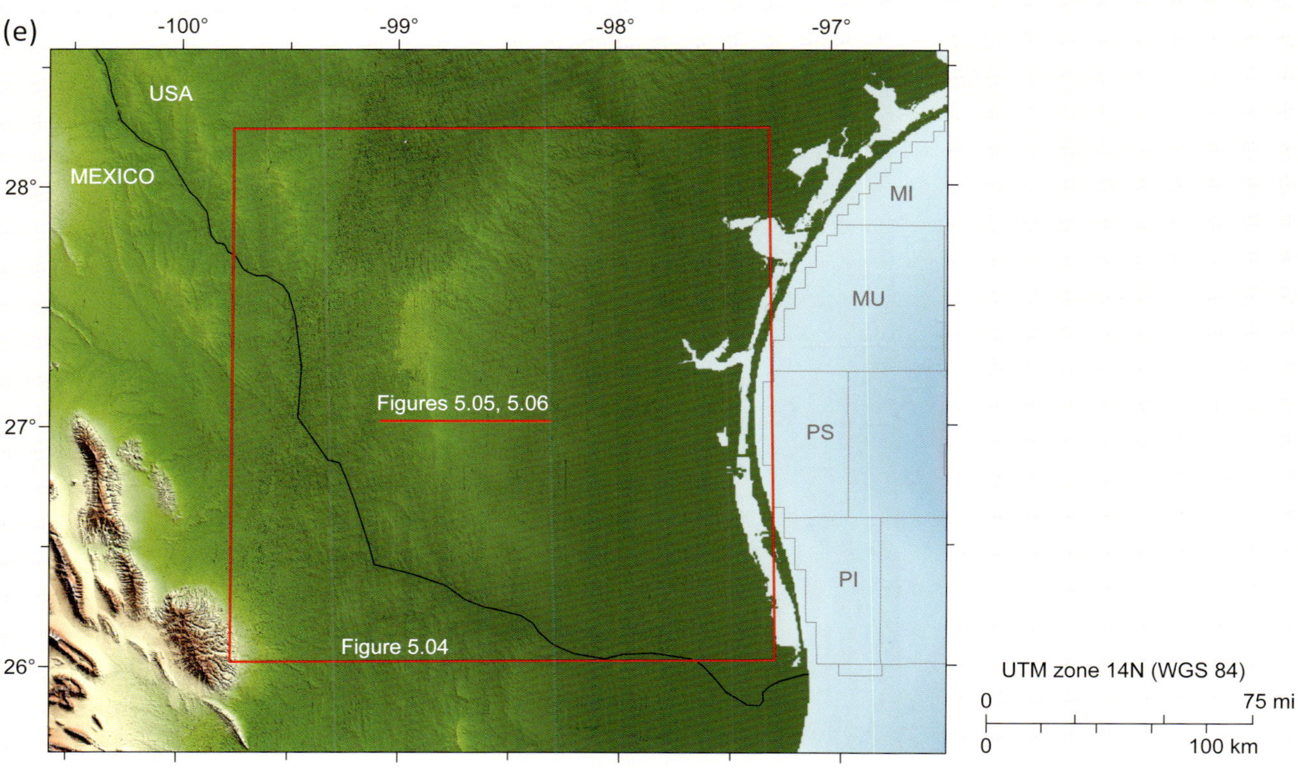

FIGURE 11.07 (CONTINUED)—Gulf of Mexico Basin, Offshore Mexico and Onshore and Offshore USA

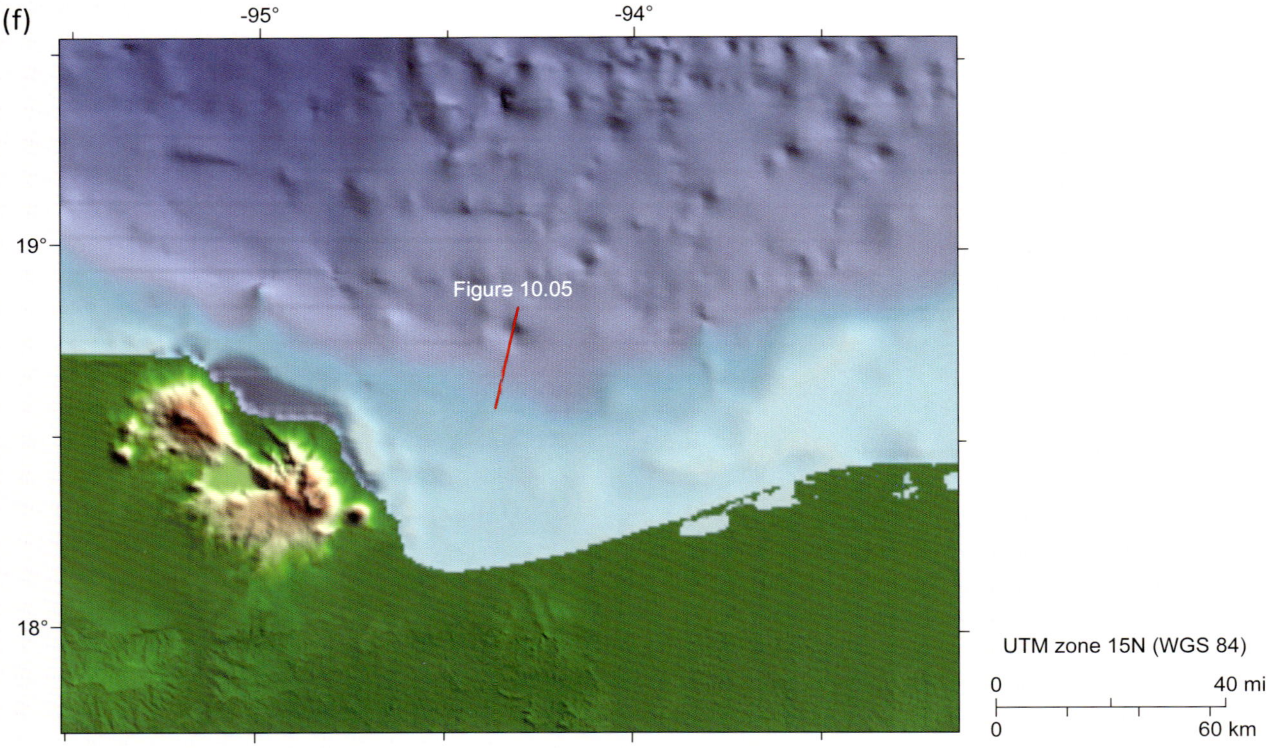

FIGURE 11.08—Jura Fold Belt, Switzerland

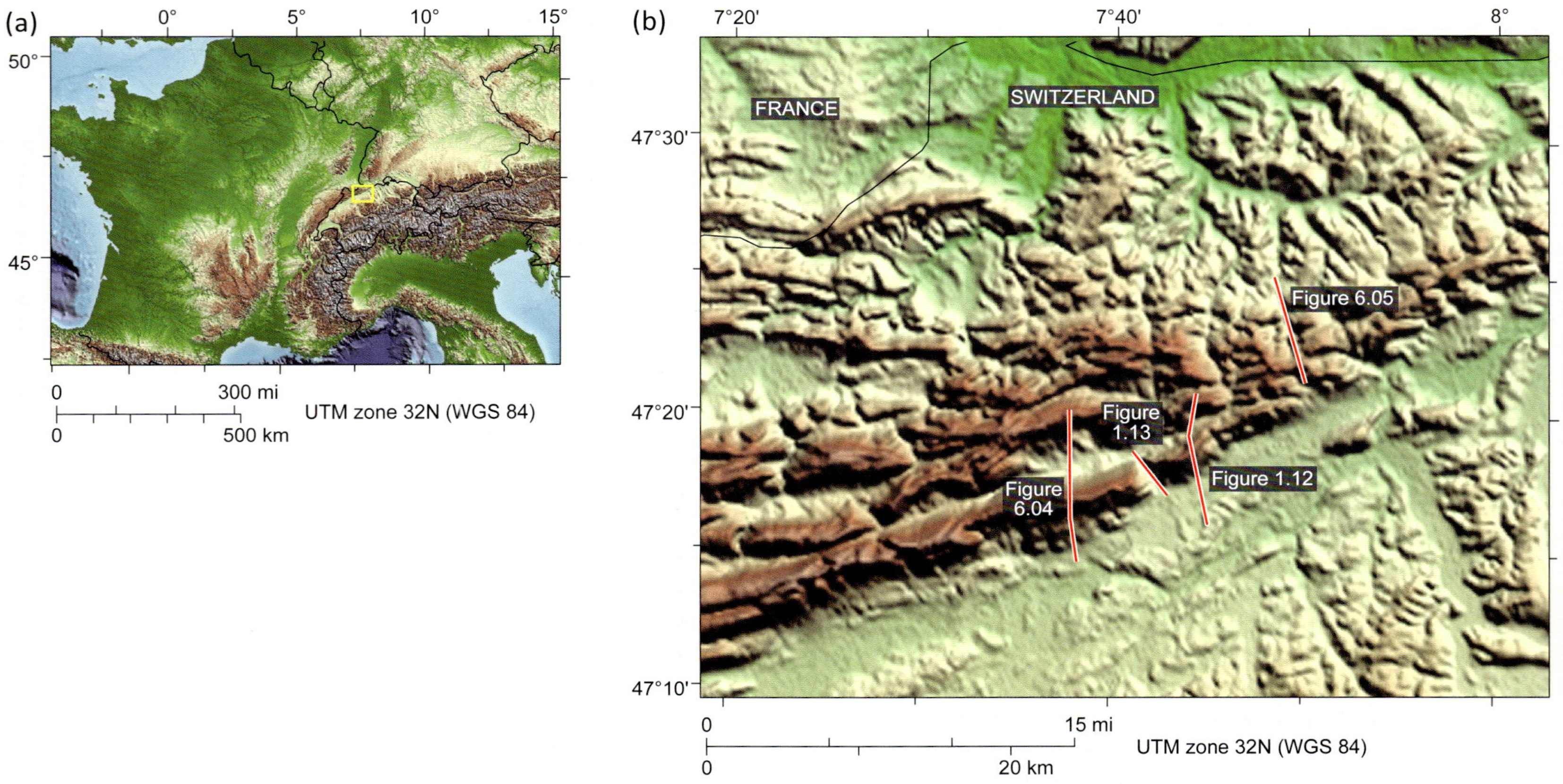

FIGURE 11.09—Katanga Fold Belt, Democratic Repulic of Congo

(a)

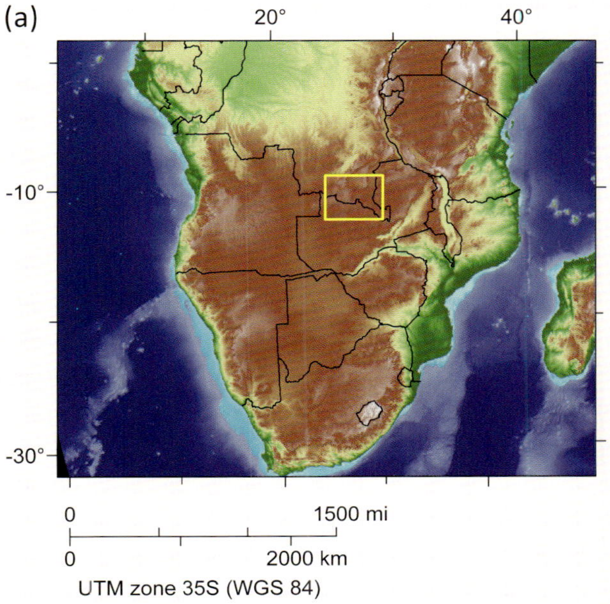

(b)

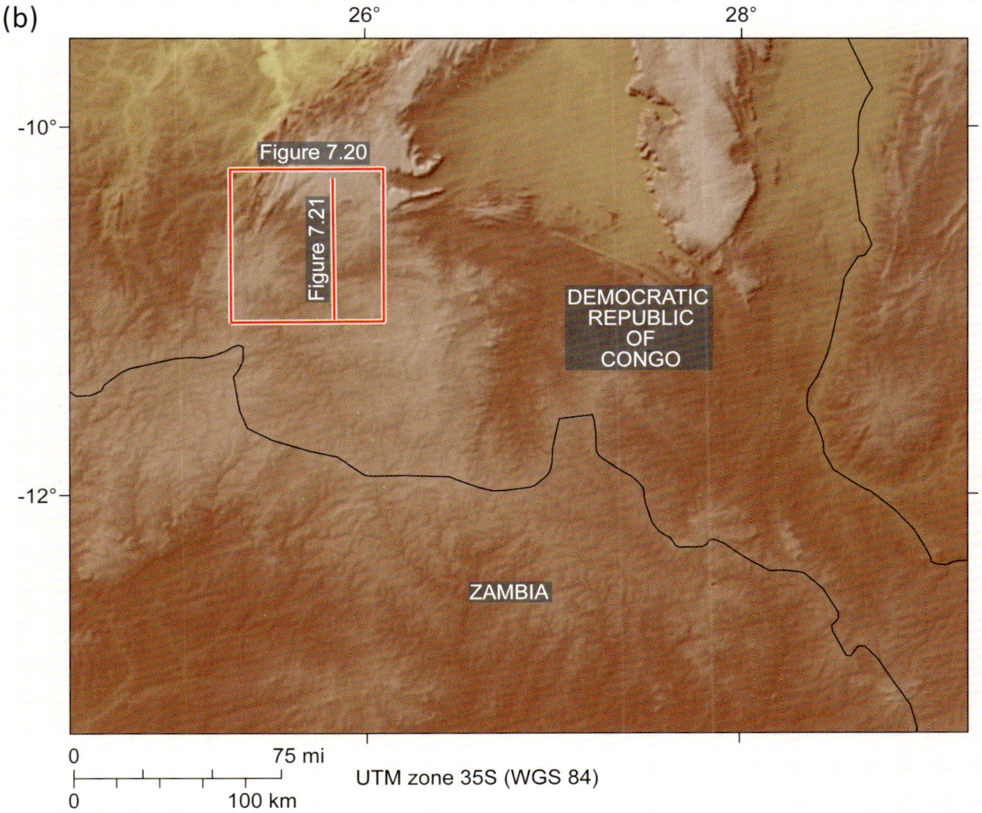

FIGURE 11.10—Kwanza Basin, Onshore and Offshore Angola

(a)

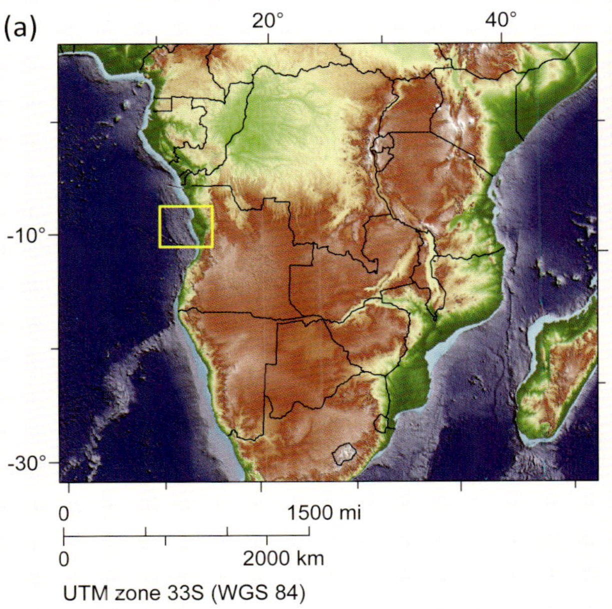

(b)

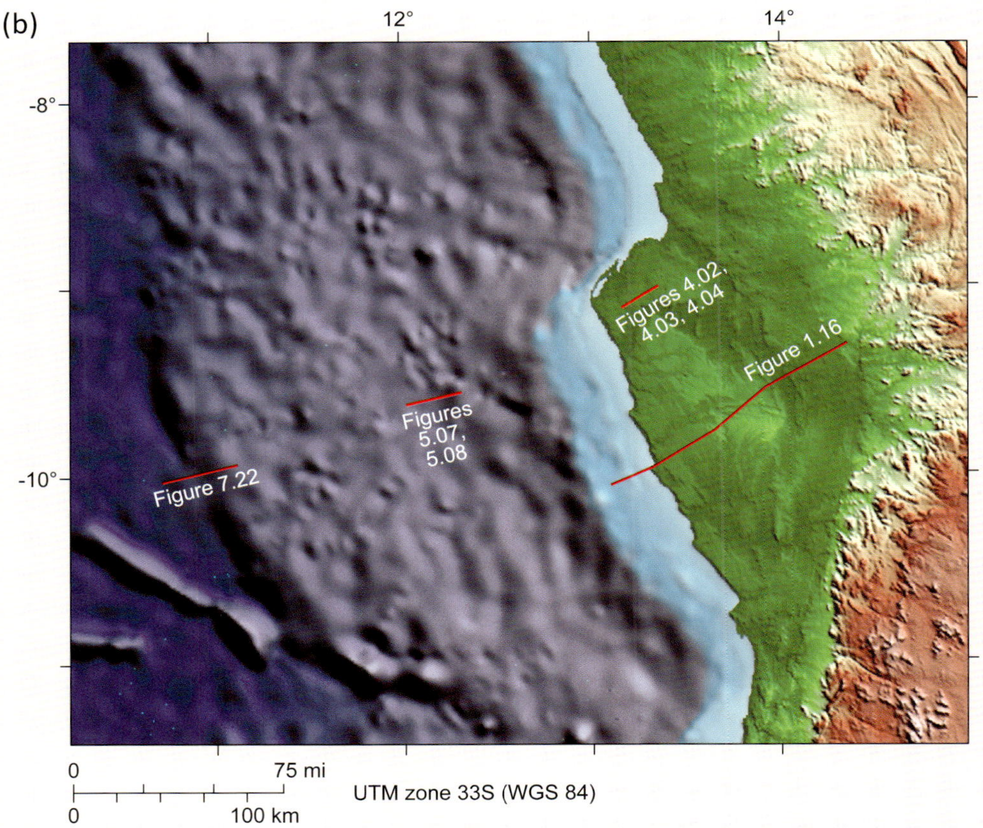

FIGURE 11.11—Lower Congo Basin, Offshore Angola, Congo, and Democratic Repulic of Congo

(a)

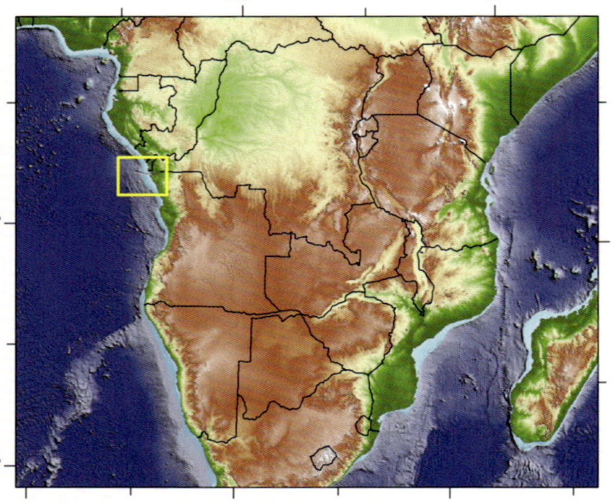

(b)

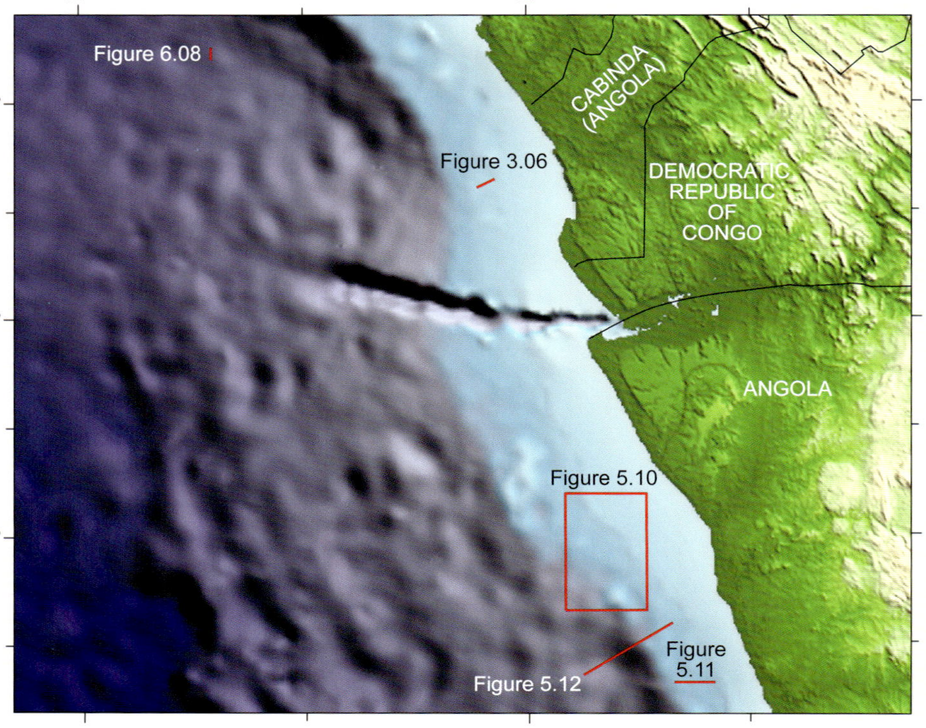

FIGURE 11.12—Maritime Alps, France

(a)

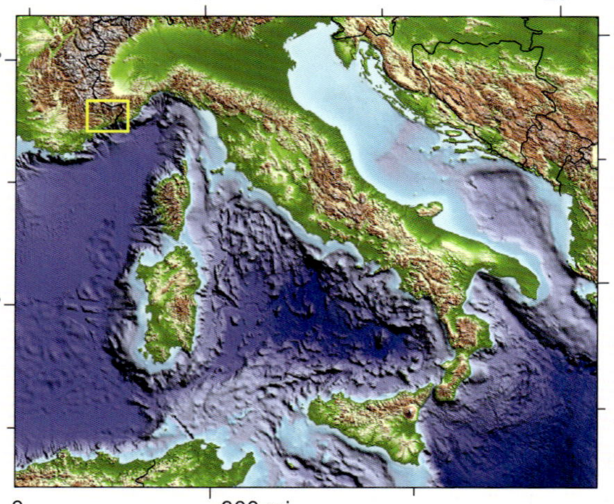

(b)

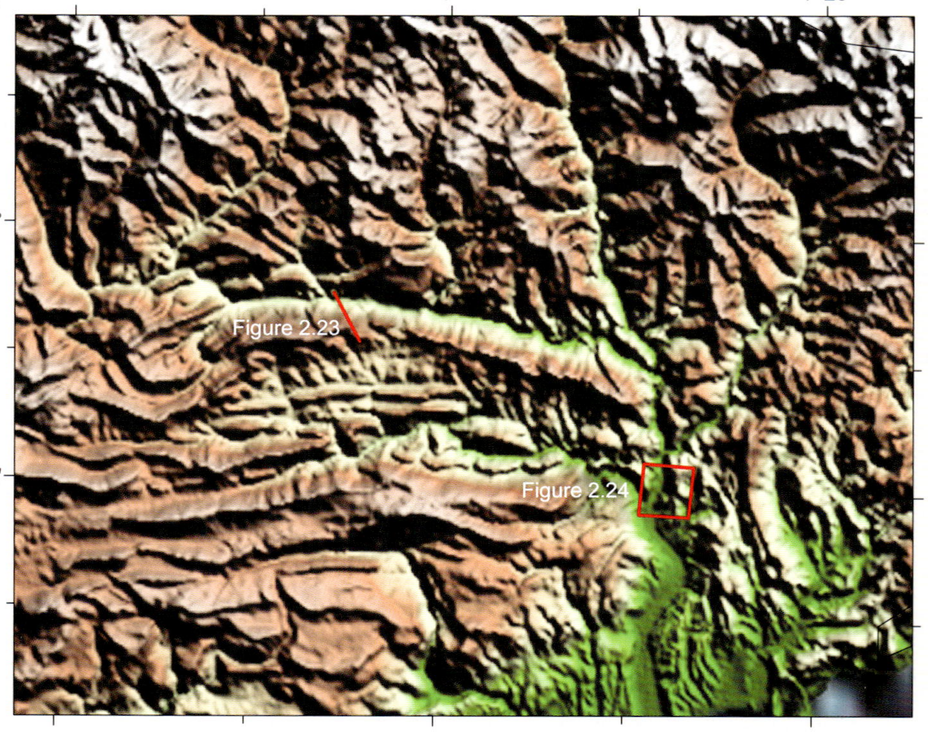

FIGURE 11.13—Nordkapp Basin, Offshore Norway

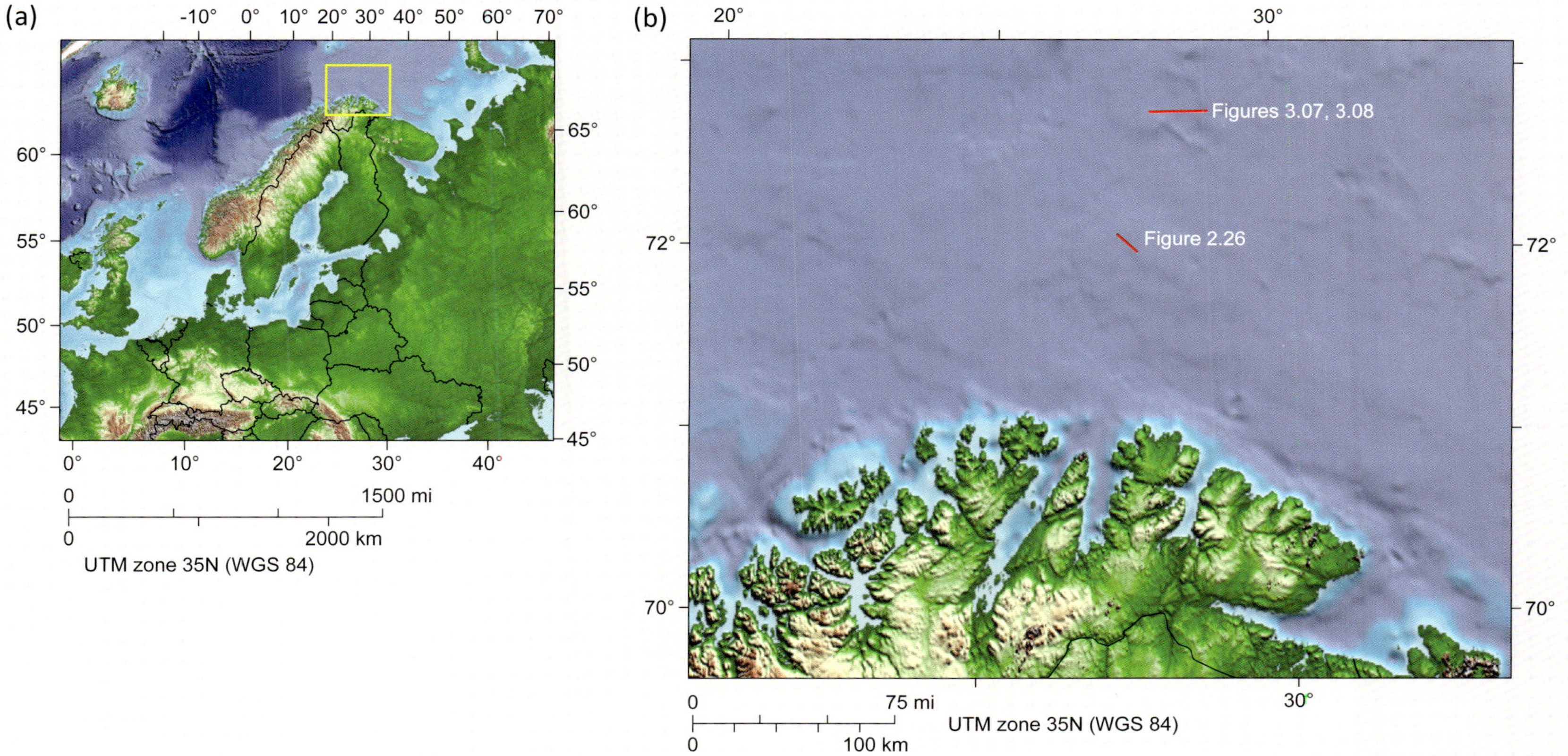

FIGURE 11.14—North German Basin, Germany

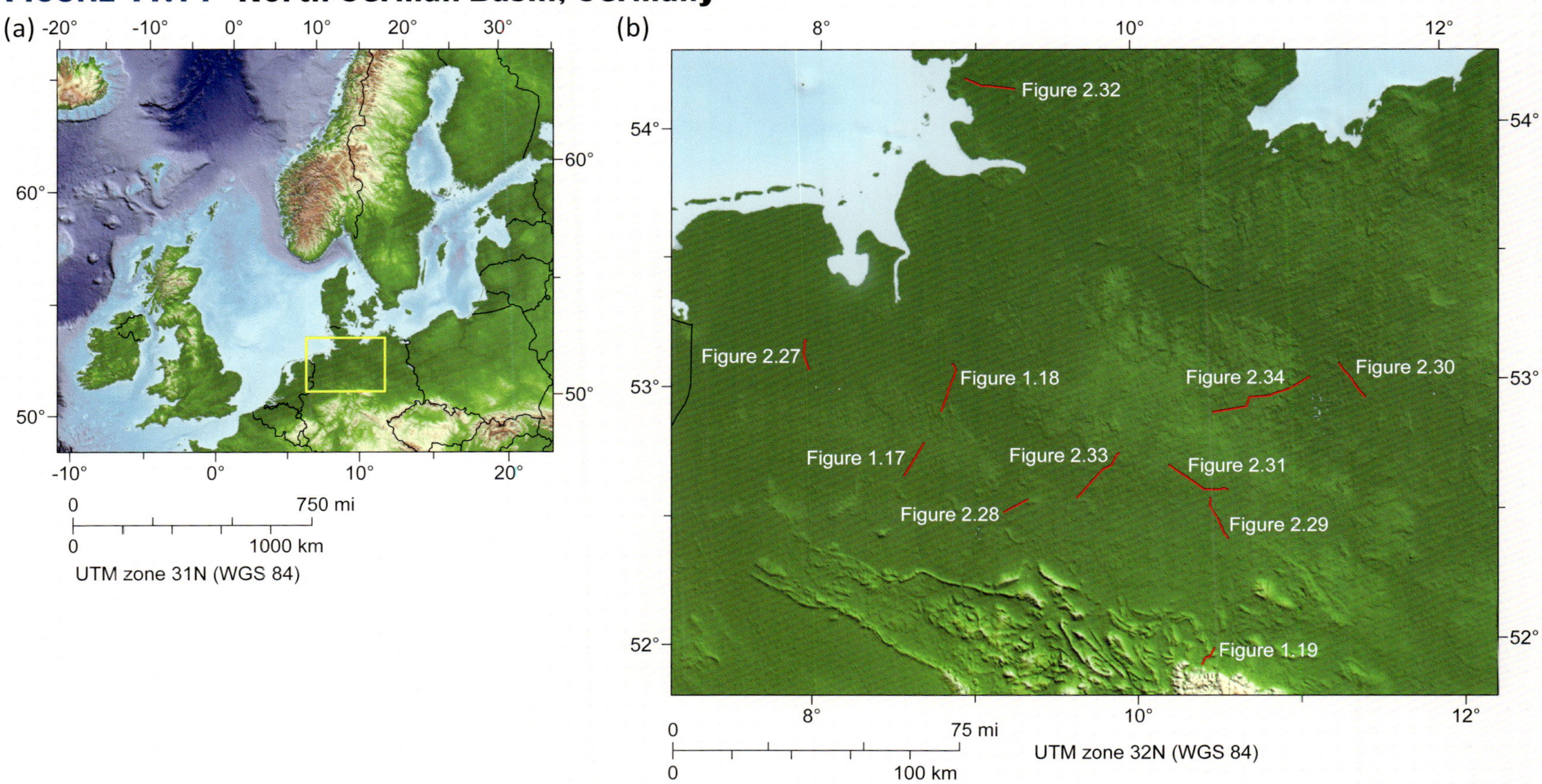

FIGURE 11.15—North Sea, Offshore UK

(a)

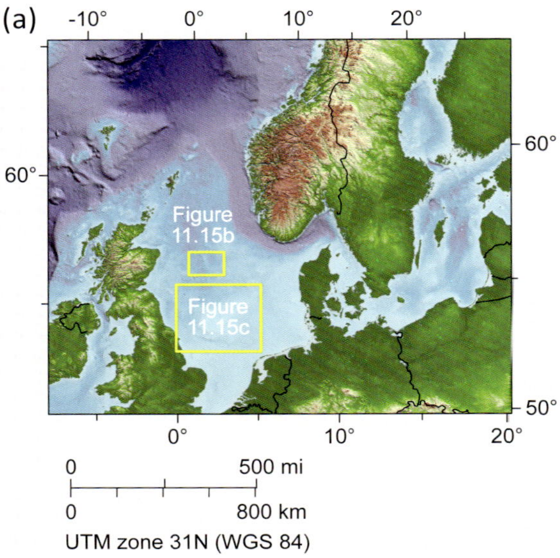

(b)

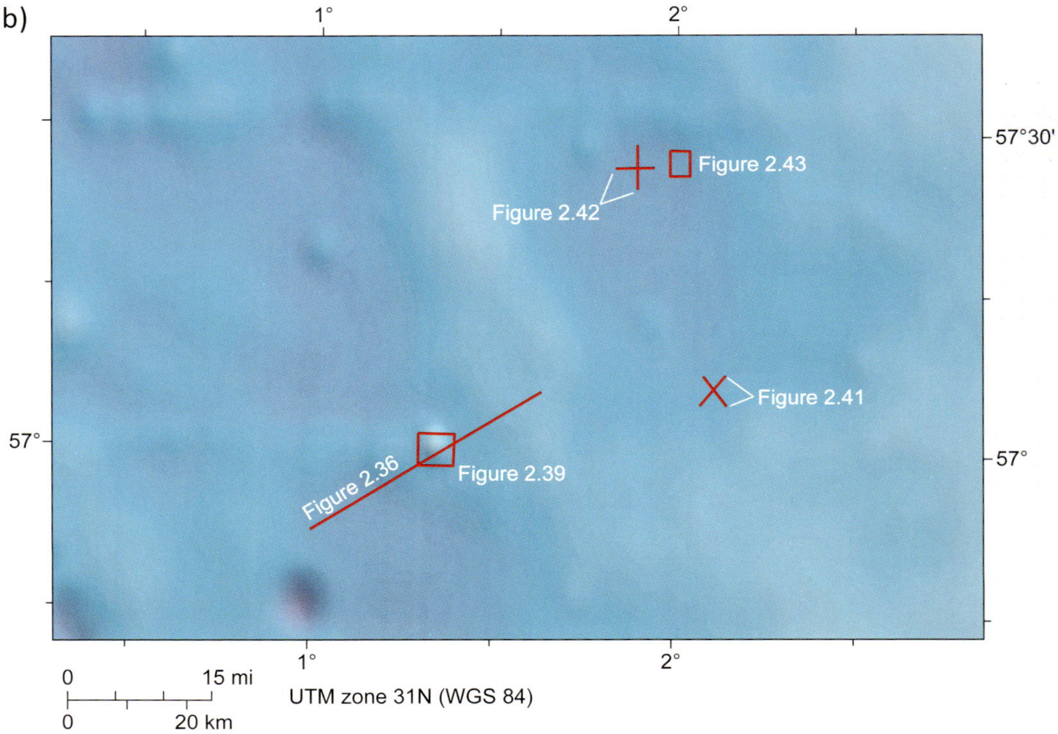

(c)

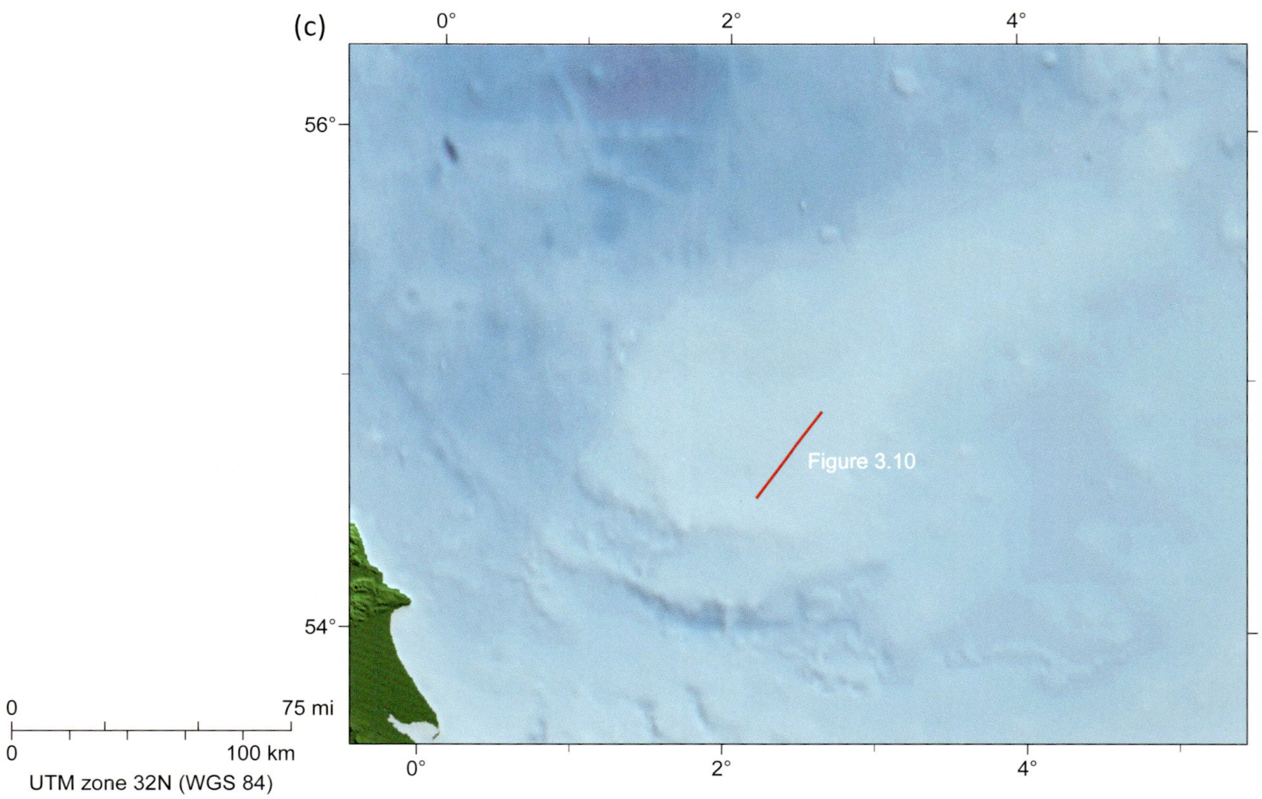

FIGURE 11.16—Norwegian Sea, Offshore Norway

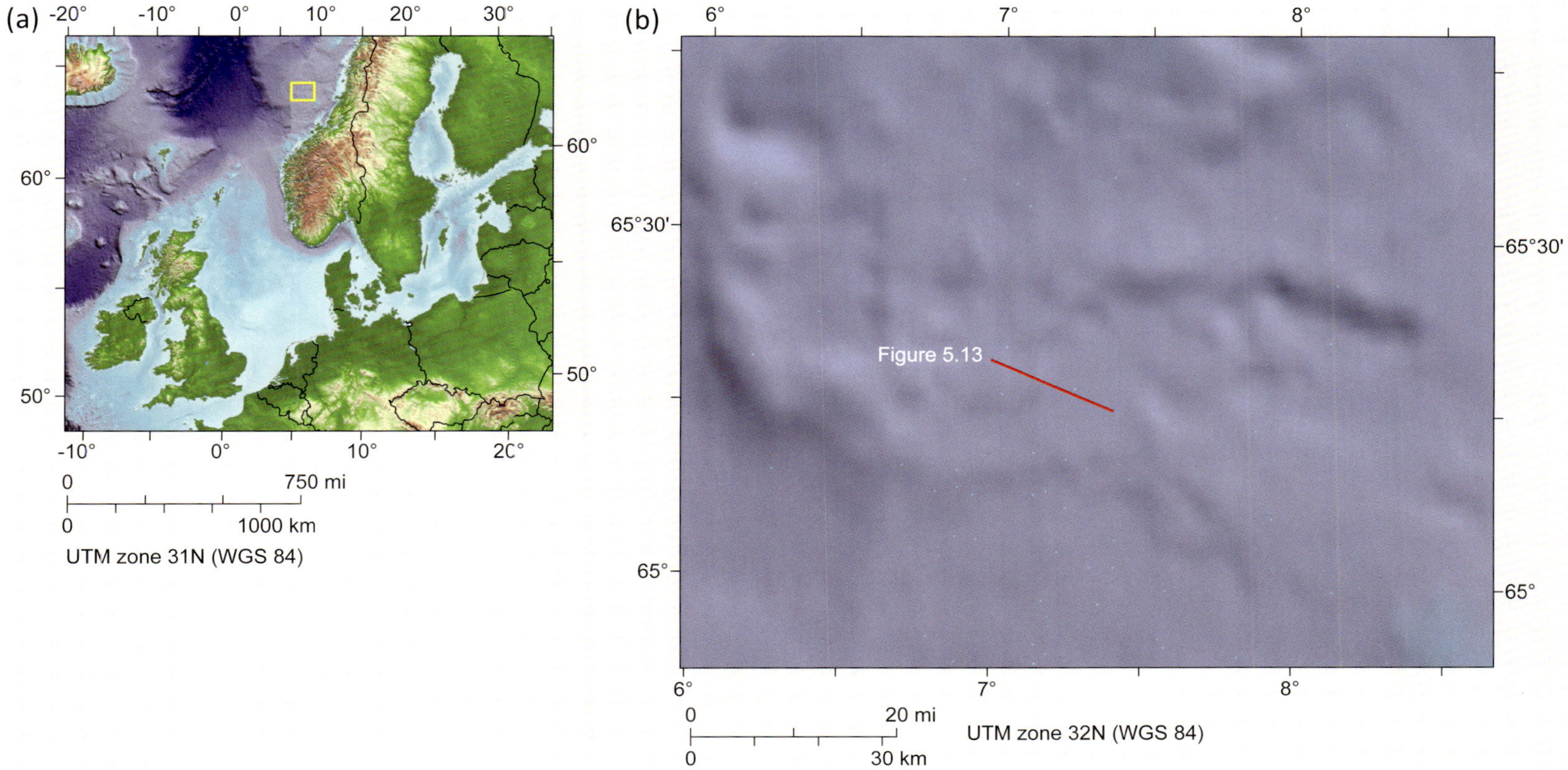

FIGURE 11.17—Paradox Basin, USA

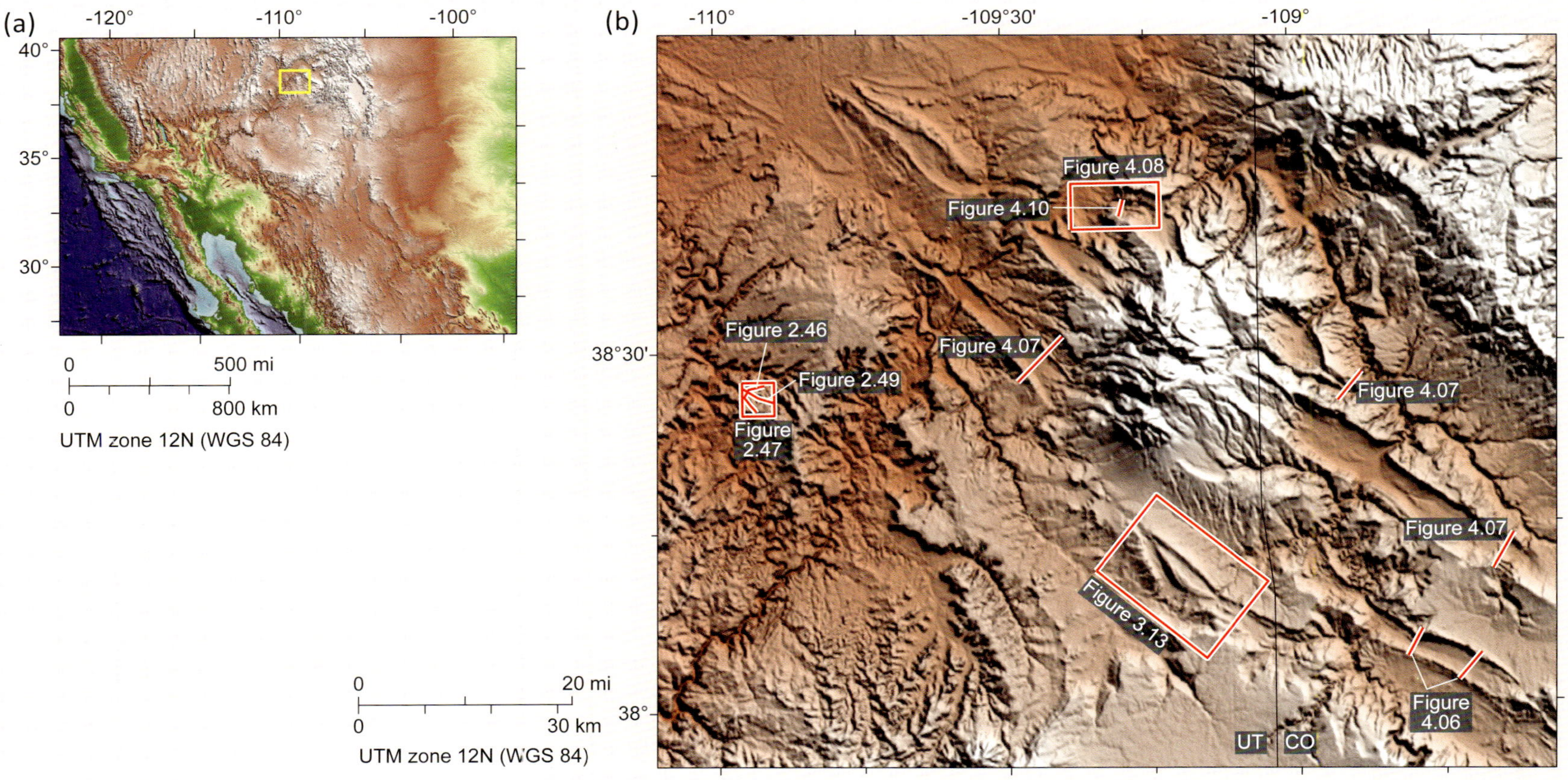

FIGURE 11.18—Potwar Plateau, Pakistan

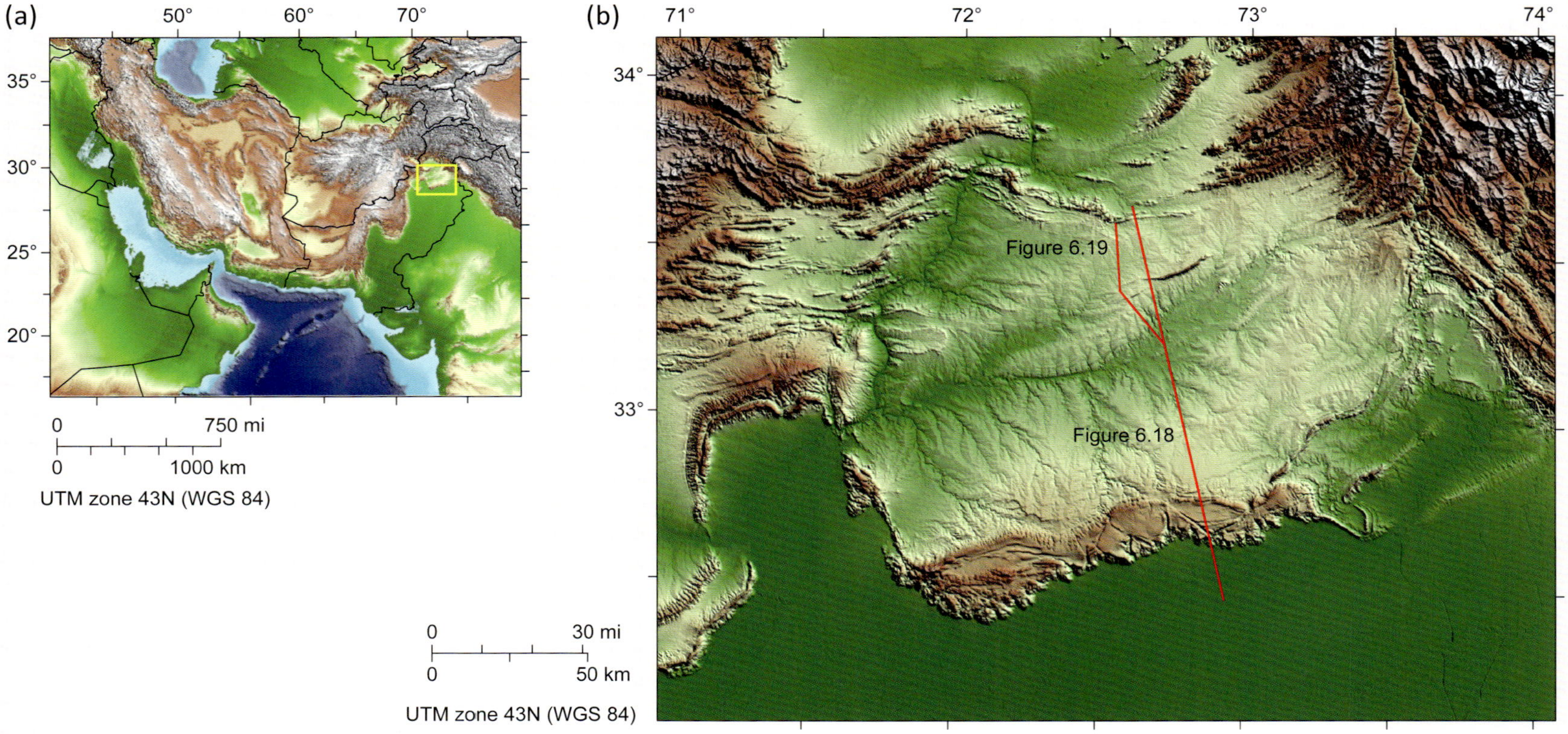

FIGURE 11.19—Precaspian Basin, Kazakhstan

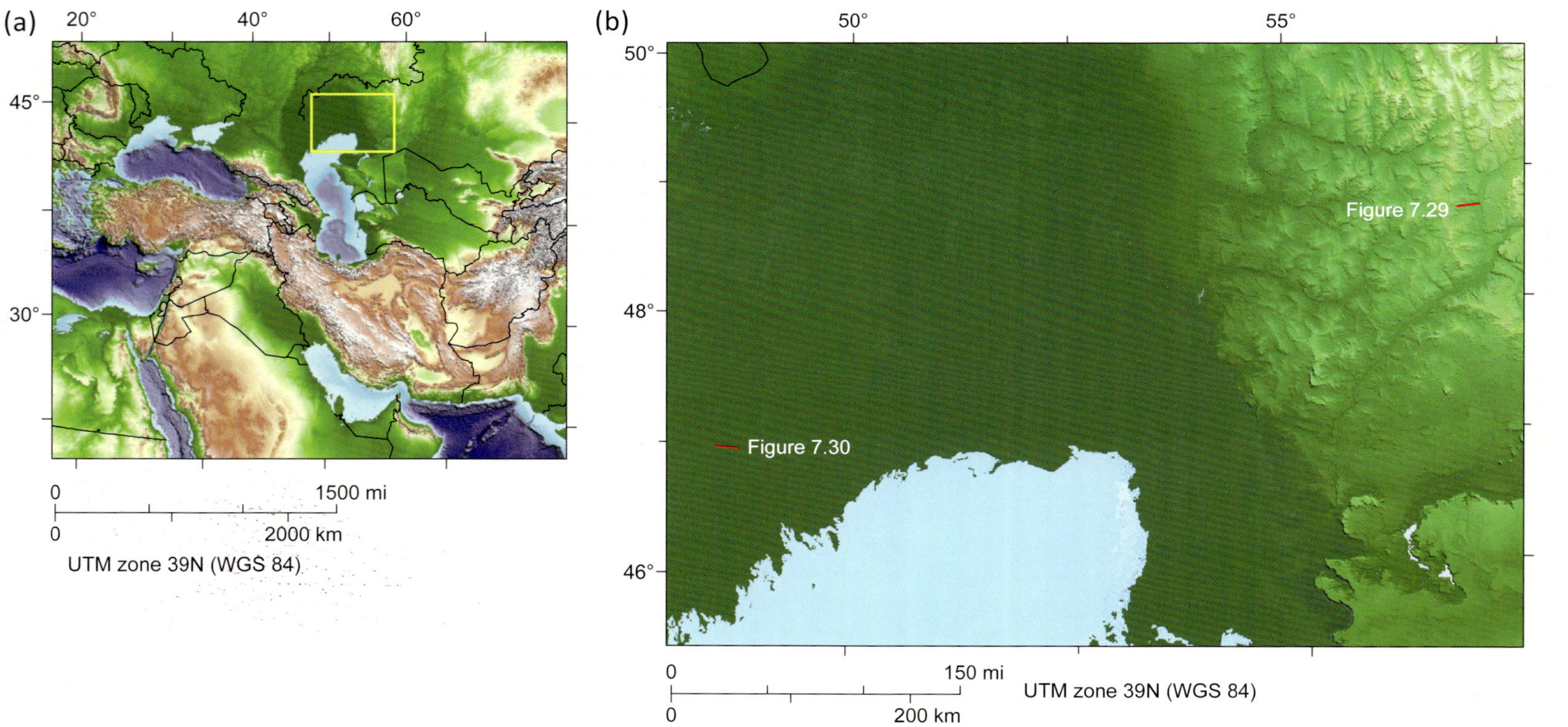

FIGURE 11.20—Pyrenees Fold Belt, France and Spain

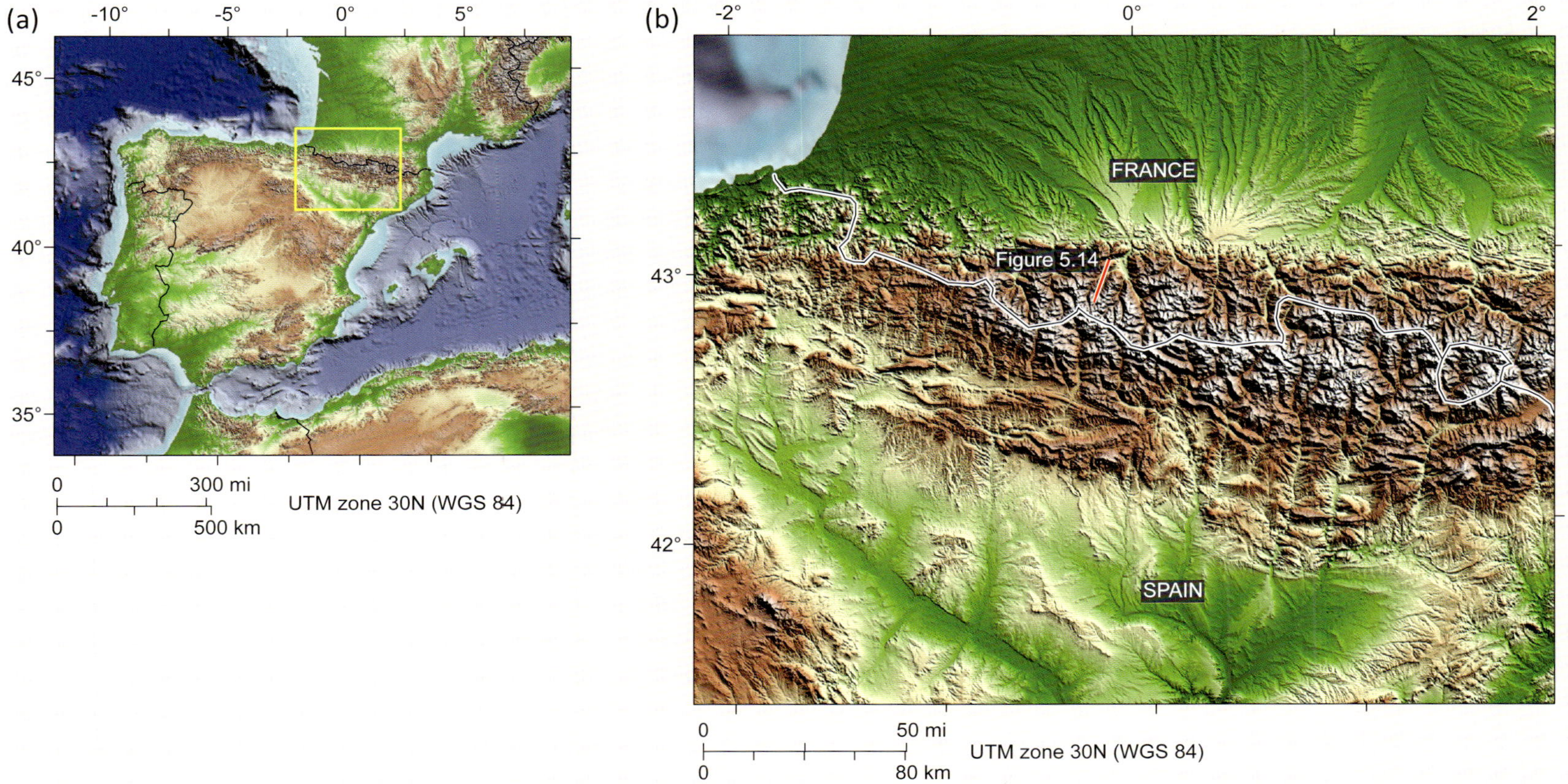

FIGURE 11.21—Sierra Madre Oriental Fold Belt, Mexico

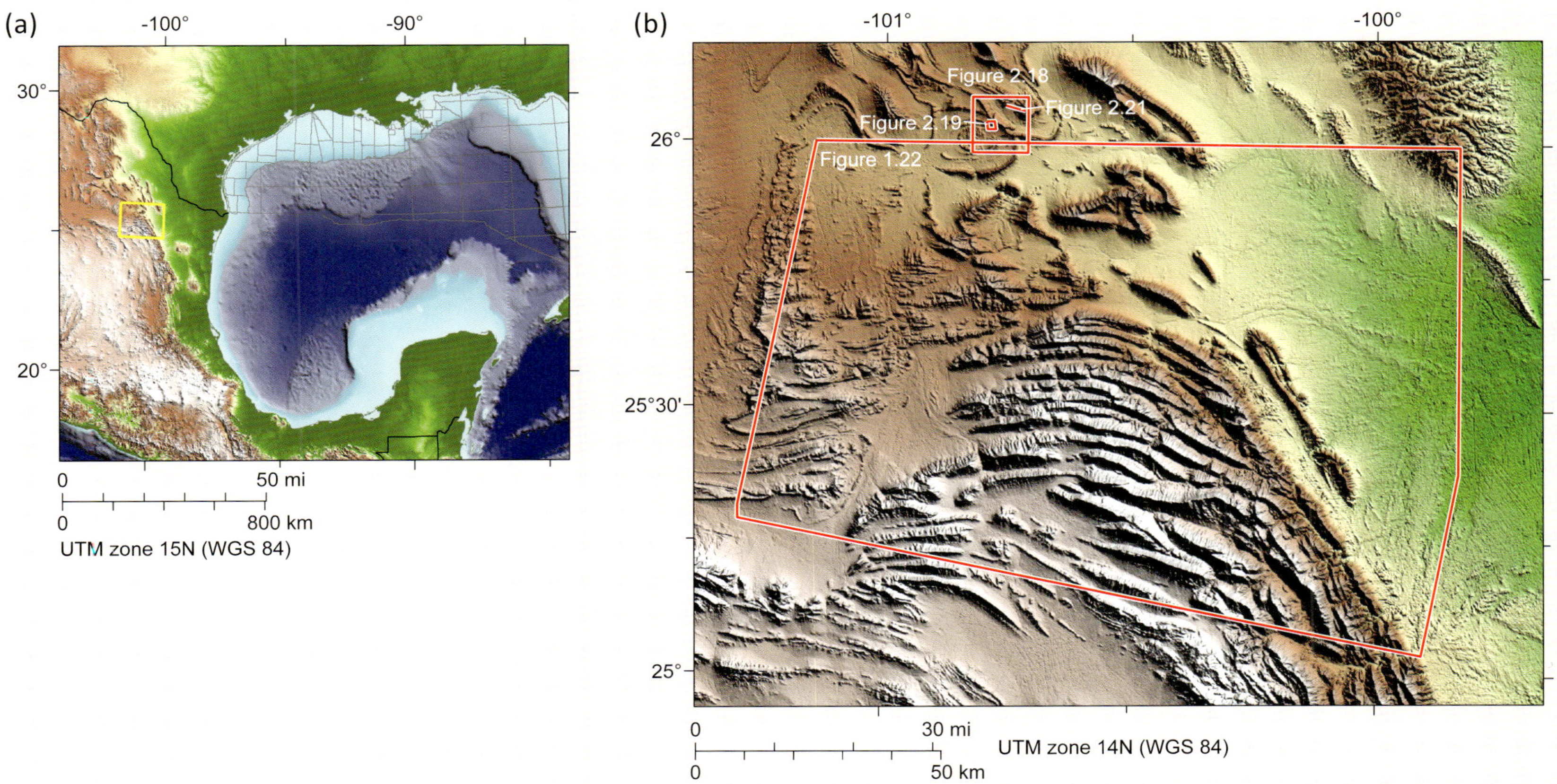

FIGURE 11.22—Zagros Fold Belt, Iran

(a)

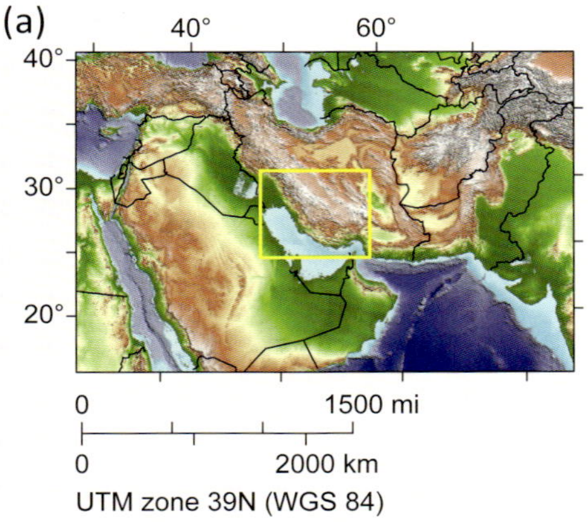

(b)

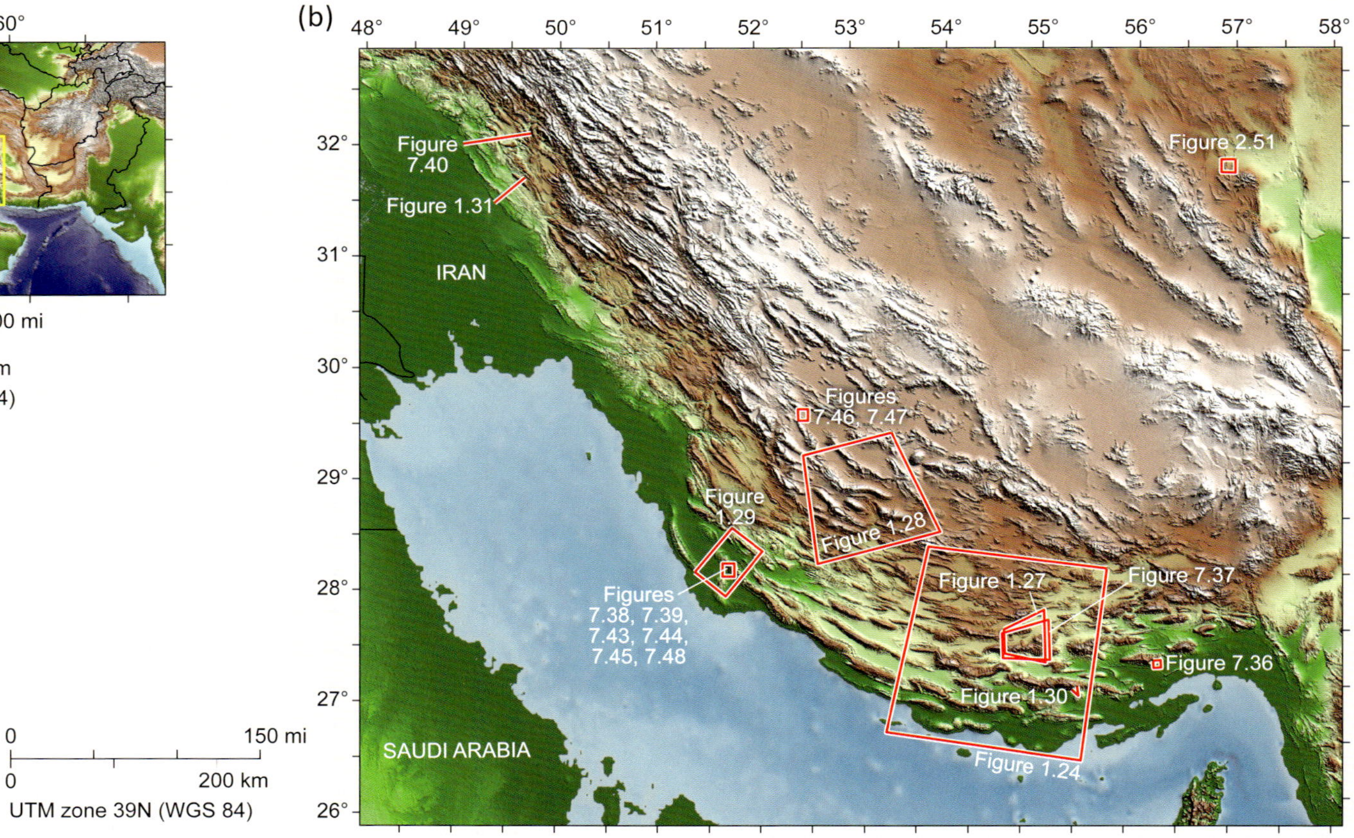

Glossary

active diapirism Diapir rise by arching, uplifting, or shouldering aside its *roof*. Active diapirism occurs in two ways. **Halokinetic active diapirism** is driven by *overburden* load on the source layer. When the average density of this *overburden* exceeds salt density, diapiric salt is pressurized and exerts an upward force on the roof. If this buoyancy force is greater than roof strength and weight, then the roof is pushed up and the diapir rises actively. Halokinetic active diapirism occurs in relatively tall diapirs overlain by relatively thin *overburden.* **Compressional active diapirism** is driven by regional compression, which squeezes the diapir, forcing salt upward to uplift, arch, or intrude the roof. Because regional tectonic forces are typically much greater than buoyancy forces, compressional active diapirism can deform a much thicker roof than can halokinetic active diapirism (Nelson, 1989).

allochthonous salt Subhorizontal or moderately dipping, sheetlike *salt diapir* emplaced at stratigraphic levels above the *autochthonous* source layer. Allochthonous salt overlies stratigraphically younger strata. The term can be applied even if the *salt sheet* remains attached to its source layer. The earliest known illustration of allochthonous salt is that of van der Fliert (1953) along the base of a thrust nappe in Algeria. Compare *autochthonous salt* and *para-autochthonous salt* (Naumann, 1858; Wilckens, 1912; Bally, 1981).

allosuture *Suture* between *salt sheets* having separate feeders.

apical suture line See *suture line*.

apical suture point See *suture point*.

apparent weld See *weld*.

autochthonous salt Salt layer that rests on its original, stratigraphically older *subsalt* strata or basement; informally known as *mother salt*. Compare *allochthonous salt* and *para-autochthonous salt*.

autosuture *Suture* between two lobes of the same *salt sheet*. There are two types of autosuture. An **encircling autosuture** forms where two lobes of a salt sheet separate to bypass an obstacle and then rejoin on the downdip side of the obstacle. An **overriding autosuture** forms where one part of a salt sheet overrides another part.

backthrust *Thrust fault* in a fold-thrust belt that dips opposite to the general dip of thrust faults and results in displacement toward the hinterland of the belt. Compare *forethrust*.

basal cutoff Oblique truncation of a *subsalt* stratigraphic layer by the lower contact of a *salt sheet*.

basal shear Advance of a *salt sheet* or *salt canopy* along a *detachment* near the salt base. The detachment may lie within the salt, along the salt contact, or within *subsalt* sediments (Chapple, 1978; Harrison and Patton, 1995).

basal shear zone See *shear zone*.

basal suture line See *suture line*.

basal suture point See *suture point*.

basement (a) Rock complex below the zone of interest. (b) In salt tectonics, basement is commonly synonymous with *presalt* rocks underlying the *source layer*. **Crystalline basement** comprises intrusive or metamorphic rocks, typically separated from the overlying *cover* by a major nonconformity.

basin-margin graben See *peripheral graben.*

boundary shear zone See *shear zone*.

bowl weld Funnel-shaped salt *weld* formed by the evacuation of salt from a funnel-shaped *salt diapir* infilled by younger clastic sediments (Rowan and others, 1999).

box fold Fold resembling three sides of a rectangle in profile. It is typically cored by a weak rock, such as evaporite or overpressured shale, which forms the detachment for folding.

breakaway graben See *peripheral graben.*

breakthrust Thrust that breaks through the steepening limb(s) of an anticline as the fold tightens (Willis, 1893).

bucket weld Structure formed by subsidence of a *minibasin* into a salt-sheet feeder until the minibasin *welds* onto the base of *autochthonous salt*. The flanks of the minibasin are welded onto the sides of the former diapir (Pilcher and others, 2011).

bulb Swollen, crestal part of a *salt diapir*. Its overhanging periphery is an **overhang**. Extremely broad bulbs grade into *salt sheets* (Jackson and Talbot, 1986).

buoyancy Instability caused by the tendency of an *overburden* to sink into a less-dense source layer. The buoyancy concept is related to *Rayleigh-Taylor instability* of a dense fluid overlying a less-dense fluid. But buoyancy is also applicable to brittle overburdens that have finite strength. The *density inversion* is due to lithologic differences (Lachmann, 1910; Arrhenius, 1913).

canopy See *salt canopy*.

canopy feeder See *salt feeder*.

canopy-margin thrust system Thrust faults forming at the leading edge of a buried, advancing *salt canopy* or *salt sheet* (Hudec and Jackson, 2009).

cap rock (a) A layer that is either more resistant to weathering or less permeable than the underlying layer. (b) In salt tectonics, a body composed predominantly of anhydrite, gypsum, or calcite above a salt body. Cap rock forms at the crest of a diapir as halite dissolves and the residue is chemically altered. Cap rock can contain economic minerals, especially sulfur and base-metal sulfides.

carapace In salt tectonics, *roof* strata comprising nearly parallel-bedded, mud-rich sediments deposited as a condensed pelagic or hemipelagic sequence on the bathymetric mound above a diapir. Carapaces can break into smaller **carapace blocks** by multidirectional spreading of underlying salt. Carapace blocks are typically older and more cemented than surrounding sediments (McGuinness and Hossack, 1993; Fletcher and others, 1995; Hart and others, 2004).

carapace block See *carapace*.

Christmas-tree structure See *salt wing*.

complete weld See *weld*.

composite halokinetic sequence A stack of *halokinetic sequences* equivalent in scale to a third-order depositional sequence and bounded by angular unconformities. Stacked *hook halokinetic sequences* form tabular composite sequences having subparallel lower and upper unconformity boundaries, a narrow zone of *drape folding* and thinning, and drape-fold axial traces stacked and subparallel to the salt-sediment interface. Stacked *wedge halokinetic sequences* form tapered composite halokinetic sequences having convergent lower and upper boundaries, a broad zone of drape folding and thinning, and stacked drape-fold axial traces inclined to the diapir. Tabular composite halokinetic sequences form if the ratio of diapir-rise rate to sediment-accumulation rate is high; tapered composite sequences form if the ratio is low (Giles and Rowan, in press).

composite suture line See *suture line*.

composite suture point See *suture point*.

compressional active diapirism See *active diapirism*.

concordant Said of a salt contact that is parallel to encasing strata. Compare *discordant*.

contraction (shortening) Occurs in a strict sense wherever an object decreases in length in any direction. As more commonly used in salt tectonics, contraction refers to a decrease in length in the horizontal direction. Even more restrictively, a **contractional fault** is a fault that accommodates layer-parallel shortening. Contractional fault *hanging walls* are typically elevated above *regional,* and strata are repeated above and below the fault. Compare *extension*.

contractional fault See *contraction*.

Couette flow Laminar flow by simple shear. Plane Couette flow is between two plates in relative motion; circular Couette flow is between two coaxial cylinders in relative motion. Couette flow applies loosely to salt flow in which maximum velocity is greatest at the top of the salt, analogous to flow in ice glaciers and ice sheets. Compare *Poiseuille flow*.

counterregional canopy *Salt canopy* fed by *counterregional systems*.

counterregional system A landward-dipping, largely welded salt system whose upper part is a weld, remnant diapir, or normal fault. Sediment wedges dip and thicken seaward. The lower part of the system may ramp continuously down to the source layer or may include flat-lying segments. If the system has more than one ramp, it is a **stepped counterregional system**. Compare *roho system* (Schuster, 1995).

cover Entire sedimentary pile above crystalline *basement*. Cover includes *presalt* strata, *salt*, and *overburden*.

crescentic fold Fold having an axial trace concentric to the surrounding diapir contact and a shallow-plunging fold hinge, common inside the *bulb* of a diapir (Jackson and Talbot, 1989a).

crestal graben Dominantly planar *normal faults* that dip toward one another and root into the crest of a salt structure. Crestal grabens include a **reactive graben** above *reactive diapirs* and a **drape graben** in the crest of a drape *monocline* above subsalt *extensional* faults (Rowan and others, 1999). Compare *keystone graben*.

crystalline basement See *basement*.

curtain fold Cylindrical fold having roughly radial axial trace and steeply plunging hinge within a diapir, possibly incorporating sheath folds originally formed in the source layer (Stier, 1915).

décollement See *detachment.*

density inversion Superposition of a denser layer above a less-dense layer. See *buoyancy*.

depotrough An elongate *minibasin.*

detachment (décollement) Horizontal or gently dipping fault, shear zone, or weak rock layer that decouples or separates deformation in overlying and underlying rocks.

diapir See *salt diapir*.

diapir fall Subsidence of a widening diapir's crest or flanks by regional *extension* or *dissolution*. In extension, the diapir subsides because salt is not imported fast enough from the adjoining source layer to fill the widening diapir (Jackson and Vendeville, 1990; Vendeville and Jackson, 1992b).

diapir family Adjoining *salt walls* or *salt stocks* comprising several generations rooted in the same *source layer*; for example, a mother diapir flanked by daughter diapirs, in turn flanked farther outward by granddaughter diapirs. The peripheral sinks of each generation overlap (Sannemann, 1968).

differential loading Application of a variable load on a salt body from any direction. All salt flow is produced by some combination of three types of differential loading: *displacement loading*, *gravitational loading*, and *thermal loading*. Which combination is most important depends on the depth of salt burial, geometry of the salt body, geologic setting, and thermal conditions of the salt (Hudec and Jackson, 2007).

discontinuous weld See *weld*.

discordant Said of a salt contact oblique to encasing strata. Compare *concordant*.

displacement loading *Differential loading* resulting from forced displacement of one boundary of a rock body relative to another. This type of loading occurs when the flanks of a salt body move toward or away from one another during regional *shortening* or *extension* (Suppe, 1985). See also *gravitational loading* and *thermal loading*.

dissolution In salt tectonics, the processes of dissolving salt into surface or subsurface water. Dissolution leaves less-soluble materials behind that can contribute to residual soil or *cap rock*. Dissolution near the surface can create karst features.

downbuilding See *passive diapirism*.

drape fold (a) Fold produced in layered rocks by movement of an underlying block at high angles to the layering; a type of forced fold. (b) In salt tectonics, a zone of upturned strata adjacent to a diapir. Upturning results from arching of the diapir *roof* above flanking strata and is an important process in the formation of *halokinetic sequences* (Rowan and others, 2003). Also known as a **flap fold** (Schultz-Ela, 2003).

drape graben See *crestal graben.*

dynamic bulge Bulge caused by pressure of flowing fluid; for example, the viscous upward flow of diapiric salt, like a slow-moving fountain. A striking example is the mountainous extrusive salt domes of southern Iran. Dynamic bulges can lift strata above *regional*; this deformation is recorded by onlapping strata or erosional truncations, which may themselves be deformed later. A diapiric bulge is likely to subside after the underlying stem pinches off or the source layer is exhausted. The record of a decaying bulge is preserved as *structural inversion* in any *synkinematic overburden* (Talbot and Jarvis, 1984).

encased minibasin Secondary minibasin covered by a higher *allochthonous salt sheet* or equivalent salt weld. Also called a **foundered minibasin** (Pilcher and others, 2011). See also *minibasin*, *primary minibasin*, *secondary minibasin*.

encircling autosuture See *autosuture*.

enterolithic fold Intestinelike fold in salt, variously ascribed to (1) volume changes, such as hydration of anhydrite to gypsum, which involves a 61% increase in volume; (2) diagenesis; (3) minor folding by gravity; or (4) larger-scale folding.

expulsion rollover See *rollover*.

extension Occurs in a strict sense wherever an object increases in length in any direction. As more commonly used in salt tectonics, extension refers to stretching in a horizontal direction. Even more restrictively, an **extensional fault** results in layer-parallel stretching. Typically, extensional fault *hanging walls* subside below *regional,* and strata are excised across the fault. Subsequent tilting does not change the class of an extensional fault, unlike a *normal fault.* Compare *contraction*.

extensional fault See *extension*.

external mushroom structure See *mushroom diapir*.

external shear zone See *shear zone*.

extrusive advance Advance of a *salt sheet* without a sedimentary *roof* or with a roof having negligible mechanical strength (McGuinness and Hossack, 1993; Fletcher and others, 1995; Hudec and Jackson, 2006).

fault weld Fault surface or fault zone joining strata originally separated by salt. Equivalent to a *weld* along which there has been significant fault slip or shear (Hossack and McGuinness, 1990).

feeder, feeder diapir, feeder stem, feeder stock See *salt feeder*.

flap fault *Normal fault* whose footwall contains an elongated *active diapir* having a *roof* that is rotated and lifted above *regional* and commonly partly eroded (Nelson, 1991).

flap fold See *drape fold*.

footwall Underlying side of a fault, orebody, or mine working. Compare *hanging wall*.

forethrust *Thrust fault* in a fold-thrust belt that dips parallel to the general dip of faults and results in displacement toward the foreland of the belt. Compare *backthrust*.

foundered minibasin See *encased minibasin*.

freeboard Height of the crest of a *salt diapir* above the surrounding sediment surface (Talbot, 1993).

frontal escarpment Gentle escarpment above the leading edge of an advancing or inflating *salt sheet*. The escarpment can be a fold scarp, fault scarp, or erosional scarp (Davis, 1903; Hudec and Jackson, 2006).

frontal rolling See *tank-tread advance*.

frontal suture See *suture*.

gravitational loading *Differential loading* produced by the weight of rocks overlying the salt and by gravitational body forces within the salt. Gravitational loading is typically caused by lateral variations in thickness or density of the *overburden*. Such variations may be sedimentary, erosional, or structural in origin. This mechanism is a factor in most salt tectonism and may initiate *halokinesis* (Harrison, 1927; Bailey, 1931; Hollingworth and others, 1945). See also *displacement loading* and *thermal loading*.

gravity gliding Downslope sliding by gravity of a rigid *overburden* sheet or block over a weak, ductile *décollement* layer (such as salt) or over a décollement fault. The base of the *overburden* dips sufficiently that gravitational body force is resolved into a downslope shear stress large enough to overcome the shear strength of the décollement layer or the frictional strength of the décollement fault (Schardt, 1893; Smoluchowski, 1909). Typically, because gravity-gliding systems are part of larger gravity-spreading systems, the term *gravity spreading* is more accurate. A rigorous distinction between gravity spreading and gravity gliding is spurious (Schultz-Ela, 2001).

gravity spreading Vertical collapse and lateral spreading of a rock mass under its own weight. In salt tectonics, gravity spreading typically operates on the salt and its *overburden*. A dipping topographic surface allows a small component of the gravitational body force to be resolved into a downslope shear stress to drive spreading. The base of the salt can be horizontal, or it can dip opposite to the surface slope. Examples are the spread of *salt glaciers* or extrusive *salt sheets* (Scrope, 1825; Nye, 1952; Bucher, 1956; Ramberg, 1981). Compare *gravity gliding*.

growth fault (synsedimentary fault) Fault that forms during sedimentation. Its throw increases with depth, and strata on the downthrown side are thicker than correlative strata on the upthrown side. Both *extension* and *contraction* can produce growth faults (Kidd, 1939; Ocamb, 1961).

growth interval See *synkinematic*.

halokinesis Type of *salt tectonics* in which salt flow is powered entirely by gravitational forces (release of gravity potential energy) in the absence of significant lateral tectonic forces (Trusheim, 1957).

halokinetic active diapirism See *active diapirism.*

halokinetic sequence Relatively conformable succession of drape-folded growth strata genetically influenced by near-surface or extrusive salt flow and equivalent in scale to parasequence-set-scale depositional sequences. Halokinetic sequences are locally bounded at the top and base by angular unconformities that become disconformable to conformable with increasing distance from the diapir (Giles and Lawton, 2002). Two geometrically distinct types of halokinetic sequences are recognized: *hook halokinetic sequences* and *wedge halokinetic sequences* (Giles and Rowan, in press). See also *composite halokinetic sequence, drape fold.*

halotectonics An obsolete term for *salt tectonics.*

hanging wall The overlying side of a fault, orebody, or mine working. Compare *footwall.*

heel-toe A stacked pair of synextensional growth wedges in which the lower wedge thickens and tilts landward and the upper wedge thickens and tilts seaward. The pair of wedges records seaward expulsion of salt by prograding sediments (concept by Worrall and Snelson, 1989; term by Chimney, 2000, and Jackson and Schultz-Ela, 2000).

hook halokinetic sequence A *halokinetic sequence* in which the basal sequence boundary curves sharply upward within 50 to 200 m of the diapir contact and is truncated at up to 90° by an overlying unconformity marking the base of the next younger sequence. Mass-wasting deposits are common at the base of the sequence, and facies change abruptly near the diapir (Giles and Rowan, in press). See also *wedge halokinetic sequence*, *composite halokinetic sequence*, *drape fold.*

imbricate wedge Zone of imbricated thrust faults forming in front of some buried advancing *salt sheets* or *salt canopies*. The wedge tapers in the direction of advance and comprises slices of sediment from the peripheral plain ahead of the advancing salt sheet (Fox, 1959; Hudec and Jackson, 2009).

incomplete weld See *weld.*

infold See *mushroom diapir.*

internal mushroom structure See *mushroom diapir.*

internal shear zone See *shear zone.*

intrasalt basin See *minibasin.*

inversion See *structural inversion.*

isopachous Refers to a sedimentary interval having uniform thickness.

keel See *salt keel.*

keystone graben *Extensional* faults in the hinge of an anticline, where strata are most strongly curved. Keystone grabens form by local stretching of the outer arcs of *turtle structures*, *rollover* anticlines, *contractional* anticlines, and domes arched above *salt diapirs* or *salt pillows*. Typically fault slip diminishes downward to zero before the faults reach salt. Compare *crestal graben* (Crosby and Crosby, 1925).

lateral suture See *suture.*

level of neutral buoyancy Depth at which the bulk density of the *overburden* equals that of salt (about 2,200 kg/m^3). At this level, a diapir spreads most rapidly if surrounded by a yielding fluid *overburden* (including air or water). Terrigenous clastics must be buried before they compact, dehydrate, or cement to reach the density of salt. This burial depth varies from basin to basin but is generally 600 to 2,200 m, depending on lithology and compaction state (Jackson and Talbot, 1989b; Hudec and others, 2009).

listric fault Curved, upward-steepening fault surface (Suess, 1909).

minibasin (intrasalt basin) Small basin largely surrounded by and subsiding into relatively thick *allochthonous* or *autochthonous* salt. Such basins can merge over time into larger, composite basins. Salt expelled from beneath a minibasin wells up around the minibasin margin, forming a network of salt walls, massifs, or welds that partly or completely surround the minibasin. See also *primary minibasin*, *secondary minibasin*, *encased minibasin*.

mock-turtle anticline Sediment-cored anticline produced by evacuation of salt from a *salt wall*. Distinguished from a *turtle structure* by missing a considerable stratigraphic section from its core (Vendeville and Jackson, 1992b).

monocline Local steepening of an otherwise uniformly dipping sequence, forming a fold having a single limb (Daly and others, 1916).

mother salt Informal equivalent of *source layer*. See *autochthonous* salt.

mushroom diapir *Salt diapir* having a broad *bulb* fringed by one or more laterally flattened, pendant **skirts** of deformed evaporites that envelop the stem of the diapir. Both the skirts and the **infolds** that they surround are downward-facing folds. With **external mushroom structure**, skirts surround antiformal, infolded synclines of *overburden*. With **internal mushroom structure**, the skirts are contained entirely within the diapir and thus surround antiformal, infolded synclines of evaporite (Jackson and Talbot, 1989a).

namakier See *salt sheet*.

normal fault Fault of any dip in which the *hanging wall* has slipped downward relative to the *footwall*. Subsequent tilting of a normal fault's dip beyond the vertical can change it to a *reverse fault*.

open-toed advance Advance of a *salt sheet* that is partly covered by a mechanically significant *roof* but that has an extrusive toe (Hudec and Jackson, 2006).

overburden Strata younger than the salt *source layer*. Overburden is generally used as a stratigraphic rather than structural term; for example, *allochthonous salt* overlies part of its overburden.

overhang See *bulb*.

overriding autosuture See *autosuture*.

para-autochthonous salt Salt that has flowed laterally beyond its depositional limit but remains in the same stratigraphic position. Para-autochthonous salt is thought to exist at the downdip ends of some passive-margin salt basins, where salt flowed seaward over new crust, such as oceanic crust or exhumed lower crust or mantle. Compare *autochthonous salt* and *allochthonous salt*.

passive diapirism (downbuilding) Syndepositional growth of a diapir whose exposed crest rises as sediments accumulate around it. The diapir's crest can be periodically buried, but the diapir repeatedly breaks through this thin, ephemeral *roof*. The diapir's base subsides, together with encasing strata, as the basin fills with sediment (Barton, 1933; Nelson, 1989; Vendeville and Jackson, 1991).

pedestal Base of a diapir, where the diapir widens to merge with the source layer.

peripheral graben (basin-margin graben, breakaway graben) Normal faults formed at the updip margin of a salt basin or *allochthonous salt sheet* to accommodate differential motion between translating *overburden* above salt and static overburden not above salt (Pierce, 1960; Hughes, 1968).

peripheral sink Locally thickened, synkinematic strata accumulating in a *rim syncline* as a result of *salt withdrawal*. Peripheral sinks may have any shape in plan and are differentiated according to thickness variations within them. A **primary peripheral sink** accumulates on the flanks of growing *salt pillows* and comprises strata that thin toward the salt structure. A **secondary peripheral sink**, which accumulates around diapirs withdrawing salt from their precursor pillow, contains strata that thicken toward the *salt diapir*. A **tertiary peripheral sink** accumulates around mature diapirs that have exhausted the salt supply from their precursor pillow and are rising by necking or thinning of the diapir's stem; coeval strata thin over the crest of the *salt diapir*. Some authors have questioned the mechanical significance of peripheral sinks (Trusheim, 1960).

pierce See *piercement*.

piercement Emplacement of a *salt diapir* to create a discordant contact, in which the diapiric margin crosscuts surrounding strata. Diapirs **pierce** by several mechanisms: see *active diapirism*, *passive diapirism*, *reactive diapirism*.

plug-fed extrusion Extrusive *salt sheet* emanating from the top of a salt stock, salt wall, or breached salt pillow (Hudec and Jackson, 2006).

plug-fed thrust *Salt sheet* emplaced by a thrust fault rooted in the crest of a preexisting *salt diapir* or pillow. Thrusting is facilitated if the salt plug is inclined and the *roof* of the diapir is thick enough to retard salt emergence (Hudec and Jackson, 2006).

Poiseuille flow Laminar flow in a pipe of circular cross section and driven by a constant pressure gradient. The term applies loosely to any salt flow in which maximum velocity is greatest in the core of the salt body. Compare *Couette flow*.

pop down Downdropped block between two reverse faults that dip away from each other.

pop up Mass of rock uplifted by reverse slip on two faults dipping toward one another.

postkinematic Occurring after salt flow, or other deformation, has ceased. The postkinematic sedimentary interval overlies the *synkinematic* interval. The postkinematic interval has regional dip, but strata show no deflection over individual structures (except compactionally induced drape). Basal postkinematic strata can onlap or truncate an underlying, uneven, deformed surface.

postsalt (a) Occurring after deposition of *autochthonous salt*. (b) Lying above autochthonous salt. Compare *presalt*.

prekinematic Occurring before salt flow, or other deformation, begins. The prekinematic sedimentary interval is *isopachous* (or changes thickness at regional rates) above local structures. The prekinematic interval underlies any *synkinematic interval* and records sedimentation before salt flow, or other visible deformation, began.

preraft Fault block resting on its original *footwall* during thin-skinned extension above a salt detachment. Compare *raft*.

presalt (a) Occurring prior to deposition of *autochthonous salt*. (b) Lying beneath autochthonous salt. Compare *postsalt*.

primary basin Sedimentary depocenter formed directly on the autochthonous salt layer with little or no stratigraphic omission between the salt and its overburden. Essentially equivalent to a *withdrawal basin* above autochthonous salt. The term is used in the deep-water Gulf of Mexico to refer to subcanopy strata as opposed to supracanopy strata (Pilcher and others, 2011).

primary minibasin Minibasin whose base rests on *autochthonous* salt or equivalent salt weld (Jamieson and others, 2000). See also *minibasin*, *secondary minibasin*, *encased minibasin*.

primary peripheral sink See *peripheral sink*.

raft Fault block that has separated from its original *footwall* and lies entirely on a *décollement*, which typically consists of thin salt. Rafts may themselves consist of smaller, older rafts, which became yoked together by later sedimentation before being ruptured again. Rafts form by extreme extension over *autochthonous* or *allochthonous* salt. Rafts are separated by troughlike depocenters of younger, *synkinematic* strata (Burollet, 1975; Duval and others, 1990).

raft tectonics Extreme extension in which deep, syndepositional grabens widen and intervening fault blocks of *overburden* separate into *rafts*, which slide downslope on a *décollement* (typically of thin salt). The décollement faults may commonly be *fault welds* if all the salt is removed or thinned beyond seismic recognition. The type area for raft tectonics is the continental margin of Angola (Burollet, 1975; Duval and others, 1990).

Rayleigh-Taylor instability Inherent instability in a layer of viscous fluid of uniform density overlying a compositionally less dense viscous fluid. Small perturbations in the horizontal interface become amplified at a rate represented by an eigenvalue that depends on the thickness, density, and viscosity of every layer, size of initial perturbation, and time elapsed.

reactive diapirism Emplacement of an elongated, sharp-crested diapir of salt or shale into the space created by regional extensional thinning during rifting or gravity spreading (Jackson and Vendeville, 1990; Vendeville and Jackson, 1992a).

reactive graben See *crestal graben*.

regional (regional datum, regional elevation) Structural elevation and dip of a marker horizon where it is unaffected by local faulting, folding, or salt tectonism. The regional is a fundamental reference surface and is commonly extrapolated and interpolated for kinematic analysis.

reverse fault Fault in which the *hanging wall* has moved upward relative to the *footwall*. Compare *thrust fault, normal fault, contraction*, and *extension*.

rim syncline Fold having an arcuate or subcircular axial trace on the outer margin of a salt upwelling. Rim syncline is a nongenetic term, but in the context of salt tectonics it typically results from salt withdrawal in the *source layer*. *Peripheral sinks* of sediments accumulate within rim synclines (Nettleton, 1934).

rock salt See *salt*.

roho system Group of listric, basinward-dipping growth faults that sole out onto an *allochthonous salt sheet,* allochthonous *salt canopy*, or allochthonous *weld*. Sediment wedges in the fault blocks dip and thicken landward but become younger seaward. Compare *counterregional system* (Schuster, 1995).

rollover Stratal interval that thickens and bends downward toward a *normal fault* or diapir. A **rollover anticline** results from rotation of *hanging-wall* strata as they slide down a *listric fault* surface (Hamblin, 1965). An **expulsion rollover** results from expulsion of underlying salt by prograding strata (Ge and others, 1997).

rollover anticline See *rollover*.

roof That part of a diapir's *overburden* that overlies the widest part of the diapir and has structure or stratigraphy affected by the diapir (Brinkmann and Lögters, 1968; Hudec and Jackson, 2006).

roof-edge thrust *Thrust fault* that cuts upsection from the tip of a *salt sheet*. The fault may reach the seafloor, typically at an inflection in the bathymetry, or the upper tip may be buried by younger sediments (Hudec and Jackson, 2009).

root Base of the *stem* of a *diapir*.

salt (a) In chemistry, a crystalline compound formed from the neutralization of an acid by a base containing a metal or group acting like a metal. (b) In geology, **rock salt** is a crystalline aggregate of halite (NaCl). Salt also loosely refers to all rocks composed mostly of any evaporite mineral.

salt anticline (salt welt) Elongated (planform axial ratio of two or more) upwelling of salt having concordant *overburden* (DeGolyer, 1925). Compare *salt pillow*. Salt anticlines forming a polygonal network connecting salt stocks have been termed **salt ridges** (Pilcher and others, 2011). Salt ridges in the Gulf of Mexico have commonly been laterally shortened, resulting in steep, concave-up flanks.

salt breakout Process or place where a buried *salt sheet* is exposed by erosion or *roof* extension and where the exposed salt advances by extrusion (Jackson and others, 1994).

salt canopy Composite diapiric structure formed by partial or complete coalescence of diapir *bulbs* or *salt sheets*. These bodies coalesce along *sutures* and may or may not be connected to their source layer by feeder stems (Jackson and Cornelius, 1985; Jackson and others, 1987). Salt canopies may be distinguished by their feeder geometries, as in *salt-stock canopies*, *salt-wall canopies*, and *counterregional canopies*, or by the shapes of the component salt sheets, as in *salt-tongue canopies* or *salt-pancake canopies*. Feeder and salt-sheet geometries tend to be related, in that salt stocks tend to produce salt-pancake canopies, whereas salt walls and counterregional systems are more likely to produce salt-tongue canopies.

salt deflation (a) The sorting out, lifting, and removal of loose, dry, fine-grained salt particles by wind erosion. (b) In salt tectonics, thinning of a salt layer in the subsurface by *salt withdrawal* (Spencer and Sharpe, 1993; Hall, 2002; Rowan, 2002).

salt diapir Mass of salt that has flowed in a ductile manner and has *discordant* contacts with the encasing *overburden*. In its broadest sense, a diapir includes intrusions or extrusions of any shape, upwelling of either buoyant or nonbuoyant rock, or emplacement by *downbuilding* or by faulting of *prekinematic overburden* (Mrazec, 1907).

salt dome (a) General term for a domal upwelling comprising a salt core and its envelope of deformed *overburden*. The salt may or may not be *discordant* to the encasing strata (Harris and Veatch, 1899). (b) A common usage is as a synonym for *salt stock*, which can cause confusion when contrasted with the original, broader definition.

salt droplet Growth stage of an emergent diapir after a *salt fountain* or *dynamic bulge* has subsided as salt flow wanes. The summit dome of the diapir dissipates by *gravity spreading* to form a flat-topped plateau of extrusive salt, equivalent to a droplet in fluid mechanics (Talbot, 1998).

salt expulsion See *salt withdrawal.*

salt feeder (canopy feeder, feeder diapir, feeder stock, feeder stem) Diapir supplying salt to an overlying *salt sheet* or *salt canopy* (Diegel and Schuster, 1990; Sumner and others, 1990).

salt flat (a) In geomorphology, a subaerial flat surface encrusted with salt. (b) In salt tectonics, a gently inclined segment of the stair-step basal contact of a *salt sheet*. Flats may form during *extrusive advance*, in which the ratio of aggradation rate to salt spreading rate is low, or during *thrust advance* as the salt moves across a flat in the frontal thrust system. Compare *salt ramp* (J. R. Hossack, written communication, 1990, in Jackson and Talbot, 1991; Hudec and Jackson, 2006).

salt fountain Growth stage of an emergent diapir in which vigorous rise of salt bulges up a parabolic summit dome, which spreads laterally under its own weight to form a surrounding apron of extrusive salt. The summit is a *dynamic bulge* supported by the pressure of salt rising from the diapir's vent (Bailey, 1931; Talbot, 1993). Compare *salt droplet*.

salt glacier (namakier) A type of *salt sheet* flowing extrusively from an exposed diapir and spreading beneath air or water (Lees, 1927; De Böckh and others, 1929; Talbot and Jarvis, 1984).

salt grounding See *salt touch-down*.

salt inflation Thickening of a salt layer by lateral flow of salt, typically expelled from beneath an *overburden* depocenter (McBride, 1998).

salt keel Downward-projecting mass of salt forming a structural low in the base of an *allochthonous salt sheet*. The keel can be the upper part of a pinched-off *salt feeder* or a *carapace block* that has subsided into an open feeder, or it can be produced by folding or faulting in the base of salt. Some apparent keels are simply oblique sections through salt feeders (Jamieson, 1995).

salt nappe Subhorizontal wedge of *allochthonous* salt tapering downslope over several tens of kilometers. The nappe includes its allochthonous *overburden*. A salt nappe advances upsection and basinward by *gravity spreading* over large distances from its feeders (Humphris, 1978; Worrall and Snelson, 1989).

salt pancake Axisymmetric variety of *salt sheet* fed by a single stem. Salt pancakes are typically thickest near the feeder and taper toward the edge of the sheet.

salt-pancake canopy *Salt canopy* formed by coalescence of *salt pancakes*.

salt pillow Subcircular (planform axial ratio of less than two) upwelling of salt having *concordant overburden* (Trusheim, 1960). Compare *salt anticline*.

salt plateau Laterally continuous, tectonically thickened *autochthonous* salt.

salt plug See *salt stock*.

salt ramp Steeply inclined segment of the stair-step basal contact of a *salt sheet*. Salt ramps cut up stratigraphic section in the direction of salt emplacement. Ramps may form during *extrusive advance*, in which the ratio of aggradation rate to salt spreading rate is high, or during thrust advance as the salt moves up a ramp in the frontal thrust system. See *basal cutoff*. Compare *salt flat* (J. R. Hossack, written communication, 1990, in Jackson and Talbot, 1991; Hudec and Jackson, 2006).

salt ridge See *salt anticline*.

salt roller Low-amplitude, asymmetric salt structure comprising two flanks: a gently dipping flank conformable with the *overburden* and a more steeply dipping flank in normal-faulted contact with the overburden. Salt rollers are an unequivocal sign of regional thin-skinned *extension* perpendicular to the strike of the salt rollers (Bally, 1981).

salt-roof thrust *Thrust fault* cutting the *roof* of a *salt sheet* and rooting into the top of salt (Hudec and Jackson, 2009).

salt sheet *Allochthonous salt* sourced from a single feeder whose breadth is several times greater than its maximum thickness; a broad, nongenetic term that includes *salt glacier*, *salt tongue*, *salt sill*, *salt nappe*, *salt pancake*, and *salt wing* (O'Brien and Lerche, 1988; Hudec and Jackson, 2006).

salt sill Intrusion of salt along a bedding plane to form a *salt sheet*. Most salt sheets were initially thought to be emplaced as sills at depths of only a few hundred meters or less. This concept for the emplacement of salt sheets has generally been replaced by that of glacial extrusion (Watkins and others, 1978; Nelson and Fairchild, 1989). However, salt sills can be intruded at depth into preexisting, thin salt beds during regional shortening, as in the Zechstein salt basin (Hudec and Jackson, 2006). These form wedge-shaped, intrusive *salt wings* from diapir flanks.

salt stock (salt plug) Pluglike *salt diapir* having subcircular planform (axial ratio of less than two) (Trusheim, 1960).

salt-stock canopy *Salt canopy* fed by *salt stocks*. Coalescence of sheets in a salt-stock canopy can produce an egg-carton geometry in the base of salt, with structural lows at the feeders and structural highs along the *basal suture lines* separating the feeders. In some areas *minibasins* have subsided into the feeders, displacing salt into interfeeder areas (Jackson and Talbot, 1989b; Rowan, 1995).

salt structure Map-scale body of salt that has changed shape by salt flow from its depositional configuration. Salt structures include *salt diapirs*, *salt pillows*, *salt anticlines*, *salt sheets*, and *salt canopies*.

salt suture See *suture*.

salt system Genetically and kinematically linked groups of salt layers, diapirs, welds, and associated subsalt and suprasalt strata and structures (Jackson and others, 1994; Rowan and others, 1999).

salt tectonics (halotectonics) Large-scale deformation involving salt or other evaporites as a *source layer*; salt tectonics includes *halokinesis* (Trusheim, 1957; Jackson and Talbot, 1986).

salt tongue Highly asymmetric variety of *salt sheet* fed by a single stem. Salt tongues are typically highly elongate in the dip direction. They are typically thickest near the feeder and taper toward the edge of the sheet (Watkins and others, 1978; D'Onfro, 1988; Wang, 1988).

salt-tongue canopy *Salt canopy* formed by coalescence of *salt tongues* (Jackson and Talbot, 1989b).

salt touch-down (salt grounding) The point in time or space at which the top of a salt layer first touches the base of the layer and salt *welding* begins. A salt weld of limited lateral extent (MacRae and Watkins, 1992; Demercian and others, 1993).

salt wall Ridge of diapiric (discordant) salt having a planform axial ratio of two or more. Walls can form sinuous, parallel rows (Trusheim, 1960).

salt-wall canopy *Salt canopy* fed by *salt walls* (Jackson and Talbot, 1989b).

salt weld See *weld*.

salt welt See *salt anticline*.

salt window A salt *weld* permeable enough to allow hydrocarbons to migrate across it (Kunz and Jones, 1994).

salt wing Subhorizontal wedge of *allochthonous* salt protruding from the steep flanks of a *salt diapir*. Most salt wings record episodes of salt extrusion over underlying strata. Repeated extrusions create stacked salt wings, forming a serrated salt contact or **Christmas-tree structure** (Ladzekpo and others, 1988; Yeilding and Travis, 1997). Under special conditions, salt wings can also be intruded (see *salt sill*).

salt withdrawal (salt expulsion) Evacuation of salt due to the weight of its *overburden*. Salt withdrawal creates accommodation space for an overlying **withdrawal basin**. Salt flows from regions of salt withdrawal into *salt structures*. This flow is typically three dimensional, so the cross-sectional area of salt can increase or decrease over time. Expulsion eventually forms salt *welds* or *fault welds*. See *salt deflation*.

secondary minibasin Minibasin whose base rests on *allochthonous* salt or equivalent salt weld. See also *minibasin*, *primary minibasin*, *encased minibasin*.

secondary peripheral sink See *peripheral sink*.

shale sheath Thin covering of overpressured bathyal shale encasing a *salt diapir*. The sheath was originally thought to have been dragged up by the diapir. However, the sheath is most likely a remnant of a condensed, muddy *roof* uplifted, rotated, and pushed aside by the rising diapir. Subsequently, both the salt and the draping shale are discordantly onlapped by younger strata during *passive diapirism* (Johnson and Bredeson, 1971; Worrall and Snelson, 1989).

shear zone Parallel-sided, comparatively thin zone of localized shearing. Ductile shear zones typically show a combination of locally strong foliation, marked reduction in grain size, and sigmoidal extension veins. Several types of ductile shear zone around or within *salt diapirs* are known. An **external shear zone,** a sheathlike collar around the diapir, is formed by the contact strain caused by rise of the diapir relative to the surrounding *overburden*. The inner part of the external shear zone typically comprises ductile evaporites. Depending on the rheology of the overburden, the external part of this shear zone can be a ductile shear zone, a brittle shear zone, a fault, or a growth fault. A **boundary shear zone**, enclosed within the diapir, contains screens or clast trains of country rock. Some boundary shear zones could result from external shear zones becoming entrained within *mushroom* structures. An **internal shear zone** is also enclosed within the diapir but comprises only sheared evaporites. A **basal shear zone** is a variety of external shear zone along the lower contact of a *salt sheet* (Ramsay and Graham, 1970; Kupfer, 1976).

sheet reactivation See *sheet segmentation.*

sheet segmentation (sheet reactivation) Deformation of a *salt sheet* acting as a second-cycle *source layer*. The salt sheet is partitioned into separate salt structures by subsidence of *minibasins* or by *growth faulting* (West, 1989).

shortening See *contraction*.

skirt See *mushroom diapir.*

source-fed thrust A *salt sheet* formed by a *thrust fault* rooted in an *autochthonous* salt layer and carrying a slice of salt in the base of its *hanging wall.* The salt slice ranges in thickness from a thin sliver in the hanging wall of a small thrust to a salt-cored nappe of *allochthonous salt* kilometers thick (Hudec and Jackson, 2006).

source layer Layer of *autochthonous* or *allochthonous* salt supplying salt for the growth of salt structures. If autochthonous, may be informally known as *mother salt.*

spine Steep-sided mass of salt within a larger salt body that moves or has moved faster or slower than neighboring spines from which it is separated by *boundary shear zones* or *internal shear zones* (Kupfer, 1976).

stem Comparatively slender part of a *salt diapir* below the *bulb.*

stepped counterregional system See *counterregional system*.

structural inversion (inversion) (a) In structural geology, a reversal of tectonics from extension to shortening or vice versa. (b) In salt tectonics, a reversal of vertical relative motion recorded by stratigraphic thickening or thinning of *synkinematic* strata. Structural inversion in *salt tectonics* is recognized by local changes in thickness or onlap directions and truncations. For example, a thickened interval (with outward onlap and truncations elsewhere than the diapir) overlying a thinned interval indicates local uplift followed by local subsidence. In contrast, a thinned interval of *synkinematic* strata (with inward onlap and truncations over the diapir) overlying a thickened interval indicates local subsidence followed by local uplift (Ziegler, 1978; Trusheim, 1960).

suballochthonous Lying beneath *allochthonous* salt.

subautochthonous Lying beneath *autochthonous* salt.

subsalt Lying beneath *autochthonous* or *allochthonous* salt. Compare *presalt*.

substrate expulsion Transfer of sediment by *gravity spreading* from an extensional zone below thick extrusive salt to a contractional zone below thin extrusive salt and immediately ahead of the salt. Concept originated for differential loading imposed by tapering glacial ice (Rotnicki, 1976; Hudec and Jackson, 2009).

suprasalt Lying above *autochthonous* or *allochthonous* salt. Compare *postsalt*.

suture The zone separating two coalesced salt sheets (*allosuture*) or two lobes of a single salt sheet (*autosuture*), including trapped sediments (Lee and others, 1989). In common usage, the term has also been applied imprecisely to a variety of individual components of a suture. A **frontal suture** is a suture whose map trace is roughly perpendicular to the dominant direction of salt flow. A **lateral suture** is a suture whose map trace is roughly parallel to the dominant direction of salt flow. An **oblique suture** is a suture whose map trace is oblique to the dominant direction of salt flow. See also *suture line*, *suture point*, and *suture surface*.

suture line Trace of a suture surface on the plane of observation (map view or cross section). A **basal suture line** is the intersection of a suture surface with the base of salt. An **apical suture line** is the intersection of a suture surface with the top of salt. A **composite suture line** is the intersection of three suture surfaces at a salt-sheet triple-junction.

suture point The end of a suture line. A **basal suture point** is the intersection of a suture line with the base of a *salt sheet* in cross section. An **apical suture point** is the intersection of a suture line with the top of a salt sheet in cross section. A **terminal suture point** is the end of a suture line in map view. A **composite suture point** is the intersection of a composite suture line with the plane of observation.

suture surface Surface separating salt and any associated sediments derived from different *allochthonous* sheets or lobes fed by the same feeder.

synkinematic Occurring during salt flow or other deformation. A synkinematic interval shows local stratigraphic thickening above subsiding structures or thinning above rising structures. Changes in thickness are also recorded by onlap or truncation. The synkinematic interval, also known as the **growth interval**, records sedimentation during salt flow or during any other type of deformation.

synsedimentary fault See *growth fault.*

tank-track advance See *tank-tread advance*.

tank-tread advance (tank-track advance, frontal rolling) Advance of a *salt sheet* whose upper surface moves faster than its base and thus rolls over and adheres to the substratum below the salt. This advance lengthens the base of the salt and forms a recumbent fold (Ramberg, 1981; Brun and Merle, 1985; Merle, 1986; Hossack, 1995; Hudec and Jackson, 2006).

terminal suture point See *suture point*.

tertiary peripheral sink See *peripheral sink*.

thermal convection Subsolidus rise of hotter salt and sinking of cooler salt because of a thermal gradient between the top and bottom of the convecting layers. A temperature-induced *density inversion* provides the instability. Thermal convection is a cyclic process, unlike the single overturn of systems having a compositional *density inversion*. Convection is promoted by increase of layer thickness, density, temperature gradient, and thermal expansivity and by decrease in thermal diffusivity and viscosity (hence, decreased grain size). Thermal convection is theoretically possible within a salt layer thicker than 2.9 km and having an unusually low viscosity of $<10^{16}$ Pa s in a geothermal gradient of 30°C/km (Talbot, 1978). See *thermal loading*.

thermal loading *Differential loading* resulting from expansion or shrinking of salt caused by temperature changes. Hot salt expands and becomes more buoyant. See also *gravitational loading* and *displacement loading*. Compare *thermal convection*.

thick-skinned deformation Deformation of both *cover* and *basement* (Rodgers, 1963). In salt tectonics, this deformation involves the *presalt* section.

thin-skinned deformation Deformation of *cover* above a subhorizontal detachment that does not involve underlying *basement* (Rodgers, 1963). In salt tectonics, typically refers to deformation detached above *autochthonous salt* or *allochthonous salt*.

thrust advance Overthrust advance of a *salt sheet* and its *roof* by means of one or more frontal *thrust faults* (Hudec and Jackson, 2006).

thrust fault Fault of any dip in which the *hanging wall* appears to have moved upward relative to the *footwall*. Originally applied only to faults dipping less than 45°, the term is now used as a synonym for *reverse fault.*

toe thrust Thrust complex at the lower or leading edge of a *gravity-spreading* system.

turtle structure (turtle-structure anticline) Mounded strata between *salt diapirs*, typically having a flat base and rounded crest. The sedimentary sequence is thick in its core and thins laterally. The anticline may or may not be cored by a low *salt pillow*. The turtle structure forms between diapirs whose flanks subside because of regional extension (see *diapir fall*) or between salt structures evolving from pillows to diapirs (Trusheim, 1960). Compare *mock-turtle anticline*.

turtle-structure anticline See *turtle structure*.

upbuilding See *active diapirism*.

vortex structure *Crescentic folds* whose lowest parts are rolled or coiled inward about fold hinges concentric to the diapir core. Vortex structures are thought to indicate *thermal convection* within a diapir (Jackson and Talbot, 1989a).

wedge halokinetic sequence *Halokinetic sequence* in which strata gradually taper toward the diapir and typically terminate against a truncating younger halokinetic sequence. The zone of drape folding is 300 to 1000 m wide, angular truncation is typically less than 30°, mass-wasting deposits are less common than in *hook halokinetic sequences*, and any facies changes are gradational over larger distances (Giles and Rowan, in press). See also *composite halokinetic sequence*, *drape fold*.

weld Surface or zone joining strata originally separated by *autochthonous* or *allochthonous* salt. The weld is a negative salt structure resulting from complete or nearly complete removal of intervening salt. The weld can consist of up to about 50 m of remnant evaporites plus insoluble material originally interbedded with the evaporites. The weld is usually but not invariably marked by a structural discordance. Another distinctive feature of welds is a *structural inversion* above them (Jackson and Cramez, 1989). A **complete weld** is a surface joining strata originally separated by autochthonous or allochthonous salt. Rocks on either side of the weld are in direct contact, not separated by evaporites or interbedded material. An **incomplete weld** is a zone joining strata originally separated by autochthonous or allochthonous salt. Rocks on either side of an incomplete weld are separated by up to about 50 m thickness of remnant evaporites and interbedded material. A **discontinuous weld** is a salt body that is welded in places but in other areas retains more than 50 m thickness of evaporites and interbedded material. An **apparent weld** is a salt body that appears welded at the scale of observation, but which may actually contain more than 50 m of evaporites and interbedded material.

withdrawal basin See *salt withdrawal*.

References and Figure Credits

The authors are grateful to the following authors and publishers, whose works are cited in, and in many cases, illustrations reprinted in, The Salt Mine. *Relevant figure numbers are listed at the end of each reference and permission statement.*

Abriel, W. L., Neale, P. S., Tissue, J. S., and Wright, R. M., 1991, Modern technology in an old area: Bay Marchand revisited: Geophysics: The Leading Edge of Exploration, v. 10, no. 6, p. 21–35.

Reproduced with permission of the Society of Exploration Geophysicists.

Figure 4.01

Allen, M. R., Griffiths, P. A., Craig, J., Fitches, W. R., and Whittington, R. J., 1994, Halokinetic initiation of Mesozoic tectonics in the southern North Sea; a regional model: Geological Magazine, v. 131, no. 4, p. 559–561.

Reproduced with permission, 1994+E663 Cambridge Journals.

Figure 3.10

Anderson, J. E., Cartwright, J., Drysdall, S. J., and Vivian, N., 2000, Controls on turbidite sand deposition during gravity-driven extension of a passive margin: examples from Miocene sediments in Block 4, Angola: Marine and Petroleum Geology, v. 17, p. 1165–1203.

Reprinted with permission from Elsevier, Copyright (2000).

Figures 5.11, 5.12

Arrhenius, S., 1913, Zur Physik der Salzlagerstätten [On the physics of salt deposits]: Meddelanden Vetensskapsakademiens Nobelinstitut, v. 2, p. 1–25.

Bailey, E. B., 1931, Salt plugs: Geological Magazine, v. 68, p. 335–336.

Baker, D. M., Lillie, R. J., Yeats, R. S., Johnson, G. D., Yousuf, M., and Zamin, A. S. H., 1988, Development of the Himalayan frontal thrust zone: Salt Range, Pakistan: Geology, v. 16, p. 3–7.

Redrawn from Baker and others (1988).

Figure 6.18

Baldschuhn, R., Binot, F., Frisch, U., and Kockel, F., 2001, Geotektonischer Atlas von Nordwest-Deutschland und dem Deutschen Nordsee-Sektor [Tectonic atlas of Northwest Germany and the German North Sea sector]: Geologisches Jahrbuch, v. A 153; 3 CD-ROM, 88 p.

http://www.schweizerbart.de

Figures 1.18, 1.19, 2.27, 2.28, 2.29, 2.31, 2.32, 2.33, 2.34

Baldschuhn, R., Frisch, U., and Kockel, F., 1996, Geotektonischer Atlas von NW-Deutschland [Tectonic atlas of NW-Germany]: Hannover, Germany, BGR (Bundesanstalt fur Geowissenschaften und Rohstoffe).

Figure 2.27

Bally, A. W., 1981, Thoughts on the tectonics of folded belts, *in* McClay, K. R., and Price, N. J., eds., Thrust and nappe tectonics: Geological Society, London, Special Publication 9, p. 13–32.

Barde, J.-P., Gralla, P., Harwijanto, J., and Marsky, J., 2002, Exploration at the eastern edge of the Precaspian basin: impact of data integration on Upper Permian and Triassic prospectivity: American Association of Petroleum Geologists Bulletin, v. 86, p. 399–415.

AAPG©2002, reprinted by permission of the AAPG whose permission is required for further use.

Figure 7.29

Barton, D. C., 1933, Mechanics of formation of salt domes with special reference to Gulf Coast salt domes of Texas and Louisiana: American Association of Petroleum Geologists Bulletin, v. 17, p. 1025–1083.

Berástegui, X., Banks, C. J., Puig, C., Taberner, C., Waltham, D., and Fernàndez, M., 1998, Lateral diapiric emplacement of Triassic evaporites at the southern margin of the Guadalquivir Basin, Spain, *in* Mascle, A., Puidefàbregas, C., Luterbacher, C., and Fernàndez, M., eds., Cenozoic foreland basins of western Europe: Geological Society, London, Special Publication 134, p. 49–68.

Reproduced with permission of The Geological Society, London.

Figures 7.01, 7.02

Bitterli, T., 1990, The kinematic evolution of a classical Jura fold: a reinterpretation based on 3-dimensional balancing techniques (Weissenstein Anticline, Jura Mountains, Switzerland): Eclogae Geologicae Helvetiae, v. 83, p. 493–511.

Springer and Birkhäuser Basel, Bitterli, T., figure 6, ©1990, reprinted with kind permission from Springer Science and Business Media.

Figures 1.12, 6.04

Boyers, W. C., 2000, Structural style and normal faulting adjacent to the Onion Creek salt diapir, Paradox Basin, Utah: Waco, Texas, Baylor University, M.S. thesis, 70 p.

Figure courtesy of Chris Boyers.

Figure 4.10

Brinkmann, R., and Lögters, H., 1968, Diapirs in western Pyrenees and foreland, Spain, *in* Braunstein, J., and O'Brien, G. D., eds., Diapirism and diapirs: American Association of Petroleum Geologists Memoir 8, p. 275–292.

Brown, J. P., 1998, Deformation and sedimentation adjacent to Carboniferous diapiric salt structures, Nova Scotia: Halifax, Nova Scotia, Dalhousie University, Ph.D. thesis, 486 p.

Figure courtesy of J. P. Brown.

Figure 2.25

Brun, J.-P., and Fort, X., 2008, Entre sel et terre: structures et mécanismes de la tectonique salifière: Paris, Société Géologique de France, Vuibert, 154 p.

Brun, J. P., and Merle, O., 1985, Strain patterns in models of spreading-gliding nappes: Tectonics, v. 4, p. 705–719.

Bucher, W. H., 1956, Role of gravity in orogenesis: Geological Society of America Bulletin, v. 67, p. 1295–1318.

Burollet, P., 1975, Technique en radeaux en Angola [Raft tectonics in Angola]: Bulletin de la Sociéte Géologique de France, v. 17, p. 503–504.

Buxtorf, A., 1916, Prognosen und Befunde beim Hauensteinbasis und Grenchenbergtunnel und die Bedeutung der letzteren für die Geologie des Jurabirges: Verhandlungen Natf, Gesellschaft Basel, v. 7, p. 184–254.

Figures 6.04, 6.05

Canérot, J., Hudec, M. R., and Rockenbauch, K., 2005, Mesozoic diapirism in the Pyrenean orogen: salt tectonics on a transform plate boundary: American Association of Petroleum Geologists Bulletin, v. 89, p. 211–229.

AAPG©2005, reprinted by permission of the AAPG whose permission is required for further use.

Figure 6.01

Chapple, W. M., 1978, Mechanics of thin-skinned fold-and-thrust belts: Geological Society of America Bulletin, v. 89, p. 1189–1198.

Chimney, P. J., 2000, Seismic evidence for "heel-toe" structural rotation, 96-G Prospect, Cabinda, offshore Angola (abs.): American Association of Petroleum Geologists Annual Meeting Program, v. 9, p. A26–A27.

Chimney, P., and Kluth, C., 2002, Evidence for low-angle, subhorizontal hanging faults in rotated fault blocks, Cabinda, offshore Angola: The Leading Edge, v. 21, no. 1084, p. 1086–1090.

Reproduced with permission of the Society of Exploration Geophysicists.

Figures 3.05, 3.06

Cobbold, P. R., and Szatmari, P., 1991, Radial gravitational gliding on passive margins: Tectonophysics, v. 188, p. 249–289.

Reprinted with permission from Elsevier, Copyright (1991).

Figure 5.01

Costa, E., and Vendeville, B. C., 2002, Experimental insights on the geometry and kinematics of fold-and-thrust belts above weak, viscous evaporitic décollement: Journal of Structural Geology, v. 24, p. 1729–1739.

Reprinted with permission from Elsevier, Copyright (2002).

Figure 6.23

Couzens-Schultz, B. A., Vendeville, B. C., and Wiltschko, D. V., 2003, Duplex style and triangle zone formation: insights from physical modeling: Journal of Structural Geology, v. 25, p. 1623–1644.

Reprinted with permission from Elsevier, Copyright (2003).

Figures 1.38, 6.31

Crosby, W. O., and Crosby, I. B., 1925, Keystone faults: Geological Society of America Bulletin, v. 36, p. 623–640.

Daly, R. A., Stose, G. W., and Hobbs, W. H., 1916, Homocline and monocline (with discussion by G. W. Stose and W. H. Hobbs): Geological Society of America Bulletin, v. 27, p. 89–92.

Dardeau, G., and De Graciansky, P. C., 1990, Halocinèse et rifting téthysien dans les Alpes-maritimes (France): Bulletin Centres Recherche Exploration-Production Elf-Aquitaine, v. 14, p. 443–464.

Figures 2.23, 2.24

Davis, W. M., 1903, An excursion to the plateau province of Utah and Arizona: Harvard College, Bulletin of the Museum of Comparative Zoology, Geological Series 6, 50 p.

Davison, I., Alsop, I., Birch, P., Elders, C., Evans, N., Nicholson, H., Rorison, P., Wade, D., Woodward, J., and Young, M., 2000a, Geometry and late-stage structural evolution of Central Graben salt diapirs, North Sea: Marine and Petroleum Geology, v. 17, p. 499–522.

Reprinted with permission from Elsevier, Copyright (2000).

Figures 2.36, 2.37, 2.38, 2.39, 2.41, 2.42, 2.43, 2.88

Davison, I., Alsop, G. I., Evans, N. G., and Safaricz, M., 2000b, Overburden deformation patterns and mechanisms of salt diapir penetration in the Central Graben, North Sea: Marine and Petroleum Geology, v. 17, p. 601–618.

Reprinted with permission from Elsevier, Copyright (2000).

Figure 2.40

Davison, I., Bosence, D. W., Alsop, G. I., and Alawi, M., 1996, Deformation and sedimentation around active Miocene salt diapirs on the Tihama Plain, northwest Yemen, *in* Alsop, G. I., Blundell, D. J., and Davison, I., eds., Salt tectonics: Geological Society Special Publications, v. 100, p. 22–39.

Figure 2.42

De Böckh, H., Lees, G. M., and Richardson, F. D. S., 1929, Contribution to the stratigraphy and tectonics of the Iranian ranges, *in* Gregory, J. W., ed., The structure of Asia: London, Methuen, p. 58–176.

Figure 7.36

DeGolyer, E., 1925, Origin of North American salt domes: American Association of Petroleum Geologists Bulletin, v. 9, p. 831–874.

Demercian, S., Szatmari, P., and Cobbold, P. R., 1993, Style and pattern of salt diapirs due to thin-skinned gravitational gliding, Campos and Santos Basins, offshore Brazil: Tectonophysics, v. 228, p. 393–433.

Reprinted with permission from Elsevier, Copyright (1993).

Diamond Geophysical, 1997, Advertisement: AAPG Explorer, February, p. 5.

AAPG©1997, reprinted by permission of the AAPG whose permission is required for further use.

Figure 8.12

Diegel, F. A., Karlo, J. F., Schuster, D. C., Shoup, R. C., and Tauvers, P. R., 1995, Cenozoic structural evolution and tectono-stratigraphic framework of the northern Gulf coast continental margin, *in* Jackson, M. P. A., Roberts, D. G., and Snelson, S., eds., Salt tectonics: a global perspective: American Association of Petroleum Geologists Memoir 65, p. 109–151.

AAPG©1995, reprinted by permission of the AAPG whose permission is required for further use.

Figures 9.05, 9.07

Diegel, F. A., and Schuster, D. C., 1990, Regional cross sections and palinspastic reconstructions, northern Gulf of Mexico (abs.): Geological Society of America, Abstracts with Programs, v. 22, p. A66.

Doelling, H. H., 1995, Interim geologic map of the Fisher Towers Quadrangle, Grand County, Utah: Utah Geological Survey Open-File Report 313.

Courtesy of the Utah Geological Survey.

Figure 4.08

D'Onfro, P., 1988, Mechanics of salt tongue formation with examples from Louisiana slope (abs.): American Association Petroleum Geologists Bulletin, v. 72, p. 175.

Dooley, T. P., Jackson, M. P. A., and Hudec, M. R., 2009, Inflation and deflation of deeply buried salt stocks during lateral shortening: Journal of Structural Geology, v. 31, p. 582–600.

Figures 2.62, 2.63, 2.64

Duval, B., Cramez, C., and Jackson, M. P. A., 1990, Raft tectonics in the Kwanza basin, Angola (abs.): Geological Society of America, Abstracts with Programs, v. 22, no. 7, p. A48.

Duval, B., Cramez, C., and Jackson, M. P. A., 1992, Raft tectonics in the Kwanza Basin, Angola: Marine and Petroleum Geology, v. 9, p. 389–404.

Figure 4.04

Duval, B., Cramez, C., Schultz-Ela, D. D., and Jackson, M. P. A., 1993, Extension, reactive diapirism, salt welding and contraction at Cegonha, Kwanza Basin, Angola (extended abs.), *in* American Association of Petroleum Geologists International Hedberg Research Conference, Abstract Volume, Bath, England, p. 41–43.

AAPG©1993, reprinted by permission of the AAPG whose permission is required for further use.

Figure 2.17

Eichenseer, H. T., Walgenwitz, F. R., and Biondi, P. J., 1999, Stratigraphic control on facies and diagenesis of dolomitized oolitic siliciclastic ramp sequences (Pinda Group, Albian, offshore Angola): American Association of Petroleum Geologists Bulletin, v. 83, p. 1729–1758.

AAPG©1999, reprinted by permission of the AAPG whose permission is required for further use.

Figure 5.10

Evans, N., Forison, P., and Sykes, G., 1999, Evaluation of steeply-dipping fractured chalk reservoir, *in* Petroleum geology of Northwest Europe: 5th Barbican Conference Proceedings, p. 975–988.

Figure 2.37

Fiduk, J. C., Anderson, L. E., and Rowan, M. G., 2004, The Wilcox Raft: an example of extensional raft tectonics in South Texas, northwestern onshore Gulf of Mexico, *in* Post, P. J., Olson, D. L., Lyons, K. T., Palmes, S. L., Harrison, P. F., and Rosen, N. C., eds., Salt-sediment interactions and hydrocarbon prospectivity: concepts, applications, and case studies for the 21st century: 24th Annual Research Conference, SEPM Foundation, p. 293–314.

Reproduced with permission of the GCSSEPM Foundation, whose permission is required for any further reproduction.

Figures 5.04, 5.05, 5.06

Fletcher, R. C., Hudec, M. R., and Watson, I. A., 1995, Salt glacier and composite sediment-salt glacier models for the emplacement and early burial of allochthonous salt sheets, *in* Jackson, M. P. A., Roberts, D. G., and Snelson, S., eds., Salt tectonics: a global perspective: American Association of Petroleum Geologists Memoir 65, p. 77–108.

AAPG©1995, reprinted by permission of the AAPG whose permission is required for further use.

Figures 7.04, 7.06, 7.79

Flinch, J. F., 1996, Accretion and extensional collapse of the external western Rif (northern Morocco), *in* Ziegler, P. A., and Horvath, F., eds., Peri-Tethys memoir 2; Structure and prospects of Alpine basins and forelands: Mémoires du Museum National d'Histoire Naturelle, v. 170, p. 61–85.

©Publications Scientifiques du Museum National d'Histoire Naturelle, Paris.

Figure 7.35

Fox, F. G., 1959, Structure and accumulation of hydrocarbons in Southern Foothills, Alberta, Canada: American Association of Petroleum Geologists Bulletin, v. 43, p. 992–1025.

Fraser, S., Robinson, A., Johnson, H., Underhill, J., Kadolsky, D., Connell, R., Johannessen, P., and Ravnas, R., 2003, Upper Jurassic, *in* Evans, I. D., Graham, C., Armour, A., and Bathurst, P., eds., The millennium atlas; petroleum geology of the central and northern North Sea: Geological Society, London, p. 157–189.

Reproduced with permission of The Geological Society, London.

Figure 3.12

Gaullier, V., and Vendeville, B. C., 2005, Salt tectonics driven by sediment progradation: Part II—Radial spreading of sedimentary lobes prograding above salt: American Association of Petroleum Geologists Bulletin, v. 89, p. 1081–1089.

AAPG©2005, reprinted by permission of the AAPG whose permission is required for further use.

Figure 5.17

Ge, H., 1996, Kinematics and dynamics of salt tectonics in the Paradox Basin, Utah and Colorado: field observations and scaled modeling: The University of Texas at Austin, Ph.D. dissertation, 317 p.

Figure courtesy of Hongxing Ge.

Figure 4.07

Ge, H., and Jackson, M. P. A., 1998, Physical modeling of structures formed by salt withdrawal: implications for deformation caused by salt dissolution: American Association of Petroleum Geologists Bulletin, v. 82, p. 228–250.

AAPG©1998, reprinted by permission of the AAPG whose permission is required for further use.

Figure 4.15

Ge, H., Jackson, M. P. A., and Vendeville, B. C., 1995, Rejuvenation and subsidence of salt diapirs by regional extension: Gulf Coast Association of Geological Societies Transactions, v. 45, p. 211–218.

Republished by permission of GCAGS, whose permission is required for further publication use.

Figures 3.13, 3.27, 3.28, 4.06

Ge, H., Jackson, M. P. A., and Vendeville, B. C., 1997, Kinematics and dynamics of salt tectonics driven by progradation: American Association of Petroleum Geologists Bulletin, v. 81, p. 398–423.

AAPG©1997, reprinted by permission of the AAPG whose permission is required for further use.

Figures 1.32, 2.30, 2.82

Geluk, M. C., 1998, Internal tectonics of salt structures: Journal of Seismic Exploration, v. 7, p. 237–251.

Figure 3.09

Giles, K. A., and Lawton, T. F., 2002, Halokinetic sequence stratigraphy adjacent to the El Papalote diapir, northeastern Mexico: American Association of Petroleum Geologists Bulletin, v. 86, p. 823–840.

AAPG©2002, reprinted by permission of the AAPG whose permission is required for further use.

Figure 2.21

Giles, K. A., and Rowan, M. G., in press, Concepts in halokinetic-sequence deformation and stratigraphy, *in* Alsop, G. I., Archer, S. Hartley, A. J., Grant, N. T., and Hodgkinson, R., eds., Salt tectonics, sediments and prospectivity: Geological Society, London, Special Publication.

Gómez-Cabrera, P. T., 2003, Stratigraphic and structural analysis of the Neogene sediments of the offshore portion of the Salina del Istmo Basin, southeastern Mexico: The University of Texas at Austin, Ph.D. dissertation, 299 p.

Figure courtesy of P. T. Gomez-Cabrera.

Figures 8.14, 8.15

Gómez-Cabrera, P. T., and Jackson, M. P. A., 2008, Regional Neogene salt tectonics in the offshore Salina del Istmo Basin, southeastern Mexico, *in* Gaxiola, J. R., Montes de Oca, R. S., Ramos, J. R. R., Bartolini, C., and Cantú Chapa, A., eds., The petroleum geology of Mexico: American Association of Petroleum Geologists Memoir 90, p. 1–28.

AAPG©2008, reprinted by permission of the AAPG whose permission is required for further use.

Figure 10.05

Guglielmo, G., Jr., Jackson, M. P. A., and Vendeville, B. C., 1997, Three-dimensional visualization of salt walls and associated fault systems: American Association of Petroleum Geologists Bulletin, v. 81, p. 46–61.

AAPG©1997, reprinted by permission of the AAPG whose permission is required for further use.

Figure 3.30

Hall, D. H., and Thies, K. J., 1995, Salt kinematics, depositional systems, and implications for hydrocarbon exploration, Eugene Island and Ship Shoal south additions, offshore Louisiana, *in* Travis, C. J., Harrison, H., Hudec, M. R., Vendeville, B. C., Peel, F. J., and Perkins, B. E., eds., Salt, sediment and hydrocarbons: 16th Annual Research Conference, GCSSEPM Foundation, Houston, Texas, p. 83–94.

Reproduced with permission of the GCSSEPM Foundation, whose permission is required for any further reproduction.

Figure 8.28

Hall, S. H., 2002, The role of autochthonous salt inflation and deflation in the northern Gulf of Mexico: Marine and Petroleum Geology, v. 19, p. 649–682.

Hamblin, W. K., 1965, Origin of "reverse drag" on the downthrown side of normal faults: Geological Society of America Bulletin, v. 76, p. 1145–1164.

Harris, G. D., and Veatch, A. C., 1899, A preliminary report on the geology of Louisiana, *in* Geological Survey of Louisiana report: Baton Rouge, Louisiana Geological Survey, Part 5, Geology and Agriculture, p. 9–138.

Harrison, H. L., and Patton, B., 1995, Translation of salt sheets by basal shear, *in* Travis, C. J., Harrison, H., Hudec, M. R., Vendeville, B. C., Peel, F. J., and Perkins, B. E., eds., Salt, sediment and hydrocarbons: 16th Annual Research Conference, GCSSEPM Foundation, Houston, Texas, p. 99–107.

Reproduced with permission of the GCSSEPM Foundation, whose permission is required for any further reproduction.

Figure 7.80

Harrison, J. C., 1995, Melville Island's salt-based fold belt, Arctic Canada: Geological Survey of Canada, Bulletin 472, 331 p.

Reproduced with permission of Natural Resources Canada 2010, courtesy of the Geological Survey of Canada.

Figures 6.12, 6.13, 6.14, 6.15, 6.16, 6.17, 6.35, 6.36

Harrison, T. S., 1927, Colorado-Utah salt domes: American Association of Petroleum Geologists Bulletin, v. 11, no. 2, p. 111–133.

Hart, W. H., Jaminski, J., and Albertin, M. L., 2004, Recognition and exploration significance of supra-salt stratal carapaces, *in* Post, P. J., Olson, D. L., Lyons, K. T., Palmes, S. L., Harrison, P. F., and Rosen, N. C., eds., Salt-sediment interactions and hydrocarbon prospectivity: 24th Annual Research Conference, GCSSEPM Foundation, Houston, Texas, p. 166–199.

Heim, A., 1921, Geologie der Schweiz: Leipzig, Tauchnitz.

Hodgkins, M. A., and O'Brien, M. J., 1994, Salt sill deformation and its implications for subsalt exploration: Geophysics: The Leading Edge of Exploration, v. 13, p. 849–851.

Reproduced with permission of the Society of Exploration Geophysicists.

Figures 8.02, 8.03, 8.04

Hollingworth, S. E., Taylor, J. M., and Kellaway, G. A., 1945, Large scale superficial structures in the Northampton ironstone field: Quarterly Journal of Geological Society, London, v. 100, p. 1–34.

Hooper, R. J., and More, C., 1995, Evaluation of some salt-related overburden structures in the U.K. Southern North Sea, *in* Jackson, M. P. A., Roberts, D. G., and Snelson, S., eds., Salt tectonics: a global perspective: American Association of Petroleum Geologists Memoir 65, p. 251–259.

AAPG©1995, reprinted by permission of the AAPG whose permission is required for further use.

Figure 3.11

Hossack, J., 1995, Structural controls on the emplacement of salt bodies, *in* Travis, C. J., Harrison, H., Hudec, M. R., Vendeville, B. C., Peel, F. J., and Perkins, B. E., eds., Salt, sediment and hydrocarbons: 16th Annual Research Conference, GCSSEPM Foundation, Houston, Texas, p. 119–120.

Hossack, J. R., and McGuinness, D. B., 1990, Balanced sections and the development of fault and salt structures in the Gulf of Mexico (abs.): Geological Society of America, Abstracts with Programs, v. 22, no. 7, p. A48.

Hudec, M. R., 2004, Salt intrusion: time for a comeback?, *in* Post, P. J., Olson, D. L., Lyons, K. T., Palmes, S. L., Harrison, P. F., and Rosen, N. C., eds., Salt-sediment interactions and hydrocarbon prospectivity: 24th Annual Research Conference, GCSSEPM Foundation, Houston, Texas, p. 119–132.

Reproduced with permission of the GCSSEPM Foundation, whose permission is required for any further reproduction.

Figures 7.27, 7.28

Hudec, M. R., and Jackson, M. P. A., 2002, Structural segmentation, inversion, and salt tectonics on a passive margin: evolution of the Inner Kwanza Basin, Angola: Geological Society of America Bulletin, v. 114, p. 1222–1244.

Figure 1.16

Hudec, M. R., and Jackson, M. P. A., 2004, Regional restoration across the Kwanza Basin, Angola: salt tectonics triggered by repeated uplift of a metastable passive margin: American Association of Petroleum Geologists Bulletin, v. 88, p. 971–990.

AAPG©2004, reprinted by permission of the AAPG whose permission is required for further use.

Figures 4.03, 4.04, 7.22

Hudec, M. R., and Jackson, M. P. A., 2006, Advance of allochthonous salt sheets in passive margins and orogens: American Association of Petroleum Geologists Bulletin, v. 90, p. 1535–1564.

AAPG©2006, reprinted by permission of the AAPG whose permission is required for further use.

Figures 7.07, 7.39, 7.72, 7.73, 7.74, 7.75, 7.76, 7.77, 7.78, 7.84

Hudec, M. R., and Jackson, M. P. A., 2007, Terra infirma: understanding salt tectonics: Earth-Science Reviews, v. 82, p. 1–28.

Figures 2.78, 2.85, 2.86, 2.89, 2.90

Hudec, M. R., and Jackson, M. P. A., 2009, Interaction between spreading salt canopies and their peripheral thrust systems: Journal of Structural Geology, v. 31, p. 114–1129.

Figures 7.08, 7.09, 7.10, 7.12, 7.13, 7.14, 7.15, 7.16, 7.81, 7.82, 7.83

Hudec, M. R., Jackson, M. P. A., and Schultz-Ela, D. D., 2009, The paradox of minibasin subsidence into salt: clues to the evolution of crustal basins: Geological Society of America Bulletin, v. 121, p. 201–221.

Figures 8.05, 8.06, 8.08, 8.09, 8.10, 8.11, 8.24, 8.25, 8.31

Hughes, D. J., 1968, Salt tectonics as related to several Smackover fields along the northeast rim of the Gulf of Mexico Basin: Gulf Coast Association of Geological Societies Transactions, v. 18, p. 320–330.

Humphris, C. C., Jr., 1978, Salt movement on continental slope, northern Gulf of Mexico, *in* Bouma, A. H., Moore, G. T., and Coleman, J. M., eds., Framework, facies, and oil-trapping characteristics of the upper continental margin: American Association of Petroleum Geologists Studies in Geology 7, p. 69–85.

Jackson, M. P. A., 1982, Fault tectonics of the East Texas Basin: The University of Texas at Austin, Bureau of Economic Geology Geological Circular 82-4, 31 p.

Figures 5.02, 5.03

Jackson, M. P. A., and Cornelius, R. R., 1985, Tertiary salt diapirs exposed at different structural levels in the Great Kavir (Dasht-i Kavir) south of Semnan, north-central Iran: a remote-sensing study of their internal structure and shape: The University of Texas at Austin, Bureau of Economic Geology, Open-File Report OF-WTWI-1985-22, 108 p.

Jackson, M. P. A., Cornelius, R. R., Craig, C. H., Gansser, A., Stöcklin, J., and Talbot, C. J., 1990, Salt diapirs of the Great Kavir, central Iran: Geological Society of America Memoir 177, 139 p.

Figures 1.03, 1.04, 1.05, 2.12, 2.52, 2.53

Jackson, M. P. A., Cornelius, R. R., Craig, C. H., and Talbot, C. J., 1987, The Great Kavir salt canopy: a major new class of salt structure (abs.): Geological Society of America, Abstracts with Programs, v. 19, p. 714.

Jackson, M. P. A., and Cramez, C., 1989, Seismic recognition of salt welds in salt tectonics regimes, *in* GCSSEPM Foundation 10th Annual Research Conference Program and Extended Abstracts, Houston, Texas, p. 66–71.

Jackson, M. P. A., and Hudec, M. R., 2005, Stratigraphic record of translation down ramps in a passive-margin detachment: Journal of Structural Geology, v. 27, p. 889–911.

Figures 5.07, 5.08, 5.09

Jackson, M. P. A., Hudec, M. R., Jennette, D. C., and Kilby, R. E., 2008, Evolution of the Cretaceous Astrid thrust belt in the ultradeep-water Lower Congo Basin, Gabon: American Association of Petroleum Geologists Bulletin, v. 92, p. 487–511.

AAPG©2008, reprinted by permission of the AAPG whose permission is required for further use.

Figures 7.24, 7.25

Jackson, M. P. A., and Schultz-Ela, D. D., 2000, Why roho, why counterregional? (abs.): American Association of Petroleum Geologists Annual Meeting Program, v. 9, p. A73.

Jackson, M. P. A., Schultz-Ela, D. D., Hudec, M. R., Watson, I. A., and Porter, M. L., 1998, Structure and evolution of Upheaval Dome: a pinched-off salt diapir: Geological Society of America Bulletin, v. 110, no. 12, p. 1547–1573.

Figures 2.46, 2.47, 2.49

Jackson, M. P. A., Schultz-Ela, M. L., Hudec, M. R., Watson, I. A., and Porter, M. L., 2001, Structure and evolution of Upheaval Dome: pinched-off salt diapir or meteoritic impact structure?: The University of Texas at Austin, Bureau of Economic Geology Report of Investigations No. 262, 93 p.

Figures 2.46, 2.48

Jackson, M. P. A., and Seni, S. J., 1983, Geometry and evolution of salt structures in a marginal rift basin of the Gulf of Mexico, East Texas: Geology, v. 11, p. 131–135.

Figure 2.04

Jackson, M. P. A., and Seni, S. J., 1984, Atlas of salt domes in the East Texas Basin: The University of Texas at Austin, Bureau of Economic Geology Report of Investigations No. 140, 102 p.

Figures 2.05, 2.07, 2.08, 2.09

Jackson, M. P. A., and Talbot, C. J., 1986, External shapes, strain rates, and dynamics of salt structures: Geological Society of America Bulletin, v. 97, p. 305–323.

Jackson, M. P. A., and Talbot, C. J., 1989a, Anatomy of mushroom-shaped diapirs: Journal of Structural Geology, v. 11, p. 211–230.

Figures 2.92, 2.93

Jackson, M. P. A., and Talbot, C. J., 1989b, Salt canopies, *in* GCSSEPM Foundation 10th Annual Research Conference Program and Extended Abstracts, Houston, Texas, p. 72–78.

Jackson, M. P. A., and Talbot, C. J., 1991, A glossary of salt tectonics: The University of Texas at Austin, Bureau of Economic Geological Circular 91-4, 44 p.

Jackson, M. P. A., and Vendeville, B. C., 1990, The rise and fall of diapirs during thin-skinned extension (abs.): American Association of Petroleum Geologists Bulletin, v. 74, p. 683.

Jackson, M. P. A., and Vendeville, B. C., 1994, Regional extension as a geologic trigger for diapirism: Geological Society of America Bulletin, v. 106, p. 57–73.

Figures 2.55, 3.32

Jackson, M. P. A., Vendeville, B. C., and Schultz-Ela, D. D., 1994, Structural dynamics of salt systems: Annual Review of Earth and Planetary Sciences, v. 22, p. 93–117.

Figures 2.79, 3.39, 3.40

Jackson, M. P. A., Warin, O. N., Woad, G. M., and Hudec, M. R., 2003, Neoproterozoic allochthonous salt tectonics during the Lufilian orogeny in the Katangan Copperbelt, central Africa: Geological Society of America Bulletin, v. 115, no. 3, p. 314–330.

Figures 7.20, 7.21

Jamieson, G. A., 1995, Subsalt and base-of-salt reflector relationships and hydrocarbon trap implications from a large 3D, depth migrated, seismic volume, *in* Offshore Technology Conference, May 1–4, Houston, 7638-MS, doi:10.4043/7638-MS.

Jamieson, G., Hannan, A., and Biles, N., 2000, Structural overview of the outer shelf, deepwater, and ultra-deepwater Gulf of Mexico, relationships between salt features and minibasin morphology, *in* Shoup, R., Watkins, J., Karlo, J., and Hall, D., eds., Integration of geologic models for understanding risk in the Gulf of Mexico: American Association of Petroleum Geologists/Datapages Discovery Series (CD-ROM) no. 1, 15 p.

Jaswal, T. M., Lillie, R. J., and Lawrence, R. D., 1997, Structure and evolution of the Northern Potwar Deformed Zone, Pakistan: American Association of Petroleum Geologists Bulletin, v. 81, p. 308–328.

AAPG©1997, reprinted by permission of the AAPG whose permission is required for further use.

Figure 6.19

Jensen, L. N., and Sørensen, K., 1992, Tectonic framework and halokinesis of the Nordkapp Basin, Barents Sea, *in* Larsen, R. M., Brekke, H., Larsen B.T., and Talleraas, E., eds., Structural and tectonic modeling and its application to petroleum geology: Amsterdam, Elsevier, Norsk Petroleum Forening (NPF) Special Publication 1, p. 109–120.

Reprinted with permission from Norsk Petroleum Forening, Copyright (1992).

Figure 3.08

Johns, M. K., and Mosher, S., 1996, Physical models of regional fold superposition: the role of competence contrast: Journal of Structural Geology, v. 18, p. 475–492.

Reprinted with permission from Elsevier, Copyright (1996).

Figure 1.34

Johnson, H. A., and Bredeson, D. H., 1971, Structural development of some shallow salt domes in Louisiana Miocene productive belt: American Association of Petroleum Geologists Bulletin, v. 55, p. 204–226.

Kidd, G., 1939, Developments in South Texas, 1938–1939: American Association of Petroleum Geologists Bulletin, v. 23, p. 860–870.

Kockel, F., 1990, Morphology and genesis of northwest-German salt structures, *in* Proceedings of Symposium on Diapirism with Special Reference to Iran: Tehran University Governery of Hormozgan, p. 225–248.

Figures 7.26, 7.27, 7.28

Kockel, F., 1998, Salt problems in Northwest Germany and the German North Sea sector, *in* Kockel, F., and Marschall, R., eds., Geology and geophysics of salt structures: Journal of Seismic Exploration, v. 7, p. 219–235.

Figures 2.27, 7.26

Kulke, H., 1991, Guide booklet for excursion to diapirs in the Atlas Mountains (Algeria): 7 October–12 October, report for BP, 29 p.

Figure 2.02

Kunz, T. J., and Jones, T. K., 1994, 3-D geological velocity model construction for 3-D depth imaging, *in* Offshore Technology Conference, May 2–5, Houston, 7398-MS, doi:10.4043/7398-MS.

Kupfer, D. H., 1976, Shear zones inside Gulf Coast stocks help delineate spines of movement: American Association of Petroleum Geologists Bulletin, v. 60, p. 1434–1447.

Lachmann, R., 1910, Über autoplaste (nichttektonische) Formelemente im Bau der Salzgesteine Norddeutschlands: Deutschen Geologischen Gesellschaft Monatsbericht, v. 62, p. 113–116.

Ladzekpo, D. H., Sekharan, K. K., and Gardner, G. H. F., 1988, Physical modeling for hydrocarbon exploration, *in* Society of Exploration Geophysicists, 58th annual meeting, SEG Abstracts, v. 1, p. 586–588.

Laubscher, H., 1977, Fold development in the Jura: Tectonophysics, v. 37, p. 337–362.

Figure 6.05

Laubscher, H., 1992, Jura kinematics and the Molasse Basin: Eclogae Geologicae Helvetiae, v. 85, p. 653–675.

Springer and Birkhäuser Basel, Laubscher, H., figure 2, ©1992 reprinted with kind permission from Springer Science and Business Media.

Figures 6.04, 6.05

Laubscher, H., and Bernoulli, D., 1980, Cross-section from the Rhine Graben to the Po Plain, *in* Schweizerische Geologische Kommission, ed., Geology of Switzerland, a guide-book, Part B: Geological Excursions: Basel, Wepf & Co., p. 183–209.

Figure 1.13

Lee, G. H., Bryant, W. R., and Watkins, J. S., 1989, Salt structures and sedimentary basins in the Keathley Canyon area, northwestern Gulf of Mexico: their development and tectonic implications, *in* GCSSEPM Foundation 10th Annual Research Conference Program and Extended Abstracts, Houston, Texas, p. 90–93.

Lees, G. M., 1927, Salzgletscher in Persien: Mitteilungen Geologische Gesellschaft Wien, v. 20, p. 29–34.

Liro, L. M., Murillas, J., Villalobos, L., Gatenby, G., and Mathur, V., 2004, Salt sutures in single- and multi-tiered allochthons, Green Canyon and Walker Ridge Areas, deep water Gulf of Mexico, *in* Post, P. J., Olson, D. L., Lyons, K. T., Palmes, S. L., Harrison, P. F., and Rosen, N. C., eds., Salt-sediment interactions and hydrocarbon prospectivity, *in* 24th Annual Research Conference, GCSSEPM Foundation, Houston, Texas, p. 200–219.

Reproduced with permission of the GCSSEPM Foundation, whose permission is required for any further reproduction.

Figures 7.18, 7.19

MacRae, G., and Watkins, J. S., 1992, Evolution of the Destin Dome, offshore Florida, north-eastern Gulf of Mexico: Marine and Petroleum Geology, v. 9, p. 501–509.

Mann, R. J., 1998, Numerical modeling of deformation around allochthonous salt tongues driven by sedimentary loading: application to the Gulf of Mexico: The University of Texas at Austin, M.S. thesis, 162 p.

Figures 7.70, 7.71

Masson, M. P., 1972, L'exploration pétrolière en Angola: Revue de l'Association Française des Techniciens du Pétrole, v. 212, p. 21–34.

Figures 4.02, 4.03

McBride, B. C., 1998, The evolution of allochthonous salt along a megaregional profile across the northern Gulf of Mexico basin: American Association of Petroleum Geologists Bulletin, v. 82, p. 1037–1054.

AAPG©1998, reprinted by permission of the AAPG whose permission is required for further use.

Figures 8.18, 9.08, 9.09

McBride, B. C., Rowan, M. G., and Weimer, P., 1998, The evolution of allochthonous salt systems, northern Green Canyon and Ewing Bank (offshore Louisiana), northern Gulf of Mexico: American Association of Petroleum Geologists Bulletin, v. 82, p. 1013–1036.

AAPG©1998, reprinted by permission of the AAPG whose permission is required for further use.

Figures 8.16, 8.17

McClay, K., Munoz, J.-A., and Garcia-Senz, J., 2004, Extensional salt tectonics in a contractional orogen: a newly identified tectonic event in the Spanish Pyrenees: Geology, v. 32, p. 737–740.

Figure 5.14

McGuinness, D. B., and Hossack, J. R., 1993, The development of allochthonous salt sheets as controlled by the rates of extension, sedimentation, and salt supply, *in* 14th Annual Research Conference, GCSSEPM Foundation, Houston, Texas, p. 127–139.

Meier, B., 1977, Zur Geologie der Klus von Balstal-Oensingen (Ostseite): Unveröffentlicht Diplomarbeit Universität Basel.

Figures 1.12, 1.13

Merle, O., 1986, Patterns of stretch trajectories and strain rates within spreading-gliding nappes: Tectonophysics, v. 124, p. 211–222.

Miralles, L., Sans, M., Pueyo, J. J., and Santanach, P., 2000, Recrystallization salt fabric in a shear zone (Cardona diapir, southern Pyrenees, Spain), *in* Vendeville, B., Mart, Y., and Vigneresse, J.-L., eds., Salt, shale and igneous diapirs in and around Europe: Geological Society, London, Special Publication 174, p. 149–167.

Reproduced with permission of The Geological Society, London.

Figure 2.10

Mohriak, W., Szatmari, P., and Couto Anjos, S. M., eds., 2008, Sal geologia e tectônica: exemplos nas Bacias Brasileiras: São Paulo, Beca Edições, 448 p.

Mount, V., Dull, K., and Mentemeier, S., 2007, Structural style and evolution of traps in the Paleogene play, deepwater Gulf of Mexico, *in* Kennan, L., Pindell, J., and Rosen, N. C., eds., The Paleogene of the Gulf of Mexico and Caribbean Basins: processes, events, and petroleum systems, *in* 27th Annual GCSSEPM Foundation Bob F. Perkins Research Conference Proceedings, p. 54–80.

Reproduced with permission of the GCSSEPM Foundation, whose permission is required for any further reproduction.

Figure 1.11

Mrazec, L., 1907, Despre cute cu simbure de strapungere [On folds with piercing cores]: Romania, Society of Stiite Bulletin, v. 16, p. 6–8.

National Iranian Oil Company, 1975a, Geological map of Iran: sheet no. 4, South-West Iran: scale 1:1,000,000.

Figure 6.20

National Iranian Oil Company, 1975b, Geological map of Iran: sheet no. 5, South-Central Iran: scale 1:1,000,000.

Figures 1.25, 1.26

Naumann, C. F., 1858, Lehrbuch der Geognosie 1 (2d ed.): Leipzig, Wilhelm Engelmann.

Nelson, T. H., 1989, Style of salt diapirs as a function of the stage of evolution and the nature of the encasing sediments, *in* GCSSEPM Foundation 10th Annual Research Conference Program and Extended Abstracts, Houston, Texas, p. 109–110.

Nelson, T. H., 1991, Salt tectonics and listric-normal faulting *in* Salvador, A., ed., The Gulf of Mexico Basin: Boulder, Colorado, Geological Society of America, The Geology of North America, v. J, p. 73–89.

Nelson, T. H., and Fairchild, L. H., 1989, Emplacement and evolution of salt sills in northern Gulf of Mexico (abs.): American Association of Petroleum Geologists Bulletin, v. 73, p. 395.

Nettleton, L. L., 1934, Fluid mechanics of salt domes: American Association of Petroleum Geologists Bulletin, v. 18, p. 1175–1204.

Nilsen, K. T., Vendeville, B. C., and Johansen, J.-T., 1995, Influence of regional tectonics on halokinesis in the Nordkapp Basin, Barents Sea, *in* Jackson., M. P. A., Roberts, D. G., and Snelson, S., eds., Salt tectonics: a global perspective: American Association of Petroleum Geologists Memoir 65, p. 413–436.

AAPG©1995, reprinted by permission of the AAPG whose permission is required for further use.

Figures 2.26, 3.07

Nye, J. F., 1952, The mechanics of glacial flow: Journal of Glaciology, v. 2, p. 82–93.

O'Brien, C. A. E., 1957, Salt diapirism in south Persia: Geologie en Mijnbouw, v. 19, p. 357–376.

Springer Netherlands, O'Brien, C. A. E., figure 4, ©1957, reprinted with kind permission from Springer Science and Business Media.

Figures 1.31, 7.40

O'Brien, J. J., and Lerche, I., 1988, Heat flow through and around salt sheets (abs.): American Association of Petroleum Geologists Bulletin, v. 72, p. 230.

Ocamb, R. D., 1961, Growth faults of South Louisiana: Gulf Coast Association of Geological Societies Transactions, v. 11, p. 139–176.

Owen, P. F., and Taylor, N. G., 1983, A salt pillow structure in the southern North Sea, *in* Bally, A. W., ed., Seismic expression of structural styles; a picture and work atlas: American Association of Petroleum Geologists Studies in Geology Series, v. 15, p. 2.3.2-7–2.3.2-10.

AAPG©1983, reprinted by permission of the AAPG whose permission is required for further use.

Figure 1.20

Pascoe, R., Hooper, R., Storhaug, K., and Harper, H., 1999, Evolution of extensional styles at the southern termination of the Nordland Ridge, Mid-Norway: a response to variations in coupling above Triassic salt, *in* Fleet, A. J., and Boldy, S. A. R., eds., Petroleum geology of Northwest Europe: Proceedings of the 5th Conference: Geological Society, London, p. 83–90.

Reproduced with permission of The Geological Society, London.

Figure 5.13

Peel, F., 2001, Emplacement, inflation and folding of an extensive allochthonous salt sheet in the Late Mesozoic (ultra-deepwater Gulf of Mexico) (abs.): American Association of Petroleum Geologists Annual Convention Official Program, v. 10, p. A155.

Figure 1.08

Peel, F. J., Travis, C. J., and Hossack, J. R., 1995, Genetic structural provinces and salt tectonics of the Cenozoic offshore U.S. Gulf of Mexico: a preliminary analysis, *in* Jackson, M. P. A., Roberts, D. G., and Snelson, S., eds., Salt tectonics: a global perspective: American Association of Petroleum Geologists Memoir 65, p. 153–175.

AAPG©1995, reprinted by permission of the AAPG whose permission is required for further use.

Figures 9.01, 9.02

Pierce, W. G., 1960, The "break-away" point of the Heart Mountain detachment fault in northwestern Wyoming, *in* U.S. Geological Survey Professional Paper, P 0400-B, p. B236–B237.

Pilcher, R. S., Kilsdonk, B., and Trude, J., 2011, Primary basins and their boundaries in the deep-water northern Gulf of Mexico: origin, trap types, and petroleum system implications: American Association of Petroleum Geologists Bulletin, v. 95, p. 219–240.

Ramberg, H., 1981, Gravity, deformation, and the Earth's crust in theory, experiments and geological application (2d ed.): London, Academic Press, 452 p.

Ramsay, J. G., and Graham, R. H., 1970, Strain variation in shear belts: Canadian Journal of Earth Sciences, v. 7, p. 786–813.

Ratcliff, D. W., Gray, S. H., and Whitmore, N. D., Jr., 1992, Seismic imaging of salt structures in the Gulf of Mexico: The Leading Edge of Exploration, v. 11, no. 4, p. 15–31.

Reproduced with permission of the Society of Exploration Geophysicists.

Figure 2.15

Rodgers, J., 1963, Mechanics of Appalachian foreland folding in Pennsylvania and West Virginia: American Association of Petroleum Geologists Bulletin, v. 47, p. 1527–1536.

Rotnicki, K., 1976, The theoretical basis for and a model of glaciotectonic deformations: Quaestiones Geographicae, v. 3, p. 103–139.

Rouby, D., Raillard, S., Guillocheau, F., Bouroullec, R., and Nalpas, T., 2002, Kinematics of a growth/raft system on the West African margin using 3-D restoration: Journal of Structural Geology, v. 24, p. 783–796.

Reprinted with permission from Elsevier, Copyright (2002).

Figure 3.04

Rowan, M. G., 1995, Structural styles and evolution of allochthonous salt, central Louisiana outer shelf and upper slope, *in* Jackson, M. P. A., Roberts, D., and Snelson, S., eds., Salt tectonics: a global perspective: American Association of Petroleum Geologists Memoir 65, p. 199–228.

AAPG©1995, reprinted by permission of the AAPG whose permission is required for further use.

Figures 8.13, 8.29, 10.10

Rowan, M. G., 1997, Three-dimensional geometry and evolution of a segmented detachment fold, Mississippi Fan foldbelt, Gulf of Mexico: Journal of Structural Geology, v. 19, p. 463–480.

Reprinted with permission from Elsevier, Copyright (1997).

Figures 6.02, 6.03

Rowan, M. G., 2002, Salt-related accommodation in the Gulf of Mexico deepwater: withdrawal or inflation, autochthonous or allochthonous?: Gulf Coast Association of Geological Societies Transactions, v. 52, p. 861–869.

Rowan, M. G., Jackson, M. P. A., and Trudgill, B. D., 1999, Salt-related fault families and fault welds in the northern Gulf of Mexico: American Association of Petroleum Geologists Bulletin, v. 83, p. 1454–1484.

AAPG©1999, reprinted by permission of the AAPG whose permission is required for further use.

Figures 3.01, 8.26, 8.27, 9.06, 9.14, 9.15, 10.03, 10.04, 10.09

Rowan, M. G., Lawton, T. F., Giles, K. A., and Ratliff, R. A., 2003, Near-salt deformation in La Popa basin, Mexico, and the northern Gulf of Mexico: a general model for passive diapirism: AAPG Bulletin, v. 87, p. 733–756.

AAPG©2003, reprinted by permission of the AAPG whose permission is required for further use.

Figures 2.18, 2.19, 2.20, 2.21, 2.50, 2.83, 2.84

Rowan, M. G., Peel, F. J., and Vendeville, B. C., 2004, Gravity-driven fold belts on passive margins, *in* McClay, K. R., ed., Thrust tectonics and hydrocarbon systems: American Association of Petroleum Geologists Memoir 82, p. 157–182.

AAPG©2004, reprinted by permission of the AAPG whose permission is required for further use.

Figures 1.08, 3.22, 3.23, 7.54

Rowan, M. G., Ratliff, R. A., Trudgill, B. D., and Duarte, J. B., 2001, Emplacement and evolution of the Mahogany salt body, central Louisiana outer shelf, northern Gulf of Mexico: American Association of Petroleum Geologists Bulletin, v. 85, p. 947–969.

AAPG©2001, reprinted by permission of the AAPG whose permission is required for further use.

Figure 9.13

Rowan, M. G., Trudgill, B. D., and Fiduk, J. C., 2000, Deep-water, salt-cored foldbelts: lessons from the Mississippi Fan and Perdido foldbelts, northern Gulf of Mexico, *in* Mohriak, W., and M. Talwani, eds., Atlantic rifts and continental margins: American Geophysical Union Geophysical Monograph 115, p. 173–191.

Redrawn from Rowan and others, *in* Mohriak and Talwani, eds., Copyright (2000) American Geophysical Union. Reproduced/modified by permission of American Geophysical Union.

Figure 1.07

Rowan, M. G., Weimer, P., McBride, B. C., Varnai, P., Acosta, Z., Budhijanto, F., Martinez, R., Navarro, A., Villamil, T., and Fiduk, J. C., 1995, Interactions between salt tectonics and sedimentation in the Louisiana offshore: preliminary results, *in* Travis, C. J., Harrison, H., Hudec, M. R., Vendeville, B. C., Peel, F. J., and Perkins, B. F., eds., Salt, sediment and hydrocarbons: Gulf Coast Section, Society of Economic Paleontologists and Mineralogists, 16th Annual Research Conference, p. 235–244.

Reproduced with permission of the GCSSEPM Foundation, whose permission is required for any further reproduction.

Figure 8.30

Sannemann, D., 1968, Salt-stock families in northwestern Germany, *in* Braunstein, J., and O'Brien, G. D., eds., Diapirism and diapirs: American Association of Petroleum Geologists Memoir 8, p. 261–270.

Schardt, H., 1893, Origine des préalpes romandes: Eclogae Geologicae Helvetiae, v. 4, p. 129–142.

Schultz-Ela, D. D., 1992, Restoration of cross-sections to constrain deformation processes of extensional terranes: Marine and Petroleum Geology, v. 9, p. 372–388.

Reprinted with permission from Elsevier, Copyright (1992).

Figures 3.02, 4.12

Schultz-Ela, D. D., 2001, Excursus on gravity gliding and gravity spreading: Journal of Structural Geology, v. 23, p. 725–731.

Schultz-Ela, D. D., 2003, Origin of drag folds bordering salt diapirs: American Association of Petroleum Geologists Bulletin, v. 87, p. 757–780.

AAPG©2003, reprinted by permission of the AAPG whose permission is required for further use.

Figures 2.72, 2.73, 2.74

Schultz-Ela, D. D., Jackson, M. P. A., and Vendeville, B. C., 1994, Mechanics of active diapirism: The University of Texas at Austin, Bureau of Economic Geology Report of Investigations No. 224, 56 p.

Figure 2.13

Schuster, D. C., 1995, Deformation of allochthonous salt and evolution of related salt/structural systems, eastern Louisiana Gulf Coast, *in* Jackson, M. P. A., Roberts, D., and Snelson, S., eds., Salt tectonics: a global perspective: American Association of Petroleum Geologists Memoir 65, p. 177–198.

AAPG©1995, reprinted by permission of the AAPG whose permission is required for further use.

Figures 9.03, 9.04, 10.01, 10.02

Scrope, G. P., 1825, Considerations on volcanoes: London, W. Phillips, 270 p.

Seni, S. J., 1994, Salt tectonics on the continental slope, northeast Green Canyon area, northern Gulf of Mexico: evolution of stocks and massifs from reactivation of salt sheets: The University of Texas at Austin, Bureau of Economic Geology Report of Investigations No. 212, 102 p.

Figures 2.14, 8.01

Seni, S. J., and Jackson, M. P. A., 1983, Evolution of salt structures, East Texas diapir province, Part 1: sedimentary record of halokinesis: American Association of Petroleum Geologists Bulletin, v. 67, p. 1219–1244.

AAPG©1983, reprinted by permission of the AAPG whose permission is required for further use.

Figure 2.06

Seni, S. J., and Jackson, M. P. A., 1984, Sedimentary record of Cretaceous and Tertiary salt movement, East Texas Basin: times, rates and volumes of salt flow and their implications to nuclear waste isolation and petroleum exploration: The University of Texas at Austin, Bureau of Economic Geology Report of Investigations No. 139, 89 p.

Figure 2.08

Smith, D. A., and Reeve, F. A. E., 1970, Salt piercement in shallow Gulf Coast salt structures: American Association of Petroleum Geologists Bulletin, v. 54, p. 1271–1289.

AAPG©1970, reprinted by permission of the AAPG whose permission is required for further use.

Figure 2.13

Smoluchowski, M. S., 1909, some remarks on the mechanics of overthrusts: Geological Magazine, v. 6, p. 204–205.

Snoke, A. W., Schamel, S., and Karasek, R. M., 1988, Structural evolution of Djebel Debadib Anticline; a clue to the regional tectonic style of the Tunisian Atlas: Tectonics, v. 7, p. 497–516.

Redrawn from Snoke and others, Copyright (1988) American Geophysical Union. Reproduced/modified by permission of American Geophysical Union.

Figure 2.03

Spencer, J. A., and Sharpe, C. L., 1993, Deep-seated salt sheet in eastern Calcasieu Parish, Louisiana: Gulf Coast Association of Geological Societies Transactions, v. 43, p. 363–371.

Stewart, S. A., and Argent, J. D., 2000, Relationship between polarity of extensional fault arrays and presence of detachments: Journal of Structural Geology, v. 22, p. 693–711.

Reprinted with permission from Elsevier, Copyright (2000).

Figure 4.05

Stewart, S. A., and Clark, A. J., 1999, Impact of salt on the structure of the Central North Sea hydrocarbon fairways, *in* Fleet, A. J., and Boldy, S. A. R., eds., Petroleum geology of Northwest Europe: Proceedings of the 5th Conference: Geological Society, London, p. 179–200.

Reproduced with permission of The Geological Society, London.

Figures 2.35, 5.30

Stier, K., 1915, Strukturbild des Benther Salzgebirges [Structural style of the Benthe salt massif]: Jahresbericht Niedersächsischer Geologischer Verein, v. 8, p. 1–14.

Stöcklin, J., 1961, Lagunäre Formationen und Salzdome in Ostiran: Eclogae Geologicae Helvetiae, v. 54, p. 1–27.

Springer and Birkhäuser Basel, Stöcklin, J., figure 8, ©1961, reprinted with kind permission from Springer Science and Business Media.

Figure 2.51

Stöcklin, J., and Nabavi, M. H., 1959, Tectonic map of Iran: National Iranian Oil Company, scale 1:2,500,000.

Figure 1.23

Stovba, S. M., and Stephenson, R. A., 2003, Style and timing of salt tectonics in the Dniepr-Donets Basin (Ukraine): implications for triggering and driving mechanisms of salt movement in sedimentary basins: Marine and Petroleum Geology, v. 19, p. 1169–1189.

Reprinted with permission from Elsevier, Copyright (2003).

Figures 1.01, 1.02

Suess, E., 1909, Das Antlitz der Erde [The face of the Earth], v. 3: Prague, Czechoslovakia, Tempsky.

Sumner, H. S., Robison, B. A., Dirks, W. K., and Holiday, J. C., 1990, Morphology and evolution of salt/mini-basin systems: lower shelf and upper slope, central offshore Louisiana (abs.): Geological Society of America, Abstracts with Programs, p. A48.

Suppe, J., 1985, Principles of structural geology: Englewood Cliffs, New Jersey, Prentice-Hall, 537 p.

Talbot, C. J., 1978, Halokinesis and thermal convection: Nature, v. 273, p. 739–741.

Talbot, C. J., 1981, Sliding and other deformation mechanisms in a glacier of salt in S. Iran, *in* McClay, K. R., and Price, N. J., eds., Thrust and nappe tectonics: Geological Society, London, Special Publication 9, p. 173–183.

Reproduced with permission of The Geological Society, London.

Figure 7.42

Talbot, C. J., 1993, Spreading of salt structures in the Gulf of Mexico: Tectonophysics, v. 228, p. 151–166.

Talbot, C. J., 1998, Extrusions of Hormuz salt in Iran, *in* Blundell, D. J., and Scott, A. C., eds., Lyell: the past is the key to the present: Geological Society, London, Special Publication 143, p. 315–334.

Reproduced with permission of The Geological Society, London.

Figure 7.41

Talbot, C. J., and Aftabi, P., 2004, Geology and models of salt extrusion at Qum Kuh, central Iran: Journal of the Geological Society, London, v. 161, p. 1–14.

Reproduced with permission of The Geological Society, London.

Figures 7.31, 7.32, 7.33, 7.34

Talbot, C. J., and Alavi, M., 1996, The past of a future syntaxis across the Zagros, *in* Alsop, G. I., Blundell, D. J., and Davison, I., eds., Salt tectonics: Geological Society, London, Special Publication 100, p. 89–109.

Figure 1.23

Talbot, C. J., and Jackson, M. P. A., 1987, Internal kinematics of salt diapirs: American Association of Petroleum Geologists Bulletin, v. 71, p. 1068–1093.

AAPG©1987, reprinted by permission of the AAPG whose permission is required for further use.

Figure 2.91

Talbot, C. J., and Jarvis, R. J., 1984, Age, budget and dynamics of an active salt extrusion in Iran: Journal of Structural Geology, v. 6, p. 521–533.

Talbot, C. J., Medvedev, S., Alavi, M., Shahrivar, H., and Heidari, E., 2000, Salt extrusion rates at Kuh-e-Jahani, Iran: June 1994 to November 1997, *in* Vendeville, B. C., Mart, Y., and Vigneresse, J. L., eds., Salt, shale and igneous diapirs in and around Europe: Geological Society, London, Special Publication 174, p. 93–110.

Reproduced with permission of The Geological Society, London.

Figures 7.46, 7.47

Trusheim, F., 1957, Über Halokinese und ihre Bedeutung für die strukturelle Entwicklung Norddeutschlands: Zeitschrift für Deutsche Geologische Gesellschaft, v. 109, p. 111–151.

Trusheim, F., 1960, Mechanism of salt migration in northern Germany: American Association of Petroleum Geologists Bulletin, v. 44, p. 1519–1540.

AAPG©1960, reprinted by permission of the AAPG whose permission is required for further use.

Figure 1.17

Valle, P. J., Gjelberg, J. G., and Helland-Hansen, W., 2001, Tectonostratigraphic development in the eastern Lower Congo Basin, offshore Angola, West Africa: Marine and Petroleum Geology, v. 18, p. 909–927.

Reprinted with permission from Elsevier, Copyright (2001).

Figure 3.03

van der Fliert, J. R., 1953, Tectonique d'écoulement et Trias diapir au Chettaaba, sud-ouest de Constantine (Algérie) [Tectonics of spreading and Triassic diapir at Chettaaba, south-west of Constantine (Algeria)]: Comptes Rendus de l'Académie des Sciences, Paris, v. 3, no. 3, p. 63–80.

Vendeville, B. C., Ge, H., and Jackson, M. P. A., 1995, Scale models of salt tectonics during basement-involved extension: Petroleum Geoscience, v. 1, p. 179–183.

Reproduced with permission of The Geological Society, London.

Figure 1.36

Vendeville, B. C., and Jackson, M. P. A., 1991, Deposition, extension, and the shape of downbuilding salt diapirs (abs.): American Association of Petroleum Geologists Bulletin, v. 75, no. 3, p. 687–688.

Vendeville, B. C., and Jackson, M. P. A., 1992a, The rise of diapirs during thin-skinned extension: Marine and Petroleum Geology, v. 9, p. 331–353.

Figures 3.14, 3.16, 3.19, 3.37

Vendeville, B. C., and Jackson, M. P. A., 1992b, The fall of diapirs during thin-skinned extension: Marine and Petroleum Geology, v. 9, p. 354–371.

Figures 3.38, 4.11, 4.14, 4.16, 4.17, 4.18

Vendeville, B. C., and Jackson, M. P. A., 1992c, The rise and fall of diapirs during thin-skinned extension: The University of Texas at Austin, Bureau of Economic Geology Report of Investigations No. 209, 60 p.

Figure 3.17

Vendeville, B. C., and Nilsen, K. T., 1995, Episodic growth of salt diapirs driven by horizontal shortening, *in* Travis, C. J., Harrison, H., Hudec, M. R., Vendeville, B. C., Peel, F. J., and Perkins, B. F., eds., Salt, sediment and hydrocarbons: Gulf Coast Section, Society of Economic Paleontologists and Mineralogists, 16th Annual Research Conference, p. 285–295.

Reproduced with permission of the GCSSEPM Foundation, whose permission is required for any further reproduction.

Figure 2.87

Ville, L., 1856, Notice géologique sur les salines des Zahrez et les gites de sel gemme de Rang el Melah et d'Ain Hadjera (Algerie): Annales des Mines, v. 15, p. 351–410.

Figures 2.01, 2.02

Volozh, Y., Talbot, C., and Ismail-Zadeh, A., 2003, Salt structures and hydrocarbons in the Pricaspian basin: American Association of Petroleum Geologists Bulletin, v. 87, p. 313–334.

AAPG©2003, reprinted by permission of the AAPG whose permission is required for further use.

Figure 7.30

Wagner, G., Mauthe, F., and Mensik, H., 1971, Der Salzstock von Cardona in Nordostspanien: Geologische Rundschau, v. 60, p. 970–966.

Figure 2.10

Wang, Y. F., 1988, Salt tongues in northern Gulf of Mexico (abs.): American Association of Petroleum Geologists Bulletin, v. 72, p. 256.

Watkins, J. S., Ladd, J. W., Buffler, R. T., Shaub, F. J., Houston, M. H., and Worzel, J. L., 1978, Occurrence and evolution of salt in deep Gulf of Mexico, *in* American Association of Petroleum Geologists Studies in Geology 7, p. 43–65.

West, D. B., 1989, Model for salt deformation on deep margin of central Gulf of Mexico basin: American Association of Petroleum Geologists Bulletin, v. 74, p. 1472–1482.

Wilckens, O., 1912, Grundzüge der tektonischen Geologie [Basic features of tectonic geology]: Jena, Fischer.

Willis, B., 1893, Mechanics of Appalachian structure, *in* U.S. Geological Survey Annual Report 13, part 2, p. 217–281.

Withjack, M. O., and Callaway, S., 2000, Active normal faulting beneath a salt layer; an experimental study of deformation patterns in the cover sequence: American Association of Petroleum Geologists Bulletin, v. 84, p. 627–651.

AAPG©2000, reprinted by permission of the AAPG whose permission is required for further use.

Figures 5.19, 5.20, 5.21

Worrall, D. M., and Snelson, S., 1989, Evolution of the northern Gulf of Mexico, with emphasis on Cenozoic growth faulting and the role of salt, *in* Bally, A. W., and Palmer, A. R., eds., The geology of North America—an overview: Boulder, Colorado, Geological Society of America, p. 97–138.

Yeilding, C. A., and Travis, C. J., 1997, Nature and significance of irregular geometries at the salt-sediment interface: examples from the deepwater Gulf of Mexico (abs.): American Association of Petroleum Geologists Annual Convention Official Program, v. 6, p. A128.

Ziegler, P. A., 1978, Northwestern Europe: tectonics and basin development: Geologie en Mijnbouw, v. 57, p. 589–626.

Subject Index

For specific locations of salt-tectonic structure, see the Geographic Index.

A

active diapir, active diapirism, 269
  compressional, 10, 12, 39–40, 42, 44, 48, 50, 52, 64–66, 68–72, 74, 77, 79–80, 89, 125–127, 269
  conceptual model, 31, 75–80, 106, 127, 204
  geologic example, 10–12, 36–40, 42, 44–45, 47–53, 89, 110, 125–127, 176–178, 221–224, 236
  halokinetic, 38, 50–51, 61–63, 74–75, 143, 157, 159, 176–178, 190, 204, 230, 269
  numerical model, 74–75, 104–105, 136–139, 159
  physical model, 60–66, 68–72, 157, 190, 230

aerial photograph—*see* outcrop photograph

allochthonous salt, 269; *see also* counterregional system, roho system, salt canopy, salt overthrust, salt sheet

allosuture, 269; *see also* suture (allosuture)

apical suture line, 281; *see also* suture (suture line)

apical suture point, 282; *see also* suture (suture point)

apparent weld, 283; *see also* weld

autochthonous salt, 269; *see also* detachment, indented diapir, salt pillow and salt anticline, salt stock and salt wall, triangular diapir

autosuture, 269; *see also* suture (autosuture)

B

backthrust, 269; see also *forethrust*

basal cutoff, 269
  counterregional system, 245–247, 249, 252, 269
  minibasin, 217–220, 222–224, 227, 230, 232–233
  roho system, 235–237, 241–243
  salt pillow and salt anticline, 6–7
  salt sheet, 161, 163–167, 170, 172, 175–177, 179, 194, 203–214

basal shear, 26–27, 73–74, 94, 159, 186, 188, 203–204, 210–212, 269

basal shear zone, 281; *see also* shear zone

basal suture line, 281; *see also* suture (suture line)

basal suture point, 282— *see also* suture (suture point)

basement, 269
  conceptual model, 77, 107–108, 140, 205, 243, 250
  extension
    contractional salt detachment, 130, 133–134, 136–140
    indented diapir, 89, 103, 107–108
    salt pillow and salt anticline, 2, 12, 23
    salt stock and salt wall, 44–48, 60–82
  geologic example
    contractional salt detachment, 142, 148–149, 152
    extensional salt detachment, 125–128, 130
    indented diapir, 112
    salt pillow and salt anticline, 2, 6, 10, 12–13
    salt sheet, 175, 178–179
    salt stock and salt wall, 40, 49
    triangular diapir, 89–91
  inversion, 10, 12–13, 46, 48, 142, 179
  numerical, 28–29, 136–139
  physical model, 23, 60–62, 99, 103, 133–136, 158, 193, 196–197
  ramp
    counterregional system, 248, 250
    extensional salt detachment, 125–129, 135–136
    indented diapir, 90, 99
    salt pillow and salt anticline, 6
    salt sheet, 175, 193, 196–197, 205
    triangular diapir, 46, 77
  shortening, 12–13, 28–29, 40, 46, 49, 142, 148–149, 152, 158

basin-margin graben, 275; *see also* graben (peripheral)

boundary shear zone, 280; *see also* shear zone

bowl weld, 270; *see also* weld

box fold, 270
  conceptual model, 80
  geologic example, 4–5, 8
  numerical model, 26
  physical model, 25, 153–156

breakaway graben, 275; *see also* graben (peripheral)

breakthrust, 1, 8–9, 24–25, 27, 29, 145, 153–156, 185, 270

bucket weld, 270

buckle fold—*see* salt pillow and salt anticline *(compressional)*

bulb, 270
  extensional salt detachment, 125–127
  salt sheet, 177, 179
  salt stock and salt wall, 31, 35–36, 38, 40–42, 44–48, 50–52, 62–63, 69, 71, 75, 77, 80–83

buoyancy, 223, 234, 269–270, 274

C

canopy, 278; *see also* salt canopy

canopy feeder, 278; *see also* salt feeder

canopy-margin thrust system, 270
  imbricate wedge, 161, 166–170, 202, 210–212
  roof-edge thrust, 161, 164–170, 202, 204, 206–213, 218, 222, 232–233, 236–238, 243

cap rock, 2, 35–36, 38, 51, 53–54, 76, 114, 186, 270

carapace, 44, 163, 175, 186, 190–191, 209, 221, 270

Christmas-tree structure, 280; *see also* salt wing

complete weld, 283; *see also* weld

composite halokinetic sequence, 270; *see also* halokinetic sequence

composite suture line, 281; *see also* suture (suture line)

composite suture point, 282; *see also* suture (suture point)

compressional active diapirism, 269; *see also* active diapirism

compressional salt anticline—*see* salt pillow and salt anticline

compressional toe
  conceptual model, 204–213, 232
  downdip end of salt basin
    Gulf of Mexico Basin, 5–8, 143
    Kwanza Basin, 9–10, 145–147, 175
    Lower Congo Basin (Gabon), 42, 145, 176–177
  numerical model, 27
  physical model, 98–99, 155, 190–191, 193–194, 198–200, 202
  salt sheet, 5, 7, 161, 164–172, 218–219, 222, 236–238, 243

conceptual model
contractional salt detachment, 160
counterregional system, 251–252
extensional salt detachment, 140
indented diapir, 118–119
minibasin, 232–234
roho system, 242–243
salt sheet, 204–215
salt stock and salt wall, 75–83
triangular diapir, 106–108

concordant, 271

constrictional strain, 55–56, 82

contraction, 271; *see also* compressional toe
above allochthonous salt
contractional salt detachment, 143
minibasin, 218, 222, 225, 227, 232
roho system, 236–238, 243
salt pillow and salt anticline, 5
salt sheet, 162–171, 173–175, 183, 185, 188, 194, 198–199, 202, 204–213
above autochthonous salt
contractional salt detachment, 141–142, 144–160
roho system, 235–243
salt pillow and salt anticline, 1–4, 6, 8–22, 24–25, 26–29
salt sheet, 176–179, 190–193
salt stock and salt wall, 33, 36, 39–40, 42–48, 50, 54–57, 59, 62–72, 75, 77, 79–81
triangular diapir, 89, 98–99
basement-detached—
see detachment (contractional)
basement-involved—
see basement (shortening)
conceptual model, 1, 31, 75, 77, 79–81, 141, 160–161, 204–213, 232, 243
diapir—*see* active diapir (compressional); plug-fed thrust
fold—*see* kink fold; monocline; salt pillow and salt anticline
geologic example
contractional salt detachment, 142–152
minibasin, 218, 222, 225, 227
roho system, 236–238
salt pillow and salt anticline, 2–6, 8–21
salt sheet, 162–171, 173–179, 183, 185, 188
salt stock and salt wall, 33, 36, 39–40, 42–48, 50, 54–57, 59
triangular diapir, 89
multidetached, 148–150, 156, 158, 160
nappe—*see* salt nappe
numerical model, 26–29, 74, 159
physical model, 22, 24–25, 62–72, 98–99, 118, 153–158, 190–194, 198–199, 202
thrust—*see* canopy-margin thrust system, salt overthrust; thrust fault
transpression—*see* transpression

contractional salt detachment, 141–160;
see also detachment (contractional)

convergent margin
Aquitaine Basin, 142
Atlas fold belt, 32–33
Betic Cordillera, 162
Ebro Basin, 36
Garmsar Basin, 163
Great Kavir, 3–4, 37
Jura fold belt, 8–9, 144
Maritime Alps, 43
Parry Islands fold belt, 148–150
Potwar Plateau, 151
Qom Basin, 181–182
Ravar Basin, 59–60
Rif Basin, 183
Sierra Madre Oriental fold belt, 14
Zagros fold belt, 15–21, 152, 183–189

Couette flow, 159, 203, 271

counterregional canopy, 271

counterregional system, 245–252, 271
conceptual model, 242–243, 245, 251–252
geologic example, 220, 224–225, 239–240, 246–248
numerical model, 204
physical model, 196–197, 201, 229, 248–250

cover, 271

crescentic fold, 82–83, 271

crestal anticline–*see* active diapir

crestal graben, 271–*see also* graben (crestal)

crystalline basement—*see* basement

curtain fold, 82, 271

D

décollement, 271;
see also detachment (décollement)

density inversion, 49, 271

depotrough, 109, 119, 121, 123–124, 217, 233, 246, 271

detachment (décollement), 271
contractional
contractional salt detachment, 141–160
minibasin, 222
roho system, 243
salt pillow and salt anticline, 1–2, 5–6, 8–13, 22, 24–27
salt sheet, 162, 166–171, 174–179, 202, 210–212
salt stock and salt wall, 39–40, 42, 44, 63–64, 69, 71, 74–75, 79–80
extensional
counterregional system, 252
extensional salt detachment, 121–140
indented diapir, 109, 111–112, 116–119
roho system, 235–243
salt stock and salt wall, 61–62, 79
triangular diapir, 85–97, 99–108
translational, 60, 193

detachment fold, 1–6, 8–22, 24–27; *see also* contraction

diapir, 278; *see also* salt diapir

diapir fall, 271; *see also* indented diapir

diapir family, 271

diapirism—*see* salt diapir

diapir shortening—*see* contraction

differential loading, 272

discontinuous weld, 283; *see also* weld

discordant, 272

displacement loading, 272

dissolution, 272
indented diapir, 109, 111, 114–115, 118
roho system, 240
salt pillow and salt anticline, 2, 16–20
salt sheet, 163, 165, 173–174, 181–182, 184–189
salt stock and salt wall, 32, 35–38, 47, 51, 53–54, 60, 73, 76

dome—*see* salt dome

downbuilding, 275; *see also* passive diapirism

drape fold, 272
basement-involved
extensional salt detachment, 133–134, 136–140
salt pillow and salt anticline, 2, 6, 13, 23, 28–29
salt stock and salt wall, 61–62
triangular diapir, 89, 103, 107–108
halokinetic
minibasin, 223
salt sheet, 203, 213
salt stock and salt wall, 39, 41–42, 58, 73, 78

drape graben, 271; *see also* graben (drape)

drape monocline—*see* drape fold; monocline (basement-involved)

duplex—*see* thrust fault

dynamic bulge, 161, 181–186, 188–189, 193, 204–206, 272

E

encased minibasin, 272

encircling autosuture—*see* suture (autosuture)

en echelon faults
conceptual model, 81
geologic example, 122, 143, 171, 236
physical model, 72, 101, 131, 135

enterolithic fold, 272

expulsion rollover, 277; *see also* rollover (expulsion)

extension, 272
basement-detached
above allochthonous salt, 86, 117, 164, 166–167, 171, 175–177, 180, 190–191, 194–197, 200–202, 204, 206, 209–213, 219, 221, 223–227, 232–233, 235–243, 245–248, 250–252
above autochthonous salt, 8–13, 23, 25, 27–29, 33, 35, 38–40, 42, 44–53, 60–62, 64, 66, 69, 71, 74–76, 79–81, 85–119, 121–140, 145, 147, 157, 246
basement-involved
contractional salt detachment, 142
extensional salt detachment, 128–130, 133–134, 136–140
salt pillow and salt anticline, 2, 12, 23, 28
salt stock and salt wall, 44–49, 60–62
triangular diapir, 87, 90–91, 107–108
conceptual model
contractional salt detachment, 141–160
counterregional system, 245, 250–252
extensional salt detachment, 140
indented diapir, 109, 118–119, 121
minibasin, 232–233
roho system, 235, 242–243, 245, 250–252
salt sheet, 204, 206, 209–213
triangular diapir, 85, 106–108
diapir fall, 271
extensional salt detachment, 133, 140
indented diapir, 109–120
salt stock and salt wall, 79, 81
triangular diapir, 86, 88, 97, 100–101, 103
diapir rise
contractional salt detachment, 141–160
counterregional system, 245–252
extensional salt detachment, 121–123, 133–140
indented diapir, 116–117, 119
minibasin, 224–227, 232
roho system, 235–236, 238–243, 251–252
salt pillow and salt anticline, 10, 23
salt sheet, 171, 176–177, 194, 197, 201–202, 204, 206, 210
salt stock and salt wall, 60–62, 64, 71, 75–76, 79, 81
triangular diapir, 85–108
fault—*see* graben; normal fault; radial fault
fold—*see* mock-turtle anticline; monocline (basement-involved); rollover (extension)
geologic example
counterregional system, 246–248
extensional salt detachment, 122–131
indented diapir, 110–115
minibasin, 219, 221, 223–227
roho system, 236–240
salt pillow and salt anticline, 2, 8–13
salt sheet, 164, 166–167, 171, 175–177, 180
salt stock and salt wall, 33, 35, 38–40, 42, 44–53
triangular diapir, 86–93
indented diapir—*see* indented diapir
numerical model, 25, 27–29, 74–75, 104–105, 136–139, 159
physical model
contractional salt detachment, 157
counterregional system, 248
extensional salt detachment, 131–136
indented diapir, 116–118
roho system, 241–242,
salt pillow and salt anticline, 23
salt sheet, 190–191, 194–197, 200–202
salt stock and salt wall, 60–62, 64, 66, 69, 71
triangular diapir, 93–103
raft—*see* raft tectonics
roho system—*see* roho system
transtension—*see* transtension
triangular diapir—*see* triangular diapir

extensional salt detachment, 121–140; *see also* detachment (extensional)

extrusive advance, 272
minibasin, 224, 226–227, 229, 233–234
salt sheet, 163, 174, 177, 180–195, 197, 200–202, 204–206, 211–212

F

falling diapir—*see* extension (diapir fall)

fault weld, 273; *see also* weld

feeder, 278; *see also* salt feeder

finite-element model—*see* numerical model

fishtail structure, 149, 160

flap fault, 273

flap fold—*see* drape fold (halokinetic)

fold belt
orogenic—*see* convergent margin
passive margin—*see* compressional toe

footwall, 273

forethrust, 273; *see also* backthrust

foundered minibasin, 272; *see also* encased minibasin

freeboard, 273

frontal escarpment, 273
counterregional system, 252
minibasin, 218, 222, 232
roho system, 236, 243
salt sheet, 161, 163–171, 175, 186, 193–196, 198, 200–202, 204–213

frontal rolling, 282; *see also* tank-tread advance

frontal suture, 281; *see also* suture

G

graben
crestal, 271
contractional salt detachment, 145, 147
counterregional system, 245–252
extensional salt detachment, 130, 133–134, 136–140
indented diapir, 110, 112–119
minibasin, 223–224, 226, 232
roho system, 242
salt pillow and salt anticline, 10–11, 23, 28–29
salt sheet, 171, 178, 194, 196, 202
salt stock and salt wall, 38–39, 48, 50, 52, 61–62, 64, 66, 70, 74–76, 79–81
triangular diapir, 85, 86, 91, 93–95, 97–99, 101, 103–108
drape, 271
extensional salt detachment, 130, 133–134, 136–139
salt pillow and salt anticline, 28–29
triangular diapir, 103, 107–109
keystone, 274
counterregional system, 243
salt pillow and salt anticline, 8–9, 11, 13, 25, 27
salt stock and salt wall, 60, 68–69, 74
triangular diapir, 99
mock turtle, 110–111, 118
peripheral (breakaway), 122–123, 135–136, 196–197, 201, 250, 275
rollover, 88, 237, 239–240, 246–248, 250

gravitational loading, 273

gravity spreading, 273

growth fault (synsedimentary fault), 273

growth interval, 282; *see also* synkinematic

H

halokinesis, 273; *see also* active diapirism (halokinetic); halokinetic sequence; passive diapir, passive diapirism

halokinetic active diapirism, 274; *see also* active diapirism (halokinetic); salt anticline (halokinetic)

halokinetic sequence, 270, 274
conceptual model, 78, 212–213
geologic example, 39, 41–42, 52, 58, 220–221
hook, 39, 41–42, 58, 78, 270, 274
numerical model, 73
physical model, 197, 201–202, 241–242, 250
wedge, 41, 52, 73, 78, 270, 274

halotectonics, 274

hanging wall, 274

heel-toe, 235, 238–241, 243, 274

hill-hole pair, 211–212

hook halokinetic sequence, 274; *see also* halokinetic sequence (hook)

I

imbricate wedge, 274; *see also* canopy-margin thrust system (imbricate wedge)

incomplete weld, 283; *see also* weld

indented diapir, 109–118
  conceptual model, 79, 106, 109, 118–119, 140
  geologic example, 86–87, 110–115
  physical model, 97, 99, 101, 103, 116–118, 230

infold, 275; *see also* mushroom diapir

intrasalt basin, 275; *see also* minibasin

inversion, 281; *see also* structural inversion

isopachous, 274

K

keel, 278–279; *see also* salt keel

keystone graben—*see* graben (keystone)

kink fold, 8–9, 22, 24–26, 28, 144, 151, 153–156, 159–160, 168–170

L

lateral suture, 281; *see also* suture

level of neutral buoyancy, 274; *see also* buoyancy

listric fault, 275

M

map
  Atlas fold belt (Djebel Debadib diapir), 33
  Atlas fold belt (Ran el Melah diapir), 32
  Betic Cordillera (Guadalquivir salt nappe), 162
  East Texas Basin, 34, 122
  Gulf of Mexico (Atwater fold belt), 5, 143
  Gulf of Mexico (counterregional systems), 246
  Gulf of Mexico (Mahogany roho), 238
  Gulf of Mexico (Mitchell and Farnella domes), 236
  Gulf of Mexico (Rum roho), 237
  Gulf of Mexico (Sigsbee Escarpment), 171
  Gulf of Mexico (Whiting Dome roho), 236
  Gulf of Mexico (Wilcox raft), 123
  Katanga Basin (West Lufilian allochthon), 173
  La Popa Basin, 40–41
  location map, 253–268
  Lower Congo Basin (Block 3 rafts), 128
  Maritime Alps, 43
  North Sea (Banff diapir), 51
  North Sea (Mungo diapir), 53
  Paradox Basin (Onion Creek diapir), 55
  Paradox Basin (Upheaval Dome), 55
  Qom Basin (Qom Kuh diapir), 181
  Ravar Basin (Diapir VII), 59
  Zagros fold belt, 15, 18
  Zagros fold belt (Jahani diapir), 188–189

minibasin (intrasalt basin), 217–234, 275
  conceptual model, 109, 118–119, 140, 217, 232–234
  encased (foundered), 229
  geologic example, 10, 49, 86, 89, 110–115, 164, 166–167, 180, 218–227, 236
  numerical model, 231
  physical model, 96, 116–118, 193–194, 197–199, 228–230
  thrusted, 222
  uplifted, 164, 221, 223

mock-turtle anticline, 79, 100, 109–111, 118, 275; *see also* turtle structure

model—*see* conceptual model, numerical model, physical model

monocline, 195–196, 275
  basement-involved, 13, 28–29, 61–62, 103, 107–108, 130, 133–134, 136–140; *see also* drape fold (basement involved)
  expulsion, 11, 21, 46, 77, 186–197, 204, 224–225, 229, 239–243, 245–252; *see also* rollover (expulsion)

mother salt, 275; *see also* autochthonous salt

multi-detachment system, multiple detachment—*see* detachment

mushroom diapir, 82–83, 275

N

namakier—*see* salt sheet

neutral buoyancy level—*see* level of neutral buoyancy

normal fault, 275
  conceptual model
    extensional salt detachment, 140
    indented diapir, 109, 118–120
    minibasin, 232–233
    roho, 235, 242–243, 245
    salt sheet, 204, 206, 209–211
    salt stock and salt wall, 31, 75–77, 79–81
    triangular diapir, 85, 106–108
  geologic example
    contractional salt detachment, 142, 145–148, 151
    extensional salt detachment, 122–125, 128–131
    indented diapir, 110–115
    minibasin, 219, 221, 223–227
    roho system, 236–240
    salt pillow and salt anticline, 2, 5, 8–13, 33–36, 38–40, 42–53
    salt sheet, 164, 166–167, 171, 175, 178–179
    triangular diapir, 86–93
  numerical model
    contractional salt detachment, 159
    extensional salt detachment, 136–139
    salt pillow and salt anticline, 25, 27–29
    salt stock and salt wall, 74–75
    triangular diapir, 104–105
  physical model
    contractional salt detachment, 157
    counterregional system, 248, 250–252
    extensional salt detachment, 131–136
    indented diapir, 116–118
    minibasin, 229
    roho system, 242–242,
    salt pillow and salt anticline, 21, 23
    salt sheet, 190–191, 194–199, 201–202
    salt stock and salt wall, 60–62, 64, 71
    triangular diapir, 93–103
numerical model
  contractional salt detachment, 159
  extensional salt detachment, 136–139
  minibasin, 231
  salt pillow and salt anticline, 25–29
  salt sheet, 203–204
  salt stock and salt wall, 72–75
  triangular diapir, 104–105

O

open-toed advance, 165, 174– 175, 194–195, 200, 203–209, 213, 275

outcrop photograph,
  Atlas fold belt (Ran el Melah diapir), 32
  Great Kavir Basin
    Diapirs 10 and 5, 37
    West Dulasian anticline, 4
  Maritimes Basin (Cape Breton Island), 44
  Paradox Basin
    Onion Creek diapir, 53–54, 114
    Upheaval Dome, 54, 56
  Qom Basin (Qom Kuh diapir), 181–182
  Ravar Basin (Diapir VII), 59–60
  Zagros fold belt
    Anguru diapir, 183
    salt diapirs and glaciers, 19–20, 183–185, 187–189

overburden, 275

overhang, 270; *see also* bulb

overriding autosuture—*see* suture (autosuture)

P

para-autochthonous salt, 275

passive diapir, passive diapirism (downbuilding), 275
  conceptual model, 31, 75–81, 86, 106, 161, 163, 204–206, 214, 232–233, 245
  geologic example, 32, 37, 39–40, 42, 44, 46–49, 57–58, 89, 177, 179–180, 193, 220, 236–237, 240, 243, 246
  numerical model, 72, 73
  physical model, 61, 64, 94, 96, 101–102, 117, 229–230, 249

passive margin (present day)
  Brazil, 122
  Gulf of Mexico (Mexico), 248
  Gulf of Mexico (United States of America), 5–8, 38–39, 86, 110, 123–124, 143, 163–172, 218–227, 236–240, 246–247
  Nordkapp Basin, 44, 89
  West Africa, 9–10, 39–40, 42, 86–88, 111, 125–129, 145–147, 175–177

pedestal, 275
- counterregional system, 245–246, 252
- extensional salt detachment, 125–126, 140
- indented diapir, 111
- minibasin, 217
- roho system, 235, 237, 242–243
- salt sheet, 161, 176–177, 180, 204–207, 214
- salt stock and salt wall, 31, 35, 40, 43–44, 46, 49, 57, 60–61, 72, 76, 77, 79–81
- triangular diapir, 89, 93–94, 99, 101

peripheral graben, 275; *see also* graben (peripheral)

peripheral sink, 275–276

physical model
- contractional salt detachment, 153–158
- counterregional system, 248–250
- extensional salt detachment, 131–136
- indented diapir, 116–118
- minibasin, 228–230
- roho system, 241–242
- salt pillow and salt anticline, 21–25
- salt sheet, 190–202
- salt stock and salt wall, 60–72
- triangular diapir, 93–103

piercement, 276; *see also* active diapirism, passive diapirism, reactive diapirism

pillow—*see* salt pillow

pinch-off—*see* salt diapir (pinched off)

plug-fed extrusion, 163–164, 177, 180–192, 194, 204–206, 276

plug-fed thrust, 207, 276

Poiseuille flow, 27, 72–74, 94, 159, 203–204, 276

pop down, 74, 150, 154–155, 276

pop up, 70–71, 74–75, 143, 148, 150–157, 159, 276

postkinematic, 276

postsalt, 276

prekinematic, 276

preraft, 276; *see also* raft tectonics

presalt, 276; *compare* subsalt
- contractional salt detachment, 142, 144–151, 156, 158
- extensional salt detachment, 122, 124–130, 133–140
- indented diapir, 112
- salt pillow and salt anticline, 2, 6–7, 9–13, 21, 23–24, 28–29
- salt sheet, 162, 173–174, 176–180, 183–185, 196–197, 205
- salt stock and salt wall, 39–40, 42, 44–49, 60–62, 77
- triangular diapir, 87–91, 103, 107–108

primary basin, 276

primary minibasin, 276

primary peripheral sink—*see* peripheral sink

R

radial fault, 31, 37, 50, 53, 55–56, 132–133, 190

raft, 276–277; *see also* raft tectonics

raft tectonics, 86–87, 102, 111, 118–119, 121, 123–124, 128–129, 193–194, 209, 277

Rayleigh-Taylor instability, 270, 277

reactive diapirism, 277
- conceptual model
  - counterregional system, 245, 251–252
  - extensional salt detachment, 121, 140
  - minibasin, 232
  - roho system, 235, 242–243
  - salt stock and salt wall, 75–76, 79
  - triangular diapir, 85–93, 106–108
- geologic example
  - extensional salt detachment, 122–123, 131, 133–134
  - minibasin, 225
  - roho system, 236–240, 246
  - salt pillow and salt anticline, 10
  - triangular diapir, 86
- numerical model, 28, 104, 136–138
- physical model
  - indented diapir, 116–119
  - roho system, 241–242
  - salt pillow and salt anticline, 21, 23
  - salt sheet, 193–197, 201–202
  - salt stock and salt wall, 61–64, 71
  - triangular diapir, 93–103

reactive graben, 271; *see also* graben (crestal)

regional datum (regional elevation), 277

rejuvenation—*see* salt diapir (rejuvenated)

restoration,
- counterregional system, 246, 250, 252
- indented diapir, 116–117
- minibasin, 224–227, 233–234
- roho system, 237–240
- salt pillow and salt anticline, 8, 10, 21, 23
- salt stock and salt wall, 41, 43, 46, 48, 57, 60–61, 76–77
- triangular diapir, 86–87, 89, 91, 96, 99, 103

reverse fault, 277—*see also* thrust fault

rim syncline, 277

rock salt, 277—*see also* salt

roho system, 235–243, 277
- conceptual model, 235, 242–243
- geologic example, 236–240
- physical model, 194, 241–242

rollover, 277
- expulsion
  - counterregional system, 245–252
  - indented diapir, 109–111, 118
  - minibasin, 225, 232–233
  - roho system, 239–242
  - salt pillow and salt anticline, 11, 21
  - salt sheet, 196, 201
  - salt stock and salt wall, 46–48, 60, 76–77
- extension
  - counterregional system, 246, 252
  - extensional salt detachment, 121–122, 124, 128–132
  - indented diapir, 116, 119,
  - roho system, 226, 235–243
  - salt sheet, 166–167, 204, 206, 209–210, 213
  - triangular diapir, 85–88, 91–93

roof, 277

roof-edge thrust, 277—*see also* canopy-margin thrust system

root, 277; *see also* pedestal

S

salt, 277

salt anticline, 277–278—*see* salt pillow and salt anticline

salt breakout, 278; *see also* salt sheet
- conceptual model, 206–208, 213
- geologic example, 162–163, 165, 175
- physical model, 196–197, 200–201

salt canopy, 278—*see also* canopy-margin thrust system; counterregional system; roho system; salt-stock canopy
- conceptual model, 206, 214–215, 217, 242–243, 245–248, 251, 252
- geologic example, 8, 164–171, 223–225, 239–240
- numerical model, 204
- physical model, 70, 193, 195, 197, 200–201, 229, 249–250

salt deflation, 278

salt detachment—*see* detachment

salt diapir, 278
- active—*see* active diapir
- allochthonous—*see* salt canopy; salt sheet
- contractional—*see* shortening
- falling—*see* extension (diapir fall)
- leaning
  - counterregional system, 245–246, 249–250, 252
  - indented diapir, 111
  - minibasin, 222, 225, 229, 232
  - roho system, 235, 237, 239, 242–243
  - salt sheet, 166–167, 174, 201
  - salt stock and salt wall, 65, 77, 79
  - triangular diapir, 102
- overthrust
  - contractional salt detachment, 141–142, 144–147, 151–155, 158, 160
  - extensional salt detachment, 125–126
  - salt pillow and salt anticline, 21, 24
  - salt sheet, 162, 163, 173–174, 176–179, 183, 185, 205, 207–208

salt stock and salt wall, 33, 42–43, 48, 59–60, 69, 71, 75, 79
passive—*see* passive diapir
pinched off
counterregional system, 246, 250, 252
extensional salt detachment, 125–126
minibasin, 225, 236, 232
roho system, 238, 242
salt sheet, 174, 176–177, 202, 207–208
salt stock and salt wall, 40, 42, 54–57, 63–64, 69, 71, 77, 79–81
reactive—*see* extension (diapir rise)
rejuvenated, 39–40, 44, 64–71, 74–75, 77, 79–80, 200
salt flow, 36, 82–83
stock—*see* salt stock and salt wall
strike slip—*see* strike slip
wall—*see* salt stock and salt wall

salt dome, 278—*see also* salt stock and salt wall

salt droplet, 278

salt expulsion—*see* rollover (expulsion); salt withdrawal

salt feeder, 278
conceptual model, 204–209, 213, 217, 232–233, 235–239, 242–243, 245–246, 251
geologic example, 8, 176–177, 179, 220, 223, 225
numerical model, 203–204
physical model, 193–194, 197, 202, 230, 248–250

salt flat, 278; *see also* salt flat-ramp pair

salt flat-ramp pair
counterregional system, 247, 249–250
minibasins, 219, 227, 230, 232–233
roho system, 236, 240, 242

salt sheet, 161, 163–169, 172, 175, 193–194, 197, 201, 204–205, 211–213

salt fountain, 278
counterregional system, 252
minibasin, 225

salt glacier (namakier), 278
conceptual model, 204–206, 208, 212–213, 242, 245
geologic example, 162–163, 167, 177, 180–189, 226–227, 238–240, 246
numerical model, 204
physical model, 190–197, 200–202, 229, 249–250

salt horn, 106, 109–116, 118

salt inflation, 278

salt keel, 278–279

salt nappe, 162–163, 175, 183, 193, 279; *see also* source-fed thrust

salt overthrust
conceptual model, 75, 79, 141, 210–212
geologic example, 8–9, 21, 24, 59, 142, 144–152, 162, 170–171, 173–177, 179, 183, 185, 194, 202, 205, 222
physical model, 71, 154–155, 158

salt pancake, 279

salt pedestal—*see* pedestal

salt pillow and salt anticline, 1–30
anticline, 1, 2, 4–11, 14–20, 22–29, 34, 45–46, 145–146, 148–152, 155–156, 158–159, 173–174, 182–185, 198, 199, 235, 239, 246, 277–278
breached, 2, 45
compressional, 1–22, 24–29, 79, 146, 148–150, 152, 154–156, 158–159, 173–175, 184–185, 193, 205
conceptual model, 1, 77, 79, 205, 235
geologic example, 2–21, 34, 45–46, 90, 122, 142, 146, 148–150, 151–152, 173–175, 184–185, 239–240, 246
halokinetic, 1, 21, 23, 34, 46, 77, 122, 142, 235
numerical model, 25–29, 159
physical model, 21–25, 154–156, 158, 193
pillow, 1, 3, 11–12, 16–18, 25, 33–34, 217, 277
transpressional, 3

salt plateau, 9–11, 21, 46, 99, 155, 175, 193, 279

salt ramp, 279; *see also* salt flat-ramp pair

salt ridge, 277–278; *see* salt anticline

salt roller, 279
conceptual model, 85, 119, 121, 140, 206, 210–211, 232, 235
geologic example
extensional salt detachment, 122–123, 129–131
minibasin, 225–226
roho system, 236–243
salt sheet, 165–167
salt stock and salt wall, 39
triangular diapir, 86–93
numerical model, 105
physical model, 99, 102, 133–134, 250

salt-roof thrust, 279
minibasin, 222, 225
roho system, 238
salt sheet, 168, 170–171, 178–179, 183, 202, 210–211

salt sheet, 161–215, 279
breakout—*see* salt breakout
conceptual model
counterregional system, 245, 251–252
minibasin, 217, 232–233
roho system, 235, 243
salt sheet, 161, 204–215
salt stock and salt wall, 77, 79
contraction above
contractional salt detachment, 151
minibasin, 222, 225
roho system, 238
salt pillow and salt anticline, 7–8, 12
salt sheet, 162, 168, 170–171, 173–175, 178–179, 183, 198, 200, 202, 210–211
counterregional system—*see* counterregional system
extension above
counterregional system, 252
indented diapir, 117
minibasin, 219, 221–227, 232–233
roho system, 235–243
salt pillow and salt anticline, 5, 11
salt sheet, 164, 166–167, 171, 175–177, 180, 194, 197–199, 201, 204, 206, 210–211
extrusive advance
counterregional system, 246, 248–250
indented diapir, 117
salt pillow and salt anticline, 6–8, 23
salt sheet, 163, 200–202, 204–206, 208–209
salt stock and salt wall, 71
triangular diapir, 60–61, 94, 99
feeder—*see* salt feeder
geologic example
contractional salt detachment, 151
counterregional system, 246–248
minibasin, 218–225
roho system, 236–240
salt pillow and salt anticline, 5–8, 11–12
salt sheet, 162–180
salt stock and salt wall, 46, 48, 54
initiation during shortening, 6, 79, 176–177, 185, 205
lineage, 206–208
numerical model, 203–204
overthrust—*see* salt overthrust
physical model
counterregional system, 248–250
indented diapir, 117
minibasin, 228–230
roho system, 241–242
salt pillow and salt anticline, 23
salt sheet, 194–202
salt stock and salt wall, 60–61, 71
triangular diapir, 94, 99
roho system—*see* roho system
salt-stock canopy—*see* salt-stock canopy
thrust advance—*see* canopy-margin thrust system

salt sill, 279

salt stock and salt wall, 31–83, 279, 280
active diapir—*see* active diapir
conceptual model
counterregional system, 245, 251–252
extensional salt detachment, 140
indented diapir 109, 119
minibasin, 217, 232–234
roho system, 235, 242–243
salt sheet, 161, 204–210, 214
salt stock and salt wall, 31, 75–83
triangular diapir, 85, 106–108

geologic example
contractional salt detachment, 142–143, 145–147
counterregional system, 246–248
extensional salt detachment, 125–127
indented diapir, 110–115
minibasin, 218–219, 221–224
roho system, 236–240
salt pillow and salt anticline, 5, 8, 10–12, 15–21
salt sheet, 163
salt stock and salt wall, 32–60
triangular diapir, 89, 92–93
indented diapir—*see* indented diapir
numerical model, 28, 72–75, 104–105, 136–139, 159, 203–204
passive diapir—*see* passive diapir
physical model
contractional salt detachment, 157
counterregional system, 248–250
extensional salt detachment, 133–136
indented diapir, 116–118
minibasin, 228–230
roho system, 241–242
salt pillow and salt anticline, 23
salt sheet, 190–194, 196–197, 201–202
salt stock and salt wall, 60–72
triangular diapir, 93–103
pinched off—*see* salt diapir (pinched off)
reactive diapir—*see* reactive diapir
salt diapir—*see* salt diapir
salt stock, 32–42, 47–57, 59–60, 65–67, 72, 76, 81–82, 279
counterregional system, 246–247, 249, 251
minibasin, 218, 224, 227, 229–230, 232–234
roho system, 236–238, 242–243, 246–247, 249, 251
salt pillow and salt anticline, 5, 15–19, 31
salt sheet, 161, 163–164, 176–177, 181–192, 205–207, 209, 214–215
triangular diapir, 102
salt wall, 31, 33–34, 43, 45–48, 60–64, 68–77, 79, 280
contractional salt detachment, 143, 146–147, 157, 159
counterregional system, 246–250
extensional salt detachment, 121–123, 130–131, 133–140
indented diapir, 109–119
minibasin, 218, 232
roho system, 237, 241–242
salt pillow and salt anticline, 10–11, 21, 23
salt sheet, 171, 173, 178–179, 193–194, 196–197, 201–202, 205, 208
triangular diapir, 85–108
salt-stock canopy, 279
counterregional system, 249
minibasin, 217–220, 223–226, 232, 234
roho system, 236–237, 242–243
salt sheet, 164, 200, 214–215

salt structure, 279

salt suture—*see* suture

salt system, 279

salt tectonics, 280

salt tongue, salt-tongue canopy, 280

salt touch-down (salt grounding), 280

salt wall—*see* salt stock and salt wall

salt-wall canopy, 280

salt weld—*see* weld

salt window, 280

salt wing, 280
counterregional system, 248, 250
indented diapir, 117
minibasin, 220, 227, 229–230
roho system, 236
salt pillow and salt anticline, 11, 12
salt sheet, 178–180 193, 196–197, 201, 202, 204
salt stock and salt wall, 47–48, 57, 60–62, 64, 76–77
triangular diapir, 94, 99

salt withdrawal, 280

satellite image,
Garmsar Basin, 163
Great Kavir, 3, 37
Parry Islands fold belt, 148
Sierra Madre Oriental fold belt, 14
Zagros fold belt, 16–20, 152, 184

secondary minibasin, 280; *see also* minibasin

seismic image
Dnieper-Donets Basin, 2
Gulf of Mexico Basin, 6–7, 38–39, 86, 110, 124, 164–166, 168–172, 218–224, 236, 247–248
Kwanza Basin, 9–10, 39–40, 125, 145
Lower Congo Basin, 42, 87–88, 145–147, 176
Nordkapp Basin, 89
North Sea Basin, 12, 13, 49–50
Precaspian Basin, 58

shale sheath, 280

shear zone, 4, 26, 27, 36, 210, 280–281

sheet segmentation (sheet reactivation), 281

shortening—*see* contraction

source-fed thrust, 281
conceptual model, 205, 208
geologic example, 151, 162–163, 173–175, 185
physical model, 153, 155–156

source layer, 281

spine, 281

squeezed diapir—*see* active diapir (compressional)

stem, 281

stepped counterregional system—*see* counterregional system

strike slip, 81, 171, 235–238, 243

structural inversion, 279
salt-controlled, 34–36, 45–47, 49, 76–77, 79, 86–87, 99–100, 109–119, 122
tectonic
contractional salt detachment, 142
extensional salt detachment, 131
salt pillow and salt anticline, 10–12, 15–21
salt sheet, 174, 178–179, 181–185, 188–192, 202, 204–205, 207–208
salt stock and salt wall, 31–33, 39–48, 50, 52, 59–60, 64–72, 77, 81
triangular diapir, 90, 103, 106

suballochthonous, 281

subautochthonous, 281

subsalt, 281; *compare* presalt
contractional salt detachment, 142, 144, 151–156, 158, 160
counterregional system, 245–252
indented diapir, 117
minibasin, 217–220, 222–227, 229–230, 232–234
roho system, 235–243
salt pillow and salt anticline, 5–8, 11–12, 21, 23
salt sheet, 161–215
salt stock and salt wall, 43, 45, 48, 57, 61–64, 71, 77, 79, 81
triangular diapir, 94, 99

substrate expulsion, 281

suprasalt, 281

suture, 281, 282
allosuture, 267
counterregional system, 249
minibasin, 217, 223, 232
roho system, 237, 239, 243
salt pillow and salt anticline, 8
salt sheet, 172, 195, 198–200, 214–215
salt stock and salt wall, 70
autosuture, 195–197, 269
suture line, 164, 172, 195, 198–200, 214–215, 281
suture point, 195, 198–200, 214–215, 282
suture surface, 215, 282

synkinematic, 282

synsedimentary fault, 282

T

tank-tread advance, 203–204, 282
teardrop diapir—*see* salt diapir (pinched off)
terminal suture point—*see* suture (suture point)
thermal convection, 282
thermal loading, 282
thick-skinned deformation, 282; *see also* basement (extension); basement (shortening)
thin-skinned deformation, 282; *see also* detachment
thrust advance, 282; *see also* canopy-margin thrust system
thrust fault, 282
- conceptual model, 1, 75–76, 79–81, 118, 141, 160–161, 204–213, 232, 243
- duplex, 148–149, 160, 162
- geologic example,
  - contractional salt detachment, 142–152
  - extensional salt detachment, 131
  - minibasin, 218, 222, 225, 227
  - roho system, 236–238
  - salt pillow and salt anticline, 5, 8–12, 15–18, 20–21
  - salt sheet, 164–171, 173–174, 176–179, 183, 185, 188
  - salt stock and salt wall, 33, 42–44, 47–48, 54–56, 59
- imbricate, 144, 148–149, 152–155, 158, 161–162, 210–212
- numerical model, 25, 28–29, 74–75, 136–139, 159
- physical model, 64–71, 103, 118, 134, 153–158, 190–192, 194, 202, 204, 241

thrust weld—*see* weld (thrust)
toe, 235–237, 241
toe thrust—*see* compressional toe
translational detachment—*see* detachment (translational)
transpression, 3, 40, 135
transtension, 131, 135, 236–237
triangular diapir, 85–108; *see also* salt roller
- conceptual model
  - counterregional systems, 251–252
  - extensional salt detachments, 121, 140
  - indented diapirs, 119
  - minibasins, 232
  - roho systems, 235, 242–243
  - salt sheets, 206, 210–211
  - salt stock and salt walls, 75, 79
  - triangular diapirs, 85, 106–109
- geologic example
  - contractional salt detachment, 146–147
  - extensional salt detachment, 121–123, 129–131
  - indented diapir, 111, 116–119
  - minibasin, 224–227
  - roho system, 236–243
  - salt pillow and salt anticline, 10
  - salt sheet, 165–167
  - salt stock and salt wall, 39, 43, 48, 62
  - triangular diapir, 86–93
- numerical model, 104–105, 136–139
- physical model
  - counterregional system, 250
  - extensional salt detachment, 133–134
  - roho system, 241–242
  - salt pillow and salt anticline, 23
  - salt sheet, 194, 186, 197, 201–202
  - salt stock and salt wall, 64, 71
  - triangular diapir, 93–103

turtle structure, turtle-structure anticline, 283; *see also* mock turtle
- geologic example, 11, 34, 47, 120
- physical model, 194, 239–240

V

vortex structure, 82–83, 283

W

wedge halokinetic sequence, 283—*see also* halokinetic sequence (wedge)
weld, 283
- allochthonous
  - counterregional system, 245–252
  - indented diapir, 117
  - minibasin, 217, 219–220, 223–230, 232–233
  - roho system, 235–243
  - salt pillow and salt anticline, 5, 7
  - salt sheet, 162, 166–167, 174, 194
- autochthonous
  - contractional salt detachment, 141–147, 156, 158
  - extensional salt detachment, 121–134, 140
  - indented diapir, 109–112, 116,
  - salt pillow and salt anticline, 1, 6, 8, 10–13, 21, 23, 24
  - salt sheet, 179–180, 193, 196–197, 201, 206–208
  - salt stock and salt wall, 31, 33, 35, 39, 40, 42, 46–48, 60, 61, 62, 76–77, 79–80
  - triangular diapir, 86–89, 99–100, 106, 118
- bowl weld, 217, 220, 223–230, 232–233, 270
- conceptual model
  - contractional salt detachment, 160
  - counterregional system, 245, 251–252
  - extensional salt detachment, 121, 140
  - indented diapir, 109
  - minibasin, 217, 232–233
  - roho system, 235, 242–243
  - salt pillow and salt anticline, 1
  - salt sheet, 206–208
  - salt stock and salt wall, 31, 76, 77, 79–80
  - triangular diapir, 85, 106
- diapir, 42, 50, 63–64, 69, 71, 80, 176–177, 202, 245–246, 248–249, 251–252
- fault weld, 273
  - contractional salt detachment, 141–147, 154, 156, 158, 160
  - extensional salt detachment, 121–132
  - indented diapir, 111–112, 118
  - roho system, 235–243
  - salt sheet, 162, 174, 176–177, 193–194
  - salt stock and salt wall, 42, 59–60, 79
  - triangular diapir, 85–89
- geologic example
  - contractional salt detachment, 142–147
  - counterregional system, 246–248
  - extensional salt detachment, 122–131
  - indented diapir, 110–112
  - minibasin, 217, 219–220, 223–227
  - roho system, 236–240
  - salt pillow and salt anticline, 5, 6–8, 10, 11–13
  - salt sheet, 162, 166–167, 174, 176–177, 179–180
  - salt stock and salt wall, 33, 35, 39, 40, 46–48
  - triangular diapir, 86–89
- physical model
  - contractional salt detachment, 154, 156, 158
  - counterregional system, 248–250
  - extensional detachment, 132–134
  - indented diapir, 116, 117, 118
  - minibasin, 228–230
  - roho system, 241–242
  - salt pillow and salt anticline, 21, 23, 24
  - salt sheet, 193–194, 196–197, 201–202
  - salt stock and salt wall, 60–64, 69, 71
  - triangular diapir, 99–100
- thrust weld, 42, 59, 79, 144, 154, 156, 158, 160, 162, 174, 176–177, 208

withdrawal basin—*see* salt withdrawal

Geographic Index

For specific types of salt-tectonic structures, see the Subject Index.

A

Algeria
- Atlas fold belt, 32, 254
- Ran el Melah diapir (Rocher de Sel de Djelfa), 32

Angola
- Kwanza Basin, 9, 39, 86, 125–127, 145, 261
  - Angola salt nappe, 175
  - Cabo Ledo uplift, 10
  - Cegonha diapir, 40
  - Quenguela mock-turtle anticline 110–111
- Lower Congo Basin, 87–88, 128–129, 262

Aptian salt basin—*see* South Atlantic salt basin

Aquitaine Basin—*see* France (Aquitaine Basin)

Atlas fold belt—*see* Algeria (Atlas fold belt); Tunisia (Atlas fold belt)

Atwater fold belt—*see* United States of America (Gulf of Mexico Basin)

B

Betic Cordillera—*see* Spain (Betic Cordillera)

Brazil (Campos Basin), 122, 255

C

Campos Basin—*see* Brazil (Campos Basin)

Canada
- Maritimes Basin, 44
- Parry Islands fold belt, 148–150

Congo—*see* Democratic Republic of Congo (Katanga fold belt)

D

Democratic Republic of Congo (Katanga fold belt), 173–174, 261

Dnieper-Donets basin—*see* Ukraine (Dnieper-Donets basin)

DRC—*see* Democratic Republic of Congo (Katanga fold belt)

E

East Texas Basin—*see* United States of America (East Texas Basin)

Ebro Basin—*see* Spain (Ebro Basin)

F

France
- Aquitaine Basin (Pyrenees fold belt), 142, 267
- Maritime Alps, 43, 262

G

Gabon (Lower Congo Basin), 42, 145–147, 176–177

Gachsaran salt, 18, 21, 185

Garmsar Basin—*see* Iran (Garmsar Basin)

Germany (North German Basin), 263
- Allertal Lineament, 46
- Barrien salt anticline, 11
- Gehlenkamp graben, 47
- Gifhorn diapir, 46
- Gorleben diapir, 46
- Gülden-Brandel diapir, 48
- Harz uplift, 12
- Heide diapir, 47
- Hohne turtle structure, 47
- Hope diapir, 48
- Hustedt diapir, 179
- Husum diapir, 45
- Jerstedt salt anticline, 12
- Meissendorf diapir, 48
- Niendorf diapir, 48
- Oppershausen diapir, 178
- Osterholz diapir, 11
- Roshe-Thondorf diapir, 48
- Salzgitter salt diapir, 12
- Thören diapir, 48
- Wolthausen diapir, 179
- Zwischenahn diapir, 45

Great Kavir—*see* Iran (Great Kavir)

Gulf of Mexico Basin (USA)—*see* Mexico (Gulf of Mexico Basin); United States of America (Gulf of Mexico Basin)

H

Hormoz salt, 15–21, 152, 183–189

I

Iran
- Garmsar Basin, 163, 257
- Great Kavir, 37, 257
  - Tapeh Zard uplift, 3
  - West Dulasian anticline, 4
- Qom Basin, 257
  - Qom Kuh diapir, 181–182
  - Shurab diapir, 182
- Ravar Basin, 59–60
- Zagros fold belt, 15–18, 268
  - Ab-e-Lashkar, 21, 185
  - Aliabad diapir, 19, 184
  - Anguru diapir, 183
  - Chahal diapir, 19, 184
  - Dezful embayment, 15
  - Do-Au diapir, 20
  - Gach diapir, 19, 184
  - Halft Kel anticline, 21
  - Jahani diapir, 188–189
  - Kazerun fault zone, 20
  - Kuh-e-Asmari, 21, 185
  - Kuh-e-Baviyun, 20
  - Kuh-e-Burkh, 19, 184
  - Kuh-e-Chachal, 19, 184
  - Kuh-e-Darang, 20
  - Kuh-e-Gach, 19, 184
  - Kuh-e-Gashu, 183
  - Kuh-e-Khonj, 152
  - Kuh-e-Kukular, 152
  - Kuh-e-Namak (Bushehr), 20, 183–189
  - Kuh-e-Sabzu, 152
  - Kuh-e-Siah, 152
  - Kuh-e-Tia, 152
  - Lahabari syncline, 21, 185
  - Ram Hormoz syncline, 21, 185
  - Siah Tagh diapir, 19, 184
  - Zagros Main Thrust, 152

J

Jura fold belt—*see* Switzerland (Jura fold belt)

K

Katanga fold belt (DRC)—*see* Democratic Republic of Congo (Katanga fold belt)

Kazakhstan (Precaspian Basin) 58, 180, 266

Keuper salt 43, 142

Kwanza Basin—*see* Angola (Kwanza Basin)

L

La Popa Basin—*see* Mexico (La Popa Basin)

Louann Salt
- East Texas Basin, 33–36, 122–123
- Gulf of Mexico Basin, 5–8, 38–39, 86, 110, 123–124, 143, 163–172, 218–227, 236–240, 246–248

Lower Congo Basin—*see* Angola (Lower Congo Basin); Gabon (Lower Congo Basin)

M

Maritimes Basin—*see* Canada (Maritimes Basin)

Mexico
- Gulf of Mexico Basin, 224–225, 248, 257, 260
- La Popa Basin, 40–42, 267
- Sierra Madre Oriental fold belt, 14, 267

Morocco (Rif Basin), 183

N

Netherlands—*see* The Netherlands (North Sea)

Nordkapp Basin—*see* Norway (Nordkapp Basin)

North German Basin—*see* Germany (North German Basin)

North Sea—*see* Norway (North Sea); The Netherlands (North Sea); United Kingdom (North Sea);

Norway
- Nordkapp Basin, 44, 89, 263
- North Sea, 13
- Norwegian Sea, 130, 265

P

Pakistan (Potwar Plateau), 151, 266

Paradox Basin—*see* United States of America (Paradox Basin)

Parry Islands fold belt—*see* Canada (Parry Islands fold belt)

Potwar Plateau—*see* Pakistan (Potwar Plateau)

Precaspian Basin (Pricaspian Basin)—*see* Kazakhstan

Pyrenees fold belt—*see* France (Aquitaine Basin); Spain (Pyrenees fold belt)

Q

Qom Basin—*see* Iran (Qom Basin)

R

Ravar Basin—*see* Iran (Ravar Basin)

Rif Basin—*see* Morocco (Rif Basin)

Röt halite
- North German Basin, 45, 48, 178–179
- North Sea, 112

Rotliegend salt (Germany), 47

S

Salina Basin—*see* Mexico (Gulf of Mexico Basin)

Sierra Madre Oriental—*see* Mexico (Sierra Madre Oriental fold belt)

South Atlantic (Aptian) salt basin
- Campos Basin, 122, 255
- Kwanza Basin, 9–10, 39–40, 86, 110–111, 125–127, 175, 261
- Lower Congo Basin, 42, 87–88, 128–129, 145–147, 176–177, 262

Spain
- Betic Cordillera, 162, 255
- Ebro Basin, 36
- Pyrenees fold belt, 131

Subbetic—*see* Spain (Betic Cordillera)

Switzerland (Jura fold belt), 8–9, 144, 260

T

The Netherlands (North Sea), 90

Tunisia (Atlas fold belt), 33, 254

U

Ukraine (Dnieper-Donets Basin), 2, 256

United Kingdom (North Sea), 91–92, 112, 264
- Banff diapir, 49–51
- Machar diapir, 51–52
- Monan diapir, 52
- Mungo diapir, 53
- North Dogger fault zone, 90
- Sole Pit, 12

United States of America
- East Texas Basin, 33, 256
  - Butler Dome, 36
  - Elkhart Graben-Mt. Enterprise Fault Zone, 34
  - Hainesville Dome, 35
  - Mexia-Talco fault zone, 34, 122–123
  - Oakwood Dome, 35
- Gulf of Mexico Basin, 39, 219–223, 257–260
  - Alaminos Canyon, 168, 170
  - Atwater fold belt, 5, 143
  - Bay Marchand, 110, 246
  - Bourbon counterregional system 246
  - Eugene Island, 231
  - Ewing Bank, 223–224
  - Farnella dome, 164
  - Garden Banks, 220
  - Grand Isle South, 223
  - Green Canyon, 38, 165–169, 172, 218, 225–227
  - Kansas anticline, 8
  - Keathley Canyon, 220
  - K2 anticline, 5
  - Mad Dog anticline, 5, 166–167
  - Mahogany salt sheet, 238
  - Mica salt sheet, 163–164
  - Mississippi Canyon, 86, 163–164
  - Mitchell dome, 163–164
  - Perdido fold belt, 6
  - Pescadores minibasin, 224
  - Queen City depotrough, 123–124
  - Rum roho system, 237
  - St. Malo anticline, 7
  - Ship Shoal, 225–226
  - Sigsbee Escarpment, 6, 164–171, 222
  - South Timbalier, 224
  - Terrebone Trough, 239–240, 246
  - Walker Ridge, 164–165, 171–172, 220
  - West Cameron, 236, 238
  - West Delta/South Pass, 247
  - White Castle dome, 38
  - Whiting dome, 236
  - Wilcox depotrough, 123–124
- Paradox Basin, 265
  - Gypsum Valley diapir, 112
  - Lisbon Valley fault, 93
  - Onion Creek diapir, 53–54, 113–115
  - Paradox Valley diapir, 113
  - Roc Creek diapir, 113
  - Spanish Valley (Moab) diapir, 113
  - Upheaval Dome, 54–57

Z

Zagros fold belt—*see* Iran (Zagros fold belt)

Zechstein salt
- North German Basin, 11–12, 45–48, 178–179
- North Sea, 12–13, 49–53, 90–92, 112
- Norwegian Sea, 130

ABOUT THE BEG

The Bureau of Economic Geology, established in 1909, is the oldest research unit at The University of Texas at Austin. The Bureau, part of the Jackson School of Geosciences, also functions as the State Geological Survey. The Bureau curates the largest volume of subsurface core and cuttings in the United States and runs a major Texas well log library. The Bureau conducts basic and applied research related to energy resources, including oil, natural gas, and coal; mineral resources; coastal processes; Earth and environmental systems; hydrogeology; carbon sequestration; nanotechnology; energy economics; and geologic mapping. The Bureau's outreach efforts involve the public, State agencies, Federal agencies, private foundations, and industry.

The Udden Book Series is aimed at a broad audience. The series comprises books that represent substantial contributions to science, address public policy, review or describe programs of national significance, or profile people who have led scientific and policy discourse. The series is named in honor of Johan A. Udden, eminent scientist and director of the Bureau of Economic Geology from 1915 through 1932.

ABOUT THE AAPG

Founded in 1917, the American Association of Petroleum Geologists is the world's largest professional geological society. The membership of the AAPG includes geologists, geophysicists, CEO's, managers, consultants, students, and academicians. The purpose of the organization is to foster scientific research, advance the science of geology, promote technology, and inspire high professional conduct. The AAPG's mission is to provide its members, other geoscientists, and nongeologists with timely and useful publications at the lowest possible cost.

The Memoir series is the premier book series of the AAPG. Created in 1962, the series was designed for broad appeal and application. The first Memoir was titled "Classification of Carbonate Rocks," and AAPG Memoirs continue to be classic, highly sought-after texts utilized for reference and research in the continuing expansion of scientific knowledge in the study of geology.